A Companion to Global Environmental History

WILEY BLACKWELL COMPANIONS TO HISTORY

This series provides sophisticated and authoritative overviews of the scholarship that has shaped our current understanding of the past. Defined by theme, period, and/or region, each volume comprises between 25 and 40 concise essays written by individual scholars within their area of specialization. The aim of each contribution is to synthesize the current state of scholarship from a variety of historical perspectives and to provide a statement on where the field is heading. The essays are written in a clear, provocative, and lively manner, designed for an international audience of scholars, students, and general readers.

A COMPANION TO GLOBAL ENVIRONMENTAL HISTORY

Edited by

J.R. McNeill
and
Erin Stewart Mauldin

WILEY Blackwell

This edition first published 2015
© 2012 John Wiley & Sons, Ltd

Edition history: Blackwell Publishing Ltd (Hardback: 2012)

Registered Office
John Wiley & Sons Ltd, The Atrium, Southern Gate, Chichester, West Sussex, PO19 8SQ, UK

Editorial Offices
350 Main Street, Malden, MA 02148-5020, USA
9600 Garsington Road, Oxford, OX4 2DQ, UK
The Atrium, Southern Gate, Chichester, West Sussex, PO19 8SQ, UK

For details of our global editorial offices, for customer services, and for information about how to apply for permission to reuse the copyright material in this book please see our website at www.wiley.com/wiley-blackwell.

The right of J.R. McNeill and Erin Stewart Mauldin to be identified as the authors of the editorial material in this work has been asserted in accordance with the UK Copyright, Designs and Patents Act 1988.

Library of Congress Cataloging-in-Publication Data

A companion to global environmental history / edited by J. R. McNeill and Erin Stewart Mauldin.
    p. cm.
  Includes index.
  ISBN 978-1-4443-3534-7 (cloth) ISBN 978-1-118-97753-8 (pbk)
1. Human ecology–History–Cross-cultural studies.   2. Global environmental change–History–Crosscultural studies.
3. Environmental policy–History–Cross-cultural studies.   4. Environmental protection–History–Cross-cultural studies.
I. McNeill, John Robert.   II. Mauldin, Erin Stewart.
  GF13.C63 2012
  304.209–dc23

                                        2012002776

A catalogue record for this book is available from the British Library.

Cover image: Philip James de Loutherbourg, *Coalbrookdale by Night*, 1801, oil on canvas.
Science Museum, London / The Bridgeman Art Library
Cover design by Richard Boxall Design Associates

Set in 10/12pt Galliard by SPi Publisher Services, Pondicherry, India

1 2015

To Julie, once more
And to Daniel

Contents

List of Maps

Notes on Contributors

Jordan Bauer is a PhD candidate in history at the University of Houston and is currently working on her dissertation, a history of urban growth and politics in post-1945 Raleigh-Durham.

Peter Boomgaard is Professor of Economic and Environmental History of Southeast Asia, University of Amsterdam, and Senior Researcher, Royal Netherlands Institute of Southeast Asian and Caribbean Studies (KITLV), Leiden. Among his publications are *Frontiers of Fear: Tigers and People in the Malay World, 1600–1950* (2001) and *Southeast Asia: An Environmental History* (2007). He is currently writing a book on the forests of Java between 1500 and 1950.

Stephen Brain is Assistant Professor of History at Mississippi State University. He is the author of *Song of the Forest: Russian Forestry and Stalinist Environmentalism, 1905–1953* (2011). He is currently conducting research for a manuscript about the environmental history of the White Sea and the fishermen who worked there, the Russian Pomor.

Jane Carruthers is Research Professor of History at the University of South Africa. Her main interests lie in environmental history, the history of national parks, and the biological sciences in South Africa, and she has published widely in these fields.

Paul D'Arcy is a Senior Fellow in the Department of Pacific and Asian History of the College of Asia and the Pacific at the Australian National University. He is author of *The People of the Sea* (2006). He has just finished editing a collection on Asian investment and engagement with Pacific Island nations, and is currently working on his next book, *Warfare and State Formation in Hawai'i: The Limits of Coercion in the Pre-Modern World*.

Daniel Headrick is Professor Emeritus of History and Social Science at Roosevelt University in Chicago. He is the author of several books, most recently *Technology: A World History* (2009) and *Power over Peoples: Technology, Environments, and Western Imperialism, 1400 to the Present* (2010). He is currently writing an environmental history of the world since the Stone Age.

J. Donald Hughes is John Evans Professor of History Emeritus at the University of Denver. He lives in Princeton, New Jersey. Author of *An Environmental History of the World* (2nd edition, 2009) and *What Is Environmental History?* (2006), he is a founding member of the American Society for Environmental History (ASEH), the European Society for Environmental History (ESEH), and the Association of East Asian Environmental Historians (AEAEH).

Paul Josephson is Professor of History at Colby College. He is a specialist in the history of big science and technology. He has written several books in environmental history including *Industrialized Nature* (2002), *Resources under Regimes* (2005), and *Motorized Obsessions* (2007). He is currently writing a history of the environmental impact of Soviet arctic conquest.

Nancy Langston is Professor in the Nelson Institute for Environmental Studies and the Department of Forest and Wildlife Ecology at the University of Wisconsin-Madison. She is author of *Forest Dreams, Forest Nightmares: The Paradox of Old Growth in the Inland West* (1995), *Where Land and Water Meet: A Western Landscape Transformed* (2003), and *Toxic Bodies: Hormone Disruptors and the Legacy of DES* (2010). She is currently editing *Environmental History* and working on a history of boreal forests.

Bao Maohong is Associate Professor of History at Peking University, China. He is the author of *Forest and Development: Deforestation in the Philippines* (2008), *Environmental Governance in China and Environmental Cooperation in Northeast Asia* (2009), and *The Origins of Environmental History and Its Development* (2012). He is currently working on the transformation of East Asia from the perspective of environmental history.

Robert B. Marks is Richard and Billie Deihl Professor of History at Whittier College. He is the author of *Tigers, Rice, Silk, and Silt: Environment and Economy in Late Imperial South China* (1998) and *The Origins of the Modern World: A Global and Ecological Narrative from the Fifteenth to the Twenty-First Century* (2007). His latest book is *China: Its Environment and History* (2012).

Joan Martinez-Alier is Professor of Economics and Economic History and Researcher at ICTA at the Autonomous University of Barcelona. He is the author of *Ecological Economics: Energy, Environment and Society* (1990) and *The Environmentalism of the Poor: A Study of Ecological Conflicts and Valuation* (2002). He is a founding member and past president of the International Society for Ecological Economics. He is also coeditor of *Getting Down to Earth: Practical Applications of Ecological Economics* (1996), *Rethinking Environmental History: World-System History and Global Environmental Change* (2007), and *Recent Developments in Ecological Economics* (2008).

Erin Stewart Mauldin is a PhD candidate in US environmental history at Georgetown University. She is currently writing her dissertation on the environmental history of the Reconstruction period in the southern United States, exploring the ecological legacies of the American Civil War and its impacts on southern agriculture and economy during the late nineteenth century.

Meredith McKittrick is Associate Professor in the Department of History and the Edmund A. Walsh School of Foreign Service at Georgetown University. She is the author of *To Dwell Secure: Generation, Christianity and Colonialism in Ovamboland* (2002) and numerous articles on the history of Namibia. She is currently writing a book about riparian farming communities in southwestern Africa.

J. R. McNeill is Professor of History and University Professor at Georgetown University. He is the author of *Mosquito Empires: Ecology and War in the Greater Caribbean, 1620–1914* (2010) and *Something New under the Sun: An Environmental History of the Twentieth-Century World* (2000), and coauthor of *The Human Web* (2003) and *A Short History of the Anthropocene* (2013). He served as president of the American Society for Environmental History from 2011 to 2013.

Martin V. Melosi is Hugh Roy and Lillie Cranz Cullen Professor and Director of the Center for Public History at the University of Houston. He is the author or editor of 19 books including his most recent, *Precious Commodity: Providing Water for America's Cities* (2011). He has also completed *Atomic Age America and the World* (2012) and has started work on *An Island Not Too Far: Fresh Kills and Staten Island*.

Alan Mikhail is Assistant Professor of History at Yale University. He is the author of *Nature and Empire in Ottoman Egypt: An Environmental History* (2011). He is currently writing a book about the changing relationships between humans and animals in Ottoman Egypt and also editing a collection of essays on Middle East environmental history.

Shawn W. Miller is Associate Professor of History at Brigham Young University. He is the author of *Fruitless Trees: Portuguese Conservation and Brazil's Colonial Timber* (2000) and *An Environmental History of Latin America* (2007). He is currently researching the environmental history of the street and the automobile in Rio de Janeiro.

David Moon is Anniversary Professor in the Department of History at York University, UK. He researches in Russian and transnational environmental history. He has published several articles, and is completing a monograph on the environmental history of the steppes. His earlier work focused on the Russian peasantry, but he also investigates connections between the Russian steppes and the North American Great Plains. He is an active member of the European Society for Environmental History.

Micah S. Muscolino is Associate Professor of History at Georgetown University. He is the author of *Fishing Wars and Environmental Change in Late Imperial and Modern China* (2009). He is currently writing a book on the environmental history of World War II in North China's Henan province.

José Augusto Pádua is Professor of Brazilian Environmental History at the Institute of History/Federal University of Rio de Janeiro, where he also coordinates the Laboratory of History and Ecology. Since 2010, he has been president of the Brazilian Association of Research and Graduate Studies on Environment and Society (ANPPAS). His most recent book, in association with J. R. McNeill and Mahesh Rangarajan, is *Environmental History: As If Nature Existed* (2010).

Liza Piper is Associate Professor of History at the University of Alberta. Her book, *The Industrial Transformation of Subarctic Canada* (2009), examines the role of industrial resource exploitation and science in the twentieth-century transformation of subarctic environments. Her current research considers how disease and climate have changed human relations to nature in the Subarctic and Arctic since the nineteenth century.

Joachim Radkau was Professor of Modern History at Bielefeld University, Germany, until his retirement in 2008. His most important publications on environmental history in English include *Nature and Power: A Global History of the Environment* (2008) and

Wood: A History (2011). In 2011 he published *Die Ära der Ökologie: Eine Weltgeschichte*, which is to be translated into English.

Libby Robin is an environmental historian at the Australian National University and at the National Museum of Australia, Canberra. She is Guest Professor at the Royal Institute of Technology, Stockholm. Her books include *How a Continent Created a Nation* (2007) and *Ecology and Empire: Environmental History of Settler Societies* (coedited with Tom Griffiths). She is currently working (with Sverker Sörlin and Paul Warde) on *Environmental Futures*, an anthology of the literature of global change.

Alan Roe is a PhD candidate in Russian environmental history at Georgetown University. He is currently researching his dissertation on outdoor recreation in the Soviet Union. He is coeditor of the *Routledge Reader for World Environmental History* (forthcoming).

Edmund Russell is a Professor in the Department of Science, Technology, and Society and the Department of History at the University of Virginia. He is the author of *Evolutionary History: Uniting History and Biology to Understand Life on Earth* (2011) and *War and Nature: Fighting Humans and Insects with Chemicals from World War I to Silent Spring* (2001), and coeditor (with Richard P. Tucker) of *Natural Enemy, Natural Ally: Toward an Environmental History of War* (2004).

Richard P. Tucker is Adjunct Professor of Environmental History at the University of Michigan. He is the author of *Insatiable Appetite: The United States and the Ecological Degradation of the Tropical World* (2000), and coeditor with Edmund Russell of *Natural Enemy, Natural Ally: Toward an Environmental History of War* (2004). He is currently writing a book on the military and the environment in the contemporary world.

Sam White is Assistant Professor at Oberlin College, where he teaches courses on global and environmental history. He is the author of *The Climate of Rebellion in the Early Modern Ottoman Empire* (2011). He is currently researching the impact of climate on the first European colonies in North America.

Acknowledgments

This book represents a team effort. As editors, we wish to register our appreciation to the platoon from Wiley-Blackwell. Tessa Harvey initiated the project and gracefully agreed to every modification of plans along the way. Gillian Kane and Isobel Bainton crisply and cheerfully helped us navigate the inevitable bumps in the road. Leah Morin, copy-editor extraordinaire, deserves our gratitude in full measure.

When editors are rash enough to agree to assemble a book that depends on original contributions from a roster of 28 authors, the probability that something will go badly wrong is considerable. Our authors defied the odds, writing their chapters promptly, answering our queries swiftly, and tolerating tweaks to their pearly prose with equanimity worthy of Marcus Aurelius or a Zen master. We thank them for that.

We also thank Peter Engelke and Yubin Shen, who translated chapters into English from German and Chinese respectively.

Finally, we wish to express our appreciation to the students and faculty of the Georgetown University History Department. Our book is a little better, and our lives a lot better, for the collegiality and support this community of scholars has offered over the years.

J. R. McNeill
Erin Stewart Mauldin

Global Environmental History: An Introduction

J. R. McNeill and Erin Stewart Mauldin

Since the 1970s, environmental history has evolved into a self-conscious and self-aware scholarly field that boasts journals, university programs, and international organizations devoted to its practice and promotion. Global environmental history, however, is much younger. Although a dynamic field with a steadily increasing number of practitioners, global environmental history remains, as yet, unclear in its structure, shape, and place within the historical profession.

This volume aims to orient readers to the fast-growing arena of scholarly inquiry known as global, or world, environmental history. It is a collection of new essays by 28 scholars from all six inhabited continents, many of whom have been instrumental in the establishment of environmental history. It surveys past developments in the field, current contours of scholarship, and possible approaches for the future. The *Companion to Global Environmental History* is intended to be useful not only to people who are coming to environmental history for the first time – serving as the equivalent of a road map to the field – but also to people who have long labored in one province of environmental history, and, for whatever reason, seek to broaden their horizons and begin to develop comparative perspectives – or deepen their existing ones.

What Is Environmental History?

Like every other subset of history, environmental history represents different things to different people. Our preferred definition of the field is the study of the relationship between human societies and the rest of nature on which they depended. Humankind has always been a part of nature, albeit a distinctive part. While the natural world has shaped and conditioned the human experience, over time, humans have made increasingly far-reaching alterations to their surroundings. Environmental history recognizes that the natural world is not merely the backdrop to human events, but evolves in its own right, both of its own accord and in response to human actions. Nature is now both natural and cultural, at least in most places on Earth. Indeed human influence upon nature has attained such proportions that some scholars maintain life on Earth has entered a new geological era, the Anthropocene. This term, while gaining acceptance, is by no means conventional yet. But its increasing use signifies growing awareness in scientific circles of the burgeoning human environmental impact.[1]

The vast scope of environmental history invites many and varied approaches. There are, we think, three chief areas of inquiry, which of course overlap and have no firm boundaries. First is the study of material environmental history, the stories of human involvement with forests and frogs, with cholera and chlorofluorocarbons. This entails the examination of human impact on the physical environment as well as nature's influence upon human affairs, each of which is always in flux and always affecting the other. This form of environmental history puts human history in a fuller context, that of the Earth and life on Earth, and recognizes that human events are part of a larger story in which humans are not the only actors. A full extension of this principle is the so-called "Big History" of David Christian and Fred Spier,[2] which places humans into the unfolding history of the universe, and finds recurrent patterns over the largest timescales. In practice, however, most of the environmental history written in the material vein stresses the economic and technological side of human actions, and thus concentrates on the last 200 years when industrialization (among other forces) greatly enhanced humankind's power to alter environments.

Second is a form of cultural and intellectual history. It concerns what humans have thought, believed, and written that treats relationships between society and nature. It emphasizes representations and images of nature in art, literature, religion, and oral traditions, how these have changed, and what they reveal about the societies that produced them.[3] The great majority of cultural environmental history is drawn from published texts, as with intellectual history, and often treats the works of influential (and sometimes not-so-influential) authors from Lucretius and Mencius to St. Francis to Mohandas K. Gandhi. This sort of environmental history tends to focus on individual thinkers, but it can also extend to the study of popular environmentalism as a cultural movement. The largest debate within this wing of environmental history, however, is the relative impact of various religio-cultural traditions on the natural world. This scholarship evaluates the texts and practices of Judeo-Christian, Islamic, East Asian, and indigenous traditions, attempting to determine their effects on the environment.[4]

The third main form is political and policy-related environmental history. This concerns the history of deliberate human efforts to regulate the relationship between society and nature, and between social groups in matters concerning nature. Although there are early examples of soil conservation, air-pollution control, and royal efforts to protect charismatic animals for a monarch's hunting pleasure, usually policy-related environmental history extends back only to the late nineteenth century. Only in the era since 1880 have states and societies mounted systematic efforts to regulate interactions with the environment generally. Between 1880 and 1965 these efforts were normally spasmodic and often modest in their impacts, so much of this sort of environmental history deals with the decades since 1965, when both states and explicitly environmental organizations grew more determined and effective in their interventions. Political environmental history is the approach that most easily dovetails with mainstream history for it uses the nation-state as its unit of analysis. Other types of environmental history tend to ignore political boundaries.

In practice, environmental history is all this and much more. More than most varieties of history, environmental history is an interdisciplinary project. Many scholars in the field trained as archeologists, geographers, or historical ecologists. In addition to the customary published and archival texts of the standard historian, environmental historians routinely use the findings culled from bio-archives (such as pollen deposits which can tell us about former vegetation patterns) and geo-archives (such as soil profiles that can tell us about past land-use practices). The subject matter of environmental history is often much the same as that in historical geography or historical ecology, although the choice of sources emphasized normally differs. An illustration is the field of climate history, which is pursued by scholars from at least half a dozen disciplines, including text-based historians. Textual historians have found useful records for climate history going back many centuries, for example, the dates of grape harvests in European vineyards. Compiling and comparing these dates over centuries allows historians to draw strong inferences about warming and cooling trends.[5]

Global Environmental History

Global environmental history has a compelling logic but presents a daunting aspect. Many ecological processes are global in scope, such as climate change or sea-level rise, and many others are found here and there around the world, such as deforestation and urban air pollution. Several of the cultural trends concerning the environment have been nearly global too, most obviously the modern, post-1960s, expression of ecological anxiety, although of course it finds different forms in different cultures.[6] But global-scale environmental history, like global and world history in general, is built upon the foundation of local work and regional surveys. No single historian can master the details of soil history or the history of water pollution around the world, just as no one can fully master the global history of wages and prices, or of women's movements. All global and world history presents this problem, and for many historians this alone suffices to make the venture illegitimate.

A moment's reflection, however, should redeem the ambition of global-scale history. Something is gained and something lost with any choice of scale. If historians required true mastery of their subjects, they could aim no more broadly than autobiography. There is no purely intellectual reason to prefer microhistory to macrohistory, whether environmental or otherwise. But it remains true that, practically speaking, bringing coherence to the subject of the global history of air pollution is much more difficult than, say, to the history of the killer fog of London in December 1952. Global environmental history, then, is often a process of stitching together scholarship from multiple geographic scales and perspectives to craft a narrative or an analysis of global ecological change.

For decades the only global environmental history syntheses came from authors who were not professional historians, and therefore less inhibited by their training and the anti-global expectations of the historical profession. British geographers and a former civil servant of the United Kingdom wrote the first notable general surveys, the former in sober style and the latter with the panache of a muckraking journalist.[7] Sociologists too joined the fray.[8] Eventually natural scientists took aim at global historical treatments of subjects such as nitrogen and soil.[9] A multidisciplinary magnum opus from 1990, B. L. Turner et al.'s *The Earth as Transformed by Human Action*, helped spur historians to try their hand at global environmental history.[10]

Professional historians began by taking slices of the whole, such as the books on global fire history by Stephen Pyne, or environmentalism by Ramachandra Guha.[11] Pyne's work, which grew out of his earlier studies of fire in American history, sought to discuss every aspect of the human relationship with fire, from cooking and the physiology of digestion to the cultural perceptions of wildfires. Guha's short treatise on modern environmentalism showed the contrasts between the social movements that go by that name in, above all, India and the US. Joachim Radkau was perhaps the first to bring the sensibilities of the historian to general global-scale environmental history in his *Natur und Macht: Eine Weltgeschichte der Umwelt*.[12] His was not a survey aiming at worldwide coverage, but a sprawling series of soundings and reflections on everything from animal domestication to contemporary tourism in the Himalaya. It reads a bit like Arnold Toynbee's *A Study of History* with its bold comparisons and juxtapositions across time and space. Unlike Toynbee, Radkau was reluctant to offer grand pronouncements, preferring to honor historians' traditional respect for the particularities of different times and places.[13]

A small platoon of professional historians brought out global-scale environmental histories of one sort or another around the same time as Radkau. Brief surveys, apparently intended for classroom use, poured forth from Europe and the US.[14] A pair of longer studies took on slices of time that, their authors claimed, exhibited some coherence: John F. Richards surveyed the early modern centuries so strongly affected by European expansion, and J. R. McNeill portrayed the twentieth century as an era of unprecedentedly tumultuous environmental change.[15] Still others presented thematic slices of global environmental history, penning accounts of deforestation or malaria over several millennia.[16] Wide-ranging anthologies added to the sudden outpouring – and sidestepped the main limitation of global history, the inability of any single author to know enough.[17] The British Empire, on which the sun famously never set, provided a framework that added coherence to global environmental history as shown in the overview by William Beinart and Lotte Hughes.[18] To date no one has chosen to follow their example with respect to any other modern empires. However, imperialism more generally served as the occasion for one of environmental history's foundational texts, Alfred W. Crosby's *Ecological Imperialism*, which, if it isn't global environmental history, surely comes very close to it. Crosby sought to explain the successes and failures of European imperial ventures from the Crusades and Greenland Norse onward in environmental terms.[19]

So global environmental history has come a long way in a brief time. The persistent presence of environmental issues in modern life has made environmental history a permanent fixture of historiography rather than a passing fancy. The growing salience of climate concerns, deforestation, water shortages, and loss of biodiversity has convinced some historians, previously working far from environmental history, that it is no longer appropriate to write history without taking the environment, especially climate change, into account.[20] Furthermore, global-scale environmental history has benefited from the rise of world or global history, an intellectual response to the recent surge of globalization and, in the US at least, a practical response to political pressures upon school curricula.[21] But, as always, further opportunities abound. Some day someone will write a global environmental history of railroads, of mining, of war, of cattle, of the oceans, of religion, of odors, of things as yet unimagined.

The *Companion*

Although there are countless ways in which one could organize the endeavor of global environmental history, this volume combines temporal, geographic, and thematic approaches. With contributions from an international roster of historians, the content of the chapters that follow is as diverse as the approaches to environmental history. Some authors emphasize natural and cultural history, while others focus on political and economic developments. Some chapters are surveys, others are historiographical, and many are a mix of the two.

Each author was given the freedom to write his or her chapter as the subject required. Consequently, there are occasional overlaps in the subjects under discussion. For instance, the impact of the first human migrations into the Americas appears in four chapters, although the authors approach the subject with differing purposes and with sometimes contrasting conclusions. Readers will also notice that some subjects recur with regularity throughout the *Companion*, such as agriculture, industrialization, and biological exchanges. That is as it should be: these are central themes for environmental history. In this volume you will find regions, themes, and time periods not as yet well represented in the historiography, new evidence for old debates, and inventive new ways of approaching the practice of environmental history.

The *Companion* is split into four parts. Part I, entitled "Times," shows how the issues and trajectories of the relationship between society and nature have evolved over time, and how they differ from one period to the next. Authors cover major milestones in human history, helping readers develop a sense of the deep past so often neglected in environmental, and indeed in all, history. Chapters in this section discuss the latest findings in the study of human origins, the methods by which environmental historians and other scholars understand ancient landscapes, and how environmental factors contributed to the rise and fall of human societies over time. The scope of this section is vast, and the authors' work demonstrates how the coevolution of humans and nature over a very *longue durée* can illuminate not only current environmental issues, but also political and economic ones.

The next section, Part II, is entitled "Places." It is a series of regional or national narratives and historiographies that show how the pieces of the global puzzle fit together. Place, although it can be defined and construed in many ways and at many scales, is usually a central concept for environmental history. In practice, most environmental history is written about specific places, some as small as a few farms, others as large as a continent. We felt that it is important to include the regional and local, for they are the foundation of the global. Many areas around the world have experienced similar historical processes that drive ecological change – biological invasions, colonialism, industrialization, conservation movements – and this section allows readers to see how geographical variations in climate, terrain, and availability of natural resources, as well as cultural patterns, political frameworks, and economic structures, have influenced the map of environmental change. Not every country or region is represented; readers will note, for example, the absence of Europe and India. In chapters that cover areas of the globe which have rich historiographies, such as the US, authors have tried to point to new issues for debate or study. Other chapters, however, provide the first surveys of areas such as the Arctic and the Middle East, regions of the globe which have yet to receive their due from environmental historians.

The third part of the *Companion* moves away from chronological and geographical organization. Here, each author examines one thematic issue across the globe and across time. There are chapters which outline the human relationship to natural elements, such as forests, rivers, and oceans, as well as chapters on how the evolution of technology, warfare, and industrial processes altered the world's environment. Authors focus partly

on the biogeophysical changes themselves, but also upon the social, economic, and political forces behind them. Some chapters present familiar themes, such as fishing or agriculture, but use an expanded temporal or geographical scope to present readers with
new, global perspectives. Other chapters, such as those on grasslands and evolution,
challenge readers with unfamiliar comparisons and unfettered imagination.

The final section of the *Companion*, Part IV, surveys different types of environmental
thought and action around the world, giving readers a sense of the variety of cultural,
intellectual, and political engagements with the environment in modern times. While the
first two parts have significant chronological depth, Parts III and IV exhibit a strong bias
toward the modern period (since 1500 CE). This is partly a reflection of the current literature in the field, and partly due to the practicalities of scholarship in global environmental history. This section contains in-depth chapters on two of the largest and
fastest-growing economies in the world – Brazil and China – while the remaining two
chapters highlight broader themes within environmentalism and environmental thought.

Global environmental history is a fast-moving field with porous boundaries and a
wide range of interdisciplinary connections. The chapters in this volume are by no means
comprehensive, and do not provide complete coverage of all themes and all places, but
the chapters help provide an understanding of how people actually work in environmental history and reflect the major approaches within the field's scholarship. We hope that
all readers of the *Companion* will find something illuminating and entrancing in global
environmental history as it is currently practiced. Consciously or unconsciously, scholars
provide scholarship for the times in which they live. Whether or not we ought to understand our own times as the Anthropocene, we live in an age of conspicuous environmental change, of pinching environmental constraints upon people's lives, and of fast-track
globalization. This *Companion* aims to offer a guide to environmental history scholarship written in, and for, our ecologically dynamic and globalizing times.

Notes

1 See for example the special issue of the *Philosophical Transactions of the Royal Society
   A: Mathematical, Physical, and Engineering Sciences* 369, 2011, which is devoted to the concept of the Anthropocene.
2 D. Christian, *Maps of Time: An Introduction to Big History*, Berkeley, University of California
   Press, 2004; F. Spier, *The Structure of Big History: From the Big Bang until Today*, Amsterdam,
   University of Amsterdam Press, 1996; F. Spier, *Big History and the Future of Humanity*,
   Oxford, Wiley-Blackwell, 2011.
3 Remarkably the most comprehensive work in this vein as regards the Western world was writ-
   ten over 40 years ago: C. Glacken's *Traces on the Rhodian Shore: Nature and Culture in Western
   Thought from Ancient Times to the End of the Eighteenth Century*, Berkeley, University of
   California Press, 1967. Glacken's massive work explored the conceptions of nature among
   several dozen prominent writers from ancient times through the European Enlightenment.
   Other examples include D. Worster's *Nature's Economy: A History of Ecological Ideas*, New
   York, Cambridge University Press, 1985, and P. Coates, *Nature: Western Attitudes since the
   Ancient Times*, Cambridge, Polity, 1998.
4 See, for example, L. White, "The Historical Roots of Our Ecologic Crisis," *Science* 155, 1967,
   pp. 1203–7; Y. Tuan, "Discrepancies between Environmental Attitude and Behaviour:
   Examples from Europe and China," *Canadian Geographer* 3, 1968, pp. 175–91; H. Amery,
   "Islam and the Environment," in N. I. Faruqui, A. K. Biswas, and M. J. Bino (eds.), *Water
   Management in Islam*, Tokyo, United Nations University Press, 2001, pp. 39–60.

5   E. Le Roy Ladurie, *Histoire du climat depuis l'an mil*, Paris, Flammarion, 1967. Le Roy Ladurie
    revised his positions substantially in *Histoire humaine et comparée du climat*, 3 vols., Paris,
    Fayard, 2004–9. A sample of more recent climate history: G. Endfield, *Climate and Society in
    Colonial Mexico: A Study in Vulnerability*, Oxford and New York, Wiley-Blackwell, 2008;
    F. Mauelshagen, *Klimageschichte der Neuzeit 1500–1900*, Darmstadt, Wissenschaftliche
    Buchgesellschaft, 2009; S. White, *Climate of Rebellion*, New York, Cambridge University Press,
    2011; W. Behringer, *A Cultural History of Climate*, Malden, MA, Polity, 2010; S. Johnson,
    Climate and Catastrophe in Cuba and the Atlantic World in the Age of Revolution, Chapel Hill,
    UNC Press, 2011; H. H. Lamb, *A History of Climate Changes*, 4 vols., London, Routledge,
    2011 (these volumes contain reprints of Lamb's pioneering work from 1966 to 1988); a
    fascinating essay that attributes great significance to climate change is R. Bulliet, *Cotton,
    Climate, and Camels in Early Islamic Iran*, New York, Columbia University Press, 2011.

6   The new standard here is J. Radkau, *Die Ära der Ökologie: Eine Weltgeschichte*, Munich, Beck,
    2011.

7   I. G. Simmons, *Changing the Face of the Earth: Culture, Environment, History*, New York,
    Blackwell, 1989; A. Mannion, *Global Environmental Change: A Natural and Cultural
    Environmental History*, Harlow, Longman, 1991; C. Ponting, *A Green History of the World*,
    Harmondsworth, Penguin, 1991. For Simmons' latest entry: *Global Environmental History*,
    Chicago, University of Chicago Press, 2008.

8   B. de Vries and J. Goudsblom, *Mappae Mundi: Humans and Their Habitats in Long-Term
    Socio-Ecological Perspective*, Amsterdam, Amsterdam University Press, 2002.

9   D. Vasey, *An Ecological History of Agriculture: 10,000 B.C. to 10,000 A.D.*, Lafayette, IN,
    Purdue University Press, 1992; V. Smil, *Energy in World History*, Boulder, CO, Westview,
    1994; G. Leigh, *The World's Greatest Fix: A History of Nitrogen and Agriculture*, Oxford,
    Oxford University Press, 2004; J. R. McNeill and V. Winiwarter (eds.), *Soils and Societies:
    Perspectives from Environmental History*, Isle of Harris, White Horse Press, 2006; A. Mannion,
    Carbon and Its Domestication, Dordrecht, Springer, 2006; D. Montgomery, *Dirt: The Erosion
    of Civilizations*, Berkeley, University of California Press, 2007.

10 B. L. Turner, W. C. Clark, R. W. Kates, et al. (eds.), *The Earth as Transformed by Human
    Action: Global and Regional Changes in the Biosphere over the Past 300 Years*, New York,
    Cambridge University Press, 1990.

11 S. J. Pyne, *World Fire: The Culture of Fire on Earth*, Seattle, University of Washington Press,
    1995; R. Guha, *Environmentalism*, New York, Longman, 2000.

12 J. Radkau, *Natur und Macht: Eine Weltgeschichte der Umwelt*, Munich, Beck, 2000; the
    English translation, by T. Dunlap, is *Nature and Power: A Global History of the Environment*,
    New York, Cambridge University Press, 2008.

13 Radkau in his youth read Toynbee with admiration. See his essay on his own work, "Nature
    and Power: An Ambiguous and Intimate Connection," *Social Science History*, forthcoming.

14 S. Sörlin and A. Öckerman, *Jorden en ö*, Stockholm, Natur och Kultur, 1998; J. D. Hughes,
    *An Environmental History of the World: Humankind's Changing Role in the Community of
    Life*, London, Routledge, 2001; J. Radkau, *Mensch und Natur in der Geschichte*, Leipzig,
    Klett, 2002; S. Mosley, *The Environment in World History*, London, Routledge, 2006;
    V. Winiwarter and M. Knoll, *Umweltgeschichte*, Cologne, Böhlau, 2007; A. Penna, *The
    Human Footprint: A Global Environmental History*, Malden, MA, Blackwell, 2009.

15 J. F. Richards, *The Unending Frontier: An Environmental History of the Early Modern World*,
    Berkeley, University of California Press, 2003; J. R. McNeill, *Something New under the Sun:
    An Environmental History of the Twentieth-Century World*, New York, Norton, 2000.

16 M. Williams, *Deforesting the Earth: From Prehistory to Global Crisis*, Chicago, University of
    Chicago Press, 2003; J. L. A. Webb, *Humanity's Burden: A Global History of Malaria*, New
    York, Cambridge University Press, 2009.

17 Three from a range of examples: A. Hornborg, J. R. McNeill, and J. Martinez-Alier (eds.),
    Rethinking Environmental History: World-System History and Global Environmental Change,

Lanham, MD, Altamira Press, 2007; E. Burke III and K. Pomeranz (eds.), *The Environment and World History*, Berkeley, University of California Press, 2009; T. Myllyntaus (ed.), *Thinking through the Environment: Green Approaches to Global History*, Cambridge, White Horse Press, 2011.

18 W. Beinart and L. Hughes, *Environment and Empire*, New York, Oxford University Press, 2007.

19 A. W. Crosby, *Ecological Imperialism: The Biological Expansion of Europe, 900–1900*, New York, Cambridge University Press, 1986.

20 D. Chakrabarty, "The Climate of History: Four Theses," *Critical Inquiry* 35, 2009, pp. 197–222.

21 In the US, where the ethnic origins of populations plays a role in the formation of political blocs, and where political blocs often interest themselves in school curricula, world history is the easiest compromise among the possible ways of presenting history to young people because – in theory at least – it leaves no one's ancestors out.

References

Amery, H., "Islam and the Environment," in N. I. Faruqui, A. K. Biswas, and M. J. Bino (eds.), *Water Management in Islam*, Tokyo, United Nations University Press, 2001, pp. 39–60.

Behringer, W., *A Cultural History of Climate*, Malden, MA, Polity, 2010.

Beinart, W., and Hughes, L., *Environment and Empire*, New York, Oxford University Press, 2007.

Bulliet, R., *Cotton, Climate, and Camels in Early Islamic Iran*, New York, Columbia University Press, 2011.

Burke, E., III, and Pomeranz, K. (eds.), *The Environment and World History*, Berkeley, University of California Press, 2009.

Chakrabarty, D., "The Climate of History: Four Theses," *Critical Inquiry* 35, 2009, pp. 197–222.

Christian, D., *Maps of Time: An Introduction to Big History*, Berkeley, University of California Press, 2004.

Coates, P., *Nature: Western Attitudes since the Ancient Times*, Cambridge, Polity, 1998.

Crosby, A. W., *Ecological Imperialism: The Biological Expansion of Europe, 900–1900*, New York, Cambridge University Press, 1986.

Endfield, G., *Climate and Society in Colonial Mexico: A Study in Vulnerability*, Oxford and New York, Wiley-Blackwell, 2008.

Glacken, C., *Traces on the Rhodian Shore: Nature and Culture in Western Thought from Ancient Times to the End of the Eighteenth Century*, Berkeley, University of California Press, 1967.

Guha, R., *Environmentalism*, New York, Longman, 2000.

Hornborg, A., McNeill, J. R. (eds.), and Martinez-Alier, Joan, *Rethinking Environmental History: World-System History and Global Environmental Change*, Lanham, MD, Altamira Press, 2007.

Hughes, J. D., *An Environmental History of the World: Humankind's Changing Role in the Community of Life*, London, Routledge, 2001.

Lamb, H. H., *A History of Climate Changes*, 4 vols., London, Routledge, 2011.

Le Roy Ladurie, E., *Histoire du climat depuis l'an mil*, Paris, Flammarion, 1967.

Le Roy Ladurie, E., *Histoire humaine et comparée du climat*, 3 vols., Paris, Fayard, 2004–9.

Leigh, G., *The World's Greatest Fix: A History of Nitrogen and Agriculture*, Oxford, Oxford University Press, 2004.

Mannion, A., *Carbon and Its Domestication*, Dordrecht, Springer, 2006.

Mannion, A., *Global Environmental Change: A Natural and Cultural Environmental History*, Harlow, Longman, 1991.

Mauelshagen, F., *Klimageschichte der Neuzeit 1500–1900*, Darmstadt, Wissenschaftliche Buchgesellschaft, 2009.

McNeill, J. R., *Something New under the Sun: An Environmental History of the Twentieth-Century World*, New York, Norton, 2000.

McNeill, J. R., and Winiwarter, V. (eds.), *Soils and Societies: Perspectives from Environmental History*, Isle of Harris, White Horse Press, 2006.

Montgomery, D., *Dirt: The Erosion of Civilizations*, Berkeley, University of California Press, 2007.

Mosley, S., *The Environment in World History*, London, Routledge, 2006.

Myllyntaus, T. (ed.), *Thinking through the Environment: Green Approaches to Global History*, Cambridge, White Horse Press, 2011.

Penna, A., *The Human Footprint: A Global Environmental History*, Malden, MA, Blackwell, 2009.

Philosophical Transactions of the Royal Society. A. Mathematical, Physical, and Engineering Sciences 369, 2011 (special issue).

Ponting, C., *A Green History of the World*, Harmondsworth, Penguin, 1991.

Pyne, S. J., *World Fire: The Culture of Fire on Earth*, Seattle, University of Washington Press, 1995.

Radkau, J., *Die Ära der Ökologie: Eine Weltgeschichte*, Munich, Beck, 2011.

Radkau, J., *Mensch und Natur in der Geschichte*, Leipzig, Klett, 2002.

Radkau, J., *Natur und Macht: Eine Weltgeschichte der Umwelt*, Munich, Beck, 2000. Translated by T. Dunlap as *Nature and Power: A Global History of the Environment*, New York, Cambridge University Press, 2008.

Radkau, J., "Nature and Power: An Ambiguous and Intimate Connection," *Social Science History*, forthcoming.

Richards, J. F., *The Unending Frontier: An Environmental History of the Early Modern World*, Berkeley, University of California Press, 2003.

Sherry, J., *Climate and Catastrophe in Cuba and the Atlantic World in the Age of Revolution*, Chapel Hill, UNC Press, 2011.

Simmons, I. G., *Changing the Face of the Earth: Culture, Environment, History*, New York, Blackwell, 1989.

Simmons, I. G., *Global Environmental History*, Chicago, University of Chicago Press, 2008.

Smil, V., *Energy in World History*, Boulder, CO, Westview, 1994.

Sörlin, S., and Öckerman, A., *Jorden en ö*, Stockholm, Natur och Kultur, 1998.

Spier, F., *Big History and the Future of Humanity*, Oxford, Wiley-Blackwell, 2011.

Spier, F., *The Structure of Big History: From the Big Bang until Today*, Amsterdam, University of Amsterdam Press, 1996.

Tuan, Y., "Discrepancies between Environmental Attitude and Behaviour: Examples from Europe and China," *Canadian Geographer* 3, 1968, pp. 175–91.

Turner, B. L., Clark, W. C., Kates, R. W., et al. (eds.), *The Earth as Transformed by Human Action: Global and Regional Changes in the Biosphere over the Past 300 Years*, New York, Cambridge University Press, 1990.

Vasey, D., *An Ecological History of Agriculture: 10,000 B.C. to 10,000 A.D.*, Lafayette, IN, Purdue University Press, 1992.

Vries, B. de, and Goudsblom, J., *Mappae Mundi: Humans and Their Habitats in Long-Term Socio-Ecological Perspective*, Amsterdam, Amsterdam University Press, 2002.

Webb, J. L. A., *Humanity's Burden: A Global History of Malaria*, New York, Cambridge University Press, 2009.

White, L., "The Historical Roots of Our Ecologic Crisis," *Science* 155, 1967, pp. 1203–7.

White, S., *Climate of Rebellion*, New York, Cambridge University Press, 2011.

Williams, M., *Deforesting the Earth: From Prehistory to Global Crisis*, Chicago, University of Chicago Press, 2003.

Winiwarter, V., and Knoll, M., *Umweltgeschichte*, Cologne, Böhlau, 2007.

Worster, D., *Nature's Economy: A History of Ecological Ideas*, New York, Cambridge University Press, 1985.

PART I

Times

Part I

Times

CHAPTER ONE

Global Environmental History:
The First 150,000 Years

J. R. MCNEILL

If, as scholars of human evolution suppose, our species emerged about 150,000 years ago, then roughly 97% of human history took place before the first cities and civilization. This chapter will briefly explore global environmental history over that very *longue durée*. It will sketch some of the ways in which the changing earthly environment affected human affairs, including almost ending them entirely about 73,000 years ago, and will outline some of the ways in which human actions changed the environment. By and large, in the 140 millennia before farming, environmental change affected human affairs more than human affairs affected the environment. But with the transition to agriculture beginning about 10,000 years ago, that began to change fundamentally: our numbers and technologies attained new levels so that, when combined with our long-standing heedlessness, we became an increasingly important force in shaping the global environment.

The Environment Shapes Paleolithic Humans and Human Affairs

About 7 million years ago our ancestors diverged, genetically speaking, from other apes. After another couple of million years, later ancestors began to walk upright (bipedalism) and develop big brains all out of proportion to their bodies. Climate change, according to prevailing interpretations, likely played a role in these fateful departures. In East Africa, where it all happened, drier conditions some 6 to 8 million years ago reduced the domain of forest and encouraged the spread of grassy savanna. This new environment rewarded upright posture and bipedalism, which allowed hominins (now the preferred term for humans plus their ancestors) to see longer distances and to move faster in open terrain. Standing upright also made it easier to dissipate heat under the tropical sun, an important task if one is obliged to keep moving to stay away from predators. East African climate also apparently became more unstable, with

A Companion to Global Environmental History, First Edition. Edited by J.R. McNeill and Erin Stewart Mauldin.
© 2012 John Wiley & Sons, Ltd. Published 2015 by John Wiley & Sons, Ltd.

rapidly alternating wet and dry phases. This instability, the thinking goes, rewarded flexible behavior and thereby big brains. So, if this line of reasoning is correct, climate change helped shape the human animal in basic ways.[1]

Climate change continued to influence human affairs in subsequent millennia. Beginning about 3 million years ago, the Earth entered a period – in which we still live – of alternating glacial and interglacial phases. In our African homeland, this oscillating climate rhythm appeared as wetter and drier phases, because it was never so cold as to encourage glaciation (outside of the highest mountains). When hominins left Africa, which some did more than a million years ago, they had to adjust to ice ages that in Eurasia involved much colder temperatures, as well as a drier, windier, and more unstable climate.

Migration

Our own species, *Homo sapiens sapiens*, evolved within Africa and by perhaps 150,000 years ago had emerged as a distinct species. A few intrepid populations walked out of Africa, perhaps 100,000 years ago. As they crossed to Arabia and Southwest Asia, they too encountered colder climate. Their migrations coincided with the early millennia of a new cold phase, an ice age that lasted from about 110,000 to 12,000 years ago. This latest ice age was not only much colder and dryer than modern climate, but in most parts of the world far more unstable. For decades or centuries comparatively sudden cooling or warming might occur, in swings of average temperatures of 5 to 10 degrees Celsius (9 to 18 degrees Fahrenheit). The slender evidence suggests these swings were smaller in Africa than on other continents. Elsewhere, the incentives to migrate, either to avoid the worst of the cold and drought or to take advantage of warming and moisture, were often strong. Staying put for centuries was usually a poor gamble because climate was too unstable.

The best aspect of ice-age conditions (for humans) was lower sea levels. This gave terrestrial species about 25 million more square kilometers to work with – the equivalent of an additional continent the size of North America. It was possible to walk across most of Indonesia, and from Australia to New Guinea, from Korea to Japan, and from Britain to France. The unfortunate part of this for historians and archeologists is that probably most people lived most of their lives in these zones, helping themselves to seafood found along the ancient shores, and all archeological remains of their existence vanished beneath the waves when sea levels rose sharply around 22,000 to 8,000 years ago.

The most challenging moment of the last ice age came around 74,000 to 70,000 years ago when a giant volcanic eruption (of Mt. Toba, on the island of Sumatra in what is now Indonesia) spewed enough dust and ash into the skies to block sunlight and lower temperatures by 5 to 15 degrees Celsius for 6 to 10 years. It may have tipped climate into another regime; the next thousand years were especially cold on average. Toba was the biggest volcanic eruption in the last 2 million years, 280 times the size of Krakatoa (1883) and about 5,000 times larger than Mt. St. Helens (1980), as measured by the quantity of tephra – rock, magma, and other material – thrown heavenward. Ash fell from the sky as far away as Arabia and the east coast of Africa. In some places in India the resulting tephra layer was, and is, six meters thick!

The Toba catastrophe played havoc with plant and animal life. The fossil pollen record shows collapses of vegetation in many parts of Asia, leaving animal species with little to munch on. DNA evidence suggests that several animals, including tigers and orangutans, suffered dramatic reductions in populations at about this time. Toba's impacts probably brought the human species close to extinction: it is possible to interpret the DNA evidence to mean that at around this time our ancestors' numbers were reduced to 10,000 or so – our closest brush with extinction so far. Quite plausibly all humans outside favorable locations in Africa were wiped out by Toba's effects. This is also the time, incidentally, when (inferred from DNA evidence in lice) humans began to wear clothing.[2]

The Toba event was of unique intensity in human experience, but the ice age contained numerous cold spells and severe droughts. Over the past 150,000 years, modern humans evolved in a time of generally cool and highly erratic global climate, and they colonized Eurasia during the colder phases of the last ice age, a circumstance that surely rewarded innovation, learning, and communication in any species. Adverse climate may well have contributed to cultural dexterity.

From their African refugia, post-Toba human populations soon migrated once more into Eurasia. Once again, lower sea levels, thanks to the buildup of ice, helped. Humans reached Australia and New Guinea, at the time united as a single continent, perhaps as early as 60,000 years ago and no later than 40,000 BP (= before the present). Getting there required a sea journey of at least 100 kilometers (60 miles). This voyage implies a considerable technological and logistical competence, as well as high tolerance for risk: the first Australians were surely a plucky lot. Other modern humans headed north into what is now China and Japan by about 30,000 BP. Distant cousins entered Europe around 40,000 years ago, to the misfortune of the indigenous Neanderthals. These new Europeans, according to genetic evidence, are the ancestors of 75–85% of contemporary Europeans. They soon encountered the depths of the last ice age and – not unlike their more affluent descendants today – headed for Spain and the south of France in search of balmier climes. Meanwhile other humans walked into the chilly expanses of Siberia, attracted by the abundance of large, tasty, naive mammals. At this time, some 40,000 to 30,000 years ago, the global population was probably only a few hundred thousand, roughly as many people as live today in Des Moines, Lubbock, or Boise (or Nottingham, Coventry, Canberra, or Christchurch). It was an uncrowded world.

Nevertheless, some people moved further afield. The last chapter in these epic migrations brought people to the Americas, possibly as early as 20,000 years ago, certainly by 13,000 BP. They crossed from Siberia to Alaska, at that time a broad land corridor because of lower sea levels. They could have come by boat or they might have walked. Once in the Americas they apparently spread out quickly, reaching Chile no later than 12,000 years ago. The archeological, linguistic, and DNA evidence concerning this discovery of America is not consistent, so arguments rage about its timing, about the size of the founding population, and about whether or not it came all at once or in two or three separate waves. It does seem that the first Americans are most closely related to peoples of southern Siberia, although rival interpretations maintain their cousins were from what is now Korea and North China.

These long, slow migrations out of Africa and throughout the world no doubt contained many setbacks. Groups guessed wrong and found themselves in deserts from time to time. Others attempted what they thought was a short sea voyage and never saw land again. The Mt. Toba deep freeze might have killed off everyone not living in warm places. But slowly, in fits and starts, humankind colonized the globe.[3]

Domestication and Farming

The deep cold of Siberia in these millennia contributed to another momentous development: the first domestication. Man's first friend was the dog. Dogs evolved from wolves over thousands of years. Just how, when, and where this first happened is unclear, but the latest (genetic) evidence suggests it occurred in Southwest Asia around 30,000 BP.

The dog–human symbiosis was a mutually profitable partnership. Dogs provided people with hunting help (compensating for our poor sense of smell), with an early-warning system against attackers, and with loyal companionship including furry warmth on cold nights. In dire circumstances, people could also eat their dogs. People provided dogs with food (or hunting help, as the dogs might see it), and sometimes protection and shelter. People with cooperative dogs enjoyed great advantages in hunting and in self-protection. People living with barking dogs would not easily fall victim to surprise attack. Dogs with cooperative people got a more reliable food supply, including access to big game such as mammoths, which dogs could scarcely bag by themselves. So, over time, a genetic selection occurred for dogs that worked well with humans – dogs that showed loyalty, barked at the appearance of strangers, accepted human commands, and could read human gestures and expressions. Meanwhile, a cultural selection took place for human groups that worked well with dogs, training them, breeding them, protecting them, and eating them only in extreme need. The Ainu, a people in Japan's northernmost island of Hokkaido, even taught their dogs to catch salmon for them. The dog–human symbiosis spread rapidly and became well-nigh universal.

The domestication of dogs was the first of many such in the human career. Dozens of animals and hundreds of plants proved susceptible to domestication. Almost all of these domestications took place in remote times before written records. But archeologists can often tell the difference between wild and domesticated species from remains of seeds and bones.

Few things in human history have mattered as much as domestication. Raising one's food as opposed to collecting or hunting it implied broad changes in the human way of life. It required people to submit to laborious routines, but allowed enormous expansions in terms of cultural richness and diversity. Mobile hunters and foragers around the world had only a few tools (often very similar ones), the same social structures just about everywhere, and – as far as we know, which is not terribly far – roughly the same sorts of ideas about nature and spirits. In the late Paleolithic some people settled down in a few choice spots, becoming at least semi-sedentary if not fully so, and a notable elaboration of culture, especially in tools and art, took place. But, by later standards, there wasn't much cultural diversity in the Paleolithic, because most people remained mobile and had to carry their culture with them.[4]

With farming, all that would change. As the ice-age cold gave way to warmer temperatures and damper conditions, plant life flourished. Forests replaced steppe and scrubland, deserts retreated, and rivers rose. In many locations, these were favorable trends for people, allowing some groups to settle down and live off newly abundant local plants and animals. In many spots in Southwest Asia, for example, there were plenty of acorns, almonds, and grasses with edible and storable seeds. Some evidence suggests people were storing seeds as long ago as 23,000 BP. Gazelles and other tasty herbivores provided meat. The scant evidence suggests that population rose across Eurasia in the millennia from 16,000 to 12,000 years ago as climate warmed and foraging became easier. In a few favored locations, it became so much easier that more people could settle down, at least for most months of the year, living off locally abundant food.

Settled populations tended to grow far faster than mobile ones, because they did not have to lug children long distances and were less inclined to try to prevent their birth, or to abandon newborns. Where people collected the seeds of wild grasses, they could mash the seeds into gruel and feed it to babies, weaning infants from their mother's milk sooner. Early weaning made mothers fertile again sooner (lactating women are much less likely to conceive). So intervals between births tended to be much shorter among settled people than among mobile ones, which meant higher fertility rates and faster population growth.

Population growth among the settled folk gradually caused difficulties. Big animals grew rare due to additional hunting. In the Levant, for example, archeologists have found that people were hunting smaller and smaller animals as time went on – fewer gazelles and more rabbits. In many parts of the world, the traditional solution for food shortage remained viable: walk somewhere else. But for settled folk with accumulated possessions, lots of small children, and perhaps spiritual and emotional commitments to preferred places, this approach was less practical. Instead, in a few circumscribed locations, such as the Levant (between a rising sea and a desert), in the Nile valley (a ribbon of oasis surrounded by desert), and in China's Yangzi valley (surrounded by hills), people intensified their quest for food. Rather than walk elsewhere, they foraged for a wider variety of plants, hunted a broader spectrum of animals, and began to work spreading the seeds of preferred plants such as wild wheat and rice. Such efforts probably took place elsewhere as well, but the archeological record is so spotty that no one knows.[5]

Not long ago scholars used to wonder why everyone did not take up farming. Now they wonder why anyone did so in the first place. Farming turns out to be more work than foraging and hunting, and usually results in worse nutrition and worse health. So why did people do it? No one knows. The archeological and, lately, genetic evidence goes some way toward illuminating where and when transitions to farming occurred. But explaining why is still a matter of ingenious guesswork.

After more than 100,000 years without bothering to farm, in the 7,000 years following the end of the ice age humans undertook at least seven transitions to farming on four continents. This seems most unlikely to be a random turn of events. It is good to bear in mind that conceivably dozens of earlier such transitions took place in environments where no archeological traces remain, either in warm humid conditions where things biodegrade quickly, or in locations now beneath the sea. So there might have been, in effect, dozens of "false starts" in agriculture. But even if so, the question merely becomes: why did at least seven *lasting* transitions take place only after the end of the ice age? Two important factors may help bring us closer to an understanding: intelligence and climate.

While turning to a way of life that brings poorer nutrition and health while requiring more work may not seem a hallmark of intelligence, it surely did take intelligence to notice which plants grow best in which sorts of places, which ones yield easily edible seeds, which seeds store well, and so forth. Developing this sort of knowledge required communication – language – as well as powers of observation, memory, and reason. So it could not have happened before modern, intelligent, language-wielding humans were on the scene. Depending on when it was that language emerged, there were either zero or very few such humans around before the last ice age. And during the ice age, conditions were too dry or too cold for agriculture outside tropical locations, sharply reducing the chances that people would make the transition successfully. Ice age conditions were too inhospitable for sedentary life, except perhaps for a few mammoth hunters in Ukraine and fisherfolk here and there, and it is hard to imagine mobile people transitioning to agriculture. Intelligent, language-bearing humans enjoying interglacial conditions, then, seem a likely prerequisite for transitions to agriculture.

But we still do not know what triggered these transitions. As usual when evidence is sparse, hypotheses abound. Some scholars think that population growth led some human groups to the brink of Malthusian crisis, so they had to engage in risky experiments including agriculture; necessity was the mother of invention. Others think that adverse climate change had the same effect. Still others prefer explanations anchored in social pressures. Perhaps there were societies that practiced competitive feast-giving, like the famous potlatches among the Amerindians of the Pacific Northwest, and so strong incentives existed to produce more food in order to achieve greater renown. Maybe there were religious reasons to try agriculture, perhaps as an effort to appease gods irritated by heavy harvesting of certain plants. There is absolutely no evidence for these social explanations, but that does not mean they must be wrong. While there is some evidence for population growth, and good evidence for climate change, that does not mean these explanations are necessarily correct. In any case, it could well be that some combination of these (and other) factors led to agriculture, and indeed the relevant combination could well have been different in each transition to agriculture – in which case no general theory would be correct.

Enduring transitions to farming took place between 11,000 and 4,000 years ago in Southwest Asia, China, Southeast Asia, Sahelian Africa, South America, Mesoamerica, and the woodlands of eastern North America. Perhaps there were others too for which we at present lack convincing evidence. From its several points of origin, agriculture slowly extended its tendrils far and wide. The cereal culture of the Levant, the rice culture of China, and the maize and beans culture of Mesoamerica diffused furthest and gave rise to the world's most influential and widespread farming systems. The Levant farming system spread fairly quickly along the northern shores of the Mediterranean, but more slowly up the Danube and into cooler climates. But between about 7000 BCE and 4000 BCE generations of farmers carried it from Anatolia to the British Isles. Others carried it eastward too, into northwestern India by perhaps 7000 BCE, but much more slowly southward toward the southern tip of India. The rice-based farming system of the Yangzi meanwhile apparently spread westward into northeastern India, into the Ganges valley, by about 6000 or 5000 BCE, although some scholars maintain rice was independently domesticated there. Much later it spread to Southeast Asia, and later still to Korea and Japan. Maize culture spread from its Mesoamerican home both north and south, and became the staple food of most of the Americas by 1000 BCE.

Wherever and whenever it happened, farming spread in two main ways. First, farmers sometimes displaced foragers, pushed them off fertile lands, and extended their own domain. Second, sometimes the *idea* of agriculture spread: people who had not formerly practiced it learned about it, and imitated the practice of others.

Farmers could often displace foragers through violence. Food production allowed greater densities of population. Bigger groups normally prevailed over smaller ones in contests of violence. Moreover, the bigger social groups became, the more likely they could support specialists in weapon-making and in the arts of violence. Foraging bands, even those that included skilled hunters, could not consistently withstand the military pressure farming peoples could bring to bear.

Farmers could achieve much the same thing without even trying, through accidental biological warfare. Farming folk lived cheek-by-jowl with herd animals. Herd animals provide suitable environments for microbes that cause diseases, trigger immune-system responses, and therefore need to get from animal to animal. Gradually, farmers' bodies came to host microbes that routinely lived in sheep, goats, cattle, or dogs. Some of these microbes caused deadly diseases among humans. Measles and tuberculosis come from cattle diseases. Influenza is a gift from pigs and ducks. Smallpox may derive from camel pox. Over many generations, the immune systems of farming folk became increasingly resistant to these microbes. Indeed, when infectious diseases were constantly present, only people with robust immune systems survived childhood. Their own children, in turn, would be likelier to have robust immune systems, or at least immune systems attuned to the risks posed by sedentary life among herd animals and animal-derived diseases. In effect, a new selection process was at work among farming peoples, for disease resistance.

But hunters and foragers did not normally face this pressure. As long as they kept their distance from farmers, they remained healthier than sons and daughters of the soil, free from the infections hosted by crowded and sedentary peoples. But if hunters and foragers came into prolonged contact with farmers, disaster routinely followed. Their immune systems lacked all resistance to the infections now becoming common among farmers, and they sickened and died in droves. This process has happened again and again in more recent times, and it is extremely likely it happened early in the history of the expansion of farming too, although the archeological evidence for it is slender. In any case, while not perhaps certain, it remains most probable that greater vulnerability to new human infections cost foragers and hunters terribly, clearing the way for further expansion of farming.[6]

In some environments farming may also have spread because it indirectly ruined landscapes for foraging. Wherever dense populations developed due to agriculture, their activities depleted nearby wild herds. Farmers' own herds and flocks munched and trampled their way over the countryside, reducing the availability of forage plants for wild animals. And farmers often set fire to vegetation, to prepare the ground for their own plantings. Burning easily got out of hand, reducing woods and scrubland to ashes, and thereby made gathering wild plants and hunting wild animals less feasible. In these ways farmers and herders by pursuing their own subsistence often made it harder for anyone to make a living without farming.

Nonetheless, not everyone took up farming. Some people either already lived in or retreated to environments too dry, too cold, or too infertile for farming. Others, like the Jomon in Japan, found farming unattractive compared to the rewards of foraging and hunting, and they persisted in their way of life for several thousand years although in contact with farmers. They learned to cultivate plants, but did so only sparingly. But such cases remained exceptional.[7]

Over the last 10,000 years the spread of farming has almost matched the earlier spread of the use of fire and of language. In all three cases, at one point in time no people had the new technology. Then some people used it, while others did not, and those who had it enjoyed great advantages against those without it. Eventually in the cases of fire and language, all people used it. This point may yet come with respect to agriculture, although to this day in the Arctic, and in several moist tropical forests, people survive who neither practice agriculture themselves nor eat its products. They now account for less than 1% of humankind – and less every day.

Paleolithic Humans Shape the Environment

People had a notable impact on their immediate environs from the time they harnessed fire. Natural fires had shaped vegetation and ecosystems long prior, probably from soon after the initial appearance of plants on Earth. Human – or, more likely, hominin – fire made cooking possible, which broadened the range of foods our ancestors could eat. It also made successful migration out of Africa into cooler climes much more likely, expanding the geographical range over which humans had environmental impacts. Humans and fire co-colonized the Earth.

Everywhere they went Paleolithic people used fire to alter landscapes to their liking. Observant humans noticed that fresh shoots of plant life attracted big herbivores that people could hunt. Hunters set fires to drive game animals into ambushes. No doubt they also caused accidental fires countless times. Specialists have reached no consensus about the scale and scope of landscape burning in the distant past, but it seems that over recent decades the weight of opinion has moved in the direction of more and more widespread burning. Scholars of the Australian Paleolithic, for example, credit the early populations there with "firestick-farming," deliberate burning that improved the browsing for kangaroos and other marsupials that could be hunted.[8] Wetter continents than Australia burned less readily, and the twists and turns in climate history, mainly precipitation history, would have affected the degree to which early humans could successfully burn vegetation. Where and when moist tropical forests prevailed, they normally resisted sparks and flame. True deserts, although dry enough for fire, lacked fuel. But, by and large, where people lived they could burn the landscape, and where the landscape could be burned, people lived. Since it is usually impossible to tell the difference between a natural fire and a human fire of tens of thousands of years ago, scholars may always remain uncertain about the pervasiveness and frequency of human fire in the Paleolithic. It does seem, however, that in most instances when humans arrived suddenly on the scene – as in the Americas around 14,000 years ago or Australia 50,000 years ago – the frequency of fire increased.[9]

A more decisive impact on the environment came with the so-called late Pleistocene extinctions, in which humans almost certainly had a hand. Some 50,000 years ago the Earth hosted at least 150 genera (families) of megafauna, defined as animals with an average body weight of more than 44 kilograms. As of 10,000 years ago, at most 43 genera remained. North America, for example, lost mastodons and mammoths, condors with 5-meter wingspans, sloths the size of hippos, horses, camels, several sorts of elephant, big cats, giant beavers, and armadillos. South America lost its entire menagerie of megafauna larger than llamas. Australia lost wombats the size of grizzly bears and indeed all its species larger than 100 kilograms. Eurasia and Africa lost smaller proportions of their megafauna.

Since the 1960s, arguments have raged concerning the relative role of human hunting and climate changes in bringing about these waves of extinctions. Other hypotheses, such as new animal diseases or storms of comets, have very few supporters. By and large, the arguments for human hunting, dubbed the "overkill hypothesis," have gained ground. For one thing, all the creatures that became extinct between 50,000 and 10,000 years ago had survived prior bouts of climate change no less severe than that of the late Pleistocene. For another, this flurry of extinctions affected big animals far more than smaller ones (and humans, when given a choice, hunt big animals first and smaller ones later). Moreover, several species of megafauna, such as woolly mammoths, held out for a few thousand years on islands without human presence. In addition, flightless birds suffered much higher rates of extinction than those able to fly.

Even the lower likelihood of extinction among African and Eurasian species suggests human agency: animals there had many millennia to grow accustomed to life amid intelligent fire- and spear-wielding bipeds and learned to be wary of humans. Or, put another way, African and Eurasian wildlife was subjected to selection pressure for wariness of humans at a time when humans were not so dangerous. By the time humans had become more dangerous, thanks to advances in weaponry and the evolution of language and complex cooperation, most species in Africa and Eurasia instinctively kept their distance from people. In the Americas and Australia, on the other hand, humans arrived suddenly and in full possession of language and of cooperative hunting techniques. Thus they could blitz their way through the big, nutritious, and naive megafauna. This blitz left too little for other big predators to eat, so some of them also became extinct.[10]

While climate change may also have helped, and conceivably other factors did too, it seems certain now that human hunting, and burning, played the largest role in bringing about the late Pleistocene extinctions. These changes in wildlife not only affected ecosystems everywhere, but in the fullness of time they proved portentous for human history. Continents with few big animals had fewer potential candidates for domestication, leaving human societies there with less to work with. The absence of horses in post-Pleistocene America, for example, meant no warhorses. When, after 1492, Europeans arrived in America, they brought cavalry. Native Americans had none.[11]

Paleolithic humans probably brought about one additional important extinction that altered the environment: that of the Neanderthals. They flourished in Europe and western Asia from roughly 500,000 years ago until 50,000 BP. They lived mainly as carnivores, hunting the deer, horses, gazelles, bison, and other wildlife that dotted western Eurasia. At their most numerous, there were perhaps 75,000 of them. But by 50,000 BP they were gone from Asia, by 30,000 BP from everywhere in Europe but its southwesternmost corner, and by 24,000 BP they were extinct.

Perhaps climate change had something to do with it. After 50,000 BP, colder and much more unstable weather set in the Neanderthal domain, so that vegetation regimes changed within a matter of decades, and therefore animal habitat and migration routes also. Perhaps the Neanderthals could not adjust to such instability. After 40,000 BP several volcanic eruptions affected the northern hemisphere, especially between Italy and the Caucasus mountains, bringing volcanic winters that were no doubt hard on vegetation, wildlife, and therefore even the cold-resistant Neanderthals.

Few specialists, however, believe that climate change alone could have wiped out a cold-resistant species that had lasted through previous glacials and interglacials. More likely, it seems, the advent of homo sapiens into western Asia and Europe spelled doom for Neanderthals. Disease might have played a role. So might violence. Or merely superior hunting skills that left the Neanderthals with less and less to eat. Modern humans were much more mobile than Neanderthals, better constructed for walking and running, which might have helped them both in military contests and in hunting. Neanderthals, stocky and powerfully built, needed more energy (more food) than modern humans. The fact that Neanderthals vanished first in western Asia and later in Europe supports the notion that modern humans had something to do with it: homo sapiens entered Europe from Southwest Asia. Probably, as with the other megafauna extinctions of the late Pleistocene, we will never know just how the extinction of the Neanderthals happened and what role in it to assign to human agency.

Unlike the other megafauna that became extinct in the Pleistocene, something of the Neanderthals apparently lives on. Recent genetic research suggests that people who are neither African nor of recent African descent derive approximately 1–4% of their DNA from Neanderthals. This strongly implies that, whether or not modern humans spread diseases to them, out-hunted them, or killed them, they definitely had sex with Neanderthals.[12]

With fire, with mobility, and with hunting prowess Paleolithic people altered environments everywhere except Antarctica. Their impacts were probably greater on continents such as Australia and the Americas where they arrived late and suddenly. In those cases they came with a full arsenal of intelligence, language, and tools and, as it were, took ecosystems by surprise. Paradoxically, their impacts on Africa were probably more modest, as African species and humans coevolved over very long spans of time. From the viewpoint of subsequent human history, the most important of these impacts was the extinction of megafauna – probably including Neanderthals.

Neolithic Farmers Shape Themselves and Their Environments

With the transitions to farming beginning about 11,000 years ago, people increasingly changed the environment around them. Human impacts manifested themselves on the smallest scales, from the microbial environment, to, perhaps, the largest, in the form of the global climate.

Health Impacts

Farming had a powerful effect on human health. Farmers ate a narrower diet than their foraging and hunting ancestors, depending heavily on a few staples, which left them at greater risk of poor nutrition due to lack of key vitamins or minerals. They ate less meat and got less protein than their ancestors. The evidence of surviving skeletons tells us that farming people were smaller than their pre-farming predecessors. Skeletal remains from various sites between Ukraine and North Africa indicate that late Paleolithic men on average stood 177 centimeters tall, and women 166 centimeters. In the last thousand years before agriculture, their average stature shrank by 5 centimeters – this was the time when game was growing scarce and climate colder and dryer. Then, in the first millennium after the origins of farming, they shrank by another 5 to 10 centimeters.[13]

Farmers were sicker as well as smaller. Hunting and foraging peoples of the Paleolithic suffered from several infections, and because they had a food-sharing ethos they shared a lot of infections with their kinfolk too. They caught diseases from eating diseased meat, and picked up plenty of bacterial infections by rooting around in the soil. In health terms, the shift to agriculture was a gigantic stride backwards for human beings.[14]

First of all, like all sedentary people, farming folk lived in the midst of their own garbage and waste. Moreover, they probably deliberately handled human feces to use it as fertilizer. It is likely, therefore, that they suffered heavily from gastrointestinal diseases carried by worms and other parasites, which one might collectively call "diseases of sedentism."

They also suffered from what one could call "diseases of domestication." Over 300 human diseases derive from domesticated animal diseases. Half of these come from dogs, cattle, sheep, and goats. Even chickens and cats have donated some of their infections to their human masters. As these infections evolved into human diseases, farmers in Eurasia (not in the Americas) increasingly were born into hazardous microbial environments, aswirl with measles, influenza, smallpox, mumps, tuberculosis, tetanus, whooping cough (pertussis), and a host of other killing diseases.

If that wasn't bad enough, farming folk also had to face what might be termed "diseases of storage." By storing grain, farmers attracted rats, mice, and other disease vectors, and put themselves at enhanced risk of bubonic plague, hemorrhagic fevers, and other nasty things. There are about 35 human diseases derived from rats or from the fleas and ticks that live on rats.

To top if off, where farmers cleared forest and especially where they allowed water to accumulate, they created new habitat for malarial mosquitoes. Malaria, which has a good claim to being one of humankind's worst scourges, seems powerfully correlated with farming environments in Africa and Asia, and seems to have taken hold among humans soon after they took up farming.[15]

What with these diseases of sedentism, of domestication, and of storage, life for farming folk – especially children – became far more hazardous than for hunting and foraging peoples – provided they could stay away from disease-bearing farmers. A large proportion of infants and toddlers in farming villages died of disease, often abetted by malnutrition. Farmers had far more babies than did hunters and foragers, but their environments also killed them much faster.

Even in the Americas, where there were almost no domesticated animals, early farming folk were far less healthy than their hunting and foraging neighbors. Careful study of some 1,500 skeletons from 22 sites in Ecuador, for example, leaves no doubt that in general farmers suffered a more severe burden of disease. If anyone had known these facts of paleopathology before adopting farming, whether in Eurasia, Africa, or the Americas, they would have avoided agriculture like the plague. But they did not know.[16]

Among the Natufians of the Levant, who left more skeletons than most pioneers of farming, life expectancy of men grew with agriculture, but that of women shortened. Two likely reasons for this are that death by violence (more likely to befall men) decreased as people took up farming, but death in the act of giving birth (much, much more likely to befall women) became more common due to higher fertility. The routines of work among farmers also left telling marks on female skeletons. They show signs of stresses from the kneeling and bending required for grinding grain and cooking: deformations of their knees, wrists, and lower backs. In general, the transition to agriculture made people shorter, sicker, and more likely to suffer from acute malnutrition.[17]

Landscape Impacts

Agriculture changed landscapes more obviously than it changed human health. In the first instance, it required the creation of new species. Wheat did not grow in the wild: it is descended from a wild progenitor, einkorn, that grew (and still grows) in southeastern Turkey and northern Syria. Maize (corn) is derived from a Mexican wild plant called teosinte. Just as dogs are not wolves, maize and wheat are different from teosinte and einkorn. Cattle, sheep, goats, camels, water buffalo, pigs, chickens, and a dozen other beasts of the field are new creations, derived from wild ancestors. In the first few millennia of farming, people created thousands of new species through patient domestication.

Beyond creating new species, early farmers created new landscapes. They used fire and ax to clear wood and scrublands to make way for their digging sticks and plows. In rice regions of East, South, and Southeast Asia, they learned to build paddies, dikes, and berms to control the flow of water for the benefit of rice. In hilly country they learned to sculpt terraces into slopes (although it is possible this skill had not developed before the first cities and civilization). They built villages. They burned woodlands to create pasture for their own flocks and herds, as their forebears had done to favor game animals. Even before the rise of cities and civilization they began to experiment with irrigation. In dozens of ways, large and small, early farmers changed their landscapes.

The scale on which early farmers wrought their changes was governed by their numbers. Prior to 3500 BCE (a reasonable date for the advent of cities and civilization), the Earth hosted only a few million people.[18] Average population density globally was only about one-thousandth of what it is today. In fact, farmers were huddled in a few choice locations as late as the fourth millennium BCE and left the greater part of the world untouched. But they had established the practices by which humankind would, over the millennia to come, alter environments almost everywhere.

Agriculture created new species through domestication. It changed landscapes, through clearing and cultivating. It changed human health. It may also have changed climate.

Possible Climate Impacts

According to a controversial hypothesis, as early as 8,000 years ago, deforestation undertaken to clear land for agriculture began to raise the concentration of carbon dioxide in the atmosphere. With the spread of agriculture, deforestation and carbon-dioxide emissions reached a scale sufficient to warm the planet. The ice cores from which scientists infer past carbon-dioxide concentrations do show modest increases beginning around that time that are not easily explicable in natural terms. Subsequently, beginning around 5,000 years ago, levels of methane in the atmosphere began to rise. Methane is an even more powerful greenhouse gas than carbon dioxide. Its emission into the atmosphere came mainly from wet rice cultivation and secondarily from livestock. The spread of rice farming from its origin in the lower Yangzi valley – perhaps – accounts for the observed upturn in methane concentrations.

Specialists remain divided about this exciting new hypothesis. Some are content to accept it and argue that human activity prevented an otherwise likely return to glacial conditions 8,000 years ago. Others, while not disputing the ice core evidence, think that the scale of early farming was too small to have the requisite effects on the atmosphere. Some embrace a rival possibility: that rising sea levels created new coastal swamps and bogs which account for the rising presence of methane in the atmosphere. If it is valid, the new hypothesis means that human action affected climate not just in the industrial era, but from the dawn (or at least the early morning) of agriculture.[19]

Conclusion

For most of the human career, our ancestors lived precariously in environments that must often have seemed hostile. They struggled to survive amid predators eager to make them into their next meal. They were at the mercy of shifts in climate that brought drought, flood, or frosts for which they were ill-prepared. When the Toba eruption blocked out the sun for years on end, the human career almost came to a sudden halt. Nonetheless, humans managed to walk all over the world and find ways to survive in Siberian forests and Australian deserts. Fire was their best tool in the Paleolithic, and with it they could alter landscapes to their advantage and roam where they could not have prospered without fire. They turned wolves into dogs.

Scholars now have strong evidence in favor of the proposition that Paleolithic people had major impacts on their environments. Archeology shows us the remains of human fire. Paleontology reveals the vast array of megafauna swept into the dustbin of prehistory with human help. Palynology indicates changes to vegetation resulting from megafauna extinctions and human burning. No one can suppose any longer that preagricultural humans lived in gentle harmony with a nature primeval.

That said, the environmental change people wrought after they developed agriculture eventually dwarfed what Paleolithic people could do. That took time, and farmers of the early Neolithic and their livestock probably had only local effects. But – despite their burden of disease – their numbers grew faster than those of hunters and foragers. Gradually their technologies became more powerful – plows and dams for example. Perhaps they even influenced climate. These two trends, of population growth and technological advance, over several millennia made Neolithic peoples shapers of their environment. The emergence of states, war, and long-distance trade combined with further growth in population and advance in technology would bestow even greater environment-shaping power upon the human race, as the remaining chapters of this book will show.

Notes

1 These ideas are reviewed on the Smithsonian Institution's Web page, accessed from http://humanorigins.si.edu/research/climate-research/effects on September 3, 2011. Bear in mind both the climate reconstruction and the interpretation of hominid evolution are based on very few data.

2 A summary of current views appears in: M. A. J. Williams, S. H. Ambrose, S. van der Kaars, et al., "Environmental Impact of the 73 ka Toba Super-Eruption in South Asia," *Palaeogeography, Palaeoclimatology, Palaeoecology* 284/3–4, 2009, pp. 295–314.

3 A useful summary is P. Manning, "*Homo sapiens* Populates the Earth: A Provisional Synthesis, Privileging Linguistic Data," *Journal of World History* 17, 2006, pp. 115–58. A recent example of the use of genetic date is H. Zhong, H. Shi, X. B. Qi, et al., "Global Distribution of Y-Chromosome Haplogroup C Reveals the Prehistoric Migration Route of African Exodus and Early Settlement in East Asia," *Journal of Human Genetics* 55, 2010, pp. 428–35.

4 S. Mithen, *After the Ice: A Global Human History, 20,000–5,000 BC*, Cambridge, MA, Harvard University Press, 2003.

5 Works that address the role of climate in transitions to agriculture include: G. Barker, *The Agricultural Revolution in Prehistory: Why Did Foragers Become Farmers?*, Oxford, Oxford University Press, 2006; P. Bellwood, *First Farmers: The Origins of Agricultural Societies*, Oxford, Blackwell, 2005; J. L. Brooke, *A Rough Journey: Human History and a Volatile Earth*, New York, Cambridge University Press, 2012; W. J. Burroughs, *Climate Change in Prehistory: The End of the Reign of Chaos*, Cambridge, Cambridge University Press, 2005.

6 These matters are well explained in J. Diamond, *Guns, Germs and Steel*, New York, Norton, 1997.

7 J. Habu, *The Ancient Jomon of Japan*, New York, Cambridge University Press, 2004.

8 The term was coined by R. Jones: "Fire Stick Farming," *Australian Natural History* 16, 1969, pp. 224–8; a recent challenge to the idea of pervasive firestick-farming appears in S. D. Mooney, S. P. Harrison, P. J. Bartlein, et al., "Late Quaternary Fire Regimes of Australasia," *Quaternary Science Reviews* 30, 2011, pp. 28–46.

9 A helpful synopsis is S. J. Pyne, *Fire: A Brief History*, Seattle, University of Washington Press, 2001, pp. 27–64.

10 A recent survey of the evidence is A. D. Barnosky, P. L. Koch, R. S. Feranec, et al., "Assessing the Causes of Late Pleistocene Extinctions on the Continents," *Science* 306, 2004, pp. 70–5.
11 This theme is elaborated in Diamond, *Guns, Germs and Steel*.
12 R. E. Green, J. Krause, A. W. Briggs, et al., "A Draft Sequence of the Neandertal Genome," *Science* 328, 2010, pp. 710–22.
13 B. Bogin, *The Growth of Humanity*, New York, Wiley, 2001.
14 The classic on this subject is M. N. Cohen, *Health and the Rise of Civilization*, New Haven, CT, Yale University Press, 1991. See also: M. N. Cohen and G. M. M. Crane-Kramer, *Ancient Health: Skeletal Indicators of Agricultural and Economic Intensification*, Gainesville, FL, University Press of Florida, 2007; J.-P. Bocquet-Appel and O. Bar-Yosef (eds.), *The Neolithic Demographic Transition and Its Consequences*, Dordrecht, Springer, 2008.
15 J. L. A. Webb, *Humanity's Burden: A Global History*, New York, Cambridge University Press, 2009, pp. 1–58.
16 On populations of the Americas, see R. Steckel and J. C. Rose (eds.), *The Backbone of History: Health and Nutrition in the Western Hemisphere*, New York, Cambridge University Press, 2002.
17 An overview is A. Mummert, E. Esche, J. Robinson, and G. J. Armelagos, "Stature and Robusticity during the Agricultural Transition: Evidence from the Bioarcheological Record," *Economics and Human Biology* 9, 2011, pp. 284–301.
18 M. Livi-Bacci (*A Concise History of World Population*, 5th ed., Oxford, Wiley-Blackwell, 2012, p. 2) says 10 million.
19 Its originator is W. F. Ruddiman, *Plows, Plagues and Petroleum: How Humans Took Control of Climate*, Princeton, NJ, Princeton University Press, 2005.

References

Barker, G., *The Agricultural Revolution in Prehistory: Why Did Foragers Become Farmers?*, Oxford, Oxford University Press, 2006.
Barnosky, A. D., Koch, P. L., Feranec, R. S., et al., "Assessing the Causes of Late Pleistocene Extinctions on the Continents," *Science* 306, 2004, pp. 70–5.
Bellwood, P., *First Farmers: The Origins of Agricultural Societies*, Oxford, Blackwell, 2005.
Bocquet-Appel, J.-P., and Bar-Yosef, O. (eds.), *The Neolithic Demographic Transition and Its Consequences*, Dordrecht, Springer, 2008.
Bogin, B., *The Growth of Humanity*, New York, Wiley, 2001.
Brooke, J. L., *A Rough Journey: Human History and a Volatile Earth*, New York, Cambridge University Press, 2012.
Burroughs, W. J., *Climate Change in Prehistory: The End of the Reign of Chaos*, Cambridge, Cambridge University Press, 2005.
Cohen, M. N., *Health and the Rise of Civilization*, New Haven, CT, Yale University Press, 1991.
Cohen, M. N., and Crane-Kramer, G. M. M., *Ancient Health: Skeletal Indicators of Agricultural and Economic Intensification*, Gainesville, FL, University Press of Florida, 2007.
Diamond, J., *Guns, Germs and Steel*, New York, Norton, 1997.
Green, R. E., Krause, J., Briggs, A. W., et al., "A Draft Sequence of the Neandertal Genome," *Science* 328, 2010, pp. 710–22.
Habu, J., *The Ancient Jomon of Japan*, New York, Cambridge University Press, 2004.
Jones, R., "Fire Stick Farming," *Australian Natural History* 16, 1969, pp. 224–8.
Livi-Bacci, M., *A Concise History of World Population*, 5th ed., Oxford, Wiley-Blackwell, 2012.
Manning, P., "*Homo sapiens* Populates the Earth: A Provisional Synthesis, Privileging Linguistic Data," *Journal of World History* 17, 2006, pp. 115–58.
Mithen, S., *After the Ice: A Global Human History, 20,000–5,000 BC*, Cambridge, MA, Harvard University Press, 2003.

Mooney, S. D., Harrison, S. P., Bartlein, P. J., et al., "Late Quaternary Fire Regimes of Australasia," *Quaternary Science Reviews* 30, 2011, pp. 28–46.

Mummert, A., Esche, E., Robinson, J., and Armelagos, G. J., "Stature and Robusticity during the Agricultural Transition: Evidence from the Bioarcheological Record," *Economics and Human Biology* 9, 2011, pp. 284–301.

Pyne, S. J., *Fire: A Brief History*, Seattle, University of Washington Press, 2001.

Ruddiman, W. F., *Plows, Plagues and Petroleum: How Humans Took Control of Climate*, Princeton, NJ, Princeton University Press, 2005.

Steckel, R., and Rose, J. C. (eds.), *The Backbone of History: Health and Nutrition in the Western Hemisphere*, New York, Cambridge University Press, 2002.

Webb, J. L. A., *Humanity's Burden: A Global History*, New York, Cambridge University Press, 2009.

Williams, M. A. J., Ambrose, S. H., Kaars, S. van der, et al., "Environmental Impact of the 73 ka Toba Super-Eruption in South Asia," *Palaeogeography, Palaeoclimatology, Palaeoecology* 284/3–4, 2009, pp. 295–314.

Zhong, H., Shi, H., Qi, X. B., et al., "Global Distribution of Y-Chromosome Haplogroup C Reveals the Prehistoric Migration Route of African Exodus and Early Settlement in East Asia," *Journal of Human Genetics* 55, 2010, pp. 428–35.

CHAPTER TWO

The Ancient World,
c. 500 BCE to 500 CE

J. DONALD HUGHES

If ancient environmental history is not always a story of origins, it does at least have a certain priority, attempting to describe the emergence and course of early civilizations and their interactions with the world of nature. In important respects, it offers the first instances of particular human impacts, and can be viewed as giving impetus and precedents for subsequent history. For this reason the term "classical" has often been applied, initially to a formative stage of Mediterranean societies, and then by analogy to many other parts of the Earth, including China and the Mayas. To use well-worn analogies, it is a foundation period, or among the first texts written on the palimpsest of the landscape, dim though the surviving traces may be under the many erasures and new inscriptions of later times. Edmund Burke calls it "deep history," offering an essential perspective to early modern and modern environmental history, as for example the way in which an understanding of ancient irrigation in the Near East and its results, both positive and damaging, is necessary to an adequate understanding of contemporary issues of water use there.[1] But a dependable knowledge of deep history is not easily attained.

Periodization

A preliminary consideration in studying ancient history is the often-vexing question of periodization. What is "ancient" history – when does it begin and end? To answer these questions is not as easy as it might look at first glance. European history has often been conventionally divided into three great chronological periods: ancient, medieval, and modern. The division points were, however, subject to continual dispute and revision, and the discussion took place within a European (with the addition, particularly in earlier times, of Mediterranean and Near Eastern) context. The change of focus from "western civilization" to "world history" inevitably challenges the Eurocentric scheme. The

A Companion to Global Environmental History, First Edition. Edited by J.R. McNeill and Erin Stewart Mauldin.
© 2012 John Wiley & Sons, Ltd. Published 2015 by John Wiley & Sons, Ltd.

"three ages" cannot serve well as a chronological base outside that narrow context without serious readjustment, and in many parts of the world they do not work at all.

The series of ages derived mainly from European archeology, and applied mostly to times before written history, that is Palaeolithic (old stone age), Neolithic (new stone age), Chalcolithic (copper-stone age), bronze age, and iron age, refer to the durable materials used in tools and weapons, and correlate with a limited range of natural resources used successively. Again, these material innovations did not occur synchronously in various parts of the world, nor is the sequence universally applicable. For example, early cities in the Near East are associated with the bronze age, but bronze tools are rare or absent in New World cities in pre-Columbian times. Iron tools are present in otherwise "Neolithic" sites in the North American Pacific Northwest, possibly deriving from contact with Asia. In the remote islands of the Pacific, a "stone-bone-shell" age persisted until the arrival of Europeans in the region.

A more basic consideration in any periodization consists of the criteria chosen to mark the transitions between periods. Jerry Bentley put it this way: "As practicing historians well know, the identification of coherent periods of history involves much more than the simple discovery of self-evident turning points in the past: it depends on prior decisions about the issues and processes that are most important for the shaping of human societies."[2] From the standpoint of environmental history, then, the issues and processes are those marking the changing modes of human interaction with the natural environment: for example, hunting and gathering, horticulture, herding and agriculture, urbanization, trade and industries. Because these occurred at different times in different regions, periods based on them must be flexible. One is tempted to say that it would be best to treat the sequence of cultural change in each locality, along with successive alterations in the landscape, as potentially unique. But that approach, though valuable as an antidote to facile overview, should not be used as an attempt to avoid useful cross-cultural comparison.

Problems of Scholarship

Readers of more than a few books on ancient environmental history will discover, perhaps rapidly, that even the major and most basic conclusions are open to serious and often passionate dispute. As Joachim Radkau remarked, "From a very early time, water and forests have been leitmotifs of environmental history … Exactly what the link between forests and water meant for humans was often controversial."[3] Taking the Mediterranean area in classical times as an example, recent historians such as Michael Williams (2003), J. R. McNeill, John Perlin, and J. Donald Hughes judge that in general there was considerable deforestation resulting in erosion, while A. T. Grove and Oliver Rackham deny both deforestation and its connection with erosion.[4] In exception to both positions, Peregrine Horden and Nicholas Purcell maintain that the amount of deforestation cannot be determined, but that most deforestation was "a good thing" because it opened land for agriculture.[5] Similar controversies rage over grazing and its effects, adequacy of food supply and its sources, hunting and fishing, mining and pollution, demographic trends, diseases such as plague and malaria, and the nature and roles of climatic change. A leading cause of these disagreements is the difficulty involved in finding dependable primary sources.

Historians like to use documents, but one of the first problems faced by environmental historians of the ancient world is the relative scantiness of written sources from the period, compared to more recent times. All Greek literature from 800 BCE to 800 CE has been published on one CD-ROM by the Thesaurus Linguae Graecae (University of California at Irvine), and there was space left over for some early medieval literature. Latin literature is more extensive, but has its limitations as well. Similar rarity exists in other ancient literatures, sometimes compounded by fragmentation; the records of Mesopotamia, for instance, were mostly written on clay, and in China much was painted on fragile bamboo strips. Deliberate destruction was often involved, as in the cases of the libraries of Alexandria and Baghdad. This problem of paucity is exacerbated when one looks for information on environmental issues. Seldom do entire books offer useful data, with some like Theophrastus' botanical works among the exceptions. Usually it is a matter of searching the surviving texts for isolated paragraphs or briefer references. One benefit is that it is possible to read all the ancient written information that survives to the present day on a chosen subject in a reasonably short time, with the added advantage for those unskilled in ancient languages that a large portion of it exists in translation. On the other hand, it is often impossible to find answers for the very questions one wants to investigate from written sources alone, so historians are required to look elsewhere as well. New texts from ancient times only rarely come to light, but other lines of inquiry, especially in the sciences, have led to discoveries that illuminate understanding of the ancient environment.

Archeology

Archeology studies past human societies and their settings through physical remains, and it often offers valuable information for ancient environmental history. Ancient historians have a special need for archeological data to help compensate for the shortcomings of other sources. From beginnings in the eighteenth century, early efforts in archeology concentrated on large, architecturally prominent sites such as temples, palaces, and fortifications, usually in urban settings, and on the collection of pieces of artistic importance for museums and collectors. This was part of a general bias toward study of the upper classes in each society. Henry Schliemann, for example, the late-nineteenth-century excavator of many sites including one in Turkey that he identified as Troy, wanted to find the remains of kings such as Priam and Agamemnon, not the huts of ordinary folk. A similar bias was predictably produced by use of written sources, since almost everywhere the only people who could write were members of the upper classes. Environmental historians have to be aware that the members of society involved in many of the most direct interactions with the environment, such as villagers, small farmers, fishermen, shepherds, and subsistence hunters, inhabited smaller and less durable structures and wrote almost nothing.

More recent archeologists, from the mid-twentieth century onward, began to remedy the aristocratic predisposition. Environmental archeology, which seeks to discover the interrelationships between societies and the environments they lived in, has broadened archeology's field of vision by looking scientifically at things once ignored or discarded, such as plant remains including seeds and pollen, animal bones, shells, soils, and microorganisms. Anatomical analyses such as those of teeth and, where possible, stomach contents can reveal diets and use of resources. Materials subjected to new laboratory methods may indicate their places of origin and the technologies used to process them.

Another expansion in a direction useful to environmental historians is provided by landscape archeology and regional archeology, which look at larger areas and study not a single site but a wider pattern of settlement and land use within the context of soils and geomorphology. This approach can establish where villages, farmsteads, roads, and water diversions existed in a district beyond a monumental center. For example, surveys of the area around the Maya site of Tikal in Guatemala, with its steep pyramids and corbel-vaulted halls, have shown that for some time it was not an isolated ceremonial site, but the center of a widespread, comprehensively utilized agricultural district with numerous dwelling sites indicating intensive land usage and a dense population.[6] Such a survey uses walkovers to find artifacts on the ground and magnetic anomalies, aerial photography to find tell-tale traces like crop marks and surface unevenness, and test pits at promising spots. One of the pioneering efforts in regional archeology, which by necessity became an interdisciplinary collaboration involving natural scientists, social scientists, and humanists, is the University of Minnesota Messenia Expedition, which surveyed 3,600 square kilometers (1,400 square miles) in southwestern Greece to elucidate the bronze age cultural and natural environment.[7] That effort, modified and including later historical periods, still continues. Technological advances such as GIS, GPS, remote sensing, geophysics, and digital photography have increased the useful information provided by several orders of magnitude.

Radiocarbon Dating

A number of scientific methodologies can augment archeology and provide valuable data for ancient environmental history. One of the most important, developed in the 1950s, is radiocarbon dating. This is based on the fact that living organisms – plants directly and animals by ingesting plants – incorporate carbon into their tissues from the carbon dioxide in the atmosphere. This carbon exists in two isotopes, Carbon 12 being by far the most prevalent, and a tiny proportion of the radioactive Carbon 14. Carbon 14 decays at a constant rate to Nitrogen 14, emitting a beta particle (electron). On the death of an organism, it no longer incorporates carbon from the atmosphere, and the Carbon 14 continues to decay, with a half-life of about 5,568 years. The age of a specimen of wood, bone, or any other organic material can therefore be dated by measuring the proportion of Carbon 14 that it contains. After about ten half-lives, the proportion is too small to be useful, so the method can be used only for materials less than 50,000 years old, which is enough for most of ancient history. The method has been tested on artifacts of known date, especially a series from ancient Egypt, where the chronology is relatively well established, and it proved to be accurate within limits. It has proved its value in dating archeological material including wood and charcoal, and notably tissues from mummies. Cross-testing with datable wood from tree rings (see below) showed, however, that the proportion of Carbon 14 in the atmosphere has varied in the past, so that the dates have had to be corrected – the older the date, the larger the correction, but not at a constant rate; there are "wiggles" in the graph that must be taken into account. Where radiocarbon dating is not useful (particularly in material older than 50,000 years), there are methods using other radioisotopes that can be used.

Paleoclimatology

Paleoclimatology is one of the most elucidating scientific methodologies. To know how the climate changed over decades and centuries could help to explain environmental trends. Often controversy rages over the question of whether a particular historical change has climatic variation or human agency as its cause. For example, abandoned Roman structures for water control exist in North Africa in locations where water supply became intermittent and inadequate.[8] Was this because the climate became drier, or because deforestation of the watersheds and attendant erosion interfered with water retention and silted up the reservoirs and canals? Certainly learning what the climatic pattern was would help to answer the question.

Since ancient written records of weather are few and far between, and rarely quantified, climatologists have turned to analysis of proxy indicators such as ice cores, pollen deposits, tree rings, and charcoal finds. One of the most useful of these is the archive of ice cores, stored in facilities around the world like the US National Ice Core Laboratory (NICL), in the Federal Center west of Denver, Colorado.[9] Climate scientists braving difficult weather drilled ice cores in Greenland, Antarctica, and mountain glaciers, revealing evidence of past temperatures and concentrations of atmospheric gases, dust, pollutants, and radioisotopes. Deposits of ice in such places represent accumulations of snowfall that fell over as much as several hundreds of thousands of years, and it is often possible to date a layer close to the very year within which the snow in it fell, allowing a quantifiable estimate of climate change during ancient centuries within what is still admittedly a statistical range wider than would be desired, although the margin of error is being reduced. Ice cores provide information in addition to climatic data. For example, studies of lead concentrations in Greenland ice show a marked increase from the sixth century BCE onward, indicating increased mining and smelting by the Greeks and Romans, and a sharp upward trend in the second century BCE, reflecting the increased efficiency and scale of Roman metallurgy, underlined by the fact that about 70 percent of the latter increase bears the isotopic signature of the great Roman mines in Rio Tinto, Spain.[10]

It is often assumed that human activities produced a very small amount of greenhouse gases and particulate pollution before the advent of the Industrial Revolution in the eighteenth and nineteenth centuries, and that climatic changes in the ancient world must therefore have been the result of natural causes. However, William F. Ruddiman has challenged this view with a hypothesis that human agriculture, through clearing land, burning and deforestation (sources of carbon dioxide), raising livestock that consume vegetation (and produce methane), and irrigating rice paddies (again a source of methane), significantly added greenhouse gases to the atmosphere and countered what would have been a natural cooling trend. This first human alteration of global climate began, he says, about 8,000 to 5,000 years ago.[11] It was not as rapid and dramatic as the global warming of the most recent centuries, but nonetheless important.

Pollen Analysis

Palynology is the study of pollen grains in stratified contexts, such as deposits in lake bottoms and caves, for evidence about historical ecosystems. Pollen grains survive well in archeological and geological situations, and by microscopic study can be differentiated, usually to genus or even species. Analysis of a pollen core, with assistance from radiocarbon dating, can show the changing abundance of the pollen rain and the relative frequency of kinds of pollen over a period of time, often centuries or millennia. An increase in the pollen of cold-loving plants might indicate a decline in average temperature, while an increase in pollen from plants that survive aridity might be evidence of declining rainfall. The number of palynological studies has increased exponentially in recent years, so keeping abreast of the information on a regional scale is daunting. Brian Huntley and John Birks have published pollen maps for Europe over the past 13,000 years.[12] Valérie Andrieu and associates have developed a computerized database covering the Mediterranean basin, emphasizing palynological evidence for human activity.[13] Pollen data are an indication not only of climate change, but also of human impacts on vegetation, and sometimes it is difficult to separate the effects of the two processes. Diana K. Davis refers to the pollen evidence in her monograph, *Resurrecting the Granary of Rome*, based on a report by Henry Lamb and associates analyzing cores taken from the Tigalmamine Lakes in the Atlas Mountains of Morocco. She says, "The most significant decrease in pollen appears to have taken place about 1,600 years ago, well before either of the 'Arab invasions.'"[14] The fall of the western Roman Empire occurred 1,600 years ago, and the chart shows a decline from the time of the Roman Republic to a low in the fifth century. It also shows a decline of oaks, with a disappearance of deciduous oaks, and a sharp rise in grass pollen. Lamb and colleagues state that this was a severe period of forest degradation, since pollen data show that all tree species declined.[15] The need for further palynological research, and for integration of the research that has been published in scattered journals, is certainly evident.

Dendrochronology

Tree-ring studies add to our knowledge of the ancient environment; dendrochronology can sometimes make dating possible with an accuracy of a single year, and dendroclimatology can give evidence of climate changes.[16] In temperate zones, the wood of many species of trees shows rings indicating annual growth, rapid in the early growing season and slower toward its end. These rings are wider when environmental conditions promote growth, narrower when they are less favorable. The limiting conditions may be temperature, rainfall, and other factors, so it may be difficult to make judgments about which kinds of climatic change are involved. Trees in a geographic area develop the same ring patterns because they experience the same climatic conditions. Cross-dating is possible between samples from living trees and older wood from archeological sites, so that sequences representing hundreds and even thousands of years have been constructed; the longest so far go back 9,000 to 11,000 years. One can therefore date ancient wood samples, and establish the age of a beam used in a building or tomb. An example of dendrochronology used to date an environmental event is a marked tree-ring growth anomaly in the seventeenth century BCE that appears to correlate with the catastrophic eruption of a volcano on the island of Thera (Santorini) in the Aegean Sea; the eruption impacted a large area with fallout of debris and tsunamis, which, among other effects, may have helped to bring about the collapse of contemporary Minoan civilization centered on Crete. The anomaly consisted of abnormally wide rings, perhaps the result of factors such as clouds, rainfall, and a new moisture-conserving layer of volcanic ash encouraging growth during the summer, when conditions are usually hot and arid. But wide and narrow rings do not constitute the only information that can be extracted from annual tree rings. Dendrochemical analysis of the wood of relevant specimens indicates the sudden absorption by the trees of atomic elements associated with volcanic activity, strengthening the correlation.[17] This can be accomplished by means of a method called neutron-activation analysis (NAA), which determines the concentration of elements in a sample that is subjected to a stream of neutrons generated by a reactor or other source. Nuclei are transformed into radioactive isotopes that emit beta particles and/or gamma rays in a predictable pattern for each element as observed in a detector. It is more easily used for heavy elements such as manganese and gold, but less easily for lighter elements like iron and sulfur. If the sample is wood taken from a single tree ring, it can give information about the elements present in the tree's environment in a particular year.

Anthracology is the analysis of charcoal evidence found in archeological contexts. It is closely related to tree-ring and other wood studies. Charcoal is often used for radiocarbon dating, but anthracology uses microscopic study of charcoal fragments to determine the species of trees or shrubs that were carbonized, and to detect changes in the presence and proportion of various species over time. The distinctive patterns of wood cells in various species are often preserved in carbonization, and larger charcoal fragments may reveal tree-ring patterns as well. Explicitly ecological interpretations of carbon analysis appeared around 1940.[18] The methodology has been applied in Mediterranean countries, with a research tradition developed since the 1960s by Jean-Louis Vernet and associates at the University of Montpellier in France and since used in other parts of the world.[19] For instance, studies of Sallèles d'Aude, near Narbonne, where a Gallo-Roman potters' workshop with 14 kilns was active from the first to the fourth century CE, showed successive use of different tree species indicating progressive deforestation and replacement of moisture-loving species by drought-resistant ones.[20] In an investigation on isolated Easter Island (Rapa Nui), charcoal from fireplace sites was found to represent a variety of tropical forest trees, indicating that the famous statue-bearing island was forested during early human occupation and only later denuded.[21] Anthracology merits application more widely in light of the pervasive use of wood and charcoal as fuel in cooking, heating, metallurgy, and the ceramic industry.

Environmental Factors in the Decline of Civilizations

An aspect of ancient history calling for explanation is the fact that a large number of societies and civilizations flourished for various periods of time and then declined and disappeared, or survived only in an utterly transformed and fragmentary manner. To give examples, the Indus valley civilization was in decline by 1800 BCE, and its major cities were abandoned by 1700 BCE. There has not been an Assyrian Empire since the seventh century BCE, and the Han Dynasty in China ended in the third century CE and was succeeded for a time by disunion and lesser achievements. The Roman Empire in the West ceased to exist as a political and economic entity sometime between 400 and 700 CE, and cannot be said to have survived even if the world today perpetuates the Roman calendar and the influence of Roman law, and many coins bear Latin mottos, such as *e pluribus unum*. The role of environmental factors in the decline, "fall," and arguably also the recovery and survival of societies is consequently a topic worthy of investigation.

Among the environmental factors considered as possible contributors to decline in ancient times are removal of vegetative land cover by deforestation and overgrazing, accelerated erosion and silt deposition, impairment of agriculture by soil depletion and accumulation of salts by evaporation in irrigated fields, changes in water supply and drainage patterns, encroachment of agriculture on marginal lands, exhaustion of accessible resources such as metallic ores, extinctions or reductions of wildlife that served as food sources or controls on fluctuations of population, dietary and environmental poisons such as lead pollution, communicable diseases including malaria and plagues, and climatic changes that exacerbated one or more of the factors just mentioned.

Understandably the pattern of factors like these varied in different times and places, and some factors were not operative in all ancient societies.

The most widely disseminated views on this general subject in recent years are those enunciated by Jared Diamond, particularly in his thought-provoking book *Collapse: How Societies Choose to Fail or Succeed*. The choice of the word "collapse" is deliberate, and he defines it as an "extreme form" of decline. He presents five contributing factors as explanations of environmental collapse in the societies he discusses, both ancient and modern: "Four of these sets of factors – environmental damage, climate change, hostile neighbors, and friendly trade partners – may or may not prove significant for a particular society. The fifth set of factors – the society's response to its environmental problems – always proves significant."[22] As to why a society's response to environmental problems may lead to collapse, Diamond gives four reasons: (1) failure to anticipate a problem, (2) failure to perceive a problem, (3) failure to try to solve the problem, and (4) failure of attempts to solve the problem.[23]

A number of historians and archeologists have criticized Diamond's work, some thoughtfully and others with barely concealed antipathy. A collection of both kinds of response, *Questioning Collapse*, appeared in 2010.[24] Some of the criticisms are: (1) Diamond does not pay enough attention to specialists on the societies he considers; (2) he offers an inadequate definition of "collapse," and/or, having offered a definition, does not adhere to it in the body of the work; (3) there are other reasons for collapse than those enumerated by Diamond; (4) he makes specific errors of fact and interpretation; and (5) his writing is too popularizing (not to mention too popular). All these issues are worth considering.

The ancient societies (in the terms used here) discussed in *Collapse* are, as failures, Easter Island and its neighbors, the Anasazi, the Maya, and, albeit briefly, Mesopotamia, and, as successes, the New Guinea highlanders and Tikopia. Below I will consider two examples that Diamond did not choose for extended treatment: Mesopotamia, where Diamond at least posits a reason for "collapse," and the famous case of the Roman Empire, to which he devotes an inconclusive paragraph and a half in a 560-page book. I will also consider the case of Easter Island (Rapa Nui), which Diamond treats at length, and which serves as one of his best examples.

To avoid some of the difficulties that appear in Diamond's definition of "collapse," it is preferable to examine the more general process of "decline," that is, a lessening of the ability of the economy and/or technology to meet the needs of the population, a loss of territory dominated by the society and/or the level of participation in common efforts by elements of the society within its sphere, and/or a deterioration of the creative arts of living, including invention, literature, visual and performing arts, music, and so on, due to the human energy necessary for sheer survival.

Mesopotamia

A factor contributing to the decline of Mesopotamian civilization given by a number of scholars is salinization – the accumulation of salts in agricultural soil due to evaporation of groundwater caused by irrigation and insufficient drainage, especially under arid or semi-arid conditions. It can reach toxic levels that stunt or prevent the growth of crop plants. Diamond says,

> Salinization is a problem today in many parts of the world besides the U.S., including India, Turkey, and especially Australia … In the past it contributed to the decline of the world's

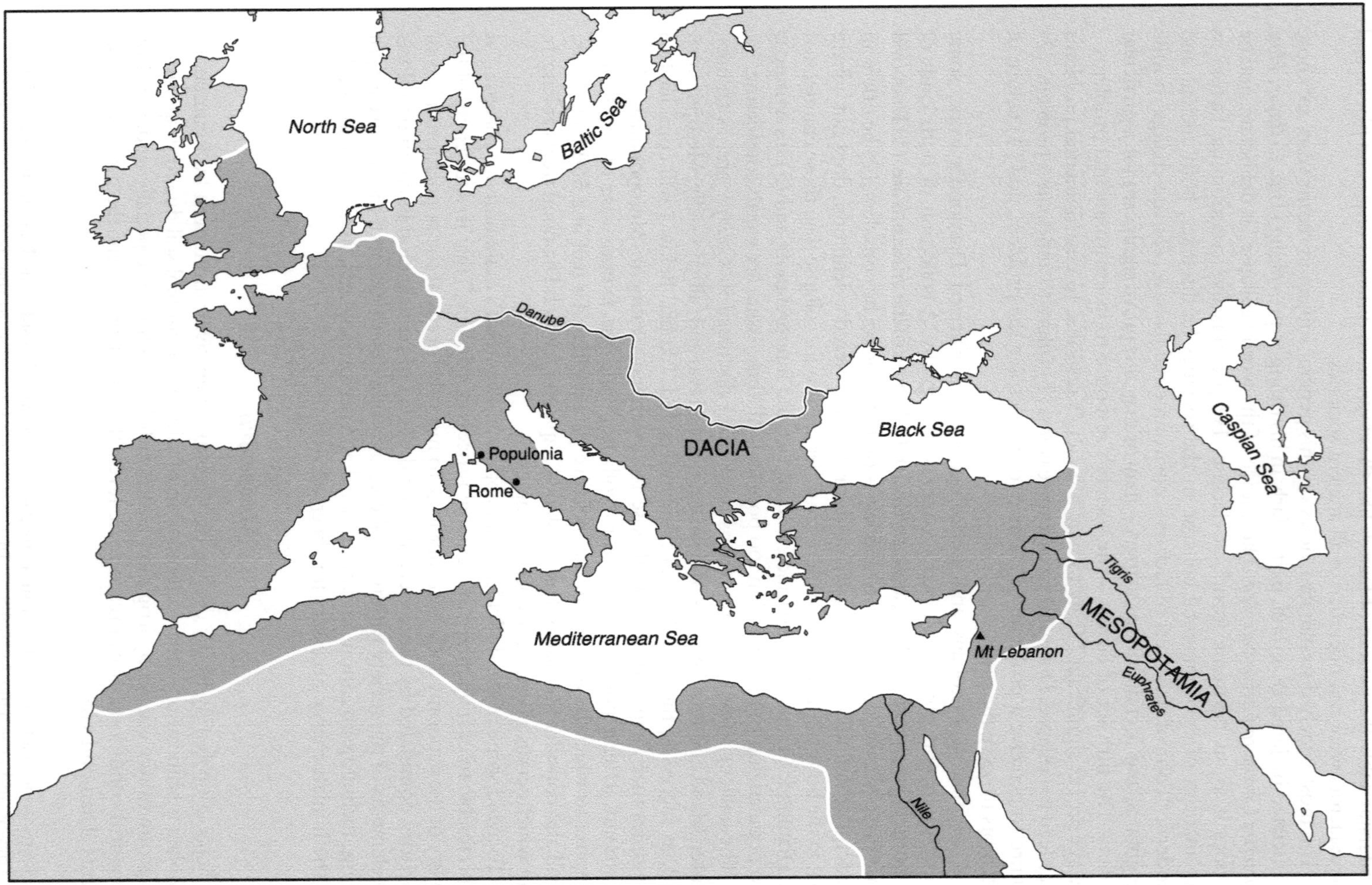

Map 2.1 Roman Empire, 395 CE

oldest civilizations, those of Mesopotamia: salinization provides a large part of the explanation for why applying the term "Fertile Crescent" today to Iraq and Syria, formerly the
leading center of world agriculture, would be a cruel joke.[25]

An important study by Jacobsen and Adams examined records of ancient temples and
found evidence of increasing salinity and declining crop yields in southern Mesopotamia
in the years between 3500 and 1700 BCE, and the authors concluded: "That growing soil
salinity played an important part in the breakup of Sumerian civilization seems beyond
question."[26] Norman Yoffee, in an article attacking environmental causation as a reason
for Mesopotamian collapse, disposes of the work of Jacobsen, Adams, and Diamond in a
footnote, saying "Powell 1985 refuted this."[27] Most of Yoffee's essay deals with Assyria,
in northern Mesopotamia, higher in the river basin, where drainage obviated salinization
as a major problem. Yoffee does not say what Powell's refutation was, leaving the reader
to consult Powell, and, if diligent, to find a later article by Artzy and Hillel that disproved
Powell. In brief, Powell claimed that the Sumerians knew how to leach and drain the salt
from their fields. Artzy and Hillel countered that drainage was virtually impossible
because the soil in southern Mesopotamia is fine-grained with a preponderance of silt
and clay, a condition that resists drainage, and that the river channel and canals had been
raised above the level of the adjacent fields by siltation, which left no place for drainage
water to flow. They conclude "that the process [of salinization] in that area was practically
unavoidable, and that it must have made irrigation-based agriculture unsustainable in
the long run," a conclusion that supports the hypothesis of Jacobsen and Adams and
vindicates the idea that salinization, an environmental factor, was a leading cause of the
decline of southern Mesopotamian civilization.[28]

The Roman Empire

In the case of the Roman Empire (see Map 2.1), more than one environmental cause was
involved in the complex process of decline and fall. The fate of the Roman Empire
constitutes one of the most argued problems of historiography. It was the subject of a
monument of English historical writing, Edward Gibbon's *The Decline and Fall of the
Roman Empire* (1776). In it, he gave reasons for the fall of Rome: two of them were that
the bureaucracy and the military were overgrown, causing the vast structure to collapse
of its own weight, and that many young men were flocking into monasteries, depriving
the government of their support. Other reasons have been offered, then and now. Even
while it was happening, some complained that the moral decadence of contemporary
Romans, or their failure to worship the gods, or the one God, was weakening society.
Later commentators cited overextension of territory, decline of population, class struggle,
the lack of individual initiative, the drain of precious metals to the East, lead poisoning,
soil exhaustion, changing climate, and so on.

Each of these explanations has been forcefully advocated, and some are supported by
good evidence. What seems certain to unbiased observers of the debate over the decline
of Rome is that no single factor is likely to have been the cause of such a complex
phenomenon. It is therefore necessary to look for a series of contributing causes. One
of these was mistreatment of the natural environment, including overexploitation of
natural resources such as forests and soils, and failure to find sustainable ways to interact
with the ecosystems of Italy and the Mediterranean. Environmental factors were
important causes of the decay of Roman economy and society, though not the only

causes, and the most important of environmental failures resulted from human activities. The consequences of the process of deterioration can be seen in the landscape. Civilizations flourish only as long as the ecosystems they depend upon. Nowhere is this clearer than in the Mediterranean basin, where the Roman Empire began and ran its course. For example, abandoned Roman olive presses have been found in desert areas of Tunisia, where today there are no olives or other trees in sight. Ancient ports in Italy and Turkey are stranded behind flats and marshes filled with the silt brought down from deforested hillsides by rivers.

Some perceptive Roman writers were aware that humans often abuse the natural world. Seneca observed, "If we evaluate the benefits of nature by the depravity of those who misuse them, there is nothing we have received that does not hurt us. You will find nothing, even of obvious usefulness, such that it does not change over into its opposite through man's fault" (*Quaestiones naturales*, 5.18.5).

It seemed to some Romans that Earth was waning: becoming less fertile and less able to sustain human beings. Some, like Lucretius, believed this was a natural process, in which Earth was growing older. Others, like Columella, maintained that environmental deterioration is due to human failures. Earth is not growing old, he said: the blame for her infertility lies in poor husbandry; declining crops are our fault, not hers. Earth was seen as responsive to human care or its lack, giving rich returns to those who treat her well and punishing those who are lazy or who weary her by trying to take from her what she is not ready to give. So environmental problems are the passionless revenge of the Earth on those who fail, through ignorance or avarice, to be attentive guardians of the land.

Among the ecological factors operating at the time, many were caused by the Romans themselves. Deforestation and its consequence, erosion, lead the list of these disasters. The Romans cut forests for many uses, such as fuel, timber for construction, and military logistics. Removal of forests to make room for farming or for grazing animals was an important process, especially in lowlands. Agricultural writers commended ways of "reducing a wooded area to an arable state" (Columella, *De re rustica*, 2.2). Roman writers also knew of the destruction caused by shepherds who set fires to burn woody vegetation and encourage the growth of grass.

Marble reliefs spiraling up Trajan's Column, an imposing Roman monument, celebrate Trajan's conquest of Dacia (Romania), and are a principal source of information about Roman military equipment and operations. They also reveal effects of the Roman army on the environment. Trees are shown being chopped down vigorously by soldiers. The axmen are clearing roads through woodland. Soldiers carry away logs to make siege terraces, catapults, battering rams, and beacon fires. Many structures demanded timber in their construction: camps, forts, palisades and other defense works, warships, boats, barges loaded with barrels, and bridges of boats. The work to supply the huge amounts of wood necessary for military operations was supplied by technical support units for the army directed by "axe masters." Transformation of the landscape by these operations was massive and intended. But was this intense use of forest resources sustainable? Study of written sources, archeological reports, scientific studies of tree rings, deposits of silt from erosion, and ancient pollen grains have led to the conclusion that it was not.

The once-flourishing cedar forests of Lebanon were valued for the quality of their wood, used in construction of ships, buildings, and fine furniture. The Emperor Hadrian proclaimed a forest reserve including the cedars for imperial use on Mount Lebanon, where more than 100 boundary markers have been found. This enlightened emperor realized the danger that they might be logged to the point of disappearance, but they continued to be cut, and today only a few protected groves survive, displaying evidence that the use of this resource was unsustainable.

Among the effects of deforestation was the vulnerability of steep slopes to torrential rains. This effect was magnified by grazing of domestic animals wherever there was vegetation, preventing regrowth of trees and shrubs and periodically destroying grass where there was overgrazing.

Farmers knew remedies for problems like siltation, salinization, and exhaustion of essential soil minerals, but could not always apply them due to political and military pressures. The tax system bore heavily on the agricultural sector. Crises of rising prices, food shortages, and labor shortages bore severely on small farmers, and many became dependents of great landowners.

The Roman economy was based on the agrarian sector. The inevitable result of a human failure to support nature was that nature could support fewer human beings. Population decline was a continuing problem. Emperors tried to counter it by edicts making marriage and childbearing mandatory. Declining population meant fewer farm workers, so that reductions in population and production operated in synergy. The Emperor Diocletian's edicts on occupations, requiring men to provide sons to fill their positions, and on prices, setting maximums particularly for food, indicate what was happening at the end of the third century CE: food was scarcer, prices were rising, and there was a labor shortage.

Industry in the Roman Empire was not as large a segment of the economy as it is in modern times, but it had significant environmental consequences. Scars of ancient mining and quarrying persist, though often eclipsed by more recent excavations. Mining and smelting operations did more than lay waste to local areas, however. The fuel needs of a large operation like the iron-making center at Populonia consumed annually the growth of wood provided by an average forest of a million acres (400,000 hectares).[29] There were many such foundries, and added to their fuel demands were the great amounts of wood and charcoal required by the pottery industry.

Pollution may not have been produced on the modern scale, but the amount of wastes from these operations was considerable. Dangerous emissions were produced by some operations; Strabo (*Geography*, 3.2.8) observed that silver-smelting furnaces in Spain were built with tall chimneys to carry deadly smoke away from the workers. Romans lacked technology to reduce effluents to the air or water. Lead can be the predominant metal in silver ore: it and other poisonous elements like mercury and arsenic were present in industrial processes such as metallurgy and pottery, leather, and textile production. Workers in these materials were notably subject to poisoning. Lead, or silver with high lead content, was used in utensils, dishes, and cooking pots, as well as coins. Studies of the Greenland ice cap have shown that lead in the atmosphere increased during Roman times. Bones from Roman burials exhibit a variable, but often high, lead content. In addition, mercury was used in gold refining, and arsenic appeared in pigments and medicines. It is likely that large numbers of people in the Roman Empire suffered from environmental poisoning produced by industrial processes.

The poor quality of the urban environment received frequent comment by poets of the early empire. Noise pollution and smoke received the most notice. Garbage and sewage disposal presented a serious health problem. Rome's main drain, the Cloaca Maxima, emptied into the Tiber River, an efficient arrangement except when floods, unfortunately frequent, backed effluents up into the city.

The trend of the Romans' actions affecting the environment over the centuries was destructive. They exploited renewable resources faster than was sustainable, and consumed nonrenewable resources as rapidly as possible. They failed to adapt their economy to the environment in sustainable ways and placed an insupportable demand on the natural resources available to them. Thus they failed to maintain the balance with nature that is necessary to the prosperity of a human community. They depleted the lands of the Mediterranean world, and in so doing undermined their own ability to survive. Environmental changes as a result of human activities must be judged to be one of the causes of the decline of the Roman Empire.

Easter Island (Rapa Nui)

Easter Island (Rapa Nui; see Map 2.2) is a celebrated historical illustration of the causes and results of environmental degradation. Clive Ponting, Jared Diamond, and others have portrayed it as a cautionary tale of a human society that destroyed its renewable resources and was consequently reduced to a fragment of the population and a shadow of the culture that had marked its zenith. A good scholarly treatment is that of John Flenley and Paul Bahn. The history of the island remains in part a mystery, since the rongo-rongo script carved on its wooden tablets, the only indigenous writing in Oceania, has not been deciphered, and oral tradition was impoverished by the death of the elders entrusted with it due to raids by slavers and epidemics in the nineteenth century. Recently, however, archeologists have found pieces of evidence that make the cultural and environmental process clearer.[30]

What was it like before the Polynesians arrived? It is a volcanic island; all the plants and animals were progeny of a few that arrived by air or sea. It was forested, but scientists have only recently understood the character of the forest ecosystem. The first Europeans to arrive found almost no trees. The native palm, the dominant species of primeval times, was by then extinct. It is known from fossil roots and tree trunks, some preserved in lava flows, and from tiny coconuts, about 2.5 to 3.3 centimeters (an inch to 1.3 inches) in diameter, found on cave floors. The Easter Island palm appears similar to the Chilean wine palm. Another tree, the sophora, a legume with yellow flowers and high-quality wood, barely survives. Until recently, evidence of few other trees or large shrubs was known, and the vegetation of the island before Polynesian settlement could be described as "Palm forest with *Sophora* and shrubs" with some areas of grassland.[31] But in the mid- to late 1990s, Catherine Orliac investigated carbonized remains of wood at several sites on Easter Island and was able to identify many woody plants. Fourteen of these had never been found on Easter Island before, including some known on other Pacific Islands as large trees. The picture that is emerging is one of a complex vegetative cover including high forests containing a variety of species, with the Easter Island palm the most numerous.[32]

There were no land mammals: no bats, no rats. No reptiles either, although insects, spiders, and snails occurred. Marine mammals such as seals, sea lions, and dolphins were present. There were a few land birds including parrots, rails, herons, and owls, all known now only from bones. Migratory sea birds found abundant nesting sites on the rocky cliffs: terns, albatrosses, seagulls, frigate birds, tropic birds, and others. A few survive today only on offshore islets. There were shellfish and crustaceans such as lobsters, but fish were not as numerous as around other Pacific islands, because Easter Island's topography and climate prevented establishment of a coral reef and lagoon. Around most of the island, cliffs fall straight into the sea.

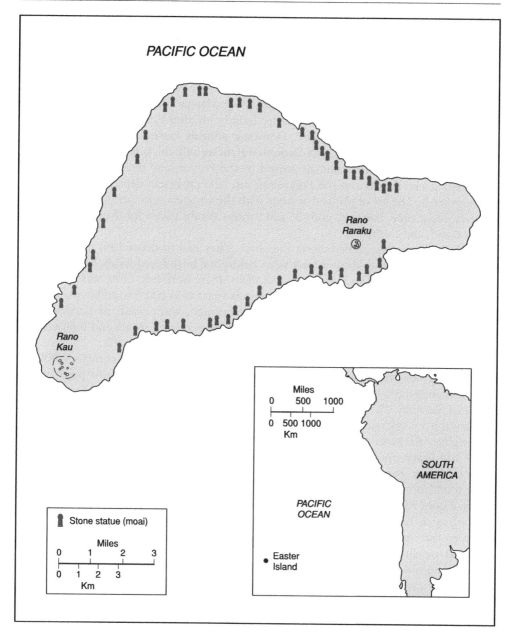

Map 2.2 Easter Island (Rapa Nui)

There is no doubt that the Easter Islanders are Polynesians: they speak a Polynesian language, and their DNA is of Polynesian type. Still, their arrival on such a tiny, distant island is a wonder of human history. Polynesians deliberately explored the Pacific, sailing eastward against the prevailing winds, so that if necessary they could return, making a colony on any suitable island they found by following up with large double-hulled canoes bearing people, domestic animals, and useful plants. One or more of these craft reached

Easter Island. We do not know the date; it has been variously estimated from 300 to 1200 CE. Radiocarbon suggests a date between 615 and 860 CE.

Where did they come from? Judging from language and material culture such as stone statues, it seems the Marquesas are most likely. Easter Island tradition calls the ancestral land "Hiva," and the largest islands of the Marquesas are Hiva Oa and Nuku Hiva. But scholars also suggest the Austral Islands, including Mangareva and Rapa Iti. The name "Rapa Nui," however, was applied to Easter Island only after European discovery.

What did they bring with them? Domestic animals carried by most Polynesian expeditions included pigs, dogs, and chickens. Of these, only chickens became established on Easter Island. Two other animals arrived by the Polynesians' deliberate choice or as stowaways: gecko lizards and the Polynesian rat. The rat began depredations on birds and palm seeds. The list of plants that came with the voyagers is longer: taro, yam, sweet potatoes, sugar cane, bananas, gourds, and various shrubs useful for dyes, paints, high-quality wood, and cloth.

The Polynesians first settled along the coast. They had to depend on the resources they found on the island, since it took years before the introduced foods could increase enough to feed the people. Fortunately for them, there were fish, birds and their eggs, and sea mammals in abundance. The indigenous vegetation had few edible plants, but they began to clear the forest by slashing and burning, and placed the familiar food plants in the soil. Like Polynesians elsewhere, they shaped large stones and built temple platforms, called *ahu* on Easter Island.

Expanding agriculture supported population increase: There were perhaps 9,000 to 15,000 inhabitants by 1500 CE. This necessitated agricultural use of most parts of the island, including the hilly interior. Erection of monumental architecture reflects development of a social hierarchy. The statues, or *moai*, figures of ancestors as high as 9 meters (30 feet) and carved from volcanic stone, were set up on the *ahu*. A progressive increase in height and weight of the *moai* may reflect a competition in size among rival communities. Many statues had heavy crowns of red stone set atop their heads, and eyes of white coral with pupils of darker stone. Meanwhile, the upper classes demanded the erection of houses whose foundations consisted of stones with holes drilled for branches. Moving all these masses of stone required the use of trunks of palm trees, a major cause of forest destruction.

Along with deforestation came soil erosion and loss of bird habitat. The native resources that had supported early expansion began to disappear, so the islanders depended on further extension and intensification of agriculture to support their increasing numbers. The technology used to support agriculture is not as startling as the ceremonial stonework, but was even more important. With few trees or none, there was little to moderate the winds, so farmers dug pits and surrounded them with stones to protect the taro and bananas. They placed stones in the fields, forming "lithic mulch" that protected plants and conserved moisture.[33] These methods were labor intensive, with the common people providing the labor.

In the latter half of the seventeenth century, a convergence of crises occurred. One day the last palm tree was cut down. Statues could no longer be moved, which helps to explain the appearance of the quarry at Rano Raraku, with sculptures in every stage of preparation, looking as if the order "tools down" had been given and all the laborers departed. The population had reached the limit of environmental support, resulting in food shortages. It was not possible for out-migration to relieve population pressure, because no materials remained for the construction of canoes large enough for interisland voyages. Conflict increased as groups attempted to seize resources. The population crashed to around 2,000. Chickens, the major source of protein, were housed in fortress-like stone coops. With starvation ever-present, the common people began to question the order of things. Direction of agriculture by the nobles had failed to provide the people with ample food. The great stone statues, guardians of safety and abundance, had also failed. There was inter-class war – a strong element in the oral tradition – and the hierarchy was overthrown. The commoners pushed down some of the statues. There were some still standing when the Europeans arrived, but by 1825 all were toppled. What role drought, crop failure, or climatic disturbances such as El Niño played in all this is a matter for further research, but surely human impact on the natural environment was the leading cause of decline.

Environmental and social disaster made a new order necessary, intellectually and economically. Worship and labor were redirected away from the veneration of ancestors and statues to the god Make-Make, creator of humankind whose image decorated Orongo, the village of the new cult of the Birdman. Carvings of men with heads of the sooty tern, a migratory sea bird that by then nested only on a rocky offshore islet, cover lava rocks. In an annual contest, young men swam out to the islet when the birds arrived. The one who brought back the first egg became the birdman of the year, endowed with political and economic privileges but kept in a special house and subjected to taboos. Also, agricultural technology began to revive and the decimated population survived, although within an impoverished landscape.

In summary, the islanders optimized their use of natural resources, and this made population growth possible. Consumption increased to the point where diminishing resources interfered with population growth. Faced with starvation, they devised new technologies to extract more production from the land. In times of crisis, social organizations collapsed and were transformed. But the bottom line was the ecosystem: the landscape with its living and non-living components. After depleting renewable resources, the Easter Islanders used clever stone-based technologies to raise sweet potatoes and sugarcane on a windswept island. But they could never bring back the palm trees and the rest of the humid high forest. The birds would never nest again in great numbers on the cliffs. And without trees for building canoes, there was no way across the sea … until the strangers came.

The Dutch admiral Jacob Roggeveen was first to arrive in 1722. A few Spanish, British, and French visits occurred during the rest of the eighteenth century. These explorers stayed briefly, killing a number of islanders and perhaps introducing venereal and other diseases.

The nineteenth century brought horrors that almost destroyed the Easter Islanders. American seal hunters and whalers, and Peruvians seeking slaves, killed or carried off half the population. The bishop of Tahiti, Tepano Jaussen, managed to save some of the captives and arranged their return to Easter Island, but they carried smallpox and tuberculosis that infected those who had remained. All but 111 died.

The islanders that are descendants of the remnant numbered about 2,275 in 2002. The birdman cult survived until about 1867, perishing from lack of young men to perform it. It has been revived as part of a festival. Chile annexed the island in 1888 and turned it over to a sheep-herding business. Today most of the sheep are gone; there are horses, cattle, and eucalyptus plantations, and some tourism. Self-rule on the island is an unresolved issue.

The decline of Easter Island society therefore is not a simple story. There were, so to speak, two "collapses," the first as a result of resource overconsumption, and the second as a result of foreign exploitation and disease. In each case, a somewhat tentative recovery followed.

Conclusion: Comparative Environmental History

An advantage of the study of ancient environmental history is that it presents the panorama of a number of societies that ran their course with the environment over an extended period of time. One may set these side by side and look for elements and processes that were similar, and, failing that, ask if there were unique factors. Were there elements that had a role in all ancient societies? Water supply, certainly, was formative in both presence and absence everywhere, but the problems it presented were different in, say, Egypt and the Maya lands, and each river-valley civilization was confronted by the idiosyncrasies of its own rivers. Deforestation appeared wherever there were forests, but varied with forest type, and some lands had little or no forest to begin with. Agriculture was distinct in North China as contrasted to South China. Natural disasters are unevenly distributed around the globe. Perhaps the insight of comparative environmental history in the ancient world is that all societies must find a *modus vivendi* with their ecological settings, but that these settings are unique, and therefore the appropriate solutions must also be to one extent or another different.

Another comparison lurks within studies of the ancient world, however much objective historians may convince themselves that they avoid it, or at least strive to avoid it, and that is the comparison between ancient and modern. We look at the past unavoidably from within our own historical setting. How was the environmental experience of the distant past like that of today? Were ruinous choices made then that are being repeated now? If so, can they be avoided? Can successful choices be imitated? Or is our situation totally other than that of the ancient world, and therefore incommeasurable? We should allow questions like these to bother us.

Notes

1   E. Burke, "The Transformation of the Middle Eastern Environment, 1500 BCE–2000 CE," in E. Burke III and K. Pomeranz (eds.), *The Environment and World History*, Berkeley, University of California Press, 2009, pp. 81–117 (p. 81).

2   J. H. Bentley, "Cross-Cultural Interaction and Periodization in World History," *The American Historical Review* 101/3, 1996, pp. 749–70 (p. 749).

3   J. Radkau, *Nature and Power*, Cambridge, Cambridge University Press, 2008, pp. 86–7.

4   M. Williams, *Deforesting the Earth: From Prehistory to Global Crisis*, Chicago, University of Chicago Press, 2003; J. R. McNeill, *The Mountains of the Mediterranean World: An Environmental History*, New York, Cambridge University Press, 1992; J. Perlin, *A Forest Journey: The Story of Wood and Civilization*, New York, W. W. Norton, 2005; J. D. Hughes, *Pan's Travail: Environmental Problems of the Ancient Greeks and Romans*, Baltimore, The Johns Hopkins University Press, 1996; A. T. Grove and O. Rackham, *The Nature of Mediterranean Europe: An Ecological History*, New Haven, CT, Yale University Press, 2001.

5   P. Horden and N. Purcell, *The Corrupting Sea: A Study of Mediterranean History*, Oxford, Blackwell, 2000, p. 743.

6   W. R. Coe and W. A. Haviland, *Tikal Report 12: Introduction to the Archaeology of Tikal, Guatemala*, Philadelphia, PA, University of Pennsylvania Museum of Archaeology and Anthropology, 1982.

7   W. A. McDonald and G. R. Rapp Jr. (eds.), *The Minnesota Messenia Expedition: Reconstructing a Bronze Age Regional Environment*, Minneapolis, MN, University of Minnesota Press, 1972.

8   B. D. Shaw, "Climate, Environment, and History: The Case of Roman North Africa," in T. M. L. Wigley, M. J. Ingram, and G. Farmer (eds.), *Climate and History: Studies in Past Climates and Their Impact on Man*, Cambridge, Cambridge University Press, 1981, pp. 379–403.

9   R. B. Alley, *The Two-Mile Time Machine: Ice Cores, Abrupt Climate Change, and Our Future*, Princeton, NJ, Princeton University Press, 2000.

10 S. Hong, J.-P. Candelone, C. C. Patterson, and C. F. Boutron, "Greenland Ice Evidence of Hemispheric Lead Pollution Two Millennia Ago by Greek and Roman Civilizations," *Science* 265/5180, 1994, pp. 1841–3.

11 W. F. Ruddiman, *Plows, Plagues, and Petroleum: How Humans Took Control of Climate*, Princeton, NJ, Princeton University Press, 2005.

12 B. J. Huntley and H. J. B. Birks, *An Atlas of Past and Present Pollen Maps for Europe: 0–13,000 Years Ago*, Cambridge, Cambridge University Press, 1983.

13 V. Andrieu, E. Brugiapaglia, R. Cheddadi, et al., "A Database for the Palynological Recording of Human Activity," in P. Leveau, F. Trément, K. Walsh, and G. Barker (eds.), *Environmental Reconstruction in Mediterranean Landscape Archaeology*, Oxford, Oxbow Books, 1999, pp. 17–24.

14 D. K. Davis, *Resurrecting the Granary of Rome: Environmental History and French Colonial Expansion in North Africa*, Athens, OH, Ohio University Press, 2007, p. 11.

15 H. G. Lamb, U. Eicher, and V. R. Switsur, "An 18.000-Year Record of Vegetation, Lake-Level, and Climatic Change from Tigalmamine, Middle Atlas, Morocco," *Journal of Biogeography* 16, 1989, pp. 65–74.

16 P. I. Kuniholm, "Dendrochronology," in D. Brothwell and A. M. Pollard (eds.), *Handbook of Archaeological Sciences*, London, Wiley, 2001; S. W. Manning, and M. J. Bruce (eds.), *Tree-Rings, Kings, and Old World Archaeology and Environment: Papers Presented in Honor of Peter Ian Kuniholm*, Oxford, Oxbow Books, 2009.

17 C. L. Pearson, D. S. Dale, P. W. Brewer, et al., "Dendrochemical Analysis of a Tree-Ring Growth Anomaly Associated with the Late Bronze Age Eruption of Thera," *Journal of Archaeological Science* 36/6, 2009, pp. 1206–14; C. L. Pearson and S. W. Manning, "Could Absolutely Dated Tree-Ring Chemistry Provide a Means to Dating the Major Volcanic Eruptions of the Holocene?," in S. W. Manning and M. J. Bruce (eds.), *Tree-Rings, Kings, and Old World Archaeology and Environment: Papers Presented in Honor of Peter Ian Kuniholm*, Oxford, Oxbow Books, 2009, pp. 97–109.

18 H. Godwin and A. G. Tansley, "Prehistoric Charcoals as Evidence of Former Vegetation, Soil and Climate," *Journal of Ecology* 29, 1941, pp. 117–26.

19 E. Asouti, *Charcoal Analysis Web*, 2009, accessed from http://pcwww.liv.ac.uk/~easouti/index.htm on February 21, 2012; G. H. Willcox, "A History of Deforestation as Indicated by Charcoal Analysis of Four Sites in Eastern Anatolia," *Anatolian Studies* 24, 1974, pp. 117–33; K. Neumann, "The Contribution of Anthracology to the Study of the Late Quaternary Vegetation History of the Mediterranean Region and Africa," *Bulletin de la Société Botanique de France* 139, 1992, pp. 421–40.

20 J.-L. Vernet, "Reconstructing Vegetation and Landscapes in the Mediterranean: The Contribution of Anthracology," in Leveau, Trément, Walsh, and Barker (eds.), *Environmental Reconstruction*, pp. 25–36; F. Laubenheimer, *20 Ans de Recherches à Sallèles d'Aude: Le Monde des Potiers Gallo-Romains*, Besançon, Presses Universitaires de Franche-Comté, 2001.

21   C. Orliac, "The Woody Vegetation of Easter Island Between the Early 14th and the Mid-17th Centuries AD," in C. M. Stevenson and W. S. Ayres (eds.), *Easter Island Archaeology: Research on Early Rapanui Culture*, Los Osos, Easter Island Foundation, 2000, pp. 211–20.

22   J. Diamond, *Collapse: How Societies Choose to Fail or Succeed*, New York, Viking, 2005, pp. 3, 11.

23   Diamond, *Collapse*, p. 421.

24   P. A. McAnany and N. Yoffee (eds.), *Questioning Collapse: Human Resilience, Ecological Vulnerability, and the Aftermath of Empire*, Cambridge, Cambridge University Press, 2010.

25   Diamond, *Collapse*, p. 48.

26   T. Jacobsen and R. M. Adams, "Salt and Silt in Ancient Mesopotamian Agriculture," *Science* 128, 1958, pp. 1251–8.

27   Yoffee, N., "Collapse in Ancient Mesopotamia: What Happened, What Didn't," in McAnany and Yoffee (eds.), *Questioning Collapse*, pp. 176–206 (p. 201).

28   Artzy, M., and Hillel, D., "A Defense of the Theory of Progressive Soil Salinization in Ancient Southern Mesopotamia," *Geoarchaeology: An International Journal* 3/3, 1988, pp. 235–8 (p. 238).

29   T. A. Wertime, "The Furnace versus the Goat: The Pyrotechnic Industries and Mediterranean Deforestation in Antiquity," *Journal of Field Archaeology* 10/4, 1983, pp. 445–52.

30   C. Ponting, *A Green History of the World: The Environment and the Collapse of Great Civilizations*, New York, St. Martin's Press, 1991; Diamond, *Collapse*; J. Flenley and P. Bahn, *The Enigmas of Easter Island: Island on the Edge*, Oxford, Oxford University Press, 2003.

31   Zizka, G., "Flowering Plants of Easter Island," *Scientific Reports PHF* 3, 1991, p. 16.

32   Orliac, "The Woody Vegetation of Easter Island."

33   C. M. Stevenson, and S. Haoa, "Diminished Agricultural Productivity and the Collapse of Ranked Society on Easter Island," in P. Wallin (ed.), *Archaeology, Agriculture and Identity*, Oslo, Kon-Tiki Museum, 1999, pp. 4–12.

References

Alley, R. B., *The Two-Mile Time Machine: Ice Cores, Abrupt Climate Change, and Our Future*, Princeton, NJ, Princeton University Press, 2000.

Andrieu, V., Brugiapaglia, E., Cheddadi, R., et al., "A Database for the Palynological Recording of Human Activity," in P. Leveau, F. Trément, K. Walsh, and G. Barker (eds.), *Environmental Reconstruction in Mediterranean Landscape Archaeology*, Oxford, Oxbow Books, 1999, pp. 17–24.

Artzy, M., and Hillel, D., "A Defense of the Theory of Progressive Soil Salinization in Ancient Southern Mesopotamia," *Geoarchaeology: An International Journal* 3/3, 1988, pp. 235–8.

Asouti, E., *Charcoal Analysis Web*, 2009, accessed from http://pcwww.liv.ac.uk/~easouti/index.htm on February 21, 2012.

Bentley, J. H., "Cross-Cultural Interaction and Periodization in World History," *The American Historical Review* 101/3, 1996, pp. 749–70.

Burke, E., "The Transformation of the Middle Eastern Environment, 1500 BCE–2000 CE," in E. Burke III and K. Pomeranz (eds.), *The Environment and World History*, Berkeley, University of California Press, 2009, pp. 81–117.

Coe, W. R., and Haviland, W. A., *Tikal Report 12: Introduction to the Archaeology of Tikal, Guatemala*, Philadelphia, PA, University of Pennsylvania Museum of Archaeology and Anthropology, 1982.

Davis, D. K., *Resurrecting the Granary of Rome: Environmental History and French Colonial Expansion in North Africa*, Athens, OH, Ohio University Press, 2007.

Diamond, J., *Collapse: How Societies Choose to Fail or Succeed*, New York, Viking, 2005.

Flenley, J., and Bahn, P., *The Enigmas of Easter Island: Island on the Edge*, Oxford, Oxford University Press, 2003.

Godwin, H., and Tansley, A. G., "Prehistoric Charcoals as Evidence of Former Vegetation, Soil and Climate," *Journal of Ecology* 29, 1941, pp. 117–26.

Grove, A. T., and Rackham, O., *The Nature of Mediterranean Europe: An Ecological History*, New Haven, CT, Yale University Press, 2001.

Hong, S., Candelone, J.-P., Patterson, C. C., and Boutron, C. F., "Greenland Ice Evidence of Hemispheric Lead Pollution Two Millennia Ago by Greek and Roman Civilizations," *Science* 265/5180, 1994, pp. 1841–3.

Horden, P., and Purcell, N., *The Corrupting Sea: A Study of Mediterranean History*, Oxford, Blackwell, 2000.

Hughes, J. D., *Pan's Travail: Environmental Problems of the Ancient Greeks and Romans*, Baltimore, The Johns Hopkins University Press, 1996.

Huntley, B. J., and Birks, H. J. B., *An Atlas of Past and Present Pollen Maps for Europe: 0–13,000 Years Ago*, Cambridge, Cambridge University Press, 1983.

Jacobsen, T., and Adams, R. M., "Salt and Silt in Ancient Mesopotamian Agriculture," *Science* 128, 1958, pp. 1251–8.

Kuniholm, P. I., "Dendrochronology," in D. Brothwell and A. M. Pollard (eds.), *Handbook of Archaeological Sciences*, London, Wiley, 2001.

Lamb, H. G., Eicher, U., and Switsur, V. R., "An 18.000-Year Record of Vegetation, Lake-Level, and Climatic Change from Tigalmamine, Middle Atlas, Morocco," *Journal of Biogeography* 16, 1989, pp. 65–74.

Laubenheimer, F., *20 Ans de Recherches à Sallèles d'Aude: Le Monde des Potiers Gallo-Romains*, Besançon, Presses Universitaires de Franche-Comté, 2001.

Manning, S. W., and Bruce, M. J. (eds.), *Tree-Rings, Kings, and Old World Archaeology and Environment: Papers Presented in Honor of Peter Ian Kuniholm*, Oxford, Oxbow Books, 2009.

McAnany, P. A., and Yoffee, N. (eds.), *Questioning Collapse: Human Resilience, Ecological Vulnerability, and the Aftermath of Empire*, Cambridge, Cambridge University Press, 2010.

McDonald, W. A., and Rapp, G. R., Jr. (eds.), *The Minnesota Messenia Expedition: Reconstructing a Bronze Age Regional Environment*, Minneapolis, MN, University of Minnesota Press, 1972.

McNeill, J. R., *The Mountains of the Mediterranean World: An Environmental History*, New York, Cambridge University Press, 1992.

Neumann, K., "The Contribution of Anthracology to the Study of the Late Quaternary Vegetation History of the Mediterranean Region and Africa," *Bulletin de la Société Botanique de France* 139, 1992, pp. 421–40.

Orliac, C., "The Woody Vegetation of Easter Island Between the Early 14th and the Mid-17th Centuries AD," in C. M. Stevenson and W. S. Ayres (eds.), *Easter Island Archaeology: Research on Early Rapanui Culture*, Los Osos, Easter Island Foundation, 2000, pp. 211–20.

Pearson, C. L., and Manning, S. W., "Could Absolutely Dated Tree-Ring Chemistry Provide a Means to Dating the Major Volcanic Eruptions of the Holocene?," in S. W. Manning and M. J. Bruce (eds.), *Tree-Rings, Kings, and Old World Archaeology and Environment: Papers Presented in Honor of Peter Ian Kuniholm*, Oxford, Oxbow Books, 2009, pp. 97–109.

Pearson, C. L., Dale, D. S., Brewer, P. W., et al., "Dendrochemical Analysis of a Tree-Ring Growth Anomaly Associated with the Late Bronze Age Eruption of Thera," *Journal of Archaeological Science* 36/6, 2009, pp. 1206–14.

Perlin, J., *A Forest Journey: The Story of Wood and Civilization*, New York, W. W. Norton, 2005.

Ponting, C., *A Green History of the World: The Environment and the Collapse of Great Civilizations*, New York, St. Martin's Press, 1991.

Radkau, J., *Nature and Power*, Cambridge, Cambridge University Press, 2008.

Ruddiman, W. F., *Plows, Plagues, and Petroleum: How Humans Took Control of Climate*, Princeton, NJ, Princeton University Press, 2005.

Shaw, B. D., "Climate, Environment, and History: The Case of Roman North Africa," in T. M. L. Wigley, M. J. Ingram, and G. Farmer (eds.), *Climate and History: Studies in Past Climates and Their Impact on Man*, Cambridge, Cambridge University Press, 1981, pp. 379–403.

Stevenson, C. M., and Haoa, S., "Diminished Agricultural Productivity and the Collapse of Ranked Society on Easter Island," in P. Wallin (ed.), *Archaeology, Agriculture and Identity*, Oslo, Kon-Tiki Museum, 1999, pp. 4–12.

Vernet, J.-L., "Reconstructing Vegetation and Landscapes in the Mediterranean: The Contribution of Anthracology," in P. Leveau, F. Trément, K. Walsh, and G. Barker (eds.), *Environmental Reconstruction in Mediterranean Landscape Archaeology*, Oxford, Oxbow Books, 1999, pp. 25–36.

Wertime, T. A., "The Furnace versus the Goat: The Pyrotechnic Industries and Mediterranean Deforestation in Antiquity," *Journal of Field Archaeology* 10/4, 1983, pp. 445–52.

Willcox, G. H., "A History of Deforestation as Indicated by Charcoal Analysis of Four Sites in Eastern Anatolia," *Anatolian Studies* 24, 1974, pp. 117–33.

Williams, M., *Deforesting the Earth: From Prehistory to Global Crisis*, Chicago, University of Chicago Press, 2003.

Yoffee, N., "Collapse in Ancient Mesopotamia: What Happened, What Didn't," in P. A. McAnany and N. Yoffee (eds.), *Questioning Collapse: Human Resilience, Ecological Vulnerability, and the Aftermath of Empire*, Cambridge, Cambridge University Press, 2010, pp. 176–206.

Zizka, G., "Flowering Plants of Easter Island," *Scientific Reports PHF* 3, 1991, p. 16.

Further Reading

Dansgaard, W., *Frozen Annals: Greenland Ice Cap Research*, Odder, Aage V. Jenkins Fonde, 2004.

Elvin, M., *The Retreat of the Elephants: An Environmental History of China*, New Haven, CT, Yale University Press, 2004.

Heyerdahl, T., *Aku-Aku: The Secret of Easter Island*, Chicago, Rand McNally, 1958.

Hillel, D., "Civilization, role of soils," in D. Hillel (ed.), *Encyclopedia of Soils in the Environment*, vol. 1, Oxford, Elsevier Academic Press, 2005, pp. 199–204.

Hughes, J. D., "Ancient Deforestation Revisited," *Journal of the History of Biology* 44/1, 2011, pp. 43–57.

Hughes, J. D., *An Environmental History of the World: Humankind's Changing Role in the Community of Life*, 2nd ed., London, Routledge, 2009.

Hughes, J. D., *Pan's Travail: Environmental Problems of the Ancient Greeks and Romans*, Baltimore, MD, and London, The Johns Hopkins University Press, 1994.

Hughes, J. D., "Warfare and Environment in the Ancient World," in B. Campbell and L. Tritle (eds.), *The Oxford Handbook of Warfare in the Classical World*, Oxford, Oxford University Press, forthcoming.

Meiggs, R., *Trees and Timber in the Ancient Mediterranean World*, Oxford, Clarendon Press, 1982.

Pinto, B., Aguiar, C., and Partidario, M., "Brief Historical Ecology of Northern Portugal during the Holocene," *Environment and History* 16/1, 2010, pp. 3–42.

Runnels, C. N., "Environmental Degradation in Ancient Greece," *Scientific American* 272/3, 1995, pp. 96–9.

Walsh, K., "Caring about Sediments: The Role of Geoarchaeology in Mediterranean Landscapes," *Journal of Mediterranean Archaeology* 17/2, 2004, pp. 223–45.

Willcox, G. H., "Charcoal Analysis and Holocene Vegetation History in Southern Syria," *Quaternary Science Reviews* 18/4–5, 1999, pp. 711–16.

CHAPTER THREE

The Medieval World, 500 to 1500 CE

DANIEL HEADRICK

Between 500 and 1500 CE, as people multiplied in numbers and pushed back the frontiers of inhabited lands, the tug-of-war between them and the rest of nature began to shift toward the human race. Human numbers grew and settlements multiplied. Forests and wetlands retreated. The tools and techniques used to squeeze more food out of the biosphere grew more sophisticated and efficient. Yet this was also a time when natural disasters demolished the achievement of once-proud civilizations. The human position within the natural world remained precarious, and the outcome of the tug-of-war still open.

Sixth-Century Disasters

For the lands around the Mediterranean, the sixth century marked the end of the classical age. Two environmental events – a dust veil in the atmosphere and an epidemic – contributed to its demise.

The first event began in 535–6.[1] As the Byzantine historian Procopius wrote: "The sun gave forth its light without brightness, like the moon, during this whole year, and it seemed exceedingly like the sun in eclipse, for the beams it shed were not clear nor such as it is accustomed to shed."[2] Writers in Constantinople, Mesopotamia, and Ireland reported a similar phenomenon. In China, *The History of the Southern Dynasties* noted that in late 535, "Yellow dust rained down like snow." The following two years, frost and snow ruined the crops, causing a major famine. Abnormally narrow tree rings – evidence of slower than normal growth – in northern Europe, Ireland, Siberia, Chile, and California, corroborate these sources.[3]

Some authors attribute this anomaly to a volcanic eruption that ejected dust into the atmosphere, thereby reducing the sunshine that reached the Earth. Others suggest that a comet could have broken up in the atmosphere. The dust veil may also have triggered

A Companion to Global Environmental History, First Edition. Edited by J.R. McNeill and Erin Stewart Mauldin.
© 2012 John Wiley & Sons, Ltd. Published 2015 by John Wiley & Sons, Ltd.

major changes in human history. In China, poor harvests caused instability and chaos, encouraging Buddhism and Daoism to flourish among the common people. Drought in southern Russia persuaded the nomadic Avars to flee westward, pushing Slavic peoples to invade the Byzantine Empire.[4]

Ominous as these events must have seemed, worse was to come. According to Procopius, in the spring of 541 thousands died every day in Constantinople. An epidemic that originated in Egypt spread from there to Palestine, Syria, and Constantinople, and then to southern Europe and Mesopotamia. It generally stayed close to maritime trade and river routes and burned itself out after four months.

Almost all authors agree that it was the bubonic plague. This disease is caused by the bacillus *Yersinia pestis*, which infects the flea that lives in the fur of rodents, especially black rats. The flea can also survive in any warm and moist environment, such as clothing. Since black rats prefer to eat grain and stay close to humans, their fleas readily jump onto people. A few days after a person is bitten by an infected flea, buboes or black swellings appear in the groin and armpits, followed, in most cases, by death. The disease can mutate into two other forms of plague – pneumonic and septicemic – that are transmitted directly from person to person and have a 100 percent death rate.[5]

Where the plague came from is uncertain. Historian William McNeill suggested that it originated in the foothills of the Himalayas, whence it traveled across India and up the Red Sea. East Africa, where rats boarded ships traveling up the coast to Yemen and Egypt, is a more likely possibility, for the epidemic bypassed Persia and did not affect China until a century later.[6]

The Byzantine Empire after Justinian (after whom the disease is named) was but a shadow of the Roman Empire at its second-century peak. In the 530s, his armies had reconquered formerly Roman territories in North Africa, Italy, and Spain, but by the end of Justinian's reign, recurrent bouts of plague and other epidemics had so reduced the empire's population – Constantinople had lost half its inhabitants – that it was too weak to hold back the renewed incursions of barbarians into the Balkans, Italy, or the coast of North Africa. Nor could it prevent Arabs from conquering most of the Middle East a century later.

The plague returned on average every 9 to 12 years. After about 650, it disappeared from Europe. In the Middle East, it reappeared in 608. During the seventh and early eighth centuries, it broke out several more times in the Middle East and perhaps China. Then it vanished for almost 700 years.[7] As the epidemic subsided, the Middle East, China, and Europe, saw a resurgence of population growth, urbanization, and economic development and a remarkable cultural flowering. From an environmental point of view, the salient story after the sixth century is the expansion of agriculture that supported this efflorescence in all three regions.

The Middle East

In the seventh and eighth centuries, Arabs conquered the Middle East and North Africa, founded new cities, extended long-distance trade, and encouraged the sciences and the arts. According to historian Andrew Watson, the opening up of trade with South and Southeast Asia stimulated an agricultural revolution throughout the Muslim world.[8]

Among the innovations were newly introduced crops, including important ones such as cotton, sugarcane, sorghum, durum wheat, and Asian rice. Some crops, previously limited to a few places, like cotton in Yemen, were more widely diffused. Others, introduced from South and Southeast Asia, were hot-weather crops that were planted in the summer in rotation with winter crops such as wheat and barley, doubling the productivity of the land. These tropical plants needed elaborate irrigation systems. When the Arabs arrived, they found the lower Tigris floodplain a marshy quagmire and Egypt and North Africa in decline since Roman times. Employing Persian engineers, they repaired old irrigation systems and introduced dams, tanks, and waterwheels. On the Iranian plateau, Arabs built irrigation tunnels known as qanats to water dry lands, and produced cotton for the booming markets of Baghdad and the Muslim world.[9]

Success came not just from crops, but from knowledge and institutions as well. During the height of the Arab Empire (the Umayyad and Abbasid caliphates) from the seventh to the eleventh century, agricultural manuals proliferated, as did works on irrigation. Experts emphasized the importance of fertilizing and crop rotation and of choosing the most appropriate crop for each type of soil, climate, and method of irrigation. Rulers built botanical gardens and encouraged irrigation, agriculture, and trade. As the land produced more crops and greater yields, incomes rose, as did population. More people meant more villages and growing cities. The agricultural revolution created a virtuous cycle of prosperity and culture.

Middle Eastern agriculture, however, rested on a delicate structure of interlocking elements. The achievements of the Umayyad caliphate (661–750) in restoring canals in Mesopotamia were undone by the epidemic of plague of 749–50 that reduced the population by 40 percent. The Abbasid caliphate (750–1258) presided, for a while, over a revival of agriculture, due as much to the use of microirrigation devices – waterwheels, cisterns, Archimedes screws, windmills – as to large reservoirs, canals, and aqueducts. With many new crops available, prosperity came from intensive gardening, specialization, and trade more than from large estates.[10]

The climate also played a part. The prosperity of the Abbasid caliphate and the Iranian cotton boom that had buttressed it correlated with a warm period throughout the ninth and tenth centuries. In the late tenth century, when the climate turned cold and the cotton boom ended, the economy and population declined. Turkish-speaking camel herders who invaded Iran damaged irrigation works or neglected to maintain them, for they had little understanding of complex agriculture or appreciation of the legal institutions and intellectual foundations that supported it. The Mongols who sacked Baghdad and overthrew the last Abbasid caliph in 1258 have often been blamed for the demise of the Arab Empire, but it had been declining long before.[11]

China

The landscape of the North China Plain is as much the work of the Chinese people over 3,000 years as of the forces of nature. In contrast, southeastern China did not achieve population densities and urbanization similar to those of North China until the period we are studying here.

The Yangzi valley and the lands south of it enjoy a warm rainy climate suitable to growing rice. Though rugged and difficult to farm, these lands tempted northern Chinese from early on to migrate south. The Sui dynasty (589–618) encouraged migration by building the Grand Canal to transport rice from the south to supplement the millet and wheat grown in the north. The first 200 years of the Tang dynasty (618–907) brought stability and prosperity to China. Under Tang rule the population rose and the lower Yangzi valley was transformed into the richest farmland in China.

In the ninth century, however, a series of droughts undermined Tang rule.[12] The time of chaos known as the "five dynasties" (907–60) accelerated the migration of people to the south.

Wet-rice agriculture was more reliable and provided larger surpluses than crops grown in the north. It also required tools and skills first perfected between the eighth and twelfth centuries. Advances in agricultural technology – waterwheels, the transplanting of rice seedlings, double cropping, and fertilizers – produced far greater yields than growing rice in naturally flooded areas. Transforming wetlands into productive rice paddies meant building canals to bring in water, ditches to remove the excess, and dikes and sluice-gates to control the flow. Farmers moved water in or out with water-lifting devices and fertilized their paddies with pond silt, ashes, and the excrement of humans and animals. Even more labor-intensive was the construction and maintenance of terraces on mountain slopes. Tunnels, aqueducts, channels, and bamboo pipes carried water to and from the terraces.

A further push to develop agriculture in the southeast came during the Song dynasty (960–1279). When northern China was conquered by the Jurchen, a pastoralist people of the steppe, in 1115, the Song rulers retreated to the south. Along with them came thousands of migrants. The Song state actively encouraged settlement and agriculture. It published instruction manuals, provided seeds, and offered low-interest loans and tax rebates. Under the direction of wealthy landowners, farmers undertook the reclamation of the Yangzi delta. Once drained and protected from tides, the delta became the most densely populated land in China, producing as much grain on less land and at lower cost than the entire North China Plain – more than ten times the size of the Yangzi delta.

The most important contribution of the Song government was the introduction of Champa rice from Vietnam in 1012. This drought-resistant variety ripened in 100 days or less, so farmers could grow two crops a year, thereby more than doubling the yield of rice culture. The population of China rose to around 100 million, almost all of the growth occurring in the south, while the north stagnated or declined. Though in later centuries China experienced vicissitudes of all sorts, the agricultural system perfected under the southern Song remained as durable as that of the ancient Egyptians.[13]

Nonetheless, there were prices to pay. One was what Sinologist Mark Elvin has called "technological lock-in."[14] In wet-rice agriculture, the yields were proportional to the amount of labor devoted to farming, keeping living standards low. Reliance on wet-rice cultivation also made Chinese agriculture vulnerable to disruption and war, as became clear when the Mongols overran China in the thirteenth century, bringing death and destruction in their wake.

What had delayed the migration of people to the south for 2,000 years after the beginning of civilization in the north? The more difficult terrain, with its swamps and steep hillsides, contributed to the delay, as did the long process of mastering the technique of growing rice in artificial ponds or paddies.[15] But diseases were an important deterrent as well, in particular malaria, dengue fever, filariasis, schistosomiasis, and other tropical diseases. As the Chinese historian Sima Qian (who died in 86 BCE) explained: "In the area south of the Yangtse the land is low and the climate humid; adult males die young."[16]

The Chinese people also suffered from epidemics that afflicted the entire country. Historians can identify smallpox, a disease known in China for centuries before the Tang and probably introduced from the north by barbarians, for the Chinese called it "Hun-pox."[17]

Another was the bubonic plague, which medical texts after 610 described as causing lumps on the body. An epidemic in 636 followed the plague that devastated Mesopotamia and Syria, regions with which China was in contact through the Silk Road.[18]

Europe

While the Chinese were migrating south, the center of gravity of Europe shifted northward. With the exception of Muslim Spain, Mediterranean Europe stagnated. In the chaos resulting from wars and epidemics, farmers abandoned the lowlands and moved into the mountains. Without them, terraces fell apart, soil eroded, and crops were replaced by tough low scrub. Sheep and goats prevented the regrowth of oak forests. Once salubrious lowland areas became infested with malaria-carrying mosquitoes, causing yet more flight to the mountains.[19]

From the early eighth to the late thirteenth century, an unusually mild climate that historians call the "medieval warm period" bathed western and central Europe. Temperatures rose 0.5 to 1 degrees Celsius (0.9 to 1.8 degrees Fahrenheit). Thanks to mild rainy winters and long hot summers, the growing seasons lasted up to three weeks longer between 800 and 1300 than they had in Roman times. Fewer May frosts threatened the growing crops. Farmers planted vineyards in regions where no grapes have grown since. Western and central Europe were usually spared the plague epidemics that periodically afflicted the Mediterranean and Middle East. Other diseases seem also to have been in abeyance. The population more than doubled from 35 million in 1000 to 80 million in 1347.[20]

The growing populations of Europe owed something to new technologies aiding in clearing the land and cultivating the thick soils of Atlantic Europe. One of these was the moldboard plow. Turning the heavy soil and cutting furrows deep enough to allow water to run off required a plow with a vertical iron coulter to break the sod, an iron plowshare at right angles to the coulter to cut it at the grass roots, and a moldboard to turn the slice of sod on its side.

The new plow demanded a radical change in both methods of farming and sources of energy. When first used on newly cleared land, it required eight oxen to pull it; even after the sod had been broken and the soil loosened, four oxen were needed. Turning such a team of oxen was difficult, so Roman-style square fields were slowly replaced by long narrow fields. Gradually moldboard plows and strip fields came to predominate in the heavier soils of western and central Europe.

Along with the new plow, European farmers introduced a balanced system of animal and crop production known as the three-field rotation. They planted one-third of their fields with winter wheat or rye, another third with spring crops like oats, barley, or legumes, and left the last third fallow to recover its fertility. Labor was distributed more evenly over the year, and diversified crops provided a more balanced diet and reduced the odds that failure of a single crop would lead to starvation.

Horse-collars, introduced in the eighth or ninth century, allowed horses to pull heavy loads such as moldboard plows or four-wheeled wagons. Iron horseshoes saved the horses' hooves from the excessive wear that occurred in wet climates. Horses could do twice as much work as oxen, but were costly to keep. Whereas oxen could survive on hay and stubble, horses needed oats, a costly fuel. The use of horses therefore spread gradually. By the end of the eleventh century, plow horses were a common sight in northern Europe, though oxen did most of the work as late as the fourteenth century.[21]

Between 800 and 1300, once-forested Europe was transformed into a land of farms, meadows, villages, and towns. Powerful lords offered lower dues and taxes to pioneers willing to settle new lands. Benedictine and Cistercian monasteries opened clearings deep inside forests and allotted forest lands to colonists. As the climate warmed, farmers and monasteries moved north or uphill onto lands once considered impossible to farm. The colonization movement was especially pronounced in central and eastern Europe. In the twelfth century, German, Flemish, and Dutch settlers moved east of the Elbe River into lands sparsely populated by Slavic speakers. In the words of agrarian historian Slicher van Bath, "It was an expansion comparable to that in North America many years later."[22]

No place in Europe was so transformed by human action as the Netherlands. In the early Middle Ages, half of that country, a 50-kilometer-wide swath between the coastal dunes and the interior, was covered with peat. Reclamation began in the ninth century. Digging ditches to drain the land required intense collective efforts. The Rijnland, a 900-square-kilometer area, was largely reclaimed by 1300. On the new land, farmers grew barley, rye, and flax and raised cattle. By 1370, the settled area of the Netherlands had more than doubled.

The effort required was huge and costly, not just once, but forever after. When peat dries out, it shrinks. Removing peat for fuel made the problem worse. Once drained, peat bogs subsided until they were below the water table. The solution was windmills to pump out the water and dikes to hold back the high tides. Though storms from the North Sea sometimes swept away dikes and villages and covered reclaimed farmland with clay and sand, after every storm the Netherlanders returned to rebuild their farms and villages. Historian Peter Hoppenbrouwers called it "a cat-and-mouse game between merciless nature and human inventiveness."[23]

Apart from the Netherlands, bogs and wetlands in several parts of Europe made tempting targets for aspiring agriculturalists. Draining and ditching required concerted efforts beyond what single families could provide, so wetlands drainage was usually an enterprise of monasteries or lords. From East Anglia to Yorkshire along England's North Sea coast, wetlands drainage began in the tenth century. Dutch experts traveled to what is now northern Poland to oversee drainage there. In southern Bohemia (Czech Republic) lords organized the transformation of wetlands into mosaics of fishponds and grainfields. As in the Netherlands, keeping water in some places and out of others required continual effort and considerable expertise everywhere it was done. The reward was more arable land, more food, and better odds of survival. In time, that translated into more mouths to feed and renewed pressure to clear more land.

Cutting trees and clearing land was never a simple matter, for forests had their uses and supporters. Peasants valued forests as sources of firewood, berries, and mushrooms, as well as places where their swine could forage for acorns. Kings and nobles hunted game in woodlands. Felling trees therefore created conflicts between people and their values. Nonetheless, deforestation proceeded apace. Peasants and monks cleared trees on the edges of their fields, then uprooted the stumps, a task requiring several oxen. By the thirteenth century, water-powered sawmills provided sawn lumber for the booming cities and villages. Forest historian Michael Williams estimated that around 500, four-fifths of western and central Europe was covered with forests; 800 years later, only half remained, with most of the clearing occurring after 1100. In his words: "By any calculation, the medieval European experience must rank as one of the great deforestation episodes in the world."[24]

The consequences of deforestation appeared as early as the eleventh century. As wild animals became rare, kings and nobles established game preserves and hired forest guards to protect their privilege. Silt filled lakes and estuaries. Near cities, shortages drove up the price of wood; by 1500, the situation had become so critical that Londoners began using coal. The flowering of medieval Europe came at the expense of forests, wetlands, and wild animals. Like the peoples of China and the Middle East, Europeans had pushed to the very limits of the carrying capacity of their land with the technology at their disposal. This made them more vulnerable to catastrophes, both natural and man-made.

Fourteenth-Century Disasters

In the late thirteenth century, summers in Europe grew cooler, autumns came sooner, and winters grew longer and harsher. Glaciers advanced down the valleys of the Alps. Off the coast of Iceland, sea ice, rare since 1000, became common. Then came years of extraordinarily bad weather. The summer of 1314 was abnormally rainy. The winter was harsh and the following spring again it rained more than usual, causing floods. The year 1316 was worse, with rains throughout the spring, summer, and fall and a bitterly cold winter that froze the Baltic Sea. After yet another rainy year, the winter of 1317–18 lasted from November to Easter. From 1318 to 1322 periodic storms battered the continent. The Baltic and parts of the North Sea froze over.

The results were disastrous. Grain production in northern Europe dropped by one-third. Molds, rusts, and mildews attacked the crops. Ergot blight caused convulsions and hallucinations among people and animals who ate the infected rye. Livestock died of "murrain," the term then used for diseases of ruminants, probably rinderpest. As there was not enough fodder for animals during the winter, their owners slaughtered them. Up to 90 percent of the livestock died, causing a shortage of manure, hence poorer crops.

Crop failures and the loss of livestock produced one of the worst famines in European history. Hungry people ate their seed corn, then roots, acorns, and nuts, even bark; some were reduced to cannibalism. Children born during the famine suffered from stunted growth and weakened immune systems. Over 30 million people in northern Europe were affected, and the population declined by 10 to 15 percent. While humans suffered, the rest of nature recovered. Land reclamation from marshes and peat bogs stopped, as did the clearing of forests. As farmers abandoned their villages and farmlands in northern Europe, trees grew back. Marginal lands that had once been farmed became pastures for the sheep and cattle that survived.[25]

Then came a more terrible calamity: the bubonic plague or Black Death (see Map 3.1). The cause of the disease in individuals is well known; that of epidemics much less so. We know that it spread wherever black rats lived, in houses with thatched roofs, in granaries and gristmills, and on board ships carrying grain. The plague was more common in towns than in small villages or large cities. It also spread most readily in warm weather.

Where the Black Death originated and how it was transmitted is a mystery. The first evidence comes from the tombs and writings of Nestorian Christians dating back to 1338–9, in what is now Kyrgyzstan. Its next appearance was in 1345 or 1346 among the Mongols who were besieging the port of Kaffa in the Crimea, then a base for Genoese merchants. Before departing, the Mongols are said to have thrown some corpses over the walls of the town; more likely, some rats made their way into the town, where they

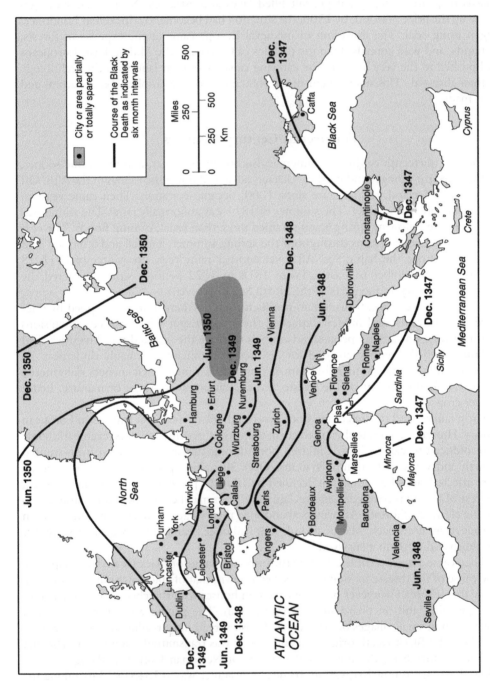

Map 3.1 The Black Death in fourteenth-century Europe

infected others. As the Genoese fled in their ships, they carried with them infected rats, bringing the disease to Constantinople in May 1347. From there it spread along the shipping lanes to Asia Minor, the Middle East, North Africa, and reached most of Europe between 1348 and 1353. It spread more rapidly and more widely than Justinian's plague, because northern and western Europe were much more densely populated and ships now sailed regularly from the Mediterranean to England and the Low Countries.

Wherever the plague appeared, it caused a demographic collapse. Recent research shows a population decline of one-half to two-thirds. Peasants abandoned the estates of their landowners to seek better wages elsewhere. One out of every four or five villages was abandoned and marginal lands reverted to forests or became pastures for sheep. Not until the mid-fifteenth century did the population of Europe begin to rise again.[26]

The plague reached Egypt and Baghdad in 1347 and Palestine and Arabia in 1348. It also spread to Morocco, Yemen, and Iran. Thereafter, it returned every few years until the end of the eighteenth century. Between 1347 and 1517, there were 55 outbreaks of plague in Egypt and 51 in Syria. During the most virulent outbreak, in 1429–30, the population of the Middle East declined by over one-third and did not recover until the nineteenth century.[27]

Egypt was particularly hard hit. The Nile and the many irrigation canals made it easy to ship grain, accompanied by rats, throughout the country. The canals, dikes, and sluices first built in the mid-thirteenth century required constant maintenance, but when the plague came, declining yields and taxes and a labor shortage prevented the government from undertaking the usual repairs after the annual Nile floods. Some fields received too much water and others not enough. Before the Black Death, Egypt and England had about the same area and population. In England, the Black Death reduced the population substantially but the arable land hardly at all, leaving more land per person and raising farm wages. In Egypt, however, the land required irrigation to be arable, hence its productivity was proportional to the labor invested in it. When the plague struck, the amount of productive land declined as much as, or more than, the population. As Egyptian agriculture entered a downward spiral, peasants fled to the cities and some of the richest farmlands were taken over by Bedouin sheepherders.[28]

In China, the fourteenth century was a time of earthquakes, floods, and rebellions. That century saw 36 years of unusually severe winters, more than any other. According to climate historian Hubert Lamb, exceptional rains in 1332 caused "one of the greatest weather disasters ever known, alleged to have taken seven million human lives in the great river valleys of China."[29] These weather events may also have forced rats to migrate in search of food. Epidemics were frequent, but the Chinese chronicles do not specify the disease(s). An epidemic in 1331 is said to have killed 9 out of 10 inhabitants in Hebei province. Possibly the first references to the plague appeared in 1353, when an epidemic broke out in eight provinces. By the end of the century, the population of China had dropped to 65 million from a peak of 123 million in 1200. As William McNeill points out, "Disease assuredly played a big part in cutting Chinese numbers in half; and bubonic plague ... is by all odds the most likely candidate for such a role."[30]

The Maya

The New World environments differed from the Old in three significant ways. The north–south orientation of the American continents delayed the spread of domesticated plants for they had to be adapted to different climates, in contrast to the domesticated plants of Eurasia that could be readily transferred between the Middle East, Europe, and China. Because Amerindians had few domesticated animals, their disease ecology was much less complex than in the eastern hemisphere; in particular, we find no evidence of pre-Columbian epidemics. And finally, while Europe was enjoying the medieval warm period, the peoples of the Americas were subject to the worst droughts in their history.[31] Yet they established complex societies in the most difficult environments, taking enormous risks. Of the hundreds of Amerindian societies, from isolated hunting bands to great empires, let us choose the most dramatic case of interaction between humans and nature: the Maya.

The Maya once occupied the highlands of southern Guatemala, the Petén or lowlands of northern Guatemala, and the Yucatán Peninsula. The plateau that slopes down from the mountains of Guatemala toward the Yucatán consists of porous limestone that absorbs rain as soon as it hits the ground, leaving few rivers or wetlands on the surface. Before the Maya transformed the land, the Petén was a forest and the Yucatán was covered with thorny bushes and savannas. A thin layer of topsoil covered the bedrock, and most of the nutrients were contained in living or recently dead organic matter. Felling trees therefore often meant turning the soil into hard laterite or exposing the bare limestone.[32]

The climate of Mesoamerica has two distinct seasons: a rainy season from May or June to November, and a dry season from December to April or May. The swamps that collect surface water during the rainy season dry out after the rains. Some years get three or four times more rain than others. In the Yucatán, the water table lies close enough to the surface that it can be reached year-round in cenotes, circular sinkholes formed by the collapse of underground caves. Though the Petén gets more rain, there are no lakes or rivers to conserve it and the water table is inaccessible.[33]

The first farmers cut and burned the vegetation during the dry season, leaving ashes to fertilize the soil. During the rains they planted and harvested maize, beans, squash, chili peppers, manioc, and cotton. When the soil lost its fertility, they burned a new patch of forest, while the earlier fields were left fallow for 4 to 20 years. Plots were dispersed throughout the countryside, and villages were often mobile.

Gradually, the population grew. At its peak in the early ninth century, when the density reached 250 people per square kilometer (650 people per square mile), as high as that of rural China in the twentieth century, slash-and-burn was no longer an option. In its place, farmers shortened the fallow period and cleared and cultivated marginal lands on hillsides. Near rivers and wetlands, they dug short canals and built raised fields with the mud dredged from the canals. In other places, they constructed terraces to prevent soil erosion and retain water. They mulched and fertilized with human wastes or composted vegetation and watered their crops from cenotes and underground cisterns.[34]

Yet the productivity of their agriculture was low, even compared with other Amerindian civilizations. As the Maya had no beasts of burden and no navigable lakes or rivers, they carried everything on their backs. Food could not be transported more than 30 kilometers or so (20 miles), beyond which distance porters would eat all they carried; thus, a food shortage in one place could not be alleviated by a surplus elsewhere. As maize could not be stored long in the humid climate, survival depended on every harvest in every location being sufficient to last a year. Continuous cultivation leached nutrients from the soil. Deforestation left the hillsides barren and eroded. The expansion of the cultivated area left fewer wild resources to fall back on. The health of the people declined, as shown by stress marks left on their teeth. By the eighth century, the Maya had reached the limit of the carrying capacity of their environment with the techniques and technologies they had.[35]

The classic Maya period began at the end of the sixth century. By the eighth century, they built cities out of local limestone, using enormous amounts of labor. At its height, 60,000 to 80,000 people lived in or near Tikal, in the central Petén. To survive the yearly dry seasons, the inhabitants paved and plastered the plazas and terraces at the city's core to catch rainwater. Water was then channeled into six major reservoirs that could store enough water for the 9,800 members of the elite who lived in the core. Around the core, densely inhabited residential neighborhoods had smaller reservoirs that could hold half as much water as the ones in the core. Beyond these neighborhoods, large basins caught the used water from the city to water and fertilize outlying fields during the dry season. These reservoirs contained enough water for up to 18 months. Tikal was an artificial oasis that thrived as long as there was sufficient rain.

Copán was the southernmost Mayan city. At its peak, it contained 4,500 structures and over 20,000 people lived in or near it. Nearby bottom lands were intensively culti-vated, but did not suffice to feed the growing population. Though located on a river, the city also stored water for the season when the river ran low and became undrinkable.[36]

Mayan kings and nobles ruled not only by their spiritual leadership and their monopoly of the means of violence, but also by controlling access to water during the dry season. In exchange, they could demand labor to build their palaces, temples, and monuments. As the population grew, so did their demands. The kings constructed monuments at an ever growing pace. Aristocrats built palaces. City-states fought one another in a crescendo of violence, seeking in conquest and human sacrifices a compensation for the decline of their economies.[37]

In the ninth century, the cities of the Petén imploded. In Copán, the last mention of a king dates from 822, and in 850 the royal palace burned. Elsewhere, the last inscription dates from 909. From a peak of 27,000, the population of the Copán valley dropped to 15,000 in 950, to no more than 8,000 in 1150, and to zero by 1250. Even as the Petén became depopulated, cities arose in the Yucatán Peninsula, until they too were abandoned around 1450. After that, there were no more cities, no more monumental stone buildings, no more inscriptions, in short, no more civilization. The Maya survived – their descendants still inhabit the land – but in greatly diminished numbers; archeologists estimate that the population declined by 90 to 99 percent.[38]

What caused such devastation? Archeologists have been debating this question since the Mayan ruins came to light a century and a half ago. Disease and malnutrition among the farmers and increasing demands on the part of the kings and elites may have triggered social unrest. According to archeologist Michael Coe: "the royalty and nobility, including the scribes … may well have been massacred by an enraged population." Another explanation emphasizes the relations between the Maya and their environment. By cutting down their forests and cultivating every scrap of land over and over, the farmers exhausted the fertility of the soil and eroded the hillsides. In Coe's words: "by the end of the eighth century, the Classic Maya population of the southern lowlands had probably increased beyond the carrying capacity of the land, no matter what system of agriculture was in use … The Maya apocalypse, for such it was, surely had ecological roots."[39]

A subset of the environmental explanation is the role of drought, which some scholars offer to the exclusion of all others.[40] Before the collapse, the Maya had survived many droughts, but after each episode, the population recovered and urban construction began anew. Then came two centuries of dry years, the driest in the last 8,000 years. The megadrought of 750 to 850 depopulated the Petén and ended the classic period of Mayan civilization. Finally, the drought of 1451–4 ruined the last Mayan cities and depopulated the Yucatán.

The combination of a rigid social hierarchy and maximum use of resources left Mayan society vulnerable to environmental changes. The power of the rulers rested on an implicit bargain between the commoners, the rulers, and the gods: in exchange for the sacrifices and labor of the commoners, the rulers performed ceremonies to ensure that the gods would provide water. When the weather (or the gods) failed them for several years and the elite controlled the last supplies of drinking water, people rebelled. It was not just hunger that drove commoners to attack their kings and lords, burn buildings, and flee; it was thirst.[41]

Conclusion

What conclusions can we draw from the example of these civilizations? In almost every case, humans multiplied in numbers and consumed natural resources to the very limits of the carrying capacity of their environments. But carrying capacity is not a constant, for it can be expanded by technological innovations: irrigation, new implements, transplanting seedlings, and introducing new plants and animals. Human actions can also reduce the carrying capacity of the environment; where once trees and grasses grew, humans have left deserts, eroded hillsides, and saline soils.

In multiplying in numbers and pushing to the limits of their environments' carrying capacities, humans behaved as though nature (or the gods) were benign, or at least controllable. But nature is fickle and indifferent to humans. By creating the densely populated and complex societies we call civilizations, humans lost the resilience of hunter-gatherers or early agricultural peoples who survived by moving on or turning to less desirable foods. Instead, they left themselves open to disasters and catastrophes, even total collapses. Climate changes that less complex societies could have survived destroyed the civilization of the Maya. New diseases, flaring into epidemics amid densely packed populations without the right immunities, ravaged the Middle East, Europe, and China. In the year 1500 it was not yet clear which side was winning the tug-of-war between humans and the rest of nature.

Notes

1 J. Gunn (ed.), *Years without Summer: Tracing A.D. 536 and Its Aftermath*, Oxford, Archaeopress, 2000;D. Keys, *Catastrophe: An Investigation into the Origins of the Modern World*, New York, Ballantine, 1999.

2 H. B. Dewing (trans.), *Procopius: History of the Wars*, vol. 4, London, Heinemann, 1916, p. 329.

3 A. T. Grove and O. Rackham, *The Nature of Mediterranean Europe: An Ecological History*, New Haven, CT, Yale University Press, 2001, p. 143; Gunn, *Years without Summer*; J. Koder, "Climate Change in the Fifth and Sixth Centuries?," in P. Allen and E. Jeffries (eds.), *The Sixth Century: End or Beginning?*, Brisbane, Australian Association for Byzantine Studies, 1996, pp. 270–85 (p. 276); M. G. L. Baillie, "Dendrochronology Raises Questions about the Nature of the A.D. 536 Dust-Veil Event," *The Holocene* 4/2, 1994, pp. 212–17 (pp. 212–13); Keys, *Catastrophe*, pp. 28, 149–53, 189, 283–4.

4 Personal communication from Dr. Dallas Abbott, Lamont-Dougherty Observatory, Columbia University, November 24, 2009 for which I am grateful. P. Farquharson, "Byzantium, Planet

Earth and the Solar System," in Allen and Jeffries (eds.), *The Sixth Century*, pp. 263–9 (p. 267); R. B. Stothers and M. R. Rampino, "Volcanic Eruptions in the Mediterranean before A.D. 630 from Written and Archaeological Sources," *Journal of Geophysical Research* 88, 1983, pp. 6357–71; M. G. L. Baillie, *Exodus to Arthur: Catastrophic Encounters with Comets*, London, B. T. Batsford, 1999, p. 128; Baillie, "Dendrochronology Raises Questions,", p. 215; Gunn, *Years without Summer*, pp. 74–5; Keys, *Catastrophe*, pp. 29–33.

5   W. Rosen, *Justinian's Flea: Plague, Empire, and the Birth of Europe*, New York, Viking, 2007, pp. 196, 200–12; O. J. Benedictow, *The Black Death, 1346–1353: A Complete History*, Woodbridge, Boydell Press, 2004, p. 39; P. Horden, "Mediterranean Plague in the Age of Justinian," in M. Maas (ed.), *The Cambridge Companion to the Age of Justinian*, Cambridge, Cambridge University Press, 2005, pp. 134–60; P. Christensen, *The Decline of Iranshahr: Irrigation and Environments in the History of the Middle East, 500 B.C. to A.D. 1500*, Copenhagen, Museum Tusculanum Press, 1993, pp. 81–2; J. N. Biraben, *Les hommes et la peste en France et dans les pays méditerranéens*, vol. 1, Paris, Mouton, 1975, pp. 25–48.

6   W. H. McNeill, *Plagues and Peoples*, Garden City, Doubleday, 1976, pp. 124–7; P. Sarris, "Bubonic Plague in Byzantium," in L. K. Little (ed.), *Plague and the End of Antiquity: The Pandemic of 541–750*, Cambridge, Cambridge University Press, 2007, pp. 119–32 (pp. 120–3); D. Twitchett, "Population and Pestilence in T'ang China," in W. Bauer (ed.), *Studia Sino-Mongolica: Festschrift für Herbert Franke*, Wiesbaden, Fritz Steiner, 1979, pp. 35–68.

7   M. Dols, *The Black Death in the Middle East*, Princeton, NJ, Princeton University Press, 1977, pp. 13–17; J. J. O'Donnell, *The Ruin of the Roman Empire*, New York, HarperCollins, 2009, pp. 257–77; Gunn, *Years without Summer*, pp. 38–9; McNeill, *Plagues and Peoples*, p. 159; Benedictow, *The Black Death*, p. 40; Twitchett, "Population and Pestilence," pp. 42–63.

8   A. Watson, "The Arab Agricultural Revolution and Its Diffusion, 700–1100," *Journal of Economic History* 34, 1974, pp. 8–35; A. Watson, *Agricultural Innovation in the Early Islamic World: The Diffusion of Crops and Farming Techniques, 700–1100*, Cambridge, Cambridge University Press, 1983. For a critique of the Watson thesis, see M. Decker, "Plants and Progress: Rethinking the Islamic Agricultural Revolution," *Journal of World History* 20/2, 2009, pp. 187–206.

9   R. W. Bulliet, *Cotton, Climate, and Camels in Early Islamic Iran: A Moment in World History*, New York, Columbia University Press, 2009, chapters 1 and 2; Watson, *Agricultural Innovation*, p. 111.

10   J. R. McNeill, *The Mountains of the Mediterranean World: An Environmental History*, Cambridge, Cambridge University Press, 1992, pp. 87–8; E. Burke III, "The Transformation of the Middle Eastern Environment, 1500 B.C.E.–2000 C.E.," in E. Burke III and K. Pomeranz (eds.), *The Environment and World History*, Berkeley, University of California Press, 2009, pp. 81–117 (pp. 85–8); Christensen, *The Decline of Iranshahr*, pp. 87–94.

11   Burke, "The Transformation of the Middle Eastern Environment," p. 85; Bulliet, *Cotton, Climate, and Camels*, p. 96; Christensen, *The Decline of Iranshahr*, p. 100. See also T. Jacobsen, and R. McAdams, "Salt and Silt in Ancient Mesopotamian Agriculture," *Science* 128/3334, 1958, pp. 1256–8.

12   M. Elvin, *The Pattern of the Chinese Past: A Social and Economic Interpretation*, Stanford, CA, Stanford University Press, 1973, p. 55; L. P. Van Slyke, *Yangtze: Nature, History and the River*, Reading, Addison-Wesley, 1988, p. 77; G. Qu and J. Li, *Population and Environment in China*, Boulder, CO, L. Rienner, 1994, p. 20; B. Fagan, *The Great Warming: Climate Change and the Rise and Fall of Civilizations*, New York, Bloomsbury Press, 2008; Gunn, *Years without Summer*, p. 75.

13   M. Elvin, "3,000 Years of Unsustainable Growth: China's Environment from Archaic Timers to the Present," *East Asian History* 6, 1993, pp. 7–46; Elvin, *The Pattern of the Chinese Past*, pp. 82, 113–26; F. Bray, *The Rice Economies: Technology and Development in Asian Societies*, Berkeley, University of California Press, 1986, pp. 203–5; C. Lamouroux, "Crise politique et

développement rizicole en Chine: La région de Jiang-Huai (VIII-Xe siècle)," *Bulletin de l'École française d'Extrême-Orient* 82, 1995, pp. 145–83; M. Cartier, "Aux origines de l'agriculture intensive du Bas Yangzi (note critique)," *Annales E.S.C.* 46/5, 1991, pp. 1013–17; C. Hsu, "Environmental History – Ancient," in L. Cheng (ed.), *Berkshire Encyclopedia of China*, 5 vols., Great Barrington, MA, Berkshire Publishing, 2009, vol. 2, pp. 731–5; P. Ho, "Early Ripening Rice in Chinese History," *Economic History Review* 2/9, 1956, pp. 200–18; J. E. Spencer, "Water Control in Terraced Rice-Field Agriculture in Southeastern Asia," in T. E. Downing and M. Gibson (eds.), *Irrigation's Impact on Society*, Tucson, AZ, University of Arizona Press, 1974, pp. 59–65; McNeill, *Plagues and Peoples*, p. 137; Twitchett, "Population and Pestilence," p. 39.

14   Elvin, "3,000 Years of Unsustainable Growth," p. 44.

15   Bray, *The Rice Economies*, pp. 9–34.

16   E. H. Schafer, *The Vermilion Bird: T'ang Images of the South*, Berkeley, University of California Press, 1967, pp. 130–1; McNeill, *Plagues and Peoples*, pp. 88–9; Elvin, *The Pattern of the Chinese Past*, pp. 185–6; Twitchett, "Population and Pestilence," p. 40.

17   V. Harden, "Smallpox," in J. Byrne (ed.), *Encyclopedia of Pestilence, Pandemics, and Plagues*, Westport, CT, Greenwood Press, 2008, pp. 647–50; D. R. Hopkins, *The Greatest Killer: Smallpox in History*, Chicago, University of Chicago Press, 2002, pp. 18, 103–5; Twitchett, "Population and Pestilence," p. 42.

18   Twitchett, "Population and Pestilence," pp. 42–52, 62–3; C. Benedict, *Bubonic Plague in Nineteenth-Century China*, Stanford, CA, Stanford University Press, 1996, p. 9.

19   McNeill, *The Mountains of the Mediterranean World*, pp. 84–7.

20   P. Acot, *Histoire du climat*, Paris, Perrin, 2003, p. 128; H. H. Lamb, *Climate, History, and the Modern World*, 2nd ed., London, Routledge, 1995, p. 159; Fagan, *The Great Warming*, pp. 12–35.

21   L. White Jr., *Medieval Technology and Social Change*, New York, Oxford University Press, 1962, pp. 54–76; T. L. Whited, J. I. Engels, R. C. Hoffmann, et al., *Northern Europe: An Environmental History*, Santa Barbara, CA, ABC-CLIO, 2005, pp. 51–2; B. H. Slicher van Bath, *The Agrarian History of Western Europe, A.D. 500–1850*, London, Edward Arnold, 1963; I. Simmons, *An Environmental History of Great Britain: From 10,000 Years Ago to the Present*, Edinburgh, Edinburgh University Press, 2001.

22   Slicher van Bath, *The Agrarian History of Western Europe*, p. 155. See also M. Williams, *Deforesting the Earth: From Prehistory to Global Crisis*, Chicago, University of Chicago Press, 2003; Whited, Engels, Hoffmann, et al., *Northern Europe*; H. C. Darby, "The Clearing of Woodland in Europe," in W. L. Thomas (ed.), *Man's Role in Changing the Face of the Earth*, Chicago, University of Chicago Press, 1956, pp. 183–216 (p. 196).

23   P. Hoppenbrouwers, "Agricultural Production and Technology in the Netherlands, c. 1000–1500," in G. Astill and J. Langdon (eds.), *Medieval Farming and Technology: The Impact of Agricultural Change in Northwest Europe*, Leiden, Brill, 1997, pp. 89–114 (p. 98); W. TeBrake, "Land Drainage and Public Environmental Policy in Medieval Holland," *Environmental Review* 12/3, 1988, pp. 75–93; G. P. Van der Ven, *Man-Made Lowlands: History of Water Management and Land Reclamation in the Netherlands*, Utrecht, Uitgeverij Matrijs, 2004, pp. 52–73, 105.

24   Williams, *Deforesting the Earth*, p. 106; R. Bechmann, *Des arbres et des hommes: La forêt au Moyen-Âge*, Paris, Flammarion, 1984, chapter 2; White, *Medieval Technology and Social Change*, pp. 82, 118; Whited, Engels, Hoffmann, et al., *Northern Europe*, p. 55; Slicher van Bath, *The Agrarian History of Western Europe*.

25   W. C. Jordan, *The Great Famine: Northern Europe in the Early 14th Century*, Princeton, NJ, Princeton University Press, 1996, pp. 7–8, 24–38, 185–6; J. R. McNeill, "Woods and Warfare in World History," *Environmental History* 9, 2004, pp. 388–410 (p. 401); N. F. Cantor, *In the Wake of the Plague: The Black Death and the World It Made*, New York, Free Press, 2001, pp. 8, 74; Lamb, *Climate, History, and the Modern World*; Slicher van Bath, *The Agrarian*

History of Western Europe, p. 142; Acot, *Histoire du climat*, p. 141; Whited, Engels, Hoffmann, et al., *Northern Europe*, p. 59.

26   R. Pollitzer, *Plague*, Geneva, World Health Organization, 1954, pp. 13–14; R. S. Gottfried, *Black Death: Natural and Human Disaster in Medieval Europe*, New York, Free Press, 1985; Benedictow, *The Black Death*.

27   S. J. Borsch, *Black Death in Egypt and England: A Comparative Study*, Austin, TX, University of Texas Press, 2005, pp. 24, 55–66; S. Tibi-Harb, "Plague in the Islamic World, 1500–1850," in J. Byrne (ed.), *Encyclopedia of Pestilence, Pandemics, and Plagues*, Westport, CT, Greenwood, 2008, pp. 516–19 (pp. 517–18); Williams, *Deforesting the Earth*, p. 117; Slicher van Bath, *The Agrarian History of Western Europe*; Christensen, *The Decline of Iranshahr*, pp. 74, 100–3; Dols, *The Black Death in the Middle East*; Benedictow, *The Black Death*, pp. 61–6, 383.

28   Borsch, *Black Death in Egypt and England*, pp. 25–53; Christensen, *The Decline of Iranshahr*, pp. 100–1.

29   Lamb, *Climate, History, and the Modern World*, p. 280.

30   McNeill, *Plagues and Peoples*, p.163; J. Dardess, "Shun-ti and the End of Yüan Rule in China," in H. Franke and D. Twitchett (eds.), *The Cambridge History of China*, vol. 6, *Alien Regimes and Border States, 907–1368*, Cambridge, Cambridge University Press, 1994, pp. 561–86 (p. 585).

31   D. A. Hodell, J. H. Curtis, and M. Brenner, "Possible Role of Climate in the Collapse of Classic Maya Civilization," *Nature* 375, 1995, pp. 391–4 (p. 391); J. Diamond, *Collapse: How Societies Choose to Fail or Succeed*, New York, Penguin, 2005; B. Fagan, *Floods, Famines and Emperors: El Niño and the Fate of Civilizations*, New York, Basic Books, 1999, p. 129; Fagan, *The Great Warming*, pp. 109–12.

32   M. Coe, *The Maya*, 6th ed., New York, Thames and Hudson, 1999, pp. 16–17; D. Webster, *The Fall of the Ancient Maya: Solving the Mystery of the Maya Collapse*, New York, Thames and Hudson, 2002, pp. 252–3.

33   R. B. Gill, *The Great Maya Drought: Water, Life, and Death*, Albuquerque, NM, University of New Mexico Press, 2000, chapter 9; A. Ford, "Critical Resource Control and the Rise of the Classic Period Maya," in S. L. Fedick (ed.), *The Managed Mosaic: Ancient Maya Agriculture and Resource Use*, Salt Lake City, UT, University of Utah Press, 1996, pp. 297–303 (pp. 299–300); M. Coe, *The Maya*, 6th ed., New York, Thames and Hudson, 1999.

34   Diamond, Collapse, pp. 163–8; Coe, *The Maya*, pp. 14–20, 35; Webster, *The Fall of the Ancient Maya*, pp. 332–4; Gill, *The Great Maya Drought*, p. 259; Williams, *Deforesting the Earth*.

35   L. Lucero, "The Collapse of the Classic Maya: A Case for the Role of Water Control," *American Anthropologist* 104, 2002, pp. 815–19; Webster, *The Fall of the Ancient Maya*, pp. 255, 330–5; Coe, *The Maya*, pp. 14–20, 166–7; Fagan, *Floods, Famines and Emperors*, pp. 147–55; Williams, *Deforesting the Earth*, pp. 49–51; Gill, *The Great Maya Drought*, pp. 318–20, 371; Diamond, *Collapse*, pp. 164–70.

36   V. L. Scarborough, "The Flow of Power: Water Reservoirs Controlled the Rise and Fall of the Ancient Maya," *The Sciences* 32/2, 1992, pp. 39–43; Ford, "Critical Resource Control and the Rise of the Classic Period Maya"; Lucero, "The Collapse of the Classic Maya"; Gill, *The Great Maya Drought*, pp. 259–66; 313–14, 371; Coe, *The Maya*, pp. 16, 93; Diamond, *Collapse*, pp. 162–70.

37   Lucero, "The Collapse of the Classic Maya," p. 815; Coe, *The Maya*, p. 21; Scarborough, "The Flow of Power," pp. 42–3; Ford, "Critical Resource Control and the Rise of the Classic Period Maya," pp. 301–2; Diamond, *Collapse*, pp. 167–9; Webster, *The Fall of the Ancient Maya*, p. 335.

38   Gill, *The Great Maya Drought*, pp. 313–14, 360; Fagan, *Floods, Famines and Emperors*, p. 154; Diamond, *Collapse*, pp. 170–5.

39 Coe, *The Maya*, p. 128. See also E. Abrams, A. Freter, D. Rue, and J. Wingard, "The Role of Deforestation in the Collapse of the Late Classic Copan Maya State," in L. E. Sponsel, T. N. Headland, and R. C. Bailey (eds.), *Tropical Deforestation: The Human Dimension*, New York, Columbia University Press, 1996, pp. 55–75.

40 Gill, *The Great Maya Drought*, p. 386.

41 Webster, *The Fall of the Ancient Maya*, pp. 239–44; Gill, *The Great Maya Drought*, pp. 313–20, 331, 360–5, 386; Coe, *The Maya*, p. 127; Diamond, *Collapse*, pp. 171–5.

References

Abrams, E., Freter, A., Rue, D., and Wingard, J., "The Role of Deforestation in the Collapse of the Late Classic Copan Maya State," in L. E. Sponsel, T. N. Headland, and R. C. Bailey (eds.), *Tropical Deforestation: The Human Dimension*, New York, Columbia University Press, 1996, pp. 55–75.

Acot, P., *Histoire du climat*, Paris, Perrin, 2003.

Baillie, M. G. L., "Dendrochronology Raises Questions about the Nature of the A.D. 536 Dust-Veil Event," *The Holocene* 4/2, 1994, pp. 212–17.

Baillie, M. G. L., *Exodus to Arthur: Catastrophic Encounters with Comets*, London, B. T. Batsford, 1999.

Bechmann, R., *Des arbres et des hommes: La forêt au Moyen-Âge*, Paris, Flammarion, 1984.

Benedict, C., *Bubonic Plague in Nineteenth-Century China*, Stanford, CA, Stanford University Press, 1996.

Benedictow, O. J., *The Black Death, 1346–1353: A Complete History*, Woodbridge, Boydell Press, 2004.

Biraben, J. N., *Les hommes et la peste en France et dans les pays méditerranéens*, vol. 1, Paris, Mouton, 1975.

Borsch, S. J., *Black Death in Egypt and England: A Comparative Study*, Austin, TX, University of Texas Press, 2005.

Bray, F., *The Rice Economies: Technology and Development in Asian Societies*, Berkeley, University of California Press, 1986.

Bulliet, R. W., *Cotton, Climate, and Camels in Early Islamic Iran: A Moment in World History*, New York, Columbia University Press, 2009.

Burke, E., III, "The Transformation of the Middle Eastern Environment, 1500 B.C.E.–2000 C.E.," in E. Burke III and K. Pomeranz (eds.), *The Environment and World History*, Berkeley, University of California Press, 2009, pp. 81–117.

Cantor, N. F., *In the Wake of the Plague: The Black Death and the World It Made*, New York, Free Press, 2001.

Cartier, M., "Aux origines de l'agriculture intensive du Bas Yangzi (note critique)," *Annales E.S.C.* 46/5, 1991, pp. 1013–17.

Christensen, P., *The Decline of Iranshahr: Irrigation and Environments in the History of the Middle East, 500 B.C. to A.D. 1500*, Copenhagen, Museum Tusculanum Press, 1993.

Coe, M., *The Maya*, 6th ed., New York, Thames and Hudson, 1999.

Darby, H. C., "The Clearing of Woodland in Europe," in W. L. Thomas (ed.), *Man's Role in Changing the Face of the Earth*, Chicago, University of Chicago Press, 1956, pp. 183–216.

Dardess, J., "Shun-ti and the End of Yüan Rule in China," in H. Franke and D. Twitchett (eds.), *The Cambridge History of China, vol. 6, Alien Regimes and Border States, 907–1368*, Cambridge, Cambridge University Press, 1994, pp. 561–86.

Decker, M., "Plants and Progress: Rethinking the Islamic Agricultural Revolution," *Journal of World History* 20/2, 2009, pp. 187–206.

Dewing, H. B. (trans.), *Procopius: History of the Wars*, vol. 4, London, Heinemann, 1916.

Diamond, J., *Collapse: How Societies Choose to Fail or Succeed*, New York, Penguin, 2005.

Dols, M., *The Black Death in the Middle East*, Princeton, NJ, Princeton University Press, 1977.

Elvin, M., *The Pattern of the Chinese Past: A Social and Economic Interpretation*, Stanford, CA, Stanford University Press, 1973.

Elvin, M., "3,000 Years of Unsustainable Growth: China's Environment from Archaic Times to the Present," *East Asian History* 6, 1993, pp. 7–46.

Fagan, B., *Floods, Famines and Emperors: El Niño and the Fate of Civilizations*, New York, Basic Books, 1999.

Fagan, B., *The Great Warming: Climate Change and the Rise and Fall of Civilizations*, New York, Bloomsbury Press, 2008.

Farquharson, P., "Byzantium, Planet Earth and the Solar System," in P. Allen and E. Jeffries (eds.), *The Sixth Century: End or Beginning?*, Brisbane, Australian Association for Byzantine Studies, 1996, pp. 263–9.

Ford, A., "Critical Resource Control and the Rise of the Classic Period Maya," in S. L. Fedick (ed.), *The Managed Mosaic: Ancient Maya Agriculture and Resource Use*, Salt Lake City, UT, University of Utah Press, 1996, pp. 297–303.

Gill, R. B., *The Great Maya Drought: Water, Life, and Death*, Albuquerque, NM, University of New Mexico Press, 2000.

Gottfried, R. S., *Black Death: Natural and Human Disaster in Medieval Europe*, New York, Free Press, 1985.

Grove, A. T., and Rackham, O., *The Nature of Mediterranean Europe: An Ecological History*, New Haven, CT, Yale University Press, 2001.

Gunn, J. (ed.), *Years without Summer: Tracing A.D. 536 and Its Aftermath*, Oxford, Archaeopress, 2000.

Harden, V., "Smallpox," in J. Byrne (ed.), *Encyclopedia of Pestilence, Pandemics, and Plagues*, Westport, CT, Greenwood Press, 2008, pp. 647–50.

Ho, P., "Early Ripening Rice in Chinese History," *Economic History Review* 2/9, 1956, pp. 200–18.

Hodell, D. A., Curtis, J. H., and Brenner, M., "Possible Role of Climate in the Collapse of Classic Maya Civilization," *Nature* 375, 1995, pp. 391–4.

Hopkins, D. R., *The Greatest Killer: Smallpox in History*, Chicago, University of Chicago Press, 2002.

Hoppenbrouwers, P., "Agricultural Production and Technology in the Netherlands, c. 1000–1500," in G. Astill and J. Langdon (eds.), *Medieval Farming and Technology: The Impact of Agricultural Change in Northwest Europe*, Leiden, Brill, 1997, pp. 89–114.

Horden, P., "Mediterranean Plague in the Age of Justinian," in M. Maas (ed.), *The Cambridge Companion to the Age of Justinian*, Cambridge, Cambridge University Press, 2005, pp. 134–60.

Hsu, C., "Environmental History – Ancient," in L. Cheng (ed.), *Berkshire Encyclopedia of China*, 5 vols., Great Barrington, MA, Berkshire Publishing, 2009, vol. 2, pp. 731–5.

Jacobsen, T., and McAdams, R., "Salt and Silt in Ancient Mesopotamian Agriculture," *Science* 128/3334, 1958, pp. 1256–8.

Jordan, W. C., *The Great Famine: Northern Europe in the Early 14th Century*, Princeton, NJ, Princeton University Press, 1996.

Keys, D., *Catastrophe: An Investigation into the Origins of the Modern World*, New York, Ballantine, 1999.

Koder, J., "Climate Change in the Fifth and Sixth Centuries?," in P. Allen and E. Jeffries (eds.), *The Sixth Century: End or Beginning?*, Brisbane, Australian Association for Byzantine Studies, 1996, pp. 270–85.

Lamb, H. H., *Climate, History, and the Modern World*, 2nd ed., London, Routledge, 1995.

Lamouroux, C., "Crise politique et développement rizicole en Chine: La région de Jiang-Huai (VIII-Xe siècle)," *Bulletin de l'École française d'Extrême-Orient* 82, 1995, pp. 145–83.

Lucero, L., "The Collapse of the Classic Maya: A Case for the Role of Water Control," *American Anthropologist* 104, 2002, pp. 815–19.

McNeill, J. R., *The Mountains of the Mediterranean World: An Environmental History*, Cambridge, Cambridge University Press, 1992.

McNeill, J. R., "Woods and Warfare in World History," *Environmental History* 9, 2004, pp. 388–410.

McNeill, W. H., *Plagues and Peoples*, Garden City, Doubleday, 1976.

O'Donnell, J. J., *The Ruin of the Roman Empire*, New York, HarperCollins, 2009.

Pollitzer, R., *Plague*, Geneva, World Health Organization, 1954.

Qu, G., and Li, J., *Population and Environment in China*, Boulder, CO, L. Rienner, 1994.

Rosen, W., *Justinian's Flea: Plague, Empire, and the Birth of Europe*, New York, Viking, 2007.

Sarris, P., "Bubonic Plague in Byzantium," in L. K. Little (ed.), *Plague and the End of Antiquity: The Pandemic of 541–750*, Cambridge, Cambridge University Press, 2007, pp. 119–32.

Scarborough, V. L., "The Flow of Power: Water Reservoirs Controlled the Rise and Fall of the Ancient Maya," *The Sciences* 32/2, 1992, pp. 39–43.

Schafer, E. H., *The Vermilion Bird: T'ang Images of the South*, Berkeley, University of California Press, 1967.

Simmons, I., *An Environmental History of Great Britain: From 10,000 Years Ago to the Present*, Edinburgh, Edinburgh University Press, 2001.

Slicher van Bath, B. H., *The Agrarian History of Western Europe, A.D. 500–1850*, London, Edward Arnold, 1963.

Spencer, J. E., "Water Control in Terraced Rice-Field Agriculture in Southeastern Asia," in T. E. Downing and M. Gibson (eds.), *Irrigation's Impact on Society*, Tucson, AZ, University of Arizona Press, 1974, pp. 59–65.

Stothers, R. B., and Rampino, M. R., "Volcanic Eruptions in the Mediterranean before A.D. 630 from Written and Archaeological Sources," *Journal of Geophysical Research* 88, 1983, pp. 6357–71.

TeBrake, W., "Land Drainage and Public Environmental Policy in Medieval Holland," *Environmental Review* 12/3, 1988, pp. 75–93.

Tibi-Harb, S., "Plague in the Islamic World, 1500–1850," in J. Byrne (ed.), *Encyclopedia of Pestilence, Pandemics, and Plagues*, Westport, CT, Greenwood, 2008, pp. 516–19.

Twitchett, D., "Population and Pestilence in T'ang China," in W. Bauer (ed.), *Studia Sino-Mongolica: Festschrift für Herbert Franke*, Wiesbaden, Fritz Steiner, 1979, pp. 35–68.

Van der Ven, G. P., *Man-Made Lowlands: History of Water Management and Land Reclamation in the Netherlands*, Utrecht, Uitgeverij Matrijs, 2004.

Van Slyke, L. P., *Yangtze: Nature, History and the River*, Reading, Addison-Wesley, 1988.

Watson, A., "The Arab Agricultural Revolution and Its Diffusion, 700–1100," *Journal of Economic History* 34, 1974, pp. 8–35.

Watson, A., *Agricultural Innovation in the Early Islamic World: The Diffusion of Crops and Farming Techniques, 700–1100*, Cambridge, Cambridge University Press, 1983.

Webster, D., *The Fall of the Ancient Maya: Solving the Mystery of the Maya Collapse*, New York, Thames and Hudson, 2002.

White, L., Jr., *Medieval Technology and Social Change*, New York, Oxford University Press, 1962.

Whited, T. L., Engels, J. I., Hoffmann, R. C., et al., *Northern Europe: An Environmental History*, Santa Barbara, CA, ABC-CLIO, 2005.

Williams, M., *Deforesting the Earth: From Prehistory to Global Crisis*, Chicago, University of Chicago Press, 2003.

CHAPTER FOUR

The (Modern) World since 1500

ROBERT B. MARKS

The modern world – the one that we live in today – can be defined in various ways, but for the purposes of this chapter we can think of it as being marked by four notable characteristics, all of which interact to produce particular kinds of environmental change and challenges: (1) industry concentrated in urban clusters; (2) sovereign territorial nation-states that order political action; (3) a large and growing gap between those living in the wealthiest parts of the world and those in the poorest; and (4) energy produced from fossil fuels. As a result, where earlier environmental impacts of human activity were more local or regional and sometimes could be reversed whenever the human footprint eased, since the beginning of the twentieth century, our impact on the environment has become increasingly global, whether seen in terms of the rate and extent of deforestation, the human interruption of global energy and chemical fluxes, or the changes to our climate.[1]

The world has not always been structured like this. There is a history of change that explains how and why the world came to be the way it is. For much of the past 200 years, those seeking to explain this "modernity" tended to focus on the history of Europe and North America because those parts of the world appeared to have been the first to make the transition to a modern world of industry and successful nation-states. The Eurocentric tale of "the rise of the West" thus structured both the narrative told about how the world got to be the way it is, and ever more historical research into why it was the West that first modernized. Some explanations focused on the unique ways in which Europe's economy developed, others on how and why European culture was different and more conducive to supporting modern economic and political forms, and still others on demographic differences between European and other states. But mostly, the explanations focused on what Europeans did, and how others in the world responded (successfully, or not). If Europe was the first to become modern, the assumption hence was that there was something special about Europe's "early modern" experience that would hold the

A Companion to Global Environmental History, First Edition. Edited by J.R. McNeill and Erin Stewart Mauldin.
© 2012 John Wiley & Sons, Ltd. Published 2015 by John Wiley & Sons, Ltd.

key to the explanation. Hence, the transition from the "early modern" world to the modern world thus also tended to center on the European experience.

More recent historical work, though, has not just called into question the Eurocentric explanations of the origins of the modern world, but provides better alternative accounts. This chapter seeks to highlight those newer (and, I think, better) explanations. In particular, I will argue that until about 1800, all the people in the world lived within boundaries set by the "biological old regime."[2] Where some peoples and states became very efficient at capturing and using the flows of solar energy to increase their numbers and powers, by the eighteenth century, most of those "successful" peoples were beginning to exhaust their natural resources (in particular forests and land), and were bumping up against a ceiling that both limited further expansion, and began to create environmental crises of various kinds and degrees of severity.[3]

What changed those global dynamics was a combination of coal and colonies that gave some in parts of western Europe the ability to escape from the constraints of the biological old regime, launching a new era in world history.[4] By 1900, fossil-fueled industrialization organized within militarily powerful nation-states set humankind on a new path that began to alter the relationship of humans to the environment. As J. R. McNeill argues, the twentieth century represents "something new under the sun," a great departure from all past human history. The most significant aspect of that century in human history, more so than two world wars, the creation of the Third World, or the growing equality and power of women, among other things, will be the impact of humans on the natural environment.[5] Indeed, the reason for the rise of environmental history in the first instance is the growing consciousness in the second half of the twentieth century of a mounting environmental crisis and the desire to find both solutions and historical explanations for how we got to this point.

The Biological Old Regime, c. 1500

Until the Industrial Revolution in the nineteenth century began to lift material constraints on food production and hence on the size of the human population, all people lived within the "biological old regime," or what others have termed the "organic economy"[6] or the "somatic energy regime," referring to human and animal muscles as being the most important prime mover.[7] All refer to the human condition of depending on the annual flows of solar energy to supply the four necessities of life: food, fuel, clothing, and housing.

Because all living things need energy to survive and increase in numbers, those that succeeded saw their numbers increase, while others saw their populations decline. In the biological old regime, humans across the planet either hunted and gathered their food, or grew it on farms or plots of land that were in effect early solar collectors. Humans cannot directly digest wood or grass, and so needed to remove those from their farms to make way for grains and vegetables that human can directly digest, or to feed animals that could be consumed or do work. The relative success of humans in concentrating and collecting solar energy for their use has led to our increasing population density and size over the past 2,000 years.

Here we look at the weight of numbers[8] to get an overall picture. By 2010, there were 7 billion people on earth. Five hundred years ago, in 1500, humankind was just 6 percent of that, or about 425 million people. By 1800, the population had almost doubled to about 800 million.[9] Moreover, in that 400-year period from 1500 to 1900, as much as 80 percent of the world's population was composed of peasants, people who lived on the land and were the direct producers of food for themselves and the rest of the population. The world was overwhelmingly rural, and the availability of land to produce food was a constant constraint on the number of people alive at any given moment.

For most of that period, the human population rose and fell in great waves lasting for centuries, even if the very long-term trend was very slightly upward and the declines came sharply and swiftly. In very broad terms, we can see three great waves of population increase and decrease over the past 1,000 years. Beginning about 900–1000 CE (probably simultaneously in China and Europe), the population rose until about 1300, then crashed precipitously around 1350 largely as a result of the Black Death. Another period of increase began about 1400 and lasted until a mid-seventeenth-century decline. The third advance, beginning around 1700, has yet to halt, reaching 6 billion by 2000, although population experts expect it to level off around 9 billion people by 2050.

Peasants not only farmed for themselves, but also most often produced a surplus that rulers and others took from them (either directly, or in the form of taxes). To that extent ruling elites were macro-parasites, living off the direct producers and reducing the food supply available for them, just as micro-parasites did, sometimes leading to the death of individuals, or, in the case of widespread virulent disease, large-scale epidemics.[10] This balancing act of people fending off or dying from both macro- and micro-parasites – elites living off peasants, civilizations fighting off or losing to nomadic invaders, and germs multiplying inside nomads and city dwellers alike and then killing them – consti-tuted humankind's "biological old regime." In this world – the world not just of 1500 but the world for millennia before and then afterward until well into the nineteenth century – the human population lived very much in the environment and had to be very mindful of the opportunities of, and limits it placed on, human activity.[11] As a result, the human population did not increase so much or so fast as to threaten the environmental basis for society, except in a few cases,[12] or until later developments shattered the bio-logical old regime and opened up new possibilities, as we will see below.

Agriculture provided not only the food for the entire society, but most of the raw materials for whatever industry there was, especially textiles for clothing. In China and India, silk and cotton reigned supreme, while wool dominated in northwestern Europe; most industrial raw materials came from farms. Fuel for processing these materials, as well as for keeping warm, also came from forests. To this extent, the biological old regime was organic, that is, it depended on solar energy to grow crops for food and trees for fuel. The biological old regime thus was one that limited the range of possibilities for people and their history because virtually all human activity drew upon *renewable* sources of energy supplied to varying degrees throughout the year by the sun.

All living things need food and energy to live, and increasing amounts of both to sustain larger populations. What agriculture allowed people to do, in effect, was to capture natural processes and to channel that energy into the human population. In the biological old regime, agriculture was the primary means by which humans altered their environment, transforming one kind of ecosystem (e.g., forest or prairie) into another (e.g., rye or wheat farms, rice paddies, fish ponds, or eel weirs) that more efficiently channeled food energy to people. The size of human populations was thus limited by the amount of land available and the ability of people to use the energy from that land for their purposes.[13]

To support a growing human population, something had to change in terms of the relationship of people to the availability of land and their efficiency in working it. On the one hand, Europeans were to encounter a whole new world, the Americas, and to populate it. Although this New World was already quite populated in 1400 and the land already used by Native Americans, a massive biological exchange would radically alter those relationships, making the Americas a relatively depopulated world by the year 1600.[14] On the other hand, global trading relationships became reestablished driven largely by economic vitality in Asia,[15] allowing a considerable increase in overall production and productivity as specialization allowed people in one part of a regional trading network to produce goods that their environment was especially suited to, and to trade via markets with countless others who were doing the same thing. Market specialization spread, thereby allowing economies throughout the world to produce more than they ever had in the past, yet without escaping the limits of the biological old regime.

In the biological old regime, some peoples were relatively more successful than others, at least in terms of how much food was produced and how large the human population could be. Those places tended to be in Asia, in particular India and China. Indian agriculture was so productive that the amount of food produced was significantly larger than in Europe, and hence its cost was far lower. In the preindustrial age, when working families spent 60–80 percent of their earnings on food, the cost of food was the primary determinant of their real wages (i.e., how much a pound, a dollar, a real, or a pagoda could buy). Agriculture in China (as well as in Japan and numerous other parts of Asia) was highly productive, yielding a harvest of 20 bushels of rice for every one sown. Rice has the unique capability of gaining nutrients not directly from the soil, but from the water (and so it is grown in "paddies"), eliminating the need for the land to lie fallow, as was the custom in Europe, to regain its fertility. Additionally, Chinese farmers had learned how to prepare the soil, to irrigate, to fertilize, and to control insect pests in order to maximize the harvest yield. Moreover, farmers in the southern half of China could get two or sometimes three harvests per year from the same plot of land, drawing the amazement of early eighteenth-century European travelers to China.[16] Such an impressively productive agriculture certainly allowed the Chinese population to grow, from 140 million in 1650, to 225 million in 1750, and then to 380–400 million by 1850, accounting for about one-third of the world's population.

In the biological old regime, productive agriculture was Asia's competitive advantage, even in industry. The causal chain went like this: high per acre yields→low-priced food → relatively low wages → comparative advantage. In England, the causal chain was like this: low per acre yields → high-priced food → relatively high wages → comparative disadvantage. Asian economies produced not only more food and condiments like peppers, spices, and tea, but also more and higher-quality manufactured goods like porcelain ("China"), silk textiles, and cotton cloth. The market exchanges among these economies of Asia, carried mostly on ships in the China seas and the Indian Ocean, created more wealth than anywhere else in the world.[17]

The Columbian Exchange, c. 1500–1600

The attractions of Asia stimulated attempts by those poorer and less well-positioned at the far western edge of the Eurasian continent (Europeans) to gain access to Asian wealth and products. Marco Polo, his father, and his uncle traveled overland across Central Asia in the thirteenth century, but the establishment of Muslim empires afterwards shut off that route. In the early fifteenth century, the Portuguese began developing the sailing skills necessary for "blue water" sailing (out of sight of land) and began working their way down the west coast of Africa, ultimately reaching the Cape of Good Hope in 1488. Then a Genoan adventurer named Christopher Columbus convinced Spain's rulers to support his attempt to get to Asia by sailing west across the Atlantic, and so he set out in 1492. We all now know that Columbus instead stumbled across the Americas.

North and South America prior to the arrival of the Europeans was populated with peoples who had constructed various kinds of social and economic systems, ranging from hunting and gathering societies to highly developed agrarian societies, in the centuries after humans first migrated into the Americas around 15,000 BCE.[18] It thus should not be too surprising that these people could also create the highest form of political organization in the biological old regime, an empire. Two in particular are important to the story of the modern world, the Aztecs in central Mexico and the Incas on the coast of Peru, because those were the ones Spanish conquistadors targeted because of the gold and silver those empires had extracted from the earth.

The conquest of the Americas led to a global exchange of people, other animals, plants, and diseases called "the Columbian exchange."[19] Maize (corn), potatoes, tomatoes, chiles, and other foods spread rapidly throughout Eurasia, enriching the diets of commoners and elites alike. Sweet potatoes, for instance, reached China by the mid-1500s, making it possible for peasants there to sell their rice rather than eat it. Certainly the spread of New World crops into the Old World made it possible for populations there to increase above what would have been possible on the basis of the existing basket of foods.

But the Columbian exchange was a two-way exchange and it seems that the native peoples of the New World were the losers, for the encounter between Old and New Worlds brought two hitherto separate disease pools into contact. The Native American ancestors had migrated into the Americas during the last ice age when a land bridge linked Alaska to Siberia, thousands of years before the agricultural revolution in Eurasia brought people and domesticated animals together in a rich recipe for the transfer of animal pathogens to humans, leading to a whole range of diseases including smallpox, chicken pox, and influenza. Eurasians contracted these diseases and over time developed some immunity to them; New Worlders did not have a chance to do the same. When the ice age ended about 12,000 years ago and the melting glaciers raised the ocean level above the Bering Strait land bridge, the peoples in the Americas were isolated from the diseases that now became an everyday part of the material world in Eurasia, rendering some of them "childhood" diseases from which most people easily recovered. The diseases for which Europeans had developed immunities over the centuries proved to be deadly to those in the Americas (and later the Pacific Islands too) without immunity.

The smallpox epidemics that weakened both the Aztecs and the Incas, paving the way for the Spanish conquest of both empires, were just the beginning of a century-long holocaust that almost wiped out Native American populations. From 1518 to 1600, 17 major epidemics were recorded in the New World, spanning a territory from what is now Argentina in the south to what is now Texas and the Carolinas in the US. Not just smallpox, but other killer diseases – measles, influenza, bubonic plague, cholera, chicken pox, whooping cough, diphtheria, and tropical malaria – ravaged American populations.[20] Disease was not the only cause of the depopulation of the Americas in the century after the Spanish conquest. The conquest itself, war among the American natives, oppression by the conquerors, the forced requisitioning of Indian labor, and lowered fertility among the surviving native population all contributed to the disaster.[21]

In Mexico alone, there had been 25 million people in 1519; 50 years later there were 2.7 million, and 100 years later there were but 750,000, or 3 percent of the original total. Similar fates befell the Incas, the inhabitants of the Caribbean islands (starting with the Arawak on Española), and the Indians of (what is now) the southeastern US, although at different rates. Whether or not European-introduced diseases ravaged the Indians of the American Northeast, the upper Mississippi, or the Northwest before the 1600s is open for scholarly debate, but after permanent European settlements were established in North America, diseases afflicted those natives too. In short, in the century after European contact with the New World, vast regions were depopulated, losing 90 percent of their pre-1500 numbers, even if we do not know with certainty what the precontact population of the Americas was. Nonetheless, it does seem certain that tens of millions of people across the Americas had vanished.[22]

European States and Colonial Empires, 1500–1800

Asian and American empires provided the political organization through which their peoples extracted resources from their environments. Until the Mughals conquered India in the sixteenth century, there was not a unified state there to coordinate large-scale water control and land "reclamation" activities. But in China, that state had been expanding into new areas inhabited by non-Chinese peoples for centuries. By the eighteenth century, the Chinese Empire had expanded to incorporate regions and peoples ranging from the tropical south, to the boreal forests on the Russian border, to the high mountain plateau of Tibet and the grasslands of Central Asia.[23] Such large land-based empires expanded to establish control of people and resources that were of strategic interest and importance.

Western Europe was differently organized. There, no single empire since the Romans was able to establish political control of the entire region, although the Spanish tried (and failed) in the sixteenth century, as did Napoleon in the early nineteenth century. Rather, much smaller political units that we now call nation-states formed out of a vast array of principalities, bishoprics, and kingdoms, among the numerous kinds of states that had emerged in the wake of the breakup of the Roman Empire.

Warfare defined the emerging European state system. Until the mid-seventeenth century, wars were mostly fought to stop the Spanish from establishing an empire or to support Protestants (in Holland and the German states) in their attempts to gain independence from the Catholic monarchs of Spain. From the 1648 Peace of Westphalia, which ended the Thirty Years War, wars mostly involved France whose fortunes had risen while Spain's declined, and then, from the late 1600s on, contests were mostly between France and England, culminating in the Seven Years War (1756–63) and leading to Britain's victory over France.

There are many things that are historically significant about wars among European states in the period from 1500 to 1800. First, the wars involved virtually all European states, tying them very clearly in a single system, especially after the Peace of Westphalia. By 1650 all European states were embroiled in a common set of entanglements defined by war. Second, wars in Europe led both to consolidation into increasingly fewer political units and to the development of a particular kind of national state as the most successful form of European state, the territorial national state. As European states dominated the world in the nineteenth and twentieth centuries, so did their form of political organization drive out other forms of political organization, in particular the large land-based empires that had dominated earlier centuries.[24]

The wars of European states drove their expenses well above the amount of silver that was left in European hands after it was used to buy Asian products, leading to standing armies and navies, taxation and state bureaucracies to collect it, representative assemblies of various kinds demanded by the taxed subjects so they could influence the level of taxation (even though Europe's "absolute monarchs" tried mightily to ignore or shut down these institutions), public indebtedness, and the institution of the national debt. All of these activities were part of a "state-building" process in seventeenth- and eighteenth-century Europe.

European states – in particular England, France, and Spain – had colonial empires, but mostly those were in the New World.[25] There, the "Great Dying" of the native peoples led to a severe need for laborers to work the new lands, a need that was met by what became the vast transatlantic slave trade. By 1800, over 10 million black Africans – by far the largest number of migrants to the New World – had been forcibly taken from their homeland and deposited in the Americas where they worked sugar and cotton plantations as slaves.

What their New World colonies provided the home countries were resources that had become scarce in the Old World. England supplied its growing navy with masts from American forests, cod to its workers, and sugar to sweeten their drinks. France took furs from northern American forests. Spain stumbled upon vast quantities of silver, which slipped out of Spanish hands and into the coffers of the English and Dutch who shipped boat loads to Asian ports in India, but increasingly in China, to pay for the spices, textiles, porcelain, and tea that they supplied to both their home and colonial markets. Indeed, the British attempt to maintain a monopoly on the sale and taxation of these commodities led the American colonists in Boston to protest in the 1765 "tea party" that contributed to the outbreak of the American revolution and independence of the British colonies there.

As the strongest and most successful European states, England and France competed not just in Europe but in the Americas and Asia as well. In the "long" eighteenth century from 1689 to 1815, Britain and France fought five wars, only one of which Britain did not initiate. Their engagement (with others) in the War of Spanish Succession was ended by the 1713 Treaty of Utrecht, which established the principle of the "balance of power" in Europe, that is, that no country should be allowed to dominate the others and to further periodic wars between the British and French.

But the most significant was the Seven Years War of 1756–63, or what Americans call the French and Indian War and interpret in terms of its impact on the American War of Independence of 1776–83 against Britain. To be sure, the spark that led to war between Britain and France came in the American colonies, and it was in fact the 22-year-old George Washington who lit it.[26] But it became a global engagement – perhaps the first real world war – with British and French troops fighting in the backwoods of the American colonies, in Canada, in Africa, in India, and in Europe. The outcome was disastrous for the French: they lost their colonial claims both in North America (the British got Canada) and in India, leading to greater British power and position in both parts of the world.[27]

By 1800, therefore, the processes of state-building in Europe had led to the creation of a system defined by war, which favored a particular kind of state exemplified by the ones built in Britain and France. Balance of power among sovereign states, not a unified empire, had become the established principle, and Britain had emerged as the strongest European state. But that does not mean that it was the strongest or richest state in the world – far from it. To be sure, Mughal power in India was declining in the early 1700s, and the British were able to begin building a colonial empire there. But the British were still too weak to be able to contest China's definition of the rules of trade in Asia. When they tried, most famously in 1793 under Lord Macartney's mission, the Chinese emperor sent them home with a stinging rebuke, and the British could do nothing about it – until, as we will see below, the Industrial Revolution gave Britain powerful new weapons forged of iron and steel and powered by steam.

The Ecological Limits of the Biological Old Regime

During the eighteenth century, there were indications that states throughout the world were reaching the ecological limits of the biological old regime. In Joachim Radkau's view, in the eighteenth century the world "entered into a new era" as resources became scarce and limits were being reached, driven by the rise in world population, market exchanges that increased consumption of resources of all kinds, and more efficient ways of transporting goods and people (and spreading diseases) around the world.[28] In China, a large-scale ecological crisis became acute:

> At that time the development in China showed in some respects a remarkable convergence with that in Europe, one of those astonishing parallels that give meaning to a *global* history of the environment. In China, as in Europe, one can detect in the eighteenth century a desire to use natural resources to their limits and to leave no more empty spaces, no quiet reserves. As in Europe, this was made easier by crops from the new world, especially the potato and maize. It was precisely the highest success of cultivation that led the country to its most severe ecological crisis: in this regard, China's fate is a warning to modern industrial society.[29]

Radkau explores the ways in which societies across the globe in the seventeenth and eighteenth centuries began exhausting and using up existing resources, and posits a global eighteenth-century "limits of nature" crisis. In his view,

> the urge to exploit the last reserves possesses an epochal character in environmental history. It led to a fundamental shift in strategies of sustainability. Until then, the sustainability of agriculture was in many cases guaranteed not only by fertilizer and fallow, but also by the fact that one could make use, as needed, of semiwild outlying areas: commons, forests, heaths, moors, and swamps.

New World crops, especially maize and the potato, "promoted population densities that led to overuse of forests and pastures": rivers were dammed and diked, irrigation extended as far as possible, and forest reserves were strained.[30] In Michael Williams' view, vast swathes of the world's temperate zones lost their forests, a major chapter in the story of "deforesting the earth."[31]

Throughout the world,

> rising demands for energy put an enormous strain on local wood, peat, and other supplies of fuel. By the late eighteenth century, there were distinct signs of increasing materials and fuel scarcities. Both Japan and the British Isles suffered an energy crisis in common with other long-steeled areas in Eurasia.[32]

In China, a growing population met the limits of the expansion of the empire, leading to an increasingly intensive use of all resources as people tried to get more out of the environment, but less was available. Living standards leveled off and began to fall, as people worked the land for less and less. China's densely populated, intensively farmed and energy-short agrarian life might have been the future for the rest of the world, Europe included, except for the discovery and use of fossil fuels as new and revolutionary sources of energy, first in a corner of England, then elsewhere in Europe, North America, and Japan by the end of the nineteenth century.

What turned out to be the Industrial Revolution was not inevitable. But for some highly specific geographic circumstances and the happenstance of a convergence of other global trends, the fossil-fuel-driven industrial world that we now inhabit might not have emerged – it was historically contingent, not inevitable.[33] In 1750, every one of the world's 720 million people, regardless of where they were or what political or economic system they had, lived and died within the biological old regime. The necessities of life – food, clothing, shelter, and fuel for heating and cooking – all came from the land, from what could be captured from annual energy flows from the sun to the earth. Industries too, such as textiles, leather, and construction, depended on products from agriculture or the forest. Even iron- and steel-making in the biological old regime, for instance, relied upon charcoal made from wood. The biological old regime thus set limits not just on the size of the human population, but on the productivity of the economy as well.

The Fossil-Fuel Escape from the Biological Old Regime, 1800–1900

Population growth and agricultural development put pressure on land resources in England as they did in China. Indeed, by 1600 much of southern England had already been deforested, largely to meet the fuel needs of the growing city of London for heating and cooking. Fortunately for the British, veins of coal were close enough to the surface of the ground and close enough to London to create both a demand for coal and the beginnings of a coal industry. By 1800, Britain was producing 10 million tons of coal a year, or 90 percent of the world's output, virtually all destined for the homes and hearths of London. As the surface deposits were depleted, mine shafts were sunk, and the deeper they went in search of coal, the more the miners encountered groundwater seeping into and flooding the mine shafts. Mine operators had a problem, and they began devising ways to get the water out of the mines.

Ultimately what they found useful was a device that used steam to push a piston. Early versions of this machine, developed first by Thomas Newcomen in 1712 and then vastly improved by James Watt in the 1760s, were so inefficient that the cost of fuel would have rendered them useless, except for one thing: at the mine head, coal was in effect free. Newcomen's (and later Watt's) steam engines thus could be used there. Between 1712 and 1800 there were 2,500 of the contraptions built, almost all of which were used at coal mines. But even that does not yet explain the Industrial Revolution, because the demand for coal (and hence steam engines) was fairly limited until new applications were devised. The one that proved most important was the idea of using the steam engine not just to draw water out of coal mines, but to move vehicles above ground.

The real breakthrough thus came with the building of the first steam-engine railway. Along with digging deeper, coal miners had to go farther from London to find coal deposits and thus had high expenses for transporting the coal overland from the pit head to water. Fixed steam engines were being used to haul coal out of the mines and to pull trams short distances. But at a mine in Durham in the north of England, the idea of putting the steam engine on the tram carriage and running it on iron rails became a reality in 1825 with an 11-kilometer (7-mile) line connecting the mine directly to the coast. The first railroad was born.

Whereas in 1830 there were a few dozen kilometers of track in England, by 1840 there were over 7,200 kilometers (4,500 miles), and by 1850 over 37,000 kilometers (23,000 miles). The prodigy of the coal mine, the railroad fueled a demand for more coal, more steam engines, and more iron and steel: each mile of railroad used 300 tons of iron just for the track. Between 1830 and 1850, the output of iron in Britain rose from 680,000 to 2,250,000 tons, and coal output trebled from 15 million to 49 million tons.[34]

A common explanation for the Industrial Revolution is that it embodied a search for more efficient *labor*-saving devices. But, as Pomeranz has argued in *The Great Divergence*, a better way to think about the Industrial Revolution is that it proceeded by finding *land*-saving mechanisms. For throughout the Old World, from China in the east to England in the west, shortages of land to produce the necessities of life were putting limits on any further growth at all, let alone allowing a leap into a different kind of economic future. This understanding of the ecological limits of the biological old regime opens a new window onto the explanation of how and why the Industrial Revolution occurred first in England.

Steam could have been produced using wood or charcoal, but that would have required vast forests and, by the end of the eighteenth century, forest covered but 5–10 percent of Britain. Under the best of circumstances, using charcoal to produce iron in 1815 would have yielded but 100,000 tons or so, a far cry from the 400,000 tons actually produced and the millions soon needed for railroads. Tens of millions of additional acres of woodland would have been needed to continue to produce iron and steel.[35] That might have been feasible, but converting land from agricultural purposes back to forest would have had other rather dire consequences. Thus without coal, and the historical accident that it was easily found and transported within England, steam, iron, and steel production would have been severely curtailed.

Similarly, Britain's New World colonies provided additional "ghost acres" beyond its borders that allowed the first part of the story of industrialization, that of cotton textiles, to unfold. To feed its textile mills, in the early 1800s Britain was importing hundreds of thousands of pounds of raw cotton from the New World, mostly from its former colonies in the US, but also from its Caribbean holdings. If the British had been forced to continue to clothe themselves with wool, linen, or hemp cloth produced from within their own borders, over 20 million acres would have been required. Similarly, Britain's sugar imports from its colonies provided substantial calories to its working population, all of which would have required more millions of acres.[36] The point is this: without coal or colonies, the dynamics of the biological old regime would have forced Britons to devote more and more of their land and labor to food production, further constraining resources for industrial production and snuffing out any hope for an industrial revolution, much as what happened to China in the nineteenth century.

Harnessing the energy captured from burning coal to make steam (no matter how inefficiently at first) enabled the British to resolve two major problems, and then to propel Britain to the top as the world's most powerful nation. First, Britain had imported such large quantities of cotton textiles from India that its native textile industry had been threatened with extinction. Import duties imposed in 1707 slowed the decline, but the application of steam engines to textile production enabled British mills in Manchester to churn out vast amounts of cheaply produced textiles with cotton bought cheaply from the slave states of the American South, and then sold to consumers in its colonies.

The Crown Colony of India centrally figured in the solution to Britain's other major international problem – finding ways to pay for the mountains of tea that it imported from China. Fortified with sugar and milk, tea stimulated tired textile workers; as that industry rapidly grew, so too did demand for tea from China. But, because Britain had nothing to trade with China, it shipped boatload after boatload of silver to China. In fact, one study estimates that one-half to two-thirds of all silver extracted from the New World from 1500 to 1800 wound up in China.[37] British trading firms discovered that, to stem the outflow, they could smuggle opium from India into China, and get the silver they needed to then pay for tea. That illegal trade flourished so much that, by the 1820s, the influx of silver into China reversed and flowed back to England. Opium addiction in China grew, and the Chinese government determined to halt both the outflow of silver and the import of opium, a decision that led to the outbreak of war with Britain in 1839 and the defeat of China in 1842.

The British might not have been interested or able to prosecute this drug war had it not been for an innovation in the use of steam power, in particular the construction of a new kind of industrial-era warship, the first of which was called the *Nemesis*.[38] England defeated China in the Opium War (1839–42) with new weapons that defied the natural limitation of the biological old regime.[39]

The Gap and the Making of the Third World, 1800–1900

The application of steam power to both industrial production and the means of war began to revolutionize the world. Fearful of falling behind Britain, of losing out on colonial opportunities, or of being colonized, first other European states (France, Germany, and Russia), then the US and Japan, entered on programs to industrialize as quickly as possible. This competition among what came to be called "the great powers" led to tensions between them in the "Scramble for Africa" in the 1870s and 1880s, and the "Scramble for Concessions" from China (1895–1900).

By the 1870s, Europeans thus had the "tools of empire" with which to engage and defeat Africans on African soil. Africans put up valiant and stiff resistance, but their technology was no match for the Maxim gun. The most famous and perhaps deadly instance was at the 1898 Battle of Omdurman where British troops confronted the 40,000-man Sudanese Dervish army. As described by Winston Churchill, the future British prime minister, the Dervish attack was quickly repulsed by Maxim guns mounted on river gunboats: "The charging Dervishes sank down in tangled heaps. The masses in the rear paused, irresolute. It was too hot even for them." On shore, the British "infantry fired steadily and stolidly, without hurry or excitement, for the enemy were far away and the officers careful." To the Sudanese "on the other side bullets were shearing through flesh, smashing and splintering bone; blood spouted from terrible wounds; valiant men struggling on through a hell of whistling metal, exploding shells, and spurting dust – suffering, despairing, dying." After five hours, the British had lost 20 soldiers; 10,000 Sudanese were killed.[40] As poet Hilaire Belloc put it:

> Whatever happens, we have got
> the Maxim gun, and they have not.[41]

With such a technological advantage, by 1900 most of Africa had been divided up among a handful of European powers, in particular Britain, France, Germany, and

Belgium, with Portugal hanging on to its seventeenth-century colonial possession in Angola. Only Ethiopia, under the extraordinary leadership of King Menelik, defeated the weakest European power, Italy, and thereby maintained its independence.[42]

Although industrialization, improvements in military technology, and strategic jockeying among "the powers" go a long way toward accounting for the dominance of Europeans, North Americans, and Japanese over Africans, Asians, and Latin Americans, there was also an ecological dimension to the making of the Third World and the gap between the industrialized and unindustrialized parts of the world.

Where the very success of China's economy in the biological old regime had begun putting stress on its forest reserves by 1800, leading to serious deforestation by the middle of the nineteenth century, other parts of Asia and Latin America were deforested by other processes. In India, forests in the peninsula were cleared long before the population began to grow around the middle of the nineteenth century. Warring Indian princes cleared forests to deny their enemies cover, a policy of "ecological warfare" that the colonizing British also carried out with gusto. Additionally, dislocated peasant farmers cleared land, and there was some commercial logging in the north as well. All of these contributed to the extensive deforestation of India by the late nineteenth century.[43]

In Latin America, different processes led to massive deforestation. There, colonial powers, intent upon extracting raw materials and transforming their Latin American holdings into sugar or coffee plantations, cleared forests. In Brazil, the great forests of the Atlantic seaboard were first cleared for sugar plantations. In the early 1800s, Brazilian landowners switched to coffee crops. As a tree (not native, but imported from Ethiopia), coffee presumably could have been planted and replanted on the same land, given adequate care to the fertility of the soil. But, as it turned out, landowners preferred to deplete the soil and, after 30 years or so, to clear another patch of virgin forest. "Thus coffee marched across the highlands, generation by generation, leaving nothing in its wake but denuded hills."[44] And on the Caribbean islands, French and British colonists in the eighteenth century removed so much forest for sugar plantations that even then observers worried that it was causing the climate of the islands to change, getting drier and drier with every stand of forest cut down.[45]

By the last quarter of the nineteenth century, then, large parts of Asia and Latin America experienced significant environmental damage caused by deforestation and the depletion of soil fertility. Of course, as these were agricultural societies, the changes put additional stresses on the biological old regime, making these regions even more susceptible to climatic shock and increasing the possibility of widespread famine.

Mostly, harvest failures were localized affairs. But, in the late nineteenth century, a climatic phenomenon we now know as El Niño (or by its more scientific name, ENSO, for El Niño–Southern Oscillation) intensified to the greatest extent in perhaps 500 years, affecting vast portions of the planet. While El Niño brings excessive rainfall to the wheat belt of North America and does not affect Europe at all, it means drought for vast portions of Asia, parts of northern and western Africa, the coast of Peru, and northeast Brazil. Three times – in 1876–9, 1889–91, and 1896–1902 – El Niño droughts afflicted what would become the Third World. The particularities of how El Niño affects Asia, Africa, Latin America, and North America, coupled with the workings of a world economy designed to benefit the industrializing parts of Europe and North America, and the aggression of the "New Imperialism" against Asians and Africans combined in a historical conjuncture of global proportions to spell famine and death for millions of people.

In all, an estimated 30 to 50 million people died horribly in famines spread across Asia and parts of Africa and Latin America. But these deaths were not just caused by the natural effects of El Niño, no matter how powerful they were in the late nineteenth century. Rather, as historian Mike Davis describes in a recent book, these massive, global famines came about as a result of El Niños working in conjunction with the new European-dominated world economy to impoverish vast swathes of the world, turning much of Asia, Africa, and Latin America into the Third World. In Asia, governments were either unwilling or unable to act to relieve the disasters. The British colonial rulers of India were more intent on ensuring the smooth workings of the "free market" and their colonial revenues than in preventing famine and death by starvation or disease. There, people died in sight of wheat being loaded onto railroads destined for consumption in Britain, and the colonial authorities spurned famine relief in the belief that it weakened "character" and promoted sloth and laziness. In China, the Manchu government, switching resources and attention from the interior to the coasts where foreign pressure was greatest, had neither the ability nor the resources to move grain to the isolated inland province of Shanxi where the drought and famine was the most severe. Likewise in Angola, Egypt, Algeria, Korea, Vietnam, Ethiopia, the Sudan, Brazil, El Niño-induced drought contributed to famines that weakened those societies and their governments, inviting new waves of imperialist expansion and consolidation.[46] The gap between industrialized countries and the future Third World had crystallized.

Although it may appear to have been a historical accident that those late-nineteenth-century El Niños hit Asians, Africans, and Latin Americans hard while improving harvest yields in the American mid-West and bypassing Europe altogether, the socioeconomic impact they had were the result of longer-run historical processes. All of these adversely affected regions either had weak states (and largely weakened because of imperialist aggression) that could act neither to industrialize nor to provide famine relief to their people, or they had colonial governments (especially the British in India) whose policies had the same results. Thus at the beginning of the twentieth century, large parts of the world and its people were condemned as best they could to fend off the worst effects of the biological old regime. It is hardly surprising that the life expectancies and life chances of people there were much less than those in the industrialized parts of the world. "The gap" was – and remains – a matter of life and death for nearly half of the world's population.

The Great Departure in the Twentieth Century

This chapter has discussed two major changes in the relationship of humans to the global environment over the past 500 years: the "Columbian exchange" of foods, pathogens, and peoples in the centuries following Europeans' 1492 encounter with the New World; and the escape from the constraints of the "biological old regime" facilitated by the use of fossil fuels (initially coal) to produce steam and power for both industry and warfare. We conclude this chapter with a brief look at another time when the relationship of humans to the global environment underwent another significant change – the twentieth century.[47]

That "great departure"[48] can be seen especially in three ways. First, the invention of an industrial means to produce nitrogen-based fertilizer has so dramatically increased the food supply that the human population has grown from 1.6 billion in 1900 to over 7 billion today. Without synthetic fertilizer, the world's farmland could support at most half that amount. Second, the post–World War II competition between the US and the Soviet Union, coupled with decolonization, unleashed demand for all kinds of natural resources to meet rapidly growing demands of consumers and states and put pressure on natural resources of all kinds, but especially global forests. And third, the energy to fuel those advances has come increasingly from oil drilled from deeper and deeper wells. Burning the oil and its refined derivatives has created a new global industry with the potential to dramatically affect the environment through spills and accidents, and the new global-circling auto industry has also poured millions of tons of carbon dioxide into the atmosphere, accelerating trends toward global warming begun by burning coal to produce electricity. In these ways (and others), in the twentieth century we created an anthroposphere that now threatens to overwhelm the natural processes not just of the atmosphere but of the hydrosphere and lithosphere as well.

One paradox of the newly emerging modern fossil-fueled industrial world was that, even as more and more power was harnessed to do work or to threaten others, in two important ways that world remained constrained by an important leftover of the bio-logical old regime. Soil scientists had begun to understand that the most significant chemical in fertilizer is nitrogen, and armament manufacturers also knew that it was critical to gun powder. But the supplies of the nitrates from which nitrogen was extracted for both uses was limited to those deposited by natural processes. A global search for the most concentrated forms of nitrates – guano (accumulated bat and bird droppings) – exhausted those supplies by the end of the nineteenth century, and Germany in particu-lar was feeling constrained both by limits on its ability to grow grain, and by its place in Central Europe surrounded by potential foes. That situation led to intense interest in, and pressure to solve, the problem of manufacturing usable nitrogen-based compounds.

By 1913, the German chemist Fritz Haber and German industrialist Carl Bosch cre-ated an industrial process for synthesizing ammonia, a nitrogen-based compound that could be easily transformed into chemically manufactured fertilizer. The Haber–Bosch process then spread slowly around the world, in part because of the world wars of the first half of the twentieth century, coming after World War II to play a central role in the vastly increased food production throughout Eurasia and the Americas, and the vast population increase as well. Perversely, solving the problem of industrially producing nitrogen-based fertilizer also proved to be a boost to armament manufacturers and the destructiveness of war: the same process produced vast quantities of substances essential for gunpowder, contributing to the bloodiness of both World War I (1914–18) and World War II (1939–45).[49]

The outcome of that combined "31-year war" of the twentieth century, which included the Great Depression era of the 1930s, set the stage for significant changes to the political ordering of the world in the second half of the twentieth century.[50] The world wars and the global crisis of capitalism and imperialism weakened or destroyed the European and Japanese colonial powers. With the European powers dismantled – regard-less of whether they "won" in World War II (e.g., Britain or France) or lost (Germany, Italy, Japan) – none could retain their colonial empires, unleashing a torrent of national independence movements in the powers' former colonies. And out of the ashes of the old colonial world order rose two new superpowers: the US and the Soviet Union. Tensions between the two countries, initially allies against Nazi Germany, soon escalated into a decades-long Cold War that lasted until 1991.

The US economic system came to be driven by stoking and satisfying a growing consumer demand for more and more stuff, especially for automobiles, while the Soviet system focused on production of ever more producer goods such as steel, railroads, and electricity, mainly to support the power of the state. The Chinese Communist revolution in 1949 added another productionist power to the globe. Both US inspired "consumerism" and Soviet-style "productionism" placed a high value on the most rapid and continuous economic development possible. In the meantime, the newly independent states of the postcolonial world too came to embrace rapid economic development as a means to raise the standard of living of their citizens and to increase their military power as well.

Like the productionism of the Soviet and Chinese models,[51] the consumerism of the US, Europe, and Japan had significant environmental consequences. Not only did refining oil and burning gasoline pollute the air in virtually all major cities in the US, Europe, and Japan, but extracting the oil and moving it around the world in tankers left spills on land and sea. Making cars also requires huge amounts of energy, creates nearly 30 tons of waste for every ton of car made, and uses charcoal burned from the Amazon rain forest, contributing directly to global warming and deforestation.[52]

Additionally, the rapid development of China and India has increased their demand for, and consumption of, oil, leading one analyst to see a "frenzied search" and "intensifying global struggle for energy" leading to competition and potential conflict among China, India, the US, Japan, and other rapidly industrializing states. Increased global demand for limited supplies of oil certainly has led to rapid increases in the price of oil; whether it will lead to increased production or armed conflict remains to be seen.[53]

The dependence of the modern world on fossil fuels is one reason a leading historian has argued that the twentieth century represents a major break with the past and the beginning of a vast, uncontrolled, and unprecedented experiment – a gamble that the fossil-fuel-consuming, twentieth-century way of organizing the globe will not undermine the ecological bases for life on Earth.[54] Twentieth-century American consumerism, Soviet productionism, and Third World developmentalism have all placed a premium on industrialized economic growth, and these approaches collectively have created an "anthroposphere" of such size and power as to rival, replace, and rechannel the natural processes of the biosphere.[55] By the end of the twentieth century, humans already were taking 40 percent of the product of the natural process of photosynthesis.[56]

In very rough terms, the size of the global economy at the beginning of the Industrial Revolution in 1800 was three times larger than it had been 300 years earlier in 1500. Since then, economic growth has accelerated, tripling from 1800 to 1900. During the twentieth century, the world economy has become 14 times as large as it was in 1900, with most of that growth coming since the end of World War II.

Economic growth is a rough indicator of our species' relationship with the environment because virtually anything that counts as "economic" results from a transformation of nature. Mining, manufacturing, and farming, all are processes which change some part of nature into something usable or consumable by humans.[57] The more economic development we have, the more we change nature. After World War II, consumerism, productionism, and developmentalism combined to push even more economic growth. While many people today are better off than their parents were 50 years ago, our pursuit of economic development has dramatically altered the environment and global ecological processes. According to the Millennium Ecosystem Assessment,

> over the past 50 years, humans have changed ecosystems more rapidly and extensively than in any comparable period of time in human history, largely to meet growing demands for

food, fresh water, timber, fiber, and fuel. This has resulted in a substantial and largely irreversible loss in the diversity of life on Earth.[58]

Nitrogen fertilizer, first manufactured in 1913 as a result of the Haber–Bosch process, now puts as much nitrogen into the environment as all global natural processes. By 1950, 3.6 million tons of nitrogen fertilizer were used, mostly in advanced industrial countries. By 1980, that had increased to 60 million tons and, by 2000, to 80 million tons. Significantly, almost 60 percent of that use is in just three countries: China (25 million tons), the US (11 million tons), and India (10 million tons); almost no synthetic nitrogen fertilizer is used in sub-Saharan Africa, the poorest place on Earth.[59] There and elsewhere, increasingly poor farmers with no money for artificial fertilizer, irrigation, or high-yielding seeds remain trapped within the biological old regime, and increasingly use up natural resources to sustain their lives, further impoverishing their environment. Poor rural populations burn forests, grasses, or anything else for cooking, heating, and lighting – survival – which further impoverishes both their environment and them. It is not just industry that transforms the global environment, but rural poverty too.

Some suggest that new technologies such as biotechnology, computers, and the Internet might foster the needed economic growth but have a lighter impact on the environment. On the one hand, manufacturing computer chips or manipulating genes uses very little raw material compared, say, to locomotives or ships. But without electricity, the computer and biotechnology worlds could not exist. Indeed, the world's energy use in the twentieth century parallels the century's economic and population explosions with 16 times as much energy being used in 1990 as in 1900. One analyst calculates that humans used more energy in the twentieth century than in all the 10,000 years from the agricultural revolution to the Industrial Revolution.[60] Coal and steam fired early industry, but the twentieth century is the oil era, fueling both the spread of the automobile and power plants for electricity. Much electricity is generated by coal, and the combination of burning oil, gas, and coal – all so-called "fossil fuels" – has led to global warming.

Global warming is caused by the release of greenhouse gases into the atmosphere (air pollution), the most commonly known being carbon dioxide from burning coal and gas, but also sulfur dioxide. Methane from irrigated rice paddies and animal herds too contributes to global warming. The effects of rising average global temperatures to date have been felt mostly in polar or cold regions with melting glaciers and thawing permafrost. A greater danger, though, is that it is not known how much more global temperatures will rise and whether the rise at some point will have a sudden, unexpected, and catastrophic consequence. Rising ocean levels are already causing problems in Alaska, low-lying Pacific islands, and Bangladesh. But rising ocean temperatures might precipitate even greater changes in ocean currents, depriving Europe, for example, of the warming effects of the Gulf Stream/North American current.

Human activity in pursuit of economic growth and development has affected other natural processes as well.[61] Besides the nitrogen cycle and global climate, humans have ripped up vast swathes of the Earth's surface for coal, copper, gold, bauxite, and other natural resources. Acid rain makes lakes in the eastern US, northeast China, and Japan unsuitable for aquatic life. Lakes and oceans filling with nitrogen-fertilizer runoff bloom with algae that suck all the oxygen from the water, creating dead zones. Additionally, consumption of beef has increased the global cattle population to 2 billion eating, drinking, defecating, and gas-emitting animals that also contribute to global warming.

Logging and farmland expansion deforest huge amounts of rain forest in Africa, Asia, and Latin America, altering local climates and driving thousands of species to extinction. To be sure, people have been burning or cutting down forest since the end of the last ice age and the beginning of agriculture 10,000 years ago. Advanced biological old regime economies in China and England, for instance, had effectively deforested their lands by 1800. The pace of deforestation quickened in the nineteenth century with the felling of North American, Russian, and Baltic forests, before slowing with war and the Great Depression in the first half of the twentieth century. But the 50 years after the end of World War II saw the most ferocious attacks on forests in human history: *half* of all deforestation in human history has occurred in the last half century, prompting historian Michael Williams to call it "The Great Onslaught."[62] Nearly all of that recent deforestation has been in the tropical world of Africa, Asia, and Latin America, where most of the world's poor live. Natural ecosystems are being radically simplified and hence made less resilient, with the loss of thousands of animal and plant species.

Will it be our fate to poison our world with the unwanted pollutants of the industrial world? Will "island Earth" go the way of the Easter Islanders who used up their natural resources, saw their population decline and descend into warring, cannibalist groups huddling in cold, dark caves?[63] If the world we live in is contingent upon what has happened in the past and the choices that people made, then so too is the future contingent upon choices and actions we make today. To be sure, the two major problems the world faces are daunting: on the one hand, providing a decent standard of living for a rapidly growing world population; and on the other halting and then reversing the degradation of the environment caused by twentieth-century-models of industrial development. Is the modern world, structured as it is, capable of addressing these global problems?[64]

The fundamental problem is the relationship of global economic growth to the environment. Economic activity, which is essential for humans to exist, has always been dependent upon and hence part of natural processes. The major difference between the twentieth century and all preceding human history is that the human impact on the environment was so small or so local before the twentieth century that modern economic theorists (beginning in the eighteenth century) developed models of how the world works that do not account for the uses or services to humans of the natural environment. In the twentieth century, and to this very day, advocates of global free trade, developmentalism, consumerism, and (until recently) productionism assume that the global economic system is separate from the global ecological system.[65] That may turn out to be a colossal mistake. The biosphere and the anthroposphere became inextricably linked during the twentieth century, with human activity increasingly driving biospheric changes in directions that can be neither known nor predicted. Albert Einstein once famously said that God does not play dice with the universe; apparently the same cannot be said about humans and the Earth.

Notes

1   The argument and much of the evidence in this chapter is drawn from R. B. Marks, *The Origins of the Modern World: A Global and Ecological Narrative from the Fifteenth to the Twenty-First Century*, 2nd ed., Lanham, MD, Rowman & Littlefield, 2007. For an overview of books dealing with global environmental history, see R. B. Marks, "World Environmental History: Nature, Modernity, and Power," *Radical History Review* 107, 2010, pp. 209–24.

2   The term is used both in F. Braudel, *Civilization and Capitalism 15th–18th Century*, vol. 1, *The Structure of Everyday Life*, New York, Harper and Row, 1981 (pp. 70–2), and in C. Ponting, *A New Green History of the World: The Environment and the Collapse of Great Civilizations*, New York, Penguin Books, 2007, chapter 12.

3   J. Radkau, *Nature and Power: A Global History of the Environment*, New York, Cambridge University Press, 2008; J. Richards, *The Unending Frontier: An Environmental History of the Early Modern World*, Berkeley, University of California Press, 2003.

4   K. Pomeranz, *The Great Divergence: China, Europe and the Making of the Modern World Economy*, Princeton, NJ, Princeton University Press, 2000.

5   J. R. McNeill, *Something New under the Sun: An Environmental History of the Twentieth-Century World*, New York, W. W. Norton, 2000.

6   E. A. Wrigley, *Continuity and Change: The Character of the Industrial Revolution in England*, Cambridge, Cambridge University Press, 1988.

7   McNeill, *Something New under the Sun*, p. 11.

8   Braudel, *The Structure of Everyday Life*, chapter 1.

9   Because no one actually took a census, these population figures are reconstructions by historical demographers, and there is much discussion and debate about all matters having to do with the size, distribution, and dynamics of human populations in the period covered by Braudel. See Braudel, *The Structure of Everyday Life*, chapter 1; C. McEvedy and R. Jones, *Atlas of World Population History*, New York, Penguin Books, 1978.

10 W. McNeill, *Plagues and Peoples*, Garden City, NY, Anchor Books, 1975.

11 A. Penna, *The Human Footprint: A Global Environmental History*, Malden, MA, Wiley-Blackwell, 2010, chapter 3.

12 Ponting, *A New Green History of the World*, chapters 1, 5, and 17.

13 V. Smil, *Energy in World History*, Boulder, CO, Westview Press, 1994; E. Burke III, "The Big Story: Human History, Energy Regimes and the Environment," in E. Burke III and K. Pomeranz (eds.), *The Environment and World History*, Berkeley, University of California Press, 2009, pp. 33–53.

14 A. Crosby, *Ecological Imperialism: The Biological Expansion of Europe, 900–1900*, New York, Cambridge University Press, 1986; K. F. Kiple and S. V. Beck, *An Expanding World: The European Impact on World History, 1450–1800*, vol. 26, *Biological Consequences of the European Expansion, 1450–1800*, Aldershot, Ashgate, 1997.

15 A. G. Frank, *ReOrient: Global Economy in the Asian Age*, Berkeley, University of California Press, 1998.

16 R. B. Marks, *Tigers, Rice, Silk, and Silt: Environment and Economy in Late Imperial South China*, New York, Cambridge University Press, 1998, pp. 284–5.

17 Frank, *ReOrient*.

18 Humans may have migrated to the Americas as early as 35,000 years ago, but the consensus among scholars is about 15,000 BCE. R. E. W. Adams and M. J. MacLeod (eds.), *The Cambridge History of the Native Peoples of the Americas*, New York, Cambridge University Press, 2000, p. 28. For a review of the debates about the size of the human populations in the pre-Columbian Americas, see C. C. Mann, *1491: New Revelations of the Americas before Columbus*, New York, Vintage Books, 2011, pp. 110–51.

19 A. Crosby, *The Columbian Exchange: Biological and Cultural Consequences of 1492*, Westport, CT, Greenwood Press, 1972.

20 B. G. Trigger and W. E. Washburn, (eds.), *The Cambridge History of the Native Peoples of the Americas*, vol. 1, Cambridge, Cambridge University Press, 1996, pp. 361–9.

21 L. Bethell (ed.), *The Cambridge History of Latin America*, vol. 2, Cambridge, Cambridge University, 1984, chapter 1.

22 The import of African slaves to fill the labor vacuum also brought malaria to the New World. J. L. A. Webb, *Humanity's Burden: A Global History of Malaria*, New York, Cambridge University Press, 2009.

23 R. B. Marks, *China: Its Environment and History*, Lanham, MD, Rowman & Littlefield, 2012, chapter 4.

24 C. Tilly, *Coercion, Capital, and European States, AD 990–1990*, Oxford, Blackwell, 1990.

25 W. Beinart and L. Hughes, *Environment and Empire*, Oxford, Oxford University Press, 2007; R. H. Grove, *Green Imperialism: Colonial Expansion, Tropical Island Edens and the Origins of Environmentalism*, New York, Cambridge University Press, 1995.

26 F. Anderson, *Crucible of War*, New York, Alfred A. Knopf, 2000.

27 A. G. Frank, *World Accumulation, 1492–1789*, New York, Monthly Review Press, 1978, p. 237.

28 D. Igler, "Diseased Goods: Global Exchanges in the Eastern Pacific Basin, 1770-1850," *American Historical Review* 109/3, 2004, pp. 693–719.

29 Radkau, *Nature and Power*, pp. 195, 111, original emphasis.

30 Radkau, *Nature and Power*, pp. 197, 198.

31 M. Williams, *Deforesting the Earth: From Prehistory to Global Crisis*, Chicago, University of Chicago Press, 2003.

32 Richards, *The Unending Frontier*, p. 12.

33 Wrigley, *Continuity and Change*.

34 E. Hobsbawm, *The Age of Revolution*, New York, Vintage Books, 1996, pp. 43–5.

35 Wrigley, *Continuity and Change*, pp. 54–5; Pomeranz, *The Great Divergence*, pp. 59–60.

36 Pomeranz, *The Great Divergence*, pp. 274–6.

37 D. O. Flynn and A. Giraldez, "Born with a 'Silver Spoon': The Origin of World Trade," *Journal of World History* 6/2, 1995, pp. 201–22.

38 W. D. Bernhard, *Narrative of the Voyages and Services of the Nemesis, from 1840 to 1843; and of the Combined Naval and Military Operations in China*, vol. 1, London, Henry Colburn, 1844.

39 D. Headrick, *The Tools of Empire: Technology and European Imperialism in the Nineteenth Century*, New York, Oxford University Press, 1981.

40 Headrick, *The Tools of Empire*, p. 118.

41 That ditty may have captured the balance of power at that moment between Africans and Europeans, but overall the balance between Europeans and others, especially those who used guerrilla tactics against European armies, was rapidly narrowing and would disappear altogether in the twentieth century. P. D. Curtin, *The World and the West: The European Challenge and the Overseas Response in the Age of Empire*, Cambridge, Cambridge University Press, 2000, chapter 2.

42 Liberia, the West African state founded by returned American slaves, was also independent, as was a small part of Morocco.

43 Bayly, C. A., *Indian Society and the Making of the British Empire*, New York, Cambridge University Press, 1988, pp. 138–9.

44 W. Dean, *With Broadax and Firebrand: The Destruction of the Brazilian Atlantic Forest*, Berkeley, University of California Press, 1995, p. 181.

45 Grove, *Green Imperialism*.

46 M. Davis, *Late Victorian Holocausts: El Niño Famines and the Making of the Third World*, London, Verso Press, 2001.

47 B. L. Turner, W. C. Clark, R. W. Kates, et al. (eds.), *The Earth as Transformed by Human Action: Global and Regional Changes in the Biosphere over the Past 300 Years*, New York, Cambridge University Press, 1990.

48 K. Polanyi, *Great Transformation: The Political and Economic Origins of Our Time*, Boston, Beacon Press, 1957; Pomeranz, *The Great Divergence*.

49 V. Smil, *Enriching the Earth: Fritz Haber, Carl Bosch, and the Transformation of World Food Production*, Cambridge, MA, MIT, 2001.

50 E. Hobsbawm, *The Age of Extremes*, New York, Pantheon Books, 1994, p. 22.

51 D. Weiner, *A Little Corner of Freedom: Russian Nature Protection from Stalin to Gorbachev*, Berkeley, University of California Press, 1999; V. Smil, *China's Environmental Crisis: An*

Inquiry into the Limits of National Development, Armonk, NY, M. E. Sharpe, 1993; J. Shapiro, *Mao's War against Nature: Politics and the Environment in Revolutionary China*, New York, Cambridge University Press, 2001.

52 McNeill, *Something New under the Sun*, chapter 7.
53 In 2005 the Bush administration began talking about "having a strategic plan for dealing with the problem." M. T. Klare, "The Intensifying Global Struggle for Energy," *JapanFocus*, May 8, 2005.
54 McNeill, *Something New under the Sun*.
55 McNeill, *Something New under the Sun*, pp. 315–23.
56 Daly, H. E., "The Perils of Free Trade," *Scientific American* 269/5, 1993, pp. 24–9.
57 Penna, *The Human Footprint*, chapter 8.
58 Millennium Ecosystem Assessment, *Ecosystems and Human Well-Being: Synthesis*, Washington, DC, Island Press, 2005, p. 1.
59 Smil, *Enriching the Earth*, chapter 7.
60 McNeill, *Something New under the Sun*, p. 15.
61 V. Smil, *Carbon Nitrogen Sulfur: Human Interference in Grand Biospheric Cycles*, New York, Plenum Press, 1985.
62 Williams, *Deforesting the Earth*, p. 420.
63 Ponting, *A New Green History of the World*, chapter 1; D. Christian, *The Maps of Time: An Introduction to Big History*, Berkeley, University of California Press, 2004; J. Diamond, *Collapse: How Societies Choose to Fail or Succeed*, New York, Penguin, 2005.
64 These are very complicated questions, and the only thing we know for sure about the future is that it cannot be predicted. Nonetheless, social and natural scientists have modeled four different scenarios to try to examine likely stresses and outcomes. B. de Vries and J. Goudsblom (eds.), *Mappae Mundi: Humans and Their Habitats in a Long-Term Socio-Ecological Perspective: Myths, Maps, and Models*, Amsterdam, Amsterdam University Press, 2003, chapter 8.
65 H. E. Daly and J. B. Cobb, *For the Common Good: Redirecting the Economy toward Community, the Environment, and Sustainable Future*, 2nd ed., Boston, Beacon Press, 1994, chapter 11.

References

Adams, R. E. W., and MacLeod, M. J. (eds.), *The Cambridge History of the Native Peoples of the Americas*, New York, Cambridge University Press, 2000.

Anderson, F., *Crucible of War*, New York, Alfred A. Knopf, 2000.

Bayly, C. A., *Indian Society and the Making of the British Empire*, New York, Cambridge University Press, 1988.

Beinart, W., and Hughes, L., *Environment and Empire*, Oxford, Oxford University Press, 2007.

Bernhard, W. D., *Narrative of the Voyages and Services of the Nemesis, from 1840 to 1843; and of the Combined Naval and Military Operations in China*, vol. 1, London, Henry Colburn, 1844.

Bethell, L. (ed.), *The Cambridge History of Latin America*, vol. 2, Cambridge, Cambridge University, 1984.

Braudel, F., *Civilization and Capitalism 15th–18th Century*, vol. 1, *The Structure of Everyday Life*, New York, Harper and Row, 1981.

Burke, E., III, "The Big Story: Human History, Energy Regimes and the Environment," in E. Burke III and K. Pomeranz (eds.), *The Environment and World History*, Berkeley, University of California Press, 2009, pp. 33–53.

Christian, D., *The Maps of Time: An Introduction to Big History*, Berkeley, University of California Press, 2004.

Crosby, A., *The Columbian Exchange: Biological and Cultural Consequences of 1492*, Westport, CT, Greenwood Press, 1972.

Crosby, A., *Ecological Imperialism: The Biological Expansion of Europe, 900–1900*, New York, Cambridge University Press, 1986.

Curtin, P. D., *The World and the West: The European Challenge and the Overseas Response in the Age of Empire*, Cambridge, Cambridge University Press, 2000.

Daly, H. E., "The Perils of Free Trade," *Scientific American* 269/5, 1993, pp. 24–9.

Daly, H. E., and Cobb, J. B., *For the Common Good: Redirecting the Economy toward Community, the Environment, and Sustainable Future*, 2nd ed., Boston, Beacon Press, 1994.

Davis, M., *Late Victorian Holocausts: El Niño Famines and the Making of the Third World*, London, Verso Press, 2001.

de Vries, B., and Goudsblom, J. (eds.), *Mappae Mundi: Humans and Their Habitats in a Long-Term Socio-Ecological Perspective: Myths, Maps, and Models*, Amsterdam, Amsterdam University Press, 2003.

Dean, W., *With Broadax and Firebrand: The Destruction of the Brazilian Atlantic Forest*, Berkeley, University of California Press, 1995.

Diamond, J., *Collapse: How Societies Choose to Fail or Succeed*, New York, Penguin, 2005.

Flynn, D. O., and Giraldez, A., "Born with a 'Silver Spoon': The Origin of World Trade," *Journal of World History* 6/2, 1995, pp. 201–22.

Frank, A. G., *ReOrient: Global Economy in the Asian Age*, Berkeley, University of California Press, 1998.

Frank, A. G., *World Accumulation, 1492–1789*, New York, Monthly Review Press, 1978.

Grove, R. H., *Green Imperialism: Colonial Expansion, Tropical Island Edens and the Origins of Environmentalism*, New York, Cambridge University Press, 1995.

Headrick, D., *The Tools of Empire: Technology and European Imperialism in the Nineteenth Century*, New York, Oxford University Press, 1981.

Hobsbawm, E., *The Age of Extremes*, New York, Pantheon Books, 1994.

Hobsbawm, E., *The Age of Revolution*, New York, Vintage Books, 1996.

Igler, D., "Diseased Goods: Global Exchanges in the Eastern Pacific Basin, 1770–1850," *American Historical Review* 109/3, 2004, pp. 693–719.

Kiple, K. F., and Beck, S. V., *An Expanding World: The European Impact on World History, 1450–1800, vol. 26, Biological Consequences of the European Expansion*, 1450–1800, Aldershot, Ashgate, 1997.

Klare, M. T., "The Intensifying Global Struggle for Energy," *JapanFocus*, May 8, 2005.

Mann, C. C., *1491: New Revelations of the Americas before Columbus*, New York, Vintage Books, 2011.

Marks, R. B., *China: Its Environment and History*, Lanham, MD, Rowman & Littlefield, 2012.

Marks, R. B., *The Origins of the Modern World: A Global and Ecological Narrative from the Fifteenth to the Twenty-First Century*, 2nd ed., Lanham, MD, Rowman & Littlefield, 2007.

Marks, R. B., *Tigers, Rice, Silk, and Silt: Environment and Economy in Late Imperial South China*, New York, Cambridge University Press, 1998.

Marks, R. B., "World Environmental History: Nature, Modernity, and Power," *Radical History Review* 107, 2010, pp. 209–24.

McEvedy, C., and Jones, R., *Atlas of World Population History*, New York, Penguin Books, 1978.

McNeill, J. R., *Something New under the Sun: An Environmental History of the Twentieth-Century World*, New York, W. W. Norton, 2000.

McNeill, W., *Plagues and Peoples*, Garden City, NY, Anchor Books, 1975.

Millennium Ecosystem Assessment, *Ecosystems and Human Well-Being: Synthesis*, Washington, DC, Island Press, 2005.

Penna, A., *The Human Footprint: A Global Environmental History*, Malden, MA, Wiley-Blackwell, 2010.

Polanyi, K., *Great Transformation: The Political and Economic Origins of Our Time*, Boston, Beacon Press, 1957.

Pomeranz, K., *The Great Divergence: China, Europe and the Making of the Modern World Economy*, Princeton, NJ, Princeton University Press, 2000.

Ponting, C., *A New Green History of the World: The Environment and the Collapse of Great Civilizations*, New York, Penguin Books, 2007.

Radkau, J., *Nature and Power: A Global History of the Environment*, New York, Cambridge University Press, 2008.

Richards, J., *The Unending Frontier: An Environmental History of the Early Modern World*, Berkeley, University of California Press, 2003.

Shapiro, J., *Mao's War against Nature: Politics and the Environment in Revolutionary China*, New York, Cambridge University Press, 2001.

Smil, V., *Carbon Nitrogen Sulfur: Human Interference in Grand Biospheric Cycles*, New York, Plenum Press, 1985.

Smil, V., *China's Environmental Crisis: An Inquiry into the Limits of National Development*, Armonk, NY, M. E. Sharpe, 1993.

Smil, V., *Energy in World History*, Boulder, CO, Westview Press, 1994.

Smil, V., *Enriching the Earth: Fritz Haber, Carl Bosch, and the Transformation of World Food Production*, Cambridge, MA, MIT, 2001.

Tilly, C., *Coercion, Capital, and European States, AD 990–1990*, Oxford, Blackwell, 1990.

Trigger, B. G., and Washburn, W. E. (eds.), *The Cambridge History of the Native Peoples of the Americas*, vol. 1, Cambridge, Cambridge University Press, 1996.

Turner, B. L., Clark, W. C., Kates, R. W., et al. (eds.), *The Earth as Transformed by Human Action: Global and Regional Changes in the Biosphere over the Past 300 Years*, New York, Cambridge University Press, 1990.

Webb, J. L. A., *Humanity's Burden: A Global History of Malaria*, New York, Cambridge University Press, 2009.

Weiner, D., *A Little Corner of Freedom: Russian Nature Protection from Stalin to Gorbachev*, Berkeley, University of California Press, 1999.

Williams, M., *Deforesting the Earth: From Prehistory to Global Crisis*, Chicago, University of Chicago Press, 2003.

Wrigley, E. A., *Continuity and Change: The Character of the Industrial Revolution in England*, Cambridge, Cambridge University Press, 1988.

CHAPTER FIVE

Southeast Asia in Global Environmental History

PETER BOOMGAARD

The Southeast Asian Region

Southeast Asia is the collection of countries wedged in between India and China (see Map 5.1). It is smaller in landmass than those two, and has fewer inhabitants by far: China had 1,275,000,000 inhabitants in the year 2000, India just over one billion, and Southeast Asia just over half a billion. Its population density is also lower.

Today, Southeast Asia consists of a number of independent states. A distinction is made between Mainland Southeast Asia and Island Southeast Asia. Burma (Myanmar), Thailand, Cambodia, Laos, and Vietnam constitute Mainland Southeast Asia, while Island Southeast Asia consists of the Philippines, Indonesia, East Timor (Timor Leste), Brunei, Malaysia, and Singapore (although Singapore and Peninsular Malaysia – which is the western part of Malaysia, the eastern part being Sarawak and Sabah, on the island of Borneo – are formally part of the mainland).

It is a region characterized by remarkable environmental contrasts. Southeast Asia claims a number of the world's biggest cities – Bangkok, Hanoi, Jakarta, Manila – responsible for some of the world's highest rates of air pollution. At the same time, and sometimes in fairly close proximity to these cities, we still find densely forested upland regions with a varied flora and fauna – "new" species of sometimes fairly large mammals are still being discovered regularly in the region – and groups of "tribal" people who so far have refused to become "modernized," even though they have been selling many of the products they collected in the forests for the international market since time immemorial, and are now producing artisanal objects – like handwoven textiles and pleated mats and hats – for the booming tourist industry, a contrast and paradox in itself.

The contrast between densely and sparsely settled areas in the same region dates back to ancient times, when many people were already living close to each other in villages and towns in river valleys like those of the Red River (Vietnam) and the Brantas (Java), usually

A Companion to Global Environmental History, First Edition. Edited by J.R. McNeill and Erin Stewart Mauldin.
© 2012 John Wiley & Sons, Ltd. Published 2015 by John Wiley & Sons, Ltd.

Map 5.1   Southeast Asia

near a capital-cum-court-city, constituting the core area of a kingdom. This core would be surrounded by sparsely populated uplands, inhabited by hunter-gatherers and shifting cultivators. These upland people – often indicated collectively with the term "tribal" – were usually connected to a core kingdom, but the links were tenuous and could easily be broken. The relationship between uplands and lowlands was often uneasy, with raids being carried out by both sides.[1] People from the lowland often tried to catch uplanders in order to enslave them, while the uplanders launched headhunting raids in the lowlands.[2]

Southeast Asia never formed an empire, in contrast to China and (parts of) India. It always consisted of a great many polities, although the number has decreased during recent centuries. Between the early sixteenth century and the decades after World War II, much of Southeast Asia was colonized by European powers – Portugal, Spain, the Netherlands, England, France – or the US, particularly during the period of modern imperialism (1870–1940), in contrast to China and Japan. This situation – not one empire, but part of many western empires – has influenced the environmental history of the region, and its historiography.

Southeast Asia, therefore, was never a unified area. In fact, up to World War II, it was not even recognized as a region by scholars. And yet the areas constituting the region had and have much in common. Most of the region was "Indianized" – with the exception of the northern part of Vietnam, which was "Sinicized." In a cultural sense, much of the region was for a long time dominated by Hinduism and Buddhism, and the monuments these religions left behind (Borobudur, Angkor Wat) can still be admired.

In this chapter, the emphasis will be on the precolonial and colonial periods, because information on the period of independence is abundantly available and well-known by most people (interested) in the region. This also means that not much will be said about "modern" phenomena like pollution, overfishing, and global warming, even though the first signs of all of these problems predate the 1950s.

The Natural Environment

The whole of Southeast Asia is located within the tropics, lying between 25 degrees north and 10 degrees south latitude. Temperatures are constantly high year round, and differences between daily maximum and minimum temperatures are small. Temperatures drop at higher elevations, and frost occurs on the highest mountain peaks – for instance in New Guinea. This means that in addition to the usual tropical crops, in many areas crops from the temperate zone can be grown in the uplands.

Within Southeast Asia, two climatic zones prevail. The zone around the equator – the equatorial zone – covers roughly the region of Island Southeast Asia, with the exception of central and eastern Java, Nusa Tenggara (Lesser Sundas), southern Sulawesi, and the western Philippines. The second zone, called the intermediate tropical zone, covers Mainland Southeast Asia and those parts of Island Southeast Asia not within the equatorial zone.

The equatorial zone is characterized by the absence of prolonged dry periods and hence by rainfall throughout the year, a pattern that is broken only in years with weather anomalies like the ENSO phenomenon (El Niño–Southern Oscillation). The type of vegetation corresponding to this rainfall pattern is that of the ever-wet and evergreen tropical forest, also called tropical rain forest. Soils are as a rule infertile clays, while nutrients are stocked in the biomass (trees) and to a much lesser degree in the soil. Cutting the forest, therefore, leaves infertile soils that will leach rapidly. Such areas appear to have been less appealing to early humans (when they arrived) than the intermediate tropical zone, probably because of a combination of the lack of edible roots and tubers and the absence of large terrestrial mammals suitable for hunting, while slash-and-burn agriculture in the ever-wet tropical rain forests is also difficult.

In the intermediate tropical zone, however, the rainfall pattern is governed by the monsoons, which means that a rainy season and a dry season alternate. In continental Southeast Asia and the Philippines, the rainy season occurs during the summer of the northern hemisphere. On the islands south of the equator it is rainy during the winter of the northern hemisphere (and therefore the summer of the southern hemisphere). The existence of a real dry season coincides with the presence of monsoon forests that include a number of deciduous (leaf-shedding) trees. Such forests are more open than the rain forest, and roots, tubers, and terrestrial mammals are available in larger quantities. Thus hunting was possible and slash-and-burn techniques could be applied more easily, while the creation of grasslands – with fire – in order to attract game and to feed livestock was more feasible. The ground here may be quite fertile, particularly in areas with alkaline volcanic soils.

In Mainland Southeast Asia, the river valleys are much more densely populated than the mountainous areas. This region is largely defined by a number of large rivers, most of which originate in the Himalayas – the Irrawaddy, Salween, Chao Phraya, Mekong, and Red River – separated by mountain chains. These mountains have their own vegetation zones, cooler and often wetter than the lowland zones. Above the 1,000-meter line, however, no prehistoric settlements are in evidence. Although, as a rule, the rivers of Island Southeast Asia are shorter, the valleys they formed are often also densely populated, in contrast to the mountain ranges from which the rivers spring. Some upland valleys – for instance in Sumatra – were also fairly densely settled at an early stage, usually because of their volcanic soils and rain shadow. Then, as now, alluvial landscapes, high population densities, monsoon regions with reliable rainfall distribution and – later – intensive (wet) rice cultivation often went together.[3]

Thus, the above-mentioned difference between densely and sparsely settled areas, now and in the past, can at least partly be explained by environmental givens. The centers of wet-rice-producing core states were often found in fertile areas with a monsoon climate along rivers or in upland valleys.[4]

The many differences in precipitation, temperature, topography, elevation, and soil types have encouraged biotic diversity in the region, which boasts of a high rate of endemicity, that is plants or animals that occur nowhere else.

Natural Change Prior to the Human Presence

It is a familiar metaphor: if we think of the age of our planet (14 billion years) expressed as the hours of a clock, humans, the usual subject matter of historians, were not present until the last second of the twelfth hour. It is beyond dispute that humans changed the face of the earth considerably, but how much change was there before humans appeared on the scene? The answer is that prehuman changes were enormous, but that they often took a long time. I will mention two types of change that influenced the region – and many other regions – profoundly.

The first major natural change that is dealt with here is that of the so-called continental drift. Most of Southeast Asia is positioned on a continental shelf called Sundaland. The only areas that are not part of this shelf are the Philippines and easternmost Indonesia (which constitute a region that has been called Wallacea, after the famous naturalist Alfred Russel Wallace), and New Guinea (both the Indonesian and the independent part) is part of another shelf, called Sahulland. Wallacea and Sahulland were originally part of a large continental mass, far to the south of their present position, drifting northward over millions of years. Proof of this development can still be found in the enormous differences between the floras and faunas of Sundaland and Sahulland, with Wallacea as a transitional zone.

Sundaland has a rich, highly diverse Asiatic flora and fauna, characterized, among other species, by large placental mammals (elephant, rhinoceros, endemic bovines, Asiatic [water] buffalo, deer, pig, monkey, gibbon, orangutan) including large predators (tiger, leopard). Sahulland, in contrast, has a relatively "poor" Australian flora and fauna, of which the marsupials (e.g., cuscus [*Phalanger*]) are typical representatives.

The second major natural change dealt with here is that of the glacial and interglacial periods, in other words the ice ages and the periods in between. During glacial periods overall temperatures were much lower – the polar ice caps grew considerably, sea levels were, therefore, much lower, and the equatorial zone contracted. During some periods, the entire part of the Sunda shelf that is now under water stood exposed, and the Malay Peninsula, Sumatra, Java, and Borneo would no longer have been separated. However, Sundaland remained separated from Wallacea, and both were still separated from Sahulland.

During these low-sea-level periods, the areas that were islands in warmer times experienced an influx of fauna and flora from the mainland. During periods of high sea levels, when the islands were isolated again, extinctions appear to have been more frequent, and some large animals underwent a process of dwarfing. In all likelihood, the area covered with tropical rain forest in warmer times was interspersed during the ice ages with large corridors of monsoon forest or even savanna vegetation, which are generally held to have been more congenial to early humans.

The last time the Sunda and Sahul shelves were exposed was a period of a few thousand years around 20,000 years ago, when sea levels dropped 100 meters or more. Prior to that episode, such drops appear to have taken place every 100,000 years or so.[5]

The Arrival of Hominids and Humans

In 1891, physician Eugene Dubois found on the island of Java part of a skull, some bones and teeth, all of which looked like those of humans or human-related beings. He thought he had discovered the "missing link" between apes and humans, and called the hypothetical being *Pithecanthropus erectus*. This being, also called "Java Man," was later renamed *Homo erectus*. Remains of a similar hominid – Peking Man – were found in China soon after the discovery of Java Man. It appears that these hominids were already present in Asia some 1.5 to 1 million years ago. Remains of *Homo erectus* were also found in Africa, and it is now assumed that this hominid started to migrate outside Africa some 1.7 million years ago. These dates, therefore, reinforce the hypothesis that hominids originated in Africa, and then spread out over Europe and Asia.[6]

The earliest traces of modern humans – *Homo sapiens* – in the Indonesian Archipelago can be dated to perhaps around 60,000 to 50,000 BP. It is not known whether they evolved locally from *H. erectus*, represented a second migration wave from Africa that overran the original population, or were a mixture of both. These dates are too early for an arrival of *H. sapiens* during the last ice age (20,000 BP), when there was no water to cross between the continent and Indonesia. As the penultimate ice age should be dated c. 100,000 years earlier, modern humans could have been present much earlier than 60,000 BP, but, if they used boats, they were not restricted to migration during a glacial period.

A spectacular discovery occurred between 2003 and 2005, when on the island of Flores, eastern Indonesia, bones of nine individual hominids of an entirely new type were found. This *Homo floresiensis* – dated between 95,000 and 12,000 BP, and therefore partly contemporary with *H. sapiens* – was much smaller than *H. erectus*. It is assumed that *H. floresiensis* was a dwarfed form of *H. erectus*, another example of the process mentioned earlier of dwarfing on an island.

Possibly around 6,000 BP a group of people that left the southern part of what today is China, and are called Southern Mongoloids accordingly, invaded Taiwan, proceeded after several centuries to the Philippines, and, again after a considerable period of time, from there migrated to the Indonesian Archipelago, the Malay Peninsula (and Oceania). The Southern Mongoloids have dominated Southeast Asia ever since, but remnants of the earlier population are still to be found in New Guinea (Papuans), and in even smaller numbers in the central and northern Philippines, in parts of Thailand and Peninsular Malaysia, and on the Andaman Islands (Negritos). The earlier inhabitants have a much darker skin coloration and shorter stature than the Southern Mongoloids. In addition, there are linguistic differences.

Foraging and Agriculture

H. erectus and the first "wave" of *H. sapiens* – the Australo-Melanesians – to come to Southeast Asia were hunters and gatherers. Evidence for artifacts of *H. erectus* is rare and weak, but it is likely that the sites they occupied are now covered by the sea, and the lack of implements, therefore, might not be meaningful. And yet their possible weaponry is a matter of some importance. All the evidence suggests that we are dealing with very small groups of people, and it is therefore, *a prima facie*, rather unlikely that these few humans have been solely responsible for the various local megafauna extinctions that have been recorded in Southeast Asia prior to the Holocene (c. 10,000 BP), like the dying out of the tiger in Borneo and the orangutan in Vietnam. It would appear likely that climatic variation during the late Pleistocene also had something to do with it. However, the jury is still out.

Nowhere in Southeast Asia proper is there any solid evidence for any form of food production prior to 3500 BCE. This is late in comparison to Southwest Asia (8500 BCE) and China (7000 BCE). The only place that can be connected to Southeast Asia, but is often considered "Pacific," where an autonomous – that is, not introduced from the mainland – form of agriculture may have started at an earlier date, is New Guinea. Here, according to some estimates, signs of cultivation could have been present as early as 8000 BCE, but this evidence appears to be rather uncertain. What does not appear to be uncertain is that New Guinea was an independent center of domestication for a number of crops, like many cultivated bananas and some taros, but no cereals and domesticated animals. It has been suggested that this early and (for Southeast Asia) different development was somehow related to the presence of the long New Guinea highland spine with its non-tropical climate.[7]

The first appearance of agriculture often went hand in hand with the presence of pottery and villages, although all three phenomena had their own time schedule. It is the time that people started to domesticate plants and animals, harnessing them to their needs, and becoming "domesticated" themselves – what we usually call sedentarized – in the process. This, clearly, was an important phase in the relationship between humans and their natural environment. Although their numbers were still fairly small, with the introduction of domesticated plants and animals, humans were now leaving their mark on the landscape in a much bigger way than before. In all likelihood, once agriculture had become successful, the number of people began to grow faster.

Crops and Livestock

Both in Mainland and in Island Southeast Asia, agriculture was introduced from China south of the Yangzi River, a primary center of plant cultivation. Here, cultivated rice was found at an early stage, as was (foxtail) millet, and there is evidence for domesticated pigs, dogs, chickens, and perhaps cattle and water buffalo.

However, that does not mean that all domesticates to be found in Southeast Asia came from China – for many crops (and some animals) a Southeast Asian origin seems plausible. Possibly the most famous original Southeast Asian crops are cloves and nutmeg/ mace, locally domesticated endemic plants, which from c. 1500 attracted European states and trading companies to the region, thus shaping its history to a considerable extent, while deeply influencing European societies as well. A case has been made for the Southeast Asian origins of staple crops like yam and taro (both root crops), sago (from a tree), and various bananas (a fruit, but for some Southeast Asian societies a staple crop). Other probable Southeast Asian domesticates include various gourds, betel nut, cardamom, and two species of cinnamon. Many fruit trees were domesticated in the region, like bilimbing, a chestnut species, some citrus species, durian, langsat, mangosteen, rambutan, snakefruit, and starfruit. Other possible Southeast Asian domesticates are coconut (but it may have come from Oceania or America), sugarcane (but perhaps from New Guinea), ginger, and melon. The list of livestock originating in Southeast Asia is much shorter, but certainly includes the banteng-type cattle (*Bos javanicus*), and possibly water buffalo, pigs, and chickens.

Prior to 1500, a number of crops came to Southeast Asia from India or Sri Lanka, or from further afield (Middle East, Iran) and only via India – pepper, eggplant, mango, mung bean, sesame, and cotton. Humped cattle, sheep, and goats had also arrived from India, and horses, much earlier, from India or China.[8] Southeast Asia, therefore, had a quite varied repertoire of domesticated plants and animals, partly locally domesticated, partly imported. Early travelers – say, prior to 1500 – almost always comment upon this fact, particularly the presence of so many types of fruit. The Southeast Asian landscape, therefore, had been undergoing almost constant change during the 5,000 years between 3500 BCE and 1500 CE, but much more was still to come.

Part of the worldwide ecological transfer of plants and animals that Alfred Crosby called the "Columbian exchange" (after the "discovery" of America by Columbus), the Europeans who went to Southeast Asia introduced a large number of important American crops there. These included maize, sweet potato, "Irish" potato, cassava, chilli peppers, peanuts, cashew nuts, pineapple, papaya, cocoa, and tobacco and rubber, while coffee was introduced from Arabia and Africa.[9] Some of the crops became the main staples in certain areas, as was the case at fairly early dates with maize and sweet potatoes; others became important export crops. Taken together, these crops changed many Southeast Asian landscapes – not to mention diets – beyond recognition. Many of these plants were upland crops, like maize, potatoes, and tobacco, and would grow where rice could not. Thus, locally the uplands acquired a denser population, which also must have led to the partial deforestation of mountains and possibly accelerated erosion. In other areas, perennials like rubber and coffee constituted "forests" (often in the form of European plantations) that took the place of the original woody cover, fulfilling some forest functions, but lacking the biodiversity of the original vegetation. It is also likely that, locally, the new food crops made for lower mortality and/or higher fertility, because food security must have increased given the increased possibility of risk spreading.[10]

Changing Agricultural Practice

Hunting and gathering is the oldest form of human land use, and the least burdensome to the environment, practiced by people at densities of below one person per square kilometer.[11] Foraging is not regarded as agriculture, although historians and archeologists are nowadays much more inclined than they used to be to argue that hunter-gatherers certainly manipulated ("managed") their natural environment, for instance by removing plants in which they had no interest in order to stimulate the growth and propagation of other plants. We should not assume that these foragers were entirely isolated, only interested in subsistence foraging, collecting for only their own livelihood. We know by now that even the people of the most "remote," seemingly untouched, "pristine" old-growth forest regions were often in contact with outsiders, with whom they exchanged (bartered) commodities. The products collected in the inland areas could and often did end up as items in long-distance trade.[12]

However, changes in the landscape became more radical and visible with shifting cultivation, also called slash-and-burn or swidden agriculture. The swiddeners were living at higher population densities, of, say, 25 to 50 per square kilometer (65 to 130 per square mile). The term "slash-and-burn," obviously the most graphic of the three given here, says it best. The cultivator cut down the vegetation growing on a certain plot, perhaps leaving the big trees standing (though possibly cutting off the foliage), let it all dry, and then set fire to it. Finally, (s)he used a pole or dibble stick to make holes in the (sometimes still hot) ash, which were then filled with seeds or cuttings. After one or two harvests, the plot was abandoned, and usually left alone until the fertility of the soil was – partly – restored. After a few years, only those who were familiar with tropical or monsoon forests would see any difference, but regrowth to the old (climax forest) situation would take at least a century.

Shifting cultivation had a bad name among western scholars at least up to the 1950s or 1960s. It was regarded as wasteful and destructive (timber that could have been traded went up in smoke), while cultivation on permanent fields was held to be less damaging to the natural environment. Since the 1960s, after the appearance of pioneering publications like those of Harold Conklin, many scholars have presented more positive views of swiddeners.[13]

However, generally speaking, the swiddener of today is unlike the swiddener of yesterday, and the past performance of shifting cultivators cannot be judged by practices observed in the present. For instance, today there is just not enough room to clear a piece of old growth forest every time a swidden has to be abandoned, and therefore fields are used in rotation, secondary forest being used again after a number of years. Historical information suggests that in some sparsely populated areas this was not the case, and that swiddeners always cleared a patch of primary forest because it was more fertile (and possibly because there would be fewer problems with weeds). When swiddeners were reusing secondary forest, the fallow period was often quite long, a decade or even much more. Nowadays, the rotation cycle is usually much shorter, which means that fewer soils revert to secondary forest, with all that this implies regarding erosion, forest resources, wildlife habitat, and carbon-dioxide uptake.

When Conklin carried out his research in the Philippines in the early 1950s, he listed the following types of plants as the 16 most prominent swidden crops in his fieldwork area: rice, maize, sweet potato, greater yam, taro, cassava, banana, sieva beans, cowpeas, pigeon peas, hyacinth beans, eggplant, squash, melon, cucumber, and kuwei.[14] Some of these crops were obviously part of the Columbian exchange.

Is slash-and-burn agriculture disappearing? Reliable figures do not appear to exist, but it would seem that, between c. 1880 and c. 1980, the area under swidden agriculture in Southeast Asia has increased slightly. However, it is unlikely that this growth will continue indefinitely, as on the one hand many swiddeners, particularly in the relatively densely populated areas, will switch to permanent field agriculture, while on the other hand it is possible that increasing forest conservation measures would limit the further expansion of slash-and-burn.

Wet-Rice Cultivation

In the meantime, permanent field cultivation, absent or unimportant during the early centuries after the arrival of agriculture in Southeast Asia, had been steadily gaining ground.[15] The type of cultivation that most readily comes to mind when thinking of the region is, of course, wet-rice cultivation. What is more Southeast Asian than an inundated rice terrace (rice paddy, *sawah*) being plowed by a man wearing a hat against the sun, steering the plow and the water buffalo pulling it?[16] Today, the two main rice exporters of the world are the Southeast Asia countries Thailand and Vietnam. Wet-rice cultivation in Southeast Asia is of a venerable age, but as the dominant type of land use it is of more recent date; at the moment it is not possible to be more precise. However, by the ninth century CE, irrigated rice cultivation in Java provided the largest single source of tax income.

The reasons why there was a shift toward wet-rice cultivation are as yet poorly understood. It can be said that the switch from swidden to *sawah* was stimulated by the state, which created incentives and the legal and administrative framework in which markets and traders could prosper (rice was locally traded in large quantities and was not merely a subsistence crop); by religious establishments, the existence of which stimulated increased social production; by the presence of "captive" factors of production like bonded labor and livestock; and by the introduction locally of new technologies. In addition to all that, population growth may have played a role locally, as did the construction of large-scale irrigation works by or on the initiative of the ruler ("hydraulic state"), although both factors have been overemphasized in the older literature.

Wet-rice cultivation and states appear to have been mutually supportive, while both often supported Hindu–Buddhist institutions, and these institutions, in turn, supported the state and wet-rice cultivation. The Southeast Asian core states – for instance Ayuthaya, Dai Viet, Majapahit – usually had a densely populated central area, with fertile volcanic or regularly flooded alluvial soils, constituting a wet-rice cultivating region. A Hindu–Buddhist temple was often part of this configuration. Thus, we come across another iconic Southeast Asian image – a temple set among rice paddies, often with a volcano and a few palms thrown in for good measure.

Irrigated wet-rice cultivation appears to have been a sustainable system par excellence. In principle it hardly required purchased inputs – at least as long as livestock could be fed from the village commons, and local rice strains were used (though strains were no doubt traded between areas). The water that flooded the fields remained on them until the crop ripened and was harvested. This practice not only guaranteed sufficient moisture for its growth but also suppressed competing weeds. In addition, irrigation water often carried many nutrients, particularly where the soil is volcanic and hence fertile. It lent cohesion to the terraced soils, thus making collapse of the terraces less likely, and it was instrumental in making soil nutrients available to the rice plants. Furthermore, nitrogen-fixing blue-green algae were growing naturally in these inundated fields, which, in combination with the dung of the buffalo that were used for plowing and that were pastured on the rice stubble after the harvest, and together with the rice straw that was plowed under, made the use of fertilizer from outside redundant.

Sometimes fish were kept in inundated rice fields, the feces of which provided additional nutrients, while, at least in relatively recent times, small outhouses were built over the inundated fields as well, thus supplying human manure, a state of affairs that nowadays is regarded by medical professionals as bad for human health.

It was a sustainable system with very little erosion, capable of feeding many people from a small acreage. In fact, in environmental terms, the only environmental problem with wet-rice cultivation was its production of methane (CH_4), a greenhouse gas. After 1870, external inputs became more important and the character of Southeast Asian agriculture grew more dynamic and less sustainable.[17]

Livestock

Whether locally domesticated or introduced from outside in domesticated form, livestock, at varying dates, became an important feature of Southeast Asian agriculture. Water buffalo and cattle were used for transport and plowing, horses were employed for riding and played an important role in wars (together with elephants); they were also used for transport. There is a lack of historical studies on livestock in Southeast Asia, but the general feeling is that domesticated animals should be regarded as important factors in environmental change. They provided the peasant-cultivator with animal manure – another understudied topic – but had to be fed, of course. In many sparsely settled areas they were let loose in the (secondary) forests, and in the wet-rice producing areas could graze on the post-harvest stubble. It would appear that supervision was not always sufficient, and in colonial sources from the seventeenth century onward damage to forests or standing crops was mentioned regularly. Sometimes, meadows appear to have been created specifically for the purpose of grazing by the larger type of livestock (particularly horses), thus making for changes in the landscape.

Although none of these animals, at least up to, say, 1970, were bred for human consumption, many of them were probably eaten in the end, slaughtered for ceremonial feasts, particularly at the death of a well-to-do person. Pigs, very important in the non-Islamic regions, and goats, were raised primarily as food. Although goats have been around in Southeast Asia for a long time, they were not as ubiquitous in the past as they are today.

In pre-1950 Southeast Asian societies, where banks were non-existent, or were shunned by the non-elite indigenous people, all these animals were also "money on the hoof," savings of those who had managed to make a profit, albeit normally a small one. From the eighteenth century, Europeans introduced various new breeds, and hybrid livestock types gradually became distributed throughout the region.[18]

Forests and Forestry

Next to the ancient monuments mentioned earlier (and, of course, the beautiful beaches), the forests of Southeast Asia are arguably what today appeals most to the visitors of the region, as they did to many travelers in the past, like Alfred Russel Wallace. One distinguishes usually two floristic regions within Southeast Asia – Indochina and Malesia (note the spelling; Malesia is formed by Malaysia, Indonesia, the Philippines, and New Guinea). Of these two regions, Malesia is the richest, consisting of the second largest block of tropical forest in the world after the Amazon. The lowland tropical forests of this region are characterized by the great abundance of tree species from the family of Dipterocarpaceae, which produces huge and straight trunks, ideal for commercial loggers with an interest in tropical hardwood.[19]

This type of logging, for large-scale long-distance trade, is relatively recent, and depends partly on modern technology like chainsaws. It was not until c. 1960 that it became important in the region. Looking at slightly older (1940) figures, we see that Southeast Asia had an estimated forest cover of between 60 and 65 percent, compared to 10 percent for China, 20 percent for India and Pakistan, and 30 percent for Europe. The all-Southeast Asian average for around 1850 or 1870 must have been somewhere between 70 and 75 percent. This, of course, is the corollary of the region's relatively low population density. Behind these average figures, quite divergent local percentages were encountered, with Java (50 percent) at one extreme, and New Guinea and Borneo (perhaps 80 to 90 percent) at the other. This is just another aspect of the two-track development pattern of densely populated (and environmentally altered) core zones coexisting with sparsely populated forest zones.

In some areas, therefore, the loss of forest cover had been considerable, as witness the figures for the teak forests of Java between 1776 and 1840, when central Java lost 40 percent of its teak-forest cover. What were the causes of such losses? The most important factor was no doubt population growth. Although low according to modern standards – probably not more than 0.2 percent annually – over time it adds up, and it was in all probability higher than the growth rate prior to, say, 1400. Population growth in such agrarian societies equals cultivation of land, which in Southeast Asia around that time equaled deforestation. Another major factor was the need for timber (houses, palaces, religious buildings, bridges, ships), while the collection and production of firewood, charcoal, and non-timber forest products increased as well. With the arrival of Europeans in larger numbers these trends were reinforced.

From the late nineteenth century, colonial forest services or departments were called into existence in order to supervise forest exploitation, which ideally would be carried out in a sustainable way, particularly regarding valuable hardwoods like teak. Another function of these institutions was the preservation of watershed-protection forests. While the latter function was in many instances carried out satisfactorily, sustainable production was a goal seldom attained, mainly because these government services were unable to hold private enterprise to its legal obligation of reforestation.[20]

The Southeast Asian forests – and other "wild" areas – were also home to many wild animals (game). The Asian elephant, the rhinoceros, wild bantengs and seladangs, tapirs, the orangutan, the tiger, and the mousedeer, to name but a few of the better known ones, could – and can still – be found in many of the wildernesses of Southeast Asia, even though it might be hard to spot any of them. Most of these animals were hunted or trapped by the indigenous population with their spears and blowguns, while the Europeans started to hunt on some scale in the nineteenth century, bringing modern firearms (breachloaders, rifles, double-barreled shotguns) to the job.[21]

In some Southeast Asian areas from the late nineteenth century attempts were made by the colonial state to stop the destruction of flora and fauna. Game laws came on the statute books, restricting the hunting of various types of wild animals. Game reserves were established, but on a very limited scale. As we have seen, some forests were protected in order to keep watersheds forested or to guarantee sustainable production, and in a small number of cases wooded areas were protected purely for scientific reasons. But it was usually too little too late, and few pre-1950 national parks have survived until today.[22]

A Two-Track Pattern

Elsewhere I have suggested a two-track demographic pattern for pre- and early modern Southeast Asia – a fast track in the wet-rice producing lowlands and mid-altitudes, and a slow track in the hunter-gatherer and shifting cultivation uplands.[23] In the wet-rice producing valleys one would find early and universal marriage and high fertility (and therefore high rates of population growth under conditions of "normal" mortality), while in the upland areas marriages would be late, celibacy relatively high, and fertility rather low (and rates of natural increase low). Generally speaking, the uplands were rather isolated, while the lowlands were commercialized and well connected to regional and long-distance trade networks.

Wet-rice production is a highly labor-intensive activity, with a high degree of female participation. It has been argued (the so-called demand-for-labor hypothesis) that this gave particularly women an incentive to bear more children, as young children, who participated in agricultural activities at quite an early age, lightened the woman's burden. In slash-and-burn areas labor requirements were lower, while in the case of hunter-gatherers, young children were a burden on the foraging mothers. Such societies, there-fore, would have incentives to keep the number of children at modest levels.

This two-track pattern was arguably reinforced by differences in religion between the uplands (Animism) and the lowlands (Buddhism, Hinduism, later Islam and Christianity), as conversion to one of the so-called world religions supposedly led to less autonomy for women, who, so the argument goes, might have wanted fewer children under less patri-archal circumstances. It was also reinforced by slave-raiding undertaken by lowland pol-ities in the uplands, and a differential sensitivity to epidemics – people in the densely settled areas having acquired a certain immunity against many diseases, whereas those in the sparsely settled zones often had not.[24]

It seems likely that Southeast Asia had more than its fair share of mortality due to epidemics and famines, perhaps partly because of the ENSO phenomenon, mentioned earlier, a frequently recurring weather pattern that causes above-average numbers of droughts and floods in some parts of Southeast Asia.[25]

This two-track demographic, agricultural, and environmental pattern would persist under (proto)colonial expansion. The wet-rice bowls were taken over, administered directly and controlled tightly by European colonizers, while the upland areas, if for-mally colonized at all, were often administered indirectly and controlled more loosely. Such lightly populated areas were only of marginal interest to the colonizers, as they were difficult to penetrate militarily, peopled by "savage tribes" (sometimes real or imag-ined headhunters or cannibals) and equally savage animals, and not easily exploited owing to the lack of a resident labor force and to transport problems. But if an area contained valuable resources like minerals, or if high-value, low-bulk crops like opium could be grown, that situation would change.

At the beginning of this chapter, I maintained that this two-track development has persisted until today, with "tribal" groups often living in forested habitats in rather close proximity to modern primate cities. Thus, the visitors of today can easily bridge the ages of Southeast Asia's environmental historical development, as long as they are aware of the fact that the people they meet are not "stone-age" people – as has sometimes been thought – but people who in several ways participate in modern society, but whose lives have been very much shaped by the environmental differences that are still visible in many areas of Southeast Asia.

Notes

1 Some of the densely populated areas were actually located at mid-altitudes, but they will be included here in the term "lowlands."

2 A recent illustration of this dichotomy is to be found in J. C. Scott, *The Art of not Being Governed: An Anarchist History of Upland Southeast Asia*, Singapore, NUS, 2010.

3 P. Bellwood, *First Farmers: The Origins of Agricultural Societies*, Oxford, Blackwell, 2005, pp. 128–45; P. Boomgaard, *Southeast Asia: An Environmental History*, Santa Barbara, CA, ABC-CLIO, 2007, pp. 18–20.

4 Scott, *The Art of not Being Governed*, pp. 50–63.

5 P. Bellwood, *Prehistory of the Indo-Malaysian Archipelago*, Honolulu, University of Hawai'i Press, 1997; C. Higham, *Early Cultures of Mainland Southeast Asia*, Bangkok, River Books, 2002; I. C. Glover and P. Bellwood (eds.), *Southeast Asia: From Prehistory to History*, London, RoutledgeCurzon, 2004.

6 Personal communication with Wil Roebroeks, University of Leiden, 2011.

7 Bellwood, *First Farmers*, pp. 128–45.

8 R. D. Hill, "Towards a Model of the History of 'Traditional' Agriculture in Southeast Asia," in P. Boomgaard and D. Henley (eds.), *Smallholders and Stockbreeders: Histories of Foodcrop and Livestock Farming in Southeast Asia*, Leiden, KITLV, 2004, pp. 19–46.

9 A. W. Crosby Jr., *The Columbian Exchange: Biological and Cultural Consequences of 1492*, Westport, CT, Greenwood Press, 1972. On rubber planting in Malaysia, see W. Beinart and L. Hughes, *Environment and Empire*, Oxford, Oxford University Press, 2007, pp. 233–50.

10 Boomgaard, *Southeast Asia*, pp. 177–80.

11 Boomgaard, *Southeast Asia*, pp. 218–41.

12 See for instance O. W. Wolters, *Early Indonesian Commerce: A Study of the Origins of Srivijaya*, Ithaca, NY, Cornell University Press, 1967, on the early long-distance trade of Sumatra; and E. Tagliacozzo, "Onto the Coasts and into the Forests: Ramifications of the China Trade on the Ecological History of Northwest Borneo, 900–1900 CE," in R. L. Wadley (ed.), *Histories of the Borneo Environment: Economic, Political and Social Dimensions of Change and Continuity*, Leiden, KITLV, 2005, pp. 25–60, on trade between China and Borneo from the tenth century CE onward.

13 H. C. Conklin, *Hanunóo Agriculture: A Report on an Integral System of Shifting Cultivation in the Philippines*, Rome, FAO, 1957.

14 Conklin, *Hanunóo Agriculture*, p. 77.

15 Boomgaard, *Southeast Asia*, pp. 33–6, 66–76; see P. Boomgaard, "From Riches to Rags? Rice Production and Trade in Asia, particularly Indonesia, 1500–1950," in G. Bankoff and P. Boomgaard (eds.), *A History of Natural Resources in Asia: The Wealth of Nature*, New York, PalgraveMacmillan, 2007, pp. 185–204, for more details. On the Mekong River basin, see T.Akimichi (ed.), *An Illustrated Eco-History of the Mekong River Basin*, Bangkok, White Lotus, 2009, pp. 16–24.

16 Though, admittedly, Southeast Asia does not have exclusive rights on this iconic image – it is also to be found elsewhere in monsoon Asia, at least insofar as the buffalo has not been replaced by a walking tractor (power tiller).

17 O. Soemarwoto, and I. Soemarwoto, "The Javanese rural ecosystem," in A. T. Rambo and P. E. Sajise (eds.), *An Introduction to Human Ecology Research on Agricultural Systems in Southeast Asia*, Los Baños, University of the Philippines, 1984, pp. 261–70; T. Whitten, R. E. Soeriaatmadja, and S. A. Afiff, *The Ecology of Java and Bali*, Singapore, Periplus Editions, 1996, pp. 563–84; Boomgaard, *Southeast Asia*, pp. 224–7.

18 P. Boomgaard and D. Henley (eds.), *Smallholders and Stockbreeders: Histories of Foodcrop and Livestock Farming in Southeast Asia*, Leiden, KITLV, 2004.

19 T. Greer and M. Perry, "Environment and Natural Resources: Towards Sustainable Development," in L. S. Chia (ed.), *Southeast Asia Transformed: A Geography of Change*, Singapore, ISEAS, 2003, pp. 143–89 (p. 147).
20 R. L. Bryant, *The Political Ecology of Forestry in Burma*, London, Hurst & Co., 1997; P. Dauvergne, *Shadows in the Forest: Japan and the Politics of Timber in Southeast Asia*, Cambridge, MA, MIT, 1997; M. L. Ross, *Timber Booms and Institutional Breakdown in Southeast Asia*, Cambridge, Cambridge University Press, 2001; T.-P. Lye, W. de Jong, and K. Abe (eds.), *The Political Ecology of Tropical Forests in Southeast Asia: Historical Perspectives*, Kyoto, Kyoto University Press, 2003; J. Kathirithamby-Wells, *Nature and Nation: Forests and Development in Peninsular Malaysia*, Copenhagen, NIAS, 2005; Boomgaard, *Southeast Asia*, pp. 167–75, 243–56, 299–306; A. D. Usher, *Thai Forestry: A Critical History*, Chiang Mai, Silkworm Books, 2009.
21 P. Boomgaard, *Frontiers of Fear: Tigers and People in the Malay World, 1600–1950*, New Haven, CT, Yale University Press, 2001.
22 E. J. Sterling, M. M. Hurley, and M. D. Le, *Vietnam: A Natural History*, New Haven, CT, Yale University Press, 2006, pp. 349–77; Boomgaard, *Southeast Asia*, pp. 256–61.
23 Boomgaard, *Southeast Asia*, pp. 132–4, 208–9.
24 L. A. Newson, *Conquest and Pestilence in the Early Spanish Philippines*, Honolulu, University of Hawai'i Press, 2009.
25 Boomgaard, *Southeast Asia*, pp. 119–31.

References

Akimichi, T. (ed.), *An Illustrated Eco-History of the Mekong River Basin*, Bangkok, White Lotus, 2009.

Beinart, W., and Hughes, L., *Environment and Empire*, Oxford, Oxford University Press, 2007.

Bellwood, P., *First Farmers: The Origins of Agricultural Societies*, Oxford, Blackwell, 2005.

Bellwood, P., *Prehistory of the Indo-Malaysian Archipelago*, Honolulu, University of Hawai'i Press, 1997.

Boomgaard, P., "From Riches to Rags? Rice Production and Trade in Asia, particularly Indonesia, 1500–1950," in G. Bankoff and P. Boomgaard (eds.), *A History of Natural Resources in Asia: The Wealth of Nature*, New York, PalgraveMacmillan, 2007, pp. 185–204.

Boomgaard, P., *Frontiers of Fear: Tigers and People in the Malay World, 1600–1950*, New Haven, CT, Yale University Press, 2001.

Boomgaard, P., *Southeast Asia: An Environmental History*, Santa Barbara, CA, ABC-CLIO, 2007.

Boomgaard, P., and Henley, D. (eds.), *Smallholders and Stockbreeders: Histories of Foodcrop and Livestock Farming in Southeast Asia*, Leiden, KITLV, 2004.

Bryant, R. L., *The Political Ecology of Forestry in Burma*, London, Hurst & Co., 1997.

Conklin, H. C., *Hanunóo Agriculture: A Report on an Integral System of Shifting Cultivation in the Philippines*, Rome, FAO, 1957.

Crosby, A. W., Jr., *The Columbian Exchange: Biological and Cultural Consequences of 1492*, Westport, CT, Greenwood Press, 1972.

Dauvergne, P., *Shadows in the Forest: Japan and the Politics of Timber in Southeast Asia*, Cambridge, MA, MIT, 1997.

Glover, I. C., and Bellwood, P. (eds.), *Southeast Asia: From Prehistory to History*, London, RoutledgeCurzon, 2004.

Greer, T., and Perry M., "Environment and Natural Resources: Towards Sustainable Development," in L. S. Chia (ed.), *Southeast Asia Transformed: A Geography of Change*, Singapore, ISEAS, 2003, pp. 143–89.

Higham, C., *Early Cultures of Mainland Southeast Asia*, Bangkok, River Books, 2002.

Hill, R. D., "Towards a Model of the History of 'Traditional' Agriculture in Southeast Asia," in P. Boomgaard and D. Henley (eds.), *Smallholders and Stockbreeders: Histories of Foodcrop and Livestock Farming in Southeast Asia*, Leiden, KITLV, 2004, pp. 19–46.

Kathirithamby-Wells, J., *Nature and Nation: Forests and Development in Peninsular Malaysia*, Copenhagen, NIAS, 2005.

Lye, T.-P., Jong, W. de, and Abe, K. (eds.), *The Political Ecology of Tropical Forests in Southeast Asia: Historical Perspectives*, Kyoto, Kyoto University Press, 2003.

Newson, L. A., *Conquest and Pestilence in the Early Spanish Philippines*, Honolulu, University of Hawai'i Press, 2009.

Ross, M. L., *Timber Booms and Institutional Breakdown in Southeast Asia*, Cambridge, Cambridge University Press, 2001.

Scott, J. C., *The Art of not Being Governed: An Anarchist History of Upland Southeast Asia*, Singapore, NUS, 2010.

Soemarwoto, O., and Soemarwoto, I., "The Javanese rural ecosystem," in A. T. Rambo and P. E. Sajise (eds.), *An Introduction to Human Ecology Research on Agricultural Systems in Southeast Asia*, Los Baños, University of the Philippines, 1984, pp. 261–70.

Sterling, E. J., Hurley, M. M., and Le, M. D., *Vietnam: A Natural History*, New Haven, CT, Yale University Press, 2006.

Tagliacozzo, E., "Onto the Coasts and into the Forests: Ramifications of the China Trade on the Ecological History of Northwest Borneo, 900–1900 CE," in R. L. Wadley (ed.), *Histories of the Borneo Environment: Economic, Political and Social Dimensions of Change and Continuity*, Leiden, KITLV, 2005, pp. 25–60.

Usher, A. D., *Thai Forestry: A Critical History*, Chiang Mai, Silkworm Books, 2009.

Whitten, T., Soeriaatmadja, R. E., and Afiff, S. A., *The Ecology of Java and Bali*, Singapore, Periplus Editions, 1996.

Wolters, O. W., *Early Indonesian Commerce: A Study of the Origins of Srivijaya*, Ithaca, NY, Cornell University Press, 1967.

PART II

Places

Chapter Six

Environmental History in Africa

Jane Carruthers

Even in our globalized era it remains generally true that issues that are regional and topical are the main drivers of historical research and that a prime responsibility for historians is to provide perspective on, and context for, pressing current problems. In this regard, the environmental history of Africa is no exception. In his article in the special issue of *History and Theory* devoted to environmental history, John McNeill writes that "environmental history is itself shaped by political and cultural concerns." Obviously, these vary in different parts of the world.[1] Africa's contemporary environmental concerns revolve around the legacy of colonialism; social, political, and environmental inequity (within states, within the continent, and within the global community); poverty and development; and different forms of understanding and knowledge. However, Africa is a very large continent with extremely diverse topography (even if one excludes the lands north of the Sahara), climate, and biota, and consequently there is no single or dominating landscape, scale, or natural-resource issue that can encapsulate the African environment as a whole. Nor does Africa have a coherent or cohesive human history; it is fragmented into many political entities, every one of which has been subject to a different historical experience. As James McCann summarizes, "Africa's physical size, the scale of its human mosaic, and its biological diversity defy both generalization and full coverage" (see Map 6.1).[2]

Continental Historiographical Challenges

As its name implies, environmental history focuses on geography and topography, climate, water resources, and biota (floral and faunal) and links these to particular human histories (social, economic, cultural, or political) in terms of how they have changed over time. Because of the variety that characterizes Africa, regional strengths, depths, gaps, and weaknesses are as evident in the environmental history literature as they are in any

A Companion to Global Environmental History, First Edition. Edited by J.R. McNeill and Erin Stewart Mauldin.
© 2012 John Wiley & Sons, Ltd. Published 2015 by John Wiley & Sons, Ltd.

Map 6.1   Contemporary Africa

other.[3] Even books with titles that suggest total treatment or synthesis – for example, McCann's *Green Land, Brown Land, Black Land: An Environmental History of Africa, 1800–1990* or *Social History and African Environments*, edited by William Beinart and JoAnn McGregor – generally present discrete case studies from different parts of Africa that often fit uneasily together and do not aid comparisons.

The environmental history coverage of Africa is uneven, and various factors account for this situation. First, researching is often inconvenient and expensive, and there are practical difficulties in gaining access to specific locations and archives. Second, when compared with scholarship in other regions of the world, there is relatively little historical ecology or environmental archeology upon which to draw. Third, there are problems in terms of recognising and surmounting cross-cultural and language barriers (more than 2,000 languages are spoken in Africa). Although some African history has appeared in languages other than English, the majority of available books and journal articles has reflected an Anglophonic dominance, and thus this perspective. In any event, much of the record of Africa's past is oral in form, not written, so there are particular challenges around orality, interviewing with the assistance of translators, and so on. Fourth, gifted individuals have inspired the growth of specific "schools" or research agendas, and thus certain topics flourish while others – equally worthy of scholarly attention – languish. In the US, McCann at Boston University has promoted the history of Ethiopia.[4] Terence Ranger, the eminent scholar of St Antony's College, Oxford, encouraged contributions to the environmental history of Zimbabwe,[5] while Beinart, Ranger's successor at Oxford, has shown a similar talent for enthusing students about southern Africa. Fifth, while many academics have made significant contributions to environmental history, few of them are indigenous black Africans. McNeill noted this absence, referring to most African environmental historians as "outsiders."[6] It is certain that more African voices would bring particular African insights into environmental history that are currently lacking and, in order to capture this vital perspective, capacity-building and encouragement for emerging and promising African scholars needs to be high on the research agenda.

Finally – and importantly – overlapping disciplinary divides compound the historian's task more in Africa than they do elsewhere. The distinctions between anthropology/history, archeology/history, and ethnography/history, for example, are often unclear and, while in some respects (as explained below) such interdisciplinarity or multidisciplinarity is a strength, traditionally trained historians are not always effectively able to incorporate these various disciplines into their own studies. Historians have been wary of inserting environmental issues into African scholarship, not only owing to competing disciplines and their lack of familiarity with matters of environmental science, but also because they have been alert to spurious arguments that purport to explain social division or physical and intellectual characteristics based on environmental determinism.[7] Moreover, although there are overlaps, synergies, and common ground, environmental history should not be equated with "environmental" writing in general, or even scholarly endeavor around "environmental or ecological issues."

Despite these challenges, and despite the variety that exists in Africa, there are certain themes in African environmental history that are more prominent than others. They are grounded in aspects of the past that relate directly to the African experience, and that also resonate with current issues, as explained above. These include imperialism and colonialism; issues around environmental justice and the environmental effects of injustice; African wildlife and national parks and protected areas; collisions between western science and indigenous knowledges; and agendas relating to developmental economics. This chapter attempts to illuminate some of these topics within a broad overview.

Contours of African Environmental History

There appears to be a general opinion that environmental history as a formally named sub-discipline originated from the environmental activism of the 1960s that manifested itself most dramatically in the US, encouraged by books such as Rachel Carson's *Silent Spring*,[8] and the first celebration of Earth Day in 1970. It was thus a response to relevant societal problems and led to the rise of ecopolitical activism in the West. There is, however, another body of opinion, championed by Richard Grove in particular, that holds that the origins of environmentalism (and thus environmental history) should be sought further back in time and in the imperial experience. Grove was adamant that environmentalism was a consequence of past imperial and colonial eras rather than of the rise of the modern environmental movement in the US. Grove promoted environmental history outside the US as being more "interesting and innovative," "more integrated, outward-looking and comparative ... in uncovering the processes and discourses of colonial expansion and cultural encounter" than the "ultra-nationalist" perspective that was embedded in the environmental history of the US.[9] Indeed, it was "to move the environmental history of the rest of the world closer to centre stage and to deliberately encourage the writing of environmental history" elsewhere, that the journal *Environment and History* was launched.[10] Its explicit aim was to prove incorrect the presumption that much non-American environmental history was derivative. While some African environmental history owes a debt to US models, more eschews them by distancing itself from many western environmental concerns and prioritizing the environmental nexus between colonialism and Africa as it played out differently in various parts of the continent.

Despite debates around its origins, as environmental history has gained ground and adherents, established academics in the more traditional areas of history have been encouraged to accept it as a valid long-term historical pursuit. Donald Worster argued that environmental history should be accepted as central to the discipline as a whole. Being the very point at which the natural and cultural intersected it should, he believed, not merely be regarded as one of many equal historical fields but one absolutely pivotal to any understanding of the past.[11]

While many historians might agree in principle with Worster, the integration of environmental history into the wider body of historical scholarship has not gone unchallenged. Amy Dalton has described McCann's *Green Land, Brown Land, Black Land* as a book in "a sub-discipline that is one of the least understood in modern academia [that claims] more inherent theoretical ambiguities and methodological dilemmas than any other area of history – and probably more than any of the numerous disciplines it butts up against."[12] In a similar vein, according to Ellen Stroud, Sverker Sörlin, and Paul Warde, a major reason why environmental history remains on the margins as an "umbrella term" is an absence of theoretical consensus or clear epistemological parameters. Stroud believes that conceptualizing how the environment should historically "be construed" is imperative. For example, neither nature nor the environment (these are not synonyms) is a category of analysis, as can be said of class, race, and gender. William Cronon would agree: "In the face of social history's classic categories of gender, race, class, and ethnicity, environmental history stands more silent than it should."[13]

However, the theoretical lacunae that are often raised as a critique against environmental history may apply less to Africa than they do to studies elsewhere. Stroud has asserted that paying attention to environmental issues can provide all historians – not only environmental ones – with fresh insights about their own work. She suggests, for example, that insights would come from studying the environment as a site for examining other axes of power. And in this respect – the environment as locus of power – African environmental history has indeed given something fresh to the discipline. It is through careful and sophisticated historical scholarship that the postcolonial trap of simplistic divides, which Aaron Sachs believes has crippled environmental history, will be avoided and fresh perspectives on colonial and other power structures unearthed.[14]

Sörlin and Warde would agree with Stroud, for they too have argued that more embedding in social and political theory is required of the field. They remind us that it is important to consider *why* we write environmental history; merely getting on with writing it is not enough. In other words, values, politics, and ideology should be explicitly involved and engaged. It is, in fact, within African environmental history that values, politics, ideology, and purpose come most strongly to the fore, arguably more so than they do in the environmental histories that are characteristic of the developed world. In Africa specifically, environmental history originated from a strong African social history and Marxist paradigm that had much to do with environmental justice relating to class and race. Like the environmental history of Africa, social history, too, had an agenda in order to broaden historical studies away from society's powerful and to consider history from "below." It would not be making too much of this point to describe the historiography of Africa more as "eco-social history" than history relating to "environmentalism."[15]

In this respect, therefore, African environmental history has stronger ties with African history generally than it does with many of the themes that arise in environmental history in other parts of the world. It has always included elements of other social sciences relevant to Africa (particularly anthropology), aimed to uncover and explain aspects of the African past, and continuously displayed both moral purpose and political purchase.[16] The moral element continues to connect environmental historians of Africa with the societies that they study, whether it be a corrective to the declensionist narrative, nuances, and clarity around African victimhood or "powerlessness," or around issues of "ownership" and exploitation of Africa's natural resources. As geographer Catherine Nash puts it, "Environmental history can offer a powerful critique of modern capitalism and colonialism but also challenge the romanticism of pre-modernity and pre-colonial societies and so counter the primitivizing claims of some environmental philosophies."[17]

William Beinart, one of the most prolific and accomplished within the small community of environmental historians of southern Africa, has – like Richard Grove – deliberately moved Africa's environmental history away from international developments and repositioned it strongly within a longer tradition of African history. He believes that African environmental history has contributed to African history more generally in refiguring colonialism in environmental terms, reevaluating dominant ideologies and outmoded tropes such as "degradation" and "decline," and questioning what is meant by precolonial environmental equilibrium.[18] A closer analysis of colonial environmental responsibility and African agency, he argues, influences our understanding of power relations and environmental transformation in Africa. For this reason, environmental history has been able to give us "more complex readings of the history of science and knowledge," and gone a long way to amplifying simplistic historical explanations to take account of local African agency and understanding. Thus, Beinart argues, environmental sensitivity is a component of a new and rich African historiography that is characterized by an innovative interdisciplinary approach, a corrective anticolonial perspective, an extension of the range of evidence in terms of new archival sources, by oral fieldwork, by incorporating nonhuman agency and African cultural constructs.[19] This close alignment between environmental history and African history in general is evident in the historiography.

Themes in African Environmental History

In their introduction to *Social History and African Environments*, Beinart and McGregor offer a trio of themes when considering the African environmental history literature as a whole: African environmental ideas and practices; colonial science, the state, and African responses; and settlers and Africans, culture, and nature.[20] To these might be added ecologies of invasion, economy, and environment, and the importance of regional peculiarity.[21] Together with the changing relationships between humans and wildlife, these are the principal issues that are discernible in African environmental history. Threaded through them all, to different degrees, is an emphasis on environmental justice and the eradication of ecoracist ideology. In terms of economy and the environment, there is practical relevance for environmental history in informing and influencing appropriate policies around economic development, including agriculture (subsistence and commercial cultivation), land restitution, ecotourism (wildlife reserves and hunting), and the appropriate extraction and export of natural resources. In some of these areas of research, organizational and social anthropologists and development economists are taking the lead with their contemporary ecopolitical agendas and future scenario-planning rather than historians per se.[22]

Science and Knowledge

Providing environmental complexity to the colonial experience has proved a fruitful avenue of historical investigation, generating nuances and uncovering hidden agencies. The cultural history of colonial and imperial science is a high-growth area, possibly because it lies in the more general domain of colonial and imperial studies where the source material is strong and accessible. One of the most exciting contributions (and a benchmark for other studies) has been the work of social anthropologists James Fairhead and Melissa Leach who have arguably transformed African environmental history by rereading the past through African agency and environmental knowledge. Their example has been followed by many other scholars eager to apply their ideas to other parts of Africa or to refine and critique their model.[23]

In their groundbreaking article "Reading Forest History Backwards: The Interaction of Policy and Local Land Use in Guinea's Forest-Savanna Mosaic" and in their subsequent book, *Misreading the African Landscape* and other writings, Fairhead and Leach creatively overturned assumptions about the African environment and the role of African people in it.[24] In 1996, Leach and Robin Mearns edited a volume that probed "science and policy based on unfounded narratives of environmental degradation," and showed that many myths sustained over many years and that had an extremely influential bearing on matters of policy, were either not historically valid or highly contestable.[25] It was John MacKenzie who first critiqued Edward Said for casting Africans merely in the role of victims and thus denying them power or agency.[26] Beinart referred to a "struggle to free historiography and social studies from narratives of dependence, victimhood and romanticism" and the research of Leach, Fairhead, McCann, and others has gone a long way to changing historical and current thinking about African issues, and to give power and agency to Africans, including women – an often ignored category.[27] While it concerns a relatively small district of central Africa, Tamara Giles-Vernick's detailed cultural history, *Cutting the Vines of the Past*, with its explorations of indigenous responses, suggests what might be achieved for other areas of Africa.[28]

Both Giles-Vernick's book and Henrietta Moore and Megan Vaughan's *Cutting Down Trees* emphasize the gender dimension inherent in natural-resource issues in Africa.[29] So does an article by Leach and Cathy Green, "Gender and Environmental History," and another by Ranger titled "Women and Environment in African Religion: The Case of Zimbabwe."[30] Traditionally, in African historiography the role of men has been central and, whether as precolonial chiefs, household heads, anticolonial insurgents, or protonationalists, men have generally dominated the narrative. Environmental history has foregrounded the role of women in sustaining African communities because their power as food providers and environmental savants is recognized in these new histories. Nancy Jacobs' *Environment, Power, and Injustice: A South African History* takes the long-term perspective from the precolonial period into the late apartheid era in the Kuruman district of South Africa. She outlines how cultivation was "a sphere of female autonomy" while men dominated political power with their wealth in livestock.[31] Over time, the relationship between women and the land led to their forceful resistance against the apartheid laws that deprived them of that land and, in the postapartheid years, has resulted in females being at the forefront of the land-restitution process. These are just a few examples of how a gender-conscious environmental history has contributed significantly to the modern historiography of many parts of Africa.

African environmental history is overtly political: it deals with human – not geological or ecological – time, as is sometimes the case in other regions. Local and imperial/colonial politics, culture, and environmental science are considered together in much of the new African environmental history and this is a vibrant area of research and writing. Beinart's *The Rise of Conservation in South Africa* is a detailed work that melds these disciplines into a historical whole.[32] This theme – imperial science – is one that can be adopted on a continental canvas.[33] Leading scholars in this regard include Peder Anker, who analyzes a number of important environmental themes linking the emerging disciplines of human and natural ecology within the international context of the British Empire.[34] Perhaps the most detailed of the twentieth century's imperial surveys was that of Lord Hailey. His comprehensive work, *An African Survey: A Study of Problems Arising in Africa South of the Sahara*, included a number of chapters dealing with the African environment – land, agriculture, forests, water, and soil erosion.[35] In many respects this book summarized the state of British knowledge about Africa at that time, its history, environment, and prospects for future development. Historians interested in the history of the environmental sciences in Africa have taken this publication as their benchmark, trawling it for information on how this western perspective intersected with other forms of knowledge. In particular, Helen Tilley analyzes this *Survey* (with which Anker's book is also concerned), with reference to the then-current bureaucracy and, in turn, there are fascinating connections with Saul Dubow's analysis of the period with reference to South Africa, "A Commonwealth of Science: The British Association in South Africa, 1905 and 1929."[36]

Case studies in the history of science are evident in the South African literature and appear to be a growth area there. Botany takes central stage in this respect with work, particularly on the Cape, by Lance van Sittert, Simon Pooley, and Elizabeth Green Musselman. Jane Carruthers has explored some aspects of wildlife management and ornithology and, in a manner similar to Dawn Nell, game-ranching. The history of plant transfers has also been attractive, and this is an area of environmental history that is presently burgeoning.[37]

Wildlife, Ecoracism, and Economic Development

Whether implicit or explicit, African environmental studies is robust when dealing with issues of economic development. Strong tropes emanated from the colonial discourse and they still affect current environmental action (and donor funding to Africa). African environments were "degrading" and "declining," and international financial interventions, related to providing subsistence for growing populations, are – even today – often unwittingly based on this discourse. Grace Carswell's "Communities in Environmental Narratives" is an excellent example. She closely interrogates why these discourses have enjoyed such a long life and concludes that "development" in different historical contexts remains externally driven.[38] Soil erosion, an important issue in African environments, is another topic that requires historical unpacking, and in this respect there is considerable scholarship relating to southern Africa, which has provoked lively discussion among soil scientists and historians alike.[39]

Ecoracism is an outgrowth of the strong social-history tradition that has dominated African historiography since the 1970s. In this regard, access to indigenous wildlife or to national parks and protected areas is a popular topic and it, too, has implications for development policy through transfrontier natural-resource conservation, community conservation, and sustainable extraction.[40] Interest in African wildlife revolves around the colonial legacy, the modern tourist industry, and also the interventions and politics of globalized conservation organizations such at the International Union for the Conservation of Nature (IUCN). As MacKenzie demonstrated so well in his groundbreaking *The Empire of Nature: Hunting, Conservation and British Imperialism*, the hunting culture is a subject that brings multiple regions of Africa together within a single environmental narrative, although there are differences between various colonial histories. To date, there is no synthetic history dealing with wildlife or its conservation that links east, west, north, and southern Africa. Within regions, however, the historiography of the colonial and modern era is relatively strong. There are well-known studies by John Mackenzie, Roderick Neumann, Gregory Maddox, James Giblin, and Isaria Kimambo, Christopher Conte, Edward Steinhart, and Jane Carruthers. Precolonial cultural attitudes toward landscape, nature, and natural resources have also been studied in Central Africa by the doyen of oral history, Jan Vansina, and also by Robert Harms and Jan Bender Shetler.[41]

Interest in wildlife and "wilderness" protection is sustained not only because it enriches national histories but also because it forms part of the global debate on what constitutes sustainable development, a buzzword in international-development policy as well as in the conservation arena – attested to by the Rio conference of 1992 and the Johannesburg Summit on Sustainable Development of 2002. Protected areas (such as national parks) are expected to deliver material benefits to local communities and regional economies as well as to entertain foreign tourists. Some issues are highly controversial and, as discussion around the Convention on International Trade in Endangered Species of Wild Fauna and Flora (CITES) indicates, often divide the "developed" from the "developing" world. Land redistribution and the forcible removal of local people add to the mix of critical historical issues around wildlife protection and land set aside for biodiversity conservation, making the environment a locus for political and economic struggles that reflect the nexus between western environmentalism and colonialism. There are

explicit claims about who best understands African environments, and who should have the right to control them – whether scientists, national governments or local people. Such arguments have become centrally important as bases for intervention, conservation and regulation. Environmentalists sometimes emphasize … responsibility to future generations for the well-being of the planet … Africanists by contrast, sometimes see access to resources as the critical issue for communities … All such approaches imply both historical investigation and historical judgment.[42]

Politics expert Rosaleen Duffy's *Killing for Conservation: Wildlife Policy in Zimbabwe* is interesting for its perspective on social equity within a global conservation arena and the challenges generated by dividing conservation management and policy objectives between rural development and poverty alleviation on the one hand, and the gratification of aesthetic enjoyment of wealthy foreign tourists on the other.[43]

The international movement regarding wildlife, national parks, and protected areas in Africa is well represented in the literature, although many of these authors are political scientists, geographers, and sociologists rather than historians. It is on this platform that the most vibrant interdisciplinary conversation is taking place and environmental history engages actively with it. While publications targeted at the general public are often laudatory or proselytizing of conservation initiatives and expanded national parks, the academic literature generally takes the form of a critique of what is regarded as a global and neoliberal approach to the environment (the commodification of nature) that is frequently detrimental to local people. In Africa, where the legacy of racial segregation and colonialism remains very much part of everyday life, and particularly in certain parts of the continent where the process of research is relatively easy (bearing in mind the constraints mentioned above), the literature is rich. In this regard, the work of Stuart Marks, Harry Wels and coauthors Marja Spierenburg and Anna Spenceley, Catherine Aubertin and Estienne Rodary, and Dan Brockington, among others, is important.[44] However, while such studies do much to inform environmental history and certainly suggest interesting lines of historical inquiry, it must also be said that they do not all belong strictly within historical studies and their methods, sources, and audiences differ from those of historians.

Because much of the focus of African history is on development and issues around poverty alleviation, there are as yet unexplored opportunities for the environmental history of Africa to gain from the "environmentalism of the poor" literature that is so characteristic of India and Latin America.[45] A good deal of this surfaces in the journal *Capitalism Nature Socialism* and is expounded upon by Ramachandra Guha and Joan Martinez-Alier, among others.[46] There was indeed a time when it seemed that this kind of environmental history was destined to take off in South Africa during the years immediately preceding the collapse of apartheid. One of the crucial policies of the new democratic South Africa was to be predicated on "apartheid divides, ecology unites" – after a divided political past, all South Africans, regardless of race, class, or age cohort, were to care for the physical environment because environmentalism was grassroots mobilization "for our future and for our children." However, this theme has not yet borne much historical fruit. Mining, carbon trading, climate change, and industrial pollution are hot political and economic topics that affect the lives of impoverished people directly, and there are no organizations such as Chipko in India or the Green Belt Movement in Kenya, or environmental champions of the poor, such as Vandana Shiva or Wangari Maathai, to mobilize people in South Africa.[47]

A Future Agenda?

Rather than decry the silences of scholarship in certain areas of study, one can regard them as rich future research opportunities and encourage work on these themes. There are, of course, exceptions, but generally speaking, African environmnenal history could be far stronger on the topics of disease, art and imagination, mineral and resource extraction, bioregional work (e.g., on transnational river basins or highlands), and disaster and crisis (so rich in Europe), to name but a few. A world-history perspective is also uncommon, although a promising avenue of linkages to global issues is occurring through studies that critique Alfred Crosby's *Ecological Imperialism* and refine our knowledge of plant transfers.[48] Given the vastness of the arid areas of Africa, it is surprising that there is a general absence of the environmental history of desert or arid areas. Water politics in Africa have generally not had the historical attention that has been devoted to them elsewhere – California and India for example – but this may be changing because of predictions of "water wars" in the future.[49]

Among the most promising areas for African environmental history is urban history. Although Africa is often still imagined as a continent with a mostly rural population, since the 1960s Africa's cities have grown exponentially and the continent is among the world's most rapidly urbanizing. With livelihood strategies maximized by way of the formal and informal sectors where there are concentrations of people, Africans from resource-poor or politically unstable rural areas have flocked to megacities such as Lagos, Nairobi, and Johannesburg to the extent that cities support the countryside economically rather than the more traditional reverse. Africa's modern cities are especially interesting because they were originally established by the colonizing powers as trading or administrative centers and they have been transformed, generally informally in terms of industrial and commercial planning, or the absence of it. There is a body of geographical, urban, and economic planning literature upon which the subfield of urban environmental history might draw as well as other specific historical studies on certain cities and areas.[50] In terms of environmental history, this form of urbanization offers fascinating historical topics because it has spawned changes in health and disease (through urban pollution, a lack of sanitation and other services, and poor housing – often shacks on the city's outskirts); social, cultural, and religious norms; migration patterns; and aspects of the informal economy by way of trade and agricultural products that are grown on the margins of densely populated areas.

Conclusion

One of the characteristics of the environmental history of Africa – and its historiography – is the rich multidisciplinary arena of environmental studies. Nonetheless, while environmental historians of Africa have certainly made use of sources and perspectives from other disciplines as explained above, environmental history in Africa has remained distinctly "history."[51] In general, it has also conformed to the dominant historiographies and present concerns of each georegion or national state, amplifying or extending arguments, challenging received wisdom, and including new lines of inquiry or unusual evidence. Its success has been to make all history more inclusive in its narratives than it had been previously. In summary: while it adheres to certain contours and topics, environmental history on African subjects is diffuse. It generally lacks the proselytizing "green" political agenda that characterized earlier literature and, in the same way as Tom Griffiths has argued for Australia, it has become more recognizably "history." Thus to discuss environmental history as being more "history" than "environmental" may – particularly in the case of Africa, where the environmental diffuses into development studies – be a real strength. Nonetheless, if African environmental history remains conceptualized as a component of modern, postcolonial African studies, it will not be what Griffiths thoughtfully calls "a distinctive endeavour ... [that] ... moves audaciously across time and space and species," that "challenges some of the conventions of history," and "questions the anthropocentric, nationalistic and documentary bases of the discipline."[52] However, if Beinart is correct that the environmental history of Africa is a thriving field closely linked with other branches of African studies, then he would probably agree with Griffiths that the strength of "the new environmental history" is that it prioritizes the "historians' traditional concerns of identity, agency, economy and politics" using the narrative form.[53] Karen Brown has made the same point: "the question [is] whether environmental history is then really 'history' if human agency is reduced to a sideline issue and the methodology is bereft of written and oral testimonies."[54] Certainly, environmental history prioritizes human agency and relates it to change over time, and it has invigorated the modern scholarship of Africa.

Notes

1 Quoted in B. Fay, "Environmental History: Nature at Work," *History and Theory* 42, 2003, pp. 1–4 (p. 1).

2 J. McCann, *Green Land, Brown Land, Black Land: An Environmental History of Africa, 1800–1990*, Portsmouth, NH, Heinemann, 1999, p. 5.

3 W. Beinart, and J. McGregor (eds.), *Social History and African Environments*, Oxford, James Currey, 2003, p. 9.

4 J. McCann, *People of the Plow: An Agricultural History of Ethiopia, 1800–1990*, Madison, University of Wisconsin Press, 1995; J. McCann, "The Plow and the Forest: Narratives of Deforestation in Ethiopia, 1840–1992," *Environmental History* 2, 1997, pp. 138–59.

5 T. O. Ranger, *Voices from the Rocks: Nature, Culture and History in the Matopos Hills of Zimbabwe*, Oxford, James Currey, 1999.

6 J. R. McNeill, "Observations on the Nature and Culture of Environmental History," *History and Theory* 42, 2003, pp. 5–43 (p. 25).

7 W. Beinart, "African History and Environmental History," *African Affairs* 99, 2000, pp. 269–302.

8 R. Carson, *Silent Spring*, New York, Houghton Mifflin, 1962.

9 R. H. Grove, *Green Imperialism: Colonial Expansion, Tropical Island Edens and the Origins of Environmentalism, 1600–1800*, Cambridge, Cambridge University Press, 1995.

10 R. Grove, "Editorial," *Environment and History* 1/1, 1995, pp. 1–2.

11 Quoted in R. White, "Environmental History, Ecology and Meaning," *A Round Table: Environmental History, special issue of Journal of American History* 76/4, 1990, pp. 1111–16 (p. 1111).

12 A. L. Dalton, "Book Review, on African Environmental History: *Green Land, Brown Land, Black Land: An Environmental History of Africa, 1800–1990*," *Current History*, 2000, pp. 231–2.

13 Quoted in C. Nash, "Environmental History, Philosophy and Difference," *Journal of Historical Geography* 26/1, 2000, pp. 23–7 (p. 24); S. Sörlin and P. Warde, "The Problem of the Problem of Environmental History: A Re-Reading of the Field," *Environmental History* 12/1, 2007, pp. 107–30.

14 E. Stroud, "Does Nature Always Matter? Following Dirt through History," *History and Theory* 42, 2003, pp. 75–81; A. Sachs, "The Ultimate Other: Post-Colonialism and Alexander von Humboldt's Ecological Relationship with Nature," *History and Theory* 42, 2003, pp. 111–35.

15 A. A. Taylor, "Unnatural Inequalities: Social and Environmental History," *Environmental History* 1/4, 1996, pp. 6–19; N. Jacobs, *Environment, Power, and Injustice: A South African History*, Cambridge, Cambridge University Press, 2003.

16 W. Cronon, "A Place for Stories: Nature, History and Narrative," *Journal of American History* 78/4, 1992, pp. 1347–76 (p. 1375).

17 Nash, "Environmental History, Philosophy and Difference," p. 25.

18 Beinart, "African History and Environmental History." See also G. Maddox, "'Degradation Narratives' and 'Population Time Bombs': Myths and Realities about African Environments," in S. Dovers, R. Edgecombe, and B. Guest (eds.), *South Africa's Environmental History: Cases and Comparisons*, Athens, OH, Ohio University Press, 2003, pp. 250–8.

19 Beinart, "African History and Environmental History," p. 292.

20 Beinart and McGregor, *Social History and African Environments*.

21 T. Griffiths and L. Robin (eds.), *Ecology and Empire: Environmental History of Settler Societies*, Edinburgh, Keele University Press, 1997; E. Burke III and K. Pomeranz (eds.), *The Environment and World History*, Berkeley, University of California Press, 2009.

22 J. Carruthers, "Past and Future Landscape Ideology: The Kalahari Gemsbok National Park," in W. Beinart and J. McGregor (eds.), *Social History and African Environments*, Oxford, James Currey, 2003, pp. 255–66; D. Bunn, "An Unnatural State: Tourism, Water and Wildlife Photography in the Early Kruger National Park," in W. Beinart and J. McGregor (eds.), *Social History and African Environments*, Oxford, James Currey, 2003, pp. 199–218. Works on economic development policy and history include C. M. Thompson, "Ranchers, Scientists, and Grass-Roots Development in the United States and Kenya," *Environmental Values* 11/3, 2002, pp. 303–26; C. Twyman, "Livelihood Opportunity and Diversity in Kalahari Wildlife Management Areas, Botswana: Re-Thinking Community Resource Management," *Journal of Southern African Studies* 26/4, 2000, pp. 783–806; D. Potts, "Worker-Peasants and Farmer-Housewives in Africa: The Debate about "Committed" Farmers, Access to Land and Agricultural Production," *Journal of Southern African Studies* 26/4, 2000, pp. 807–32; J. D. Hackel, "A Case Study of Conservation and Development Conflicts: Swaziland's Hlane Road," *Journal of Southern African Studies* 27/4, 2001, pp. 813–32; G. Carswell, "African Farmers in Colonial Kigezi, Uganda, 1930–1962: Opportunity, Constraint and Sustainability," PhD thesis, University of London, 1997.

23 J. Fairhead and M. Leach, "Reading Forest History Backwards: The Interaction of Policy and Local Land Use in Guinea's Forest-Savanna Mosaic, 1893–1993," *Environment and History* 1/1, 1995, pp. 55–91; J. Fairhead and M. Leach, *Misreading the African Landscape: Society and Ecology in a Forest-Savannah Mosaic*, Cambridge, Cambridge University Press, 1996; J. Fairhead and M. Leach, *Reframing Deforestation: Global Analyses and Local Realities, Studies in West Africa*, London, Routledge, 1998. Other examples of this work include A. F. Clark, "Environmental Decline and Ecological Response in the Upper Senegal Valley, West Africa, from the Late Nineteenth Century to World War I," *Journal of African History* 36/2, 1995, pp. 197–218.

24 Fairhead and Leach, *Reframing Deforestation*.

25 M. Leach and R. Mearns (eds.), *The Lie of the Land: Challenging Received Wisdom on the African Environment*, Portsmouth, NH, Heinemann, 1996.

26 J. M. MacKenzie, "Edward Said and the Historians," *Nineteenth Century Contexts* 18, 1994, pp. 9–25.

27 Beinart, "African History and Environmental History," p. 302.

28 T. Giles-Vernick, *Cutting the Vines of the Past: Environmental Histories of the Central African Rain Forest*, Charlottesville, VA, University Press of Virginia, 2002. See also T. Giles-Vernick, "Leaving a Person Behind: History, Personhood, and Struggles of Forest Resources in the Sangha Basin of Equatorial Africa," *International Journal of African Historical Studies* 32–3/2, 1999, pp. 311–38, and T. Giles-Vernick, "*Doli*: Translating an African Environmental History of Loss in the Sangha River Basin of Equatorial Africa," *Journal of African History* 41/3, 2000, pp. 373–94.

29 H. L. Moore and M. Vaughan, *Cutting Down Trees: Gender, Nutrition, and Agricultural Change in the Northern Province of Zambia, 1890–1990*, Portsmouth, NH, Heinemann, 1994.

30   M. Leach and C. Green, "Gender and Environmental History: From Representation of Women and Nature to Gender Analysis of Ecology and Politics," *Environment and History* 3, 1997, pp. 343–70; T. O. Ranger, "Women and Environment in African Religion: The Case of Zimbabwe," in W. Beinart and J. McGregor (eds.), *Social History and African Environments*, Oxford, James Currey, 2003, pp. 72–86.

31   Jacobs, *Environment, Power, and Injustice*, p. 52.

32   W. Beinart, *The Rise of Conservation in South Africa: Settlers, Livestock and the Environment 1770–1950*, Oxford, Oxford University Press, 2003.

33   Climate is another: see J. McCann, "Climate and Causation in African History," *International Journal of African Historical Studies* 32/2–3, 1999, pp. 261–79; I. Pikirayi, "Environmental Data and Historical Process: Historical Climatic Reconstruction and the Mutapa State 1450–1862," in W. Beinart and J. McGregor (eds.), *Social History and African Environments*, Oxford, James Currey, 2003, pp. 60–71; J.-B. Gewald, "El Negro, El Niño, Witchcraft and the Absence of Rain in Botswana," *African Affairs* 100/2–3, 2001, pp. 555–80.

34   P. Anker, *Imperial Ecology: Environmental Order in the British Empire, 1895–1945*, Cambridge, MA, Harvard University Press, 2001.

35   Lord Hailey, *An African Survey: A Study of Problems Arising in Africa South of the Sahara*, London, Oxford University Press, 1938.

36   H. Tilley, "Africa as a Living Laboratory," PhD thesis, University of Oxford, 2001; H. Tilley, "African Environments and Environmental Sciences: The African Research Survey, Ecological Paradigms and British Colonial Development, 1920–1940," in W. Beinart and J. McGregor (eds.), *Social History and African Environments*, Oxford, James Currey, 2003, pp. 109–30; S. Dubow, "A Commonwealth of Science: The British Association in South Africa, 1905 and 1929," in S. Dubow (ed.), *Science and Society in Southern Africa*, Manchester, Manchester University Press, 2000, pp. 66–99.

37   L. Van Sittert, "Making the Cape Floral Kingdom: The Discovery and Defence of Indigenous Flora at the Cape ca.1890–1939," *Landscape Research* 28, 2003, pp. 113–29; S. Pooley, "Pressed Flowers: Nations of Indigenous and Alien Vegetation in South Africa's Western Cape, c. 1902-1945," *Journal of Southern African Studies* 36, 2010, pp. 599–618; E. Green Musselman, "Plant Knowledge at the Cape: A Study in African and European Collaboration," *International Journal of African Historical Studies* 36, 2003, pp. 367–92; J. Carruthers, "Conservation and Wildlife Management in South African National Parks, 1930s–1960s," *Journal of the History of Biology* 41/2, 2007, pp. 203–36; J. Carruthers, "Influences on Wildlife Management and Conservation Biology in South Africa c.1900 to c.1940," *South African Historical Journal* 58/1, 2007, pp. 65–90; J. Carruthers, "Our Beautiful and Useful Allies: Aspects of Ornithology in Twentieth Century South Africa," *Historia* 49/1, 2004, pp. 89–109; J. Carruthers, "Wilding the Farm or Farming the Wild: The Evolution of Scientific Game Ranching in South Africa from the 1960s to the Present," *Transactions of the Royal Society of South Africa* 63/2, 2008, pp. 160–81; D. D'A. Nell, "The Development of Wildlife Utilisation in South Africa and Kenya, c.1950–1990," PhD thesis, University of Oxford, 2003; W. Beinart and K. Middleton, "Plant Transfers in Historical Perspective: A Review Article," *Environment and History*, 10/1, 2004, pp. 3–30; J. Carruthers and L. Robin, "Taxonomic Imperialism in the Battles for *Acacia*: Identity and Science in South Africa and Australia," *Transactions of the Royal Society of South Africa* 65/1, 2010, pp. 48–64.

38   G. Carswell, "Communities in Environmental Narratives," *Environment and History* 9/1, 2003, pp. 3–30.

39   S. Schirmer, and P. Delius, "Soil Conservation in a Racially Divided Society," *Journal of Southern African Studies* 26/4, 2000, pp. 719–42; W. Beinart, "Soil Erosion, Conservationism and Ideas about Development: A Southern African Exploration, 1900–1960," *Journal of Southern African Studies* 11/1, 1984, pp. 52–83; McCann, *Green Land, Brown Land, Black Land*, pp. 141–74; K. Showers, *Imperial Gullies: Soil Erosion and Conservation in Lesotho*, Athens, OH, Ohio University Press, 2005.

40 See E. Garland, "The Elephant in the Room: Confronting the Colonial Character of Wildlife Conservation in Africa," *African Studies Review* 51/3, 2008, pp. 51–74.

41 J. M. MacKenzie, *The Empire of Nature: Hunting, Conservation and British Imperialism*, Manchester, Manchester University Press, 1988; J. M. MacKenzie (ed.), *Imperialism and the Natural World*, Manchester, Manchester University Press, 1990; R. P. Neumann, *Imposing Wilderness: Struggles over Livelihood and Nature Preservation in Africa*, Berkeley, University of California Press, 1998; G. Maddox, J. L. Giblin, and I. Kimambo, *Custodians of the Land: Ecology and Culture in the History of Tanzania*, London, James Currey, 1996; C. Conte, *Highland Sanctuary: Environmental History in Tanzania's Usambara Mountains*, Athens, OH, Ohio University Press, 2004; E. Steinhart, *Black Poachers, White Hunters: A Social History of Hunting in Colonial Kenya*, Oxford, James Currey, 2006; J. Carruthers, *The Kruger National Park: A Social and Political History*, Pietermaritzburg, University of Natal Press, 1995; J. Vansina, *Paths in the Rainforests*, Madison, University of Wisconsin Press, 1994; R. Harms, *Games against Nature: An Eco-Cultural History of the Nunu of Equatorial Africa*, Cambridge, Camssbridge University Press, 1987; J. B. Shetler, *Imagining Serengeti: A History of Landscape Memory in Tanzania from Earliest Times to the Present*, Athens, OH, Ohio University Press, 2007.

42 Beinart and McGregor, *Social History and African Environments*, p. 2.

43 R. Duffy, *Killing for Conservation: Wildlife Policy in Zimbabwe*, Oxford, James Currey, 2000.

44 S. A. Marks, "Back to the Future: Some Unintended Consequences of Zambia's Community-Based Wildlife Program (ADMADE)," *Africa Today* 48/1, 2001, pp. 120–41; H. Wels, *Private Wildlife Conservation in Zimbabwe: Joint Ventures and Reciprocity*, Leiden, Brill, 2003; H. Wels and M. J. Spierenburg, "Securing Space: Mapping and Fencing in Transfrontier Conservation in Southern Africa," *Space and Culture* 9/3, 2006, pp. 294–312; H. Wels, B. Wishitemi, and A. Spenceley (eds.), *Culture and Community: Tourism Studies in Eastern and Southern Africa*, Amsterdam, Rozenberg Publishers, 2007; C. Aubertin and E. Rodary, *Protected Areas, Sustainable Land?*, Aldershot, Ashgate, 2011; D. Brockington, *Fortress Conservation: The Preservation of the Mkomazi Game Reserve Tanzania*, Oxford, James Currey, 2002; D. Brockington, R. Duffy, and J. Igoe, *Nature Unbound: Conservation, Capitalism and the Future of Protected Areas*, London, Earthscan, 2008; D. Brockington, *Celebrity and the Environment: Fame, Wealth and Power in Conservation*, London, Zed, 2009.

45 M. Carey, "Latin American Environmental History: Current Trends, Interdisciplinary Insights, and Future Directions," *Environmental History* 14/2, 2009, pp. 221–52.

46 R. Guha and M. Gadgil, *This Fissured Land: An Ecological History of India*, Delhi, Oxford University Press, 1992; R. Guha, *The Unquiet Woods: Ecological Change and Peasant Resistance in the Himalaya*, Berkeley, University of California Press, 1989; D. Arnold and R. Guha, *Nature, Culture, Imperialism: Essays on the Environmental History of South Asia*, Delhi, Oxford University Press, 1995; J. Martinez-Alier, *The Environmentalism of the Poor: A Study of Ecological Conflicts and Valuation*, Northampton, MA, Edward Elgar, 2002.

47 J. Cock and E. Koch (eds.), *Going Green: People, Politics and the Environment in South Africa*, Cape Town, Oxford University Press, 1991, p. 15; E. Koch, D. Cooper, and H. Coetzee, *Water, Waste and Wildlife: The Politics of Ecology in South Africa*, London, Penguin, 1990; T. Le Quesne, "The Divorce of Environmental and Economic Policy under the First ANC Government, 1994–1999," *Journal of Environmental Law and Policy* 7, 2000, pp. 1–20; P. Steyn, "Environmental Management in South Africa: Twenty Years of Governmental Response to the Global Challenge," *Historia* 46, 2002, pp. 25–53; P. Steyn, "Popular Environmental Struggles in South Africa, 1972–1992," *Historia* 47, 2002, pp. 125–58; P. Steyn, "The Lingering Environmental Impact of Repressive Governance: The Environmental Legacy of the Apartheid Era for the New South Africa," *Globalizations* 2/3, 2005, pp. 391–403; P., Steyn and A. Wessels, "The Emergence of New Environmentalism in South Africa, 1988–1992," *South African Historical Journal* 42, 2000, pp. 210–31.

48   A. W. Crosby, *Ecological Imperialism: The Biological Expansion of Europe, 900–1900*, Cambridge, Cambridge University Press, 1986; Beinart and Middleton, "Plant Transfers in Historical Perspective"; K. Middleton, "Who Killed Malagasy Cactus? Science, Environment and Colonialism in Southern Madagascar (1924–1930)," *Journal of Southern African Studies* 25/2, 1999, pp. 215–48.

49   A. Isaacman and C. Sneddon, "Toward a Social and Environmental History of the Building of Cahora Bassa Dam," *Journal of Southern African Studies* 26/4, 2000, pp. 597–632; M. Thabane, "Shifts from Old to New Social and Ecological Environments in the Lesotho Highlands Water Scheme: Relocating Residents of the Mohale Dam Area," *Journal of Southern African Studies* 26/4, 2000, pp. 633–54.

50   D. M. Anderson and R. Rathbone (eds.), *Africa's Urban Past*, Portsmouth, NH, Heinemann, 2000; B. Raftopoulos and T. Yoshikuni (eds), *Sites of Struggle: Essays in Zimbabwe's Urban History*, Harare, Weaver Press, 1999; T. A. Barnes, *We Women Worked So Hard: Gender, Urbanisation and Social Reproduction in Colonial Harare, Zimbabwe, 1930–1956*, Portsmouth, NH, Heinemann, 1999; B. Freund, *The African City: A History*, Cambridge, Cambridge University Press, 2007; Howard, A. M., "Contesting Commercial Space in Freetown, 1860–1930: Traders, Merchants, and Officials," *Canadian Journal of African Studies* 37/2–3, 2003, pp. 236–61; J. Iliffe, *Honour in African History*, Cambridge, Cambridge University Press, 2005; D. Jeater, "No Place for a Woman: Gwelo Town, Southern Rhodesia, 1894–1920," *Journal of Southern African Studies* 26, 2000, pp. 29–43; G. A. Myers, *Disposable Cities: Garbage, Governance and Sustainable Development in Urban Africa*, Aldershot, Ashgate, 2005; K. Sheldon, "Markets and Gardens: Placing Women in the History of Urban Mozambique," *Canadian Journal of African Studies*, 37/2–3, 2003, pp. 359–95; A. Simone, *For the City yet to Come: Changing African Life in Four Cities*, Durham, NC, Duke University Press, 2004; Penvenne (1995).

51   S. Dovers, "On the Contribution of Environmental History to Current Debate and Policy," *Environment and History* 6/2, 2000, pp. 131–50.

52   T. Griffiths, "How Many Trees Make a Forest? Cultural Debates about Vegetation Change in Australia," *Australian Journal of Botany* 50/4, 2002, pp. 375–89 (pp. 376–7).

53   Beinart, "African History and Environmental History," pp. 269–302; Griffiths, "How Many Trees Make a Forest?," pp. 376–8.

54   K. Brown, "Book Review: S. Dovers, R. Edgecombe, and B. Guest, *South Africa's Environmental History: Cases and Comparisons*," *African Affairs* 103/411, 2004, pp. 311–12 (p. 311).

References

Anderson, D. M., and Rathbone, R. (eds.), *Africa's Urban Past*, Portsmouth, NH, Heinemann, 2000.

Anker, P., *Imperial Ecology: Environmental Order in the British Empire, 1895–1945*, Cambridge, MA, Harvard University Press, 2001.

Arnold, D., and Guha, R., *Nature, Culture, Imperialism: Essays on the Environmental History of South Asia*, Delhi, Oxford University Press, 1995.

Aubertin, C., and Rodary, E., *Protected Areas, Sustainable Land?*, Aldershot, Ashgate, 2011.

Barnes, T. A., *We Women Worked So Hard: Gender, Urbanisation and Social Reproduction in Colonial Harare, Zimbabwe, 1930–1956*, Portsmouth, NH, Heinemann, 1999.

Beinart, W., "African History and Environmental History," *African Affairs* 99, 2000, pp. 269–302.

Beinart, W., *The Rise of Conservation in South Africa: Settlers, Livestock and the Environment 1770–1950*, Oxford, Oxford University Press, 2003.

Beinart, W., "Soil Erosion, Conservationism and Ideas about Development: A Southern African Exploration, 1900–1960," *Journal of Southern African Studies* 11/1, 1984, pp. 52–83.

Beinart, W., and McGregor, J. (eds.), *Social History and African Environments*, Oxford, James Currey, 2003.

Beinart, W., and Middleton, K., "Plant Transfers in Historical Perspective: A Review Article," *Environment and History*, 10/1, 2004, pp. 3–30.

Brockington, D., *Celebrity and the Environment: Fame, Wealth and Power in Conservation*, London, Zed, 2009.

Brockington, D., *Fortress Conservation: The Preservation of the Mkomazi Game Reserve Tanzania*, Oxford, James Currey, 2002.

Brockington, D., Duffy, R., and Igoe, J., *Nature Unbound: Conservation, Capitalism and the Future of Protected Areas*, London, Earthscan, 2008.

Brown, K., "Book Review: S. Dovers, R. Edgecombe, and B. Guest, *South Africa's Environmental History: Cases and Comparisons,*" *African Affairs* 103/411, 2004, pp. 311–12.

Bunn, D., "An Unnatural State: Tourism, Water and Wildlife Photography in the Early Kruger National Park," in W. Beinart and J. McGregor (eds.), *Social History and African Environments*, Oxford, James Currey, 2003, pp. 199–218.

Burke, E., III, and Pomeranz, K. (eds.), *The Environment and World History*, Berkeley, University of California Press, 2009.

Carey, M., "Latin American Environmental History: Current Trends, Interdisciplinary Insights, and Future Directions," *Environmental History* 14/2, 2009, pp. 221–52.

Carruthers, J., "Conservation and Wildlife Management in South African National Parks, 1930s–1960s," *Journal of the History of Biology* 41/2, 2007, pp. 203–36.

Carruthers, J., "Influences on Wildlife Management and Conservation Biology in South Africa c.1900 to c.1940," *South African Historical Journal* 58/1, 2007, pp. 65–90.

Carruthers, J., *The Kruger National Park: A Social and Political History*, Pietermaritzburg, University of Natal Press, 1995.

Carruthers, J., "Our Beautiful and Useful Allies: Aspects of Ornithology in Twentieth Century South Africa," *Historia* 49/1, 2004, pp. 89–109.

Carruthers, J., "Past and Future Landscape Ideology: The Kalahari Gemsbok National Park," in W. Beinart and J. McGregor (eds.), *Social History and African Environments*, Oxford, James Currey, 2003, pp. 255–66.

Carruthers, J., "Wilding the Farm or Farming the Wild: The Evolution of Scientific Game Ranching in South Africa from the 1960s to the Present," *Transactions of the Royal Society of South Africa* 63/2, 2008, pp. 160–81.

Carruthers, J., and Robin, L., "Taxonomic Imperialism in the Battles for *Acacia*: Identity and Science in South Africa and Australia," *Transactions of the Royal Society of South Africa* 65/1, 2010, pp. 48–64.

Carson, R., *Silent Spring*, New York, Houghton Mifflin, 1962.

Carswell, G., "African Farmers in Colonial Kigezi, Uganda, 1930–1962: Opportunity, Constraint and Sustainability," PhD thesis, University of London, 1997.

Carswell, G., "Communities in Environmental Narratives," *Environment and History* 9/1, 2003, pp. 3–30.

Clark, A. F., "Environmental Decline and Ecological Response in the Upper Senegal Valley, West Africa, from the Late Nineteenth Century to World War I," *Journal of African History* 36/2, 1995, pp. 197–218.

Cock, J., and Koch, E. (eds.), *Going Green: People, Politics and the Environment in South Africa*, Cape Town, Oxford University Press, 1991.

Conte, C., *Highland Sanctuary: Environmental History in Tanzania's Usambara Mountains*, Athens, OH, Ohio University Press, 2004.

Cronon, W., "A Place for Stories: Nature, History and Narrative," *Journal of American History* 78/4, 1992, pp. 1347–76.

Crosby, A. W., *Ecological Imperialism: The Biological Expansion of Europe, 900–1900*, Cambridge, Cambridge University Press, 1986.

Dalton, A. L., "Book Review, on African Environmental History: *Green Land, Brown Land, Black Land: An Environmental History of Africa, 1800–1990*," *Current History*, 2000, pp. 231–2.

Dovers, S., "On the Contribution of Environmental History to Current Debate and Policy," *Environment and History* 6/2, 2000, pp. 131–50.

Dubow, S., "A Commonwealth of Science: The British Association in South Africa, 1905 and 1929," in S. Dubow (ed.), *Science and Society in Southern Africa*, Manchester, Manchester University Press, 2000, pp. 66–99.

Duffy, R., *Killing for Conservation: Wildlife Policy in Zimbabwe*, Oxford, James Currey, 2000.

Fairhead, J., and Leach, M., *Misreading the African Landscape: Society and Ecology in a Forest-Savannah Mosaic*, Cambridge, Cambridge University Press, 1996.

Fairhead, J., and Leach, M., "Reading Forest History Backwards: The Interaction of Policy and Local Land Use in Guinea's Forest-Savanna Mosaic, 1893–1993," *Environment and History* 1/1, 1995, pp. 55–91.

Fairhead, J., and Leach, M., *Reframing Deforestation: Global Analyses and Local Realities, Studies in West Africa*, London, Routledge, 1998.

Fay, B., "Environmental History: Nature at Work," *History and Theory* 42, 2003, pp. 1–4.

Freund, B., *The African City: A History*, Cambridge, Cambridge University Press, 2007.

Garland, E., "The Elephant in the Room: Confronting the Colonial Character of Wildlife Conservation in Africa," *African Studies Review* 51/3, 2008, pp. 51–74.

Gewald, J.-B., "El Negro, El Niño, Witchcraft and the Absence of Rain in Botswana," *African Affairs* 100, 2001, pp. 555–80.

Giles-Vernick, T., *Cutting the Vines of the Past: Environmental Histories of the Central African Rain Forest*, Charlottesville, VA, University Press of Virginia, 2002.

Giles-Vernick, T., "*Doli*: Translating an African Environmental History of Loss in the Sangha River Basin of Equatorial Africa," *Journal of African History* 41/3, 2000, pp. 373–94.

Giles-Vernick, T., "Leaving a Person Behind: History, Personhood, and Struggles of Forest Resources in the Sangha Basin of Equatorial Africa," *International Journal of African Historical Studies* 32/2–3, 1999, pp. 311–38.

Green Musselman, E., "Plant Knowledge at the Cape: A Study in African and European Collaboration," *International Journal of African Historical Studies* 36, 2003, pp. 367–92.

Griffiths, T., "How Many Trees Make a Forest? Cultural Debates about Vegetation Change in Australia," *Australian Journal of Botany* 50/4, 2002, pp. 375–89.

Griffiths, T., and Robin, L. (eds.), *Ecology and Empire: Environmental History of Settler Societies*, Edinburgh, Keele University Press, 1997.

Grove, R., "Editorial," *Environment and History* 1/1, 1995a, pp. 1–2.

Grove, R. H., *Green Imperialism: Colonial Expansion, Tropical Island Edens and the Origins of Environmentalism, 1600–1800*, Cambridge, Cambridge University Press, 1995b.

Guha, R., *The Unquiet Woods: Ecological Change and Peasant Resistance in the Himalaya*, Berkeley, University of California Press, 1989.

Guha, R., and Gadgil, M., *This Fissured Land: An Ecological History of India*, Delhi, Oxford University Press, 1992.

Hackel, J. D., "A Case Study of Conservation and Development Conflicts: Swaziland's Hlane Road," *Journal of Southern African Studies* 27/4, 2001, pp. 813–32.

Hailey, Lord, *An African Survey: A Study of Problems Arising in Africa South of the Sahara*, London, Oxford University Press, 1938.

Harms, R., *Games against Nature: An Eco-Cultural History of the Nunu of Equatorial Africa*, Cambridge, Cambridge University Press, 1987.

Howard, A. M., "Contesting Commercial Space in Freetown, 1860–1930: Traders, Merchants, and Officials," *Canadian Journal of African Studies* 37/2–3, 2003, pp. 236–61.

Iliffe, J., *Honour in African History*, Cambridge, Cambridge University Press, 2005.

Isaacman, A., and Sneddon, C., "Toward a Social and Environmental History of the Building of Cahora Bassa Dam," *Journal of Southern African Studies* 26/4, 2000, pp. 597–632.

Jacobs, N., *Environment, Power, and Injustice: A South African History*, Cambridge, Cambridge University Press, 2003.

Jeater, D., "No Place for a Woman: Gwelo Town, Southern Rhodesia, 1894–1920," *Journal of Southern African Studies* 26, 2000, pp. 29–43.

Koch, E., Cooper, D., and Coetzee, H., *Water, Waste and Wildlife: The Politics of Ecology in South Africa*, London, Penguin, 1990.

Le Quesne, T., "The Divorce of Environmental and Economic Policy under the First ANC Government, 1994–1999," *Journal of Environmental Law and Policy* 7, 2000, pp. 1–20.

Leach, M., and Green, C., "Gender and Environmental History: From Representation of Women and Nature to Gender Analysis of Ecology and Politics," *Environment and History* 3, 1997, pp. 343–70.

Leach, M., and Mearns, R. (eds.), *The Lie of the Land: Challenging Received Wisdom on the African Environment*, Portsmouth, NH, Heinemann, 1996.

MacKenzie, J. M., "Edward Said and the Historians," *Nineteenth Century Contexts* 18, 1994, pp. 9–25.

MacKenzie, J. M., *The Empire of Nature: Hunting, Conservation and British Imperialism*, Manchester, Manchester University Press, 1988.

MacKenzie, J. M. (ed.), *Imperialism and the Natural World*, Manchester, Manchester University Press, 1990.

Maddox, G., "'Degradation Narratives' and 'Population Time Bombs': Myths and Realities about African Environments," in S. Dovers, R. Edgecombe, and B. Guest (eds.), *South Africa's Environmental History: Cases and Comparisons*, Athens, OH, Ohio University Press, 2003, pp. 250–8.

Maddox, G., Giblin, J. L., and Kimambo, I., *Custodians of the Land: Ecology and Culture in the History of Tanzania*, London, James Currey, 1996.

Marks, S. A., "Back to the Future: Some Unintended Consequences of Zambia's Community-Based Wildlife Program (ADMADE)," *Africa Today* 48/1, 2001, pp. 120–41.

Martinez-Alier, J., *The Environmentalism of the Poor: A Study of Ecological Conflicts and Valuation*, Northampton, MA, Edward Elgar, 2002.

McCann, J., "Climate and Causation in African History," *International Journal of African Historical Studies* 32/2–3, 1999, pp. 261–79.

McCann, J., *Green Land, Brown Land, Black Land: An Environmental History of Africa, 1800–1990*, Portsmouth, NH, Heinemann, 1999.

McCann, J., *People of the Plow: An Agricultural History of Ethiopia, 1800–1990*, Madison, University of Wisconsin Press, 1995.

McCann, J., "The Plow and the Forest: Narratives of Deforestation in Ethiopia, 1840–1992," *Environmental History* 2, 1997, pp. 138–59.

McNeill, J. R., "Observations on the Nature and Culture of Environmental History," *History and Theory* 42, 2003, pp. 5–43.

Middleton, K., "Who Killed Malagasy Cactus? Science, Environment and Colonialism in Southern Madagascar (1924–1930)," *Journal of Southern African Studies* 25/2, 1999, pp. 215–48.

Moore, H. L., and Vaughan, M., *Cutting Down Trees: Gender, Nutrition, and Agricultural Change in the Northern Province of Zambia, 1890–1990*, Portsmouth, NH, Heinemann, 1994.

Myers, G. A., *Disposable Cities: Garbage, Governance and Sustainable Development in Urban Africa*, Aldershot, Ashgate, 2005.

Nash, C., "Environmental History, Philosophy and Difference," *Journal of Historical Geography* 26/1, 2000, pp. 23–7.

Nell, D. D'A., "The Development of Wildlife Utilisation in South Africa and Kenya, c.1950–1990," PhD thesis, University of Oxford, 2003.

Neumann, R. P., *Imposing Wilderness: Struggles over Livelihood and Nature Preservation in Africa*, Berkeley, University of California Press, 1998.

Penvenne, J. M., *African Workers and Colonial Racism: Mozambican Strategies and Struggles in Lourenço Marques, 1877–1962*, Portsmouth, NH, Heinemann, 1995.

Pikirayi, I., "Environmental Data and Historical Process: Historical Climatic Reconstruction and the Mutapa State 1450–1862," in W. Beinart and J. McGregor (eds.), *Social History and African Environments*, Oxford, James Currey, 2003, pp. 60–71.

Pooley, S., "Pressed Flowers: Nations of Indigenous and Alien Vegetation in South Africa's Western Cape, c. 1902–1945," *Journal of Southern African Studies* 36, 2010, pp. 599–618.

Potts, D., "Worker-Peasants and Farmer-Housewives in Africa: The Debate about "Committed" Farmers, Access to Land and Agricultural Production," *Journal of Southern African Studies* 26/4, 2000, pp. 807–32.

Raftopoulos, B., and Yoshikuni, T. *(eds), Sites of Struggle: Essays in Zimbabwe's Urban History*, Harare, Weaver Press, 1999.

Ranger, T. O., *Voices from the Rocks: Nature, Culture and History in the Matopos Hills of Zimbabwe*, Oxford, James Currey, 1999.

Ranger, T. O., "Women and Environment in African Religion: The Case of Zimbabwe," in W. Beinart and J. McGregor (eds.), *Social History and African Environments*, Oxford, James Currey, 2003, pp. 72–86.

Sachs, A., "The Ultimate Other: Post-Colonialism and Alexander von Humboldt's Ecological Relationship with Nature," *History and Theory* 42, 2003, pp. 111–35.

Schirmer, S., and Delius, P., "Soil Conservation in a Racially Divided Society," *Journal of Southern African Studies* 26/4, 2000, pp. 719–42.

Sheldon, K., "Markets and Gardens: Placing Women in the History of Urban Mozambique," *Canadian Journal of African Studies*, 37/2–3, 2003, pp. 359–95.

Shetler, J. B., *Imagining Serengeti: A History of Landscape Memory in Tanzania from Earliest Times to the Present*, Athens, OH, Ohio University Press, 2007.

Showers, K., *Imperial Gullies: Soil Erosion and Conservation in Lesotho*, Athens, OH, Ohio University Press, 2005.

Simone, A., *For the City yet to Come: Changing African Life in Four Cities*, Durham, NC, Duke University Press, 2004.

Sörlin, S., and Warde, P., "The Problem of the Problem of Environmental History: A Re-Reading of the Field," *Environmental History* 12/1, 2007, pp. 107–30.

Steinhart, E., *Black Poachers, White Hunters: A Social History of Hunting in Colonial Kenya*, Oxford, James Currey, 2006.

Steyn, P., "Environmental Management in South Africa: Twenty Years of Governmental Response to the Global Challenge," *Historia* 46, 2002a, pp. 25–53.

Steyn, P., "The Lingering Environmental Impact of Repressive Governance: The Environmental Legacy of the Apartheid Era for the New South Africa," *Globalizations* 2/3, 2005, pp. 391–403.

Steyn, P., "Popular Environmental Struggles in South Africa, 1972–1992," *Historia* 47, 2002b, pp. 125–58.

Steyn, P., and Wessels, A., "The Emergence of New Environmentalism in South Africa, 1988–1992," *South African Historical Journal* 42, 2000, pp. 210–31.

Stroud, E., "Does Nature Always Matter? Following Dirt through History," *History and Theory* 42, 2003, pp. 75–81.

Taylor, A. A., "Unnatural Inequalities: Social and Environmental History," *Environmental History* 1/4, 1996, pp. 6–19.

Thabane, M., "Shifts from Old to New Social and Ecological Environments in the Lesotho Highlands Water Scheme: Relocating Residents of the Mohale Dam Area," *Journal of Southern African Studies* 26/4, 2000, pp. 633–54.

Thompson, C. M., "Ranchers, Scientists, and Grass-Roots Development in the United States and Kenya," *Environmental Values* 11/3, 2002, pp. 303–26.

Tilley, H., "Africa as a Living Laboratory," PhD thesis, University of Oxford, 2001.

Tilley, H., "African Environments and Environmental Sciences: The African Research Survey, Ecological Paradigms and British Colonial Development, 1920–1940," in W. Beinart and J. McGregor (eds.), *Social History and African Environments*, Oxford, James Currey, 2003, pp. 109–30.

Twyman, C., "Livelihood Opportunity and Diversity in Kalahari Wildlife Management Areas, Botswana: Re-Thinking Community Resource Management," *Journal of Southern African Studies* 26/4, 2000, pp. 783–806.

Van Sittert, L., "Making the Cape Floral Kingdom: The Discovery and Defence of Indigenous Flora at the Cape ca.1890–1939," *Landscape Research* 28, 2003, pp. 113–29.

Vansina, J., *Paths in the Rainforests*, Madison, University of Wisconsin Press, 1994.

Wels, H., *Private Wildlife Conservation in Zimbabwe: Joint Ventures and Reciprocity*, Leiden, Brill, 2003.

Wels, H., and Spierenburg, M. J., "Securing Space: Mapping and Fencing in Transfrontier Conservation in Southern Africa," *Space and Culture* 9/3, 2006, pp. 294–312.

Wels, H., Wishitemi, B., and Spenceley, A. (eds.), *Culture and Community: Tourism Studies in Eastern and Southern Africa*, Amsterdam, Rozenberg Publishers, 2007.

White, R., "Environmental History, Ecology and Meaning," *A Round Table: Environmental History, special issue of Journal of American History* 76/4, 1990, pp. 1111–16.

Further Reading

Allman, J., Geiger S., and Musisi, N. (eds.), *Women in African Colonial Histories*, Indianapolis, Indiana University Press, 2002.

Neely, A., "Blame It on the Weeds: Politics, Poverty and Ecology in the New South Africa," *Journal of Southern African Studies* 36, 2010, pp. 869–87.

Worster, D. (ed.), *The Ends of the Earth*, New York, Cambridge University Press, 1988.

CHAPTER SEVEN

Latin America in Global Environmental History

SHAWN W. MILLER

Latin America, with its startling ecological diversity, is still in its infancy as a place of focused interest for environmental historians. The field's growth, however, has been rapid since the 1990s, and its practitioners share a reasonable hope for the topic's precocious development. For being often perceived as a global backwater, the region has had an outsized global influence: New World tropical plantations gave birth to modern, monocultural agriculture; the region's ecological and topographical diversity challenged classical and early modern science, inspiring new views of nature's functions and of life's origins; and a constant stream of indigenous and adopted products (sugar, silver, maize, chocolate, coffee, rubber, guano, bananas, copper, oil, etc.) helped feed, finance, and build the modern world. Historians have only just begun to investigate the physical realities, let alone the cultural and social expressions, of the region's environmental past (see Map 7.1).

The Region, Precontact

Some of the most developed research in the field has demonstrated the substantial impacts of indigenous peoples on the region's varied landscapes, millennia before the arrival of Europeans. Contrary to the pristine myth, which still strongly influences our collective memory of the pre-Columbian past as a place before time, indigenous peoples substantially reshaped American landscapes to meet their physical and cultural needs. Both regional[1] and insular[2] studies of the pre-Columbian populations show that indigenous numbers were large. Some American cities and regions rivaled or surpassed those of Eurasia in population density, and it appears that the entire region was inhabited, offering no empty landscapes and, by 1200 CE, no significant frontiers. Moreover, civilizations and their cities began more than 3,000 years ago: indigenous impacts on the New World landscape were not recent events in 1492, but processes of significant longevity.

A Companion to Global Environmental History, First Edition. Edited by J.R. McNeill and Erin Stewart Mauldin.
© 2012 John Wiley & Sons, Ltd. Published 2015 by John Wiley & Sons, Ltd.

Map 7.1 Latin America

The best studied of these processes to date is agriculture, whose vestiges are discernible to archeologists. In a variety of forms, indigenous people left substantial marks on the landscape and, due in part to their rapid depopulation during a century of conquest, many such evidences are still intact. Andean peoples entirely refashioned their mountains and valleys with sophisticated works of irrigation and terracing that today are evident along great stretches. In Peru alone, there are some 6,000 square kilometers (2,300 square miles) of known terraces, nearly all of them precolonial in origin, most of which required substantial construction in stone. Rather than simple cut-and-fill, Andean terraces were carefully constructed and integrated with complex irrigation systems; their graded surfaces drained excess water, took advantage of the high tropical sun at high altitudes, and improved the texture of steep rocky soils.[3] Terraces were, in a sense, reclamation projects, making farming possible in a landscape that in its natural form was too steep, too arid, or too cold to produce food.

Among the Aztecs and their cultural neighbors, *chinampa* agriculture reshaped the high valley lakes, supporting numerous large cities. Various forms of raised-field agriculture were practiced by Mexican, Central American, and Andean cultures to maximize the use of wetland landscapes, but the cultures of the Valley of Mexico refined the practice, creating one of the world's most productive agricultural regimes. Often referred to as floating gardens, the *chinampas* were in reality fields built up of soil in the midst of the valley's shallow lakes. The lakes, in addition to water for irrigation, provided fertile sediments that could be applied to fields annually. The lakes also imparted a warmer microclimate in high valleys prone to frost. With the further addition of recycled plant materials and urban human waste, central Mexico's farmers could harvest three to four crops per year and never had to let the land lie fallow. In order to make the system work, a strong, centralized, technically capable authority was required to manage water levels in various districts and also limit salinity, for these high basin lakes had no outlets other than evaporation. The farms and lakes were riddled with canals, sluices, and dikes, many of great length and size, that together managed the lakes' levels and their salinity. The largest such construction, Nezahualcóyotl's Dike, ran 16 kilometers (10 miles), dividing Lake Texcoco's saline north, which was fed by a few rivers, from its fresh-water south, fed largely by springs, some of which were tapped as well for urban drinking water and urban garden irrigation. Hence, most *chinampas* were located in the southern portions of the Valley of Mexico. The Aztecs' complex hydrological system, which largely fell into disrepair after the Spanish conquest, had drastically altered the lake's ecology, which itself was intensively utilized for fish, fowl, and a variety of other comestibles, such as spirulina.[4]

Human impacts are also evident in heavily forested ecologies such as the Amazon basin. There a number of pre-Columbian Amazonian cultures, some of whom built cities, have left evidence of significant soil amendment. Across sections of the Amazon, researchers have uncovered what locals have long referred to as black soils, which had been assumed to be natural. But digging has revealed them to be human creations. By centuries of careful amendment, using waste, vegetation, offal, and charcoal, these cultures created deep, rich soils – in stark contrast to the region's typically thin, leached varieties – that farmers still prize today and current entrepreneurs excavate to sell as potting soil. Some confirmed black soils cover as many as 100 hectares and reach 2 meters (6.5 feet) in depth.

By the practice of agroforestry, forest peoples also selected, to some extent, the forests they lived in. When felling forests for agriculture, they protected those trees that produced fruits, nuts, waxes, fibers, and other goods, and worked to disseminate them beyond their natural ranges. As much as 12 percent of the Amazon basin's forest species were heavily shaped by ancient, rather than modern, human interventions.[5] Rather than a pristine wilderness, it is more accurate to see the New World at contact as a cultivated garden. And even among groups that did not build cities, hunting, fishing, and repeated deforestation through slash-and-burn agriculture had significant impacts. In some marine regions, shell middens, piled over centuries as refuse heaps, reach astonishing sizes, and the accumulated impact of the hunt over millennia resulted in the comparatively small number of large game species throughout the New World tropics.

Beyond these rather material interests in indigenous relations with nature, we still know little about their cultural aspects. The landscape provides a variety of physical evidences that scientists have examined with great creativity and sophisticated tools, but substantially less evidence for the humanities which is needed to answer questions about indigenous perspectives of and relations with nature and cosmology, attitudes toward consumption, and efforts at conservation. Some persuasive work has been done among the indigenous of North America[6] where evidence is yet thinner, but investigations of pre-Columbian Latin America on these points have only just begun to tap the sources.[7] The relative abundance of pre-Columbian materials and of ethnographic documentation produced by sixteenth-century colonists, in addition to the cautious use of ethnographic projection, may provide the gate to an important path of research.

Conquest and Colonization

New World conquest left many environmental legacies. Foremost stands the indigenous demographic catastrophe. While scholars still disagree about exact numbers and specific infections, the consensus holds that the European conquest of the Americas was a human disaster of unparalleled proportions. Within about a century and a half, tens of millions of people died, the majority due to introduced infectious diseases whose likely microbial suspects were, among others, the common cold, influenza, measles, and smallpox. Combined with direct military campaigns and the often harsh conditions of subjugation, entire cities and cultures disappeared. To date, scholars have debated the demographic aspects of the calamity, but few have looked at other environmental consequences.[8] If indigenous peoples, who were populous and widespread, had made pervasive impacts on the natural world before the conquest, how did human relations with the environment change after the total human population had been diminished to a remnant? Europeans came to dominate the land politically, but the region's total 1491 population, based on current estimates, would not be achieved again in most areas until the early twentieth century.

Europeans, with a limited number of new technologies, the introduction of long-distance trade, and the help of African slaves imported to repopulate the landscape with laborers, began to have environmental impacts that were notable if not notorious. Sugar devastated forests in Brazil and numerous Caribbean islands, such as Barbados, wreaking ecological devastation that led to local extinctions for animals and diminishing returns for settlers. Declines in soil fertility, unmistakable evidence of soil erosion, and the disappearance of lumber and fuel would, over time, destroy fortunes and force the enterprise to constantly seek new land beyond the new colonial frontiers. The smaller islands of the Caribbean were in numerous cases deforested within a couple of generations, their formerly fertile resources squeezed dry, like sugarcane at the mill, leaving them a husk.[9] While the Indians who previously inhabited these islands had left their mark, European colonization brought human environmental impacts of an entirely different order.

Likewise, European introductions of domesticated animals and weed seeds transformed many landscapes. Unlike human populations which seemed to struggle in order to recover, cattle, sheep, pigs, and some introduced plants reproduced massively, usually in temperate areas conducive to their growth. Due to a shortage of human labor, ranching became the most widespread economic activity throughout Spanish America. Sheep have been characterized as an ecological plague, erupting numerically and transforming formerly fertile highland valleys into desert scrublands, whereas studies of colonial cattle have found a more benign impact due to the imported adaptation of rotating seasonal grazing (a practice that had been common among Andean pastoralists).[10]

Europeans also sought treasure more directly, mining silver and gold ores in quantities so large as to transform the global monetary order. Mountain ranges in Mexico, Peru, and Brazil even today bear the scars of colonial pits, shafts, washings, and tailings. While some indigenous peoples had mined as well, colonization developed mining on a new scale with significant long-term consequences. The patio process, which employed mercury to process silver ores, introduced a dangerous toxin to local environments: miners at Huancavelica's mercury mines and workers who handled mercury at silver-processing sites died of poisoning in appalling numbers.[11] Yet, while on the map the conquest's political reach was estimable, the reach of the settlers, who were a miserable fraction of the former indigenous populations, did not penetrate far. Sugar-planting, silver-mining, and sheep-ranching all had documented consequences, but in much of the rest of the region, human activities and pressures shaped nature less dramatically than they had in the period before the European conquest.

For those who survived war and disease in the sixteenth century, Latin America seems to have been something of a cornucopia. Agricultural landscapes continued to provide maize, potatoes, and manioc, to which were added Eurasian and African crops, such as wheat, rice, and sweet potatoes, and a large variety of vegetables and fruits, in particular the banana, which would become the staple food for much of the lowlands. With the decline of the indigenous population, it is probable that wild fish and game stocks increased. And to the Indians' turkeys, edible dogs, and llamas, the conquerors introduced cows, sheep, pigs, and chickens. In the Columbian exchange, it was the peoples living in the Americas who were the first beneficiaries in terms of crops and livestock. Europe would not see the caloric benefits of the exchange for a number of generations. Famines and food crises remained common features in Europe at the time; there are no such events, however, recorded for the major regions colonized in Latin America. The surviving indigenous populations themselves took note. When Philip II's officials surveyed hundreds of indigenous communities in the late sixteenth century, among the questions they asked was what they believed had caused so many of them to die of disease. They pointed to a number of changes associated with the Christian conquest, including declines in the practice of bathing, the use of excess clothing, and the inferiority of European medicine. But by far the most common explanation proffered was food, not its kind or scarcity, but its excess. Before the conquest, the Indians of Mexico said their more spartan diets enhanced their health. Now there was too much food, they argued, and Indians ate more meat than the Spaniards.[12]

Another factor that may need greater consideration in the assessment of the colonial period is the extent to which mercantilism, with its extensive barriers to free trade and its restrictive monopolies, limited economic growth and, hence, environmental change. Iberian powers, while promoting the forced immigration of Africans, established rather effective barriers to most other immigrants, despite an ongoing interest by many, including non-Catholics and non-Iberians, to come to the New World, thus limiting a potential source of population growth and economic expansion in the region. In Brazil, the crown monopolized to itself the profits of such commodities as southern right whales for oil, high-grade timber for shipbuilding and construction, brazilwood for dyeing textiles, and also diamonds. Such monopolies, which had the promise of becoming major industries, even if not sustainable ones, remained relatively small operations that in effect helped conserve natural resources.[13]

Hence, the conquest entailed many environmental processes, with relative gains and losses, depending upon one's perspective. Indians, their domesticates, their pathogens, and a variety of native plants and animals suffered badly as a result of the invasion of Europeans, their domesticates, and new ways of interacting with nature. Latin America became, after the conquest, a unique cultural hybrid made up of diverse elements loosely held together by a dominant Iberian Christian tradition. Environmentally, the Americas did not hybridize, as species do not mix and mingle as do cultures, but Latin America did become something very different, an ecological amalgam. While the Americas' original biodiversity may have declined, with the tragic loss of some species, overall, the region's biodiversity increased with the substantial importation of foreign genetic materials. As a result of the conquest, the New World really did become something entirely new, both culturally and environmentally.

The transformation of nature had significant geopolitical consequences. As we have seen, introduced microbes played a central role in the process of the conquest, "favoring" the invaders in the casualty count, although taking many European lives as well. The conquest, however, created a new disease regime, both by microbial introductions and by the modification of local ecologies. The old regime privileged European outsiders over indigenous insiders. A new disease regime emerging in the seventeenth century was created by the introduction of African diseases, and it would greatly advantage locals over outsiders. Yellow fever, long endemic in Africa, became epidemic in the Caribbean in the mid-seventeenth century. Fatalities, especially among Europeans, were catastrophic. However, yellow fever, like chicken pox, quickly became a disease of childhood in those areas it visited frequently. When acquired as a child, the course of the disease was milder, morbidity was lower, and those who survived acquired life-long immunity. Over time, an ever larger fraction of Latin America's lowland populations acquired immunity to yellow fever. However, any new immigrants from Europe, especially young males, were vulnerable and died in the formerly high percentages.

This reality played a significant geopolitical role in the lowlands of Latin America. From the beginning, northern Europeans, particularly the Dutch, French, and British, all of whom had colonies in North America, coveted the far more valuable colonies of the Spanish and Portuguese in Latin America and the Caribbean. And they attacked them repeatedly, hoping to secure their own tropical colonies. What is striking is that, despite their greater military power, greater wealth in the period after 1650, and larger populations, northern Europeans failed markedly in their grand designs. They did take a few scraps of land, mostly small, poorly settled, or little defended Caribbean islands, but their attempts to take Havana, Santo Domingo, Port-au-Prince, and Cartagena, among other prizes, or settle remote, uninhabited coasts, were either aborted before victory or miserably short-lived. The realities were two. In case after case, the invaders learned they had to conquer Iberian fortifications quickly, or local diseases, such as yellow fever and malaria, took such a toll on their soldiers that they could no longer fight. Second, even if they succeeded in a military campaign, they had to constantly supply new bodies to replace the dead and keep the territorial gain viable. They had few successes.

The same challenge was a central feature of independence movements. Haitian slaves, the first peoples in Latin America to declare their independence, faced two successive invasions by the British and the French, who sent massive fleets to recapture Haiti. But in both cases the invaders lost tens of thousands of men to tropical diseases and soon desisted. Like the Spanish, the Haitian slaves knew how to use their immunities to their advantage. With so much to lose, they were bold fighters, but they were also willing to avoid conflict while microbes took the vanguard. In Spanish America's wars of independence, which carried on for more than a decade, time and again local soldiers, with immunities to disease acquired in childhood, had a notable advantage over their metropolitan opponents.[14]

Export Orientations

With independence, many of Latin America's new republics broke the chains of monopoly and the barriers to free trade. With more liberal political and economic policies, many hoped to be able to follow the lead of the advancing industrial nations of the north, and to commence manufacturing. But they found that they were not yet ready to compete with the industrializing nations and that their place in the world economy would be one of supplying calories to northern workers and raw materials to northern factories. With the barriers to immigration and trade down, populations began to grow rapidly, and exports expanded through foreign and domestic investments, with telling consequences.[15]

But Latin America's economies and exports did not expand as states and producers wished. In addition to human factors such as foreign competition and a poorly developed infrastructure, environmental factors such as natural disasters, plant diseases, and environmental degradation shaped the region's economic course. A number of Latin American regions made the profitable shift from sugar production to coffee, a product that would enrich some states, driving incipient industrialization. However, in Cuba's case, where the transition to coffee was in full swing, a series of devastating hurricanes, each passing over or near Havana, changed the economic course. Hurricanes had long plagued the Caribbean basin. Over the centuries, hurricanes had destroyed entire cities with heavy winds, flooding rains, and storm surges. In the middle of the nineteenth century, hurricanes came with such frequency to western Cuba that planters gave up on coffee, making Cuba a monocultural sugar producer for much of the rest of its history, creating an economic homogeneity that made Cuba vulnerable to the global pricing of a single commodity. It was simply too expensive to constantly replant coffee trees destroyed by high winds, trees that took several years to mature and bear fruit. Sugarcane, which was less susceptible to high winds, even if destroyed, could be replanted and return to profitability in a single year. Sugar, however, was more detrimental to the Cuban landscape than coffee would have been: sugar consumed forests for space and for firewood, depleted soils, and even placed heavier burdens on the slaves than did the arboricultural regime of coffee.[16]

Rubber production in the Amazon faced a different kind of natural obstacle. From the late nineteenth century, first with the appearance of the bicycle and then that of the automobile, global rubber demand swelled causing rubber's price to rise dramatically. Brazil, which had by this time become quite dependent on coffee, found in natural rubber an opportunity for economic diversification. Brazil dominated world coffee production, but in rubber it essentially held a global monopoly. The price of rubber, however, was extraordinarily high not due to any artificial limits on its production, but because only so much could be gathered in the wild by tappers, who had to walk great distances from one tree to another, scattered throughout the forests. The British, Dutch, and French, who no longer had direct designs on Latin American territory, took rubber-tree seeds and planted them in their Asian colonies, which by the second decade of the twentieth century outproduced Brazil's primitive system by large margins and brought the global price of rubber lower than Brazil's cost of production. To compete, Brazil would also have to plant rubber in plantations, but every attempt to do so was defeated by South American leaf blight, a fungus that had long evolved with the rubber tree, but which only had a limited impact on the species until it was gathered into plantations. Then, destruction was near total. Henry Ford, who invested millions in Amazonian rubber plantations, hiring the best technical help, failed miserably. Asian plantations, on the other hand, thrived as the fungus never found its way to Asian shores. Rubber's demise in the Amazon left few viable economic alternatives in the broad basin for decades, and many who had migrated to the Amazon during the rubber boom returned to their regions of origin.[17]

In Peru, export revenues declined due to the exhaustion of an ancient, yet renewable resource. Global agriculture had always faced a fertility barrier that had forced farmers to resort to fallowing and rotating crops. Farmers also continued to fortify soils with a variety of materials, from manure and fish to blood and bone meal, but such efforts were labor-intensive and costly, when they were possible at all. From the 1840s, Europeans rediscovered Peruvian guano, deep, rich piles of bird droppings that the Incas had utilized centuries before. With its extremely high concentrations of nitrogen, guano was the best accessible fertilizer ever available, and Peru, while fighting wars to protect the resource, mined the product from coastal islands and promontories where birds had nested and defecated for millions of years. Peru exported guano to every corner of the world that could afford it, and earned a substantial income. Crops in Europe, China, and Australia not only exhibited yields never before seen, but with consistent supplies, guano made it possible to plant crops year after year without fallowing and also to plant crops in soils that had previously been judged too infertile to be worth the effort. However, by the 1870s, Peru ran out of guano, quite unexpectedly. Exporters had simply mined out guano's ancient bank accounts. Bird populations continued to make hourly deposits, but global farmers spent the assets on their fields far faster than the birds could replenish the stocks. Farmers had come to depend on powerful external fertility inputs. Searching drastically, they found chemical substitutes, primarily mineral nitrates which also came from South America's Pacific coastal nations and also caused military conflict. Then, in the early twentieth century, German scientists discovered how to capture nitrogen from the air with sophisticated machinery and massive inputs of energy. But it was lowly guano that set global agriculture on the path to dependency on external sources of fertility.[18]

Conservation

There is evidence that indigenous peoples engaged in various forms of conservation. More commonly, this took the form of various local practices, such as not harvesting the females of hunted species, but there are reported instances of conservation as policy, particularly regarding forests and large game. Europeans generally neglected conservation in their colonial domains until almost the period of independence, when they became more acutely aware of the costs of some of their colonial endeavors. Deforestation in particular had resulted in erosion and the disappearance of firewood, the region's main source of energy. But it was also argued in the eighteenth century that deforestation had altered the climate, decreasing essential rainfall and increasing local temperatures. Changing perceptions of nature, of nature's centrality in economic development, and even of the place of slaves and peasants on the landscape began to shape state policies. A number of efforts at forest protection and agricultural reform were discussed and legislated, although their geographic and temporal reach was limited.[19] With independence and the rise of liberalism in the region, conservation was neglected again for much of the nineteenth century, although some forested areas were converted to reserves and there were even some urban cases of reforestation.[20]

Toward the turn of the twentieth century, states began to give the issue greater attention. In Peru, after the guano bust, the state began to manage and protect guano birds, often by killing predatory birds. Guano was now to be used not to support exports but to enhance national agriculture, and did so with some success until a boom in fisheries depleted fishing stocks, causing bird populations and guano production to suffer.[21] In Venezuela, egrets were protected after 1917, after hunters had destroyed them in their millions for decades in order to harvest the breeding plumage, popular for adorning women's hats.[22]

Yet many such efforts at conservation were too late, were contradicted by later developmental legislation, or were poorly enforced. From Argentina and Chile's Andean slopes, chinchilla pelts had been exported to North America and Europe from the 1820s to manufacture fur clothing. Named by the Spanish after the Chincha Indians who used to wear their fur, chinchilla became a major component in northern fur markets. By the last decade of the nineteenth century, Southern Cone exports averaged nearly half a million pelts per year – a long fur coat required as many as 150 pelts. Unlike that of most rodents, chinchilla reproduction is limited by small litters, usually one set of twins, and populations began to crash at the turn of the twentieth century. In 1905, Argentina exported about 200,000 pelts. By 1909, exports had fallen to 28,000. Despite an international agreement in 1910 between Argentina, Chile, Bolivia, and Peru, which outlawed the trade, one species became extinct, and the two remaining, one of which was not rediscovered until the 1970s, remain endangered with declining populations despite careful protection in national reserves in Chile. For some who wanted to Europeanize their national territories, the demise of native species was considered progress. The native deer, the huemul, was driven to near extinction by overhunting and by the introduction of the larger, more aggressive European red deer. And in 1946, Argentina imported beaver from Canada, a species that has thrived and recently expanded north from Tierra del Fuego despite efforts to prevent it. Over the course of the early twentieth century, Argentina, for example, legislated away colonial and nineteenth-century hunting regulations, and the international trade in wild species grew rapidly. Between 1976 and 1984, Argentina officially exported 11.6 million iguana skins and 24 million nutria pelts, in addition to tens of thousands of boa constrictor and crocodile skins, and nearly 1 million live birds, all to satisfy the demands of northern consumers and pet owners. After signing the Convention on International Trade in Endangered Species of Wild Fauna and Flora (CITES) in 1980, Argentina's official export numbers declined substantially, but there was little domestic enforcement of the convention for lack of political will, and agencies remained hopelessly understaffed.[23] Despite reported declines in the illegal commerce in wild species, it remains the third most valuable illicit trade in the region, after drugs and weapons, and poses a significant threat to a variety of species, particularly colorful birds.

Not long after the US established the first national park, a number of Latin American nations followed suit. Mexico established Desierto de Leones as a forest preserve in 1876 (which became a national park in 1901), and by 1940 had 40 parks, all but two of which were designated in the 1930s by President Lázaro Cárdenas. The US, in 1940, only had 30 national parks, although altogether they were larger in total extent than Mexico's. But Mexico's justification for parks was different from that of the US, where parks embodied closed areas ruled by elitist visions of nature, frequently excluding poor and indigenous peoples who had long subsisted on the same lands. By the 1930s, Mexico saw national parks as tools of the revolution, and attempted to incorporate within them the needs of farmers, ranchers, loggers, scientists, and the urban dweller, which explains why most were located near large cities in the central valleys. Unlike the US, which focused its park attentions on the frontier, on preserving what was deemed pristine, and on protecting what was considered aesthetically or monumentally beautiful, Mexico's interests focused on reclaiming already degraded landscapes, places of important cultural heritage, and locations that were accessible to large sectors of the population which needed spaces for recreation.[24] While the creation of parks cooled in Mexico after Cárdenas left office in 1940, other nations continued apace, establishing reserves with various aims, most notably in Costa Rica although most of the nations of the region participated, and special emphasis was given to the Amazon basin. Today, many nations in the region have national parks, reserves, and monuments whose total protected area exceeds that of the US as a total percentage of their landscapes. Unfortunately, with some notable and hopeful exceptions, many national parks exist only on paper due to the lack of state funding and sometimes local support.[25] Despite the formation of several reserves in Mexico to protect the essential wintering grounds of the monarch butterfly, most have now been destroyed due to deforestation by local people seeking a simple living. And even at the state level, the goals of development and poverty eradication are often at odds with those of land conservation and species protection.

Among the most significant, state-sponsored projects are hydroelectric dams. Despite a few exceptional or recent cases, Latin America has historically had a limited quantity of the fuels that drove modernization elsewhere. Lacking coal and oil, nations invested heavily in dam construction, and as a result have become highly dependent on hydropower to produce the electricity necessary to industrialize. After a century of dam construction, many nations rely on hydropower for more than 90 percent of their electrical energy production, and hydroelectric projects, some of the most expensive public works ever built, are also a major contributor to the region's indebtedness. In addition to displacing tens of thousands, many of them indigenous peoples whose cultures have been intimately tied to place, natural features, and burial grounds, which have all been inundated, dams have also modified the region's landscapes on a vast scale. Some single cases submerged forests the size of small European nations. Despite initial plans to incorporate locks, navigation, fish ladders, and recreational features, most dams produced only electricity at the expense of fisheries, farms, and forests. In the Amazon some dams altered local concepts of time and season, and farmers, who relied on the region's annual flooding not only to fertilize their fields but to indicate the time to plant, stood confused by rivers that now ran clear and with little deviation in volume. In lowland regions, new reservoirs sometimes bred mosquito populations so large as to make their shorelines uninhabitable. Yet much of the region's hydroelectric capacity remains untapped, and nations are pushing forward ambitious projects despite continuing local and increasing international concern.

Current Trends and New Directions

The majority of environmental research by historians to date in Latin America is rather focused on the material aspects of human relations with nature, from the physical impacts of colonialism and capitalism to the opposing efforts of conservation. However, even under these familiar themes, many questions have yet to be asked, and among those that have, the answers have not been much contested, nor has there been sufficient time to reach consensus. Nevertheless, the field has recently begun to branch out in new directions, sometimes taking a cultural turn or connecting environmental aspects of Latin America's past with traditional historical fields such as social, labor, and indigenous history, and the histories of science, colonialism, and nation-building. An important contribution of some recent studies is the emphasis on expressing the many discordant human voices in the debates about nature and our place in it, beyond the better-studied views of political elites, extractive capitalists, developmentalist technocrats, and foreign scientists. Oil production in Mexico, for example, not only degraded landscapes and depleted resources: it also harmed the physical and cultural human ecologies of the Tampico region. Oil transformed landscapes, social hierarchies, and national agendas, but only did so in the context of a significant response from local indigenous peoples and eventually the state.[26] Likewise in the Andes, global warming, melting glaciers, and natural disasters exposed competing interpretations of natural processes, conflicting solutions, and entrenched social, racial, and national divisions, evidence that nature and the discourses it inspires reach into nearly every sphere of human society.[27] In Brazil, in addition to state policies and environmental change associated with sugar production in the twentieth century, class-distinct interpretations of sugar landscapes themselves shaped human work routines and labor militancy.[28]

New work has also shown the value of examining environmental issues that have transnational or global implications. By the late nineteenth century, Latin American banana production was linked by a long commodity chain to banana consumption in the US and Europe. Southern banana production shaped a new mass market, changed northern eating habits, and enhanced consumer health. The mass market, in turn, transformed Latin American landscapes, altered social relations, and facilitated the spread of plant diseases, the fight against which compromised the health of banana workers.[29] Also, in a less direct form of northern consumption, henequen fibers, used to bind such temperate commodities as wheat, enslaved indigenous Mexicans who were uprooted to the Yucatan's tropics.[30] Such studies demonstrate that to fully understand historical environmental processes, we often must take a global view and, more importantly, emphasize the importance of the oft-neglected consumption side of environmental relations. Global consumption, as well as domestic attitudes toward consumption and acquisitiveness, are at the heart of much of environmental history, and must be examined in their biological, cultural, psychological, and economic aspects. Extraction and production remain at the center of most of our studies, but choices about consumption, and how much people choose to consume, will be a useful key to understanding human civilization's weight on the landscape.[31]

Some recent studies have also helped blur what had been a stark line dividing nature and culture. Cultural studies have done so by examining nature as a discourse, as much as a physical reality, which connects and interacts with a myriad of other human ideas and concepts. Such an approach has the power to integrate and connect ideas about nature with those of race, gender, class, society, geography, politics, and economics, and above all to highlight cultural differences among various national, regional, and global discourses. There is a tendency to see Latin America through North American and European lenses, and understanding local cultures is an important corrective. Those historians who focus on the actual materiality of nature have helped blur the distinction by refusing to see human relationships with nature as a monolithic battle between "culture" and "nature." Humans can be profitably characterized as one important species among many, all seeking to feed themselves, reproduce, and, in many cases, find shelter, create suitable habitats, and get around. Humans are part of nature too, and biological perspectives can help ground our analysis, noting that physically, if not culturally, all humans have much in common. There is a still a lot of human hubris in our histories. To lose sight of humans altogether is to slip into natural history, but to see man and culture as the measure of all things runs the risk of destroying what environmental history has to offer that is original and instructive.

There are many areas of obvious neglect in the field, gaping blanks that historians have not seen or have only begun to fill with exciting new dissertations. Entire regions, such as the Southern Cone, need more attention, and many of the region's smaller republics have been entirely ignored. The basic outlines of environmental legislation and popular environmental movements in some nations remain poorly studied. Too few historians have taken a decisively environmental approach to the Latin American city, one of the most urbanized regions of the world. We have yet to examine the historical meanings and uses of urban space, the temporal implications of urban life's rapid pace on human biology, the environmental fallout of a species in constant movement, or even much on such traditional issues as demography, sanitation, hygiene, health, water, and urban planning. In agriculture we are just beginning to scratch the soil and examine fertility and sustainability, especially in the latter centuries. Energy, both its production and its consumption, is central to the development and degradation of the region, and we need more studies of oil, offshore drilling, hydroelectric dams, and alternative fuels, and particularly the state's role in their promotion. The automobile, possibly the most transformative technology of the twentieth century, needs attention. It has altered the built and unbuilt landscape, changed the chemical composition of the air, and broadened substantially the human footprint. Moreover, from its introduction, the car became a major public-health issue, killing millions of passengers and pedestrians over the course of the following century.

To date, topics that have been of most interest to historians include forests and other extractive activities, conservation (primarily of forests), agriculture, commodity chains, human and plant diseases, domesticated animals, and natural disasters, although in all these areas, one must say, research is still at its beginning. Meteorology, climate, and, just as importantly, weather would seem to offer topics of original promise in the tropics. Marine, estuarine, and riverine environments, which have been so critical economically and culturally, are also only just beginning to merit consideration. Beyond domesticated animals, we have thus far paid little attention to wild species of animals, fish, birds, and plants, or to biodiversity more generally. What has been unique about cultural attitudes toward wild things and wilderness in the region, the rates of extinction, the status of the endangered and efforts at wildlife conservation, and what of the long international trade in species as diverse as rain-forest birds and aquarium fish? While in North America, historical studies of such icons as beaver, bison, pigeons, California condors, and wolves are available, in Latin America, similar species, such as the broad-ranged jaguar, the monarch butterfly, the Andean condor, tapir, scores of primates, as well as many exotic or predatory birds, are almost unknown historically. Tourism, the consumption of nature by foreigners in situ rather than through exports, needs a historian's perspective. Beach tourism and cruising offer unique approaches to what is an unusually transnational affair in which people, as much as goods, are in movement.

Due to the region's unique ecologies and histories, the field holds much promise and probably many surprises. Happily, there are many environmental aspects of Latin America's history about which we have not even thought to inquire. Because of the region's many exceptional qualities, we should not lean too heavily on the creativity of scholars of other regions to formulate what are likely to be the most profitable and pressing questions.

Notes

1   W. M. Denevan, *The Native Population of the Americas in 1492*, Madison, University of Wisconsin Press, 1976.
2   M. Livi-Bacci, "Return to Hispaniola: Reassessing a Demographic Catastrophe," *Hispanic American Historical Review* 83/1, 2003, pp. 3–51.
3   W. M. Denevan, *Cultivated Landscapes of Native Amazonia and the Andes*, Oxford, Oxford University Press, 2001.
4   T. M. Whitmore and B. L. Turner, *Cultivated Landscapes of Middle America on the Eve of Conquest*, Oxford, Oxford University Press, 2002. Called tecuitlatl by the Aztecs, spirulina is a blue-green algae rich in protein, vitamins, and minerals that could be gathered from lakes with fine nets.
5   Denevan, *Cultivated Landscapes*.
6   S. Krech, *The Ecological Indian: Myth and History*, New York, W. W. Norton, 1999.
7   L. Simonian, *Defending the Land of the Jaguar: A History of Conservation in Mexico*, Austin, TX, University of Texas Press, 1995, pp. 10–26.
8   N. D. Cook, *Born to Die: Disease and New World Conquest, 1492–1650*, New York, Cambridge University Press, 1998.
9   D. Watts, *The West Indies: Patterns of Development, Culture, and Environmental Change since 1492*, Cambridge, Cambridge University Press, 1987.
10 E. G. K. Melville, *A Plague of Sheep: Environmental Consequences of the Conquest in Mexico*, Cambridge, Cambridge University Press, 1994; A. S. Sluyter, *Colonialism and Landscape: Postcolonial Theory and Applications*, Lanham, MD, Rowman & Littlefield, 2002.
11 K. W. Brown, "Workers' Health and Colonial Mercury Mining at Huancavelica, Peru," *The Americas* 57/4, 2001, pp. 467–96.
12 J. C. Super, *Food, Conquest, and Colonization in Sixteenth-Century Spanish America*, Albuquerque, NM, University of New Mexico Press, 1988.
13 S. W. Miller, *An Environmental History of Latin America*, New York, Cambridge University Press, 2007.
14 J. R. McNeill, *Mosquito Empires: Ecology and War in the Greater Caribbean, 1620–1914*, Cambridge, Cambridge University Press, 2010.
15 R. P. Tucker, *Insatiable Appetite: The United States and the Ecological Degradation of the Tropical World*, Berkeley, University of California Press, 2000.
16 L. A. Pérez, *Winds of Change: Hurricanes and the Transformation of Nineteenth-Century Cuba*, Chapel Hill, NC, University of North Carolina Press, 2001.
17 W. Dean, *Brazil and the Struggle for Rubber: A Study in Environmental History*, Cambridge, Cambridge University Press, 1987; G. Grandin, *Fordlandia: The Rise and Fall of Henry Ford's Forgotten Jungle City*, New York, Metropolitan Books, 2009.

18   G. T. Cushman, "The Most Valuable Birds in the World: International Conservation Science and the Revival of Peru's Guano Industry, 1909–1965," *Environmental History* 10, 2005, pp. 477–509.

19   R. H. Grove, *Green Imperialism: Colonial Expansion, Tropical Island Edens and the Origins of Environmentalism, 1600–1860*, New York, Cambridge University Press, 1995; J. A. Pádua, *Um sopro de destruição: pensamento político e crítica ambiental no Brasil escravista,1786–1888*, 2nd ed., Rio de Janeiro, Jorge Zahar, 2004.

20   J. A. Drummond, *Devastação e preservação ambiental: os parques nacionais do Estado do Rio de Janeiro*, Niterói, EdUFF, 1997.

21   Cushman, "The Most Valuable Birds in the World."

22   A. Zerpa Mirabal, *Explotación y comercio de plumas de garza en Venezuela: fines del siglo XIX – principios del siglo XX*, Caracas, Ediciones del Congreso de la República, 1998.

23   A. Brailovsky and D. Foguelman, *Memoria verde: historia ecológica de la Argentina*, Buenos Aires, Editorial Sudamericana S.A., 1991.

24   E. Wakild, "Border Chasm: International Boundary Parks and Mexican Conservation, 1935–1945," *Environmental History* 14/3, 2009, pp. 453–75.

25   S. D. Evans, "Historiografía verde: estado de la historia sobre la conservación de la naturaleza en América Latina," in R. Funes Monzote (ed.), *Naturaleza en declive: miradas a la historia ambiental de América Latina y el Caribe*, Valencia, Centro Francisco Tomás y Valiente UNED Alzira-Valencia, 2008, pp. 81–96.

26   M. I. Santiago, *The Ecology of Oil: Environment, Labor, and the Mexican Revolution, 1900–1938*, New York, Cambridge University Press, 2006.

27   M. Carey, *In the Shadow of Melting Glaciers: Climate Change and Andean Society*, Oxford, Oxford University Press, 2010.

28   T. D. Rogers, *The Deepest Wounds: A Labor and Environmental History of Sugar in Northeast Brazil*, Chapel Hill, NC, University of North Carolina Press, 2010.

29   J. Soluri, *Banana Cultures: Agriculture, Consumption, and Environmental Change in Honduras and the United States*, Austin, TX, University of Texas Press, 2005; S. Marquardt, "Green Havoc: Panama Disease, Environmental Change, and Labor Process in the Central American Banana Industry," *American Historical Review* 106/1, 2001, pp. 49–80; S. Marquardt, "Pesticides, Parakeets, and Unions in the Costa Rican Banana Industry, 1938–1962," *Latin American Research Review* 37/2, 2002, pp. 3–36.

30   S. D. Evans, *Bound in Twine: The History and Ecology of the Henequen-Wheat Complex for Mexico and the American and Canadian Plains, 1880–1950*, College Station, TX, Texas A&M University Press, 2007.

31   A. J. Bauer, *Goods, Power, History: Latin America's Material Culture*, New York, Cambridge University Press, 2001.

References

Bauer, A. J., *Goods, Power, History: Latin America's Material Culture*, New York, Cambridge University Press, 2001.

Brailovsky, A., and Foguelman, D., *Memoria verde: historia ecológica de la Argentina*, Buenos Aires, Editorial Sudamericana S.A., 1991.

Brown, K. W., "Workers' Health and Colonial Mercury Mining at Huancavelica, Peru," *The Americas* 57/4, 2001, pp. 467–96.

Carey, M., *In the Shadow of Melting Glaciers: Climate Change and Andean Society*, Oxford, Oxford University Press, 2010.

Cook, N. D., *Born to Die: Disease and New World Conquest, 1492–1650*, New York, Cambridge University Press, 1998.

Cushman, G. T., "The Most Valuable Birds in the World: International Conservation Science and the Revival of Peru's Guano Industry, 1909–1965," *Environmental History* 10, 2005, pp. 477–509.

Dean, W., *Brazil and the Struggle for Rubber: A Study in Environmental History*, Cambridge, Cambridge University Press, 1987.

Denevan, W. M., *Cultivated Landscapes of Native Amazonia and the Andes*, Oxford, Oxford University Press, 2001.

Denevan, W. M., *The Native Population of the Americas in 1492*, Madison, University of Wisconsin Press, 1976.

Drummond, J. A., *Devastação e preservação ambiental: os parques nacionais do Estado do Rio de Janeiro*, Niterói, EdUFF, 1997.

Evans, S. D., *Bound in Twine: The History and Ecology of the Henequen-Wheat Complex for Mexico and the American and Canadian Plains, 1880–1950*, College Station, TX, Texas A&M University Press, 2007.

Evans, S. D., "Historiografía verde: estado de la historia sobre la conservación de la naturaleza en América Latina," in R. Funes Monzote (ed.), *Naturaleza en declive: miradas a la historia ambiental de América Latina y el Caribe*, Valencia, Centro Francisco Tomás y Valiente UNED Alzira-Valencia, 2008, pp. 81–96.

Grandin, G., *Fordlandia: The Rise and Fall of Henry Ford's Forgotten Jungle City*, New York, Metropolitan Books, 2009.

Grove, R. H., *Green Imperialism: Colonial Expansion, Tropical Island Edens and the Origins of Environmentalism, 1600–1860*, New York, Cambridge University Press, 1995.

Krech, S., *The Ecological Indian: Myth and History*, New York, W. W. Norton, 1999.

Livi-Bacci, M., "Return to Hispaniola: Reassessing a Demographic Catastrophe," *Hispanic American Historical Review* 83/1, 2003, pp. 3–51.

Marquardt, S., "Green Havoc: Panama Disease, Environmental Change, and Labor Process in the Central American Banana Industry," *American Historical Review* 106/1, 2001, pp. 49–80.

Marquardt, S., "Pesticides, Parakeets, and Unions in the Costa Rican Banana Industry, 1938–1962," *Latin American Research Review* 37/2, 2002, pp. 3–36.

McNeill, J. R., *Mosquito Empires: Ecology and War in the Greater Caribbean, 1620–1914*, Cambridge, Cambridge University Press, 2010.

Melville, E. G. K., *A Plague of Sheep: Environmental Consequences of the Conquest in Mexico*, Cambridge, Cambridge University Press, 1994.

Miller, S. W., *An Environmental History of Latin America*, New York, Cambridge University Press, 2007.

Pádua, J. A., *Um sopro de destruição: pensamento político e crítica ambiental no Brasil escravista,1786–1888*, 2nd ed., Rio de Janeiro, Jorge Zahar, 2004.

Pérez, L. A., *Winds of Change: Hurricanes and the Transformation of Nineteenth-Century Cuba*, Chapel Hill, NC, University of North Carolina Press, 2001.

Rogers, T. D., *The Deepest Wounds: A Labor and Environmental History of Sugar in Northeast Brazil*, Chapel Hill, NC, University of North Carolina Press, 2010.

Santiago, M. I., *The Ecology of Oil: Environment, Labor, and the Mexican Revolution, 1900–1938*, New York, Cambridge University Press, 2006.

Simonian, L., *Defending the Land of the Jaguar: A History of Conservation in Mexico*, Austin, TX, University of Texas Press, 1995.

Sluyter, A. S., *Colonialism and Landscape: Postcolonial Theory and Applications*, Lanham, MD, Rowman & Littlefield, 2002.

Soluri, J., *Banana Cultures: Agriculture, Consumption, and Environmental Change in Honduras and the United States*, Austin, TX, University of Texas Press, 2005.

Super, J. C., *Food, Conquest, and Colonization in Sixteenth-Century Spanish America*, Albuquerque, NM, University of New Mexico Press, 1988.

Tucker, R. P., *Insatiable Appetite: The United States and the Ecological Degradation of the Tropical World*, Berkeley, University of California Press, 2000.

Wakild, E., "Border Chasm: International Boundary Parks and Mexican Conservation, 1935–1945," *Environmental History* 14/3, 2009, pp. 453–75.

Watts, D., *The West Indies: Patterns of Development, Culture, and Environmental Change since 1492*, Cambridge, Cambridge University Press, 1987.

Whitmore, T. M., and Turner, B. L., *Cultivated Landscapes of Middle America on the Eve of Conquest*, Oxford, Oxford University Press, 2002.

Zerpa Mirabal, A., *Explotación y comercio de plumas de garza en Venezuela: fines del siglo XIX – principios del siglo XX*, Caracas, Ediciones del Congreso de la República, 1998.

Further Reading

Dean, W., *With Broadax and Firebrand: The Destruction of the Brazilian Atlantic Forest*, Berkeley, University of California Press, 1995.

Evans, S. D., *The Green Republic: A Conservation History of Costa Rica*, Austin, TX, University of Texas Press, 1999.

Gallini, S., *Una historia ambiental del café en Guatemala: la Costa Cuca entre 1830 y 1902*, Ciudade de Guatemala, Asociación para el Avance de las Ciencias Sociales en Guatemala, 2009.

Gerbi, A., *Nature in the New World: From Christopher Columbus to Gonzalo Fernández de Oviedo*, trans. Jeremy Moyle, Pittsburg, University of Pittsburgh Press, 1986.

Lipsett-Rivera, S., *To Defend Our Water with the Blood of Our Veins: The Struggle for Resources in Colonial Puebla*, Albuquerque, NM, University of New Mexico Press, 1999.

McCook, S. G., *States of Nature: Science, Agriculture, and Environment in the Spanish Caribbean, 1760–1940*, Austin, TX, University of Texas Press, 2002.

Miller, S. W., *Fruitless Trees: Portuguese Conservation and Brazil's Colonial Timber*, Stanford, CA, Stanford University Press, 2000.

Sonnenfeld, D., "Mexico's 'Green Revolution,' 1940–1980: Towards an Environmental History," *Environmental Review* 16/4, 1992, pp. 29–52.

CHAPTER EIGHT

The United States in Global Environmental History

ERIN STEWART MAULDIN

The borders of the United States of America encompass a wide range of landscapes – from grasslands and deciduous forests to tropical marshland and arid deserts. The breakup of Pangaea over 200 million years ago pushed North America into its current position, and the first Paleoindians traveled from Asia into the Americas perhaps as early as 14,000 years ago.[1] Later Native American groups supported themselves by managing and extracting the natural resources found within the continent's varied regions, but it was the biological exchanges of the fifteenth and sixteenth centuries that created the ecosystems and economies we now associate with the modern Americas. At the dawn of the nineteenth century, the US was a "supplier" nation, providing raw materials, such as timber and cotton, for other countries' consumption. For the last century, however, the US has possessed what Richard Tucker calls a "global ecological reach," using the natural resources of other countries – as well as its own – to sustain its power and influence.[2]

Unlike some other areas of the globe, the US has not suffered for attention from environmental historians. At least three decades of essays, edited volumes, and monographs exist on the subject; even the textbook industry has paid its respects. Many scholars have presented a far more comprehensive treatment of the nation's environmental history than could be achieved in the space allotted here. Therefore, this chapter will provide an overview of US environmental history with an emphasis on transnational subjects and areas of study that have lately attracted attention.

The Pre-Columbian Period

It is tempting to begin a history of the US with the arrival of humans on the continent. But natural processes formed the landscape of the North American continent over tens of millions of years, and factors such as plate tectonics, volcanic activity, and glaciations

A Companion to Global Environmental History, First Edition. Edited by J.R. McNeill and Erin Stewart Mauldin.
© 2012 John Wiley & Sons, Ltd. Published 2015 by John Wiley & Sons, Ltd.

were – and still are – an active influence on the region's history. The Appalachian Mountains are the result of a fold-and-thrust system of plate collision; shockwaves from the submersion of the Pacific Ocean floor beneath the continent generated the Rocky Mountains; and plate movements welded microcontinents and ancient islands to North America to form what is now the Pacific Northwest and New England (see Map 8.1).[3] Glaciation bestowed its own natural advantages onto the continent, sculpting the Great Lakes and creating the mounds of glacial till that form New England pastures. The southern US escaped the last glaciation, leaving its soils older and more weathered. Volcanic activity helped to shape the western edge of the continent; basalt lava flows formed the Columbia Plateau between 17 and 11 million years ago, and many younger volcanoes exist within the Cascade Mountain Range (including Mt. St. Helens).[4]

Long before the last ice age (the Pleistocene epoch) ended and the glaciers began to retreat, the first Paleoindians arrived. The continent was filled with cold-adapted megafauna such as mammoths, ground sloths, giant beavers, and tapirs. Historians and scientists disagree as to whether the extinction of these species toward the end of the Pleistocene was the work of humans or that of climate change; currently, the verdict seems to be that climate change altered wildlife's habitats and food supply, making them more vulnerable to human predators. Regardless of the cause, the effects were the same: North America had no large domesticable animals to provide food, transportation, and clothing. This not only kept human population numbers in check, but also gave later European invaders a decided biological advantage.[5] Although the megafaunal extinctions limited large-game hunting and livestock herding, hunting remained a significant part of native subsistence, with deer, bear, elk, bison, turkey, caribou (in Alaska), and moose replacing vanished megafauna as food sources. Fishing, too, was a large part of many tribes' subsistence, especially along the coastlines where over half of the yearly food supply came from rivers, lakes, bays, and oceans. Cultivation was difficult in some areas, such as northern New England, but eventually most Indians engaged in some type of agriculture in conjunction with hunting, fishing, and gathering.

The combination of these strategies varied greatly among the continent's tribes. Those in the southeastern US planted maize, beans, and squash – adopted from Mexico – in addition to cultivating indigenous plants such as sunflower for their edible seeds; hunting was a secondary food source. Conversely, the Indians who lived in the modern Midwest used farming as insurance against famine: most of their food came from buffalo meat.[6] Although they often exploited the landscape in ways that either mimicked natural patterns or encouraged species diversity, natives transformed the landscape to a remarkable degree. Native Americans used fire to clear land for crops, as a method in hunting, and to encourage game-animal populations in certain areas. Agricultural production necessitated the conversion of bottomland forest to cropland, increasing soil erosion. In present-day Arkansas, archeologists have discovered evidence that pre-Columbian salt production led to the clearance of hardwood forest in areas near salt brines – the wood was used to burn off the briny water, leaving behind the salt. The Cahokia site in Illinois, which supported tens of thousands of people, collapsed partly due to the overexploitation of forest resources.[7] There was no "ecological Indian" – Native Americans had a complex relationship with the natural world, sometimes characterized by unsustainable actions. The wilderness that Europeans "discovered" was largely a human construct.[8]

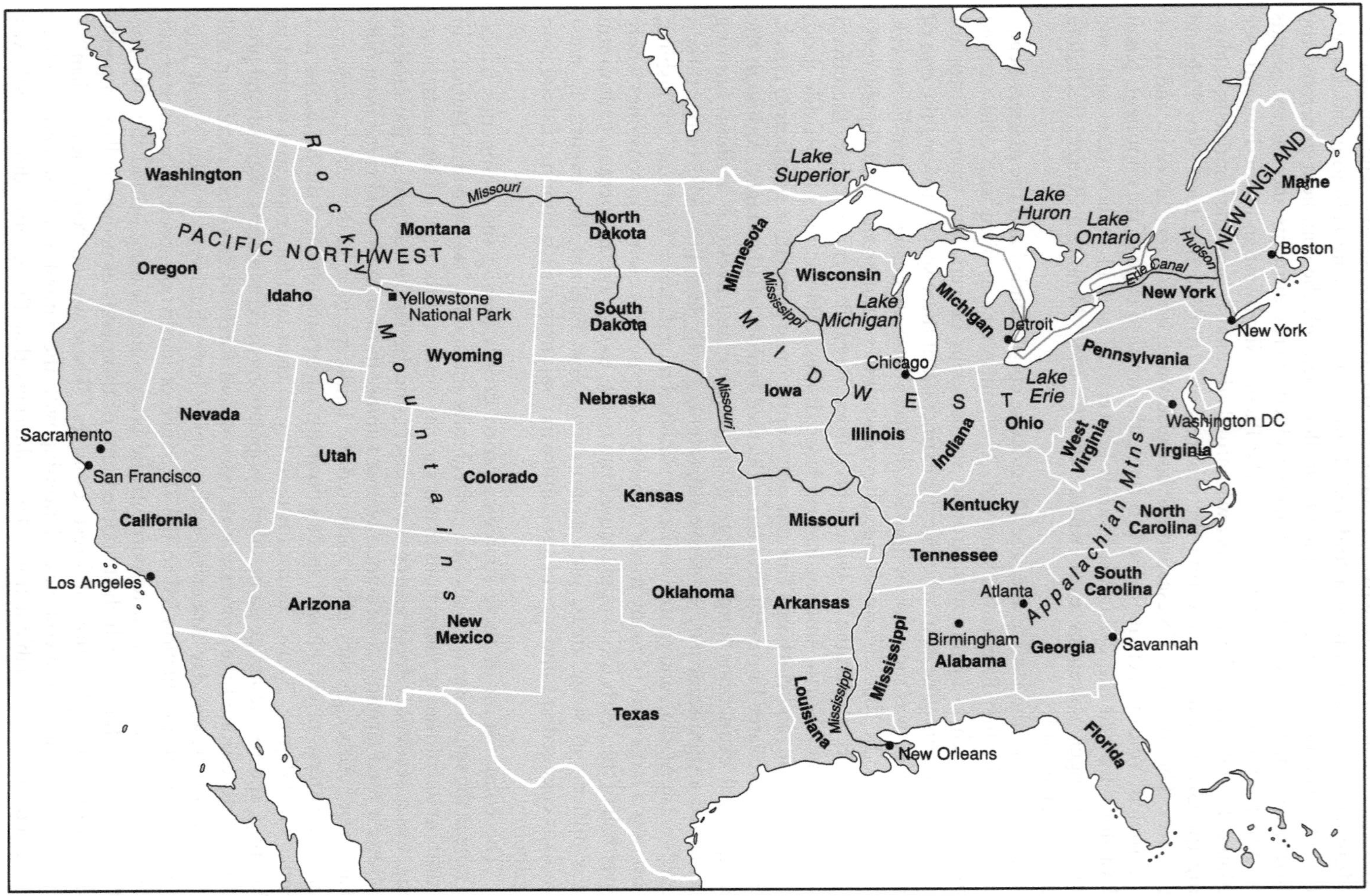

Map 8.1 The continental United States

The Ecological Atlantic

When Columbus sailed the ocean blue in 1492, he unknowingly created a link between hemispheres, civilizations, and environments which until that time had evolved in isolation. The vast networks created by expanding transatlantic empires enabled the transfer of plants, animals, and diseases from the Old World to the New, and vice versa. This process, now known as the Columbian exchange, enabled Europeans to occupy and settle the Americas more quickly. It also gave birth to an economy based on cash crops, slave labor, the fur trade, and the transfer of raw materials such as timber, hides, and fish to Europe. This new economy transformed the American landscape long after the initial wave of biological invasions.

Between the mid-sixteenth and late eighteenth centuries, Europeans from Spain, Britain, France, the Netherlands, and elsewhere carved out territories on the continent. Although North America appeared to lack the vast sources of gold and silver that Europeans hoped to find, it contained plants and animals that could easily be turned into marketable products. On the east coast, hunters coveted beavers, muskrat, raccoon, mink, fox, moose, and deer for their furs or hides. White pine, cedar, and oaks were felled and sent to Europe. The removal of these species impacted many aspects of the landscape: forest composition shifted, rivers cut deeper and curvier beds, and scavenger animals flourished while large predators decreased. But the export of the region's flora and fauna also severely limited the Indians' diet, made them more susceptible to disease, and shifted traditional gender divisions – since there were fewer animals for men to hunt, they joined native women in the fields. The English who settled New England and the Mid-Atlantic region encouraged natives to acquire livestock and fence their crops, a policy which brought other unintended consequences. American grasses and perennials, not evolved to handle livestock, gave way to European grasses and weeds. Predators, especially wolves, likely experienced a population boom thanks to the presence of livestock, an additional (and largely slow-footed) food source provided by the English. Virginia DeJohn Anderson describes how cattle and hogs became "weapons" in the colonial struggle for land, an advance guard for English settlers. Indeed, livestock did become a significant source of friction between Native Americans and European settlers, and, in at least one instance, led to war.[9]

In the Great Lakes region and upper Mississippi Valley, French traders worked alongside Iroquois and Algonquin peoples in a lucrative fur-trading enterprise, one that depleted stocks of beavers and other wildlife from the northern US and Canada. The French did spread disease among natives which led to cultural and religious shifts within the Indian populace, but their presence was not as transformative (in the short term) to the landscape as that of their British counterparts.[10] The Spanish, who claimed the lands which make up California and the southwestern states, were experienced pastoralists. Confronted with arid lands, many of which were not suited for Mediterranean crops, the Spanish who settled in America used ranching in combination with agriculture to reduce reliance on imported food. The Spanish did not settle in large numbers, but the destruction of the land and native plant species through overgrazing pushed Indians into smaller and smaller parcels of uncultivable land, and, ultimately, forced them to rely on Spanish missions and haciendas for employment and food.[11]

The triumph of the European way of life over that of the Native Americans was often in doubt, and the timeline of the transition from Indian dominance to European dominance was very different from region to region. Sometimes the introduction of an Old World species strengthened Native American influence and power. The adoption of the horse by some of the Plains Indians made running down buffalo and enemies alike much easier, and they eclipsed native agricultural tribes as the dominant force in the modern Midwest. The horse also buoyed the fortunes of the Choctaw, who lived in the lower Mississippi Valley. When the deerskin trade collapsed, the horse allowed the Choctaw to conduct highly profitable raids on neighboring tribes and European settlements.[12] In other instances, the European rejection of native practices endangered settlers' lives. Carville Earle argues that mortality in the Jamestown settlement was largely due to water-borne illnesses, which local Indian tribes avoided by moving upland during the summer. Europeans considered this semi-nomadism barbaric, and they suffered for it. For the most part, however, the biological invasions of the New World by the Old gradually eroded the traditional lifestyle of native peoples.[13]

The shift from Indian to European dominance of the American landscape and the subsequent commodification and, in some respects, biological impoverishment of that landscape are two of the major themes in US environmental historiography. But the expansion of empires during this period engendered similar transformations around the Atlantic basin, and too often environmental histories of the US ignore the nation's place within this larger "ecological Atlantic." Some global environmental histories, such as John Richards' *The Unending Frontier*, and more specific studies, such as Marcy Norton's *Sacred Gifts, Profane Pleasures*, explore the Atlantic connections that impacted the colonial American landscape.[14] Environmental historians of the US, however, have been slow to take up these threads of transatlantic inquiry.

The Early Republic and Antebellum Periods

The American Revolution, fought from 1775 to 1781, and the subsequent adoption of the Constitution of the United States of America receive few mentions by environmental historians. This is perhaps because these events, while major watersheds in US, Atlantic, and world history, did little to alter the ways in which the landscape of the North American continent was viewed, manipulated, or managed. The 70 years that followed, however, transformed the US from a nation of farmers on the eastern edge of a continent to an industrial power that controlled what is now the contiguous 48 states. These immense changes were generated by new technologies, urbanization, aggressive wars of expansion, and the labor of immigrants and enslaved Africans.

In the 1790s, the nation was very new, but many towns and villages were over 100 years old and suffering from population pressure. The problem was especially acute in New England. Average farm size plummeted due to the scarcity of land, population growth, and the vagaries of inheritance law. Soil exhaustion and the disappearance of meadowland forced down agricultural yields. At the same time, the climate became particularly unpredictable: this period marked the transition away from the Little Ice Age to warmer temperatures, and volcanic activity around the world impacted US harvests in 1789, 1811, 1816, and 1818. In response, many New Englanders turned to market-oriented agriculture to protect themselves against famine. They focused less on products that would feed their families, and more on products that would make them money – cattle, hay, wood, and wheat.

This shift toward commercial agriculture in the northeastern US is often termed the "market revolution" by US historians, although farmers who lived near port towns had been producing products for international markets since the seventeenth century. Other labels for this critical period might include the "transportation revolution," or even an "agricultural revolution." The completion of the Erie Canal in 1829, the invention of the cast-steel plow in 1837, and the spread of railroads enabled grain to travel from the fertile farms in the Northwest Territory (now the upper Midwest) to the population of the Northeast who were experiencing the first stirrings of the American Industrial Revolution. Without revolutions in agriculture and transportation, there would have been no revolution in industry.[15]

The industrialization of US manufacturing began around the 1820s. Spurred on by the removal of British trade restrictions and the exodus of many young people from farming, textile mills and blast furnaces popped up along the rivers of New England. Textile factories needed river water to run their machines, and constructed elaborate systems of dams and canals to harness it. Not only did the dams block the migration of fish upstream, destroying their spawning grounds, but the appetite of factories for energy led to the packaging and selling of (free-flowing) water for the exclusive use of a corporation. Both outcomes put textile companies at odds with local farmers and fishermen who depended on rivers for income and food, and timber companies who needed the rivers for the transport of lumber. The disputes that ensued over the ownership of river water in New England played out again and again as industrial capitalists claimed the rights to natural resources, displacing subsistence farmers and smaller landowners.[16]

The burgeoning industrial system was also fueled by the further acquisition of territory west of the original 13 states. But acquisition and occupation were separate processes: in order to settle its new lands, the US had to extinguish competing claims from both Indian and European nations. The Creek War of 1812 and the three Seminole Wars (1814–56) secured the areas that would become the Deep South and eliminated the Spanish presence from Florida. The Indian Removal Act of 1832 forced the migration of thousands of American Indians from their homes in the southeastern US across the Mississippi River. And the Texas Revolution (1836) and subsequent US–Mexican War (1846) saw the Mexican cession of the far west. These territorial conflicts receive attention from historians because of their political and social ramifications, but there were ecological components as well. Environmental factors driving expansion included increased population pressure, the scarcity of new land, depletion of natural resources, soil exhaustion in older areas, and the persistent belief that Native Americans had failed to subdue or properly improve the land.

Environmental factors also contributed to the outcome or timelines of these wars. The Second Seminole War, for instance, has been called "America's first Vietnam" because the swampy terrain of Florida – with its razor-like sawgrass, endemic malaria, and deadly reptiles – served to both weaken US troops and shelter the opposing Seminoles, who essentially fought a guerilla war. The conflict dragged on from 1835 to 1842, costing more than any other Indian war undertaken by the US, and led many soldiers to wonder why their government fought so hard for a such a marshy, "God-abandoned hellscape."[17] Another example comes from the 1830s, when buffalo hunting in the southern plains came under increased pressure because of drought (a normal part of life in the plains) and the influx of other Indian groups into the area, who had been pushed west by US removal policies. This constituted an ecological and economic crisis for the Comanche, who responded by raiding and plundering Mexican villages in present-day Texas. These raids caused Mexicans to largely abandon the territory and invite American settlers to move in – thus sparking the Texas Revolution and, later, the US–Mexican War.[18]

Slavery and African-American Environmental History

While industrialization and the depletion of natural resources drove expansion in the North, the growth of slavery and the exhaustion of soil was the motive for territorial acquisition in the South. Although slavery existed in most regions of colonial America, by the late eighteenth century, the institution was largely limited to the coastline south of Maryland. Slaves lived and worked on rice and tobacco plantations, and the economic decline of these areas led many to believe that slavery would "naturally" die out as an economic system in the US. But the Haitian Revolution, which created a demand for US sugar, and the invention of the cotton gin, breathed new life into the institution of slavery, powering the economic and territorial growth of the South throughout the antebellum period.[19] Although most southerners did not own slaves, the slave-based plantation complex became a dominant ecological regime in that region.

Slaves made biological, cultural, and, of course, economic contributions to American society. It was slaves' labor that supported the agricultural and shipping sectors of the southern economy, in addition to supplying northern and British textile industries with cotton. Their agricultural knowledge proved indispensable to the success of certain crop regimes, rice in particular. Slaves also contributed to the region's foodways by planting West African food crops such as okra, benne seed, yams, and guinea corn (sorghum). They also inadvertently introduced West African diseases such as malaria, yellow fever, yaws, and dengue.[20] While slaves helped shape the natural landscape, the landscape, in turn, helped to shape their lives. It left marks on their bodies, provided them with sustenance and a measure of independence, and dictated their daily tasks. Despite the wealth of subject matter, slavery occupies a peculiar place in US environmental historiography. It is mentioned often, but almost never treated on its own, although it influenced human interaction with the environment for hundreds of years and provides historians with a transnational issue that binds the US economy with plantation systems throughout the Americas. There are exceptions, but in many works on this period, slaves simply move about in the background, clearing forests and building fences, planting cotton and stripping tobacco.

Recently, scholars such as Mart Stewart and Dianne Glave have begun to raise interesting questions as to the place of African-Americans in US environmental history. If there are only a handful of works on slavery and environmental history, even fewer have been published about blacks after the Civil War.[21] But African-Americans comprised the bulk of agricultural workers and sharecroppers between the Civil War and the 1960s, contributed to the growth of industrial cities such as Chicago and Detroit, and participated in Progressive-era sanitation and beautification movements. They have sometimes experienced the forces of American history differently from whites, and perhaps African-American environmental history might have different contours from the basic narrative used in overviews like this one. Would their story start in Africa, or colonial America? What events would be emphasized, since things like the market revolution matter less for African-Americans? Would the story of the late-nineteenth-century preservation movements disappear altogether? It is important to consider the ways in which the story of US environmental history changes when we shift our focus to other groups – not only blacks, but women and immigrants as well.

The Civil War Era

The plantation system needed vast amounts of land to be profitable, and politicians of both the North and the South were avid expansionists – they feared that, without the continual addition of new lands, slavery would collapse. Ironically, it was the incorporation of new states in the West that brought sectional conflict to a head, caused 11 states to secede from the Union, and sparked a four-year civil war that led to the demise of slavery. The Civil War (1861–5) and years of Reconstruction (the military occupation of the South by the Union army ending in 1877) left social, political, and ecological legacies that drastically altered the landscapes of the US. Although some historians have argued that the war had few long-term consequences,[22] the processes initiated during the Civil War set the stage for many of the major environmental shifts of the late nineteenth century: federal irrigation projects, the cattle boom, urban sanitation efforts, development of the southern timber and mining industries, and the spread of sharecropping.

Most of the Civil War was fought in southern states, thus, the Confederacy suffered more physical destruction than other regions. Soldiers needed wood to cook with, to erect their tents, and to keep themselves warm, so they used whatever timber was at hand – fences, outbuildings, woodland, shrubs, even houses. After the war, when the Bureau of Refugees, Freedmen, and Abandoned Lands (Freedmen's Bureau) had to account for abandoned or confiscated plantations throughout the South, many of their entries note, "timber cut by troops" in the margin.[23] Timber was also needed for the construction of defenses and fortifications, and heavy artillery occasionally sparked wildfires that consumed entire forests.[24] The war also accelerated land abandonment in many areas. Of the thousands of acres of farmland abandoned during the war, most were reforested with pine trees or their soil was eroded away – both outcomes were inconvenient and costly to remedy for postwar farmers. Forage raids severely reduced the region's stocks of animals and foodstuffs. Some campaigns – such as Sherman's March to the Sea and the Shenandoah Valley campaign – were particularly destructive, making provisions and livestock very expensive.[25]

In the years immediately following the war, military occupation, a heavy tax burden, and the emancipation of slaves jarred the region's traditional economic systems, and the ecological processes initiated by the war – such as increased soil erosion and the removal of livestock – only amplified these problems. Many farmers – white or black – could not afford basic agricultural necessities, so they turned increasingly to the most profitable option: cotton. But growing cotton is a risky business, and the land often did not produce the profits that farmers had gambled on; subsequently, they became entangled in the cycle of debt and tenancy that became a hallmark of the late-nineteenth-century South.

Northern states suffered little physical damage.[26] Instead, the war greatly expanded the importance of manufacturing to the northern economy, fueling the twin forces of industrialization and urbanization. Over 1.5 million men served in the Union army during the Civil War, and they needed rations, uniforms, shoes, and guns. Timber, especially, was in great demand for everything from railroad ties to coffins. The South, of course, needed the same supplies, but many of the factories, iron foundries, and textile mills there were destroyed by the end of the war. Technological innovations of the 1850s, such as the sewing machine or mechanical harvesters and binders, helped northern industry and agriculture maintain output despite the severe labor shortages created by conscription and casualties. Changes in the packaged-food industry, too, aided the war effort. Canned foods were introduced during the war, and condensed milk in particular became a valuable ration in the Union army because of its high caloric content.[27] The manufacture and transport of the food required by the federal war machine also greatly increased the economic importance of the Midwest, especially Chicago. The Union blockade of southern ports prevented western farmers and Texas cattle breeders from reaching their usual markets in southern states, so they sent their pigs and steers to Chicago to be slaughtered. At the same time, the demand for meat skyrocketed as a result of orders from the Union army. Due to the confluence of railroads in Chicago and the disruption of Mississippi River trade by warships, Chicago's stockyards and slaughterhouses became the most popular destination for Midwestern pigs and steers. Between 1860 and 1863, Chicago's meat-packing industry grew sixfold and, by 1870, Chicago had surpassed Cincinnati and St. Louis to become the largest meat-packing city in the world.[28]

The logistics of the Union war effort required a government and infrastructure strong enough to quickly and efficiently command the natural resources of the land. Thus, legislators passed a series of acts which aimed to consolidate power within the federal government and create a cohesive national economy. Between 1861 and 1866, Congress established a national currency, authorized the backing of war bonds with paper money, freed the slaves, and established the first national social-welfare bureau. It also ordered the building of a transcontinental railroad on government-owned land, and promised 160-acre homesteads to any settler who would settle on public lands for five years. Gone was the old antebellum republic comprised of a loose confederation of states – the America that emerged from the Civil War was one of heavy industry, big business, and increasingly capital-intensive agriculture.[29]

Industrialization and Ecological Conflict

In the decades following the Civil War, the US experienced remarkable economic growth fueled by an abundance of domestic natural resources and the influx of foreign capital and labor. Railroad transportation transformed every region of the country, agricultural innovation helped to improve the quality and variety of peoples' diets, and new technologies such as barbed wire, the typewriter, and the telephone heralded vast changes in American culture. "Progress," however, came at a cost to the American landscape. By the turn of the twentieth century, the nation's forests, wildlife, and supply of arable land – once seen as limitless – now looked decidedly limited. Governments, corporations, and citizens fought for access to a fast-dwindling set of natural resources. Concerned for the continued growth of the US economy and fearful over future shortages, some Americans began to advocate the conservation and preservation of domestic resources. Others, however, looked to the acquisition of foreign ones.

After the war ended in 1865, southern entrepreneurs desired to remake their economy in the image of the victorious industrial North. But the South was capital-poor – its antebellum wealth had been tied up in land and slaves – so industrialists searched for ways to extract natural resources for quick economic gains. Timber was one of the few products in which investors had any interest: the white-pine stocks of northern and Great Lakes forests were depleted by this point, and southern pine made an adequate substitute. To reach the timber, however, additional railroads had to be built, and to build the railroads, new supplies of coal and iron ore were needed. The subsequent expansion of industry in the South – created to remove, transport, and process natural resources – redrew the region's urban landscape. Small towns such as Atlanta, Georgia, and Birmingham, Alabama blossomed into thriving metropolises because of their proximity to mines or railroads. Other cities, such as Savannah, Georgia, whose industry once thrived on the plantation complex, were repurposed to process and ship the vast amounts of longleaf pine being cut from southern states.[30] The timber and mining industries also had a transformative effect on the far West. Lumbermen swarmed the forests of the Northwest where they harvested the enormous Douglas fir and the nearby (still more enormous) redwoods of California using mechanical donkeys and railroad cars. And between the 1849 California gold rush and the Black Hills gold rush of 1875, the discovery of gold, silver, lead, copper, and iron-ore deposits drew thousands of hopeful prospectors and transformed the outposts of the "wild West" into major urban centers.[31]

Industrialization touched rural areas as well, in the form of technological and organizational innovations. In states such as Illinois and Iowa, livestock farmers realized that feeding corn to their animals would maximize the efficiency of meat production. So instead of grazing in pastures, cows and hogs were kept in enclosures and fed a diet of corn. This method, dubbed the feedlot system, provided an outlet for excess grain production and solved the problem of maintaining grazing land. At the same time, assembly- (or disassembly-)line techniques for breaking down animals spread through meat-packing plants: animals could now be separated into their constituent parts within seconds, and the refrigerated railroad car could transport them already dressed and packaged to consumers.[32]

In California, the development of the irrigation pump gave farmers access to immense amounts of water stored in underground reservoirs and helped turn the state's Central Valley into the largest fruit-growing region in the country. Railroads enabled growers to transport fragile produce quickly and effectively, and the arrival of legions of East Asian (and, later, Hispanic) agricultural laborers turned small orchards into large agribusinesses. By the 1930s, an 800-mile "fruit belt" stretched the length of the state, growing plums, peaches, pears, grapes, apricots, oranges, and cherries, all expected by the urban consumer to fit a standard of quality and appearance. Thanks to the railroad, feedlot, and irrigation systems, the old seasonal calendar of production was broken: consumers could buy fruit, vegetables, and meat at any time of year, from anywhere in the country. Many Americans began to think of their food not as something they made, but as something they bought in a store.[33]

The US' resource-intensive industrial expansion inevitably produced significant amounts of waste and pollution. In 1870, 60 percent of Sacramento, California's deaths were from tuberculosis and other lung diseases caused by the "heavy clouds of poisonous metals" emitted from the city's 221 iron foundries. Birmingham residents could not hang their clothes outside to dry because they would collect coal dust from steel plants. In their hurry to "cut out and get out," timber companies left vast areas of clear-cut land, increasing erosion and flooding. Hydraulic mining required billions of gallons of water to expose mountain deposits, while the slag and solid waste produced by this process fouled rivers, flooded farmlands, and destroyed fish-spawning grounds. And despite their efforts to utilize the whole animal, meat-packers continued to dump animal waste and unused parts into cities' water supplies.[34] The obviously wasteful practices of timber and mining companies, the rapid disappearance of regional resources, and the state of urban living conditions galvanized interest at both local and federal levels for the conservation, regulation, and preservation of the American landscape.

Regulation of the use of the environment had existed in the US as far back as the seventeenth century. Most early conservation laws sought to regulate hunting and fishing – for instance, through the creation of a hunting "season" – although many restrictions were widely ignored. Agricultural reforms such as contour plowing, terracing, and replenishing soil nutrients also experienced brief periods of popularity throughout the eighteenth and nineteenth centuries, even among slave-holding plantation owners (although less out of respect for the land and more out of respect for profit).[35] At the end of the nineteenth century, however, those local conservation measures were institutionalized at the federal level. Yellowstone, created in 1872, was the first national park in the world. During the 1890s, Congress passed the Forest Reserve Act and the Forest Management Act, allowing the government to set aside forest land and manage grazing and timber operations on public lands. The Newlands Act, passed in 1902, gave the federal government the authority to control and conserve water resources in the West. Cities passed "anti-smoke" ordinances aimed at controlling the quantity and timing of industrial plants' emissions.[36]

These acts were designed not to prevent the extraction of natural resources or the growth of industry, but to increase the efficiency of these processes. The conservation movement sought to apply scientific expertise to the long-term maintenance of forest, water, and wildlife resources, exemplified by Gifford Pinchot's efforts with the US Forest Service. There was, however, a contingent of Americans led by John Muir and the Sierra Club who advocated the preservation of nature for its innate, spiritual value. Preservationists believed that "wilderness" – usually thought of as spectacular, sparsely populated mountain landscapes in the West – should not be used at all for human ends.[37]

Growing awareness and action regarding environmental issues around the turn of the twentieth century was part of a larger reform movement called Progressivism. Responding to the problems created by industrialization and urbanization, Progressive initiatives included settlement houses, temperance, women's suffrage, putting an end to child labor, and combating political corruption, in addition to causes placed under the umbrella of environmentalism: conservation, municipal garbage collection, the creation of urban parks, street paving, and the expansion of sewage systems and waterworks. Although conservation is usually associated with Theodore Roosevelt's masculine "strenuous life" philosophy, both white and black women found a voice in the Progressive conservation movement, joining gardening clubs and fighting for better urban living conditions.[38]

Interestingly, environmental historians have not yet made the connection – if there is one – between Progressive causes such as the development of forestry management and the revival of wildlife-protection laws on the one hand, and Populist agitation on the other. During the late nineteenth century, farmers in the West and South fomented a protest movement resulting in the founding of a political party, the Populists, which enjoyed considerable, albeit brief, success at local and national levels. But the social and economic transformations historians use to explain this widespread agrarian protest had ecological components as well. Steve Hahn argues that, in the South, fence laws and restrictions on hunting and gaming made it more difficult for land-poor people to keep livestock and to supplement their diets with game. Although Hahn's argument hinges on these farmers' adherence to a "republican producer ideology," the legislation they were fighting had very specific environmental causes. For instance, farmers in the West joined Populist groups not only because they disliked railroad "middle men" who overcharged them on freight, but also because of a drop in agricultural prices, competition with ranchers, feedlots for land, resentment over federal control of water and timber, and the increasingly poor soils they had to plant. Environmental historians of Africa have noted that, in several instances, government or colonial imposition of conservation schemes led to peasant resistance. More might be done to investigate the links between Populism and the ecological conflicts which resulted in the first wave of US environmentalism, and how they compare to similar responses across the globe.[39]

The American Century

While the conservation and preservation movements in the US sought to set aside natural resources for future generations, the environmental and economic shocks of the 1890s led others to look overseas for sources of natural capital. Beginning with the acquisition of overseas territories during the 1890s and accelerating greatly after World War II, the American military–industrial complex, consumer culture, and insatiable appetite for energy have exacted an enormous environmental price around the world. Those same forces, however, gave birth to the modern environmental movement, culminating in the founding of Earth Day, the establishment of the Environmental Protection Agency, and legislation such as the Clean Air and Clean Water Acts. The greening of American culture, however, has done little to slow the nation's exploitation of foreign natural resources, despite the country's declining political influence in the early twenty-first century.

In the 1890s, with the continental US secured and settled, Americans turned outward to find new markets for industrial products and sources of raw materials. The "opening" of Japan in 1853 had failed to give the US unrestricted access to Asian consumers, and the 1867 purchase of Alaska had only recently begun to show economic potential.[40] After its victory in the Spanish–American War of 1898, however, the US acquired an overseas empire: it controlled Guam, Puerto Rico, the Philippines, Hawai'i, and Cuba (briefly). American capital, technology, and military began to transform the economies and ecologies of these islands through the expansion of cash-crop farming and the construction of new ports, bases, and fortifications. US investment was not limited to Pacific and Caribbean islands, however, and interest in developing the resources of Latin America for US markets soared after 1898. American companies became involved in the production and marketing of sugar, coffee, and bananas, encouraging the expansion of capital-intensive plantations, deforestation in tropical lowlands, and the displacement of indigenous agriculture. The logging of Central American hardwoods also increased, as did the demand for rubber, copper, and, perhaps most importantly, petroleum.[41]

When oil was first discovered in the US in the 1850s, petroleum was mostly used as a lubricant for railroad cars and as an illuminant, in the form of kerosene. Although the US possessed enormous domestic reserves of petroleum, the refinement of the internal combustion engine and the invention of the automobile and airplane around the turn of the twentieth century made petroleum a key strategic resource. American firms began to seek out foreign oil fields in Latin America and, later, in the Middle East, Southeast Asia, and Canada. The environmental costs of the reliance on oil-based products have been well covered in environmental history scholarship, but the process of extracting oil was a dirty business too, and inflicted horrific wounds on local landscapes. Camps, pipelines, and roads had to be constructed, displacing people from their homes and causing deforestation. Millions of gallons of fresh water were used to keep crude moving through the pipes. Immense fires broke out, emitting toxic fumes that killed plant life for miles around, and, and, of course, oil spills often occurred, harming wildlife and polluting water-ways.[42] Although US oil companies, led by Rockefeller's Standard Oil, began to exploit foreign oil reserves as early as 1900, it was not until the second and third decades of the century that the demand for oil caught up with the supply. During these decades, Henry Ford's mass-production techniques made the automobile affordable to working-class Americans. By 1929, 50 percent of American families owned an automobile. Meanwhile, the US military began the transition from coal to oil as a fuel source, and the integration of oil-fueled machines into World War I – such as the submarine, airplane, and tank – intensified the military's reliance on petroleum.

The acceleration of American industrial production and demand for foreign natural resources created by World War I was significantly dampened by the global depression of the 1930s. Environmental historians tend to focus discussions of this period on the spec-tacular natural (and man-made) disasters such as the decade-long drought in the West and the Dust Bowl, as well as scientists' and governments' responses to them.[43] However, the Great Depression and Franklin Roosevelt's efforts to alleviate it also created vast, federally funded projects such as electrification of the rural South through the Tennessee Valley Authority and the erection of dams in the far West. These projects were intended to improve the lives of Americans, provide employment, and prevent drought and flood-ing, but unsurprisingly carried unforeseen environmental consequences. For instance, in 1937, Congress authorized the Central Valley Project in California, resulting in the construction of four dams and four canal systems. Intended to insulate farmers from drought and alleviate pressure on the state's aquifers, the project actually accelerated the depletion of water reserves: the availability of inexpensive water fueled agricultural expansion, so demand soon outstripped supply.[44] Another example comes from the South, where national concern over soil erosion encouraged the importation of exotic plants to help restore vegetative cover on farms – the most infamous of these exotics is kudzu, a perennial vine native to Japan and China. The federal government distributed millions of kudzu seedlings throughout the southeastern US, but they grew far faster than expected. Called "the vine that ate the South," kudzu expands its botanical empire by thousands of acres annually, smothering native plants, draping buildings, and cover-ing roadsides.[45] There were, however, Depression-era programs which had more amelio-rative effects on the American landscape: the Civilian Conservation Corps built terraces for erosion control, planted trees, and created public parks and recreational spaces.

The outbreak of World War II reignited the US race for raw materials both at home and across the globe. Although the US war machine consumed oil, rubber, timber, min-erals, and food products at an alarming rate, the war effort had other, more long-term ecological impacts. For example, World War II drained Midwestern farms of laborers while demand for food spiked, so farmers began to use the newly developed hybrid corn seeds which eliminated the need for pollination and dramatically improved crop yields. In the decades after the war, the US exported hybrid seeds of all types to countries such as Mexico, India, and the Philippines, sparking green revolutions around the world and increasing the world's food supply. The export of seeds, however, also entailed the export of American-style industrial agriculture, which relied heavily on inputs of water, technol-ogy, and artificial fertilizers to succeed.[46]

But perhaps the legacy of World War II with the most far-reaching impact was the development of the atomic bomb and the nuclear arms race it sparked. The production, testing, and use of these bombs have released a multitude of radioactive, cancer-causing elements into biomes, landscapes, and communities around the world beginning in

1945. In the US alone, it will take billions (possibly trillions) of dollars and almost a century to contain the immediate contamination from bomb production and testing. The radioactive by-products of the arms race will last for tens of thousands of years, and its true cost is impossible to estimate.[47]

Environmentalism and Environmental Justice

Although the immediate post-World War II years were difficult for the US economy, the 1950s and 1960s were a time of exuberant economic expansion and growing affluence. Government spending, which had increased dramatically during the Great Depression, remained high through the postwar period; public funds built schools, provided benefits to GIs, and constructed the nationwide interstate system of highways. Military spending also remained elevated due to American involvement in the Korean War and Cold War. The birth rate skyrocketed, generating a demand for new homes, automobiles, consumer products, roads, and public services – all of which stimulated the economy.[48] The new prosperity fueled a culture of consumption and convenience, one in which the ability to buy things – cars, TVs, hamburgers – was increasingly associated with the quality of life. Americans eagerly adopted new products such as garbage disposals and dishwashers. Disposable paper towels and napkins became essential household items. Developments in the petroleum-based plastics industry encouraged the use of plastics in everything from grocery bags to furniture. The rise of suburban sprawl and intercity highways meant that almost every family needed at least one car to get around. This led to a car-centered infrastructure of shopping centers, fast-food restaurants, and drive-through bank-teller windows. During these same decades, however, concern grew over the gap between the affluence enjoyed by Americans and the impoverishment of the environment, the "imbalance between private wealth and public poverty."[49] A vast array of issues – the nuclear arms race, Americans' rampant consumption of materials, the environmental costs of suburban development, urban pollution, the degradation of national parkland – coalesced within the countercultural movement of the 1960s, giving rise to the modern environmental movement.

There were many points of origin for the environmental movement involving all sectors of society. The science of ecology, which came of age during this period, provided hard data as to the impact of human development on landscapes. Politicians and public thinkers such as Lyndon Johnson and John Kenneth Galbraith sought to safeguard Americans' quality of life by using government programs to protect and improve the environment. The 1962 publication of Rachel Carson's *Silent Spring* shocked readers with its descriptions of the dangers of DDT and other pesticides; the book showed Americans how their desire for perfect fruit, green lawns, and pest-free gardens had endangered not only the landscape and its wildlife, but their own lives. Housewives and homeowners became increasingly vocal about the dangers of suburban development, such as the lack of open space, increased soil erosion, and the disappearance of woodlands and wildlife populations. The Sierra Club and other wilderness advocates mobilized resistance against the damming of the Green River and flooding of Echo Park in Colorado and Utah. News coverage of America's involvement in the Vietnam War produced haunting images of widespread environmental destruction, galvanizing interest in environmental causes among student war protesters.

These threads of environmental awareness, protest, and activism culminated in the first national Earth Day teach-in on April 22, 1970, in which over 20 million people participated.[50] This outpouring of public concern put legislators under enormous pressure to craft policies to tackle environmental issues. As a result, the 1970s were a landmark decade in environmental legislation. Beginning with the National Environmental Policy Act of January 1970, and the establishment of the Environmental Protection Agency, more than a dozen environmental laws were passed during that decade: the Clean Air Act, the Water Pollution Control Act, the Endangered Species Act, Energy Policy and Conservation Act, the Superfund Law, and others transformed national policy toward the environment.

The political and legal victories of the environmental movement were significant, improving water and air quality across the country and regulating corporate activity and practices. But, as Andrew Hurley argues, "the age of ecology was also the age of environmental inequality."[51] The mainstream environmental movement, characterized primarily by white middle-class members of organizations such as the Sierra Club, largely failed to address the issues affecting working-class and minority Americans, especially those in the decaying urban areas of the country. Scholars debate the reasons as to why so few African-Americans joined the environmental movement: Robert Bullard notes that blacks were busy fighting for voting rights and social equality at the time and felt that environmentalists were leaching political will away from "important" issues, while Cassandra Johnson and Josh McDaniel contend blacks' "collective memory of land-based labor" reduced their interest in nature outdoor activities and, thus, environmentalism.[52] Jeffrey Stine suggests that often corporations obliged poor Americans to take on additional environmental risks – air pollution, industrial waste, superfunds – in exchange for the promise of jobs. Environmental inequality is an issue in rural areas as well. In Appalachia and other areas, strip-mining destroyed farms and forests, and polluted waterways. Chad Montrie explains how those affected by mining, working-class whites, took coal companies to court, staged protests, and even engaged in ecoterrorism – simply to protect their private property. Once the call to ban strip-mining went national, however, environmental groups based far from the areas affected were too quick to compromise, trading the health, livelihoods, and land of Appalachia residents for a political victory.[53]

Conclusion

Since the 1970s, environmental regulations have positively impacted American lives and the American landscape: air quality is measurably better, municipal water supplies are cleaner, industrial waste has been reduced, and harmful substances have largely been eliminated from a wide range of food and manufactured products. Technology has made everything from household appliances to cars to new buildings more energy efficient. The organic- and local-food movements – although decidedly limited in their impact and practicality – have helped to raise awareness as to the environmental costs of the American food industry, and, for better or for worse, "green" and "eco" are trendy, marketable adjectives for almost anything.

Despite whatever progress has been made, however, the US is no closer to solving the big questions of our global ecological future. Climate change is, of course, one of the biggest questions, and both the US government and a portion of the US population have refused to confront the issue, or even admit it is an issue. The fear that increased emissions regulations will hinder economic growth is pervasive and unlikely to dissipate. The energy sector is still dominated by oil and coal, and, although rising gas prices hint at the end of cheap energy, too much of the US economy is tied to fossil fuels for alternatives to gain traction. Water supply is an increasingly contentious issue, not only in western states, but all over the country; even states such as Alabama and Georgia, which normally receive more than adequate rainfall, are currently in legal battles over reservoir and lake access. And it is increasingly clear that, as the economy becomes more globalized, the environmental impacts of US corporations and agribusinesses are merely being shifted to countries across the border, across the ocean, or simply downstream.

Notes

1 The arrival date of humans in the Americas is still debated; most agree that a group of humans called the Clovis people established themselves on the continent between 13,000 and 11,000 years ago. See T. Silver, *A New Face on the Countryside: Indians, Colonists, and Slaves in the South Atlantic Forests, 1500–1800*, Cambridge, Cambridge University Press, 1990, and T. Steinberg, *Down to Earth: Nature's Role in American History*, Oxford, Oxford University Press, 2002.

2 R. P. Tucker, "The Global Ecological Reach of the United States: Exporting Capital and Importing Commodities," in D. C. Sackman (ed.), *A Companion to American Environmental History*, Oxford, Wiley-Blackwell, 2010, pp. 505–28.

3 G. G. Whitney, *From Coastal Wilderness to Fruited Plain: A History of Environmental Change in Temperate North America from 1500 to the Present*, Cambridge, Cambridge University Press, 1994, pp. 40–1; A. M. Sowards, *United States West Coast: An Environmental History*, Santa Barbara, CA, ABC-CLIO, 2007, p. 2.

4 A. Cowdrey, *This Land, This South: An Environmental History*, Lexington, KY, University Press of Kentucky, 1983, pp. 2–3; Steinberg, *Down to Earth*, pp. 5–6; Whitney, *From Coastal Wilderness to Fruited Plain*, pp. 42–6; Sowards, *United States West Coast*, pp. 3–5.

5 Steinberg, *Down to Earth*, p. 13; A. Crosby, *Ecological Imperialism: The Biological Expansion of Europe, 900–1900*, Cambridge, Cambridge University Press, 1986. See also McNeill (Chapter 2) in this volume.

6 Silver, *A New Face on the Countryside*, pp. 36–9; A. Isenberg, *The Destruction of the Bison: An Environmental History, 1750–1920*, Cambridge, Cambridge University Press, 2000.

7 E. Peacock, "Historical and Applied Perspectives on Prehistoric Land Use in Eastern North America," *Environment and History* 4, 1998, pp. 1–29 (pp. 11–13). See also N. H. Lopinot and W. I. Woods, "Wood Overexploitation and the Collapse of Cahokia," in C. M. Scarry (ed.), *Foraging and Farming in the Eastern Woodlands*, Gainesville, FL, University Press of Florida, 1993, pp. 206–31.

8 Sowards, *United States West Coast*, pp. 24, 27; S. Krech, *The Ecological Indian: Myth and History*, New York, W. W. Norton, 1999, p. 212. Peacock, in "Historical and Applied Perspectives," also provides an excellent summary of both archeological and historical literature debating the environmental impacts of pre-Columbian American Indians.

9 King Philip's War (1675–6), fought around Massachusetts Bay. See V. D. Anderson, *Creatures of Empire: How Domestic Animals Transformed Early America*, New York, Oxford University Press, 2004. For a strong parallel in colonial Mexico, see E. G. K. Melville, *A Plague of Sheep: Environmental Consequences of the Conquest of Mexico*, Cambridge, Cambridge University Press, 1994; and for a similar occurrence in New Zealand, see Crosby, *Ecological Imperialism*.

10 R. White, *The Middle Ground: Indians, Empires, and Republics in the Great Lakes Region, 1650–1815*, Cambridge, Cambridge University Press, 1991.

11 Melville, *A Plague of Sheep*, chapter 5; S. W. Hackel, *Children of Coyote, Missionaries of Saint Francis: Indian–Spanish Relations in Colonial California, 1769–1850*, Chapel Hill, NC, University of North Carolina Press, 2005, chapter 3.

12 Isenberg, *The Destruction of the Bison*, pp. 39, 61; J. T. Carson, "Horses and the Economy and Culture of the Choctaw Indians, 1690–1840", in P. Sutter and C. Manganiello (eds.), *Environmental History and the American South: A Reader*, Athens, GA, University of Georgia Press, 2009, pp. 61–79.

13 C. Earle, *Geographical Inquiry and American Historical Problems*, Stanford, CA, Stanford University Press, 1992, chapter 2. European settlers, of course, faced difficulties in taming their new environment; see A. Taylor, "'Wasty Ways': Stories of American Settlement," *Environmental History* 3/3, 1998, pp. 291–310.

14 J. F. Richards, *The Unending Frontier: An Environmental History of the Early Modern World*, Berkeley, University of California Press, 2003; M. Norton, *Sacred Gifts, Profane Pleasures: The History of Tobacco and Chocolate in the Atlantic World*, Ithaca, NY, Cornell University Press, 2008.

15 T. Steinberg, *Nature Incorporated: Industrialization and the Waters of New England*, Cambridge, Cambridge University Press, 1991.

16 Steinberg, *Nature Incorporated*. For a very similar story set in Italy, see S. Barca, *Enclosing Water: Nature and Political Economy in a Mediterranean Valley, 1796–1916*, Cambridge, White Horse Press, 2010.

17 M. Grunwald, *The Swamp: The Everglades, Florida, and the Politics of Paradise*, New York, Simon and Schuster, 2006, pp. 41–3.

18 B. DeLay, *War of a Thousand Deserts: Indian Raids and the US–Mexican War*, New Haven, CT, Yale University Press, 2008. See also P. Hamalainen, *The Comanche Empire*, New Haven, CT, Yale University Press, 2009.

19 A. Rothman, *Slave Country: American Expansion and the Origins of the Deep South*, Cambridge, MA, Harvard University Press, 2005.

20 M. A. Stewart, *"What Nature Suffers to Groe": Life, Labor and Landscape on the Georgia Coast, 1680–1920*, Athens, GA, University of Georgia Press, 1996, p. 135; Silver, *A New Face on the Countryside*.

21 D. D. Glave and M. Stoll,(eds.), *"To Love the Wind and the Rain": African Americans and Environmental History*, Pittsburgh, University of Pittsburgh Press, 2006.

22 J. T. Kirby, *Mockingbird Song: Ecological Landscapes of the South*, Chapel Hill, NC, University of North Carolina Press, 2008.

23 "Report Submitted by Capt. Horace James for the months of July–September of 1865," Quarterly Reports of Abandoned or Confiscated Lands, Records of the Assistant Commissioner for the State of North Carolina Bureau of Refugees, Freedmen, and Abandoned Lands, M843, Roll 36, National Archives.

24 D. E. Davis (ed.), *Southern United States: An Environmental History*, Santa Barbara, CA, ABC-CLIO, 2006, pp. 149–50.

25 For more on these campaigns, see L. Brady, *War Upon the Land: Union Operations against the Southern Landscape, 1862–1865*, Athens, GA, University of Georgia Press, 2012. Brady's work is the first, and so far only, book on the environmental history of the Civil War.

26 A few battles were fought in Maryland and Pennsylvania, both of which were on the side of the Union during the Civil War.

27 J. P. McPherson, *Battle Cry of Freedom: The Civil War Era*, New York, Oxford University Press, 1988, p. 449; R. F. Selcer, *Civil War America, 1850 to 1875 (Almanacs of American Life)*, New York, Facts on File, 2006, pp. 265, 420. For more on coffins as a wartime industry, see D. G. Faust, *This Republic of Suffering: Death and the American Civil War*, New York, Alfred A. Knopf, 2008.

28 W. Cronon, *Nature's Metropolis: Chicago and the Great West*, New York, W. W. Norton, 1991, p. 211.

29 McPherson, *Battle Cry of Freedom*, p. 452.

30 Cowdrey, *This Land, This South*; Stewart, *"What Nature Suffers to Groe."*

31 A. Isenberg, *Mining California: An Ecological History*, New York, Hill and Wang, 2005, pp. 37, 78.

32 Cronon, *Nature's Metropolis*, chapter 5.

33 Cronon, *Nature's Metropolis*; Steinberg, *Down to Earth*.

34 Isenberg, *Mining California*, p. 37; D. Stradling (ed.), *Conservation in the Progressive Era: Classic Texts*, Seattle, University of Washington Press, 2004; Cronon, *Nature's Metropolis*.

35 Silver, *A New Face on the Countryside*, p. 195; Kirby, *Mockingbird Song*.

36 For an interesting comparison of US and British efforts to control air pollution, see D. Stradling and P. Thorsheim, "The Smoke of Great Cities: British and American Efforts to Control Air Pollution, 1860–1914," *Environmental History* 4/1, 1999, pp. 6–31.

37 Stradling, *Conservation in the Progressive Era*, introduction; S. P. Hays, *Conservation and the Gospel of Efficiency: The Progressive Movement, 1890–1920*, Cambridge, MA, Harvard University Press, 1959.

38 Stradling, *Conservation in the Progressive Era*; Glave and Stoll, "*To Love the Wind and the Rain.*"

39 S. Hahn, *The Roots of Southern Populism: Yeoman Farmers and the Transformation of the Georgia Upcountry, 1850–1890*, New York, Oxford University Press, 1993. Phillips discusses compulsory terracing schemes and other soil erosion interventions by the British in South Africa, and how these measures led to peasant rebellions and independent African farmer's movements: S. T. Phillips, "Lessons from the Dust Bowl: Dryland Agriculture and Soil Erosion in the United States and South Africa, 1900–1950," *Environmental History* 4/2, 1999, pp. 245–66. Sunseri shows how the introduction of scientific forestry in East Africa by the German colonial state sparked peasant resistance: T. Sunseri, *Wielding the Ax: State Forestry and Social Conflict in Tanzania, 1820–2000*, Athens, OH, Ohio University Press, 2009.

40 The Klondike gold rush began in 1897.

41 Tucker, "The Global Ecological Reach of the United States," pp. 508–12.

42 B. Black, *Petrolia: The Landscape of America's First Oil Boom*, Baltimore, MD, The Johns Hopkins University Press, 2000; J. R. McNeill and D. S. Painter, "The Global Environmental Footprint of the US Military, 1789–2003," in C. Closmann (ed.), *War and the Environment*, College Station, TX, Texas A&M University Press, 2009, pp. 10–31; M. L. Santiago, *The Ecology of Oil: Environment, Labor, and the Mexican Revolution, 1900–1938*, Cambridge, Cambridge University Press, 2006.

43 For some of the global implications of the American Dust Bowl, see Phillips, "Lessons from the Dust Bowl."

44 M. Reisner, *Cadillac Desert: The American West and Its Disappearing Water*, New York, Penguin Books, 1993; D. Worster, *Rivers of Empire: Water, Aridity and the Growth of the American West*, New York, Oxford University Press, 1985, pp. 242–50, 262.

45 Stewart, "*What Nature Suffers to Groe.*"

46 Tucker, "The Global Ecological Reach of the United States"; Steinberg, *Down to Earth*, pp. 268–70.

47 McNeill and Painter, "The Global Environmental Footprint of the US Military," p. 26. Wills points out that recently atomic test sites have begun to attract rare species of flora and fauna, and some have even become nature reserves: J. Wills, "Welcome to the Atomic Park: American Nuclear Landscapes and the 'Unnaturally Natural'," *Environment and History* 7, 2001, pp. 449–72.

48 A. Brinkley, *American History: A Survey*, New York, McGraw-Hill, 2003, pp. 800, 809.

49 A. Rome, "Give Earth a Chance: The Environmental Movement and the Sixties," *The Journal of American History* 90/2, 2003, pp. 525–54 (p. 530); S. Strasser, *Waste and Want: A Social History of Trash*, New York, Metropolitan Books, 2000, chapter 4.

50 S. Stoll, *U.S. Environmentalism since 1945: A Brief History with Documents*, New York, PalgraveMacmillan, 2007, introduction; A. Rome, *Bulldozer in the Countryside: Suburban Sprawl and the Rise of American Environmentalism*, Cambridge, Cambridge University Press, 2001, and Rome, "Give Earth a Chance."

51 A. Hurley, *Environmental Inequalities: Class, Race, and Industrial Pollution in Gary, Indiana, 1945–1980*, Chapel Hill, NC, UNC Press, 1995, pp. xiii–xiv, 2; H. K. Rothman, *The Greening of a Nation?: Environmentalism in the US since 1945*, New York, Harcourt Brace Publishing, 1998.

52 R. D. Bullard, *Dumping in Dixie: Race, Class, and Environmental Quality*, Boulder, CO, Westview Press, 1990; C. Johnson and J. McDaniel, "Turpentine Negro," in D. D. Glave and M. Stoll (eds.), *"To Love the Wind and the Rain": African Americans and Environmental History*, Pittsburgh, University of Pittsburgh Press, 2006, pp. 51–62.

53 J. K. Stine, *Mixing the Waters: Environment, Politics, and the Building of the Tennessee–Tombigbee Waterway*, Akron, OH, University of Akron Press, 1993; C. Montrie, *To Save the Land and People: A History of Opposition to Surface Coal Mining in Appalachia*, Chapel Hill, NC, University of North Carolina Press, 2003.

References

Anderson, V. D., *Creatures of Empire: How Domestic Animals Transformed Early America*, New York, Oxford University Press, 2004.

Barca, S., *Enclosing Water: Nature and Political Economy in a Mediterranean Valley, 1796–1916*, Cambridge, White Horse Press, 2010.

Black, B., *Petrolia: The Landscape of America's First Oil Boom*, Baltimore, MD, The Johns Hopkins University Press, 2000.

Brady, L., *War Upon the Land: Union Operations against the Southern Landscape, 1862–1865*, Athens, GA, University of Georgia Press, 2012.

Brinkley, A., *American History: A Survey*, New York, McGraw-Hill, 2003.

Bullard, R. D., *Dumping in Dixie: Race, Class, and Environmental Quality*, Boulder, CO, Westview Press, 1990.

Carson, J. T., "Horses and the Economy and Culture of the Choctaw Indians, 1690–1840", in P. Sutter and C. Manganiello (eds.), *Environmental History and the American South: A Reader*, Athens, GA, University of Georgia Press, 2009, pp. 61–79.

Cowdrey, A., *This Land, This South: An Environmental History*, Lexington, KY, University Press of Kentucky, 1983.

Cronon, W., *Nature's Metropolis: Chicago and the Great West*, New York, W. W. Norton, 1991.

Crosby, A., *Ecological Imperialism: The Biological Expansion of Europe, 900–1900*, Cambridge, Cambridge University Press, 1986.

Davis, D. E. (ed.), *Southern United States: An Environmental History*, Santa Barbara, CA, ABC-CLIO, 2006.

DeLay, B., *War of a Thousand Deserts: Indian Raids and the US–Mexican War*, New Haven, CT, Yale University Press, 2008.

Earle, C., *Geographical Inquiry and American Historical Problems*, Stanford, CA, Stanford University Press, 1992.

Faust, D. G., *This Republic of Suffering: Death and the American Civil War*, New York, Alfred A. Knopf, 2008.

Glave, D. D., and Stoll, M. (eds.), *"To Love the Wind and the Rain": African Americans and Environmental History*, Pittsburgh, University of Pittsburgh Press, 2006.

Grunwald, M., *The Swamp: The Everglades, Florida, and the Politics of Paradise*, New York, Simon and Schuster, 2006.

Hackel, S. W., *Children of Coyote, Missionaries of Saint Francis: Indian–Spanish Relations in Colonial California, 1769–1850*, Chapel Hill, NC, University of North Carolina Press, 2005.

Hahn, S., *The Roots of Southern Populism: Yeoman Farmers and the Transformation of the Georgia Upcountry, 1850–1890*, New York, Oxford University Press, 1993.

Hamalainen, P., *The Comanche Empire*, New Haven, CT, Yale University Press, 2009.

Hays, S. P., *Conservation and the Gospel of Efficiency: The Progressive Movement, 1890–1920*, Cambridge, MA, Harvard University Press, 1959.

Hurley, A., *Environmental Inequalities: Class, Race, and Industrial Pollution in Gary, Indiana, 1945–1980*, Chapel Hill, NC, UNC Press, 1995.

Isenberg, A., *The Destruction of the Bison: An Environmental History, 1750–1920*, Cambridge, Cambridge University Press, 2000.

Isenberg, A., *Mining California: An Ecological History*, New York, Hill and Wang, 2005.

Johnson, C., and McDaniel, J., "Turpentine Negro," in D. D. Glave and M. Stoll (eds.), *"To Love the Wind and the Rain": African Americans and Environmental History*, Pittsburgh, University of Pittsburgh Press, 2006, pp. 51–62.

Kirby, J. T., *Mockingbird Song: Ecological Landscapes of the South*, Chapel Hill, NC, University of North Carolina Press, 2008.

Krech, S., *The Ecological Indian: Myth and History*, New York, W. W. Norton, 1999.

Lopinot, N. H., and Woods, W. I., "Wood Overexploitation and the Collapse of Cahokia," in C. M. Scarry (ed.), *Foraging and Farming in the Eastern Woodlands*, Gainesville, FL, University Press of Florida, 1993, pp. 206–31.

McNeill, J. R., and Painter, D. S., "The Global Environmental Footprint of the US Military, 1789–2003," in C. Closmann (ed.), *War and the Environment*, College Station, TX, Texas A&M University Press, 2009, pp. 10–31.

McPherson, J. P., *Battle Cry of Freedom: The Civil War Era*, New York, Oxford University Press, 1988.

Melville, E. G. K., *A Plague of Sheep: Environmental Consequences of the Conquest of Mexico*, Cambridge, Cambridge University Press, 1994.

Montrie, C., *To Save the Land and People: A History of Opposition to Surface Coal Mining in Appalachia*, Chapel Hill, NC, University of North Carolina Press, 2003.

Norton, M., *Sacred Gifts, Profane Pleasures: The History of Tobacco and Chocolate in the Atlantic World*, Ithaca, NY, Cornell University Press, 2008.

Peacock, E., "Historical and Applied Perspectives on Prehistoric Land Use in Eastern North America," *Environment and History* 4, 1998, pp. 1–29.

Phillips, S. T., "Lessons from the Dust Bowl: Dryland Agriculture and Soil Erosion in the United States and South Africa, 1900–1950," *Environmental History* 4/2, 1999, pp. 245–66.

Reisner, M., *Cadillac Desert: The American West and Its Disappearing Water*, New York, Penguin Books, 1993.

Richards, J. F., *The Unending Frontier: An Environmental History of the Early Modern World*, Berkeley, University of California Press, 2003.

Rome, A., *Bulldozer in the Countryside: Suburban Sprawl and the Rise of American Environmentalism*, Cambridge, Cambridge University Press, 2001.

Rome, A., "Give Earth a Chance: The Environmental Movement and the Sixties," *The Journal of American History* 90/2, 2003, pp. 525–54.

Rothman, A., *Slave Country: American Expansion and the Origins of the Deep South*, Cambridge, MA, Harvard University Press, 2005.

Rothman, H. K., *The Greening of a Nation?: Environmentalism in the US since 1945*, New York, Harcourt Brace Publishing, 1998.

Santiago, M. L., *The Ecology of Oil: Environment, Labor, and the Mexican Revolution, 1900–1938*, Cambridge, Cambridge University Press, 2006.

Selcer, R. F., *Civil War America, 1850 to 1875 (Almanacs of American Life)*, New York, Facts on File, 2006.

Silver, T., *A New Face on the Countryside: Indians, Colonists, and Slaves in the South Atlantic Forests, 1500–1800*, Cambridge, Cambridge University Press, 1990.

Sowards, A. M., *United States West Coast: An Environmental History*, Santa Barbara, CA, ABC-CLIO, 2007.

Steinberg, T., *Down to Earth: Nature's Role in American History*, Oxford, Oxford University Press, 2002.

Steinberg, T., *Nature Incorporated: Industrialization and the Waters of New England*, Cambridge, Cambridge University Press, 1991.

Stewart, M. A., *"What Nature Suffers to Groe": Life, Labor and Landscape on the Georgia Coast, 1680–1920*, Athens, GA, University of Georgia Press, 1996.

Stine, J. K., *Mixing the Waters: Environment, Politics, and the Building of the Tennessee–Tombigbee Waterway*, Akron, OH, University of Akron Press, 1993.

Stoll, S., *U.S. Environmentalism since 1945: A Brief History with Documents*, New York, PalgraveMacmillan, 2007.

Stradling, D. (ed.), *Conservation in the Progressive Era: Classic Texts*, Seattle, University of Washington Press, 2004.

Stradling, D., and Thorsheim, P., "The Smoke of Great Cities: British and American Efforts to Control Air Pollution, 1860–1914," *Environmental History* 4/1, 1999, pp. 6–31.

Strasser, S., *Waste and Want: A Social History of Trash*, New York, Metropolitan Books, 2000.

Sunseri, T., *Wielding the Ax: State Forestry and Social Conflict in Tanzania, 1820–2000*, Athens, OH, Ohio University Press, 2009.

Taylor, A., "'Wasty Ways': Stories of American Settlement," *Environmental History* 3/3, 1998, pp. 291–310.

Tucker, R. P., "The Global Ecological Reach of the United States: Exporting Capital and Importing Commodities," in D. C. Sackman (ed.), *A Companion to American Environmental History*, Oxford, Wiley-Blackwell, 2010, pp. 505–28.

White, R., *The Middle Ground: Indians, Empires, and Republics in the Great Lakes Region, 1650–1815*, Cambridge, Cambridge University Press, 1991.

Whitney, G. G., *From Coastal Wilderness to Fruited Plain: A History of Environmental Change in Temperate North America from 1500 to the Present*, Cambridge, Cambridge University Press, 1994.

Wills, J., "Welcome to the Atomic Park: American Nuclear Landscapes and the 'Unnaturally Natural'," *Environment and History* 7, 2001, pp. 449–72.

Worster, D., *Rivers of Empire: Water, Aridity and the Growth of the American West*, New York, Oxford University Press, 1985.

Further Reading

Carney, J. A., *Black Rice: The African Origins of Rice Cultivation in the Americas*, Cambridge, MA, Harvard University Press, 2001.

Cronon, W., *Changes in the Land: Indians, Colonists and the Ecology of New England*, New York, Hill and Wang, 1983.

Hays, S. P., *Beauty, Health, and Permanence: Environmental Politics in the United States, 1955–1985*, Cambridge, Cambridge University Press, 1987.

Stewart, M. A., "Slavery and African American Environmentalism," in D. D. Glave and M. Stoll (eds.), *"To Love the Wind and the Rain": African Americans and Environmental History*, Pittsburgh, University of Pittsburgh Press, 2006, pp. 9–20.

Williams, M., *Americans and Their Forests: A Historical Geography*, Cambridge, Cambridge University Press, 1989.

CHAPTER NINE

The Arctic and Subarctic in Global Environmental History

Liza Piper

This chapter examines the environmental history of arctic and subarctic regions, here referred to collectively as the Circumpolar North, and including Iceland, northern Scandinavia, Siberia, Alaska, northern Canada, and Greenland (see Map 9.1).[1] From an ecological perspective, regional unity is apparent in the extent of tundra, taiga (boreal forest), and the polar ice cap, yet the potential for homogeneity on a broad regional scale should not obscure the considerable physical differences between the diverse mixed-wood forests that line the Slave River of Canada, with its dry interior climate and expansive freshwater delta, the sea ice that dominates along the northwest coast of Greenland, and the urban conglomeration of Norilsk, Russia. The objective of this chapter is to examine the environmental history of the region as a whole by focusing upon those episodes that best illuminate human relations with the rest of nature at high latitudes.

Migration to the Arctic

The most significant arrival of people in the Circumpolar North and their movements across the region historically correspond to climatic changes. These include the climatic oscillations that accompanied the movement of ice sheets and the transgression or regression of oceans before deglaciation, followed by later, less dramatic climatic shifts in the Holocene. Across northern Europe, the ice sheets created a dramatically different coastline. The Fennoscandian ice sheet blocked lands to the west and only opened to human colonization at the end of the last glaciation, although the northern part of the Russian plain was accessible. The Siberian far north was blocked by a major transgression of the Arctic Ocean, which likewise only retreated at the outset of the Holocene. Beringia, the land bridge connecting eastern Siberia to the Mackenzie River, opened as a result of a regression of the Pacific Ocean during

A Companion to Global Environmental History, First Edition. Edited by J.R. McNeill and Erin Stewart Mauldin.
© 2012 John Wiley & Sons, Ltd. Published 2015 by John Wiley & Sons, Ltd.

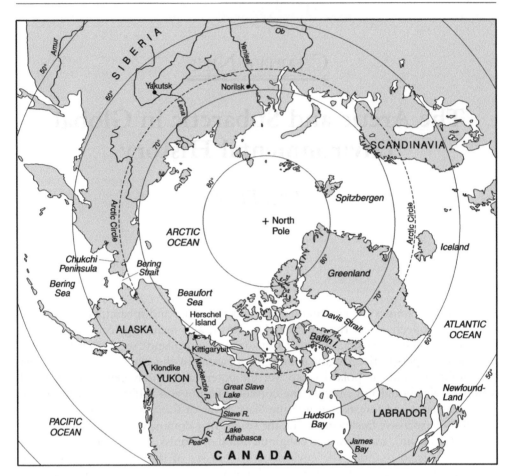

Map 9.1　The Circumpolar North

the last glaciation. With the retreat of the ice sheets, people, fish, plants, and animals came flooding back into deglaciated or newly ice-edge territories. From the end of the last glaciation to 1200 CE in Arctic North America, climatic stability tended to encourage the growth of settlements and wider communities, which then moved or disappeared altogether as climate deteriorated.[2] Where conditions were more favorable, as across much of Scandinavia moderated by the Gulf Stream, permanent settlements had long since emerged and, by the ninth century, the Normans were already pursuing expansion.[3] Nevertheless, for all circumpolar populations in the short and longer term, mobility remained key. This is not the story of remote and isolated communities establishing themselves in close adaptation to immediate environments, but rather the history of peoples moving across large physical stretches and, when settling, remaining tied to expansive areas that provided the opportunities for culture and life.

Mobile responses to climatic oscillations shaped the history of the medieval warm period or little climatic optimum (c. 900–1200 CE) across the Circumpolar North. This warming period provided opportunities for expansion in northern climates: large quantities of pack ice, the main hazard to navigation, shifted northward and thinned. Both the Norse and the Thule took advantage of relatively advanced maritime technologies and the prolonged warming of the little climatic optimum (the Norse also had the benefit of favorable currents in the North Atlantic) to make their long-distance journeys. Iceland was well-known in Anglo-Saxon England and among Irish monks, but the first permanent settler, a Norwegian called Ingólfur Arnarson, arrived in 870 CE. The settlement in Iceland was the first step in a much greater westward movement that brought the Norse to Greenland and then Vinland where they encountered Thule people moving eastward from the west coast of Alaska along the path that is today thought of as the Northwest Passage. The Thule had developed techniques to efficiently harvest seal and walrus along the coast and to hunt whales in open seas. With the warming conditions, habitats changed for seals, walrus, and whales. As the ice pack moved north, whale habitat enlarged, and the west-coast Alaskans pursued them eastward. The Thule reached Greenland, likely around the eleventh century, and for 300 years overlapped with their Norse counterparts.

The Norse came to Greenland as Iceland faced pressures from overpopulation as early as 975 CE. In many respects the communities in Greenland were extensions of those in Iceland. Settlers in Greenland practiced much the same medieval dairying economy as they had in Iceland previously, combined with harvesting of seal and caribou in the spring and fall.[4] From Greenland, Leif Eiriksson traveled again westward along a coast he described as Helluland, Markland (Forest Land, probably southern Labrador), and Vinland (Wine/Pasture Land). This last name has been assigned to the Norse settlement found in northern Newfoundland, where a handful of people lived for a few years before abandoning the settlement.

There were two settlements on Greenland, the larger Eastern Settlement in the Julianehab area with 190 or so farms, and the smaller Western Settlement near Godthaab with 90 farms. With the onset of the Little Ice Age (LIA) around 1350 and the poorer environmental conditions it brought to high latitudes, the Greenland settlements, as essentially colonies of Iceland and more distantly Scandinavia, now faced more hazardous sailing conditions back across the North Atlantic. Moreover, the shortened growing season had dramatic impacts upon the ability of the Greenland colonists to derive subsistence from the land. The Norse settlements in Greenland collapsed relatively quickly, and were abandoned in the fifteenth century.[5] The Thule similarly retreated as a consequence of the poorer environmental conditions of the LIA. They had already adapted to the changed circumstances they found as they moved eastward. In the relatively shallow straits of the Coronation Gulf area, for instance, neither whales nor walrus thrived, and the Thule had turned instead to harvesting caribou, fish, and ringed seal.[6] Further west, once ocean temperatures fell, the sea ice advanced and persisted longer through the year, and whales became relatively more scarce. Communities responded by turning to other food sources (again fish, caribou, and seal), as they had in the Coronation Gulf area. Thule and their descendants also abandoned sites such as Somerset Island and south Baffin in the 1300s. The movement of the Thule to the west and the Norse to the east, while sometimes classed as explorations, are more importantly understood as movements in search of homes, even if the villages established did not necessarily endure. The new environmental hazards after 1300 led not only to the abandonment and collapse of villages and settlements, but also to disrupted trade connections that had emerged during the more moderate times.

The Fur Trade and Whaling

The period from the fourteenth to the nineteenth century saw the overharvesting, depletion, extirpation, and in some instances extinction of arctic and subarctic species – in particular fur bearers such as sable and beaver, and marine mammals such as walrus, sea otters, and eastern Arctic bowhead whales – as a direct consequence of the rise of global commerce in products made from these creatures and their pursuit across the full extent of their ranges. On land, the opportunities for fur harvesting drove Russian colonial expansion across Siberia. Moscow controlled a network that supplied furs to much of the medieval world, and after that to Europe and the Ottoman Empire. Furs were luxury items, in high demand across the cooling northern hemisphere. However, by the seventeenth century, while demand continued to increase, yields from the White Sea region could no longer keep pace and hence Moscow's rulers, as well as private trappers, turned their attention east to Siberia. Trappers sought sable (*Martes zibellina*), followed in importance by ermine, marten, beaver, and squirrel. As Russians moved into Siberia they subjugated indigenous peoples across the region, and forced them to pay a tax, the *iasak*, levied in furs (preferably sable), on every "fit, native Siberian male" aged 18 to 50.[7] This tax ensured a continuous supply of furs to Moscow. The other route for furs to market was through the hands of private traders who hired commercial Russian trappers and whose harvests greatly exceeded those collected to pay the *iasak*.

Continuous increases in demand not only led Russians to move further and further east, across central Siberia to the Chukchi Peninsula and the Pacific Ocean, they also ensured that rising prices caused sables and other small fur bearers to be hunted across the full extent of their ranges.[8] That sable populations were systematically depleted by sustained predation can be discerned from the changes in the number of sable pelts that were required to pay the *iasak*. By the mid-seventeenth century, the rate had dropped from 10–12 sables per man per year, to just three pelts per man in western Siberia. Moreover, the requirement for indigenous Siberians to hunt the animal, as well as the broader expansion of the fur trade, led to shifts in harvesting practices and new pressures on a wide range of species, not only sable but also other fur bearers large and small, including bears, wolves, wolverines, reindeer, otters, minks, ermines, and weasels.

A similar pattern of colonial expansion in pursuit of fur bearers brought the English and French westward across Subarctic and into Arctic North America. The Hudson's Bay Company (HBC) emerged in 1670 when the English government granted a royal charter giving the company wide powers, including exclusive trading rights in the watershed of Hudson Bay, known as Rupert's Land. Predicating the company's control upon a watershed reflected the centrality of rivers and other inland waterways both to transportation for trade and to beaver (*Castor canadensis*) habitat. In 1763, following the Treaty of Paris and the defeat of France, the distinct, more loosely structured fur trade operating out of Montreal was reorganized under the direction of Scottish, English, American, and a few French Canadian managers. They formed the North West Company (NWC) in 1784, which significantly rivalled the HBC until the two merged in 1821.

The environmental consequences of the Canadian fur trade intensified in this 40-year period as a direct consequence of trade rivalries. Aboriginal populations were complicit in overharvesting as they trapped the beaver for trade. Intense competition in the hinterland of York Fort and Albany Fort led the HBC to raise prices. Higher prices encouraged increased harvesting, which in turn placed excessive pressure on beaver populations and ultimately led to declines. Overharvesting also ensued as both the NWC and the HBC expanded their hunting territories, increased the number of trade posts, and thus pressured fur-bearing animals across their habitats. Fur bearers were not alone in dealing with the impacts of competition. Game animals, which supplied the food for the trade, faced serious depletion by the 1820s in the most competitive areas, such as east of Lake Winnipeg in the Petit Nord, and as far north as Lake Athabasca and Great Slave Lake by the 1830s.

Pressure on Canadian subarctic fauna eased after the 1821 merger under the banner of the HBC, as the intense competition subsided and as the new governor, George Simpson, introduced conservation measures to ensure the long-term health of the trade.[9] What saved the Canadian beaver in the long run were not the self-imposed conservation efforts of the HBC, however, but rather a change in fashion. The shift from beaver felt to silk hats in the 1860s truncated the external demand that had been the principal factor driving the overharvesting of the beaver. With populations under less pressure, the beaver subsequently staged a remarkable recovery across subarctic Canada.

The devastation of the eastern Arctic bowhead (or Greenland right whale) (*Balaena mysticetus*) and Atlantic walrus (*Odobaenus rosmarus rosmarus*) from the seventeenth through the twentieth century was more profound than the overharvesting of fur bearers on land. Indigenous harvesting of arctic whale stocks long predated commercial whale operations. Indigenous communities, whether along the Chukchi or Labrador coasts, likely harvested fewer than 15 and 5 bowhead whales each year respectively, given that the size of these animals ensured considerable resources to the community with each catch. For indigenous communities, whaling was just one part of the harvest of sea mammals, which, more importantly, also included walruses and seals. For the Inuit on Baffin Island, whales provided abundant oil, baleen, and bones, while seal meat was the dietary staple supplemented by whale meat, blubber, fish, and caribou.[10] The commercial hunt of arctic sea mammals that began in the sixteenth century sought bowhead whales for their oil and baleen, and hunted walruses or gray whales to supplement the oil harvest as needed. Neither on land nor on water did newcomers seek foodstuffs in the North, rather they came to exploit the environment for fashion and for fuel.

The Basque moved into northern waters off the Labrador coast from the Bay of Biscay by 1536. They conducted an exclusively shore-based whale hunt, which eventually extended into the Gulf of St. Lawrence. Intensified commercial hunting of the bowhead whale began later in the sixteenth century, shortly after the discovery of Spitsbergen (part of the Svalbard archipelago). By 1611 the Dutch began commercial whaling operations that harvested the Greenland–Spitsbergen and Davis Strait stocks for the next 300 years. Initially dominated by the Dutch, German, French, English, and American whalers all participated over the coming centuries. The geography of the whale hunt shifted as local populations of whales were depleted or avoided heavily harvested areas. Depletion likely brought the Basques north to Labrador in the first instance. By 1670, Dutch whaling ships off Spitsbergen had been pushed offshore into the open waters, as whales stayed away from sheltered bays. The Dutch also moved to the western coasts of Greenland, into the Davis Strait area, after 1719, in search of improved hunting, although catches from Davis Strait never equalled those from the Greenland–Spitsbergen waters. Climatic changes also influenced the geography of whaling. Colder climatic conditions in the Arctic during the LIA meant that the winter pack ice extended farther south and the Greenland whales that followed the edge of the pack ice traveled farther south too.[11]

The colder climate further influenced the availability of plankton as food for the whales, ice thickness, and the character of drift ice – one of the main hazards to early modern whale ships – with varying impacts on whales and their predators.[12]

By the 1780s the Dutch and German whaling industries were in decline, but the British, impressed not only by the Dutch profits but also by the role of the whaling industry in serving as a nursery for seamen, had encouraged the growth of its own hunt, and not only outcompeted the Dutch, but also effectively hunted both the Greenland–Spitsbergen and the Davis Strait stocks to extinction.[13] The rise of British and American commercial whaling moved the focus of harvesting efforts to North American waters. By the early nineteenth century, British whalers had moved into Baffin Bay and the Arctic archipelago. Ice was a much greater hazard in these waters, particularly during the tail end of the LIA, as compared to the relatively open seas between Greenland and Spitsbergen, yet the depletion of whale populations elsewhere made this move worthwhile. In these same years, American vessels moved into the Bering Strait whaling grounds, sending over 2,700 cruises to the Bering, Chukchi, and Beaufort seas between 1849 and 1914.[14]

Assessing the impact of commercial harvesting by reconstructing whale populations from the seventeenth century has been much more precise than any equivalent reconstructions for arctic land animals. Louwrens Hacquebord accounted for 122,000 whales killed by English, Dutch, and German whalers in the seas between Jan Mayen and Spitsbergen from 1669 onwards. The Greenland bowhead whale stock that was thus exterminated would have been about 46,000 whales. John Bockstoce and colleagues' reconstruction of the western American Arctic bowhead populations indicate that 16,594 bowhead whales were killed between 1849 and 1914. These stocks were smaller and depleted more rapidly than those to the east. The ecological impacts of the hunt for marine mammals were greater than the impacts of the terrestrial fur trade. Whales, in particular, played a more important role in arctic food relationships than did either beaver or sable. The destruction of the Greenland bowhead effectively removed 46,000 whales from that ecosystem, adding 3.5 million tons of plankton food annually for seabirds and fish.[15]

For the human populations that had previously relied upon small annual catches of whales to sustain their communities, the devastation brought by commercial harvesting was considerable. In the Mackenzie Delta, for instance, Inuvialuit (western Inuit) harvested bowhead and beluga, most notably at the village of Kittigaryuit, the site of a natural beluga whale trap. Although not interested in beluga (much faster swimmers than bowhead, which tended to sink when killed), the American whalers moved into the delta after 1889 as these were also the summer feeding grounds of the remaining bowhead in the western American Arctic. From their base on Herschel Island, American whalers extirpated bowhead stocks, thereby undermining Inuvialuit livelihoods in the delta. As with the depletion of game populations to provision the land-based fur trade, so too the rise and demise of the commercial whale harvest affected a wider range of species. The presence of commercial whalers in Hudson Bay up to 1915 led Inuit hunters to focus more exclusively upon whales in this period, rather than continuing to diversify their harvests with other sea mammals. Many Inuit, as well, became involved in provisioning the whale crews, thus intensifying their harvests of caribou in particular. Near Baffin Island, Hudson Bay, and in the northern Yukon, whalers' subsistence demands contributed to the decline of caribou populations while recreational hunting of polar bears and musk-ox also undermined these wildlife populations.

Colonization and the Destruction of Indigenous Economies

Animals were domesticated in the Circumpolar Arctic, most notably reindeer in Siberia and northern Scandinavia (Sápmi) as early as the first millennium CE.[16] The transition to pastoral economies based upon reindeer herding did not end other harvesting practices but rather integrated herding (which enabled Sami to obtain meat, skins, and milk, and to manufacture milk products such as cheese) with hunting, fishing, and gathering plants. As with other indigenous economies across the Circumpolar North, reindeer herding demanded extensive rather than intensive land use. Sami herders relied on large areas, moving seasonally between different sites to access a range of resources at optimal times of year. Herding families returned to the same sites year after year (hence the problems of using "nomadism" to describe such harvesting practices, a term which carries with it connotations of impermanence). The Swedish state recognized and taxed the lands required by the Sami for such subsistence activities.[17] Samoyeds in western Siberia also herded reindeer and kept domesticated dogs. Many Yakuts kept large herds of cattle and horses, which they moved seasonally to suitable grazing lands found in the pine and large taiga of central Siberia. In northeastern Siberia there were both so-called "settled" and "reindeer" Yukagir.[18] The latter followed the migratory reindeer, while the former kept domesticated reindeer in herds as well as fishing and hunting elk. Chukchi and Korak peoples on the Chukchi Peninsula practiced seasonal transhumance and the herding of reindeer. The domestication of animals was thus widespread across the Circumpolar North prior to the arrival of settlers from more temperate latitudes who introduced new domesticated species into arctic and subarctic regions.

The colonization of the Circumpolar Arctic and Subarctic, as part of the larger expansion of populations from Europe after the sixteenth century, met with mixed success. European portmanteau biota, most notably introduced diseases, played a role in transforming indigenous relations with the natural environment across the Circumpolar North.[19] A simple model of virgin-soil epidemics arriving at first contact and decimating arctic and subarctic populations does not hold. Sami and Swedes, for instance, had long been territorially linked, but low population densities meant that Sami and new settlers remained susceptible to crowd infections.[20] Similarly, although joined by land to continental Europe, Siberia was nevertheless sufficiently remote to constitute a sort of New World. Acute infectious disease epidemics arrived in Siberia in conjunction with the expansion of the fur trade. There was a severe outbreak of smallpox in western Siberia in the 1630s, and by the 1690s smallpox epidemics had significantly affected the Yukagir population.[21] The impact of disease epidemics facilitated the introduction of the *iasak* taxation system, subjugating Siberian indigenous populations to the authority of Russia and contributing to new hunting and harvesting relationships. In Subarctic and Arctic North America, introduced diseases caused epidemic outbreaks among local populations (both indigenous and newcomer) well into the twentieth century. The toll taken by infectious disease epidemics was only accentuated by the corrosive impact of tuberculosis, which arrived in the Aleutians by the eighteenth century and was a leading cause of death for Aboriginal people in Arctic North America from the late nineteenth through the mid-twentieth century. The transition that occurred in the mid-twentieth century was not one of population, which remained small and relatively dispersed, but rather one of health care.

In spite of the impacts of introduced infectious diseases in reducing indigenous populations across the Circumpolar North, settlers from more southern areas did not flock north into depopulated spaces to establish viable agricultural economies. One of the great challenges within the Siberian fur trade was provisioning; difficulty obtaining sufficient provisions through the extension of agriculture "intensified in direct proportion to the Russians' eastward advance."[22] There were hopes that the Amur River in eastern Siberia would provide rich grain lands to provision the main settlement at Yakutsk, rather than having to send food from western Siberia. Ultimately, though, the region failed to provide in the hoped-for fashion as the obstacles to agriculture, and particularly large-scale export agriculture, were too great. Russian settlers nevertheless moved into all parts of Siberia; while Cossacks primarily moved into the Amur region, elsewhere peasant migrants arrived in more southerly taiga regions, outside the permafrost zone, and established communities around grain cultivation and livestock grazing.[23]

In northern Scandinavia, agricultural settlement encouraged by late-seventeenth-century legislation pressed at the edges of lands long occupied by indigenous peoples engaged in more diverse harvesting activities.[24] Agriculture and the opening up of grazing lands for domesticated reindeer led to the destruction of old-growth woodlands. Nevertheless, preindustrial clearance was much more modest across the Circumpolar North than in more temperate environments because the pressure for agricultural settlement was relatively subdued.[25] In North America, the Peace River district elicited similar hopes to the Amur River region in Siberia and met with more modest failures, as displaced grain farmers did move north to the region and to northern Saskatchewan in the face of southern droughts in the 1920s and 1930s. Other attempts at grain farming and animal domestication (including reindeer herding in the Mackenzie Delta) were made, particularly in some of the climatically more moderate parts of the region and in the immediate vicinity of trade and mission posts. Ultimately, the more profound environmental impacts were a direct consequence of harvesting the land not for food, but for minerals, timber, oil, and gas.

The Klondike gold rush epitomises the character of interest in, and the transformative capacity of, mineral exploitation in arctic and subarctic regions.[26] In that instance, tens of thousands of southerners – Americans in particular but by no means exclusively – traveled over land and water into Canada's Yukon Territory after 1896 in the hopes of making their individual fortunes. The Klondike gold rush was, in many respects, one of the last international "poor man's" mining rushes. It offered an opportunity for men and women willing to move into an unfamiliar and relatively challenging environment to potentially strike it rich through hard labor and some good luck. These men and women needed neither scientific training nor vast capital resources to partake in the disassembly of the Yukon and Klondike watersheds in their search for gold. Nevertheless their labor would, soon enough, be supplanted by machine technology, as dredges in short order took over the work of extracting gold from Yukon and Alaskan gravels.

External demand for specialized minerals and fuel sources drove large-scale exploitation. The Soviet Union was very interested in the resource potential of its own Siberian and Far Eastern territories from the 1920s forward. The Gulag system facilitated major mine developments, such as those that arose centered on Norilsk, by forcing much-needed labor northward. The USSR also looked farther afield, participating significantly along with Norway in the long-term exploitation of coal from Spitsbergen.[27] Industrialization drove the extraction of coal, oil, and gas whether on Spitsbergen or in the Mackenzie Valley where oil extraction was tied to regional radium and hard-rock gold-mine development. Commercial forestry across the boreal region took off in much the same period as mineral exploitation and in some places, such as northern Sweden, had the same transformative effect. During the nineteenth century, the timber frontier moved steadily northward in Scandinavia, generally preceding the agricultural frontier. Timber harvesting proceeded along the major river valleys and thence moved inland focused upon large-diameter timber trees. Ultimately, the growth of lumbering not only transformed northern Sweden, but also contributed directly to the industrial development of the nation as a whole.[28]

Nationalism and Conservation

From the late nineteenth into the early twentieth century, northern territories (whether lying within or beyond national boundaries) became essential to national identities in Europe and Canada. Those who linked national identity with arctic and subarctic territories drew upon prevailing late-nineteenth-century beliefs in the environmental roots of racial and cultural superiority. Where previously arctic environments were characterized as harsh and threatening, by the late nineteenth century, the difficulties of life thought to be inherent in these places (the cold, the short to non-existent daylight hours in winter, the encounters with wildlife, the remoteness) became challenges to be overcome, proving ground for individual and national life, and opportunities for the extension of commerce and civilization.[29] The arctic wilderness became a wilderness sublime, with the concurrent aesthetic erasure of Aboriginal peoples and scientific erasure of indigenous and other forms of competing, typically local, knowledge.

With these erasures, northern wilderness became important terrain for emerging conservationist concern. In Canada, the Dominion government implemented new hunting legislation directed at both game and migratory birds, in close cooperation with southern American neighbors rather than northerners. The first northern national park, Wood Buffalo, was created in 1922, followed shortly afterwards by the Thelon Game Sanctuary in 1927.[30] To the west, the American naturalist Olaus Murie spent much time in the early 1920s conducting field research in Alaska and the Yukon. This early work directly influenced his advocacy in the 1950s for the creation of the Arctic National Wildlife Refuge, which he saw as a wilderness where predator–prey relationships could be studied "with minimal human interference."[31] Spitsbergen became the focus of attention in the first decade of the twentieth century as a possible European national park, or at the very least protected by new conservationist legislation in the interests of science and tourism. Indeed, as with Murie's interest in preserving a natural history wilderness, so too Swedish scientists in the 1910s sought to preserve Spitsbergen as "a reserve not so much for wildlife but for the scientists themselves."[32]

The heyday of the North to national identities was shortlived. Northern nationalisms persist into the present, with continued debates over northern sovereignty in the context of a melting Arctic figuring prominently among the concerns of northern nations. Nevertheless, over the course of the twentieth century, Northern national identities became less stable while international perspectives grew in significance. At mid-century, the Cold War drew international attention northward and arctic and subarctic environments became key strategic and "hostile" environments.[33] Although ideologically and latitudinally distant, North America and the Soviet Union were closest across the North Pole. Northern Canada was, moreover, a key site of uranium mining.[34] Northern North America was thus a critical flank in the Cold War and concern about nuclear attack drove the construction of radar stations across Canada and Alaska as part of the Distant Early Warning (DEW), Pinetree, and Mid-Canada lines. American military interest was not confined to Canada, but extended to Greenland, which scholars have argued was effectively colonized by US military science in this period.[35] Concerns about Cold War security led to considerable new research and surveying across the Circumpolar world, producing countless new maps, photographs, and data about climate and terrain in particular, as well as human responses to these environments – key questions should the Arctic become the next battleground.

In addition to the role played by the Cold War, the other two significant developments that contributed to the internationalization of the Arctic and Subarctic were the rising power of indigenous peoples who, while situated within national contexts, generally found greater political power and solidarity through connections across the Circumpolar North; and growing awareness of distant impacts on northern environments. The rise of indigenous political movements came in the 1970s within the longer context of twentieth-century decolonization and advocacy for civil rights. In Sweden, the Reindeer Husbandry Act of 1971 helped to establish land-use rights for certain Sami and thus gave them additional freedom to regulate their own affairs. In Canada, northern indigenous people played a significant role in the Berger Inquiry (1974–7), which delayed the development of the Mackenzie Valley Pipeline to allow the resolution of outstanding land-claim issues and to ensure environmental protection.[36] The James Bay and Northern Quebec Agreement in 1975, which enabled a major hydroelectric development on the shores of James Bay, played a significant role in reshaping Cree relationships with the land and the federal government.[37] In each of these instances, as in the Sami hunger strikes in Norway in the late 1970s, indigenous and land rights were indivisible and in this fashion articulated relationships with the land that effectively transcended nationalist narratives of development.

The internationalization of the North in the twentieth century made it "a distinctive realm, inseparable from the rest of the planet."[38] This was particularly apparent, from the 1960s on, with new awareness of the impacts of distant activities on northern ecosystems. By 1961 scientific research detailed radioactive fallout at high latitudes and described its impacts upon humans and animals in northern places.[39] The effects of fallout at high latitudes reflected two divergent developments in this period. One was the growing debate about military and scientific uses of "pristine" or "fragile" – or alternatively "barren" and "wasted" – arctic and subarctic environments.[40] Nuclear tests on Amchitka in the 1960s, for instance, played a major role in the rise of Greenpeace as one of the foremost international environmentalist organizations. In part, then, concern about the arctic impacts of the atomic age reflected actual uses of northern spaces. But they also pointed to the ways in which northern environments were intimately and profoundly connected with more distant, southern ones. The fallout in the circumpolar region came less from high-latitude testing than from testing across the globe that entered the upper atmosphere and then traveled via atmospheric circulation to concentrate at the poles, where it precipitated out of the atmosphere into polar ice and oceans. Through biomagnification, radioactive fallout and a range of other pollutants become highly concentrated in animals such as caribou or reindeer, seals, and polar bears, which made up the country foods of many northerners – whether Sami, Inuit, Dene, or others.

Over the coming decades it would become clear that radioactive fallout was far from the only such distant impact. From 1967 onwards, there was recognition in Sweden of the role played by distant sources of pollution in harming fish populations and surface waters through acidification, although the debate did not extend to northern Sweden until the 1990s.[41] In the 1980s, Eric Dewailly, a university researcher in Quebec City, wished to use Inuit from the North as a control group in studying polychlorinated biphenyl (PCB) contamination. What he learned was that, far from being pristine, arctic residents and in particular Aboriginal northerners comprised one of the most exposed populations when it came to PCBs and other persistent organic pollutants (POPs).[42] The impacts of climate change are similarly understood within the context of a Circumpolar North influenced by outside developments. This is apparent from the Arctic Climate Impact Assessment, the foremost scientific summary of ongoing climate change, which eschewed examinations at a national scale, employing instead a "subregional" approach that divided the Arctic into quadrants that express ecological and social integrity.[43] Sheila Watt Cloutier, International Chair of the Inuit Circumpolar Council (2002–6), worked to represent international Inuit interests in banning POPs and filed a petition asserting that the arctic impacts of greenhouse gas emissions violated Inuit human rights. These most recent developments have yet to receive sustained historical analysis. They clearly lie within longer shifts in perceptions of northern environments and space, the role of scientists in creating knowledge about the North, the impacts of industrialization, and the respective power of indigenous and nonindigenous northerners – all key themes in the long term for Circumpolar Arctic and Subarctic environmental history.

Notes

1   Northern Mongolia also falls within the boundaries of the Subarctic; it is not included here, however.
2   R. Fossett, *In Order to Live Untroubled: Inuit of the Central Arctic, 1550–1940*, Winnipeg, University of Manitoba Press, 2001, p. 21.
3   M. Bravo and S. Sörlin (eds.), *Narrating the Arctic: A Cultural History of Nordic Scientific Practices*, Canton, Watson Publishing International, 2002, pp. 77–9.
4   P. C. Buckland, T. Amorosi, L. K. Barlow, et al., "Bioarchaeological and Climatological Evidence for the Fate of Norse Farmers in Medieval Greenland," *Antiquity* 70, 1996, pp. 88–96 (p. 89).
5   Buckland, Amorosi, Barlow, et al., "Bioarchaeological and Climatological Evidence."
6   Fossett, *In Order to Live Untroubled*, pp. 22–3, 26.
7   J. F. Richards, *The Unending Frontier: An Environmental History of the Early Modern World*, Berkeley, University of California Press, 2003, p. 531.
8   Richards, *The Unending Frontier*, p. 536.
9   A. J. Ray, "Some Conservation Schemes of the Hudson's Bay Company, 1821–1870: An Examination of the Problems of Resource Management in the Fur Trade," *Journal of Historical Geography* 1, 1975, pp. 49–68.
10 Fossett, *In Order to Live Untroubled*, p. 168.
11 L. Hacquebord, "Three Centuries of Whaling and Walrus Hunting in Svalbard and Its Impact on the Arctic Ecosystem," *Environment and History* 7, 2001, pp. 169–85 (p. 174).
12 J. R. Bockstoce, D. B. Botkin, A. Philp, et al., "The Geographic Distribution of Bowhead Whales, *Balaena mysticetus*, in the Bering, Chukchi, and Beaufort Seas: Evidence from Whaleship Records, 1849–1914," *Marine Fisheries Review* 67/3, 2005, pp. 1–43 (p. 7).
13 R. C. Allan and I. Keay, "Saving the Whales: Lessons from the Extinction of the Eastern Arctic Bowhead," *Journal of Economic History* 64/2, 2004, pp. 400–32 (p. 401). Keeping in mind as well that Dutch and German whalers caught between 95 and 98 percent of Greenland bowheads between 1661 and 1719.
14 Bockstoce, Botkin, Philp et al., "The Geographic Distribution of Bowhead Whales."
15 Hacquebord, "Three Centuries of Whaling and Walrus Hunting," p. 178.

16 T. Josefsson, I. Bergamn, and L. Östlund, "Quantifying Sami Settlement and Movement Patterns in Northern Sweden 1700–1900," *Arctic* 63/2, 2010, pp. 141–54.

17 Josefsson, Bergman, and Östlund, "Quantifying Sami Settlement," pp. 144, 151. See also Bravo and Sörlin, *Narrating the Arctic*, p. 78.

18 Richards, *The Unending Frontier*, p. 529.

19 L. Piper and J. Sandlos, "A Broken Frontier: Ecological Imperialism in the Canadian North," *Environmental History* 12/4, 2007, pp. 759–95.

20 L. Koerner, *Linneaus: Nature and Nation*, Cambridge, MA, Harvard University Press, 1999, p. 69; Bravo and Sörlin, *Narrating the Arctic*, p. 78.

21 Richards, *The Unending Frontier*, p. 538.

22 M. Bassin, "Expansion and Colonialism on the Eastern Frontier: Views of Siberia and the Far East in pre-Petrine Russia," *Journal of Historical Geography*, 14/1, 1988, pp. 3–21 (p. 12).

23 Richards, *The Unending Frontier*, p. 539.

24 L. Östlund, O. Zackrisson, and G. Hörnberg, "Trees on the Border between Nature and Culture: Culturally Modified Trees in Boreal Sweden," *Environmental History* 7/1, 2002, pp. 48–68.

25 For other preindustrial impacts upon the boreal treescape in northern Scandinavia see Östlund, Zackrisson, and Hörnberg, "Trees on the Border between Nature and Culture."

26 K. Morse, *The Nature of Gold: An Environmental History of the Klondike Gold Rush*, Seattle, University of Washington Press, 2003.

27 D. Avango, L. Hacquebord, Y. Aalders, et al., Between Markets and Geopolitics: Natural Resource Exploitation on Spitsbergen from 1600 to the Present Day," *Polar Record* 47/240, 2011, pp. 29–39.

28 Bravo and Sörlin, *Narrating the Arctic*, p. 86.

29 J. Sandlos, "From the Outside Looking In: Aesthetics, Politics, and Wildlife Conservation in the Canadian North," *Environmental History* 6/1, 2001, pp. 6–31; Bravo and Sörlin, *Narrating the Arctic*; U. Wråkberg, "Nature Conservationism and the Arctic Commons of Spitsbergen 1900–1920," *Acta Borealia* 23/1, 2006, pp. 1–23.

30 Sandlos, "From the Outside Looking In," p. 13.

31 J. M. Glover, "Sweet Days of a Naturalist: Olaus Murie in Alaska, 1920–26," *Forest & Conservation History* 36/3, 1992, pp. 132–40 (p. 140).

32 Wråkberg, "Nature Conservationism," p. 19; Bravo and Sörlin, *Narrating the Arctic*, pp. 99–100.

33 M. Farish, "Frontier Engineering: From the Globe to the Body in the Cold War Arctic," *The Canadian Geographer* 50/2, 2006, pp. 177–96 (p. 179); P. W. Lackenbauer and M. Farish, "The Cold War on Canadian Soil: Militarizing a Northern Environment," *Environmental History* 12, 2007, pp. 920–50.

34 L. Piper, *The Industrial Transformation of Subarctic Canada*, Vancouver, UBC Press, 2009.

35 M. Heymann, H. Knudsen, M. L. Lolck, et al., "Exploring Greenland: Science and Technology in Cold War Settings," *Scientia Canadensis* 33/2, 2010, pp. 11–42.

36 P. Sabin, "Voices from the Hydrocarbon Frontier: Canada's Mackenzie Valley Pipeline Inquiry, 1974–1977," *Environmental History Review* 19/1, 1995, pp. 17–48.

37 H. M. Carlson, "A Watershed of Words: Litigating and Negotiating Nature in Eastern James Bay, 1971–75," *Canadian Historical Review* 85/1, 2004, pp. 63–84.

38 S. Bocking, "Science and Spaces in the Northern Environment," *Environmental History* 12, 2007, pp. 867–94 (p. 887).

39 C. L. Smith, "Radiation Hazard to Man and Animals from Fallout in the Arctic," *Polar Record* 12/81, 1965, pp. 709–16.

40 P. Coates, "Amchitka, Alaska: Toward the Bio-Biography of an Island," *Environmental History* 1/4, 1996, pp. 20–45.

41 M. Cioc, B. O. Linnér, and M. Osborn, "Environmental History Writing in Northern Europe," *Environmental History* 5/3, 2000, pp. 396–406; H. Laudon, "Separating Natural Acidity from Anthropogenic Acidification in the Spring Flood of Northern Sweden," PhD thesis, Swedish University of Agricultural Sciences, 2000, p. 9.

42   H. Hung, R. Kallenborn, K. Breivik, et al., "Atmospheric Monitoring of Organic Pollutants in the Arctic under the Arctic Monitoring and Assessment Programme (AMAP): 1993–2006," *Science of the Total Environment* 408/15, 2010, pp. 2854–73.

43   Arctic Climate Impact Assessment, *Impacts of a Warming Arctic*, Cambridge, Cambridge University Press, 2004.

References

Allan, R. C., and Keay, I., "Saving the Whales: Lessons from the Extinction of the Eastern Arctic Bowhead," *Journal of Economic History* 64/2, 2004, pp. 400–32.

Arctic Climate Impact Assessment, *Impacts of a Warming Arctic*, Cambridge, Cambridge University Press, 2004.

Avango, D., Hacquebord, L., Aalders, Y., et al., "Between Markets and Geopolitics: Natural Resource Exploitation on Spitsbergen from 1600 to the Present Day," *Polar Record* 47/240, 2011, pp. 29–39.

Bassin, M., "Expansion and Colonialism on the Eastern Frontier: Views of Siberia and the Far East in pre-Petrine Russia," *Journal of Historical Geography*, 14/1, 1988, pp. 3–21.

Bocking, S., "Science and Spaces in the Northern Environment," *Environmental History* 12, 2007, pp. 867–94.

Bockstoce, J. R., Botkin, D. B., Philp, A., et al., "The Geographic Distribution of Bowhead Whales, *Balaena mysticetus*, in the Bering, Chukchi, and Beaufort Seas: Evidence from Whaleship Records, 1849–1914," *Marine Fisheries Review* 67/3, 2005, pp. 1–43.

Bravo, M., and Sörlin, S. (eds.), *Narrating the Arctic: A Cultural History of Nordic Scientific Practices*, Canton, Watson Publishing International, 2002.

Buckland, P. C., Amorosi, T., Barlow, L. K., et al., "Bioarchaeological and Climatological Evidence for the Fate of Norse Farmers in Medieval Greenland," *Antiquity* 70, 1996, pp. 88–96.

Carlson, H. M., "A Watershed of Words: Litigating and Negotiating Nature in Eastern James Bay, 1971–75," *Canadian Historical Review* 85/1, 2004, pp. 63–84.

Cioc, M., Linnér, B. O., and Osborn, M., "Environmental History Writing in Northern Europe," *Environmental History* 5/3, 2000, pp. 396–406.

Coates, P., "Amchitka, Alaska: Toward the Bio-Biography of an Island," *Environmental History* 1/4, 1996, pp. 20–45.

Farish, M., "Frontier Engineering: From the Globe to the Body in the Cold War Arctic," *The Canadian Geographer* 50/2, 2006, pp. 177–96.

Fossett, R., *In Order to Live Untroubled: Inuit of the Central Arctic, 1550–1940*, Winnipeg, University of Manitoba Press, 2001.

Glover, J. M., "Sweet Days of a Naturalist: Olaus Murie in Alaska, 1920–26," *Forest & Conservation History* 36/3, 1992, pp. 132–40.

Hacquebord, L., "Three Centuries of Whaling and Walrus Hunting in Svalbard and Its Impact on the Arctic Ecosystem," *Environment and History* 7, 2001, pp. 169–85.

Heymann, M., Knudsen, H., Lolck, M. L., et al., "Exploring Greenland: Science and Technology in Cold War Settings," *Scientia Canadensis* 33/2, 2010, pp. 11–42.

Hung, H., Kallenborn, R., Breivik, K., et al., "Atmospheric Monitoring of Organic Pollutants in the Arctic under the Arctic Monitoring and Assessment Programme (AMAP): 1993–2006," *Science of the Total Environment* 408/15, 2010, pp. 2854–73.

Josefsson, T., Bergamn, I., and Östlund, L., "Quantifying Sami Settlement and Movement Patterns in Northern Sweden 1700–1900," *Arctic* 63/2, 2010, pp. 141–54.

Koerner, L., *Linneaus: Nature and Nation*, Cambridge, MA, Harvard University Press, 1999.

Lackenbauer, P. W., and Farish, M., "The Cold War on Canadian Soil: Militarizing a Northern Environment," *Environmental History* 12, 2007, pp. 920–50.

Laudon, H., "Separating Natural Acidity from Anthropogenic Acidification in the Spring Flood of Northern Sweden," PhD thesis, Swedish University of Agricultural Sciences, 2000.

Morse, K., *The Nature of Gold: An Environmental History of the Klondike Gold Rush*, Seattle, University of Washington Press, 2003.

Östlund, L., Zackrisson, O., and Hörnberg, G., "Trees on the Border between Nature and Culture: Culturally Modified Trees in Boreal Sweden," *Environmental History* 7/1, 2002, pp. 48–68.

Piper, L., *The Industrial Transformation of Subarctic Canada*, Vancouver, UBC Press, 2009.

Piper, L., and Sandlos, J., "A Broken Frontier: Ecological Imperialism in the Canadian North," *Environmental History* 12/4, 2007, pp. 759–95.

Ray, A. J., "Some Conservation Schemes of the Hudson's Bay Company, 1821–1870: An Examination of the Problems of Resource Management in the Fur Trade," *Journal of Historical Geography* 1, 1975, pp. 49–68.

Richards, J. F., *The Unending Frontier: An Environmental History of the Early Modern World*, Berkeley, University of California Press, 2003.

Sabin, P., "Voices from the Hydrocarbon Frontier: Canada's Mackenzie Valley Pipeline Inquiry, 1974–1977," *Environmental History Review* 19/1, 1995, pp. 17–48.

Sandlos, J., "From the Outside Looking In: Aesthetics, Politics, and Wildlife Conservation in the Canadian North," *Environmental History* 6/1, 2001, pp. 6–31.

Smith, C. L., "Radiation Hazard to Man and Animals from Fallout in the Arctic," *Polar Record* 12/81, 1965, pp. 709–16.

Wråkberg, U., "Nature Conservationism and the Arctic Commons of Spitsbergen 1900–1920," *Acta Borealia* 23/1, 2006, pp. 1–23.

Further Reading

Berg, R., and Jakobsson, E., "Nature and Diplomacy: The Struggle over the Scandinavian Border Rivers in 1905," *Scandinavian Journal of History* 31/3–4, 2006, pp. 270–89.

Bolotova, A., "Colonization of Nature in the Soviet Union: State Ideology, Public Discourse, and the Experience of Geologists," *Historical Social Research* 29/3, 2004, pp. 104–23.

Carlos, A. M., and Lewis, F. D., "Indians, the Beaver and the Bay: The Economics of Depletion in the Lands of the Hudson's Bay Company, 1700–1763," *Journal of Economic History* 53/3, 1993, pp. 465–94.

Hewitt, G., "The Genetic Legacy of the Quaternary Ice Ages," *Nature* 405, 2000, pp. 907–13.

Östlund, L., "Logging the Virgin Forest: Northern Sweden in the Early-Nineteenth Century," *Forest & Conservation History* 39/4, 1995, pp. 160–71.

CHAPTER TEN

The Middle East in Global Environmental History

ALAN MIKHAIL

One of the gaping holes in the global story of the environment has so far been the history of the Middle East. This is a major problem for our understanding of global environmental history since, from at least antiquity, the Middle East has been the crucial zone of connection between, on the one hand, Europe, the Mediterranean, and Africa and, on the other, East, Central, South, and Southeast Asia.[1] It was an arena of contact where European and South Asian merchants traveled and conducted commerce, where each year religious pilgrims from Mali to Malaysia came not only to visit Mecca and Medina but also to buy goods in the many bazaars of cities throughout the Middle East. Pastoral nomads moved from northern India and Central Asia through Iran and into Anatolia (see Map 10.1). Ships carrying goods, people, vermin, and ideas sailed from India, China, and Southeast Asia to ports on the Red Sea and Persian Gulf. Beginning in the late medieval and early modern periods, the world from the western Mediterranean to China was one characterized by intense circulation, interconnection, and movement.[2] All the threads of these connections ran through the Middle East, with important environmental implications.[3]

Just as global trends in trade were deeply intertwined with what was happening in the Middle East, so too large-scale trends in climate, disease, and crop diffusion impacted and were impacted by the histories of the Middle East's many environments. For example, one of the results of the rise and spread of Islam in the seventh and eighth centuries was the movement of new kinds of crops and agricultural technologies across Eurasia. In what one scholar has described as the "Islamic green revolution," the unity provided by the new religion for the first time brought disparate parts of the world together into a unified ecological zone.[4] In the medieval period and later, it was most likely through the Middle East that plague moved westward to the Mediterranean basin.[5] In another medieval example of environmental exchange, knowledge of irrigation technologies and waterworks was transferred from the Muslim world to Spain and Italy.[6] In the fifteenth

A Companion to Global Environmental History, First Edition. Edited by J.R. McNeill and Erin Stewart Mauldin.
© 2012 John Wiley & Sons, Ltd. Published 2015 by John Wiley & Sons, Ltd.

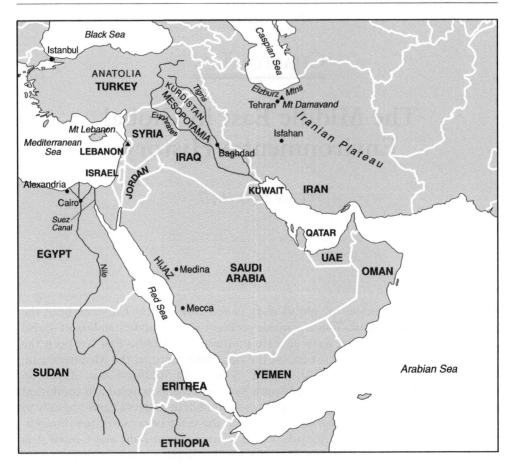

Map 10.1   The Middle East

and sixteenth centuries, coffee grown in the soils of Yemen, using techniques borrowed from East Africa, soon filled cups in Istanbul, Vienna, and Isfahan.[7] Trans-Saharan caravan networks ensured a steady circulation of animals, salt, paper, disease, and human slaves between the Middle East, North Africa, and West Africa.[8] And it was most likely through the Ottoman Empire that maize and other New World crops moved east.[9] In the twentieth century, clearly the extraction, refinement, and distribution of Middle Eastern oil were of enormous global environmental and geopolitical significance.

The history of climate change as well was deeply affected by what was happening in various parts of the Middle East. Whether the Little Ice Age in the early modern period that greatly reduced agricultural yields in the Ottoman Empire, Icelandic volcano eruptions at the end of the eighteenth century that affected Nile flood levels, El Niño-induced famines at the end of the nineteenth century in Anatolia and Iran, or today's warnings of global warming, the Middle East has clearly been very deeply involved in these instances of climatic alteration, and studying these cases can offer us much as we try to understand the history of climate change.[10]

In other words, the Middle East was a region that served to diffuse agricultural and natural products, knowledge of environmental manipulation techniques, climatic trends, and disease agents across large swathes of Eurasia and the world. Therefore, to understand the history of these crops, diseases, commodities, weather patterns, and technologies, we must necessarily understand the Middle Eastern components of their global histories. This is perhaps clear and unsurprising, but it remains the fact that much of this history continues to be underresearched.

In what follows, I first review the historiographical possibilities of writing Middle East environmental histories with an eye toward the question of available sources for these histories. I then discuss four areas of research on environmental topics in the Middle East: climate, energy, disease, and culture. My goal is both to review some of the work that has already been done on these subjects and to suggest their significance to future considerations of the global importance of Middle East environmental history.

Sources for Middle East Environmental History: Seasons of Want or Plenty?

One of the most commonly cited reasons for the lack of environmental knowledge of the region is a paucity of sources. As much new work is showing, however, it is simply untrue that there are not enough sources to narrate environmental histories of the Middle East and North Africa. In fact just the opposite is the case. Like China and South Asia, the Middle East offers a wealth of continuous source materials from antiquity to the present.[11] In the Nile Valley for instance, we have an unbroken documentary record of Egyptians' interactions with the river from at least as early as 3000 BCE to the present. Similarly, archeological and written sources from antiquity to today exist on the environmental manipulation techniques of populations in Anatolia, the Iranian Plateau, the Mesopotamian lands of Iraq, and elsewhere. Thus, the available published and unpublished sources of the Middle East offer environmental historians nearly unparalleled empirical depth and detail to be able to track environmental change, landscape manipulation, and human–environment interactions over millennia. Having the sources to operate over this timescale and at such a level of detail are luxuries not enjoyed by historians of most other regions of the globe.

Thus in debates over the long-term effects of human manipulation of the natural world, the Middle East offers us perhaps the most complete and longest available documentary record. Scholars have so far used these sources in many different ways. There is a sizeable geographical, palynological, biological, and geoarcheological literature that gives us a historic and cultural geographic picture of ecological systems and their operations.[12] These works can be roughly divided into two major categories: descriptions of the continuities of various delicate equilibriums of human settlements and environments, and studies of human groups inevitably doomed to collapse because of the inherent contradictions of their ecological systems. Both of these modes of narrating the environmental past are more concerned with painting a picture of ecological systems and their function than they are with either how humans were integrally connected to environments or how the many relationships between humans and nature changed over time. In the first story of continuity, the picture we get is one of dynamic stasis – of human communities constantly working and struggling to survive in the face of enormous environmental challenges. In this story of society constantly running only to stay in the same place, there is very little attention given to human agency's effects on ecological systems. Likewise, the second narrative of eventual and inevitable decline also gives primacy of place to ecological limits while sidelining historical specificity and contingency. If certain human communities were predetermined to decline because of a fundamental flaw or contradiction inherent in their local world, then what, other than documenting this downward spiral, is left for the historian to do?

Two examples from this literature are Karl Butzer's work on Egypt and Peter Christensen's study of Iran.[13] Both accounts sketch a picture of the ecological limits within which human communities were made to live and survive. In both of these cases and in others as well, the picture is one of systems in overall equilibrium on the timescale of generations despite massive fluctuations from year to year. It is no coincidence that the examples of both Egypt and the Iranian plains are areas reliant on the yearly floods of massively complex river systems. Floods could be bad or good from year to year, but overall, through the development of various technologies of environmental manipulation, chiefly irrigation and agriculture, communities were able to ride out these vagaries to achieve some semblance of ecological balance that either continued for millennia in the case of Egypt or eventually broke down as in the case of Iran. Other examples of this kind of description of ecological systems that prevailed throughout the Middle East before 1500 include J. M. Wagstaff's general study, Carlos Cordova's work on landscape change in Jordan, Russell Meiggs' and J. V. Thirgood's examinations of wood supplies around the Mediterranean, and Robert M. Adams' examination of southern Iraq.[14]

Historians have also successfully used the wealth of available evidence to examine how and with what techniques humans in extremely ecologically fragile environments were able to survive for millennia and what perturbations to those environments made them no longer viable.[15] In this vein, Peter Christensen shows how various enclave communities on the Iranian Plateau were able to manage their environmental vulnerability for centuries through the resilient and careful use of irrigation technologies, agriculture, and nomadic pastoralism. Similarly, J. R. McNeill paints a picture of Mediterranean mountain environments in Turkey and Morocco that shows just how fragile these environments were over time. With a long documentary record at his disposal, he is able to show that from year to year the fortunes of any one environment might rise and fall drastically, but, seen over the course of centuries, these environments were actually quite resilient and able to deal with intense change. Thus, it is only the presence of such a long record that allows us to see how these environments functioned over time. Without these sources, we might mistakenly judge a year of acute food shortages to be the norm in the mountains.

The presence of such a long record of human interaction with the environment in the Middle East also allows environmental historians plenty of opportunity to offer new perspective on some long-debated issues in the field of environmental history. One of the most obvious and discussed of these historiographical interventions is that revolving around Karl Wittfogel's thesis of oriental despotism.[16] Several of his examples concerning the emergence of despotic forms of authoritarian government come from the Middle East – Mesopotamia and the Nile Valley. But it is precisely these examples that refute his point. An examination of the documentary record that Wittfogel fails to consult shows that it was the very people he claims were exploited by sultans, emperors, and shahs who were actually the ones in control of the day-to-day function and maintenance of the large-scale irrigation networks that are so crucial to his argument.[17]

Environmental historians interested in the history of natural disasters and their effects on human communities will likewise find much material in the Middle East to test their hypotheses. We have records and accounts of earthquakes and other events from antiquity to the present.[18] The history of public works is also similarly usefully addressed by considering the record of such projects in the Middle East. As it was elsewhere in the world (India and the American Southeast for instance), the middle of the twentieth century was a period of intense dam-building throughout the Middle East.[19] What was the connection between these major public-works projects and earlier instances of this kind of environmental work (the Suez Canal, the Ottoman Don–Volga canal-building project, or Safavid plans to build a dam to divert the course of the Karun River toward the new capital at Isfahan) or current ones (the Red to Dead pipeline project in Jordan or Turkey's dam projects in its southeast for example)?[20] Again, because we have a record of such projects and social, economic, cultural, and political commentaries on them, they offer a rich source base for thinking about the environmental history of public works in the Middle East and beyond.

Climate

Historians of climate change stand to gain much from recent work done on the environmental history of the Middle East that challenges various propositions put forth in the literature on climate change.[21] One of these is the idea that the Middle Ages were a period of global warming. Evidence for this worldwide weather trend comes mostly from European and Chinese sources. Using Persian and Arabic sources from medieval Iran, however, Richard W. Bulliet has shown that in fact the opposite seems to have been occurring in Iran in this period.[22] That is, an intense period of cold – what he calls the "Big Chill" – gripped Iran in the eleventh and early twelfth centuries leading to reductions in cotton cultivation levels. The Big Chill facilitated Turkic migrations west into Iran and eventually Anatolia and perhaps even contributed to the emergence of Shi'ism as the dominant form of religiosity in Iran. This period of cooling decimated large landholdings and much of the economic and social base of Iran, which likely made the Iranian Plateau vulnerable to Mongol and other incursions. After this period of cold, the Middle East and parts of Central Asia enjoyed a few centuries of temperate weather patterns that likely contributed to the emergence of the great empires of the early modern period – the Ottomans, Safavids, Mughals, Uzbeks, and Mamluks.

In the sixteenth and seventeenth centuries, however, another more well-known period of cool set into the Middle East and elsewhere. The Little Ice Age contributed to various economic and political troubles for the Ottoman Empire forcing a realignment of resources and sovereignty and was also a contributing factor to the general crisis of the seventeenth century.[23] A consideration of this climatic event puts the historiographically vexed issue of the decline, decentralization, or realignment of the Ottoman Empire in the seventeenth and eighteenth centuries in a wholly new light. One of the most important and discussed facets of this period in the Ottoman Empire are the Celali peasant revolts at the end of the sixteenth and the beginning of the seventeenth centuries. As Sam White's research shows, there is now ample evidence linking these rural rebellions to ecological pressures brought about by the Little Ice Age. The manifestations of this climatic change in central Anatolia – drought, famine, plagues of livestock and people, and eventually increasing rates of mortality among Ottoman peasants – led to widespread violence and banditry that finally coalesced into an organized rebellion against the Ottoman state. Thus, the causes of the Celali revolts are not only, or perhaps even most importantly, to be found in the realms of the economic and the political, but rather must also be seen as results of ecological factors. This middle period of Ottoman history that so many historians have interpreted as the crucial hinge between the rise and sixteenth-century efflorescence of the empire and its eventual efforts at reform in the nineteenth century is perhaps more usefully thought of as a time when the empire was attempting to manage massive climatic fluctuations and their attendant effects on agricultural production, human and animal populations, and flood levels.

Likewise, the last two decades of the eighteenth century were also a period of, albeit less dramatic, but no less crucial, weather fluctuations. The eruption in 1783 and 1784 of the Laki fissure in Iceland began a climate event of enormous proportions.[24] Monsoon levels decreased over the Indian Ocean, temperatures plunged around the Mediterranean, flood levels of various riverine systems in the Middle East were dramatically reduced, and a 20-year period of famine, drought, disease, and hardship began. These climatic features were no doubt contributing factors to the political turmoil of the end of the eighteenth century that resulted in, among other things, various European incursions into the Middle East and eventually the period of Ottoman reform known as the Tanzimat. In the second half of the nineteenth century, a similar period of climatically induced famine and food shortages contributed to discontent among provincial notables in the Ottoman Empire and to calls for political reform in the empire. In short, there is plenty of good reason to conceive of the last millennium (and perhaps earlier periods as well) of Middle Eastern history along the lines of climate and weather patterns rather than political states. Indeed, it seems that periods of cooling or climatic fluctuation set off by events like the Laki eruption in Iceland often had much more impact on the vast majority of the rural peoples of the Middle East than did the political powers ostensibly ruling over them.

Energy

The Middle East also has much to contribute to understandings of the history of energy around the globe. Accepting that there has been only one major transition in energy regimes over the course of human history – from solar energy to fossil fuels – the Middle East, with the largest known petroleum supplies in the world, clearly played a central role in this story.[25] Understanding how humans cultivated and then exploited their energy supplies must be at the heart of any history of human communities. Since the latter half of the nineteenth century, much of the world's energy supplies have come from the Middle East, and the region must therefore be included in any history of energy since this period. The Middle East, however, was clearly also very important to the global story of energy long before oil was ever discovered in the region. As in most of the pre-industrial world, energy utilization in the Middle East before the nineteenth century was characterized by near-total reliance on human and animal power. Water and wind power never made up a large portion of the energy sources utilized by people from Morocco to Iran. This was a result not of a lack of running-water resources, but rather of the availability of abundant quantities of cheap animal resources. Waterwheels were thus pushed by oxen; goods were very often transported by donkey rather than by river; and fields were tilled by animal-drawn plow. Thus, the history of animal and human power in the Middle East helps us understand this most vital of preindustrial environmental energy stories as it played out elsewhere in the world as well.

The more recent global transition to a fossil-fuel economy is one that can therefore be thought of as the replacement of animals with oil as the region's most abundant and cheapest source of energy.[26] This energy transition is one that has largely occurred throughout the world, increasing the geopolitical importance of Middle Eastern oil reserves. Putting petroleum utilization and consumption in the larger historical context of transitions between energy regimes shows us that oil and the supposed need to seek out alternative sources of energy in the Middle East are the latest iterations of a much longer story of low-cost energy.[27] Because animal power was (and in many places still is) so effective and cheap, there was no incentive to find and use other sources of energy, like water or wind. Thus, animals remained the primary means of transport and power in the Middle East and Central Asia while Europeans began to experiment with water-powered mills. Oil, like animals, therefore, represents a new phase of energy regime in the Middle East, and as long as petroleum remains abundant and cheap, there will be little incentive for Middle Eastern oil states to seek out other forms of energy.[28] Thus, by making energy outputs, whether caloric or carbon organic, our unit of analysis, the politics of oil in the Middle East no longer comes to be seen as only a problem of modern industrial economies, but rather as a much older problem of energy and society.

Disease

One of the ripest areas for research on the global import of the environmental history of the Middle East is the epidemiology and etiology of various diseases. Because so many disease epidemics in the Middle East, whether plague, cholera, or something else, deeply impacted populations and almost always traveled between the region and elsewhere, we have ample documentary evidence of these outbreaks, which historians have used to great effect. Some of this work is beginning to put disease in a wider ecological context as one of various other environmental phenomena.[29] This is partly a result of the realization of how disease functioned in Middle Eastern environments and of how people in those environments conceived of disease. In other words, people who experienced plague and other diseases in the Middle East considered them to be one in a series of environmental forces that included famine, rain, drought, flood, and wind.

For example, in the case of Egypt, plague was just one component of a regular biophysical pathology of the environment.[30] The disease came and went as one iteration in a cycle that included famine, flood, drought, price inflation, wind, and revolt. Climatic factors, the level of the Nile's inundation, and relative populations of rats, fleas, and humans contributed to the incidence and severity of plague for any given year in Egypt. As such, like a low Nile flood, famine, or other hardships, plague was an accepted and expected environmental reality. Egyptians understood it to be one pathological element of the Egyptian environment. Part of the reason for this acceptance of plague as part of the Egyptian environment was its regularity and frequency. A new plague epidemic visited Egypt on average every nine years for the entire period from 1347 to 1894,[31] and plague was reported in Egypt in 193 of these 547 years.[32] This high incidence of plague suggests that the disease and its epidemics were regular and expected occurrences in the lives of most Egyptians and that historians should pay closer attention to the role of the disease in the shaping of Egyptian and Middle Eastern history.

Plague in Egypt owed the regularity of its outbreaks to Egypt's strategic location as a center of trade and to the many thousands of people – and rats – that came to Egypt every year from elsewhere. In strict epidemiological terms, plague was not endemic to Egypt. The endemic foci of plague include areas of Central Asia, Kurdistan, Central Africa, and northwestern India, where the disease is permanently maintained among rodent populations. Plague came to Egypt from these and other places through the movement of goods, rats, fleas, and people.[33] Indeed, the arrival of hundreds of ships and caravans every week from places like Istanbul, India, Yemen, the Sudan, China, Central Africa, and Iraq ensured a consistent flow and constantly replenished supply of disease carriers, hosts, and vectors. The two major entry points of plague were through the main trading areas of Egypt: its Mediterranean ports and the southern route from the Sudan. Thus, plague either entered Egypt through the ports of Alexandria and Rosetta before making its way further inland, or it traveled from Central Africa to the Sudan and then into Egypt. Most Egyptians, however, did not concern themselves with whether or not plague was endemic to Egypt. What mattered most was that plague functioned as though it were endemic. Whether a particular epidemic's origins lay in Kurdistan or the Sudan did not affect the ultimate outcome of the disease on the Egyptian population.

Another example of the centrality of the Middle East to the global environmental history of disease is the role of the annual Islamic pilgrimage to Mecca (*ḥajj*) in the spread of various epidemic diseases. As perhaps the largest single annual gathering of people anywhere in the early modern and modern worlds, the *ḥajj* had important epidemiological implications. For a few days every year, hundreds of vessels, thousands of people and caravan animals, and millions of germs from all over the world gathered and jostled around in the Arabian Peninsula. Upon returning home, many people found that they had contracted some form of epidemic disease. In the nineteenth century, this was most usually cholera. All over the world, but most especially in South Asia, North America, and Europe, in the decades after the 1830s, cholera struck major cities with great force, leading to the application of various kinds of protective measures.[34]

The pilgrimage was one of the main avenues through which cholera moved between South Asia and the Mediterranean and between areas of Europe and the Middle East. Both before and after the opening of the Suez Canal in 1869, Egypt was one of the main transit points for *ḥajj* pilgrims – and hence disease – moving to and from the Mediterranean. Egyptian ports thus became especially important as chokepoints to prevent the movement of cholera to and from the Hijaz during the pilgrimage season. In 1833, for example, 2,000 pilgrims from various parts of the Ottoman Empire and several hundred Russian Tatars from around the Black Sea were quarantined in a lazaretto in Alexandria on their way to Mecca.[35] Once it was determined that they were not infected with cholera, plague, or any other disease, they were allowed to join the caravan to the Hijaz. The Alexandria lazaretto could accommodate up to 2,500 people, and food, clean quarters, and various other provisions were supplied by the government of Egypt to those in quarantine. The British consul general who visited the Alexandria lazaretto in 1833 made special note of the facility's cleanliness, convenience, and comfort.

Such quarantine efforts notwithstanding, cholera epidemics originating in the yearly pilgrimage would at various other points in the nineteenth century overwhelm stations charged with controlling the disease's diffusion. For example, in 1881, there was a particularly virulent outbreak of cholera in the Hijaz.[36] In response, the quarantine board of Alexandria ordered all pilgrims returning to Egypt from the Hijaz to spend at least 28 days in one of three quarantine stations in the city. These three facilities, however, were soon overwhelmed by the enormous number of pilgrims that came to the city. Overcrowding, a lack of food and water, and a decision by the quarantine board to extend the mandatory detention made an already bad situation even worse. A contained outbreak of cholera thus became a full-blown epidemic throughout these camps and in surrounding areas. Many of the pilgrims being held in these facilities soon began to riot and some even succeeded in burning down one of the lazarettos before escaping altogether. Such experiences of diseases associated with the pilgrimage led various governments at the end of the nineteenth century to institute medical regulations for pilgrims going on and coming from the yearly pilgrimage.[37] These measures were thus attempts to manage the centrality of the *ḥajj* to the global ecology and etiology of disease.

Culture

The Middle East is the birthplace and home of many of the world's cultural, linguistic, and religious traditions. From the earliest documentary evidence until today, Middle Eastern religions and cultures have shown an interest in the interactions between humans and the rest of nature. There is thus much work to be done in the general realm of relationships between the environment and Middle Eastern cultures and religions. Foremost among these religious traditions in the last 1,500 years are the cultures of Islam in the region. Thus to understand how the people of one of the world's largest religious traditions thought about, dealt with, and otherwise related to the natural world, the Middle East offers much material.

The subject of Islam and nature is of the utmost importance to advancing our understanding of how humans in the past have thought about the environment.[38] Islam was and is a cultural idiom stretching from Africa and Europe in the west to China and Indonesia in the east. And given the massive numbers of Muslims living all over the world, Islam is now a religious practice of global import. As such, both what various Islamic texts from different places and times say about the natural world, environmental manipulation, the nonhuman living realm, and so forth and the history of how Muslims have actually put their religious ideas about nature into practice in the real world should be a major area of scholarly research. From Clarence Glacken's *Traces on the Rhodian Shore* to Ramachandra Guha's and others' writings on environmentalism in India, much work exists on Jewish, Buddhist, Christian, Hindu, and Shamanist views of nature.[39] Islam and its various traditions deserve no less.

The Middle East's extreme demographic heterogeneity also affords environmental historians ample opportunity to examine how different cultural groups have thought about the same natural spaces and environments. In Mt. Lebanon, for instance, Shi'ite, Druze, Maronite, Sunni Muslim, Orthodox Christian, Armenian, and other cultural and religious groups all live on the same mountains and in the same valleys. Similarly in southeastern Turkey, communities of Turks, Kurds, Arabs, Armenians, and others share the same rivers, mountains, and plains.[40] Is there a difference in access to environmental resources or in environmental manipulation techniques between groups? What is unique or similar about how each of these groups thinks about and interacts with their shared environments? In general, both in the past and today, what are the different ways distinct cultural and religious groups engage nature?

Conclusion

Given the Middle East's enormous cultural traditions and diversity, its centrality to the global traffic of economics, politics, and ecology, and the vast quantities of source materials available from various regions and from all periods, it should be no surprise that the region has much to teach us about various facets of global environmental history. Add to these facts the presence of the world's longest river along with other important watersheds; the global significance of the region's oil; the Suez Canal and other crucial past and present public works; the historic importance of cotton, coffee, and other agricultural commodities from the region; and the presence of a vast diversity of landscapes and ecologies and it becomes clear that there are countless topics to address in the environmental history of the Middle East and that these histories are crucial to any study of the global environment.

I have focused here on only a few examples from the existing literature on Middle East environmental history and have suggested some future areas of research. Others include the environmental impacts of war, gender and the environment, pollution, the extraction of resources other than oil, and the role of environmentalism and other kinds of environmental thought in the Middle East. In the years ahead, all these topics and more will surely find their historians.

Notes

1   This chapter is based on a text previously published in the online journal *History Compass*. The author thanks Wiley-Blackwell for permission to republish.

2   For works that discuss the Middle East as a Eurasian contact zone, see J. L. Abu-Lughod, *Before European Hegemony: The World System* AD *1250–1350*, New York, Oxford University Press, 1989; R. M. Eaton, "Islamic History as Global History," in M. Adas (ed.), *Islamic and European Expansion: The Forging of a Global Order*, Philadelphia, Temple University Press, 1993, pp. 1–36. For a study of these connections in the early modern Muslim world, see M. Alam and S. Subrahmanyam, *Indo-Persian Travels in the Age of Discoveries, 1400–1800*, Cambridge, Cambridge University Press, 2007.

3   The most ambitious work to tackle this subject is J. F. Richards, *The Unending Frontier: An Environmental History of the Early Modern World*, Berkeley, University of California Press, 2003. In a telling indication of the absence of the Middle East from global environmental history, this magisterial work has very little to say about the Middle East.

4   A. Watson, *Agricultural Innovation in the Early Islamic World: The Diffusion of Crops and Farming Techniques, 700–1100*, Cambridge, Cambridge University Press, 1983.

5   M. W. Dols, *The Black Death in the Middle East*, Princeton, NJ, Princeton University Press, 1977; W. H. McNeill, *Plagues and Peoples*, Garden City, NY, Anchor Press, 1976; D. Panzac, *La peste dans l'Empire Ottoman, 1700–1850*, Louvain, Association pour le développement des études turques, 1985.

6   T. F. Glick, *Irrigation and Society in Medieval Valencia*, Cambridge, MA, Harvard University Press, 1970; T. F. Glick, *Irrigation and Hydraulic Technology: Medieval Spain and Its Legacy*, Aldershot, Variorum, 1996.

7   R. Hattox, *Coffee and Coffeehouses: The Origins of a Social Beverage in the Medieval Near East*, Seattle, University of Washington Press, 1985; M. Tuchscherer (ed.), *Le commerce du café avant l'ère des plantations coloniales: Espaces, réseaux, sociétés (XVᵉ-XIXᵉ siècle)*, Cairo, Institut français d'archéologie orientale, 2001.

8   G. Lydon, *On Trans-Saharan Trails: Islamic Law, Trade Networks, and Cross-Cultural Exchange in Nineteenth-Century Western Africa*, Cambridge, Cambridge University Press, 2009.

9   F. Tabak, *The Waning of the Mediterranean, 1550–1870: A Geohistorical Approach*, Baltimore, The Johns Hopkins University Press, 2008, pp. 255–69. On the diffusion of maize from Egypt and North Africa to other parts of Africa, see J. C. McCann, *Maize and Grace: Africa's Encounter with a New World Crop, 1500–2000*, Cambridge, MA, Harvard University Press, 2007.

10 S. White, *The Climate of Rebellion in the Early Modern Ottoman Empire*, Cambridge, Cambridge University Press, 2011; M. Davis, *Late Victorian Holocausts: El Niño Famines and the Making of the Third World*, London, Verso, 2001.

11 On the environmental histories of China from antiquity to the present, see M. Elvin, *The Retreat of the Elephants: An Environmental History of China*, New Haven, CT, Yale University Press, 2004; M. Elvin and L. Ts'ui-Jung (eds.), *Sediments of Time: Environment and Society in Chinese History*, Cambridge, Cambridge University Press, 1998. On South Asia, see R. H. Grove, V. Damodaran, and S. Sangwan (eds.), *Nature and the Orient: The Environmental History of South and Southeast Asia*, Delhi, Oxford University Press, 1998.

12 For a representative sampling of some of this work, see W. C. Brice (ed.), *The Environmental History of the Near and Middle East since the Last Ice Age*, London, Academic Press, 1978.

13 K. W. Butzer, *Early Hydraulic Civilization in Egypt: A Study in Cultural Ecology*, Chicago, University of Chicago Press, 1976; P. Christensen, *The Decline of Iranshahr: Irrigation and Environments in the History of the Middle East, 500 BC to AD 1500*, Copenhagen, Museum Tusculanum Press, 1993.

14 J. M. Wagstaff, *The Evolution of Middle Eastern Landscapes: An Outline to AD 1840*, London, Croon Helm, 1985. To be accurate, Wagstaff's book does address the post-1500 period, but the bulk of the work (over two-thirds) concerns the earlier period. C. E. Cordova, *Millennial Landscape Change in Jordan: Geoarchaeology and Cultural Ecology*, Tucson, AZ, University of Arizona Press, 2007; R. Meiggs, *Trees and Timber in the Ancient Mediterranean World*, Oxford, Clarendon Press, 1982; J. V. Thirgood, *Man and the Mediterranean Forest: A History of Resource Depletion*, London, Academic Press, 1981; R. McC. Adams, *Land behind Baghdad: A History of Settlement on the Diyala Plains*, Chicago, University of Chicago Press, 1965.

15 Christensen, *The Decline of Iranshahr*; J. R. McNeill, *The Mountains of the Mediterranean World: An Environmental History*, Cambridge, Cambridge University Press, 1992.

16 K. A. Wittfogel, *Oriental Despotism: A Comparative Study of Total Power*, New Haven, CT, Yale University Press, 1957.

17 This argument is advanced more thoroughly in the case of Egypt in A. Mikhail, *Nature and Empire in Ottoman Egypt: An Environmental History*, Cambridge, Cambridge University Press, 2011.

18 M. R. Sbeinati, R. Darawcheh, and M. Mouty, "The Historical Earthquakes of Syria: An Analysis of Large and Moderate Earthquakes from 1365 BC to 1900 AD," *Annals of Geophysics* 48, 2005, pp. 347–435; E. Zachariadou (ed.), *Natural Disasters in the Ottoman Empire*, Rethymnon, Crete University Press, 1999; N. N. Ambraseys and C. F. Finkel, *The Seismicity of Turkey and Adjacent Areas: A Historical Review, 1500–1800*, Istanbul, Eren, 1995.

19 On the Aswan Dam, see Y. A. Shibl, *The Aswan High Dam*, Beirut, The Arab Institute for Research and Publishing, 1971. See also the relevant sections of J. Waterbury, *Hydropolitics of the Nile Valley*, Syracuse, NY, Syracuse University Press, 1979; R. O. Collins, *The Nile*, New Haven, CT, Yale University Press, 2002. On the construction of dams during what came to be known as Iran's First Seven-Year Plan from 1948 to 1955, see P. Beaumont, "Water Resource Development in Iran," *The Geographical Journal* 140, 1974, pp. 418–31; G. R. Clapp, "Iran: A TVA for the Khuzestan Region," *Middle East Journal* 11, 1957, pp. 1–11.

20 On the Suez Canal, see D. A. Farnie, *East and West of Suez: The Suez Canal in History, 1854–1956*, Oxford, Clarendon Press, 1969. On the Don–Volga Canal, see H. İnalcık, "The Origins of the Ottoman–Russian Rivalry and the Don–Volga Canal 1569," *Les annales de l'Université d'Ankara* 1, 1947, pp. 47–106; A. N. Kurat, "The Turkish Expedition to Astrakhan and the Problem of the Don–Volga Canal," *Slavonic and East European Review* 40, 1961, pp. 7–23. On the Karun River project, see A. Khazeni, *Tribes and Empire on the Margins of Nineteenth-Century Iran*, Seattle, University of Washington Press, 2009, pp. 23–5. On the construction of a pipeline between the Red and Dead seas, see B. N. Asmar, "The Science and Politics of the Dead Sea: Red Sea Canal or Pipeline," *Journal of Environment and Development* 12, 2003, pp. 325–39. On Turkey's Southeast Anatolia Development Project (GAP), see A. Çarkoğlu, and M. Eder, "Development *alla Turca*: The Southeastern Anatolia Development Project (GAP)," in F. Adaman and M. Arsel (eds.), *Environmentalism in Turkey: Between Democracy and Development?*, Aldershot, Ashgate, 2005, pp. 167–84;

L. Harris, "Postcolonialism, Postdevelopment, and Ambivalent Spaces of Difference in Southeastern Turkey," *Geoforum* 39, 2008, pp. 1698–708.

21 Some of the earliest attempts to integrate climate change into the history of the Middle East are R. Murphey, "The Decline of North Africa since the Roman Occupation: Climatic or Human?," *Annals of the Association of American Geographers* 41, 1951, pp. 116–32; W. Griswold, "Climatic Change: A Possible Factor in the Social Unrest of Seventeenth Century Anatolia," in H. W. Lowry and D. Quataert (eds.), *Humanist and Scholar: Essays in Honor of Andreas Tietze*, Istanbul, Isis Press, 1993, pp. 37–57. The best recent studies of the impacts of climate change in the Middle East are R. W. Bulliet, *Cotton, Climate, and Camels in Early Islamic Iran: A Moment in World History*, New York, Columbia University Press, 2009; White, *The Climate of Rebellion*.

22 Bulliet, *Cotton, Climate, and Camels in Early Islamic Iran*, pp. 69–95.

23 White, *The Climate of Rebellion*.

24 L. Oman, A. Robock, G. L. Stenchikov, and T. Thordarson, "High-Latitude Eruptions Cast Shadow over the African Monsoon and the Flow of the Nile," *Geophysical Research Letters* 33, 2006, p. L18711.

25 E. Burke III, "The Big Story: Human History, Energy Regimes, and the Environment," in E. Burke III and K. Pomeranz (eds.), *The Environment and World History*, Berkeley, University of California Press, 2009, pp. 33–53 (p. 35).

26 R. W. Bulliet, "The Camel and the Watermill," *International Journal of Middle East Studies* 42, 2010, pp. 666–8.

27 Burke, "The Big Story," pp. 33–53.

28 Bulliet, "The Camel and the Watermill."

29 For a recent review of this literature, see S. White, "Rethinking Disease in Ottoman History," *International Journal of Middle East Studies* 42, 2010, pp. 549–67.

30 For a more thorough discussion of this subject, see A. Mikhail, "The Nature of Plague in Late Eighteenth-Century Egypt," *Bulletin of the History of Medicine* 82, 2008, pp. 249–75.

31 M. W. Dols, "The Second Plague Pandemic and Its Recurrences in the Middle East: 1347–1894," *Journal of the Economic and Social History of the Orient* 22, 1979, pp. 162–89 (pp. 169 and 176).

32 Dols, "The Second Plague Pandemic," pp. 168–9 and 175–6; A. Raymond, "Les grandes épidémies de peste au Caire aux XVIIᵉ et XVIIIᵉ siècles," *Bulletin d'études orientales* 25, 1973, pp. 203–10.

33 M. W. Dols, "Plague in Early Islamic History," *Journal of the American Oriental Society* 94, 1974, pp. 371–83 (p. 381); Raymond, "Les grandes épidémies," pp. 208–9.

34 On cholera in India, see D. Arnold, *Colonizing the Body: State Medicine and Epidemic Disease in Nineteenth-Century India*, Berkeley, University of California Press, 1993, pp. 159–99. On the disease in the US, see C. E. Rosenberg, *The Cholera Years: The United States in 1832, 1849, and 1866*, Chicago, University of Chicago Press, 1987. On cholera in Europe, see R. J. Evans, *Death in Hamburg: Society and Politics in the Cholera Years, 1830–1910*, New York, Oxford University Press, 1987; C. J. Kudlick, *Cholera in Post-Revolutionary Paris: A Cultural History*, Berkeley, University of California Press, 1996.

35 L. Kuhnke, *Lives at Risk: Public Health in Nineteenth-Century Egypt*, Berkeley, University of California Press, 1990, p. 95.

36 Kuhnke, *Lives at Risk*, pp. 107–8.

37 J. R. McNeill, *Something New under the Sun: An Environmental History of the Twentieth-Century World*, New York, W. W. Norton, 2000, p. 196.

38 M. I. Dien, *The Environmental Dimensions of Islam*, Cambridge, Lutterworth Press, 2000; H. A. Haleem (ed.), *Islam and the Environment*, London, Ta-Ha Publishers, 1998; R. C. Foltz, F. M. Denny, and A. Baharuddin (eds.), *Islam and Ecology: A Bestowed Trust*, Cambridge, MA, Harvard University Press, 2003; R. C. Foltz (ed.), *Environmentalism in the Muslim World*, New York, Nova Science Publishers, 2005.

39 C. J. Glacken, *Traces on the Rhodian Shore: Nature and Culture in Western Thought from Ancient Times to the End of the Eighteenth Century*, Berkeley, University of California Press, 1967; M. Gadgil and R. Guha, *This Fissured Land: An Ecological History of India*, Berkeley, University of California Press, 1993. On Buddhism, see D. E. Cooper and S. P. James, *Buddhism, Virtue and Environment*, Aldershot, Ashgate, 2005. On Judaism, see M. D. Yaffe (ed.), *Judaism and Environmental Ethics: A Reader*, Lanham, MD, Lexington Books, 2001. For a useful beginning on the relationships between various religious traditions and the environment, see B. R. Taylor (ed.), *The Encyclopedia of Religion and Nature*, 2 vols., London, Thoemmes Continuum, 2005.

40 L. Harris, "Water and Conflict Geographies of the Southeastern Anatolia Project," *Society and Natural Resources* 15, 2002, pp. 743–59; A. I. Bagis, "Turkey's Hydropolitics of the Euphrates–Tigris Basin," *International Journal of Water Resources Development* 13, 1997, pp. 567–82.

References

Abu-Lughod, J. L., *Before European Hegemony: The World System* AD *1250–1350*, New York, Oxford University Press, 1989.

Adams, R. McC., *Land behind Baghdad: A History of Settlement on the Diyala Plains*, Chicago, University of Chicago Press, 1965.

Alam, M., and Subrahmanyam, S., *Indo-Persian Travels in the Age of Discoveries, 1400–1800*, Cambridge, Cambridge University Press, 2007.

Ambraseys, N. N., and Finkel, C. F., *The Seismicity of Turkey and Adjacent Areas: A Historical Review, 1500–1800*, Istanbul, Eren, 1995.

Arnold, D., *Colonizing the Body: State Medicine and Epidemic Disease in Nineteenth-Century India*, Berkeley, University of California Press, 1993.

Asmar, B. N., "The Science and Politics of the Dead Sea: Red Sea Canal or Pipeline," *Journal of Environment and Development* 12, 2003, pp. 325–39.

Bagis, A. I., "Turkey's Hydropolitics of the Euphrates–Tigris Basin," *International Journal of Water Resources Development* 13, 1997, pp. 567–82.

Beaumont, P., "Water Resource Development in Iran," *The Geographical Journal* 140, 1974, pp. 418–31.

Brice, W. C. (ed.), *The Environmental History of the Near and Middle East since the Last Ice Age*, London, Academic Press, 1978.

Bulliet, R. W., "The Camel and the Watermill," *International Journal of Middle East Studies* 42, 2010, pp. 666–8.

Bulliet, R. W., *Cotton, Climate, and Camels in Early Islamic Iran: A Moment in World History*, New York, Columbia University Press, 2009.

Burke, E., III, "The Big Story: Human History, Energy Regimes, and the Environment," in E. Burke III and K. Pomeranz (eds.), *The Environment and World History*, Berkeley, University of California Press, 2009, pp. 33–53.

Butzer, K. W., *Early Hydraulic Civilization in Egypt: A Study in Cultural Ecology*, Chicago, University of Chicago Press, 1976.

Çarkoğlu, A., and Eder, M., "Development *alla Turca*: The Southeastern Anatolia Development Project (GAP)," in F. Adaman and M. Arsel (eds.), *Environmentalism in Turkey: Between Democracy and Development?*, Aldershot, Ashgate, 2005, pp. 167–84.

Christensen, P., *The Decline of Iranshahr: Irrigation and Environments in the History of the Middle East, 500* BC *to* AD *1500*, Copenhagen, Museum Tusculanum Press, 1993.

Clapp, G. R., "Iran: A TVA for the Khuzestan Region," *Middle East Journal* 11, 1957, pp. 1–11.

Collins, R. O., *The Nile*, New Haven, CT, Yale University Press, 2002.

Cooper, D. E., and James, S. P., *Buddhism, Virtue and Environment*, Aldershot, Ashgate, 2005.

Cordova, C. E., *Millennial Landscape Change in Jordan: Geoarchaeology and Cultural Ecology*, Tucson, AZ, University of Arizona Press, 2007.

Davis, M., *Late Victorian Holocausts: El Niño Famines and the Making of the Third World*, London, Verso, 2001.

Dien, M. I., *The Environmental Dimensions of Islam*, Cambridge, Lutterworth Press, 2000.

Dols, M. W., *The Black Death in the Middle East*, Princeton, NJ, Princeton University Press, 1977.

Dols, M. W., "Plague in Early Islamic History," *Journal of the American Oriental Society* 94, 1974, pp. 371–83.

Dols, M. W., "The Second Plague Pandemic and Its Recurrences in the Middle East: 1347–1894," *Journal of the Economic and Social History of the Orient* 22, 1979, pp. 162–89.

Eaton, R. M., "Islamic History as Global History," in M. Adas (ed.), *Islamic and European Expansion: The Forging of a Global Order*, Philadelphia, Temple University Press, 1993, pp. 1–36.

Elvin, M., *The Retreat of the Elephants: An Environmental History of China*, New Haven, CT, Yale University Press, 2004.

Elvin, M., and Ts'ui-Jung, L. (eds.), *Sediments of Time: Environment and Society in Chinese History*, Cambridge, Cambridge University Press, 1998.

Evans, R. J., *Death in Hamburg: Society and Politics in the Cholera Years, 1830–1910*, New York, Oxford University Press, 1987.

Farnie, D. A., *East and West of Suez: The Suez Canal in History, 1854–1956*, Oxford, Clarendon Press, 1969.

Foltz, R. C. (ed.), *Environmentalism in the Muslim World*, New York, Nova Science Publishers, 2005.

Foltz, R. C., Denny, F. M., and Baharuddin, A. (eds.), *Islam and Ecology: A Bestowed Trust*, Cambridge, MA, Harvard University Press, 2003.

Gadgil, M., and Guha, R., *This Fissured Land: An Ecological History of India*, Berkeley, University of California Press, 1993.

Glacken, C. J., *Traces on the Rhodian Shore: Nature and Culture in Western Thought from Ancient Times to the End of the Eighteenth Century*, Berkeley, University of California Press, 1967.

Glick, T. F., *Irrigation and Hydraulic Technology: Medieval Spain and Its Legacy*, Aldershot, Variorum, 1996.

Glick, T. F., *Irrigation and Society in Medieval Valencia*, Cambridge, MA, Harvard University Press, 1970.

Griswold, W., "Climatic Change: A Possible Factor in the Social Unrest of Seventeenth Century Anatolia," in H. W. Lowry and D. Quataert (eds.), *Humanist and Scholar: Essays in Honor of Andreas Tietze*, Istanbul, Isis Press, 1993, pp. 37–57.

Grove, R. H., Damodaran, V., and Sangwan, S. (eds.), *Nature and the Orient: The Environmental History of South and Southeast Asia*, Delhi, Oxford University Press, 1998.

Haleem, H. A. (ed.), *Islam and the Environment*, London, Ta-Ha Publishers, 1998.

Harris, L., "Postcolonialism, Postdevelopment, and Ambivalent Spaces of Difference in Southeastern Turkey," *Geoforum* 39, 2008, pp. 1698–708.

Harris, L., "Water and Conflict Geographies of the Southeastern Anatolia Project," *Society and Natural Resources* 15, 2002, pp. 743–59.

Hattox, R., *Coffee and Coffeehouses: The Origins of a Social Beverage in the Medieval Near East*, Seattle, University of Washington Press, 1985.

İnalcık, H., "The Origins of the Ottoman–Russian Rivalry and the Don–Volga Canal 1569," *Les annales de l'Université d'Ankara* 1, 1947, pp. 47–106.

Khazeni, A., *Tribes and Empire on the Margins of Nineteenth-Century Iran*, Seattle, University of Washington Press, 2009.

Kudlick, C. J., *Cholera in Post-Revolutionary Paris: A Cultural History*, Berkeley, University of California Press, 1996.

Kuhnke, L., *Lives at Risk: Public Health in Nineteenth-Century Egypt*, Berkeley, University of California Press, 1990.

Kurat, A. N., "The Turkish Expedition to Astrakhan and the Problem of the Don–Volga Canal," *Slavonic and East European Review* 40, 1961, pp. 7–23.

Lydon, G., *On Trans-Saharan Trails: Islamic Law, Trade Networks, and Cross-Cultural Exchange in Nineteenth-Century Western Africa*, Cambridge, Cambridge University Press, 2009.

McCann, J. C., *Maize and Grace: Africa's Encounter with a New World Crop, 1500–2000*, Cambridge, MA, Harvard University Press, 2007.

McNeill, J. R., *The Mountains of the Mediterranean World: An Environmental History*, Cambridge, Cambridge University Press, 1992.

McNeill, J. R., *Something New under the Sun: An Environmental History of the Twentieth-Century World*, New York, W. W. Norton, 2000.

McNeill, W. H., *Plagues and Peoples*, Garden City, NY, Anchor Press, 1976.

Meiggs, R., *Trees and Timber in the Ancient Mediterranean World*, Oxford, Clarendon Press, 1982.

Mikhail, A., *Nature and Empire in Ottoman Egypt: An Environmental History*, Cambridge, Cambridge University Press, 2011.

Mikhail, A., "The Nature of Plague in Late Eighteenth-Century Egypt," *Bulletin of the History of Medicine* 82, 2008, pp. 249–75.

Murphey, R., "The Decline of North Africa since the Roman Occupation: Climatic or Human?," *Annals of the Association of American Geographers* 41, 1951, pp. 116–32.

Oman, L., Robock, A., Stenchikov, G. L., and Thordarson, T., "High-Latitude Eruptions Cast Shadow over the African Monsoon and the Flow of the Nile," *Geophysical Research Letters* 33, 2006, p. L18711.

Panzac, D., *La peste dans l'Empire Ottoman, 1700–1850*, Louvain, Association pour le développement des études turques, 1985.

Raymond, A., "Les grandes épidémies de peste au Caire aux XVIIᵉ et XVIIIᵉ siècles," *Bulletin d'études orientales* 25, 1973, pp. 203–10.

Richards, J. F., *The Unending Frontier: An Environmental History of the Early Modern World*, Berkeley, University of California Press, 2003.

Rosenberg, C. E., *The Cholera Years: The United States in 1832, 1849, and 1866*, Chicago, University of Chicago Press, 1987.

Sbeinati, M. R., Darawcheh, R., and Mouty, M., "The Historical Earthquakes of Syria: An Analysis of Large and Moderate Earthquakes from 1365 BC to 1900 AD," *Annals of Geophysics* 48, 2005, pp. 347–435.

Shibl, Y. A., *The Aswan High Dam*, Beirut, The Arab Institute for Research and Publishing, 1971.

Tabak, F., *The Waning of the Mediterranean, 1550–1870: A Geohistorical Approach*, Baltimore, The Johns Hopkins University Press, 2008.

Taylor, B. R. (ed.), *The Encyclopedia of Religion and Nature*, 2 vols., London, Thoemmes Continuum, 2005.

Thirgood, J. V., *Man and the Mediterranean Forest: A History of Resource Depletion*, London, Academic Press, 1981.

Tuchscherer, M. (ed.), *Le commerce du café avant l'ère des plantations coloniales: Espaces, réseaux, sociétés (XVᵉ-XIXᵉ siècle)*, Cairo, Institut français d'archéologie orientale, 2001.

Wagstaff, J. M., *The Evolution of Middle Eastern Landscapes: An Outline to AD 1840*, London, Croon Helm, 1985.

Waterbury, J., *Hydropolitics of the Nile Valley*, Syracuse, NY, Syracuse University Press, 1979.

Watson, A., *Agricultural Innovation in the Early Islamic World: The Diffusion of Crops and Farming Techniques, 700–1100*, Cambridge, Cambridge University Press, 1983.

White, S., *The Climate of Rebellion in the Early Modern Ottoman Empire*, Cambridge, Cambridge University Press, 2011.

White, S., "Rethinking Disease in Ottoman History," *International Journal of Middle East Studies* 42, 2010, pp. 549–67.

Wittfogel, K. A., *Oriental Despotism: A Comparative Study of Total Power*, New Haven, CT, Yale University Press, 1957.

Yaffe, M. D. (ed.), *Judaism and Environmental Ethics: A Reader*, Lanham, MD, Lexington Books, 2001.

Zachariadou, E. (ed.), *Natural Disasters in the Ottoman Empire*, Rethymnon, Crete University Press, 1999.

CHAPTER ELEVEN

Australia in Global Environmental History

LIBBY ROBIN

Origins of Environmental History in Australia

The first of the environmental history professional groups, the American Society for Environmental History, commenced its journal in 1976 (now known as *Environmental History*). It began as a subdiscipline within (national) history in the US, but this has not been true for environmental history in other places. In Britain an independent journal, *Environment and History*, commenced in 1995, reaching beyond a North American focus to include, most notably, the comparative environmental histories of Africa and India, and also Australia and New Zealand. *Environment and History* drew authors and readers from beyond history departments, particularly geography and social anthropology, where transnational comparative work is more common.[1] The newest and most explicitly global of the international environmental history journals is *Global Environment: Journal of History and Natural and Social Sciences*, published in Naples, Italy, by the Global Environment Society since 2008. It is the first to include explicitly natural sciences in its definition of environmental history. Its organization most closely resembles the largely natural-science community that first promoted environmental history in Australia.

Natural science has been a strong force shaping environmental history and policy in Australia, and the interplay between them.[2] Agricultural, ecological, and forest sciences, in particular, have been closely allied with environmental policy-making in Australia, and they have emerged as forces in environmental history, too. It is no coincidence that an early (1991) synthetic overview of the emergence of environmental history in Australia, although published in a history journal, was undertaken by authors originally trained in ecological sciences, Stephen Dovers and John Dargavel.[3] They traced a confluence of disciplines in the field, and picked out themes such as landscape frontiers, biological invasion and forest history as defining Australian environmental history. These themes are all of major concern to ecologists, but not commonly found in history curricula.

A Companion to Global Environmental History, First Edition. Edited by J.R. McNeill and Erin Stewart Mauldin.
© 2012 John Wiley & Sons, Ltd. Published 2015 by John Wiley & Sons, Ltd.

Geography and economic history nurtured environmental history in New Zealand, Canada, and Britain, but Australia's trajectory was rather different, not least because, since the mid-1990s, Australian geographers have most commonly worked in inter-disciplinary departments of environmental studies (predominantly natural sciences), while economic history has been systematically phased out of commerce and business programs, its main stronghold in Australian universities until the 1990s. The history of science, sponsored by both arts and science faculties, however, survived until the twenty-first century, and there are many environmental historians (including myself), who come from an interdisciplinary background in history, philosophy, and sociology of science (HPS). HPS in Australia extended the strong tradition of valuing science and building cultural aspirations on scientific foundations.[4] Anthropology and under-standings of Indigenous traditional knowledge have flourished in Australia in the period since the Indigenous Revolution of the 1970s, when Aboriginal people demanded to move from being "objects of scientific study" to being the subjects and authors of their own histories. Traditional ecological knowledge is increasingly part of environmental management, but is seldom treated historically, while Indigenous histories are seldom identified as environmental history. Meanwhile research into the history of environmentalism, an important element of US and German environmental history, for example, has most often been undertaken in departments of political science in Australia.[5]

A Confluence of Disciplines on a Regional Environmental Scale

The natural scale of ecological studies is a bioregion or ecosystem. This scale of history appealed to Dovers and Dargavel, but they also noted work at state level, the default scale for environmental policy-makers. Australia's state boundaries (drawn as colonial boundaries in the nineteenth century in London surveyors' offices; see Map 11.1) have little in common with natural ecosystem barriers, but archival sources about environ-mental change are held by the states, because lands and forests were the principal sources of revenue for the Australian states after the colonies federated in 1901. Agriculture and forestry ministries were very much the domain of state bureaucracies until about the 1990s, when in most states they were amalgamated into "superministries" (usually of natural resources, environment, and more recently, sustainability). From the mid-nine-teenth century, the social organisation of space was the colonial or state bureaucrat's domain. Officials became landscape authors, harnessing the land's environmental possibilities to meet the revenue needs of the state, the basic needs of the British settler society, and, some would argue, forging its moral virtue.[6] There was no national depart-ment for land management equivalent to the United States Department of Agriculture (USDA), although, in international scientific relations, the Council for Scientific and Industrial Research (CSIR; after 1949, the Commonwealth Scientific and Industrial Research Organisation [CSIRO]) sometimes played this role, and worked closely with USDA at times.

Frontier history, the precursor of environmental history in the US, was most notable as a *silence* in Australian history. There is no frontier (or frontier "heroism") if warfare with Indigenous peoples is denied. The doctrine of *terra nullius*, the notion that the land was legally "empty," was a major myth of British settlement in Australia. The land was "empty" because there was no evidence of agriculture (fences) and the people did

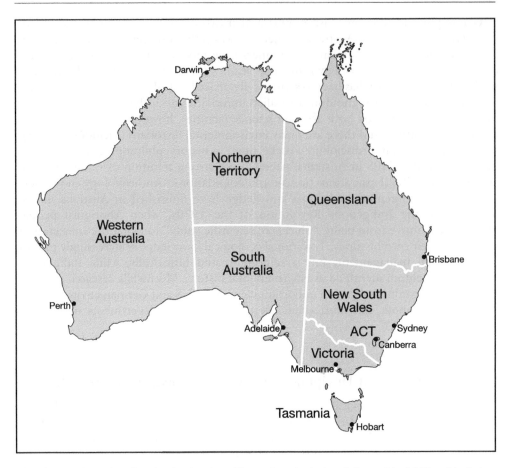

Map 11.1 Australia, showing its six states (formerly colonies) as federated in 1901, with their capital cities. Until 1911, the Northern Territory was part of South Australia. The Australian Capital Territory (ACT) was excised from New South Wales the same year, and Canberra was established as the national capital in 1913. Map drawn by Clive Hilliker, The Australian National University. Used by kind permission

not defend it with recognized military technologies (they used other forms of warfare). In 1969, archeologist Rhys Jones defined "fire-stick farming," the concept that Aboriginal people managed the land with fire to increase the green pick for hunting prey.[7] Revisionist historians, particularly Henry Reynolds, criticized the basis for *terra nullius* in Australia and native title was finally recognized by the High Court in the Mabo judgment of 1992.[8]

Much earlier, the frontier concept in its western American form shaped an early geography of pastoral expansion in the South Australian wheat belt by visiting American geographer Donald Meinig, *On the Margins of the Good Earth*, published in Chicago in 1962.[9] Another geographer, Michael Williams, expanded on the theme in *The Making of the South Australian Landscape*, bringing training in British historical geography, and publishing in London in 1974.[10] The landscape frontier in South Australia was clearly defined spatially (unlike settlement in other colonies). In the period from 1869 to 1884 a wheat belt expanded into the desert interior, beyond Goyder's Line, "the line of reliable rainfall."[11] There were wheat belts in other states too, but South Australia's was the iconic and tragic frontier of the nineteenth century. The full horror of the cartographic transgression of settlement in districts of unreliable rainfall became apparent in the Federation drought of 1895–1902. The same drought decimated the internal pastoral expansion of western New South Wales, but its wheat belt was not as vulnerable, while Western Australia's expansion and wheat-belt development was a twentieth-century story. In Victoria, as late as 1969, the expansion of the wheat belt was stopped by a popular environmental movement.[12] These environmental histories came later and emerged from other disciplinary traditions.

Biological invasions were part of the tragedy of nineteenth-century land settlement. *They All Ran Wild* is the title of the 1969 work of farmer-historian and nature writer Eric Rolls.[13] Most famously, the spread of rabbits affected all states. The 1930s sandstorms of inland Australia, at the same time as the US Dust Bowl, were greatly aggravated by the rabbit plague. *Flying Fox and Drifting Sand* was written by Francis Ratcliffe, who studied sand drift for CSIR. His popular book, subtitled "the adventures of a biologist in Australia," was on the reading lists of schools throughout the 1950s and 1960s.[14] Perhaps the most destructive biological invasion was the cloven hooves of introduced cattle and sheep to country where, as Sam Clayton, a New South Wales soil scientist, put it in 1938, "our destiny is wrapped up with the surface six-inches of soil." As historian Tom Griffiths writes, before pastoral expansion into the "outside country" in the 1860s and 1870s (during uncharacteristic good years), there were only soft-footed hopping kangaroos in scattered company: they did not travel in "damaging single file" like the cattle and sheep.[15]

Forest history was the other strength identified by Dovers and Dargavel in their 1991 overview, along with pastoral expansion and biological invasion. Eric Rolls' important and popular book *A Million Wild Acres* (1981) was a major regional environmental history portraying a forest under siege from agricultural settlement. But Australian forest historians were also influenced by international trends in scientific forestry. The International Union of Forest Research Organisations (IUFRO) established a subject group on forest history in 1965, before environmental history was known as such anywhere.[16] IUFRO also has strong links with the *Global Environment* journal. Its global focus comes from long-established connections and networks between forestry and forest-history scholars in Europe, North America, and also Asia and Australia and, since the late 1980s, has included a tropical forest history subgroup from Oceania, Latin America, and Africa. The strength of IUFRO, an international organization devoted to forests and related sciences promoting scientifically grounded policy-making, shaped the discipline of environmental history even in Australia, where much of the continent is treeless or sparsely treed, and almost all unsuited to forestry. Forests, however, are close to major cities and became the subject of environmental dispute in the 1980s and 1990s at the time when the disciplinary field of environmental history emerged.

Despite the strength in forest history, other exploited resources remain understudied. Mining history is, extraordinarily, one of the neglected aspects of Australian environmental history, despite the present importance of mining to the economy. The gap in economic history has proved a limit on Australian environmental history. Much of what has been written about mining has been commissioned accounts of single mines and settlements, often paid for by the mining companies themselves, or by governments in the guise of heritage projects.[17] There is considerably more reflective writing about nineteenth-century gold rushes than there is about twenty-first-century gold mines, which like coal and iron-ore mines are transforming the landscape in our own era with huge open-cut operations. Much of the action is in the desert and remote northern Australia, but in the case of the Newcrest goldmine, one of the largest in the southern hemisphere, the arsenic-affected mountains of soil left over after the gold is extracted are vast – and right in the middle of some of New South Wales' prime pastoral country, not far from Canberra, the national capital. Some 26 major properties – each of which was for many generations a family's livelihood – have been razed already. The different dreams of the Australian economy are sometimes at odds with each other, and these contradictions play out dramatically in environmental history.

Australian Environmental Identity – a National Question?

Tom Griffiths has commented that "environmental history often makes best sense on a regional or global scale, rarely on a national one."[18] This is particularly true for the environment in Australia, where the ecological and the political perspectives both function best on levels other than the national. The national environment should have been part of a "continent for a nation," but, despite the fact that first Prime Minister Edward Barton used this phrase rhetorically at the time of federation in 1901, the foundational Australian Constitution focused on the nation, leaving environment for the states.

In *Spoils and Spoilers*, an early national view of environmental history in Australia first published in 1981, historian Geoffrey Bolton suggested that European land settlement could be seen as a tension between economic and cultural aspirations. It was "a conflict between those who exploited the country to serve preconceived economic goals and imported attitudes of mind, and those on the other hand who sought to create a civilisation where human use of resources was compatible with a sense of identity with the land."[19] "They hated trees" was the title of one chapter. Environment became entangled with identity, as British settlers struggled to "improve" the land by clearing forest culture for pasture. This was a civilizing and national mission: the Bible and the plough were invoked together in the rhetoric of the Protestant work ethic, while the regulatory authorities resumed properties where owners had failed to clear trees as required by their land-deed entitlements.[20]

The disciplinary divides exacerbated this division between nature and nation: historians dealt with nation, and ecologists with nature. Environmental history has had to fight blind spots on both sides. It was a conscious decision that gave the National Museum of Australia a brief to use environmental history to bridge the divide between people and nature. The committee proposing the content for the prospective museum in 1975 declared: "to divorce man from nature in the new museum would be to perpetuate a schism which the nineteenth century, in the interests of science, did much to foster."[21]

Several national stories did emerge early in Australia, despite the fact that policy and settlement patterns tended to colonial scales. Significantly these syntheses mostly came from those environmental historians (like Geoffrey Bolton) with a strong historical training. The national is, as David Christian has commented, the conventional (or default) scale of modern historiography.[22]

George Seddon, who identified as an environmental historian, a landscape architect, a historian of science, and a geologist, and was also a professor of English literature, was interested in Australia's national environmental sensibility. His book *The Old Country* and his collected essays *Landprints*, are wide-ranging studies of what it means to be Australian and the slow adaptation of European sensibility to a strange land.[23] Australia has been accused of lacking a nature-writing tradition like the transcendentalists of the US, but, as Tom Griffiths has shown, there is a long history of nature writing associated with the (more British) field naturalists' tradition. Some of it consciously modeled itself on Thoreau. "The Woodlanders" (Charles Barrett, Claude Kinane, and Brooke Nicholls) named their bush hut in the hills near Melbourne "Walden" in 1905.[24] George Seddon and Eric Rolls, in more recent years, have been among Australia's distinguished nature writers, as well as historians. Seddon's other passion was water. He observed shrewdly that "domestic lawn is one of the major irrigated crops in Australia."[25] Others, including J. M. Powell, Michael Cathcart, and Heather Goodall and Allison Cadzow, have written important histories of water and its management in the driest inhabited continent in the world.[26] Seddon's epigrammatic summary of suburban water sensibility in Australia is very different from Powell's scholarly historical study of water policy, but both are considered environmental histories. Seddon's 1994 book, *Searching for the Snowy*, is more a literary meditation, but it is the only one of the water histories to include "environmental history" in its subtitle.[27]

Environmental health science is often omitted from Australia's environmental history canon, yet both David Walker's *Anxious Nation* and Warwick Anderson's *The Cultivation of Whiteness* have strong environmental as well as racial dimensions.[28] Their debates center on Australia's tropical north, tropical country, and desert, where questions of suitability for white settlers raged in the early years of the twentieth century as part of the White Australia policy. Geographer Griffith Taylor, in the end, left Australia in 1928 to work in the US and Canada, at least in part because his recommendations against European–Australian settlement in environmentally limited regions met such opposition from Australian governments.[29]

The tropical north is only one of the silences of national history. Histories of the sea have been rare in both the national and the environmental canon. The nation is generally about the land, but in 1998 Frank Broeze sought to shake this assertion with *Island Nation: A History of Australians and the Sea*.[30] James and Margarita Bowen's magisterial 2002 history, *The Great Barrier Reef: History, Science, Heritage*, was perhaps too much about science to be seen as "environmental" even by environmental historians, who regularly comment on the dearth of marine historiography. The history of science in Australia has been written outside history departments, and seldom constructs itself on the national scale favored by history curricula. One exception is Libby Robin's *How a Continent Created a Nation*, which explicitly considers the ways scientific endeavor was integrated in nation-building, and how the European scientific vision for agriculture clashed with the natural ecology of the continent.

The most surprising silence in environmental history is in urban history. Although people live in large cities, Australia's national identity has, since the years around federation, focused on the bush (agricultural and pastoral Australia) as captured by the writings of Henry Lawson and others, who deplored the evils of the city. The bush, rather than wilderness or desert, was where Australian citizens proved their mettle.[31] The civic emphasis on the rural left the built environment bypassed by environmental history. Despite the fact that Australia is an overwhelmingly urban nation (or, perhaps more accurately, a suburban nation), urban histories have generally not been perceived as either national or environmental.

One exception was Perth. George Seddon's classic histories of Perth – *Sense of Place* and *A City and Its Setting* – were, from their publication in the 1980s, unequivocally environmental history because of their landscape emphasis.[32] But important urban histories such as Graeme Davison's *The Rise and Fall of Marvellous Melbourne* (1978, republished 2004) were not environmental. Newer histories, such as Andrea Gaynor's 2006 *Harvest of the Suburbs* (an environmental history of growing food in both Perth and Melbourne) and Grace Karskens' 2009 *The Colony: A History of Early Sydney* are more explicitly environmental. As the environmental history field itself matures, there is less need to have the word "environmental" in the title of the book for it to be recognized as environmental history. Dovers and Dargavel's survey included urban planner Dan Coward's *Out of Sight: Sydney's Environmental History 1851–1981* (1988), but excluded historian Alan Mayne's *Fever Squalor and Vice: Sanitation in Victorian Sydney* (1982), also on waste and waste-management in the same city at the same time, perhaps simply because the title of the latter did not declare itself as "environmental" history. While the historical community was shy of science, scientifically strong histories were included by Dovers and Dargavel, who were both confident readers and writers of technical and scientific language. They included biohistory, for example, even though Stephen Boyden's subject was Hong Kong.[33] In this case Boyden was Australian, and Hong Kong was a human ecology of western civilization, so it was environmental history.

Scholars trained as historians often did not ask the "environmental" question about their sources before the turn of the twenty-first century, perhaps because the "national" scale of their work precluded it. Environmental history was not part of the undergraduate history curriculum, although world history was taught as an undergraduate course at several Australian universities from the 1980s, and it included environmental and scientific history. It took the move to a global scale to bring environment into focus for historians.

Global Environmental Stories from Australia

Tom Griffiths' essay "Ecology and Empire: Towards an Australian History of the World" synthesised the confluence of historical ideas as the last millennium drew to a close.[34] Deep time is now essential to Australian environmental history, and this opens up new questions of how the Australian (continental) story fits into global history. Thermoluminescence dating techniques show Aboriginal people present in the land 55,000 years ago, about twice as long as earlier techniques could date reliably.[35] A long Indigenous history – including the use of firestick-farming – had become part of Australia's environmental history, and Australian history thereby became an important part of global environmental history.

The continent was biogeographically isolated, and its biota had evolved independently of other places. Then "it had a radically new technology imposed upon it, suddenly, twice."[36] The two waves of human arrivals each brought major technological shocks to the ecosystems. Aboriginal people hunted and modified the landscape with fire. The British settlement brought simultaneous agricultural and industrial revolutions, something that only New Zealand shared. Australia is no longer just a new land with a settler history and an Indigenous "prehistory." It is world history: an ancient continent beyond Wallace's Line (between Bali and Lombok in modern Indonesia), with an environmental exceptionalism that is important to a world history dominated by tropes from far northern lands trapped under ice until about 10,000 years ago, when the people arrived.

Australia's human story is longer than Europe's or North America's, spanning some 55,000 years. It is an "old" land, not a new one, and it was populated throughout the ice age when northern lands were frozen. The whole hominid history of Australia and New Guinea (which were one continent, Sahul, comprising the present Australian continent and the land bridges joining it to New Guinea in the north and Tasmania in the south until after the last glacial period) is about fully modern humans. *Homo erectus* never made it across Wallace's Line. In a new twist to the "last of lands" idea, the animals and plants of Sahul first encountered hominid species in their most potent form, as *Homo sapiens*, fully modern people, unlike almost everywhere else, and this, biologists argue, made the biota more vulnerable to extinction.[37] The role of hominid species in shaping evolutionary adaptation and extinctions is an emerging field of study that is surely global environmental history.

Another great force in this continent is fire. Australian ecosystems evolved to adapt to infrequent hot lightning fires before people arrived. As climatic changes dried the landscape, the number and ferocity of fires increased, and gradually eucalypts replaced the rainforest species that had clad the land in Gondwanan times. In the long Aboriginal period, the land was recultivated by fire through cool burns that are still normal practice over much of northern Australia. Fire is a mark of people in a landscape, and many of the early European explorers arriving by ship saw smoke before they saw land. When British settlers arrived, they also fired the land to clear it for farms and in pastoral areas to encourage green pick for sheep and cattle.

World fire historian Stephen J. Pyne has taken a particular interest in Australia, naming its southeastern corner the "fire flume," the most fire-prone region in the world.[38] The juxtaposition of hot, dry summers and flammable vegetation with fierce winds and sudden wind changes makes for a potent mix. The eucalypt with its high oil content (rendering distant hills a very distinct blue) adds tremendous heat to fires, as has been discovered in places like California and Portugal, where there are major plantations of imported eucalypts and fire ferocity has increased.[39]

There is another factor in the southeastern fire flume, and that is the mountain ash (*Eucalyptus regnans*), which has evolved with a greater fire-dependence than other eucalypts. These tall straight trees, prized for their timber, depend solely on their seed supply to regenerate (while other eucalypts resprout, or coppice from lignotubers underground), and they need a really hot fire to crack open their seeds. Ash-type eucalypts have to renew themselves *en masse*. As Tom Griffiths puts it, these "grand and magnificent trees have evolved to commit mass suicide once every few hundred years – and in European times, more frequently."[40] Living with fire in and near mountain-ash forests is a major Australian experience, and fire policy needs both history and ecology to understand the ferocity of the problem.

Place Studies: A Return to the Regional and Ecological in the Face of the Global

Many of the newest environmental histories have returned to "human-sized" places, teasing out ecological specificity and community sensibility. George Main draws strongly on Indigenous histories of place in his study of his home region, the southwest tablelands of central New South Wales.[41] Ian Lunt is one of many ecologists using history to reconstruct and understand the influence of land-use history in creating past, current, and future patterns of biodiversity in fragmented agricultural landscapes. Here history has a practical purpose: it is integral to understanding landscape processes.[42] Reunification of Indigenous and settler histories of environmental understandings on a place-based scale is urgently needed. Rebe Taylor's prize-winning history *Unearthed* (2008), for example, is primarily a study of Tasmanian Aboriginal people on Kangaroo Island, but includes much detail about environment and the sealing and farming livelihoods on the island for people of all heritages.[43] Perhaps the most innovative social history, with an environmental flavor, is Mike Smith's 2005 *Peopling the Cleland Hills*. Smith is better known as an archeologist, but this book is a social and spatial history of the Aboriginal people who lived in remote outback Australia during the nineteenth and twentieth centuries, pieced together from wide-ranging sources. It focuses on how people lived (in good times) in the Central Australian desert country around the Cleland Hills, and where they withdrew to when resources declined during long drought periods.[44]

Conclusion: Scaling Australia into Global Environmental History

Australian environmental history works on a number of scales. If we follow the scaling framework outlined by David Christian, from microhistorical to Big History, we can see that Australia's first consciously environmental histories were (in common with many other places) local or regional.[45] The first environmental histories were of wheat-belt landscapes, forests, and stories of individual pastoral enterprises and fisheries. The national came later. Frontier histories, a theme of such national importance in the US, were not part of Australian nationalist historiography until late in the twentieth century. Environmental histories were the domain of geographers and ecologists, and some exceptional polymaths such as Eric Rolls and George Seddon, rather than mainstream historians. But since the emergence of ideas about "deep time" in Australian history, and a widespread reimagining of the Australian country through Indigenous perspectives, Australian stories have been finding their way into global-scale environmental history, not least because world-history scales are challenged by the patterns in Australia and its earlier form, Sahul.

Christian's scales are shaped by history – the scale of enlightenment thinking (500 years), the scale of the agricultural revolution (5,000 years), the scale of the Big Bang (4.6 billion years). Costanza and colleagues in their 2007 global history, *Sustainability or Collapse?*, predominantly written by scientists, organized their chapters using scales of number – the millennial scale (10,000 years), the centennial scale (1,000 years), the decadal scale (100 years). They also add chapters on "the future" – an explicit element of the Integrated History and future of People on Earth (IHOPE) project, and a typical concern of the scientifically trained. Both the historical and the scientific scales work attractively for histories of North America and Europe. The people arrive at the millennial scale as the ice melts, the Vikings find the Americas about 1,000 years ago (and 1066 is a powerful date in British historical thinking), then the exceptional twentieth century is treated as a separate subject. But Australia's industrial and agricultural revolutions are just 200 years old, and simultaneous. Further back, the "big moment" was the arrival from Asia of the dingo, the first placental mammal about 4,000 years ago.[46] The next important story was the rising of sea levels 21,000 to 7,000 BP, which separated New Guinea and Tasmania from the Australian continent – and how Aboriginal people coped with this. Only now is archeological evidence emerging that casts light on a group of societies that came to terms with a temperature increase of 10–12 degrees Celsius (18–22 degrees Fahrenheit) and a sea-level rise of 136 meters (446 feet) as the ice sheets melted, reducing the size of the continent by a third. Fourteen centuries is a long time, but some of these changes happened very rapidly, and surely have interest for a global environmental sensibility today as we debate temperature changes and sea-level rises again, albeit on much smaller scales. Global scales are equivalent to the agricultural revolution for Christian, the historian, but are planetary for global-change scientists. We need more history to decide what scales are appropriate for environmental histories. The Australian case suggests that environmental history is becoming an "interdisciplinary metadiscipline," rather than a subdiscipline of history. It is a space where people with expertise at the decadal level (roughly, the national) need to have conversations with the millennial and global scale, and some of these conversations need to come from places where there are significant silences in the European and North American record, because the frozen lands supported no people.

Notes

1 The founding editor of *Environment and History*, Richard Grove, is a graduate of the geography department of University College London, and the most recent past editor, Georgina Endfield, was trained as and works as a geographer.

2 L. Robin, *How a Continent Created a Nation*, Sydney, University of New South Wales Press, 2007; L. Robin and T. Griffiths, "Environmental History in Australasia," *Environment and History* 10/4, 2004, pp. 439–74.

3 S. Dovers and J. Dargavel, "Environmental History – A Confluence of Disciplines," *Australian Historical Association Bulletin* 66/67, 1991, pp. 25–30.

4 The first HPS department in the world was established at the University of Melbourne in 1948. It has been absorbed into other "mega-departments" since 2008. On Australia's culture of science, see Robin, *How a Continent Created a Nation*.

5 T. Doyle, *Green Power: The Environment Movement in Australia*, Sydney, University of New South Wales Press, 2001. Earlier, Australian political scientist Elim Papadakis had set the disciplinary pattern; see E. Papadakis, *The Green Movement in West Germany*, London, St. Martin's Press, 1984.

6 R. Wright, *The Bureaucrat's Domain: Space and the Public Interest in Victoria 1836–1884*, Melbourne, Oxford University Press, 1989, p. xiv; J. M. Powell, *An Historical Geography of Modern Australia: The Restive Fringe*, New York, Cambridge University Press, 1988.

7 R. Jones, "Fire-Stick Farming," *Australian Natural History* 16, 1969, pp. 224–8.

8 H. Reynolds, *The Law of the Land*, Ringwood, Penguin Press, 1987. The High Court upheld native title for Eddie Mabo (on the island of Mer in the Torres Strait) on June 3, 1992 ("The Mabo judgment"), and the Commonwealth Native Title Act was passed in December 1993. The Aboriginal Land Rights (Northern Territory) Act (1976) also recognized native title, but only in the Northern Territory.

9 D. Meinig, *On the Margins of the Good Earth*, Chicago, Rand McNally, 1962.

10 M. Williams, *The Making of the South Australian Landscape*, London, Academic Press, 1974.

11 J. Sheldrick, "Goyder's Line: The Unreliable History of the Line of Reliable Rainfall," in T. Sherratt, T. Griffiths, and L. Robin (eds.), *A Change in the Weather: Climate and Culture in Australia*, Canberra, National Museum of Australia Press, 2005, pp. 56–65.

12 L. Robin, *Defending the Little Desert: The Rise of Ecological Consciousness in Australia*, Melbourne, Melbourne University Press, 1998.

13 E. Rolls, *They All Ran Wild: The Animals and Plants that Plague Australia*, Sydney and London, Angus and Robertson, 1969.

14 L. Robin and T. Griffiths, "Francis Noble Ratcliffe, 1904–1970," in N. Koertge (ed.), *New Dictionary of Scientific Biography*, Farmington Hills, MI, Charles Scribner's Sons, 2007, pp. 207–11.

15 T. Griffiths, "The Outside Country," in T. Bonyhady and T. Griffiths (eds.), *Words for Country*, Sydney, University of New South Wales Press, 2002, pp. 222–43 (p. 228).

16 The Australian Forest History Society (AFHS) was established in 1988, but there is still no Australian Environmental History Society. An Australian Environmental History Network was commenced in 1997, founded by Stephen Dovers, Robert Wasson, and Richard Grove, and convened since 1999 by Libby Robin. It now includes New Zealand.

17 R. J. Donovan et al., "A Mining History of Australia: Part 1 of the National Mining Heritage Research Project for the Australian Council of National Trusts," unpublished report to the Commonwealth Government, 1995.

18 T. Griffiths, "Ecology and Empire: Towards an Australian History of the World," in T. Griffiths and L. Robin (eds.), *Ecology and Empire: Environmental History of Settler Societies*, Edinburgh, Keele University Press, 1997, pp. 1–16.

19 G. Bolton, *Spoils and Spoilers: Australians Make Their Environment, 1788–1980*, Sydney, Allen & Unwin, 1981, p. 23.

20 "The Bible and the Plough" was the title of Tom Stannage's environmental history undergraduate course at the University of Western Australia in the 1990s, which inspired an exhibition on wheat at the National Museum of Australia (*Tangled Destinies*, 2001). See L. Robin, "The Love–Hate Relationship with Land in Australia: Presenting 'Exploitation and Sustainability' in Museums," *Nova Acta Leopoldina*, 2012.

21 Committee of Inquiry on Museums and National Collections, *Museums in Australia 1975: Report of the Committee of Inquiry on Museums and National Collections including the Report of the Planning Committee on the Gallery of Aboriginal Australia*, Canberra, Australian Government Publishing Service, 1975, pp. 70–1.

22 D. Christian, "Scales," in M. Hughes-Warrington (ed.), *Palgrave Advances in World Histories*, New York, PalgraveMacmillan, 2005, pp. 64–89.

23 G. Seddon, *The Old Country: Landscapes, Places and People*, Melbourne, Cambridge University Press, 2006; G. Seddon, *Landprints: Reflections on Place and Landscape*, Melbourne, Cambridge University Press, 1997; T. Griffiths,"The Man from Snowy River," *Thesis Eleven* 74, 2003, pp. 7–20.

24 T. Griffiths, "The Natural History of Melbourne," in *Hunters and Collectors: The Antiquarian Imagination in Australia*, Melbourne, Cambridge University Press, 1996, pp. 121–49 (pp. 128–9).

25 Seddon, *Landprints*, p. 183.

26 J. M. Powell, *Watering the Garden State: Water, Land and Community in Victoria 1834–1988*, Sydney, Allen & Unwin, 1989; M. Cathcart, *The Water Dreamers: The Remarkable History of Our Dry Continent*, Melbourne, Text Publishing, 2009; H. Goodall and A. Cadzow, *Rivers and Resilience: Aboriginal People on Sydney's Georges River*, Sydney, University of New South Wales Press, 2009.

27 G. Seddon, *Searching for the Snowy: An Environmental History*, St. Leonards, Allen & Unwin, 1994.

28 D. Walker, *Anxious Nation: Australia and the Rise of Asia 1850–1939*, St. Lucia, University of Queensland Press, 1999; W. Anderson, *The Cultivation of Whiteness: Science, Health and Racial Destiny in Australia*, Carlton, Melbourne University Press, 2002.

29 C. Strange and A. Bashford, *Griffith Taylor: Visionary, Environmentalist, Explorer*, Canberra, National Library of Australia Press, 2008; C. Strange,"The Personality of Environmental Prediction: Griffith Taylor as 'Latter-day Prophet'," *Historical Records of Australian Science* 21/2, 2010, pp. 133–48.

30 Other omissions were Nonie Sharp's 2002 *Saltwater People: The Waves of Memory*, about Indigenous understandings of sea-country and *The Pearl Fishers of Torres Strait*, by Regina Ganter in 1994, a study of a declining environmental resource categorized as race relations rather than environmental history.

31 Robin, *How a Continent Created a Nation*.

32 G. Seddon and D. Ravine, *A City and Its Setting*, Fremantle, Fremantle Arts Centre Press, 1986; G. Seddon, *Sense of Place*, Nedlands, University of New South Wales Press, 1972, includes an ecological appraisal of the Swan coastal plan as well as the city.

33 S. Boyden, *The Ecology of a City and Its People: The Case of Hong Kong*, Canberra, ANU Press, 1981.

34 Griffiths, "Ecology and Empire."

35 For example, R. G. Roberts, R. Jones, N. A. Spooner, et al., "The Human Colonisation of Australia: Optical Dates of 53,000 and 60,000 Years Bracket Human Arrival at Deaf Adder Gorge, Northern Territory," *Quaternary Science Reviews* 13/5–7, 1994, pp. 575–83; and R. G. Roberts, R. Jones, and M. A. Smith,"Thermoluminescence Dating of a 50,000 Year-Old Human Occupation Site in Northern Australia," *Nature* 345, 1990, pp. 153–6.

36 G. Seddon,"The Man-Modified Environment," in J. McLaren (ed.), *A Nation Apart*, Melbourne, Longman Cheshire, 1983, pp. 9–31 (p. 10).

37 T. Flannery, *Here on Earth: An Argument for Hope*, Melbourne, Text Publishing, 2010.

38 S. Pyne, *World Fire: The Culture of Fire on Earth*, New York, Holt, 1995; and S. Pyne, *Burning Bush: A Fire History of Australia*, New York, Holt, 1991.

39 I. Tyrrell, *True Gardens of the Gods: Californian–Australian Environmental Reform 1860–1930*, Berkeley, University of California Press, 1999.

40 T. Griffiths,"We Have Still Not Lived Long Enough," *Inside Story*, February 16, 2009, accessed from http://inside.org.au/we-have-still-not-lived-long-enough/ on March 2, 2012; T. Griffiths, *Forests of Ash: An Environmental History*, Melbourne, Cambridge University Press, 2001.

41 G. Main, *Heartland: The Regeneration of Rural Place*, Sydney, University of New South Wales Press, 2005.

42 I. Lunt and P. Spooner, "Using Historical Ecology to Understand Patterns of Biodiversity in Fragmented Agricultural Landscapes," *Journal of Biogeography* 32, 2005, pp. 1859–73.

43 R. Taylor, *Unearthed: The Aboriginal Tasmanians of Kangaroo Island*, Kent Town, Wakefield Press, 2008.

44 M. A. Smith, *Peopling the Cleland Hills: Aboriginal History in Western Central Australia, 1850–1980*, Canberra, Aboriginal History, 2005.

45 Christian, "Scales."

46 This was the key story of T. Flannery, "The Trajectory of Human Evolution in Australia," in R. Costanza, L. J. Graumlich, and W. Steffen (eds.), *Sustainability or Collapse?: An Integrated History and Future of People on Earth*, Cambridge, MA, MIT, 2007, pp. 89–94.

References

Anderson, W., *The Cultivation of Whiteness: Science, Health and Racial Destiny in Australia*, Carlton, Melbourne University Press, 2002.

Bolton, G., *Spoils and Spoilers: Australians Make Their Environment, 1788–1980*, Sydney, Allen & Unwin, 1981.

Boyden, S., *The Ecology of a City and Its People: The Case of Hong Kong*, Canberra, ANU Press, 1981.

Cathcart, M., *The Water Dreamers: The Remarkable History of Our Dry Continent*, Melbourne, Text Publishing, 2009.

Christian, D., "Scales," in M. Hughes-Warrington (ed.), *Palgrave Advances in World Histories*, New York, PalgraveMacmillan, 2005, pp. 64–89.

Committee of Inquiry on Museums and National Collections, *Museums in Australia 1975: Report of the Committee of Inquiry on Museums and National Collections including the Report of the Planning Committee on the Gallery of Aboriginal Australia*, Canberra, Australian Government Publishing Service, 1975.

Costanza, R., Graumlich, L. J., and Steffen, W. (eds.), *Sustainability or Collapse?: An Integrated History and Future of People on Earth*, Cambridge, MA, MIT, 2007.

Donovan, R. J., et al., "A Mining History of Australia: Part 1 of the National Mining Heritage Research Project for the Australian Council of National Trusts," unpublished report to the Commonwealth Government, 1995.

Dovers, S., and Dargavel, J., "Environmental History – A Confluence of Disciplines," *Australian Historical Association Bulletin* 66/67, 1991, pp. 25–30.

Doyle, T., *Green Power: The Environment Movement in Australia*, Sydney, University of New South Wales Press, 2001.

Flannery, T., *Here on Earth: An Argument for Hope*, Melbourne, Text Publishing, 2010.

Flannery, T., "The Trajectory of Human Evolution in Australia," in R. Costanza, L. J. Graumlich, and W. Steffen (eds.), *Sustainability or Collapse?: An Integrated History and Future of People on Earth*, Cambridge, MA, MIT, 2007, pp. 89–94.

Goodall, H., and Cadzow, A., *Rivers and Resilience: Aboriginal People on Sydney's Georges River*, Sydney, University of New South Wales Press, 2009.

Griffiths, T., "Ecology and Empire: Towards an Australian History of the World," in T. Griffiths and L. Robin (eds.), *Ecology and Empire: Environmental History of Settler Societies*, Edinburgh, Keele University Press, 1997, pp. 1–16.

Griffiths, T., *Forests of Ash: An Environmental History*, Melbourne, Cambridge University Press, 2001.

Griffiths, T., "The Man from Snowy River," *Thesis Eleven* 74, 2003, pp. 7–20.

Griffiths, T., "The Natural History of Melbourne," in *Hunters and Collectors: The Antiquarian Imagination in Australia*, Melbourne, Cambridge University Press, 1996, pp. 121–49.

Griffiths, T., "The Outside Country," in T. Bonyhady and T. Griffiths (eds.), *Words for Country*, Sydney, University of New South Wales Press, 2002, pp. 222–43.

Griffiths, T., "We Have Still Not Lived Long Enough," *Inside Story*, February 16, 2009, accessed from http://inside.org.au/we-have-still-not-lived-long-enough/ on March 2, 2012.

Jones, R., "Fire-Stick Farming," *Australian Natural History* 16, 1969, pp. 224–8.

Lunt, I., and Spooner, P., "Using Historical Ecology to Understand Patterns of Biodiversity in Fragmented Agricultural Landscapes," *Journal of Biogeography* 32, 2005, pp. 1859–73.

Main, G., *Heartland: The Regeneration of Rural Place*, Sydney, University of New South Wales Press, 2005.

Meinig, D., *On the Margins of the Good Earth*, Chicago, Rand McNally, 1962.

Papadakis, E., *The Green Movement in West Germany*, London, St. Martin's Press, 1984.

Powell, J. M., *An Historical Geography of Modern Australia: The Restive Fringe*, New York, Cambridge University Press, 1988.

Powell, J. M., *Watering the Garden State: Water, Land and Community in Victoria 1834–1988*, Sydney, Allen & Unwin, 1989.

Pyne, S., *Burning Bush: A Fire History of Australia*, New York, Holt, 1991.

Pyne, S., *World Fire: The Culture of Fire on Earth*, New York, Holt, 1995.

Reynolds, H., *The Law of the Land*, Ringwood, Penguin Press, 1987.

Roberts, R. G., Jones, R., and Smith, M. A., "Thermoluminescence Dating of a 50,000 Year-Old Human Occupation Site in Northern Australia," *Nature* 345, 1990, pp. 153–6.

Roberts, R. G., Jones, R., Spooner, N. A., et al., "The Human Colonisation of Australia: Optical Dates of 53,000 and 60,000 Years Bracket Human Arrival at Deaf Adder Gorge, Northern Territory," *Quaternary Science Reviews* 13/5–7, 1994, pp. 575–83.

Robin, L., *Defending the Little Desert: The Rise of Ecological Consciousness in Australia*, Melbourne, Melbourne University Press, 1998.

Robin, L., *How a Continent Created a Nation*, Sydney, University of New South Wales Press, 2007.

Robin, L., "The Love–Hate Relationship with Land in Australia: Presenting 'Exploitation and Sustainability' in Museums," *Nova Acta Leopoldina*, 2012.

Robin, L., and Griffiths, T., "Environmental History in Australasia," *Environment and History* 10/4, 2004, pp. 439–74.

Robin, L., and Griffiths, T., "Francis Noble Ratcliffe, 1904–1970," in N. Koertge (ed.), *New Dictionary of Scientific Biography*, Farmington Hills, MI, Charles Scribner's Sons, 2007, pp. 201–11.

Rolls, E., *They All Ran Wild: The Animals and Plants that Plague Australia*, Sydney and London, Angus and Robertson, 1969.

Seddon, G., *Landprints: Reflections on Place and Landscape*, Melbourne, Cambridge University Press, 1997.

Seddon, G., "The Man-Modified Environment," in J. McLaren (ed.), *A Nation Apart*, Melbourne, Longman Cheshire, 1983, pp. 9–31.

Seddon, G., *The Old Country: Landscapes, Places and People*, Melbourne, Cambridge University Press, 2006.

Seddon, G., *Searching for the Snowy: An Environmental History*, St. Leonards, Allen & Unwin, 1994.

Seddon, G., *Sense of Place*, Nedlands, University of New South Wales Press, 1972.

Seddon, G., and Ravine, D., *A City and Its Setting*, Fremantle, Fremantle Arts Centre Press, 1986.

Sheldrick, J., "Goyder's Line: The Unreliable History of the Line of Reliable Rainfall," in T. Sherratt, T. Griffiths, and L. Robin (eds.), *A Change in the Weather: Climate and Culture in Australia*, Canberra, National Museum of Australia Press, 2005, pp. 56–65.

Smith, M. A., *Peopling the Cleland Hills: Aboriginal History in Western Central Australia, 1850–1980*, Canberra, Aboriginal History, 2005.

Strange, C., "The Personality of Environmental Prediction: Griffith Taylor as 'Latter-day Prophet'," *Historical Records of Australian Science* 21/2, 2010, pp. 133–48.

Strange, C., and Bashford, A., *Griffith Taylor: Visionary, Environmentalist, Explorer*, Canberra, National Library of Australia Press, 2008.

Taylor, R., *Unearthed: The Aboriginal Tasmanians of Kangaroo Island*, Kent Town, Wakefield Press, 2008.

Tyrrell, I., *True Gardens of the Gods: Californian–Australian Environmental Reform 1860–1930*, Berkeley, University of California Press, 1999.

Walker, D., *Anxious Nation: Australia and the Rise of Asia 1850–1939*, St. Lucia, University of Queensland Press, 1999.

Williams, M., *The Making of the South Australian Landscape*, London, Academic Press, 1974.

Wright, R., *The Bureaucrat's Domain: Space and the Public Interest in Victoria 1836–1884*, Melbourne, Oxford University Press, 1989.

CHAPTER TWELVE

Oceania: The Environmental History of One-Third of the Globe

PAUL D'ARCY

Oceania, the region encompassing the Pacific Islands, is the longest continually settled oceanic environment (one much more aquatic than terrestrial) settled by humans on our predominantly blue aquatic planet – for this reason alone it deserves far greater attention than it receives. The Pacific Ocean spans almost one-third of the earth's surface – all the world's other oceans could fit into it, as could all landmasses of the world. Map 12.1 illustrates the dominance of maritime spaces in the Pacific hemisphere. The Pacific is also home to many of the world's islands with terrestrial and marine fauna and flora that generally derived from broadly similar origins and diverged in relative isolation in these islands – comparative biologists' dream laboratories. The same concepts have been applied somewhat inappropriately to Pacific Island societies, diminishing their cultural achievements and interactions. Hence, Pacific environmental history is a potential tool for rectifying Eurocentric misrepresentation of their cultures as isolated and vulnerable to external influences.

Oceania makes up the eastern part of the wedge of islands bounded by Tokyo, Jakarta, and Rapanui (Easter Island) that extends southeast from Asia into the Pacific. These islands are conventionally divided into three geographically discrete regions: Melanesia, Micronesia, and Polynesia. Melanesia consists of large, continental islands stretching from New Guinea to Fiji in the southwest Pacific. Micronesia lies to the north, across the equator. It consists of a number of "oceanic" islands – smaller, volcanic high islands and numerous atolls stretching west beyond the Philippines into the central Pacific. Polynesia also consists of oceanic islands and extends west from Fiji to Rapanui. These regions are often also portrayed as culturally and linguistically distinct, although great diversity within each is now also generally acknowledged.[1] Map 12.2 illustrates these conceptual divisions alongside biogeographical ones.

An early work on the environmental history of the Pacific Islands is J. R. McNeill's 1994 "Of Rats and Men."[2] McNeill surveys the entire history of human settlement in the

A Companion to Global Environmental History, First Edition. Edited by J.R. McNeill and Erin Stewart Mauldin.

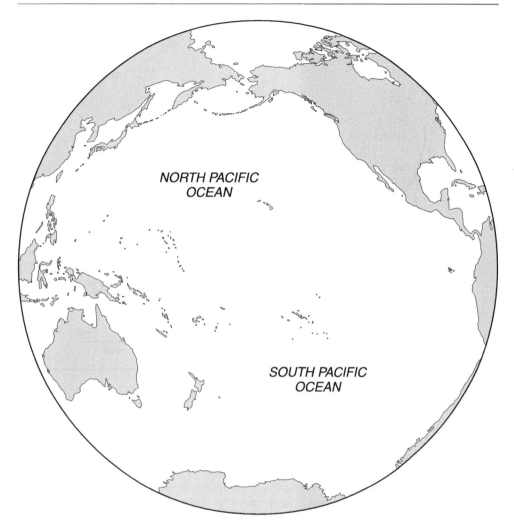

Map 12.1   The Pacific. © Paul D'Arcy and ANU Cartography. Used by kind permission

Pacific Islands, dividing this history into eras based on the type of technology that allowed the introduction of exotic flora and fauna into the islands, and the pervasive economic approach to resource management from indigenous tenure and sustainable harvest practices, to plantation agriculture to modern-day conservation and sustainable-growth paradigms. The more efficient the transport technology, the more intrusive exotic elements became in islands relatively isolated for millennia from competing continental species. McNeill thus integrates the Pacific world into world history as no one has done since or before, and focuses most clearly and comprehensively on environmental impacts. He makes a compelling case for further study of the myriad of island labs by environmental historians to investigate the subtleties of native–exotic interaction in specific and divergent contexts.

Few historians have taken up this challenge in the two decades since McNeill published. Studies that have done so have generally supported McNeill's contention, but

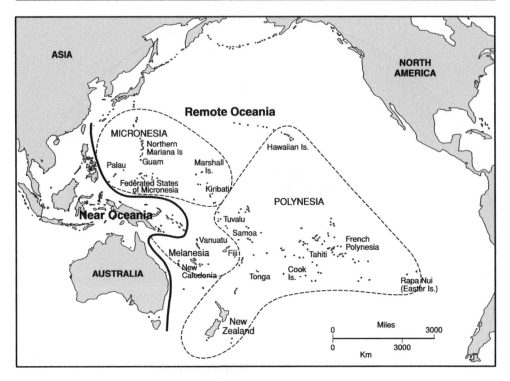

Map 12.2 Remote Oceania and Near Oceania. © Paul D'Arcy and ANU Cartography. Used by kind permission

placed more emphasis on the importance of cultural perceptions in shaping human–environment relations. McNeill's vision is also shared by most biological scientists and social scientists, particularly archeologists, although a revisionist stream has emerged questioning the degree to which islands were as isolated as many scholars assert. McNeill subsequently compiled a collection of previously published environmental history articles that combined Pacific Rim and Oceanic perspectives, but only one general environmental history has been published since "Of Rats and Men," and this largely focuses on Australia and New Zealand, while two general environmental overviews have been published by scientists that focus on natural processes more than human–environment interactions. The best general environmental study since 1993 is geographer Moshe Rapaport's edited collection of thematic essays.[3] Little wonder then that the Pacific Islands merit less than a page of text in Donald Hughes' *What Is Environmental History?*, in which he notes this neglect of such a huge expanse of the earth's surface.[4]

Most Pacific Island history has focused on intercultural relations between Pacific Islanders and Europeans over the last two and a half centuries.[5] Pacific historians have largely focused on the impact of western products, peoples, and ideas on Pacific Islanders, with increasing emphasis placed on presenting Pacific Islanders as rational, active agents in this process. The majority of Pacific Islanders' millennia of history was recorded and conveyed orally, which has meant that much Pacific history has become multidisciplinary to incorporate nonliterate sources such as oral traditions, linguistic patterns, and material remains. For this reason, Pacific history cannot be viewed in isolation, but must been seen as part of a larger body of work produced by historians' Pacific-focused colleagues in anthropology, archeology, linguistics, and geography. This has enhanced rather than diluted the nature of Pacific history. However, this cultural and multidisciplinary approach is muted within the environmental history community because there are relatively few Pacific historians focused on environmental history.

Exploring and Colonizing Oceania

The original exploration of the Pacific Islands, begun thousands of years ago, was one of the greatest seafaring achievements of early humankind and perhaps one of the greatest feats of empirical science in the ancient world. Encountering subtle and substantial differences in environments each time they pushed further out into the Pacific, Pacific Islanders had to observe their new surroundings, plot their position relative to their home island, memorize new environmental information instantly, and interpret, apply, or modify existing navigational techniques and maxims as they went.

Two distinct biogeographical zones are recognized within Oceania: Near Oceania and Remote Oceania (see Map 12.2).[6] These divisions map the progressive diminution of terrestrial and marine species' diversity eastward from their dispersal point into the Pacific as the gaps between islands increase.[7] The boundary between Near and Remote Oceania lies east and south of the present-day Solomon Islands.[8] It then passes east and north of the Bismarck archipelago, extending westward off the north coast of New Guinea before turning north to pass east of the Philippines. Near Oceania is located in the western Pacific, and encompasses most of the islands designated as Melanesia under the conventional classification. This area demonstrates a great deal of environmental continuity with Island Southeast Asia in terms of its large "continental" islands, and small gaps between islands. In contrast, Remote Oceania, broadly coinciding with Micronesia and Polynesia, is characterized by large gaps between smaller Oceanic islands and archipelagos. It is also notable for its very limited land area relative to ocean area. To populate this area, humans had to bring many if not most food crops and domestic birds and animals with them to maintain their usual diet and supplement the food sources they found in their new homes. In other words, Pacific landscapes are largely landscapes created by humans for humans from first settlement.[9]

Linguistic and archeological evidence suggests the ancestors of the region's indigenous inhabitants came to the southwest Pacific from southeast China and Taiwan. Oceania was first settled by Papuan-speaking peoples more than 35,000 years ago whose penetration was limited to New Guinea and perhaps some nearby Melanesian islands. They were followed by Austronesian-speaking peoples 3,500 years ago whose main route from Southeast Asia was along the northern coast of New Guinea into Island Melanesia, and from there east to Polynesia or north to eastern Micronesia. Others sailed to the western islands of Micronesia directly from Southeast Asia. They were skilled seafarers and explorers who evolved sophisticated navigational techniques involving waves, swells, stars, and signs of land such as land-based seabirds to map the oceans as well as the lands they came across. Linguistic reconstructions of their protolanguage suggest they came into the Pacific with canoes and food-storage techniques capable of supporting long open sea voyages into the unknown which then evolved in situ over time.[10]

Long-discredited claims of links between Oceania and South America first suggested by Thor Heyerdahl in the 1950s are now being reviewed in light of mounting biological and botanical evidence. Detailed published examinations of the historical and botanical record pertaining to this topic by historian Robert Langdon have largely been overlooked by nonhistorians.[11] The implications of this body of evidence for theories of Pacific colonization remain highly contested and uncertain. The means of distribution of the South American cultivar, the sweet potato (*ipomoea batatas*), across much of Oceania by the time of European contact also remains a major unresolved issue in Pacific history.[12]

The Oceanic Environment as a Human Habitat

Short-term variation in atmospheric circulation patterns, temperature, and rainfall were the most important environmental influences on Pacific Islanders as tillers of the land and travelers and fishers on the sea. Seasonal and annual climatic variations necessitated economies able to cope with times of scarcity through maximizing the potential benefits of times of surplus. There is great variability within Oceania's seasonal climatic patterns. Typhoons can occur during any month of the year. Similarly, storms with gale-force winds may occur during any month over much of the region.[13] Rainfall also varies widely. Generally areas near the equator experience high rainfall with limited seasonal variation. Further from the equator annual rainfall diminishes substantially, and is subject to marked seasonal variation.[14] Climatic conditions and weather patterns can also vary dramatically from year to year. Onotoa Atoll in Kiribati recorded 216 centimeters (85 inches) of rain in 1946, but only 16.8 centimeters (6.6 inches) in 1950.[15] In such variable conditions, the worst-case scenario rather than the overall average determines the viability of a community. During El Niño conditions, zones of convergence that cause heavy rainfall in this part of the Pacific move toward the equator. The usual areas of convergence such as the Carolines and Fiji then experience a decline in rainfall. Variation even occurs within the Caroline Islands. Islands that normally have high rainfall like Palau and Pohnpei may experience drought, while drier areas further east such as the Marshalls may experience high rainfall.[16]

Climatic variation and natural hazards are often deeply intrusive and disruptive of human activity. Few Islanders reached old age without experiencing the destructive forces of nature. The western Pacific is dominated by monsoon and typhoon weather patterns arising from the periodic heating and cooling of the Asian landmass. The monsoon winds blow from the northwest away from Asia in the northern-hemisphere winter, and from the southeast toward Asia in the northern-hemisphere summer. Typhoons or hurricanes occur in much of the western tropical Pacific. These spiral storms begin as areas of slowly circulating clouds that gather energy from the warm ocean waters that they pass over. They develop into giant mobile whirlwinds that can last for weeks. Their high winds and torrential rains carve a path of destruction, while the accompanying storm waves devastate coastal areas and low-lying coral islands. The Pacific Rim of Fire where plates of the earth's crust meet bisects Oceania. Activity along these plate boundaries gives rise to volcanic eruptions, earthquakes, and tsunami. While the overall death toll from natural disasters was probably well below that caused by disease, they could be equally devastating on specific localities. Rarely a decade passed without some community being leveled. Although these episodes are important indicators of Islanders' capacity to deal with disruption and population loss, they have received little attention from scholars.[17]

Volcanic eruptions are the best-covered natural hazard in Pacific Islands scholarship. Geomorphologist R. J. Blong led the way with his 1982 study *The Time of Darkness*, which correlates geological evidence of a large eruption in the mid-seventeenth century on Long Island in the Bismarck Sea off the northern coast of New Guinea with numerous oral traditions across Papua New Guinea of a time of darkness when ash fell over an area of more than 80,000 square kilometers (31,000 square miles) for up to three days, blocking the sun and smothering houses and crops. Blong noted considerable variation in the generational dating of the time of darkness among the traditions, although the distinct ash in the stratigraphy allowed him to correlate these tales with this eruption.[18]

An even larger eruption occurred in 1452 on the island of Kuwae in what is now central Vanuatu, shattering the 75-kilometer-(46.6-mile-)long island into eight islands between present-day Epi and Tongariki. Kuwae was one of the eight largest volcanic events in the past 10,000 years, hurling at least 30 million cubic meters (1 billion cubic feet) of rock, earth and magma into the atmosphere and creating enough dust to circle the world, remain in the atmosphere for three years, and block enough sunlight to create unseasonal and prolonged winters that stunted crop and vegetation growth in China and Europe, resulting in thousands of deaths from freezing and starvation. Upheaval and forced migrations are also recorded in this part of the Pacific, including the cessation of the trade with distant Tonga of local kava, a narcotic root beverage.[19]

Typhoons are a more regular destructive force in the Pacific. Even if people survive their high winds and storm waves, they face starvation in the wake of the destruction of flora and fauna. High winds blow over trees, and defoliate those that remain standing. Breadfruit trees are particularly susceptible to wind damage. Salt from sea spray and storm waves also destroy crops and delay regeneration by altering the soil chemistry and biology.[20] Severe typhoons could alter the course of history. The vibrant chieftaincy of Leluh on Kosrae seems to have been crippled by a super-typhoon in the 1770s that also all but wiped out the populations of the nearby atolls of Pingelap and Mwaekil.[21] Faka'ofa's domination of the Tokelau chain of atolls ended abruptly in 1846, when strong winds destroyed most of its coconut trees. Faced with starvation, a number of residents set off for neighboring atolls, but many were dispersed en route by adverse winds.[22] Geographer P. D. Nunn's argument for substantial natural environmental change, both geomorphologic and climatic, after Pacific Islander colonization has been fiercely contested by archeologists such as Matthew Spriggs. Nunn's 2009 *Vanished Islands and Hidden Continents of the Pacific* argues for far greater geological change and instability in Oceania than is conventionally recognized.[23]

Islands and Relative Isolation

Cultural development on Pacific Islands is usually depicted as being driven by the interaction of internal processes related to environmental factors. These include: adaptation of the founding culture to a new environment; population growth in a limited land area; environmental change, both natural and human-induced; and cultural emphasis on competition for status channeled into warfare, or the intensification of production for redistribution to forge social and political obligations. The possibility of new arrivals introducing cultural innovations was not dismissed, but it was almost always considered of secondary importance.[24] As the most influential archeologist of the Pacific Islands today P. V. Kirch asserts, "Little worlds unto themselves, circumscribed and frequently isolated, islands are natural history's best shot at something approaching the controlled experiment."[25]

This image of isolated, fragile Pacific communities pervades global, comparative studies. Isolated Rapanui is portrayed by scientist Jared Diamond as a harbinger for the planet, an ecological morality tale in which human neglect and abuse of the environment lead to deforestation in the process of social and political competition, isolation without canoe wood, and population pressure on a now isolated, denuded, and degraded island landscape. Trapped on their island, the Rapanui became locked in an increasing bitter struggle for resources and a downward spiral of cultural decline.[26] Archeologist Terry Hunt disputes this vision. He notes that evidence suggests a more complex interaction of factors, and primarily the introduction of rats (*Rattus exulans*) before the arrival of the first Europeans in 1722, which led to the destruction of the main native plant, the palm (*Jubaea chiliensis*), through rats eating palm seeds and as a result preventing palm reproduction.[27]

A new generation of scholarship questions this vision of isolated, fragile outposts. Island communities are increasingly now portrayed as connected "in a wider social world of moving items and ideas."[28] Local traditions, the distribution of cultural traits, and observations by literate outsiders all attest to interisland voyaging within most archipelagos in the first generation to experience sustained and widespread European contact from the 1760s onward. Historian Ian Campbell notes that archeological, linguistic, and traditional evidence all suggest that the period from c. 1100 to 1500 CE was an era of significant upheaval and interisland movement through much of Oceania.[29] This was a world of long-term cultural continuities, punctuated by periods of rapid, often externally influenced change.[30] Expectations of forces from beyond the horizon were deeply embedded in their worldviews. Centuries of experience had taught them that new lands, new opportunities and new and old threats hovered just beyond the walls of heaven where the sky touched the horizon.

This revisionist image of more interconnected Islander communities poses a major challenge to the fundamental assumptions of much archeology and environmental biogeography theory. If this more fluid and interlinked reality of island life is more common than not, then are indigenous systems and biota more able to resist external influences and colonizers less penetrating than hitherto believed? Exploring such issues of cultural construction and cultural resilience requires collaboration between biological scientists, archeologists, anthropologists, and historians.[31]

It is important to remember that the majority of the region's peoples lived away from the sea, however, practicing intensive agriculture and living much more geographically restricted lives in the fertile highlands of the island of New Guinea. Archeological excavations in this area have revealed the beginnings of agriculture there some 7,000 years ago and a remarkable process of agricultural intensification to support large populations. This process of agricultural intensification in the highlands was not associated with the concentration of political power to mobilize a large workforce as was often the case. The political organization of the highlands revolved around the distribution of pigs and valuables by political contenders known as big men to recruit and retain supporters from within their kin groups and beyond.

Cultural Determinism versus Environmental Determinism

Explanations of the evolution of social and political institutions in Oceania vacillate between environmental and cultural influences in attempting to explain why some developed complex hierarchical societies in which power was increasingly centralized under one ruler as in Tonga and Hawai'i, and others exhibited localized, more segmented sociopolitical organization such as in the New Guinea Highlands. Environmental explanations emphasize the need for sufficient surpluses to support the specialist administrators and warriors associated with the sustained concentration of power, and cultural ones the driving force of collective and kin-group status rivalry.

In *Social Stratification in Polynesia,* anthropologist Marshall Sahlins proposed that atolls and smaller high islands lacked the resource bases either to sustain a large enough population or to provide enough natural resources to allow or require distribution of surplus beyond immediate subsistence needs. In contrast, larger high islands like Tahiti provided the opportunity to produce substantial surpluses that ambitious chiefs could gain control of and redistribute to support and develop bureaucracies and coercive forces.[32]

Most archeologists and anthropologists focus on economic modes of production as the key to understanding political power in Pacific Island history. Kirch noted that intensive dry-land agriculture gave rise to inherently unstable and expansive chiefdoms. According to him, variable rainfall reduced food security and necessitated the conquest of other agricultural land.[33] Anthropologist William Alkire divides the coral islands of the Pacific into four groups on the basis of the size of their accessible resource base within their regular spheres of voyaging and uncontested user rights. He argues that relative isolation and the resource base within regular spheres of interaction and movement played a significant role in determining cultural patterns.[34] Essentially Alkire's argument is that the larger the environmental system, the greater the cultural development, until the population reaches the limits of its environment's carrying capacity.

Irving Goldman presented the other side of the case in arguing that internal competition for political influence was the driving force behind political evolution. He classified Polynesian societies into three categories on the basis of their political organization: traditional, open, and stratified. Goldman notes that their historical evolution was generally characterized by the development of stronger political controls, more exploitative relations between the rulers and the ruled, more violence and conflict, and greater overall insecurity.[35]

In *Nature, Culture, and History,* historian Kerry Howe demonstrates that the vacillation between environmental and cultural explanations of behavior in Pacific scholarship has a long history, with periods of ascendancy and decline that are measured in decades rather than years. Howe notes that the dominant sentiment among European intellectuals in the nineteenth century was the power of nature over humans, and the dominant themes in European descriptions of the Pacific were "paradise lost" and the dangers Pacific environments and cultures posed to Europeans. It was only with technical and medical advances in the late nineteenth and early twentieth centuries that Europeans began to believe that humans could control nature. The importance of culture was revived and linked to the idea of progress. By the early twentieth century, social environments were considered alongside natural environments as shapers of human affairs. Howe believes that culture is now the main variable used by those seeking to explain human behavior and history in Oceania.[36] He concludes that the common thread linking the varied paradigms is that, in all, European and Pacific Island cultures meet in what is essentially a "morality tale"[37] – a contest between various manifestations of good and evil in which Europeans confront their demons on exotic Pacific shores. Pacific "otherness" is rooted in European definitions of normality.

Cultural Ecology

One strand of Pacific scholarship has recently moved toward seeing culture and nature as intertwined and mutually influential. All physical environments contain culturally specific, symbolic markers that are only apparent to people versed in those cultures and known as cultural landscapes and seascapes. In other words, landscapes and seascapes can be read as history by those sufficiently versed in the material and conceptual worlds of the cultures occupying them.

In *Guardians of Marovo Lagoon*, anthropologist Edvard Hviding suggests that the inhabitants of the lagoon construct their perception of nature through their day-to-day use of that environment. The cultural system is not merely transferred from the mind of the elders to the memory of the young, but rather emerges from engagements with the environment in multiple forms of knowledge and practice.[38] Few scholars have questioned whether cultural perceptions of the environment correlate with western scientific explanations, or at least tried to ascertain whether there are phenomena in nature that might account for these perceptions. Archeologist Simon Best's survey of historical and traditional records about mythical lizard-like taniwha and salt-water crocodiles across Oceania revealed a close geographical correlation between the two phenomena. His argument that Pacific Island legends of taniwha might refer to salt-water crocodiles provides a superb model for reconciling environmental and cultural frameworks.[39] Similarly, recent environment-focused studies have documented that the food production and collection capacity of women equaled or exceeded that of men in many communities, and documented a prominent role for women in navigation in a number of island groups which challenged long-held academic beliefs that the sea was a predominantly male space.[40]

Culture Contact and the Impact of Precolonial European Influences

European encounters with the South Pacific began in 1520 when Magellan sailed westward across Oceania from South America. Other Spanish voyages of discovery followed in his wake.[41] European colonial rule over most of the Pacific Islands only came late in the nineteenth century, however. Sustained European contact only began from the 1770s for certain archipelagos, particularly in malaria-free Micronesia and Polynesia. Colonial rule was generally preceded by decades of interaction between indigenous peoples and European explorers, traders, beachcombers, missionaries, and settlers.[42]

New ideas and experiences, goods, and pathogens are generally regarded as the most significant influences on indigenous cultures during the first century of sustained western contact. Scholars are divided over the extent to which indigenous peoples exercised control over these influences. Most scholars portray Pacific Islanders as vibrant peoples who actively engaged these new elements from beyond their local worlds. The counterargument that local communities could not respond and that European contacts had a "fatal impact" on Pacific societies remains popular in the public imagination, but today has few academic defenders. It is seen to have the most resonance in the impact of introduced diseases.[43] Another strand of Pacific-focused environmental history for this period considers how Europe's discovery of the Pacific influenced western scientific thought about the physical and biological worlds as well as spurring major advances in the fields of cartography and navigation.[44]

Isolation from Eurasian infectious diseases made Pacific populations vulnerable to any epidemics that managed to reach their shores.[45] European contact introduced exotic diseases and took a heavy toll. The magnitude of the ensuing depopulation remains controversial. This issue rose to prominence in 1989 when David Stannard, professor of American studies at the University of Hawai'i, published a book arguing that the indigenous Hawai'ian population at European contact was at least double former estimates, so that postcontact depopulation until the first accurate census was truly catastrophic. Debate raged within the world of Pacific scholarship about Stannard's method for calculating contact populations and whether depopulation primarily resulted from epidemics or from disease-induced infertility.[46]

Epidemiologist Stephen Kunitz argues that indigenous populations were much more robust than has been allowed for, and only failed to bounce back from disease-induced depopulation when their political, cultural, and economic institutions were disrupted by the imposition of European rule. He contrasts the rapid decline of Hawai'ian and Māori populations in situations of widespread colonial settlement and land alienation, with Tonga, Samoa, and Tahiti where settler populations were outnumbered by the indigenous population and the latter retained possession of most of their lands.[47]

As historian Donald Denoon notes, "Depopulation was for Stannard a cause, for Kunitz an effect, of dispossession."[48] Denoon also makes the important point that specific localized circumstances such as local diet and the presence or absence of malaria led to variation in the degrees of western contact and depopulation.[49] Depopulation reduced stress on terrestrial and marine environments by reducing the need and capacity to harvest, although most stressed marine populations in this period were the result of European whaling and harvesting rather than enhanced indigenous capacity due to introduced western technology.

A number of works have explored the unsustainable commercial extraction of flora and fauna by Europeans – sandalwood, bêche-de-mer, whale oil, and pearl shell, particularly whaling and sandalwood. Heritage specialist Anita Smith notes that none of these industries left much of a human footprint in the land or sea beyond species depletion, as none required major permanent settlement or infrastructure.[50] Dorothy Shineberg's publications on the sandalwood trade remain models of scholarship on resource extraction and its cultural and ecological implications, despite being written over 40 years ago. Her work demonstrates that the trade was mutually beneficial, and Islanders were astute and skilled traders. The price obtained for the wood tended to rise over time as did the quality of goods sought. Shineberg also notes that once demand for western goods reached saturation point, many indigenous societies involved in the trade reverted to their preference for traditional valuables. Western goods enhanced traditional activities, but did little to alter them, although they did favor male activities such as land clearance through metal axes rather than female activities such as crop cultivation.[51]

The impact of western food crops has also received some attention, particularly concerning Hawai'i and Aotearoa/New Zealand.[52] The devastating impact of introduced flora and fauna such as cattle, rats, possums, and rabbits on local species with few competitors or predators and fragile ecosystems has been detailed in a number of studies. While indigenous peoples changed their landscapes through burning to facilitate agriculture and hunting, the degree of deforestation and conversion of landscapes into pasture or plantations escalated dramatically after the arrival of Europeans, especially in Aotearoa/New Zealand, Hawai'i, Fiji, and New Caledonia.[53]

Ecological exchanges in the early contact period were addressed in the 1980s and early 1990s and then neglected until Jenny Newell's 2010 book *Trading Nature*. Newell presents a historical account of the exchanges of plants and animals that first occurred from the first recognized European contact with Tahiti in the 1760s. Newell argues for the need to combine environmental history, cultural history, and contact history to better understand the dynamic of exchange taking place. She concludes that culture was more influential than ecological factors and thus the simple arrival of fauna and flora on a shore did not necessarily make its introduction and colonization inevitable. Prior to Newell's book, most works on the environmental consequences of European contact were heavily skewed toward biological aspects to the neglect of cultural aspects. The best by far in this vein are Linda Cuddihy and Charles P. Stone's *Alteration of Native Hawaiian Vegetation* and John Culliney's *Islands in a Far Sea*, which is also about Hawai'i, but much more comprehensive in its temporal coverage.[54]

European Settler Societies and Plantation Colonies

With the exception of the Spanish in the Mariana Islands in the seventeenth century, the French in the Society and Marquesas Islands and New Caledonia in the 1840s and 1850s, and the British in New Zealand in the 1840s, most Pacific Islands did not experience European attempts at military conquest and settlement until the last decades of the nineteenth century. European settlement was restricted across most of the tropical Pacific, however, with the exception of the French convict settlement of New Caledonia, and to a lesser extent the American settlement in Hawai'i. Elsewhere, European discomfort with tropical conditions, particularly malaria, resulted in small European populations who supported themselves through plantation economies supplying European and North American markets with copra and sugar especially, and relying on indentured labor. The largest labor flows were from Japan to Hawai'i, the Solomon Islands and New Hebrides (modern Vanuatu) to Queensland, India to Fiji, and New Guineans within German New Guinea.[55]

Pacific Islanders remained the majority of the population in most plantation colonies in tropical Oceania despite the ravages of introduced disease. The restricted European presence of European-run plantation enclaves, resident traders, and by and large poorly resourced and thinly stretched colonial administrations had a limited impact on most local communities beyond the collection of annual head taxes and perhaps periods of annual labor on plantations to earn taxes if the administration insisted on payment in cash rather than kind.[56] The incomplete nature of conquest and consolidation meant that indigenous tenure and use of the land and sea continued to varying degrees alongside the rules and practices of the new colonizers. Pockets of Islander seafaring survived the colonial era reasonably intact – especially on small islands remote from European centers of administration.[57]

Phosphate, gold, and nickel mining had larger but more localized impacts on Pacific ecosystems and indigenous communities than plantations The mining of guano or phosphate on small islands including Peleliu, Makatea, and Nauru, nickel mining on Grand Terre in New Caledonia, and gold mining in Aotearoa/New Zealand and New Guinea were all large-scale enterprises that involved the construction of significant infrastructure, the displacement and/or dislocation of local indigenous communities, and significant degradation of land and waterways. The New Caledonian economy profited from its nickel wealth from the 1890s until the present, while New Zealand's gold rush which began in the 1860s was less sustained and farming was soon restored as the national economic mainstay.[58] The most commonly cited Pacific example of mining degradation is Nauru, which was stripped of guano and much of it left a desolate wasteland of underlying rock pinnacles by colonial powers in the twentieth century to add fertility to pastures in New Zealand and Australia. One of the world's smallest nations, Nauru struggles to survive with most of its guano mined out and compensatory payments also exhausted, but never sufficient to rehabilitate the island's core to its former habitable self.[59]

Europeans built far more in stone than did Pacific Islanders, so that their houses, stores, barracks, and other infrastructure left an enduring imprint on the cultural landscapes of the Pacific Islands. The most common European structures of influence became churches after the conversion of most of the population. Most villages built their own Christian churches, and often in stone even when funded and built by local communities. In her study of colonial architecture in the New Hebrides, anthropologist Margaret Rodman argues that individual houses reflect cultural, historical, and highly personal histories of their occupants, especially when read as historical "texts" alongside archival records, diaries, and ethnographic literature. The same is true for the indigenous Pacific where cultural traits and material manifestations have long been the staple of archeological reconstructions of cultural histories and interactions.[60] Colonial communities overcame barriers of funding, distance, and isolation to develop land and sea transport that often incorporated best-practice construction, especially in Hawai'i.[61]

The Pacific War and the Nuclear Age

The colonial era was interrupted by the outbreak of war in the Pacific in December 1941. World War II in Oceania posed significant logistical challenges for all combatants. Moving and supplying large numbers of combatants in tropical conditions on small undeveloped islands scattered across vast stretches of ocean meant that logistical efficiency and technological superiority were crucial for both survival and victory. The industrial power of the US realized its true potential in the Pacific, and modern, total war ushered in the new, nuclear age in the Pacific. The oceanic environment, vast distances involved, and ferocity of Japanese resistance led to many Japanese garrisons simply being bypassed and isolated through American aerial dominance, and their resistance was ultimately broken by the use of a new super-weapon of unprecedented killing power – the atomic bomb.

Judith Bennett's *Natives and Exotics* is the first environmental history of the Pacific War. Bennett focuses on the logistical and health challenges that remote, scattered undeveloped tropical islands posed for combatants and those provisioning them. She shows how this demanding environment spurred the invention or refinement of insecticides, portable refrigeration units, and the like, as well as technologically advanced weaponry in an environment where control of the air and sea lanes determined the fate of troops on land. Transport also enhanced the movement of exotic species into the Pacific, most noticeably with the introduction of the brown tree snake into Guam, perhaps on US transport planes. In the absence of competitors and predators, and an abundance of prey, the brown snake multiplied to unprecedented levels and feasted on birds' eggs with acute consequences for some avian populations. The refuse of war remains today as silent witness to a violent past and also as environmental hazards in the form of unexploded munitions and fuel. Pacific War vintage firearms have been recycled and reused in recent conflicts on Bougainville. Bennett also discusses how the interaction of preconceptions and perceptions of the environment influenced and were in turn influenced by the physical experience of the land and sea – a theme she pursues in a section on the evolving memories of war landscapes.[62]

The Pacific continued to be a focus for nuclear weapons after the war. The vast and relatively underpopulated expanses of Oceania proved ideal for nuclear testing by the Americans, British, and French, especially as Islanders were politically marginal in essentially colonial relationships in American Micronesia, the British Gilbert Islands, and French Polynesia. Atmospheric testing of nuclear warheads was eventually replaced by first underground testing and then computer simulations and delivery-system testing as protests mounted within the region. The South Pacific Nuclear Free Zone (SPNFZ) was declared in 1985 with only Australia qualifying its support out of concern for its military alliance with the US, which used nuclear-powered and nuclear-armed vessels.[63] SPNFZ bordered the South American nuclear-free zone, making it the largest contiguous nuclear-free zone in the world, covering well over half of the southern hemisphere.[64]

Environmental historians have an important opportunity and as yet unfulfilled role to play in analyzing this era. Most histories of the nuclear era in the Pacific have been either anthropological studies of communities displaced by nuclear testing or largely focused on the region-wide political machinations and implications of the nuclear Cold War. Smith has noted the unique, but relatively unexplored, landscapes of infrastructure and abandonment that resulted.[65] Detailed environmental histories are beginning to emerge for affected atolls using the vast array of ethnographic data on marine and land tenure and subsistence prior to evacuation for testing, and scientific data on environmental degradation and recovery due to nuclear contamination, human resilience, ecosystem tolerance, and adaptation of nuclear fallout as Islanders now move toward recolonizing the atolls of their parents and grandparents. These islands are now also landscapes of harm and neglect, with a culture of fear and enduring sense of place for the dispossessed.[66]

Independence, Economic Viability, and Sustainable Development

Two fears delayed independence until the 1970s for most Pacific nations. The first concerned economic viability as microeconomies in a global world; the second was of possible political incoherence because arbitrary colonial boundaries aggregated diverse peoples. Environmental research on this post-independence era has been dominated by themes of resource depletion, environmental degradation, and questions about sustainable futures for island states. Many now doubt the ability of Oceania's sovereign states to make significant economic advances and secure political stability without external assistance. While many policy-makers subscribe to this analysis, the majority of academic specialists on the Pacific have rejected it or called for serious modification.[67] The latter urge more acknowledgment of Pacific nations' highly inadequate preparation for independence as a fundamental reason for their current problems. They are also more cautious about the efficacy of applying foreign models to Pacific problems.

Independent Pacific Island states generally inherited limited infrastructure from their colonial rulers and since independence have been unable to provide the transport, educational, health, and economic facilities needed to make their citizens willing and able to operate as citizens of a modern state. Most people still practice essentially highly localized subsistence lifestyles occasionally supplemented by cash crops. Despite poor communications and at times tense relations between social groups, sizable minorities of the population now travel beyond their kin-group area to work in the modern economy, especially in national capitals or large multinational undertakings such as mine sites. Such gathering places are sources of both identity formation and tension.[68]

The dominant development paradigm since independence has been the so-called MIRAB economy used to describe many of the smaller nations of contemporary Oceania. MIRAB refers to the main funding sources of these economies: migration, remittances, aid, and bureaucracy. The MIRAB image is one of tiny, nonviable economies forever condemned to dependency on aid from former colonial powers. As most aid is absorbed in civil-service salaries, remittances dependent on variable access to Pacific Rim labor markets become the main income source for those outside of government. The late Tongan scholar Epeli Hau'ofa argued that these MIRAB "basket cases" were the result of barriers created by colonial boundaries and policies that imposed an artificial sense of isolation and separation upon Islanders. They must now decolonize their minds, and recast their sense of identity by rediscovering the vision of their ancestors for whom the Pacific was a boundless sea of possibilities and opportunities.[69]

The resource-rich and larger Pacific nations of Melanesia display a different economic pattern of large-scale and foreign-controlled multinational companies extracting minerals, timber, and fish and exporting them largely unprocessed so that host nations develop comparatively few ancillary and value-added industries and receive only a fraction of the potential value of their raw materials. Limited resources for environmental policing of resource-extraction sites, the political marginalization of land owners in the immediate vicinity of these operations because of their small numbers and therefore limited parliamentary representation, and the large percentage of national income accounted for by taxes on multinationals have given rise to the resource–curse paradigm, where resources are considered at best a double-edged sword. They attract outsiders with no commitment to local sustainability, and divide local communities, pitting these same communities against what is often perceived as an unholy alliance between big business and national governments. A lack of domestic resources to monitor offshore waters and develop effective fishing fleets has forced Pacific nations into fishing-access agreements with wealthier Pacific Rim nations that have returned a mere fraction of the value of the catch at market to Island nations.[70]

Resource-rich areas are often lightly populated, leading tragically to common scenarios where resources perceived to be owned by no one are poorly protected and sustained, especially on the high seas and away from national capitals where they are designated *terra nugax*, a sort of unclaimed frontier ripe for harvesting. The term "terra nugax" was first used in Papua New Guinea in relation to the Ok Tedi copper and gold mine in the remote and sparsely populated far western mountainous interior of the country near the border with Indonesia. The small, scattered population around the mine, and the mine's huge contribution to government revenue, resulted in weak local representation in the national parliament, poor environmental safeguards, and ecological catastrophe downstream as a result of the dumping of toxic waste into local waterways.[71]

Most academic analysis of modern resource disputes has been conducted by anthropologists, economists, and political scientists, all of whom focus almost exclusively on the last two decades and see the problems as having arisen relatively recently. A welcome historical perspective on contemporary unsustainable resource development is provided by Judy Bennett's *Pacific Forest*.[72] This long-term history of forest use in the Solomon Islands stretches from the creation of forests through indigenous use, colonial forestry, into the era of rapid deforestation in the postcolonial era. This longer-term environmental history perspective is vital for demonstrating the historical roots of many contemporary resource disputes over resource access and ownership, and unsustainable practices that result. It also serves to place current practices in perspective, making poor practices seem all the more counterproductive by comparison, even if they are logical for those locked in contemporary contexts of insecure, uncertain, and conflicting user rights. In keeping with current Pacific historiography, Bennett also makes a compelling case that changing perceptions and beliefs about forests and their uses are as influential as developments in forestry technology.

Conclusion

All environmental history must embrace many disciplinary perspectives and methods to succeed, but Pacific Island environmental history perhaps especially so, because of the diverse range of sources scholars must rely on to fill gaps and address distortions in the written record about local cultures that recorded and communicated their history orally in chant, traditions, and cultural land- and seascapes. Pacific historians emphasize the need to consider cultural differences in perceiving the world, and in interacting with others and with their environments.

Certain crossdisciplinary exchanges offer great promise for further refining the core debates and issues facing environmental historians of the Pacific Islands: insularity and biological and cultural atrophy/endemism, the interaction and influence of cultural and environmental influences in human history, and the nature, consistency, and rapidity of change in island contexts.

Environmental historians must work more closely with archeologists and linguists on the vast majority of Pacific history that occurred prior to western contact. Historians usually adopt more short-term perspectives than archeologists and linguists and have documented numerous examples of rapid and dramatic changes induced by external elements in the early post-European-contact era. Such rapid change is difficult to detect in the archeological record in the absence of oral traditions, while material remains and landscape transformations detectable in the archeological record bear witness to slower moving changes better than to intragenerational change. Archeologists have generally made more limited use of recorded traditions than have historians.[73] Scholars such as Luders, Gunson, and Berg have demonstrated that there are sufficient sources and methodologies to locate and date periods of rapid change well before European contact, when human forces and natural hazards such as typhoons had dramatic and far-reaching impacts upon historical processes, as has been documented for the era of European contact.[74]

Historians must also focus more on human–environment relations before and after European contact through applying anthropologically influenced cultural ecology perspectives and archeological perspectives, as well as environmental history perspectives. Geographers have also made important contributions to the literature on the exploration and colonization of Oceania by Pacific Islanders, and especially to colonial and postcolonial economic development and spatial relations. Their methods and perspectives have perhaps the most direct application to environmental histories of the region.[75] It is especially important to document changes brought about by the introduction of new flora, fauna, and attitudes to the environment in a variety of different island ecosystems and contexts. Environmental history has really only begun its journey of exploration into the Pacific.

Notes

1 Discussions of these three areas as cultural entities can be found in K. R. Howe, *Where the Waves Fall: A New South Sea Islands History from First Settlement to Colonial Rule*, Honolulu, University of Hawai'i Press, 1984, and I. C. Campbell, *A History of the Pacific Islands*, Christchurch, Canterbury University Press, 1989, pp. 11–27.

2 J. R. McNeill, "Of Rats and Men: A Synoptic Environmental History of the Island Pacific," *Journal of World History* 5/2, 1994, pp. 299–349.

3 J. R. McNeill (ed.), *Environmental History in the Pacific World*, Aldershot, Ashgate, 2001; M. Rapaport (ed.), *The Pacific Islands: Environment and Society*, Honolulu, Bess, 1999; D. Garden, *Australia, New Zealand and the Pacific: An Environmental History*, Santa Barbara, ABC-CLIO, 2005; C. S. Lobban and M. Schefter, *Tropical Pacific Island Environments*, Mangilao, University of Guam Press, 1997; P. D. Nunn, *Oceanic Islands*, Oxford, Blackwell, 1994.

4 J. D. Hughes, *What Is Environmental History?*, Cambridge, Polity, 2006, pp. 69–71.

5 See D. Munro and B. Lal (eds.), *Texts and Contexts: Reflections in Pacific Islands Historiography*, Honolulu, University of Hawai'i Press, 2006. The terms "Pacific Islander" and "Islander" are used here interchangeably to refer to Pacific Islanders in general, while the terms "European" and "western" are used interchangeably to refer to influences and people emanating from European and North American Caucasian societies.

6 R. C. Green, "Near and Remote Oceania: Disestablishing Melanesia in Culture History," in A. Pawley (ed.), *Man and a Half: Essays in Pacific Anthropology and Ethnobiology in Honour of Ralph Bulmer*, Aukland, Polynesian Society, 1991, pp. 491–502 (pp. 493–5).

7 See E. A. Kay, *Little Worlds of the Pacific: An Essay on Pacific Basin Biogeography, Harold L. Lyon Arboretum Lecture 9*, Honolulu, University of Hawai'i Press, 1980, pp. 25, 33.

8 Excluding the Santa Cruz Group, 352 kilometers (219 miles) to the east of the main chain.

9 Smith and Jones call them "transported landscapes": A. Smith and K. L. Jones, *Cultural Landscapes of the Pacific Islands*, Paris, ICOMOS Thematic Study, 2007, p. 30.

10 See K. R. Howe (ed.), *Waka Moana: Voyages of the Ancestors: the Discovery and Settlement of the Pacific*, Honolulu, University of Hawai'i Press, 2007.

11 See, for example, R. C. Green, "A Range of Disciplines Support a Dual Origin for the Bottle Gourd in the Pacific," *Journal of the Polynesian Society* 110/2, 2000, pp. 191–7, and M. Burtenshaw, "Maori Gourds: An American Connection?," *Journal of the Polynesian Society* 108/4, 1999, pp. 427–33, versus R. Langdon, "The Secret History of the Papaw in the South Pacific: An Essay in Reconstruction," *Journal of Pacific History* 24/1, 1989, pp. 11–20; R. Langdon, "When the Blue-Egg Chickens Come Home to Roost: New Thoughts on the Prehistory of the Domestic Fowl in Asia, America and the Pacific Islands," *Journal of Pacific History* 24/2, 1989, pp. 164–92; R. Langdon, "The Banana as a Key to Early American and Polynesian History," *Journal of Pacific History* 28/1, 1993, pp. 15–35; R. Langdon, "The Soapberry, a Neglected Clue to Polynesia's Prehistoric Past," *Journal of the Polynesian Society* 105/2, 1996, pp. 185–200.

12 D. E. Yen, *The Sweet Potato and Oceania, Bernice P. Bishop Museum Bulletin* 236, Honolulu, Bishop Museum Press, 1974; C. Ballard, P. Brown, R. M. Bourke, and T. Harwood (eds.), *The Sweet Potato in Oceania: A Reappraisal, Oceania Monograph* 56, Sydney, University of Sydney, 2005.

13 On seasonal wind patterns and their variability see D. L. Oliver, *Oceania: The Native Cultures of Australia and the Pacific Islands*, 2 vols., Honolulu, University of Hawai'i Press, 1989, pp. 1, 14, and B. R. Finney, R. Rhodes, P. Frost, and N. Thompson, "Wait for the West Wind," *Journal of the Polynesian Society* 98/3, 1989, pp. 261–302 (pp. 265–7, 272–3).

14 Oliver, *Oceania*, pp. 1, 15–17.

15 H. J. Wiens, *Atoll Environment and Ecology*, New Haven, CT, Yale University Press, 1962, p. 155.

16   Lobban and Schefter, *Tropical Pacific Island Environments*, pp. 104–6.

17   On natural hazards in the Pacific see G. Hicks and H. Campbell (eds.), *Awesome Forces: The Natural Hazards that Threaten New Zealand*, Wellington, Te Papa Press, 2003; W. C. Dudley and M. Lee, *Tsunami!*, Honolulu, University of Hawai'i Press, 1988; and W. A. Lessa, "The Social Effects of Typhoon Ophelia (1960) on Ulithi," in A. P. Vayda (ed.), *Peoples and Cultures of the Pacific: An Anthropological Reader*, Garden City, NY, The Natural History Press, 1968, pp. 330–79.

18   R. J. Blong, *The Time of Darkness: Local Legends and Volcanic Reality in Papua New Guinea*, Seattle, University of Washington Press, 1982.

19   D. Luders, "Legend and History: Did the Vanuatu-Tonga Kava Trade cease in AD 1447?," *Journal of the Polynesian Society* 105/3, 1996, pp. 287–310 (pp. 289–92, 303–5); M. Monzier, C. Robin, and J. P. Eissen, "Kuwae, 1425 AD: The Forgotten Caldera," *Journal of Volcanology and Geothermal Research* 59, 1994, pp. 207–18. For another example, see P. W. Taylor, "Myths, Legends and Volcanic Activity: An Example from Northern Tonga," *Journal of the Polynesian Society*, 104/3, 1995, pp. 323–46. The best account of the impact of a modern eruption on a Pacific community is K. Neumann, *Rabaul Yu Swit Moa Yet: Surviving the 1994 Volcanic Eruption*, Oxford, Oxford University Press, 1996.

20   Lobban, and Schefter, *Tropical Pacific Island Environments*, p. 98, and W. H. Alkire, *Coral Islanders*, Arlington Heights, IL, AHM, 1978, p. 37.

21   P. D'Arcy, *The People of the Sea: Environment, Identity and History in Oceania*, Honolulu, University of Hawai'i Press, 2006, chapter 7.

22   Alkire, *Coral Islanders*, p. 102.

23   See P. D. Nunn, *Climate, Environment and Society in the Pacific during the Last Millennium*, Amsterdam, Elsevier, 2007; P. D. Nunn, *Vanished Islands and Hidden Continents of the Pacific*, Honolulu, University of Hawai'i Press, 2009; M. Spriggs, "Geomorphic and Archaeological Consequences of Human Arrival and Agricultural Expansion on Pacific Islands: A Reconsideration after 30 Years of Debate," in S. Haberle, J. Stevenson, and M. Prebble (eds.), *Altered Ecologies: Fire, Climate and Human Influence on Terrestrial Landscapes, Terra Australis* 32, Canberra, ANU E-Press, 2009, pp. 239–52; M. S. Allen, "New Ideas about Late Holocene Climate Variability in the Central Pacific," *Current Anthropology* 47/3, 2006, pp. 521–35.

24   This scheme is most elegantly argued in P. V. Kirch, *The Evolution of the Polynesian Chiefdoms*, Cambridge, Cambridge University Press, 1984. See also P. V. Kirch, "Microcosmic Histories: Island Perspectives on 'Global' Change," *American Anthropologist* 99/1, 1997, pp. 30–42 (p. 30).

25   Kirch, "Microcosmic Histories," p. 30.

26   J. Diamond, "Twilight at Easter," *New York Review of Books*, March 25, 2004, pp.6–10; J. Diamond, *Collapse: How Societies Choose to Fail or Succeed*, New York, Viking, 2005.

27   Hunt and Lipo dispute the vision of scholars such as Bahn, Flenley, and Diamond. T. L. Hunt, "Rethinking Easter Island's Ecological Catastrophe," *Journal of Archaeological Science* 34/3, 2007, pp. 485–94; T. L. Hunt and C. P. Lipo, "Ecological Catastrophe, Collapse, and the Myth of 'Ecocide' on Rapanui (Easter Island)," in P. A. McAnany and N. Yoffee (eds.), *Questioning Collapse: Human Resilience, Ecological Vulnerability, and the Aftermath of Empire*, Cambridge, Cambridge University Press, 2010, pp. 21–40; P. Bahn and J. Flenley, *Easter Island, Earth Island*, London, Thames and Hudson, 1992; J. Flenley, *The Enigmas of Easter Island: Island on the Edge*, Oxford, Oxford University Press, 2003; and Diamond, *Collapse*, p. 118–19.

28   G. Irwin, *The Prehistoric Exploration and Colonisation of the Pacific*, Cambridge, Cambridge University Press, 1992, p. 204; D'Arcy, *The People of the Sea*, chapters 3 and 7, and N. Gunson, "Great Families of Polynesia: Inter-Island Links and Marriage Patterns," *Journal of Pacific History* 32/2, 1997, pp. 139–52. For an early exploration of this concept see P. V. Kirch, "Exchange Systems and Inter-Island Contact in the Transformation of an Island Society: The

Tikopia Case," in P. V. Kirch (ed.), *Island Societies: Archaeological Approaches to Evolution and Transformation*, Cambridge, Cambridge University Press, 1986, pp. 33–41. While noting that certain small islands benefited from external contacts, Kirch does not see it as a widespread trend.

29   Campbell, *A History of the Pacific Islands*, p. 36; D. Lewis, *From Maui to Cook: The Discovery and Settlement of the Pacific*, Lane Cove, Doubleday, 1977, p. 29; B. R. Finney, "Voyaging," in J. D. Jennings (ed.), *The Prehistory of Polynesia*, Canberra, ANU, 1979, pp. 324–51 (pp. 349–50); and Irwin, *The Prehistoric Exploration and Colonisation of the Pacific*, pp. 213–14.

30   D'Arcy, *The People of the Sea*, chapter 7.

31   These issues are discussed in more detail in P. D'Arcy, "Cultural Divisions and Island Environments since the Time of Dumont d'Urville," *The Journal of Pacific History* 38/2, 2003, pp. 217–35.

32   M. Sahlins, *Social Stratification in Polynesia*, Seattle, University of Washington Press, 1958.

33   P. V. Kirch, *The Wet and the Dry: Irrigation and Agricultural Intensification in Polynesia*, Chicago, University of Chicago Press, 1994, p. 8.

34   Alkire, *Coral Islanders* pp. 65–9.

35   I. Goldman, "Status Rivalry and Cultural Evolution in Polynesia," *American Anthropologist* 57/4, 1955, pp. 680–97 (p. 694).

36   K. R. Howe, *Nature, Culture, and History: The "Knowing" of Oceania*, Honolulu, University of Hawai'i Press, 2000, pp. 41–2.

37   Howe, *Nature, Culture, and History*, p. 58.

38   E. Hviding, *Guardians of Marovo Lagoon: Practice, Place, and Politics in Maritime Melanesia*, Honolulu, University of Hawai'i Press, 1996, pp. 27, 369, 371; Smith and Jones, *Cultural Landscapes of the Pacific Islands*, p. 11.

39   S. Best, "Here Be Dragons," *Journal of the Polynesian Society* 97/3, 1988, pp. 239–59.

40   On the role of women in agriculture, see J. Linnekin, "Gender Division of Labour," in D. Denoon (ed.), *The Cambridge History of the Pacific Islanders*, Cambridge, Cambridge University Press, 1997, pp. 105–13. For women and marine subsistence and seafaring, see M. D. Chapman, "Women's Fishing in Oceania," *Human Ecology* 15/3, 1987, pp. 267–88; E. Huffer, "Women and Navigation: Does the Exception Confirm the Rule?," *International Journal of Maritime History* 20/2, 2008, pp. 259–324; K. L. N. Wilson, "Nā Wāhine Kanaka Maoli Holowa'a: Native Hawaiian Women Voyagers," in P. D'Arcy (ed.), *Women and the Sea in the Pacific, special three-article forum plus introduction in the International Journal of Maritime History* 20/2, 2008, pp. 259–324.

41   O. H. K. Spate, *The Pacific since Magellan*, vol. 1, *The Spanish Lake*, Canberra, Australian National University Press, 1979.

42   P. Hempenstall, "Imperial Manoeuvres," in K. R. Howe, R. C. Kiste, and B. V. Lal (eds.), *The Pacific Islands in the Twentieth Century*, St. Leonards, Allen & Unwin, 1994, pp. 3–28.

43   On "fatal impact" versus the revisionists, see J. Linnekin, "Contending Approaches," in D. Denoon (ed.), *The Cambridge History of the Pacific Islanders*, Cambridge, Cambridge University Press, 1997, pp. 3–36 (pp. 24–5).

44   H. Liebersohn, *The Traveler's World: Europe to the Pacific*, Cambridge, MA, Harvard University Press, 2006; B. Douglas and C. Ballard (eds.), *Foreign Bodies: Oceania and the Science of Race 1750–1940*, Canberra, ANU E-Press, 2008.

45   On pre-European health see Howe, *Where the Waves Fall*, pp. 44–50; and D. Denoon, "Pacific Island Depopulation: Natural or Un-Natural History?," in L. Bryder and D. A. Dow (eds.), *New Countries and Old Medicine: Proceedings of an International Conference on the History of Medicine and Health*, Auckland, Pyramid, 1994, pp. 324–39 (pp. 329–31).

46   D. E. Stannard, *Before the Horror: The Population of Hawai'i on the Eve of Western Contact*, Honolulu, University of Hawai'i Press, 1989, and A. F. Bushnell, "The 'Horror' Reconsidered: An Evaluation of the Historical Evidence for Population Decline in Hawai'i, 1778–1803," *Pacific Studies* 16/3, 1993, pp. 115–61.

47 S. Kunitz, *Disease and Social Diversity: The European Impact on the Health of Non-Europeans*, Cambridge, Cambridge University Press, 1994, chapter 3.

48 Denoon, "Pacific Island Depopulation," p. 325.

49 Denoon, "Pacific Island Depopulation," pp. 332–4; D. E. Stannard, "Disease and Infertility: A New Look at the Demographic Collapse of Native Populations in the Wake of Western Contact," *Journal of American Studies* 24/3, 1990, pp. 325–50 (pp. 331–5); Kunitz, *Disease and Social Diversity*, p. 51.

50 Smith and Jones, *Cultural Landscapes of the Pacific Islands*, pp. 55–6.

51 D. Shineberg, "The Sandalwood Trade in Melanesian Economics, 1841–65," *Journal of Pacific History* 1, 1966, pp. 129–46; D. Shineberg, *They Came for Sandalwood: A Study of the Sandalwood Trade in the South-West Pacific, 1830–1865*, Melbourne, Melbourne University Press, 1967.

52 On Hawai'i, see R. Cordy, "The Effects of European Contact on Hawai'ian Agricultural Systems: 1778–1819," *Ethnohistory* 19/4, 1972, pp. 393–418; on New Zealand, see J. Belich, *Making Peoples: A History of the New Zealanders from Polynesian Settlement to the End of the Nineteenth Century*, Auckland, Penguin Books, 1996, p. 152.

53 On New Zealand, see A. Crosby, *Ecological Imperialism: The Biological Expansion of Europe, 900–1900*, New York, Cambridge University Press, 1986, pp. 217–68; on Oceania, see McNeill, "Of Rats and Men."

54 J. Newell, *Trading Nature: Tahitians, Europeans, and Ecological Exchange*, Honolulu, University of Hawai'i Press, 2010; L. Cuddihy, and C. P. Stone, *Alteration of Native Hawaiian Vegetation: Effects of Humans, Their Activities, and Introductions*, Honolulu, University of Hawai'i Press, 1990; J. Culliney, *Islands in a Far Sea: Nature and Man in Hawaii*, San Francisco, Sierra Club, 1988.

55 P. Corris, *Passage, Port and Plantation: A History of Solomon Islands Labour Migration 1870–1914*, Melbourne, Melbourne University Press, 1973.

56 Campbell, *A History of the Pacific Islands*, pp. 156–85. Pacific Islanders became minorities in their own lands in the colonial era in Hawai'i, New Caledonia, the Mariana Islands, and New Zealand, although Hawai'i was the only plantation colony among these.

57 D. Lewis, *The Voyaging Stars: Secrets of the Pacific Island Navigators*, Sydney, Collins, 1978, pp. 108–10; A. Couper, *Sailors and Traders: A Maritime History of Pacific Peoples*, Honolulu, University of Hawai'i Press, 2009; M. Thomas, *Schooner from Windward: Two Centuries of Hawaiian Interisland Shipping*, Honolulu, University of Hawai'i Press, 1983; P. D'Arcy, "Variable Rights and Diminishing Control: The Evolution of Indigenous Maritime Sovereignty in Oceania," in D. Ghosh, H. Goodall, and S. H. Donald (eds.), *Water, Sovereignty and Borders: Fresh and Salt in Asia and Oceania*, London, Routledge, 2009, pp. 20–37 (pp. 23–5, 28–31); R. Teiwaki, *Management of Marine Resources in Kiribati*, Suva, Univelrsity of the South Pacific, 1988, p. 40.

58 Smith, and Jones, *Cultural Landscapes of the Pacific Islands*, p. 56; J. Salmon, *A History of Goldmining in New Zealand*, Wellington, Government Printer, 1963; M. Lyons, *The Totem and the Tricolour: A Short History of New Caledonia since 1774*, Kensington, New South Wales University Press, 1986, pp. 109–20; H. Nelson, *Black, White and Gold: Goldmining in Papua New Guinea, 1878–1930*, Canberra, Australian National University Press, 1976.

59 C. Weeramantry, *Nauru: Environmental Damage under International Trusteeship*, Melbourne, Oxford University Press, 1992.

60 Smith, and Jones, *Cultural Landscapes of the Pacific Islands*, pp. 55–6; A. Smith and K. Buckley, "Convict Landscapes: Shared Heritage in New Caledonia," *Historic Environment* 20/2, 2007, pp. 27–32; S. Bedford, "Post-Contact Maori: The Ignored Component in New Zealand Archaeology," *Journal of the Polynesian Society* 105/4, 1996, pp. 411–40; M. Sahlins and P. V. Kirch, *Anahulu: the Anthropology of History in the Kingdom of Hawai'i*, 2 vols., Chicago, University of Chicago Press, 1992; M. C. Rodman, *Houses Far from Home: British Colonial Space in the New Hebrides*, Honolulu, University of Hawai'i Press, 2001.

61 D. E. Duensing, "The Hana Belt Road: Paving the Way for Tourism," *Hawaiian Journal of History* 41, 2007, pp. 119–48; Thomas, *Schooner from Windward*.

62 J. Bennett, *Natives and Exotics: World War II and Environment in the Southern Pacific*, Honolulu, University of Hawai'i Press, 2009.

63 S. Firth, *Nuclear Playground*, Honolulu, University of Hawai'i Press, 1987. Algeria's independence from France in 1962 removed the desert interior of Algeria as a nuclear testing ground option for France.

64 Firth, *Nuclear Playground*, pp. 128–31, 137–43.

65 M. Merlin and R. Gonzalez, "Environmental Impacts of Nuclear Testing in Remote Oceania," in J. R. McNeill and C. Unger (eds.), *Environmental Histories of the Cold War*, New York, Cambridge University Press, 2010, pp. 167–202; B. R. Johnston and H. Barker, *Consequential Damages of Nuclear War: The Rongelap Report*, Walnut Creek, CA, Left Coast Press, 2008; A. Smith, "Colonialism and the Bomb in the Pacific," in J. Schofield and W. Cocroft (eds.), *A Fearsome Heritage: Diverse Legacies of the Cold War*, California, Left Coast Press, 2007, pp. 51–72.

66 See, for example, R. C. Kiste, *The Bikinians*, Menlo Park, CA, Cummings Publishing, 1974; F. R. Fosberg, "Vegetation of Bikini Atoll, 1985," *Atoll Research Bulletin* 315, National Museum of Natural History, Smithsonian Institution, Washington, DC, 1988; Z. T. Richards, M. Beger, S. Pinca, and C. C. Wallace, "Bikini Atoll Coral Biodiversity Resilience Five Decades after Nuclear Testing," *Marine Pollution Bulletin* 56, 2008, pp. 503–15.

67 See B. Reilly, "The Africanisation of the South Pacific," *Australian Journal of International Affairs* 54/3, 2000, pp. 261–8, versus J. Fraenkel, "The Coming Anarchy in Oceania? A Critique of the 'Africanisation of the South Pacific'," *Journal of Commonwealth and Comparative Politics*, 42/1, 2004, pp.1–34.

68 On contemporary social and economic circumstances, see K. Nero, "The Material World Remade," in D. Denoon (ed.), *The Cambridge History of the Pacific Islanders*, Cambridge, Cambridge University Press, 1997, pp. 359–96.

69 G. Bertram and R. Watters, "The MIRAB Economy in Pacific Microstates," *Pacific Viewpoint* 26/3, 1985, pp. 497–519; G. Bertram, "The MIRAB Model Twelve Years On," *The Contemporary Pacific* 22/2, 1999, pp. 105–38; E. Hau'ofa, "Our Sea of Islands," *The Contemporary Pacific* 6/1, 1994, pp. 148–61.

70 There is a vast array of literature on post-independence resource disputes. For general overviews of resource conflict paradigms, see M. Allen, "Greed and Grievance: The Role of Economic Agendas in the Conflict in Solomon Islands," *Pacific Economic Bulletin* 20/2, 2005, pp. 56–71; on mining see G. Banks, "Mining and the Environment in Melanesia: Contemporary Debates Reviewed," *The Contemporary Pacific* 14/1, 2002, pp. 39–67; on forestry see T. T. Kabutaulaka, "Rumble in the Jungle: Land, Culture and (Un)sustainable Logging in the Solomon Islands," in A. Hooper (ed.), *Culture and Sustainable Development in the Pacific*, Canberra, Asia Pacific Press, 2000, pp. 88–97; for fisheries see S. Chand, R. Q. Grafton, and E. Petersen, "Multilateral Governance of Fisheries: Management and Cooperation in the Western and Central Pacific Tuna Fisheries," *Marine Resource Economics* 18, 2003, pp. 329–48, and Teiwaki, *Management of Marine Resources in Kiribati*; and on community responses see C. Filer and M. Macintyre, "Grass Roots and Deep Holes: Community Responses to Mining in Melanesia," *The Contemporary Pacific* 18/2, 2006, pp. 215–31; R. Scheyvens and L. Lagisa, "Women, Disempowerment and Resistance: An Analysis of Logging and Mining Activities in the Pacific," *Singapore Journal of Tropical Geography* 19/1, 1998, pp. 51–71.

71 G. Banks and C. Ballard, (eds.), *The Ok Tedi Settlement: Issues, Outcomes and Implications*, Canberra, National Center for Development Studies, 1997, and especially J. Burton, "Terra Nugax and the Discovery Paradigm: How Ok Tedi Was Shaped by the Way It Was Found and How the Rise of Political Process in the North Fly Took the Company by Surprise," in Banks and Ballard (eds.), *The Ok Tedi Settlement*, pp. 27–55.

72   J. Bennett, *Pacific Forest: A History of Resource Control and Contest in Solomon Islands, 1800–1997*, Cambridge and Leiden, White Horse Press and Brill, 2000.

73   There are a few notable exceptions, however. See R. J. Hommon, "Social Evolution in Ancient Hawai'i," in P. V. Kirch (ed.), *Island Societies: Archaeological Approaches to Evolution and Transformation*, Cambridge, Cambridge University Press, 1987, pp. 55–68; A. J. Anderson, *The Welcome of Strangers: An Ethnohistory of Southern Māori ad 1650–1850*, Dunedin, Otago University Press, 1998; P. V. Kirch, *How Chiefs Become Kings: Divine Kingship and the Rise of Archaic States in Ancient Hawai'i*, Berkeley and Los Angeles, University of California Press, 2010.

74   D'Arcy, *The People of the Sea*, pp. 128–33; Luders, "Legend and History"; Gunson, "Great Families of Polynesia"; M. Berg, "Yapese Politics, Yapese Money and the Sawei Tribute Network before World War 1," *Journal of Pacific History* 27/2, 1992, pp. 150–64.

75   See particularly M. Levison, R. G. Ward, and J. W. Webb, *The Settlement of Polynesia: A Computer Simulation*, Minneapolis, University of Minnesota Press, 1973; R. G. Ward and E. Kingdom (eds.), *Land, Custom and Practice in the South Pacific*, Cambridge, Cambridge University Press, 1995; M. Bourke and T. Harwood (eds.), *Food and Agriculture in Papua New Guinea*, Canberra, ANU E-Press, 2009. Ray Watters, coauthor of the MIRAB articles cited above, is also a geographer, as is Rapaport; see Rapaport, *The Pacific Islands*. Two particularly good articles on spatial relations are R. G. Ward, "Remote Runways: Air Transport and Distance in Tonga," *Australian Geographical Studies* 36/2, 1998, pp. 177–86, and S. G. Britton, "The Evolution of a Colonial Space-Economy: The Case of Fiji," *Journal of Historical Geography* 6/3, 1980, pp. 251–74.

References

Alkire, W. H., *Coral Islanders*, Arlington Heights, IL, AHM, 1978.

Allen, M., "Greed and Grievance: The Role of Economic Agendas in the Conflict in Solomon Islands," *Pacific Economic Bulletin* 20/2, 2005, pp. 56–71.

Allen, M. S., "New Ideas about Late Holocene Climate Variability in the Central Pacific," *Current Anthropology* 47/3, 2006, pp. 521–35.

Anderson, A. J., *The Welcome of Strangers: An Ethnohistory of Southern Māori AD 1650–1850*, Dunedin, Otago University Press, 1998.

Bahn, P., and Flenley, J., *Easter Island, Earth Island*, London, Thames and Hudson, 1992.

Ballard, C., Brown, P., Bourke, R. M., and Harwood, T. (eds.), *The Sweet Potato in Oceania: A Reappraisal*, Oceania Monograph 56, Sydney, University of Sydney, 2005.

Banks, G., "Mining and the Environment in Melanesia: Contemporary Debates Reviewed," *The Contemporary Pacific* 14/1, 2002, pp. 39–67.

Banks, G., and Ballard, C. (eds.), *The Ok Tedi Settlement: Issues, Outcomes and Implications*, Canberra, National Center for Development Studies, 1997.

Bedford, S., "Post-Contact Maori: The Ignored Component in New Zealand Archaeology," *Journal of the Polynesian Society* 105/4, 1996, pp. 411–40.

Belich, J., *Making Peoples: A History of the New Zealanders from Polynesian Settlement to the End of the Nineteenth Century*, Auckland, Penguin Books, 1996.

Bennett, J., *Natives and Exotics: World War II and Environment in the Southern Pacific*, Honolulu, University of Hawai'i Press, 2009.

Bennett, J., *Pacific Forest: A History of Resource Control and Contest in Solomon Islands, 1800–1997*, Cambridge and Leiden, White Horse Press and Brill, 2000.

Berg, M., "Yapese Politics, Yapese Money and the Sawei Tribute Network before World War 1," *Journal of Pacific History* 27/2, 1992, pp. 150–64.

Bertram, G., "The MIRAB Model Twelve Years On," *The Contemporary Pacific* 22/2, 1999, pp. 105–38.

Bertram, G., and Watters, R., "The MIRAB Economy in Pacific Microstates," *Pacific Viewpoint* 26/3, 1985, pp. 497–519.

Best, S., "Here Be Dragons," *Journal of the Polynesian Society* 97/3, 1988, pp. 239–59.

Blong, R. J., *The Time of Darkness: Local Legends and Volcanic Reality in Papua New Guinea*, Seattle, University of Washington Press, 1982.

Bourke, M., and Harwood, T. (eds.), *Food and Agriculture in Papua New Guinea*, Canberra, ANU E-Press, 2009.

Britton, S. G., "The Evolution of a Colonial Space-Economy: The Case of Fiji," *Journal of Historical Geography* 6/3, 1980, pp. 251–74.

Burtenshaw, M., "Maori Gourds: An American Connection?," *Journal of the Polynesian Society* 108/4, 1999, pp. 427–33.

Burton, J., "Terra Nugax and the Discovery Paradigm: How Ok Tedi Was Shaped by the Way It Was Found and How the Rise of Political Process in the North Fly Took the Company by Surprise," in G. Banks and C. Ballard (eds.), *The Ok Tedi Settlement: Issues, Outcomes and Implications*, Canberra, National Center for Development Studies, 1997, pp. 27–55.

Bushnell, A. F., "The 'Horror' Reconsidered: An Evaluation of the Historical Evidence for Population Decline in Hawai'i, 1778–1803," *Pacific Studies* 16/3, 1993, pp. 115–61.

Campbell, I. C., *A History of the Pacific Islands*, Christchurch, Canterbury University Press, 1989.

Chand, S., Grafton, R. Q., and Petersen, E., "Multilateral Governance of Fisheries: Management and Cooperation in the Western and Central Pacific Tuna Fisheries," *Marine Resource Economics* 18, 2003, pp. 329–48.

Chapman, M. D., "Women's Fishing in Oceania," *Human Ecology* 15/3, 1987, pp. 267–88.

Cordy, R., "The Effects of European Contact on Hawai'ian Agricultural Systems: 1778–1819," *Ethnohistory* 19/4, 1972, pp. 393–418.

Corris, P., *Passage, Port and Plantation: A History of Solomon Islands Labour Migration 1870–1914*, Melbourne, Melbourne University Press, 1973.

Couper, A., *Sailors and Traders: A Maritime History of Pacific Peoples*, Honolulu, University of Hawai'i Press, 2009.

Crosby, A., *Ecological Imperialism: The Biological Expansion of Europe, 900–1900*, New York, Cambridge University Press, 1986.

Cuddihy, L., and Stone, C. P., *Alteration of Native Hawaiian Vegetation: Effects of Humans, Their Activities, and Introductions*, Honolulu, University of Hawai'i Press, 1990.

Culliney, J., *Islands in a Far Sea: Nature and Man in Hawaii*, San Francisco, Sierra Club, 1988.

D'Arcy, P., "Cultural Divisions and Island Environments since the Time of Dumont d'Urville," *The Journal of Pacific History* 38/2, 2003, pp. 217–35.

D'Arcy, P., *The People of the Sea: Environment, Identity and History in Oceania*, Honolulu, University of Hawai'i Press, 2006.

D'Arcy, P., "Variable Rights and Diminishing Control: The Evolution of Indigenous Maritime Sovereignty in Oceania," in D. Ghosh, H. Goodall, and S. H. Donald (eds.), *Water, Sovereignty and Borders: Fresh and Salt in Asia and Oceania*, London, Routledge, 2009, pp. 20–37.

Denoon, D., "Pacific Island Depopulation: Natural or Un-Natural History?," in L. Bryder and D. A. Dow (eds.), *New Countries and Old Medicine: Proceedings of an International Conference on the History of Medicine and Health*, Auckland, Pyramid, 1994, pp. 324–39.

Diamond, J., *Collapse: How Societies Choose to Fail or Succeed*, New York, Viking, 2005.

Diamond, J., "Twilight at Easter," *New York Review of Books*, March 25, 2004, pp.6–10.

Douglas, B., and Ballard, C. (eds.), *Foreign Bodies: Oceania and the Science of Race 1750–1940*, Canberra, ANU E-Press, 2008.

Dudley, W. C., and Lee, M., *Tsunami!*, Honolulu, University of Hawai'i Press, 1988.

Duensing, D. E., "The Hana Belt Road: Paving the Way for Tourism," *Hawaiian Journal of History* 41, 2007, pp. 119–48.

Filer, C., and Macintyre, M., "Grass Roots and Deep Holes: Community Responses to Mining in Melanesia," *The Contemporary Pacific* 18/2, 2006, pp. 215–31.

Finney, B. R., "Voyaging," in J. D. Jennings (ed.), *The Prehistory of Polynesia*, Canberra, ANU, 1979, pp. 324–51.

Finney, B. R., Rhodes, R., Frost, P., and Thompson, N., "Wait for the West Wind," *Journal of the Polynesian Society* 98/3, 1989, pp. 261–302.

Firth, S., *Nuclear Playground*, Honolulu, University of Hawai'i Press, 1987.

Flenley, J., *The Enigmas of Easter Island: Island on the Edge*, Oxford, Oxford University Press, 2003.

Fosberg, F. R., "Vegetation of Bikini Atoll, 1985," *Atoll Research Bulletin* 315, National Museum of Natural History, Smithsonian Institution, Washington, DC, 1988.

Fraenkel, J., "The Coming Anarchy in Oceania? A Critique of the 'Africanisation of the South Pacific'," *Journal of Commonwealth and Comparative Politics*, 42/1, 2004, pp.1–34.

Garden, D., *Australia, New Zealand and the Pacific: An Environmental History*, Santa Barbara, ABC-CLIO, 2005.

Goldman, I., "Status Rivalry and Cultural Evolution in Polynesia," *American Anthropologist* 57/4, 1955, pp. 680–97.

Green, R. C., "Near and Remote Oceania: Disestablishing Melanesia in Culture History," in A. Pawley (ed.), *Man and a Half: Essays in Pacific Anthropology and Ethnobiology in Honour of Ralph Bulmer*, Aukland, Polynesian Society, 1991, pp. 491–502.

Green, R. C., "A Range of Disciplines Support a Dual Origin for the Bottle Gourd in the Pacific," *Journal of the Polynesian Society* 110/2, 2000, pp. 191–7.

Gunson, N., "Great Families of Polynesia: Inter-Island Links and Marriage Patterns," *Journal of Pacific History* 32/2, 1997, pp. 139–52.

Hau'ofa, E., "Our Sea of Islands," *The Contemporary Pacific* 6/1, 1994, pp. 148–61.

Hempenstall, P., "Imperial Manoeuvres," in K. R. Howe, R. C. Kiste, and B. V. Lal (eds.), *The Pacific Islands in the Twentieth Century*, St. Leonards, Allen & Unwin, 1994, pp. 3–28.

Hicks, G., and Campbell, H. (eds.), *Awesome Forces: The Natural Hazards that Threaten New Zealand*, Wellington, Te Papa Press, 2003.

Hommon, R. J., "Social Evolution in Ancient Hawai'i," in P. V. Kirch (ed.), *Island Societies: Archaeological Approaches to Evolution and Transformation*, Cambridge, Cambridge University Press, 1987, pp. 55–68.

Howe, K. R., *Nature, Culture, and History: The "Knowing" of Oceania*, Honolulu, University of Hawai'i Press, 2000.

Howe, K. R. (ed.), *Waka Moana: Voyages of the Ancestors: the Discovery and Settlement of the Pacific*, Honolulu, University of Hawai'i Press, 2007.

Howe, K. R., *Where the Waves Fall: A New South Sea Islands History from First Settlement to Colonial Rule*, Honolulu, University of Hawai'i Press, 1984.

Huffer, E., "Women and Navigation: Does the Exception Confirm the Rule?," *International Journal of Maritime History* 20/2, 2008, pp. 259–324.

Hughes, J. D., *What Is Environmental History?*, Cambridge, Polity, 2006.

Hunt, T. L., "Rethinking Easter Island's Ecological Catastrophe," *Journal of Archaeological Science* 34/3, 2007, pp. 485–94.

Hunt, T. L., and Lipo, C. P., "Ecological Catastrophe, Collapse, and the Myth of 'Ecocide' on Rapanui (Easter Island)," in P. A. McAnany and N. Yoffee (eds.), *Questioning Collapse: Human Resilience, Ecological Vulnerability, and the Aftermath of Empire*, Cambridge, Cambridge University Press, 2010, pp. 21–40.

Hviding, E., *Guardians of Marovo Lagoon: Practice, Place, and Politics in Maritime Melanesia*, Honolulu, University of Hawai'i Press, 1996.

Irwin, G., *The Prehistoric Exploration and Colonisation of the Pacific*, Cambridge, Cambridge University Press, 1992.

Johnston, B. R., and Barker, H., *Consequential Damages of Nuclear War: The Rongelap Report*, Walnut Creek, CA, Left Coast Press, 2008.

Kabutaulaka, T. T., "Rumble in the Jungle: Land, Culture and (Un)sustainable Logging in the Solomon Islands," in A. Hooper (ed.), *Culture and Sustainable Development in the Pacific*, Canberra, Asia Pacific Press, 2000, pp. 88–97.

Kay, E. A., *Little Worlds of the Pacific: An Essay on Pacific Basin Biogeography, Harold L. Lyon Arboretum Lecture 9* , Honolulu, University of Hawai'i Press, 1980.

Kirch, P. V., *The Evolution of the Polynesian Chiefdoms*, Cambridge, Cambridge University Press, 1984.

Kirch, P. V., "Exchange Systems and Inter-Island Contact in the Transformation of an Island Society: The Tikopia Case," in P. V. Kirch (ed.), *Island Societies: Archaeological Approaches to Evolution and Transformation*, Cambridge, Cambridge University Press, 1986, pp. 33–41.

Kirch, P. V., *How Chiefs Become Kings: Divine Kingship and the Rise of Archaic States in Ancient Hawai'i*, Berkeley and Los Angeles, University of California Press, 2010.

Kirch, P. V., "Microcosmic Histories: Island Perspectives on 'Global' Change," *American Anthropologist* 99/1, 1997, pp. 30–42.

Kirch, P. V., *The Wet and the Dry: Irrigation and Agricultural Intensification in Polynesia*, Chicago, University of Chicago Press, 1994.

Kiste, R. C., *The Bikinians*, Menlo Park, CA, Cummings Publishing, 1974.

Kunitz, S., *Disease and Social Diversity: The European Impact on the Health of Non-Europeans*, Cambridge, Cambridge University Press, 1994.

Langdon, R., "The Banana as a Key to Early American and Polynesian History," *Journal of Pacific History* 28/1, 1993, pp. 15–35.

Langdon, R., "The Secret History of the Papaw in the South Pacific: An Essay in Reconstruction," *Journal of Pacific History* 24/1, 1989a, pp. 11–20.

Langdon, R., "The Soapberry, a Neglected Clue to Polynesia's Prehistoric Past," *Journal of the Polynesian Society* 105/2, 1996, pp. 185–200.

Langdon, R., "When the Blue-Egg Chickens Come Home to Roost: New Thoughts on the Prehistory of the Domestic Fowl in Asia, America and the Pacific Islands," *Journal of Pacific History* 24/2, 1989b, pp. 164–92.

Lessa, W. A., "The Social Effects of Typhoon Ophelia (1960) on Ulithi," in A. P. Vayda (ed.), *Peoples and Cultures of the Pacific: An Anthropological Reader*, Garden City, NY, The Natural History Press, 1968, pp. 330–79.

Levison, M., Ward, R. G., and Webb, J. W., *The Settlement of Polynesia: A Computer Simulation*, Minneapolis, University of Minnesota Press, 1973.

Lewis, D., *From Maui to Cook: The Discovery and Settlement of the Pacific*, Lane Cove, Doubleday, 1977.

Lewis, D., *The Voyaging Stars: Secrets of the Pacific Island Navigators*, Sydney, Collins, 1978.

Liebersohn, H., *The Traveler's World: Europe to the Pacific*, Cambridge, MA, Harvard University Press, 2006.

Linnekin, J., "Contending Approaches," in D. Denoon (ed.), *The Cambridge History of the Pacific Islanders*, Cambridge, Cambridge University Press, 1997a, pp. 3–36.

Linnekin, J., "Gender Division of Labour," in D. Denoon (ed.), *The Cambridge History of the Pacific Islanders*, Cambridge, Cambridge University Press, 1997b, pp. 105–13.

Lobban, C. S., and Schefter, M., *Tropical Pacific Island Environments*, Mangilao, University of Guam Press, 1997.

Luders, D., "Legend and History: Did the Vanuatu-Tonga Kava Trade cease in AD 1447?," *Journal of the Polynesian Society* 105/3, 1996, pp. 287–310.

Lyons, M., *The Totem and the Tricolour: A Short History of New Caledonia since 1774*, Kensington, New South Wales University Press, 1986.

McNeill, J. R. (ed.), *Environmental History in the Pacific World*, Aldershot, Ashgate, 2001.

McNeill, J. R., "Of Rats and Men: A Synoptic Environmental History of the Island Pacific," *Journal of World History* 5/2, 1994, pp. 299–349.

Merlin, M., and Gonzalez, R., "Environmental Impacts of Nuclear Testing in Remote Oceania," in J. R. McNeill and C. Unger (eds.), *Environmental Histories of the Cold War*, New York, Cambridge University Press, 2010, pp. 167–202.

Monzier, M., Robin, C., and Eissen, J. P., "Kuwae, 1425 AD: The Forgotten Caldera," *Journal of Volcanology and Geothermal Research* 59, 1994, pp. 207–18.

Munro, D., and Lal, B. (eds.), *Texts and Contexts: Reflections in Pacific Islands Historiography*, Honolulu, University of Hawai'i Press, 2006.

Nelson, H., *Black, White and Gold: Goldmining in Papua New Guinea, 1878–1930*, Canberra, Australian National University Press, 1976.

Nero, K., "The Material World Remade," in D. Denoon (ed.), *The Cambridge History of the Pacific Islanders*, Cambridge, Cambridge University Press, 1997, pp. 359–96.

Neumann, K., *Rabaul Yu Swit Moa Yet: Surviving the 1994 Volcanic Eruption*, Oxford, Oxford University Press, 1996.

Newell, J., *Trading Nature: Tahitians, Europeans, and Ecological Exchange*, Honolulu, University of Hawai'i Press, 2010.

Nunn, P. D., *Climate, Environment and Society in the Pacific during the Last Millennium*, Amsterdam, Elsevier, 2007.

Nunn, P. D., *Oceanic Islands*, Oxford, Blackwell, 1994.

Nunn, P. D., *Vanished Islands and Hidden Continents of the Pacific*, Honolulu, University of Hawai'i Press, 2009.

Oliver, D. L., *Oceania: The Native Cultures of Australia and the Pacific Islands*, 2 vols.,Honolulu, University of Hawai'i Press, 1989.

Rapaport, M. (ed.), *The Pacific Islands: Environment and Society*, Honolulu, Bess, 1999.

Reilly, B., "The Africanisation of the South Pacific," *Australian Journal of International Affairs* 54/3, 2000, pp. 261–8.

Richards, Z. T., Beger, M., Pinca, S., and Wallace, C. C., "Bikini Atoll Coral Biodiversity Resilience Five Decades after Nuclear Testing," *Marine Pollution Bulletin* 56, 2008, pp. 503–15.

Rodman, M. C., *Houses Far from Home: British Colonial Space in the New Hebrides*, Honolulu, University of Hawai'i Press, 2001.

Sahlins, M., *Social Stratification in Polynesia*, Seattle, University of Washington Press, 1958.

Sahlins, M., and Kirch, P. V., *Anahulu: the Anthropology of History in the Kingdom of Hawai'i*, 2 vols., Chicago, University of Chicago Press, 1992.

Salmon, J., *A History of Goldmining in New Zealand*, Wellington, Government Printer, 1963.

Scheyvens, R., and Lagisa, L., "Women, Disempowerment and Resistance: An Analysis of Logging and Mining Activities in the Pacific," *Singapore Journal of Tropical Geography* 19/1, 1998, pp. 51–71.

Shineberg, D., "The Sandalwood Trade in Melanesian Economics, 1841–65," *Journal of Pacific History* 1, 1966, pp. 129–46.

Shineberg, D., *They Came for Sandalwood: A Study of the Sandalwood Trade in the South-West Pacific, 1830–1865*, Melbourne, Melbourne University Press, 1967.

Smith, A., "Colonialism and the Bomb in the Pacific," in J. Schofield and W. Cocroft (eds.), *A Fearsome Heritage: Diverse Legacies of the Cold War*, California, Left Coast Press, 2007, pp. 51–72.

Smith, A., and Buckley, K., "Convict Landscapes: Shared Heritage in New Caledonia," *Historic Environment* 20/2, 2007, pp. 27–32.

Smith, A., and Jones, K. L., *Cultural Landscapes of the Pacific Islands*, Paris, ICOMOS Thematic Study, 2007.

Spate, O. H. K., *The Pacific since Magellan*, vol. 1, *The Spanish Lake*, Canberra, Australian National University Press, 1979.

Spriggs, M., "Geomorphic and Archaeological Consequences of Human Arrival and Agricultural Expansion on Pacific Islands: A Reconsideration after 30 Years of Debate," in S. Haberle,

J. Stevenson, and M. Prebble (eds.), *Altered Ecologies: Fire, Climate and Human Influence on Terrestrial Landscapes, Terra Australis 32*, Canberra, ANU E-Press, 2009, pp. 239–52.

Stannard, D. E., *Before the Horror: The Population of Hawai'i on the Eve of Western Contact*, Honolulu, University of Hawai'i Press, 1989.

Stannard, D. E., "Disease and Infertility: A New Look at the Demographic Collapse of Native Populations in the Wake of Western Contact," *Journal of American Studies* 24/3, 1990, pp. 325–50.

Taylor, P. W., "Myths, Legends and Volcanic Activity: An Example from Northern Tonga," *Journal of the Polynesian Society*, 104/3, 1995, pp. 323–46.

Teiwaki, R., *Management of Marine Resources in Kiribati*, Suva, University of the South Pacific, 1988.

Thomas, M., *Schooner from Windward: Two Centuries of Hawaiian Interisland Shipping*, Honolulu, University of Hawai'i Press, 1983.

Ward, R. G., "Remote Runways: Air Transport and Distance in Tonga," *Australian Geographical Studies* 36/2, 1998, pp. 177–86.

Ward, R. G., and Kingdom, E. (eds.), *Land, Custom and Practice in the South Pacific*, Cambridge, Cambridge University Press, 1995.

Weeramantry, C., *Nauru: Environmental Damage under International Trusteeship*, Melbourne, Oxford University Press, 1992.

Wiens, H. J., *Atoll Environment and Ecology*, New Haven, CT, Yale University Press, 1962.

Wilson, K. L. N., "Nā Wāhine Kanaka Maoli Holowa'a: Native Hawaiian Women Voyagers," in P. D'Arcy (ed.), *Women and the Sea in the Pacific, special three-article forum plus introduction in the International Journal of Maritime History* 20/2, 2008, pp. 259–324.

Yen, D. E., *The Sweet Potato and Oceania, Bernice P. Bishop Museum Bulletin 236*, Honolulu, Bishop Museum Press, 1974.

Further Reading

Brookfield, H. C., and Hart, D., *Melanesia: A Geographical Interpretation of an Island World*, London, Methuen & Co., 1971.

Doulman, D., *Tuna Issues and Perspectives in the Pacific Islands Region*, Honolulu, East–West Center, 1987.

Kirch, P. V., and Green, R. C., *Hawaiki, Ancestral Polynesia: An Essay in Historical Anthropology*, Cambridge, Cambridge University Press, 2001.

Terrell, J. E., Hunt, T. L., and Gosden, C., "The Dimensions of Social Life in the Pacific: Human Diversity and the Myth of the Primitive Isolate," *Current Anthropology* 38, 1997, pp. 155–96.

CHAPTER THIRTEEN

The Environmental History of the Soviet Union

STEPHEN BRAIN

On the morning of September 30, 1957, Soviet citizens living near the town of Chelyabinsk, in the southern Ural Mountains (see Map 13.1), should have awakened to updates on some alarming news. The previous afternoon, a tank containing 70 tons of liquid radioactive waste at the nearby Mayak nuclear fuel reprocessing plant overheated and exploded with the force of a 75-kiloton bomb, throwing its 160-ton concrete roof 30 meters (100 feet) into the air and releasing a cloud of high-level radioactivity into the atmosphere. An eerie yellow fog was seen to float away from the site, and people began to complain of odd symptoms, including hair loss and sloughing skin. According to the International Nuclear and Radiological Event scale, the explosion ranked as the second most serious nuclear accident until 2011, when an earthquake and tsunami struck the Fukushima nuclear power plant in Japan.

But, despite the danger, the official Soviet response was complete silence. A full week passed before any evacuation took place. When it came, with no explanation provided, the resettlement applied only to a small percentage of the affected population. Ultimately, more than a quarter of a million people were exposed to radioactive pollution, with the number of resulting illnesses and deaths still unknown because of the government cover-up. Worse still, the 1957 explosion merely punctuated a much more chronic and arguably worse pattern of pollution at the Chelyabinsk-40 nuclear facility since its foundation in 1945. For the first seven years of the plant's operation, the main reactor employed an open-cycle cooling system; water from a nearby lake was circulated through the reactor to cool it. Additionally, from 1949 to 1956, more than 57 million cubic meters (2 billion cubic feet) of nuclear waste was dumped directly into the Techa River, the only source of drinking water for 24 surrounding villages. When radiation was detected in the region of the Arctic Ocean where the waters of the Techa ultimately flow, the waste was thereafter diverted into a nearby lake, which dried up in 1967 and became the source for an enormous cloud of blowing radioactive dust. As a result of these events, the area around the

A Companion to Global Environmental History, First Edition. Edited by J.R. McNeill and Erin Stewart Mauldin.
© 2012 John Wiley & Sons, Ltd. Published 2015 by John Wiley & Sons, Ltd.

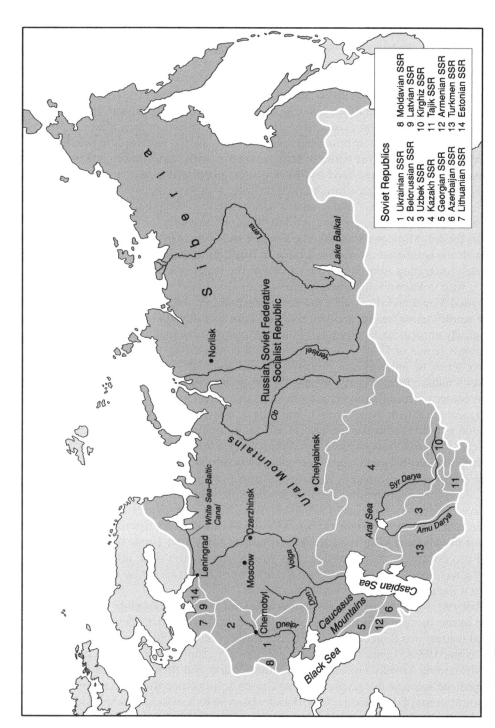

Soviet Republics

1 Ukrainian SSR
2 Belorussian SSR
3 Uzbek SSR
4 Kazakh SSR
5 Georgian SSR
6 Azerbaijan SSR
7 Lithuanian SSR
8 Moldavian SSR
9 Latvian SSR
10 Kirghiz SSR
11 Tajik SSR
12 Armenian SSR
13 Turkmen SSR
14 Estonian SSR

Map 13.1   The USSR

Mayak facility is known as the most polluted place on earth, with a radiation level still 50 times higher than is normal. It stands as a vast uninhabitable zone, first repurposed as a radiological training ground for civil defense troops and since 1966 as the East Ural Nature Reserve, stretching across southwestern Siberia.[1]

The nuclear contamination of the southern Urals may represent the most dramatic case of Soviet environmental mismanagement, but, as a number of authors have documented, the list of environmental catastrophes in the former Soviet Union is distressingly long.[2] For instance, the Aral Sea, once the fourth largest lake in the world, is now mostly a desert called the Aralkum, reduced to 10 percent of its former area, after the Soviets diverted its tributaries to cotton plantations in Central Asia. Because the lake bed has become a major source of windborne carcinogens, with agricultural chemicals once washed into the lake aerosolized and blown across populated areas, the government of Kazakhstan has undertaken efforts to refill the lake, with some preliminary success in the northern section. However, the southern part of the sea, which borders Turkmenistan, continues to desiccate because the river flowing from the south, the Syr Darya, no longer reaches the sea. More ominous still, the Soviet Union's most important biological weapons facility was situated on what was once an island in the South Aral Sea, but is now a peninsula; migrating animals have easy access to the stores of weapons buried around the facility, with unknown long-term effects.[3]

Beyond Mayak and the Aral lie the environmental catastrophes of Chernobyl (the site of the world's worst nuclear accident to date), Lake Baikal (the world's deepest and most biologically unusual lake, polluted by industrial plants whose end products are obsolete), the Virgin Lands of Central Asia (where an ill-conceived agricultural program led to massive levels of soil erosion), Norilsk (a nickel smelting city in the far north which releases 4 million tons of heavy metals into the air annually), and Dzerzhinsk (the home of the Soviet chemical weapons industry and the most chemically polluted city in the world, according to some accounts). And outside the critical zones, millions of Russians continue to suffer from the prosaic yet serious consequences of Soviet environmental neglect: widespread water pollution and air pollution leading to chronic illness and contributing to lower life expectancy. According to Feshbach and Friendly, at the end of the Soviet period even privileged Moscow suffered from soil "saturated with zinc, lead, molybdenum and chromium,"[4] and the average life span of a Muscovite had decreased by 10 years between 1970 and 1990 as a result. Soviet secrecy greatly exacerbated the human cost of all these calamities, especially the Mayak accident described at the beginning of this chapter. With the partial exception of the Virgin Lands Campaign, whose initial successes were well publicized but whose later failures drew little attention, the Soviet government sought to cover up the severity of the problems associated with these places, if the problems were acknowledged at all.

What can account for this callous disregard for human well-being? This question, posed in the context of human-rights abuses rather than in strict reference to environmental matters, has long been central to the study of Soviet politics and history. Even as early as the 1930s, Lenin's lieutenant Leon Trotsky claimed that his rival Josef Stalin, after drumming Trotsky out of the Communist Party and then the Soviet Union, had betrayed the spirit of the revolution and chosen to prioritize the interests of the state over the interests of the workers.[5] Two decades later, Stalin's successor, Nikita Khrushchev, shocked the Soviet leadership by choosing to echo Trotsky's arguments at the 1956 party conference, accusing Stalin of constructing a cult of personality that disregarded the needs of ordinary citizens while allowing grave errors to go unpunished. In the years since Trotsky and Khrushchev leveled their charges, numerous dividing lines have been devised, each attempt seeking to pinpoint the moment when the Soviet government departed from the democratic, utopian ideals of the revolutionary period in favor of *raisons d'état*. The Soviet government, unsurprisingly, always denied that any permanent departure had taken place, while other analysts claimed that abusive despotism simply inhered in Marx's ideas or, alternatively, in Bolshevik ideology, and especially in the decision of a small band of intellectuals to seize power in the name of the working class.[6] Yet another approach points to the Russian Civil War of 1918–21 as a formative experience, one that hardened the Bolshevik Party and helped to make instrumentalist methods of statecraft appealing.[7] However, Stalin's consolidation of power in 1929, because of the massive economic and social changes that his dictatorship brought with it, is generally recognized as the key turning point in Soviet history, for it was at this time that, according to the dominant interpretation, the various revolutionary dreams of 1917 were whittled down to the single, illiberal, and statist variant that Stalin preferred.[8]

A similar discussion has recently developed focusing specifically on Soviet environmental history, and on identifying what force or actor in Soviet history is to blame for catastrophes like the Mayak explosion and the vanishing Aral Sea. As is the case for the debate about the Soviet project as a whole, there is (or was) an official position espoused by the Soviet government, corresponding poorly to the historical record, claiming that communist societies did not and could not do significant social or environmental harm, as capitalist societies often did. A second parallel between the debate about Soviet environmental history and the discussion of Soviet history in general is the broad consensus among western scholars that it was Stalin's "revolution from above," and specifically its philosophical orientation, that deserves the blame for Russia's subsequent problems. In both cases, the root cause of Soviet abuses is Stalinism's inherent mistrust of elemental natural forces, a hostility that targeted (and twinned) undisciplined human nature and the chaos of the nonhuman world. The most recent research on the Soviet environmental history adds nuance to this consensus, however, and suggests that the search for the moment when the Soviet Union embraced an environmentally hostile stance may be misbegotten, because no such fall from grace ever took place. Instead, the historical record shows that the Soviet Union pursued simultaneously, from the outset, astonishingly progressive and cataclysmically destructive policies, and continued to do so until its fall in 1991. Throughout its history, the Soviet Union acted as a hyperbolically exaggerated version of a capitalist society or, more specifically, an exaggerated version of prerevolutionary Russia, with tsarist-era problems magnified and multiplied, but also with the protoenvironmentalist ethos present in prerevolutionary Russia, including its flaws and virtues, developed and transformed into real and meaningful policies.

It is worthy of remark that the first assertions about Soviet environmental management came from the Soviets themselves, some years before environmentalism became a mass social and political movement in the West. The Soviets began portraying Lenin as an environmentalist of the romantic variety, but also as an active force for environmental legislation, as early as the mid-1950s.[9] The trend reached its zenith in the late 1960s, with the publication of scores of books, articles, and pamphlets including *V. I. Lenin on the Protection of Nature* (1969), *V. I. Lenin on Forests and Forest Management* (1967), and *Leninist Principles for the Protection of Nature* (1969).[10] Brian Bonhomme reviewed the content of these publications, and found a suspiciously hagiographical tone in many of the anecdotes; in one story, Lenin received a gift of 15 cherry trees and was so touched by the gift that he made sure that the trees "were planted in a good place and that their delicate trunks were protected from the cold."[11] Bonhomme expresses doubt, given the paucity of references to nature in Lenin's own work, that Lenin considered himself a true friend of nature, and concludes that Lenin was likely most interested in the utilitarian side of conservation, if he was interested at all. However, leaving aside momentarily the question of whether these works accurately portrayed the motivations and achievements of Soviet environmentalism in the 1920s, the publications do tell us a number of noteworthy things about Soviet environmentalism of the 1950s, 1960s, and 1970s: first, that there were authors who *wanted* Lenin to be an environmentalist, and second, that the Soviet authorities saw no reason to object. Holding Lenin up as an environmentalist hero helped support a narrative, generated both for domestic and for foreign consumption, that communist societies accorded the natural world more respect than did capitalist societies, because the elimination of private property had also eliminated the causes for environmental destruction.[12] For a time in the 1960s, this message was the only one available about environmental conditions in the Soviet Union and, given the lack of countervailing evidence, as well as the conspicuousness of American environmental crises at the time, the superiority of Soviet environmental management seemed plausible – and even inspirational for those concerned about environmental mismanagement in the capitalist world.[13]

Soviet claims about environmental harmony raised the suspicion of a number of western analysts, who responded in the early 1970s with a series of books exposing the widespread public-health problems plaguing the Warsaw Bloc countries. The first to emerge were Marshall Goldman's *The Spoils of Progress: Environmental Misuse in the Soviet Union* and Philip Pryde's *Conservation in the Soviet Union*, both published in 1972. Both refused to accept Soviet assertions about environmental protection at face value. Pryde acknowledged that a dictatorial system could theoretically accelerate the drafting and promulgation of environmental legislation but, in reality, Soviet environmental laws were "much too generalized, and the procedures set up for enforcing them … almost totally ineffective."[14] Although he lauded the Soviet rejection of the West's wasteful consumer culture, Pryde concluded that Soviet efforts at environmental protection had proven that "the centralized planning of an economy per se provides no necessary guarantee that … pollution of the environment will not occur."[15] Goldman was more strident in his critiques, having discovered the details about the Aral Sea as well as the degraded state of the Volga River, which had caught fire in 1970 just as Cleveland's Cuyahoga River famously had in 1969. He contended that communist systems actually posed a greater threat to the environment than did capitalist systems, since "the highest potential for increased destructiveness comes when the state monopolizes all the productive powers of the country … as often happens in a communist country."[16] Both books, then, cast doubt on the possibility of effective environmental protection under a communist regime, regardless of good intentions and well-trained technicians, because the very functioning of centralized political systems prevented meaningful laws from taking effect. Six years later, Ze'ev Vol'fson, writing under the pseudonym of Boris Komarov, published perhaps the most influential book exploring the failures of communist environmentalism, *The Destruction of Nature in the Soviet Union*.[17]

The flaw inherent in dictatorial command economies identified by Pryde, Goldman, and Vol'fson, significant enough to prevent even timely and well-intended environmental legislation from effecting real change, can help explain the failure of Soviet comprehensive environmental laws. On June 7, 1957, Estonia was the first Soviet republic to adopt a legal code setting standards for environmental quality and resource conservation, and on October 27, 1960, the Russian republic adopted a similar measure, "On the Conservation of Nature in the R.S.F.S.R."[18] These laws established standards for the permissible use of land, minerals, waters, forests, air, and parklands, established the required amount of suburban greenbelts near cities, and mandated compulsory environmental education for schoolchildren, "with a view to instilling in youth a feeling for a considerate attitude toward natural wealth and habits of the correct use of natural resources."[19] In addition, the state required itself to popularize environmental problems, marshaling "publishing houses, museums, theaters, radio, television, and the editorial staffs of newspapers and magazines to publicize the goals of conservation of nature."[20] Undoubtedly ahead of their time when compared with the environmental efforts of other countries, the comprehensive laws nevertheless failed to protect water or air quality for Soviet urbanites, because ultimately, as Pryde and Goldman both allege, the Soviet dictatorship was unwilling to prosecute itself for environmental infractions.[21] Hence, in the estimation of authors like Pryde and Goldman, the root cause for the Soviet environmental crisis was built into the Bolshevik revolution, and specifically into Lenin's decision to implement a dictatorship – temporary in theory but permanent in practice – for dictatorships lack the pluralist structures required to enforce environmental standards.

Douglas Weiner, likely the most influential historian of Soviet environmental politics, challenged this view by providing evidence in *Models of Nature* (1988) and *A Little Corner of Freedom* (1999) that the Bolshevik dictatorship did not in fact make environmental safeguards impossible, but instead created space for unprecedented and meaningful levels of nature protection.[22] Weiner examined the origins and development of the *zapovedniki* (a word with religious connotations due to its relationship with the Russian word *zapoved'*, or sacred commandment), a network of nature preserves begun during the tsarist era but greatly expanded in the first years after the Bolshevik revolution. As Weiner explains, the *zapovedniki* were an unusual form of nature preserve, conceived to advance scientific understanding of ecological questions rather than provide leisure activities, or enhance economic performance. Each reserve was dedicated to a single ecological problem, and each was designed to be strictly inviolable, so as to facilitate the study of nature's functioning without human intervention. Scientists at one reserve, for instance, studied the ecosystemic function of the Ukrainian steppe; at another they examined the growing conditions of the Khosta forest in the Caucasus Mountains. It is clear that the *zapovedniki* were neither a Soviet idea nor a communist ideal, since the law first authorizing them was passed in 1916, and the idea behind them had been circulating among experts in the Russian Geographical Society for almost a decade. But the Soviet dictatorship made its mark on the endeavor by expanding into an organized system of preserves what had previously been only an embryonic program, such that by 1925 there were 24 preserves and by 1929 61, covering almost 4 million hectares.[23] (This, although only about 0.2% of the gigantic Soviet territory, is equal to roughly the size of Kentucky, or of the six largest national parks in the US – Death Valley, Yellowstone, Everglades, Grand Canyon, Glacier, and Olympic – combined.)

The *zapovednik* concept was able to flourish in the early years of the Soviet Union due to a fortuitous alliance of dictatorial rule and support among the Bolshevik leadership for expertise and what today would be called "wise use." Lenin himself, as Weiner relates, valued socialism most of all for its efficiency, so it follows logically that he endorsed conservation policies. Recognizing that human effort could not fully replace the forces of nature, but rather could only work within natural limits, Lenin sought to make the

communist economy function in accordance with natural laws, and accordingly, within a
few months of taking power, the Bolsheviks passed laws codifying forest management
and hunting, as well as authorizing the expansion of the *zapovedniki*. The *zapovedniki*
would act as "models of nature" where the possibilities and limits of natural function
could be determined, after careful investigation by trained experts.

But also important for the acceptance of environmental protection during Lenin's
rule was his personal enthusiasm for creating a "cultured" society. The impetus to create
a Soviet network of *zapovedniki* came not from the agricultural bureaucracy or the industrial authorities, but rather from the People's Commissariat for Enlightenment – the
bureau charged with education and cultural matters. As Weiner relates, Lenin's minister
of education, Anatolii Lunacharskii, found the proposal to create a network of nature
preserves personally intriguing for their educational potential, and made sure that the
idea gained Lenin's attention. Lenin heard the proposal on January 16, 1919, despite
the civil war still raging in the provinces, and was favorably disposed, stating that the
cause of conservation was important "for the whole republic, and that he considered it
an urgent priority."[24] Bureaucratic red tape prevented speedy adoption of the law, but on
May 4, 1920, Lenin signed the legislation, thereby making the Soviet Union the first
country to set aside parcels of land specifically for the scientific study of ecological function. The scientists seeking to test the innovative concept of the "biogeocenosis," their
name for the sum total of biological and geological interactions on a given territory,
received sanction to conduct their investigations, although never as much financial support as they would have liked, even in the best of times. For Weiner, then, the advent of
communist rule did not represent the crucial turning point in Soviet environmental history, for the creation of the *zapovednik* system gave proof of the environmental potential
of socialist society.

In Weiner's view, the moment of "original sin" in Soviet environmental politics came
after Lenin had passed from the scene, when his successor, Josef Stalin, took command
and began to pursue very different political and economic goals. Beginning in 1928,
Stalin embarked upon a program of hyperindustrialism via the First Five-Year Plan, which
in short order transformed the Soviet Union into an industrial superpower. Throughout
the country, often in previously uninhabited lands, enormous industrial combines were
built, vast new mining operations established, and electric plants of all varieties constructed, resulting in dramatically increased industrial production after only a few short
years.[25] Accompanying the five-year plan was an explosion of Promethean propaganda,
summarized well by Maksim Gorkii's pronouncement, made in reference to the convict
labor used to dig the White Sea–Baltic Canal, that "Man, in transforming nature, transforms himself."[26] Also accompanying the First Five-Year Plan was much stricter criminal
enforcement of economic policies than had been the case before Stalin consolidated
power; engineers were arrested for failing to meet quotas, accused of intentionally
"wrecking" the Soviet economy, and sentenced to lengthy prison terms, while those who
doubted the wisdom of Stalin's breakneck tempos were hounded from their jobs and
frequently punished.[27]

Beyond its human cost, the industrial effort took its toll on environmental conditions
in the Soviet Union, and undermined the stability of the *zapovedniki*. When Lunacharskii,
the patron and protector of the network in the 1920s, resigned as minister of education
in September 1929, his emphasis on cultural pluralism and public education evaporated,
and the *zapovedniki* entered a period of slow decline. The preserves were shuttled
from one ministry to another beginning in 1933, and were explicitly reoriented toward

economic goals, including increasing the numbers of valuable domestic species and acclimatizing potentially useful foreign species to Russian conditions – both missions clearly opposed to the original intentions of the founding ecologists. As Stalin's tenure progressed, the fortunes of the preserves fell further. Perhaps the most important *zapovednik*, the Askania-Nova preserve in southern Ukraine, saw not only its research institute defunded but its staff arrested in 1934. In 1935, the authorities compiled a 100-page report documenting the anti-Soviet tendencies of the people who worked in the reserves, although no direct action was taken as a result. And finally, at the end of Stalin's reign, the *zapovedniki* were nearly eliminated entirely, when in 1951 the 128 reserves were reduced to 40, and their total area cut by 85 percent.[28]

Weiner explains the drastic change in the fortunes of the *zapovedniki* by pointing to a cultural shift accompanying the start of the First Five-Year Plan, presided over by Stalin, and wrought by the state's enforcement of an ethical code that discounted the importance of the natural world. The primary responsibility for the Soviet citizen individual, according to this code, was to assist the state in its goal of developing the economy so as to create the preconditions for socialism. Similarly, the worth of the natural world – to the extent it had worth – lay in its capacity to be transformed according to human reason. The personal preferences of the individual, as well as the traditional processes of nature, were not important when compared with the state's mission of forging a path toward the perfect human society. Set against this backdrop, the struggle between the ecologists and the central authorities over the fate of the *zapovedniki* was in fact a proxy fight about liberal individualism and the right to live apart from smothering state control. "Common to many of these cases," Weiner writes when referring to the struggles of Soviet environmental activists, was the fact that "nature protection served as a surrogate for politics, as actual political discourse was prohibited and punished."[29] Advocating nature protection was, at its root, the scientists' way of preserving their "professional identity and esprit de corps in the face of adverse and dangerous conditions," as well as their social autonomy.[30] Environmental activism functioned as a way "for Soviet people to forge or affirm various independent, unofficial defining identities for themselves."[31]

The impact of Weiner's research helped to solidify a developing consensus that the Soviet Union's environmental troubles could be traced directly to Stalin, the onset of rapid industrialization, and the adoption of repressive social measures to enforce a belief in anti-environmentalist Prometheanism. Earlier writers had hinted in this direction; Goldman wrote in *The Spoils of Progress* that in the "three decades after Lenin's death in 1924, slight attention was paid to preserving the country's natural resources,"[32] thereby implying that matters had been better before Lenin's death, but Weiner provided concrete evidence of a visionary environmental initiative favored by Lenin but then crushed by Stalin. Within a few years, this viewpoint had found expression in general textbooks about Soviet history. For instance, in *The Soviet Experiment*, Ronald G. Suny summarizes the environmental impact of the First Five-Year Plan in this way:

> The rush to modernity ... meant that attention was paid almost exclusively to output and productivity and almost no notice was taken of the impact of rapid industrialization on the natural environment ... In the Soviet Union, ecological ignorance was compounded by the bravado of the Communists, who looked upon nature simply as an obstacle to be overcome on the road to progress.[33]

Brian Bonhomme refers to Stalin as "the undefended villain of environmentalism," and even contrary evidence has been interpreted to maintain Stalin in this villainous role.[34] When William Husband found that Stalin-era children's literature sometimes presented nature in a sympathetic light, he concluded this literature "demonstrated an important limit to political control in the USSR," rather than a conscious decision to allow environmentalist themes to appear in children's books.[35]

The identification of Stalinism as the root cause of Soviet environmental problems gained popularity in part because of its logical appeal – it follows that the political movement that sent thousands of falsely accused victims to the Gulag would also have turned a blind eye to environmental destruction – but also because Stalinist industrialization did significantly decrease the environmental quality of life for millions of Soviet citizens. Donald Filtzer documents the depth of the public-health problems in his recent work, *The Hazards of Urban Life in Late Stalinist Russia*, and relates the horrific conditions that Soviet urbanites were forced to confront in their daily lives: "Soviet cities," Filtzer writes, "were filthy places, covered for most of the year in piles of garbage, mounds of human excrement, and torrents of raw sewage flowing through open gutters or simply spilling out onto streets and sidewalks."[36] The consequences of the failure to adequately control waste, human and industrial alike, created a public-health nightmare in the Soviet Union. Infant mortality, rates of respiratory diseases, and gastrointestinal infections all far exceeded western norms. At the heart of these problems, Filtzer asserts, stood the inhumane values embodied by the political and economic system that Stalin constructed: "Stalin's personal distrust of, and indeed contempt for, the ordinary producers of Soviet society meant that in questions over the allocation of resources he naturally gravitated toward solutions that involved the suppression of consumption."[37] The connection between environmental issues and resource allocation is direct and explicit: in Filtzer's view, Stalin did not want to expend valuable resources on expensive programs that would benefit individuals rather than the state, and this misanthropy precluded the financing of programs that would have protected the environment. Here Filtzer sounds a note nearly identical to Weiner's description of Stalin's government, especially in light of its environmental neglect, as a "kleptocracy."[38] Supporters of the Stalin-as-villain consensus can also point to Stalin's ecologically irresponsible "hero projects," as Paul Josephson does in reference especially to hydroelectric installations.[39]

But a close analysis of the evidence used to buttress the consensus reveals that the sharp discontinuity assigned to Stalin's ascendance has been exaggerated. Put another way, the changes ascribed to Stalin were not as great as have been described, and those great changes that did occur were common to other industrial societies. For instance, when reviewing the history of the *zapovedniki*, the record of Stalin's hostility to the preserves is ambiguous. The network actually grew steadily throughout Stalin's tenure, from 7 preserves in 1920, to 9 in 1929, 15 in 1933, 37 in 1937, 91 in 1947, and 128 in 1951.[40] While it is true that there was a sharp reduction in the number of preserves in 1951, at the very end of Stalin's life, the number began to increase again in 1957, indicating that Stalin had put in place no permanent philosophical antagonism toward the preserves. Furthermore, the full reason for the 1951 reorganization has only recently come to light. The opponents of the *zapovedniki*, including officials in the Ministry of Forest Management, did not wish to abolish the preserves or economically exploit them, but rather to manage them more actively, using methods that the under the existing rules of strict inviolability were impermissible. Throughout the Soviet Union, the *zapovedniki*, because they could not be logged, thinned, or even cleaned, were by the later 1940s becoming flashpoints for forest fires and insect outbreaks, and foresters wanted approval to clear away underbrush and cut firebreaks, actions which would require a change in designation. However, once the reassignment of the *zapovedniki* was completed in 1952, the forests were not opened to logging, but rather were reassigned as "Class I" protected forests, where absolutely no commercial use was allowed.[41] Furthermore, the political repression of preserve researchers on scientific grounds, discussed by Weiner, is difficult to substantiate, since the reasons for the arrest of the Askania-Nova staff is unknown, and the report drawn up in 1935 attacking the *zapovednik* workers did not result in any state action. The preserve system did undergo a scientific reorientation in the early 1930s, toward an emphasis on practical results, yet they persisted nevertheless. Should a network of preserves dedicated to science, unduplicated anywhere else in the world, be considered evidence of an anti-environmental attitude because the original mission of the founders was changed from one kind of scientific inquiry to another?

A review of public-health statistics raises similar questions. According to Filtzer's statistics, for instance, infant mortality in Russia actually declined throughout the Soviet period, and sharply so, except during the tragic, state-generated famine that accompanied agricultural collectivization in 1932–3, during World War II, and during a postwar famine in 1947. In 1901–5, the average infant mortality in Russia was 253 deaths per 1,000 live births, and in 1911–15, this number increased to 273. However, under Soviet power, the death rate steadily decreased. The data for the early Soviet period are incomplete, but the numbers that exist show a fairly regular pattern (see Table 13.1).

By 1956, the infant mortality rate in the Russian republic was 50 deaths per 1,000 live births – much higher than in Western Europe, but 82 percent lower than the rate had been 40 years prior. This improvement in public health stemmed from several causes, but among them were successful efforts at checking infectious disease, including waterborne disease. Stalin killed millions, but clean-up of the water supply under his rule also saved millions from early death.

Indeed, there is evidence supporting the contention that Stalin deserves recognition as a pioneer in environmental protection. Throughout his reign, Stalin intentionally and consistently supported remarkably stringent levels of environmental protection for the forests of the Russian heartland. Ultimately he authorized the world's largest forest preserve, not situated in the hinterlands, but embracing Russia's most productive and most centrally located woodlands.[42]

Table 13.1 Soviet infant mortality, 1926–55

1926–30	183	1948	95
1936–40	193	1949	86
1941–5	183	1950	89
1946	81	1951	91
1947	132	1952	78
1953	73	1955	62
1954	71		

Note: All figures in deaths per 1,000 live births. Compiled from Filtzer, D., *The Hazards of Urban Life in Late Stalinist Russia: Health, Hygiene and Living Standards, 1943–1953*, Cambridge, Cambridge University Press, 2010, pp. 258, 283, 303.

Stalin's sweeping forest-protection program evolved in response to a dilemma resulting from the decision, made in the first years after the Bolshevik revolution, to divide forest management into two spheres, forest cultivation and forest industry, with a different bureaucracy in charge of each. The rationale provided in 1918, at the time of the division, held that the work of planting and nurturing new forests differed so significantly from the work of felling and removing timber that different bureaucratic entities should exist for each task. Ideally, the forest-cultivation bureau, under the auspices of the People's Commissariat for Agriculture, was supposed to tend the new forests from sowing to maturity, and each year, in response to the requests of the logging bureaus under the aegis of the Supreme Soviet of the Economy, would indicate which parcels should be cut. However, when this division of labor was actually implemented, the People's Commissariat for Agriculture and the Supreme Soviet of the Economy began to quarrel almost immediately, complaining that each made the other's work impossible, and frequently petitioning for each other's elimination. The logging trusts alleged that the People's Commissariat of Agriculture impeded them in their goal of meeting their quotas, while the cultivation bureaus complained that the loggers exploited the forest inefficiently, felled forests not assigned to them, and left the logged forests in such rough shape that regeneration followed only with extreme difficulty. Throughout the 1920s, the two agencies waged a relentless bureaucratic war against one another, and although the Supreme Soviet of the Economy might seem to have held the advantage of ideological propinquity to the communist leadership, the upper echelons of the government repeatedly upheld forest conservation as an ideal, and denied the requests of the logging bureaus. For instance, in 1929, Stalin's lieutenant Lazar Kaganovich expressed his support of forest conservationism at a meeting of the Party Central Committee:

> When we approach the question about who should be master of the forest ... then we arrive at a sticking point between two agencies – the People's Commissariat for Agriculture and the Supreme Soviet of the Economy. The Supreme Soviet has the larger appetite – they say "I will take it all and never be satisfied. I am afraid they will gobble up the entire forest.[43]

On more than one instance in the 1920s, the Supreme Soviet of the Economy petitioned the government for more access to land and was not only denied, but had its allotments decreased.

The predominance of the conservationists, albeit a temporary one, came to an end with the adoption of the First Five-Year Plan in 1929. After a poor timber harvest in the 1928–9 logging season, the Supreme Soviet of the Economy complained that the existing system did not and could not provide the amount of timber required for rapid industrialization. Armed with indisputable numbers, the Supreme Soviet was able to wrest control of the forest away from the People's Commissariat of Agriculture. By the end of 1931, the logging bureaus had won the right to determine logging levels and patterns throughout the USSR, and after doing so, went about installing a remorselessly aggressive felling regime. The 1930 timber harvest exceeded annual growth in Leningrad province by 47 percent, in Western province by 125 percent, in Moscow province by 129 percent, and in Ivanovo-Voznesenk province by 104 percent.[44] Forests in Riazan district scheduled for logging in 1976 were chopped down 46 years ahead of schedule.[45]

According to the prevailing consensus about Stalinist anti-environmentalism, the story of Soviet forest management should end here, with the elimination of forest conservation as an effective force in Soviet politics. And yet the domination of the industrialists was, in the event, extremely short-lived. Within a matter of months, in response to the protests of the academic foresters, the Soviet government soon reversed itself and not only returned to the conservationism of the 1920s, but strengthened forest protection far beyond the previous levels, consistently increasing the rigor of the conservation laws for almost 25 years. In 1931, pursuant to Stalin's personal order and reflecting the foresters' concerns that increased logging rates were causing ruinous erosion that threatened the nation's hydroelectric plants, new forest legislation was adopted. Control over the forest was once again divided, and the forests in a 6-kilometer (4-mile) belt on either side of the major rivers of Russia were returned to the People's Commissariat of Agriculture, with special instructions to maintain forest cover there. The Supreme Soviet of the Economy protested vehemently, but to no avail. In 1936, the protection regime grew stricter still, with a special agency, the Main Administration of Forest Protection and Afforestation, created to oversee the protected forests, after it was determined that the Commissariat of Agriculture could not police the preserves effectively. In addition, the width of the protective zones was increased, from 6 to 20 kilometers (4 to 12.5 miles). Finally, in 1947, the Main Administration of Forest Protection and Afforestation was transformed into an agency of the highest governmental designation, the Ministry of Forest Management, and charged with the management not merely of the protected forests, but of all the forests of the Soviet Union, with the industrial bureaus placed in an explicitly subordinate position. This arrangement lasted until March 1953, when Stalin died. In the aftermath of his death, the Ministry of Forest Management was disbanded and the forests returned to industrial control. The wide protective forest belts paralleling the Volga, the Dnieper, and the Don maintained their protective status throughout the Soviet period, however, and as such constituted the world's largest forest preserve, encompassing an area the size of Mexico.

The story of Stalinist forest management suggests that the implacable antagonism to environmentalism commonly ascribed to the Soviet leadership may be exaggerated, if not entirely illusory. This is not to say that Stalin did not prioritize the rapid industrialization of the Soviet Union above all other initiatives, or that hyperindustrialism did not exact a serious environmental toll. However, Stalin's forest policy (which, it must be noted, allowed for completely irresponsible levels of logging outside the protected zones) provides a reason to believe that Soviet leaders, including Stalin, wanted industrialization to proceed rapidly but without doing irreparable harm. When foresters began to publicize the real dangers that overlogging posed to the country's hydrology and future economic growth, the Soviet government responded forcefully and effectively. (It is true that Stalin's government failed to introduce antipollution legislation, but such laws were nearly unheard of anywhere in the world in the 1930s.) Stalin's rapid adoption of legislation in defense of the forest may provide the answer to the vexing question, posed by Douglas Weiner in *A Little Corner of Freedom*: why did the Soviet authorities so often turn a blind eye to environmental activism, given the ideological hostility of the regime?[46] It may be the case that this hostility did not exist.

Returning, then, to a question posed earlier in this chapter, what can explain the many environmental failures of the Soviet Union, if not a structural antipathy toward nature inherent in communist dictatorship, or a conscious animus built into Stalinism? In light of the analysis provided above, perhaps a better question would be to ask how it is possible that the country that built the *zapovednik* system, created the largest forest preserve in world history, and reduced infant mortality by 82 percent in 40 years despite a civil war (1918–21) and the Nazi invasion (1941) also allowed such ruinous levels of air pollution, the cover-up of the Mayak nuclear accidents, and the Aral Sea disaster. Posing the question this way does not represent an attempt to deny Soviet environmental problems, but rather reflects an effort to characterize the Soviet environmental record more accurately, as a mixture of remarkable achievement and distressing dysfunction, so as to find explanations for its shortcomings.

Some of the environmental tragedies that befell the Soviet Union were not distinctively Soviet. The story of the evaporated Aral Sea, for instance, parallels very strongly that of the Owens Valley Dry Lake, near Death Valley in southeastern California. As recently as 1924, the Owens Lake was 260 square kilometers (100 square miles) in size and, because of its location in the high desert near the Mojave and the Great Basin, presented an important oasis for birds migrating from Canada to Mexico.[47] The city of Los Angeles, however, began in 1913 to divert the water for urban use, and soon the lake disappeared entirely. The dry lake bed then became the single greatest source in the world of small particulate matter or PM10 – an aerosolized mixture of arsenic, cadmium, nickel, and sulfates with particles smaller than 10 microns in diameter – that due to its extremely fine-grained quality was easily inhaled and thus caused serious health problems for nearby residents. The Los Angeles *Times* reported that, in the 1990s, the air in the towns nearest the lake regularly contained particulate matter at 23 times the federal acceptable norm.[48] Both the Aral Sea and Owens Lake have been partially refilled in the very recent past, but the common underlying cause for the desiccated lakes – water diversion for economic purposes – remains.

Similarly, the dreadful nuclear accidents at the Mayak facility and at Chernobyl also have analogues – albeit less serious ones – in other countries. The nuclear industry, as is commonly accepted and yet perhaps still insufficiently documented, poses a serious environmental threat wherever its installations are sited, and the high levels of security and secrecy surrounding them has resulted in an alarming number of underreported incidents around the world. For instance, the Nevada Test Site, located 105 kilometers (65 miles) northwest of Las Vegas, served as the location for more than 900 nuclear detonations between 1951 and 1992, 100 of which were above ground. These tests together released more than 300 million curies of radiation into the atmosphere. The most critical, the Baneberry blast of December 18, 1970, produced a more violent reaction than expected and released 6.7 million curies into the air. (In comparison, the Three Mile Island nuclear accident released 20 curies.) To take another example, weapons production at the nuclear facility at Hanford Site in the state of Washington has severely contaminated the surrounding area, and only public pressure on the Department of Energy forced the publication of government documents testifying to the release of plutonium and other nuclear materials directly into the air and into the Columbia River.[49] Secrecy concerning radiation pollution, to one degree or another, has been common among all nuclear powers. The American nuclear accidents and mismanagement do not in any way exonerate Soviet malfeasance, but they do highlight the inherent dangers of nuclear energy and especially nuclear weapons technology which, when coupled with military secrecy, create greatly increased chances for environmental hazards, and for insufficient warnings to the public.

If water diversion and nuclear contamination are international environmental trends, and the Soviet cases represent only particularly serious manifestations of the trends, the same cannot be said for the failure of the Soviets to enforce their own environmental laws, and this disconnect between intentions and results requires explanation. The phenomenon results from a combination of two factors: first, the extreme influence exerted by the industrial bureaus over the Soviet political system, a consequence of the fact that the Soviet project largely amounted to an effort to rapidly modernize a country that had fallen far behind that of its neighbors, and second, the source of the Russian environmental ethos in elite prerevolutionary culture, which led the mostly proletarian Soviet leadership to respect, but not entirely understand, environmentalist values. The first factor accords with a very prominent interpretation of Soviet history, presented most succinctly by Theodore von Laue in *Why Lenin? Why Stalin? Why Gorbachev?*, which holds that the Soviet experiment was not actually intended to create a socialist utopia. According to von Laue, the Bolshevik revolution initiated a concerted effort to transform backward Russia into a country that would not lose European wars, as it had reliably been doing since the mid-nineteenth century.[50] Lenin emphasized Russia's cultural and infrastructural backwardness, Stalin Russia's military and industrial backwardness, and Khrushchev emphasized Russia's backwardness in standard of living, but each of the early Soviet leaders allowed the managers of economic industrial bureaus a nearly free hand over policy, even when this led to conflict with the party leadership.[51] One result of this conflict was the acceptance of generally admirable environmental legislation at the highest governmental levels, but the concomitant inability to enforce the legislation on the ground.[52]

The second factor, the socially remote origins of Russian environmentalism, does not help explain the failure to enforce curbs on air and water pollution, but does help explain why the Soviets continued to put environmental legislation in place despite the difficulties in enforcement. As a number of scholars have demonstrated, prerevolutionary Russian culture featured a well-developed artistic, literary, and scientific consciousness with a strong environmentalist component, but this consciousness developed in a social stratum that wielded only indirect authority – mostly moral – after the Bolshevik revolution. Jane Costlow, for example, has discussed the prevalence of environmental themes in Russian literature, finding a concern about deforestation in poetry and prose expressed throughout the nineteenth century in the works of the most illustrious Russian writers. Leo Tolstoy, himself an enthusiastic novice in forest management, notably employed forest conservation as a recurrent theme in *Anna Karenina*, with responsible forest management symbolizing ethically responsible behavior toward the greater world; "Tolstoy's defense of the forest," Costlow writes, "is grounded not in economics or legislation, but in religious ethics and spiritual transformation."[53] In *Anna Karenina*, the choice of the hero, Levin, to forgo selling off his timberland for much-needed cash underscores the moral meaning of nature: the forest is a place of communion with larger forces – historical, social, spiritual – and to cut it down thoughtlessly is to destroy a link to the transcendent and to the past. Other Russian authors of note, including Fyodor Dostoevsky, Anton Chekhov, Ivan Turgenev, and Mikhail Lermontov, made similar connections between the integrity of the natural world and the integrity of the human world.[54] Christopher Ely has highlighted a similarly powerful connection in Russian painting, between natural beauty and nationalism. The beloved paintings of Ivan Shishkin, according to Ely, "offered his audiences of city-dwellers a chance to take in appreciation of the rural values and spirit of the nation"; Ely stresses that "by creating numerous realistic scenes of simple Russian forests and fields that stood as symbols of Russian nationality, Shishkin invited urban Russians to imagine a profound connection between themselves and their natural surroundings."[55] The list of artists who made environmentalist statements with their art, like the list of authors, stretches far beyond Ivan Shishkin, featuring Mikhail Nesterov, Mikhail Klodt, and Isaac Levitan.

The search for transcendent meaning in nature also found its way into other manifestations of Russian high culture. One of Russia's most influential historians, Vasilii Kliuchevskii (1841–1911), for example, linked the Russian national spirit with the forest, in contrast with the geographically alien steppe peoples of Central Asia. And in addition to the artistic and academic appreciation for the greater significance of nature in Russian cultural life, there was, as has already been noted, a strong scientific emphasis on nature protection in prerevolutionary Russia, expressed in the support for the *zapovedniki*, as well as a nationalist form of forest conservation whose followers eventually formed the core of Stalin's Ministry of Forest Management.[56]

These approaches to nature protection all share common roots in a social class that lost a great deal of standing after the Bolshevik revolution. Although the Soviet government frequently failed to enact policies that improved the material conditions of the working class, especially in its first decades, it did, as Sheila Fitzpatrick has argued, create a new ruling class by elevating workers and peasants into positions of prominence.[57] Especially during the Cultural Revolution of the late 1920s and early 1930s, the old tsarist intelligentsia was largely coerced from power, although not destroyed completely.[58] The cultural values of the old Russian ruling class survived, and even returned to prominence in the later 1930s as part of a trend identified by Nicholas Timasheff as a "Great Retreat," but they generally served to reinforce the philosophical materialism and state capitalism championed by the Bolsheviks.[59] As Mark Bassin shows in "'I Object to Rain that Is Cheerless': Landscape Art and the Stalinist Aesthetic Imagination," the conflict of the old and new value systems did not destroy one or the other in the short term, but rather created a puzzling ambiguity in official pronouncements on art.[60] Katerina Clark finds the same ambiguity in Stalinist literature, noting that an "ambivalent attitude toward nature is built into the very structure of the Stalinist novel" because, for most Soviet writers, nature was "more wholesome and pure, more vital and captivating" than industry.[61] It is this deep ambiguity about the meaning of nature in Soviet ideology and among Soviet leaders, exaggerated by the winner-take-all dictatorial decision-making process, that explains the sharp distinction between Soviet environmental accomplishments and failures.

In the long run, the values of the new Soviet elite won out. Russia urbanized during the Soviet period, and the old, primarily rural, Russian society that had generated the prerevolutionary environmental ethos disappeared permanently. The upsurge of environmental activism accompanying the fall of communism has largely subsided,[62] and the forest preserves created at Stalin's behest in the 1930s were eliminated by Vladimir Putin in 2007 with scarcely a murmur of public protest. For more than a decade, the post-Soviet intelligentsia has mourned the disappearance of the "Russian Atlantis," the lost world where writers and composers such as Tolstoy, Dostoevsky, Akhmatova, Tchaikovsky, and Shostakovich once lived and worked. In time, the environmental component of the Russian Atlantis may gain greater recognition, as well as a greater comprehension of how it vanished, under the waves of economic modernization.

Notes

1 The accident took place in the town of Ozyorsk, a closed city whose name was unknown in the West until the fall of the Soviet Union. As a result, the accident is known outside Russia as the Kyshtym accident. The Mayak reactors are currently shut down, although not dismantled, and the facility disposes of nuclear waste for Bulgaria and Ukraine. For discussions of the Kyshtym incident, see Z. Medvedev, *Nuclear Disaster in the Urals*, New York, Vintage Books, 1980; P. Josephson, *Red Atom*, New York, W. H. Freeman, 2000; D. Soran and D. B. Stillman, *An Analysis of the Alleged Kyshtym Disaster*, Los Alamos, NM, Los Alamos National Laboratory, 1982; V. Larin, *Russkie atomnye akuly: razmyshleniia s elementami sistemizatskii i analiza*, Moscow, KMK, 2005; M. R. Edelstein, M. Tysiachniouk, and L. V. Smimova, *Cultures of Contamination: Legacies of Pollution in Russia and the U.S.*, Amsterdam, Elsevier, 2007; A. M. Kellerer, "The Southern Urals Radiation Studies: A Reappraisal of the Current Status," *Radiation and Environmental Biophysics* 41/4, 2002, pp. 307–16.

2 Authors who have documented Soviet environmental problems include D. J. Peterson, *Troubled Lands: The Legacy of Soviet Environmental Destruction*, Boulder, CO, Westview Press, 1993; M. I. Goldman, *The Spoils of Progress: Environmental Misuse in the Soviet Union*, Cambridge, MA, MIT, 1972; B. Komarov, *The Destruction of Nature in the Soviet Union*, White Plains, NY, M. E. Sharpe, 1980; M. Feshbach and A. Friendly, *Ecocide in the USSR*, New York, Basic Books, 1992. Works in the Russian and Ukrainian languages include V. Boreiko, *Belye piatna istorii prirodookhrany*, Kiev, Kievskii ekologo-kul'turnyi tsentr, 1996, and F. Shtil'mark, *Otchet o prozhitom: zapiski ekologa-okhotoveda*, Moscow, Logata, 2006.

3 For more on the Aral Sea, see P. Micklin and N. V. Aladin, "Reclaiming the Aral Sea," *Scientific American* 298, 2008, pp. 64–71; M. L. Glantz, *Creeping Environmental Problems and Sustainable Development in the Aral Sea Basin*, Cambridge, Cambridge University Press, 1999; R. W. Ferguson, *The Devil and the Disappearing Sea*, Vancouver, Raincoast Books, 2003; G. F. S. Wiggs, S. L. O'Hara, J. Wegerdt, et al., "The Dynamics and Characteristics of Aeolian Dust in Dryland Central Asia: Possible Impacts on Human Exposure and Respiratory Health in the Aral Sea Basin," *The Geographical Journal* 169/2, 2003, pp. 142–57.

4 Feshbach and Friendly, *Ecocide in the USSR*, p. 9.

5 L. Trotsky, *The Revolution Betrayed*, Garden City, NY, Doubleday, 1937.

6 Malia argues that communism leads inextricably to oppression: M. Malia, *The Soviet Tragedy: A History of Socialism in Russia, 1917–1991*, New York, Free Press, 1994.

7 S. Fitzpatrick, "Civil War as Formative Experience," in A. Gleason, P. Kenez, and R. Stites (eds.), *Bolshevik Culture: Experiment and Order in the Russian Revolution*, Bloomington, IN, Indiana University Press, 1985, pp. 57–76.

8 Perhaps foremost among scholars portraying the Stalin era as a period of tragic destruction is Lewin: M. Lewin, *Russian Peasants and Soviet Power*, Evanston, IL, Northwestern University Press, 1968, and M. Lewin, *The Making of the Soviet System*, London, Methuen, 1985. A possible alternative path to Stalinism is presented in S. Cohen, *Bukharin and the Bolshevik Revolution*, New York, Alfred A. Knopf, 1973.

9 Z. L. Zile, "Lenin's Contribution to Law: The Case of Protection and Preservation of the Natural Environment," in B. Eissenstat (ed.), *Lenin and Leninism*, Lexington, KY, Lexington Books, 1971, pp. 83–100.

10 A fuller, but by no means comprehensive, list of works would include the aforementioned I. I. Kurortsov, *V. I. Lenin ob okhrane prirody*, Maikop, Adgeiskoe otdelenie krasnodarskogo knizhnogo izdatel'stva, 1969; P. V. Vasil'ev, *V. I. Lenin o lesakh i lesnom khoziaistve*, Irkutsk, Izdatel'stvo "Lesnaia promyshlennost," 1967; P. I. Zhukov, *Leninskie printsipy okhrany prirody*, Minsk, Gosudarstvennyi komitet soveta ministrov Belorusskoi SSR po okhrane prirody, 1969; P. I. Zhukov, *Vladimir Il'ich Lenin i priroda: Besda dlia shkol'nikov*, Irkutsk, Irkutskaia oblastnaia stantsiia iunykh naturalistov, 1980; N. A. Gladkov, *Okhrana prirody v pervye gody*

Sovetskoi vlasti, Moscow, Izdatel'stvo Moskovskogo universiteta, 1972; I. N. Kurazhkovskii, *Iz istorii organizatsii okhrane prirody v Astrakhanskom krae*, Astrakhan, Nizhe-volzhskoe knizhnoe izdatel'stvo, 1959; I. N. Kurazhkovskii, *Vladimir Il'ich Lenin i priroda*, Astrakhan, Nizhe-volzhskoe knizhnoe izdatel'stvo, 1969.

11 Quoted in B. Bonhomme, *Forests, Peasants, and Revolutionaries: Forest Conservation and Organization in Soviet Russia, 1917–1929*, New York, East European Monographs, 2005.

12 DeBardeleben discusses the environmental attitudes inherent in Marxism: J. DeBardeleben, *The Environment and Marxism-Leninism: The Soviet and East German Experience*, Boulder, CO, Westview, 1985.

13 The appeal of Marxism in remedying capitalist environmental excess is given voice in A. Gare, "Soviet Environmentalism: The Path Not Taken," in T. Benton (ed.), *The Greening of Marxism*, New York, Guilford Press, 1996, pp. 111–28.

14 P. R. Pryde, *Conservation in the Soviet Union*, Cambridge, Cambridge University Press, 1972, p. 163.

15 Pryde, *Conservation in the Soviet Union*, p. 177.

16 Goldman, *The Spoils of Progress*, p. 214.

17 In addition to the three books mentioned here, other works addressing the failures of Soviet environmental practices include D. E. Powell, "The Social Costs of Modernization: Ecological Problems in the USSR," *World Politics* 23/4, 1971, pp. 618–24; K. Bush, "Environmental Problems in the USSR," *Problems of Communism* 21, 1972, pp. 21–31; I. Volyges, *Environmental Deterioration in the Soviet Union and Eastern Europe*, New York, Praeger, 1974; F. Singleton (ed.), *Environmental Misuse in the Soviet Union*, New York, Praeger, 1976; C. Ziegler, *Environmental Policy in the USSR*, Amherst, MA, University of Massachusetts Press, 1987; J. Massey-Stewart (ed.), *The Soviet Environment*, Cambridge, Cambridge University Press, 1992; M. Feshbach, *Ecological Disaster: Cleaning Up the Hidden Legacy of the Soviet Regime*, New York, Twentieth Century Fund Press, 1995; see also the works mentioned at the outset of this chapter.

18 The Russian republic represented only one of 15 republics in the Soviet Union, although by far the largest, comprising three-quarters of the territory of the USSR as a whole.

19 Quoted in Pryde, *Conservation in the Soviet Union*, p. 182.

20 Pryde, *Conservation in the Soviet Union*, p. 182.

21 For more information about the comprehensive laws, see W. E. Butler, "Soviet Environmental Law as a Model for Other Countries," *Connecticut Journal of International Law* 4/2, 1989, pp. 279–86; A. G. Tarnavskii, "Law and Voluntary Nature Conservation in the USSR," *Connecticut Journal of International Law* 4/2, 1989, pp. 369–78.

22 See D. R. Weiner, *Models of Nature*, Pittsburgh, University of Pittsburgh Press, 1988; D. R. Weiner, *A Little Corner of Freedom*, Berkeley, University of California Press, 1999.

23 Weiner, *Models of Nature*, p. 61.

24 Weiner, *Models of Nature*, p. 27.

25 For a vivid example of the First Five-Year Plan in operation, see S. Kotkin, *Magnetic Mountain*, Berkeley, University of California Press, 1997.

26 M. Gorkii, *The White Sea Canal: Being an Account of the Construction of the New Canal between the White Sea and the Baltic Sea*, London, John Lane, 1935; this work was banned in 1937, perhaps for putting matters a bit too bluntly. See also D. R. Weiner, "Man of Plastic: Gor'kii's Visions of Humans in Nature," *The Soviet and Post-Soviet Review* 22/1, 1995, pp. 65–88.

27 See L. Graham, *Science in Russia and the Soviet Union: A Short History*, Cambridge, Cambridge University Press, 1993, chapter 8 for a discussion of the related "Promparty" trial.

28 Pryde, *Conservation in the Soviet Union*, p. 51.

29 Weiner, *A Little Corner of Freedom*, p. 444.

30 Weiner, *A Little Corner of Freedom*, p. 445.

31 Weiner, *A Little Corner of Freedom*, p. 20, citing sociologist Oleg Ianitskii's unpublished address, "Ekologicheskaia politika i ekologicheskoe dvizhenie v Rossii," presented at the Working Group of Eco-Sociologists and Leaders of Ecological Organizations, Moscow, May 19–21, 1995.

32 Goldman, *The Spoils of Progress*, p. 28.

33 R. G. Suny, *The Soviet Experiment*, New York, Oxford University Press, 1998, pp. 238–9.

34 B. Bonhomme, "Forests, Peasants, and Revolutionaries: Forest Conservation in Soviet Russia, 1917–1925," PhD thesis, *The City University of New York*, 2000, p. 291.

35 W. Husband, "'Correcting Nature's Mistakes': Transforming the Environment and Soviet Children's Literature, 1928–1941," *Environmental History* 11/2, 2006, pp. 300–18, paragraph 38.

36 D. Filtzer, *The Hazards of Urban Life in Late Stalinist Russia: Health, Hygiene and Living Standards, 1943–1953*, Cambridge, Cambridge University Press, 2010, p. 22.

37 Filtzer, *The Hazards of Urban Life in Late Stalinist Russia*, pp. 20–1.

38 D. R. Weiner, "Environmental Activism in the Soviet Context," in C. Mauch and N. Stoltzfus (eds.), *Shades of Green: Environmental Activism around the Globe*, Lanham, MD, Rowman & Littlefield, 2006, pp. 101–34 (p. 129).

39 Josephson has explored the failures of Soviet environmental policy in *Totalitarian Science and Technology*, Atlantic Highlands, NJ, Humanities Press, 1996; *Industrialized Nature*, Washington, DC, Island Press, 2002; and *Resources under Regimes: Technology, Environment, and the State*, Cambridge, MA, Harvard University Press, 2004.

40 Pryde, *Conservation in the Soviet Union*, p. 51.

41 For more information about the redesignation of the *zapovedniki*, see S. C. Brain, *Song of the Forest*, Pittsburgh, University of Pittsburgh Press, 2011, pp. 135–6.

42 See S. C. Brain, "Stalin's Environmentalism," *Russian Review* 69/1, 2010, pp. 93–118, and Brain, *Song of the Forest*.

43 "Voprosy lesnogo khoziaistva na II plenume," *Lesovod* 7–8, July–August 1929, pp. 12–38 (p. 18).

44 "Sostoianie lesnoi promyshlennosti," *Lesnoe khoziaistvo i lesnaia promyshlennost'* 84–5, September–October 1930, pp. 77–9 (p. 77).

45 M. Tkachenko, "Zadachi lesnogo khoziaistva i 'Den' Lesa' v 1930 godu," *Lesnoi spetsialist* 7–8, April 1930, pp. 10–13 (p. 10).

46 Weiner provides numerous examples of actions that would have been suicidal in intent had the activists who carried them out suspected that the Stalinist state would consider them traitorous: requests to travel abroad to attend a conservation conference in Vienna in 1937; public praise for American conservationism in 1938; public protests against illegal tree removal in 1948. In response to his own question, asking how it is that such actions failed to incur the wrath of the secret police, Weiner writes: "We have no answer to the riddle of the regime's failure to snuff out nature protection activism." Weiner, *A Little Corner of Freedom*, p. 444.

47 Joseph Grinnell, a biologist from the Museum of Vertebrate Zoology in Berkeley, visited the lake in 1917 and found that "Great numbers of water birds are in sight along the lake shore – avocets, phalaropes, ducks. Large flocks of shorebirds in flight over the water in the distance, wheeling about show in mass, now silvery now dark, against the gray-blue of the water. There must be literally thousands of birds within sight of this one spot. En route around the south end of Owens Lake to Olancha I saw water birds almost continuously." J. Grinnell, *Owen's Lake*, 2001, accessed from http://www.ovcweb.org/owensvalley/owenslake.html on March 7, 2012.

48 M. Cone, "Owens Valley Plan Seeks L.A. Water to Curb Pollution," *Los Angeles Times*, December 17, 1996, accessed from http://articles.latimes.com/1996-12-17/news/mn-9875_1_owens-valley on March 7, 2012.

49 *A Brief History of Hanford*, accessed from http://www.downwinders.com/hanford_hist.php on March 7, 2012.

50 T. von Laue, *Why Lenin? Why Stalin? Why Gorbachev?*, New York, HarperCollins, 1993.

51 The tension between the timber trusts and the central government cited earlier is only one instance when the party leadership expressed uneasiness about the decisions that the industrial bureaus tended to make when given freedom to do so. More discussion of this tension can be found in R. W. Davies, M. Harrison, and S. G. Wheatcroft (eds.), *The Economic Transformation of the Soviet Union, 1913–1945*, Cambridge, Cambridge University Press, 1994.

52   For more about the conflict of state interests with those of the industrial bureaucracies, see
     D. Kelley, "Environmental Policy-Making in the USSR: The Role of Industrial and
     Environmental Interest Groups," *Soviet Studies* 28/4, 1976, pp. 570–89; B. Jancar,
     *Environmental Management in the Soviet Union and Yugoslavia: Structure and Regulation in
     Federal Communist States*, Durham, NC, Duke University Press, 1987.
53   J. Costlow, "Imaginations of Destruction: The 'Forest Question' in Nineteenth-Century
     Russian Culture," *Russian Review* 62, 2003, pp. 91–118 (p. 113).
54   Still other writers who used nature to explore moral and ethical issues include Ivan Bunin and
     Sergei Esenin, and even the early Maksim Gorky.
55   C. Ely, *This Meager Nature*, DeKalb, IL, Northern Illinois University Press, 2002, p. 205.
56   For more information about Russian romantic forestry in the prerevolutionary period, see
     Brain, *Song of the Forest*.
57   See S. Fitzpatrick, *Education and Social Mobility in the Soviet Union*, Cambridge, Cambridge
     University Press, 1979, and S. Fitzpatrick, *The Russian Revolution*, Oxford, Oxford University
     Press, 2007.
58   S. Fitzpatrick (ed.), *Cultural Revolution in Russia, 1928–1931*, Bloomington, IN, Indiana
     University Press, 1978; S. Fitzpatrick, *The Cultural Front: Power and Culture in Revolutionary
     Russia*, Ithaca, NY, Cornell University Press, 1992.
59   N. Timasheff, *The Great Retreat: The Growth and Decline of Communism in Russia*, New
     York, E. P. Dutton, 1946. David Hoffman has argued that the importance of a great retreat
     in Soviet politics has been overstated: D. Hoffman, "Was There a 'Great Retreat' from Soviet
     Socialism? Stalinist Culture Reconsidered," *Kritika: Explorations in Russian and Eurasian
     History* 5/4, 2004, pp. 651–74.
60   M. Bassin, "'I Object to Rain that Is Cheerless': Landscape Art and the Stalinist Aesthetic
     Imagination," *Cultural Geographies* 7/3, 2000, pp. 313–36.
61   K. Clark, *The Soviet Novel: History as Ritual*, Chicago, University of Chicago Press, 1985,
     p. 111.
62   For an analysis of post-Soviet environmental activism, see P. Jacques, "A Political Economy of
     Russian Nature Conservation Policy: Why Scientists Have Taken a Back Seat," *Global
     Environmental Politics* 2/4, 2002, pp. 102–24; L. Henry, "Thinking Globally, Acting Locally:
     The Russian Environmental Movement and Sustainable Development," in J. Agyeman and
     Y. Ogneva-Himmelberger (eds.), *Environmental Justice and Sustainability in the Former
     Soviet Union*, Cambridge, MA, MIT, 2009, pp. 47–70; and L. Henry, *Red to Green:
     Environmental Activism in Post-Soviet Russia*, Ithaca, NY, Cornell University Press, 2010.

References

Bassin, M., "'I Object to Rain that Is Cheerless': Landscape Art and the Stalinist Aesthetic
    Imagination," *Cultural Geographies* 7/3, 2000, pp. 313–36.
Bonhomme, B., *Forests, Peasants, and Revolutionaries: Forest Conservation and Organization in
    Soviet Russia, 1917–1929*, New York, East European Monographs, 2005.
Bonhomme, B., "*Forests, Peasants, and Revolutionaries: Forest Conservation in Soviet Russia,
    1917–1925*," PhD thesis, The City University of New York, 2000.
Boreiko, V., *Belye piatna istorii prirodookhrany*, Kiev, Kievskii ekologo-kul'turnyi tsentr, 1996.
Brain, S. C., *Song of the Forest*, Pittsburgh, University of Pittsburgh Press, 2011.
Brain, S. C., "Stalin's Environmentalism," *Russian Review* 69/1, 2010, pp. 93–118.
A Brief History of Hanford, accessed from http://www.downwinders.com/hanford_hist.php on
    March 7, 2012.
Bush, K., "Environmental Problems in the USSR," *Problems of Communism* 21, 1972,
    pp. 21–31.
Butler, W. E., "Soviet Environmental Law as a Model for Other Countries," *Connecticut Journal
    of International Law* 4/2, 1989, pp. 279–86.

Clark, K., *The Soviet Novel: History as Ritual*, Chicago, University of Chicago Press, 1985.

Cohen, S., *Bukharin and the Bolshevik Revolution*, New York, Alfred A. Knopf, 1973.

Cone, M., "Owens Valley Plan Seeks L.A. Water to Curb Pollution," *Los Angeles Times*, December 17, 1996, accessed from http://articles.latimes.com/1996-12-17/news/mn-9875_1_owens-valley on March 7, 2012.

Costlow, J., "Imaginations of Destruction: The 'Forest Question' in Nineteenth-Century Russian Culture," *Russian Review* 62, 2003, pp. 91–118.

Davies, R. W., Harrison, M., and Wheatcroft, S. G. (eds.), *The Economic Transformation of the Soviet Union, 1913–1945*, Cambridge, Cambridge University Press, 1994.

DeBardeleben, J., *The Environment and Marxism-Leninism: The Soviet and East German Experience*, Boulder, CO, Westview, 1985.

Edelstein, M. R., Tysiachniouk, M., and Smimova, L. V., *Cultures of Contamination: Legacies of Pollution in Russia and the U.S.*, Amsterdam, Elsevier, 2007.

Ely, C., *This Meager Nature*, DeKalb, IL, Northern Illinois University Press, 2002.

Ferguson, R. W., *The Devil and the Disappearing Sea*, Vancouver, Raincoast Books, 2003.

Feshbach, M., *Ecological Disaster: Cleaning Up the Hidden Legacy of the Soviet Regime*, New York, Twentieth Century Fund Press, 1995.

Feshbach, M., and Friendly, A., *Ecocide in the USSR*, New York, Basic Books, 1992.

Filtzer, D., *The Hazards of Urban Life in Late Stalinist Russia: Health, Hygiene and Living Standards, 1943–1953*, Cambridge, Cambridge University Press, 2010.

Fitzpatrick, S., "Civil War as Formative Experience," in A. Gleason, P. Kenez, and R. Stites (eds.), *Bolshevik Culture: Experiment and Order in the Russian Revolution*, Bloomington, IN, Indiana University Press, 1985, pp. 57–76.

Fitzpatrick, S., *The Cultural Front: Power and Culture in Revolutionary Russia*, Ithaca, NY, Cornell University Press, 1992.

Fitzpatrick, S. (ed.), *Cultural Revolution in Russia, 1928–1931*, Bloomington, IN, Indiana University Press, 1978.

Fitzpatrick, S., *Education and Social Mobility in the Soviet Union*, Cambridge, Cambridge University Press, 1979.

Fitzpatrick, S., *The Russian Revolution*, Oxford, Oxford University Press, 2007.

Gare, A., "Soviet Environmentalism: The Path Not Taken," in T. Benton (ed.), *The Greening of Marxism*, New York, Guilford Press, 1996, pp. 111–28.

Gladkov, N. A., *Okhrana prirody v pervye gody Sovetskoi vlasti*, Moscow, Izdatel'stvo Moskovskogo universiteta, 1972.

Glantz, M. L., *Creeping Environmental Problems and Sustainable Development in the Aral Sea Basin*, Cambridge, Cambridge University Press, 1999.

Goldman, M. I., *The Spoils of Progress: Environmental Misuse in the Soviet Union*, Cambridge, MA, MIT, 1972.

Gorkii, M., *The White Sea Canal: Being an Account of the Construction of the New Canal between the White Sea and the Baltic Sea*, London, John Lane, 1935.

Graham, L., *Science in Russia and the Soviet Union: A Short History*, Cambridge, Cambridge University Press, 1993.

Grinnell, J., Owen's Lake, 2001, accessed from http://www.ovcweb.org/owensvalley/owenslake.html on March 7, 2012.

Henry, L., *Red to Green: Environmental Activism in Post-Soviet Russia*, Ithaca, NY, Cornell University Press, 2010.

Henry, L., "Thinking Globally, Acting Locally: The Russian Environmental Movement and Sustainable Development," in J. Agyeman and Y. Ogneva-Himmelberger (eds.), *Environmental Justice and Sustainability in the Former Soviet Union*, Cambridge, MA, MIT, 2009, pp. 47–70.

Hoffman, D., "Was There a 'Great Retreat' from Soviet Socialism? Stalinist Culture Reconsidered," *Kritika: Explorations in Russian and Eurasian History* 5/4, 2004, pp. 651–74.

Husband, W., "'Correcting Nature's Mistakes': Transforming the Environment and Soviet Children's Literature, 1928–1941," *Environmental History* 11/2, 2006, pp. 300–18.

Jacques, P., "A Political Economy of Russian Nature Conservation Policy: Why Scientists Have Taken a Back Seat," *Global Environmental Politics* 2/4, 2002, pp. 102–24.

Jancar, B., *Environmental Management in the Soviet Union and Yugoslavia: Structure and Regulation in Federal Communist States*, Durham, NC, Duke University Press, 1987.

Josephson, P., *Industrialized Nature*, Washington, DC, Island Press, 2002.

Josephson, P., *Red Atom*, New York, W. H. Freeman, 2000.

Josephson, P., *Resources under Regimes: Technology, Environment, and the State*, Cambridge, MA, Harvard University Press, 2004.

Josephson, P., *Totalitarian Science and Technology*, Atlantic Highlands, NJ, Humanities Press, 1996.

Kellerer, A. M., "The Southern Urals Radiation Studies: A Reappraisal of the Current Status," *Radiation and Environmental Biophysics* 41/4, 2002, pp. 307–16.

Kelley, D., "Environmental Policy-Making in the USSR: The Role of Industrial and Environmental Interest Groups," *Soviet Studies* 28/4, 1976, pp. 570–89.

Komarov, B., *The Destruction of Nature in the Soviet Union*, White Plains, NY, M. E. Sharpe, 1980.

Kotkin, S., *Magnetic Mountain*, Berkeley, University of California Press, 1997.

Kurazhkovskii, I. N., *Iz istorii organizatsii okhrane prirody v Astrakhanskom krae*, Astrakhan, Nizhe-volzhskoe knizhnoe izdatel'stvo, 1959.

Kurazhkovskii, I. N., *Vladimir Il'ich Lenin i priroda*, Astrakhan, Nizhe-volzhskoe knizhnoe izdatel'stvo, 1969.

Kurortsov, I. I., *V. I. Lenin ob okhrane prirody*, Maikop, Adgeiskoe otdelenie krasnodarskogo knizhnogo izdatel'stva, 1969.

Larin, V., *Russkie atomnye akuly: razmyshleniia s elementami sistemizatskii i analiza*, Moscow, KMK, 2005.

Lewin, M., *The Making of the Soviet System*, London, Methuen, 1985.

Lewin, M., *Russian Peasants and Soviet Power*, Evanston, IL, Northwestern University Press, 1968.

Malia, M., *The Soviet Tragedy: A History of Socialism in Russia, 1917–1991*, New York, Free Press, 1994.

Massey-Stewart, J. (ed.), *The Soviet Environment*, Cambridge, Cambridge University Press, 1992.

Medvedev, Z., *Nuclear Disaster in the Urals*, New York, Vintage Books, 1980.

Micklin, P., and Aladin, N. V., "Reclaiming the Aral Sea," *Scientific American* 298, 2008, pp. 64–71.

Peterson, D. J., *Troubled Lands: The Legacy of Soviet Environmental Destruction*, Boulder, CO, Westview Press, 1993.

Powell, D. E., "The Social Costs of Modernization: Ecological Problems in the USSR," *World Politics* 23/4, 1971, pp. 618–24.

Pryde, P. R., *Conservation in the Soviet Union*, Cambridge, Cambridge University Press, 1972.

Shtil'mark, F., *Otchet o prozhitom: zapiski ekologa-okhotoveda*, Moscow, Logata, 2006.

Singleton, F. (ed.), *Environmental Misuse in the Soviet Union*, New York, Praeger, 1976.

Soran, D., and Stillman, D. B., *An Analysis of the Alleged Kyshtym Disaster*, Los Alamos, NM, Los Alamos National Laboratory, 1982.

"Sostoianie lesnoi promyshlennosti," *Lesnoe khoziaistvo i lesnaia promyshlennost'* 84–5, September–October 1930, pp. 77–9.

Suny, R. G., *The Soviet Experiment*, New York, Oxford University Press, 1998.

Tarnavskii, A. G., "Law and Voluntary Nature Conservation in the USSR," *Connecticut Journal of International Law* 4/2, 1989, pp. 369–78.

Timasheff, N., *The Great Retreat: The Growth and Decline of Communism in Russia*, New York, E. P. Dutton, 1946.

Tkachenko, M., "Zadachi lesnogo khoziaistva i 'Den' Lesa' v 1930 godu," *Lesnoi spetsialist* 7–8, April 1930, pp. 10–13.

Trotsky, L., *The Revolution Betrayed*, Garden City, NY, Doubleday, 1937.

Vasil'ev, P. V., *V. I. Lenin o lesakh i lesnom khoziaistve*, Irkutsk, Izdatel'stvo "Lesnaia promyshlennost," 1967.

Volyges, I., *Environmental Deterioration in the Soviet Union and Eastern Europe*, New York, Praeger, 1974.

von Laue, T., *Why Lenin? Why Stalin? Why Gorbachev?*, New York, HarperCollins, 1993.

"Voprosy lesnogo khoziaistva na II plenume," *Lesovod* 7–8, July–August 1929, pp. 12–38.

Weiner, D. R., "Environmental Activism in the Soviet Context," in C. Mauch and N. Stoltzfus (eds.), *Shades of Green: Environmental Activism around the Globe*, Lanham, MD, Rowman & Littlefield, 2006, pp. 101–34.

Weiner, D. R., *A Little Corner of Freedom*, Berkeley, University of California Press, 1999.

Weiner, D. R., "Man of Plastic: Gor'kii's Visions of Humans in Nature," *The Soviet and Post-Soviet Review* 22/1, 1995, pp. 65–88.

Weiner, D. R., *Models of Nature*, Pittsburgh, University of Pittsburgh Press, 1988.

Wiggs, G. F. S., O'Hara, S. L., Wegerdt, J., et al., "The Dynamics and Characteristics of Aeolian Dust in Dryland Central Asia: Possible Impacts on Human Exposure and Respiratory Health in the Aral Sea Basin," *The Geographical Journal* 169/2, 2003, pp. 142–57.

Zhukov, P. I., *Leninskie printsipy okhrany prirody*, Minsk, Gosudarstvennyi komitet soveta ministrov Belorusskoi SSR po okhrane prirody, 1969.

Zhukov, P. I., *Vladimir Il'ich Lenin i priroda: Besda dlia shkol'nikov*, Irkutsk, Irkutskaia oblastnaia stantsiia iunykh naturalistov, 1980.

Ziegler, C., *Environmental Policy in the USSR*, Amherst, MA, University of Massachusetts Press, 1987.

Zile, Z. L., "Lenin's Contribution to Law: The Case of Protection and Preservation of the Natural Environment," in B. Eissenstat (ed.), *Lenin and Leninism*, Lexington, KY, Lexington Books, 1971, pp. 83–100.

Further Reading

Tarnavskii, A. G., "Law and Voluntary Nature Conservation in the USSR," *Connecticut Journal of International Law* 4/2, 1989, pp. 369–78.

PART III

Drivers of Change and Environmental Transformations

CHAPTER FOURTEEN

The Grasslands of North America and Russia

DAVID MOON

The Great Plains of North America and the steppes of Russia share similarities in their natural environments and environmental histories. Both are semi-arid grasslands, with low and unreliable rainfall, but very fertile soil (chernozems, mollisols). Both were settled and plowed up by migrants from regions with higher rainfall and more trees. The settlers displaced smaller populations – nomadic pastoralists and Plains Indians – who herded or hunted animals that grazed on the grass. On both, farmers reaped bumper harvests in good years, but also experienced recurring droughts that caused soil erosion and crop failures. The steppes, however, have a slightly more continental and arid climate.[1]

Both the Russian steppes and the North American Great Plains are part of the vast expanse of grasslands found around the world in lands as far-flung as the Argentine Pampas, the South African veld, and the rolling hills of Mongolia. The grass was in some places fully natural vegetation, but in others sustained by periodic burning and grazing that prevented the emergence of other plant life. Prior to the nineteenth century, these lands had hosted scant settlement and often supported mobile pastoralists rather than farmers. But in the nineteenth and twentieth centuries, often abetted by both railways and government action, a globe-girdling movement of farming folk spilled out onto these grasslands. These farmers converted tens of millions of hectares of grass into fields of grain in one of the grand environmental transformations of modern history. How are these epic environmental changes to be understood? As heroic conquests of unforgiving nature? As foolhardy ventures into biomes unsuited to farming?

The main focus of this chapter is the "Dust Bowl" on the southern plains of the US in the 1930s (see Map 14.1) and the drought, crop failure, and famine on the Russian steppes in 1891–2. Two Russian scientists, Aleksandr Chibilev and S. V. Levykin, have recently argued that, by the turn of the twentieth century, the "barbaric" plowing up of the steppes and the prairies had created the preconditions for the world's first ecological

A Companion to Global Environmental History, First Edition. Edited by J.R. McNeill and Erin Stewart Mauldin.
© 2012 John Wiley & Sons, Ltd. Published 2015 by John Wiley & Sons, Ltd.

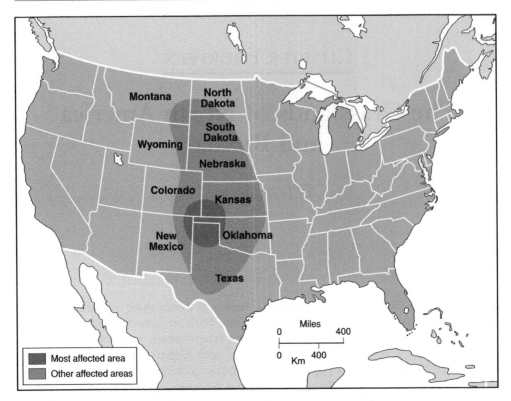

Map 14.1   The Dust Bowl of the 1930s

crisis caused by plowing up virgin land.[2] This chapter analyzes the responses to these two disasters in order to explore competing interpretations of the environmental histories of these two grasslands so important to modern world history.

The Great Plains

In 1936, during the "Dust Bowl," the US Resettlement Administration made a documentary film entitled *The Plow that Broke the Plains*. The documentary's narrator presented one version of the environmental history of the Great Plains from the start of Euro-American settlement after the Civil War to the 1930s:

> This is a record of land ... of soil, rather than people – a story of the Great Plains: the 400 million acres of wind-swept grass lands that spread up from the Texas Panhandle to Canada ... A high, treeless continent, without rivers, without streams ... A country of high winds, and sun ... and of little rain ... By 1880 we had cleared the Indian, and with him, the buffalo, from the Great Plains ... First came the cattle ... which grazed on the unfenced range ... an uncharted ocean of grass. [Then came the railroad, which] brought the world into the plains ... new populations, new needs ... the plowman followed the herder and the pioneer came to the plains ... Progress came to the plains. Two hundred miles from water, two hundred miles from home, but the land is new. High winds and sun ... a country without rivers and with little rain. Settler, plow at your peril!

The earlier settlers were disappointed: "The rains failed ... and the sun baked the light soil. Many left ..." The Great War brought new hopes:

> "Wheat will win the war!" "Plant wheat ..." The boom continued after the war. Then we reaped the golden harvest ... then we really plowed the plains ... We had the man-power ... we invented new machinery ... the world was our market. By 1933 the old grass lands had become the new wheat lands.

Once again, however, "The rains held off and the sun baked the earth. This time no grass held moisture against the winds and the sun ... this time millions of acres of plowed land lay open to the sun." Then came the film's final, tragic, refrain:

> Baked out – blown out – and broke! Year in, year out, uncomplaining they fought the worst drought in history ... their stock choked to death on the barren land ... their homes were nightmares of swirling dust night and day. Many went ahead of it – but many stayed until stock, machinery, homes, credit, food, and even hope were gone. On to the West! Once again they headed for the setting sun. Once again they headed West. Last year [1935] in every summer month 50,000 people left the Great Plains and hit the highways for the Pacific Coast ...[3]

This interpretation was similar to the findings of the federal government's Great Plains Committee, which submitted its report, entitled *The Future of the Great Plains*, in 1937. It stated: "In 1934 and ... 1936 drought conditions in the Great Plains area ... became so severe that it was necessary for the Federal Government to take emergency steps to rescue dying cattle, relieve destitute families, and safeguard human life." The cause was described as follows: "Current methods of cultivation were so injuring the land that large areas were decreasingly productive even in good years, while in bad years they tended more and more to lapse into desert." The committee contrasted "human modification of natural conditions" since "the coming of the white man" after 1866 with the impact of the Plains Indians. The Indians had hunted buffalo and set fire to the grass, but had not reduced the numbers of buffalo or destroyed the grass cover.

> There is no evidence – the Committee asserted – that in historic times there was ever a severe enough drought to destroy the grass roots and cause wind erosion comparable to that which took place in 1934 and 1936; that phenomenon was due to the plowing and over cropping of comparatively recent years. ... Nature has established a balance in the Great Plains ... The white man has disturbed this balance.

The dust storms of recent years, the report stated, were a result of "the cultivation of unsuitable lands, and of suitable lands by improper methods." Overgrazing had also made the land susceptible to wind erosion, which had "ruined or impaired the productivity of thousands of acres." The settlers' "destructive tendencies" were explained by their "lack of understanding concerning the critical differences between the physical conditions of the Great Plains ... and those ... east of the Mississippi whence they had come."[4]

This narrative was followed by the prominent environmental historian, and native of Kansas, Donald Worster. In 1979 he argued: "Some environmental catastrophes are nature's work, others are the slowly accumulating effects of ignorance or poverty. The Dust Bowl, in contrast, was the inevitable outcome of a culture that deliberately, self-consciously, set itself that task of dominating and exploiting the land for all it was worth." Worster blamed capitalism. "The expansionary energy of the United States," he continued, "had finally encountered a volatile marginal land, destroying the delicate ecological balance that had evolved there." He also contrasted the impact of the white settlers with the culture of the Plains Indians, who had a "conservation ethic" and had been "part and parcel of nature." At the heart of Worster's account were case studies of one of the worst affected counties, Cimarron County in the Oklahoma panhandle, and one in the "second tier," Haskell County in southwest Kansas. On the eve of the Dust Bowl, Cimarron County had been converted to a "wheat empire" as a result of the arrival of tractors to plow up the land and railroads to carry the produce to markets. In Haskell County, cattle-ranching had been replaced by "mono-culture factory wheat farming." "Virtually none of [these farmers] was interested in the land as a home for themselves and their children; to them it was solely an instrument to make cash," Worster writes.[5]

This interpretation presents a narrative of humans coming from outside and destroying nature: the settlers plowed up too much land and used farming methods that were destructive to the fragile grassland environment. The fertile soils yielded bumper harvests in wet years, but the settlers reaped only misery, and choked on the dust, in bad years when the rains failed. Geoff Cunfer, the author of a major recent study of the Great Plains, has termed this the "declensionist narrative."[6]

On the eve of the Dust Bowl, however, the prevailing interpretation of the history of the Great Plains was a progressive one in which humans – in particular white humans – conquered nature with the aid of science and technology. One classic use of this progressive narrative is *The Great Plains* by Walter Prescott Webb, who was a professor at the University of Texas at Austin. His narrative stops short of the Dust Bowl for the simple reason that it was published in 1931. Webb began with the formation of the Great Plains, concentrating on the land, the climate (which was "deficient in water"), the vegetation (predominantly grasses), and the wildlife. Webb then overlaid the human history of the region – the Plains Indians, the Spanish, and then American settlers – onto this environment. When he came to the American settlers, he noted that their way of life in the east had stood on three legs: land, water, and timber. When they crossed the 98th parallel onto the Plains, however, "two of these legs were withdrawn, – water and timber, – and civilization was left on one leg – land." Webb traced the adjustments the white man made to adapt to the Great Plains environment. Following the example of the Plains Indians, white men rode on horses. They needed new weapons to fight on horseback, and by the end of the 1830s, the Texas Rangers were using revolvers to fight against the formidable Comanches. Beginning in the 1860s, railroads brought large numbers of settlers who could not farm the fertile land until they could fence their fields to keep out cattle. The shortage of trees meant that they could not use timber. The solution was barbed wire, which was invented in the early 1870s. The farmers needed water. The solution, widely adopted from the 1870s, was windmills to pump water from wells. Irrigation, where it was practicable, dry-farming, to conserve moisture in the soil, and drought-resistant crops all followed.[7]

The notion that, armed with science and technology, humans could triumph over nature was not choked by the clouds of dust rolling off the plains in the 1930s. In the spirit of Franklin Roosevelt's New Deal, the federal government intervened. In 1935, it set up the Soil Conservation Service, whose scientists surveyed the types of soil and the extent of erosion. Some land was taken out of cultivation and families resettled. On land that remained in cultivation, irrigation was expanded and extension services advised farmers how to cultivate land in ways that conserved moisture and protected land from erosion (e.g., plowing parallel to the contours in the land).[8] Shelterbelts of trees were planted to reduce wind erosion. Roosevelt's Prairie States Forestry Project, launched in 1934, was initially intended to change the climate of the region by planting a vast belt of trees from the Texas panhandle to the Canadian border. It was later scaled down to smaller windbreaks.[9] It was based, in part, on Russian experience of forestry on the steppes.[10]

Historians have offered different assessments of both the causes of the Dust Bowl and the conservationists' efforts to mitigate its effects. Worster, for example, criticized government initiatives for failing to change the attitudes of farmers who treated the land as a resource to be exploited.[11] Geoff Cunfer, on the other hand, pointed out that Great Plains farmers spoke out against the conservationists' view that they had farmed land unfit for cultivation and should reduce the area they sowed.[12] Cunfer presented a more balanced view of agriculture and environment on the Great Plains. He called his interpretation a "middle ground" between the declensionist and progressive narratives. Insisting that people are "fully part of the natural world," Cunfer used Geographic Information Systems (GIS) to analyze data on the entire Great Plains region. In seeking the causes of the Dust Bowl, he investigated the precise region affected; the types of soil; the proportion of the land that had been ploughed up and how this had changed over the preceding period; the amount of rainfall over the 1930s in comparison with regional averages; and the difference in temperatures compared with the average. His conclusion was that the drought of the 1930s was the main cause of the "Dust Bowl." Plowing up land may have "tipped the balance" in the southern Texas panhandle, but overall, the weather was a more important cause than farming. He pointed out, significantly, that there were dust storms on land that had never been plowed up.[13] A recent scientific study, moreover, has shown that the drought which caused the dust storms was a result of "anomalous tropical sea surface temperatures."[14]

Cunfer was interested not just in the Dust Bowl, but in the longer history of interaction between people and the natural world since the start of Euro-American settlement on the Plains in the 1870s. Surveying data on land use, he argued that patterns of cropping and pasture had been surprisingly consistent over the entire period as farmers made choices on how best to use the land in the environmental conditions of the region with the technology and resources they had available. By the end of the 1920s, the proportion of the total area of the Great Plains that had been converted to cropland was approaching 30 percent, and it remained at around that level for the rest of the twentieth century. The percentage of the land plowed up was higher in the eastern tall-grass prairies, where it exceeded 50 percent, and lower in the more arid west. From the mid-1930s, crucially, farmers began to tap the Ogallala Aquifer, which extends from north Texas to South Dakota, to provide water to irrigate their fields. Cunfer concluded that American farmers achieved a "sequence of periods of temporary equilibrium" and that agriculture on the plains is sustainable. The Dust Bowl was "a temporary disruption in a stable system."[15] Cunfer drew on the work of Kansas historian James Malin, who had vehemently attacked the New Dealers' interpretation of the "Dust Bowl," and the implication that the settlers' plows had broken the plains. Malin drew on his considerable knowledge of Kansas history and made pioneering use of ecology in presenting his interpretation of the ability of farmers to adapt to the conditions on the Great Plains.[16] The key to the interpretations of both Cunfer and Malin is that they focus not just on the 1930s, but on the longer-term history of farming in the region.

Nevertheless, it is the declensionist interpretation of human culpability, supported by dramatic images of clouds of dust blotting out the sun, which has been the dominant one since the 1930s. Perhaps this is because the human tragedy of the Dust Bowl became an enduring theme in American culture. In Texas in the mid-1930s, Cindy Walker wrote the song "Dusty Skies," which was recorded by Bob Wills and his Texas Playboys in 1941.[17] The previous year Woody Guthrie, a native of Oklahoma, recorded his "Dust Bowl Ballads," including lines such as "You could see that dust storm comin', the cloud looked deathlike black."[18] The best-known portrayal is John Steinbeck's novel *The Grapes of Wrath*, published in 1939, which brought the author Pulitzer and Nobel prizes, and in 1940 was made into a movie. The human story of Tom Joad and his family escaping the dust storms in Oklahoma and seeking a new life in California, in a camp run by the Resettlement Administration, has become etched in the popular consciousness in a way that tales of "anomalous tropical sea surface temperatures" can never be.

The Steppes

Halfway round the northern hemisphere is another semi-arid grassland: the steppes that extend across the south of present-day Ukraine and Russia, through Kazakhstan and Mongolia, to northern China.[19] Steinbeck visited the steppes, while on a tour of the Soviet Union, in the late 1940s. He described the view as his party flew into Kiev:

> We were over the flat grainlands of the Ukraine, as flat as our Middle West, and almost as fruitful. The huge bread basket of Europe ... the endless fields lay below us, yellow with wheat and rye ... There was no hill ... of any kind. The flat stretched away to a round unbroken horizon.

When they visited collective farms, their hosts, eager to impress, "put on the same kind of show a Kansas farmer would." Steinbeck related how they were asked about American farming, tractors, crop varieties, agricultural education, and extension services. Later they visited Stalingrad, where they saw the famous tractor factory, the site of some of the fiercest fighting in World War II and part of Soviet iconography thereafter.[20] Nowhere, however, did Steinbeck write about dust storms or droughts. Woody Guthrie, who was known for his Communist sympathies,[21] wrote no ballads about dust storms or people who had fallen on hard times on the steppes. Nevertheless, there had recently been a disaster far greater than the Dust Bowl: the catastrophic famine of 1932–3 (known as the "Holodomor" in Ukraine), caused when Stalin's forced collectivization of agriculture (launched in 1930) combined with drought.[22] The Soviet authorities tried to cover it up. Most of the foreign press corps, including Walter Duranty of the *New York Times*, underreported or even denied the famine.[23]

Rather than agonize over the famine, in which millions died, Soviet culture celebrated the alleged successes of collectivization, in the face of opposition from "kulaks," and the epic story of bringing new land in the steppe region into cultivation. The best-known example is Mikhail Sholokhov's turgid novel *Virgin Soil Upturned*, set in his native Don region, published in two volumes in 1932 and 1959. Like *The Grapes of Wrath*, it was made into a movie, in 1959. Like Steinbeck, Sholokhov won prizes, a Lenin prize, and a Nobel prize for his earlier work. But, while Steinbeck's novel presented a declensionist narrative of the experiences of farmers on the grasslands, Sholokhov's depicted a progressive one. In private, however, Sholokhov wrote to Stalin in 1933, drawing his attention to the plight of the population of the Don region during the famine and asking for aid.[24]

The famine of 1932–3 was the latest, and worst, of a series of human and environmental disasters to hit the steppes. Better known in the US at the time were the droughts, crop failures, and famines of 1921–2 and 1891–2. On both occasions, Americans were involved in famine relief.[25] The disaster of 1891–2, moreover, prompted outbursts of anxiety in Russia about the state of agriculture and the human impact on the environment, which offer a parallel to American reactions to the Dust Bowl four decades later, and provide evidence for a declensionist narrative of the environmental history of the steppes.[26] In 1891, the governor of the province of Samara, on the mid-Volga, reported:

> As a result of the complete failure of the harvest of grain and grass, Samara province was one of the worst victims. The main reason for the harvest failure was extremely unfavorable climatic phenomena – the lack of rain from the spring stopped the winter grain from growing ... and the scorching winds from the southeast in the middle of the summer finally destroyed the spring grain.

He also noted, however, that harvest yields were falling year by year due to the fairly careless cultivation of the soil and absence of regular crop rotations. Peasants were interested more in plowing up as much land as possible, and sowing only wheat, in the hope of enriching themselves after one or two good harvests. Their economy was, the governor wrote, on a very unsound basis. The harvest failure had seriously damaged the peasant way of life, the area of land sown was declining, and peasants were in arrears with their taxes.[27] The story was similar across much of the steppes in 1891: drought in the spring, when the crops most needed water, heat waves and scorching winds in the summer, which killed off crops that managed to grow. The harvest was around half the average.[28] Dust storms, similar to those on the southern plains in the 1930s, were reported. As one eyewitness related:

> From the morning a strong, violent, easterly wind began to blow, which at times raised a significant amount of dust ...; the air became dry, far off there was a haze in the air, which foreshadowed an abrupt change. By midday ... the entire horizon was covered with a very fine dust; the sun ... was obscured as if by a light cloud; all that could be seen was a red blotch ... The house trembled under the force of the strong wind ... All living things hid, [and] kept quiet, as if in anticipation of something even more terrible.[29]

The causes of the disaster, and how to deal with it, were discussed by officials, scientists, and the public. There was debate over the degree to which the disaster was the result of the anomalous climatic conditions, and whether speculative plowing up of marginal land was a factor.[30] Some commentators suggested that the drought had been caused by deforestation, a view that echoed one widely held earlier in the nineteenth century.[31] The pioneering soil scientist and steppe ecologist Vasilii Dokuchaev argued instead that plowing up the steppes and removing native grasses was causing the land to dry out. Plowed land was less able to absorb moisture, which evaporated or drained off the land through ravines that were expanding as cultivation had made the land more liable to erosion. As a result, the water table was falling and, in times of drought, crops dried up because they were not able to draw on moisture in the ground.[32] The philosopher Vladimir Solov'ev feared that the destruction of the steppe environment by "rapacious" agriculture was threatening the region with desertification, and the loss of Russian civilization.[33] Thus, as in the US four decades later, the immediate intellectual response to the disaster was a declensionist interpretation in which human activity was held responsible.

If the focus is broadened, however, a progressive narrative can be found also for the steppe region. In the wake of the disaster of 1891–2, elements in the Russian government and scientific community believed that science and technology could address the ecological problem. With financial support from the government, Dokuchaev led a scientific expedition to the steppes. The scientists identified what they saw as samples of the steppe environment that had not been altered by human activity, and studied their hydrology, soil and geology, meteorological conditions, and flora and fauna. They aimed to learn from the "virgin" environment in order to devise more sustainable ways of cultivating the land. Dokuchaev's plan included planting trees to protect the soil from the drying effects of the wind and to reduce erosion, accumulating water in reservoirs, regulating the use of scarce water, and working out norms for the relative areas of arable land, meadows, forest, and water in conformity with environmental conditions.[34] The Russian government had been involved in engineering solutions to the shortage of water on the steppes before 1891. In 1880, General Zhilinskii was charged with leading an expedition to carry out experimental work on irrigation. The expedition's work was expanded in 1892.[35]

This progressive narrative goes back at least as far as the completion of the Russian political and military conquest of the steppes under Catherine the Great (reigned 1762–96). In the late 1760s, the Russian Academy of Sciences sent expeditions to explore the region. The leaders were instructed to investigate, among other things: the nature of the land and water; any uncultivated or unpopulated land that could be used for growing grain or other crops; the economic activities of populated places and how they could be improved. The volumes they produced documented the wealth and potential of the region, in particular the fertility of the soil.[36] Catherine visited the steppes in 1787 during a tour organized by Grigorii Potemkin. Catherine and her entourage delighted in the sights of the "Asiatic" steppes that awaited "civilization."[37] There were doubters. At the end of the 1830s, the Swiss manager of farms attached to the military settlements in the region wrote that, due to the periodic droughts (and locusts), the steppes were destined to be a "land of pastures." Such views were challenged by officials in the Ministry of State Domains that had responsibility for agriculture and worked hard to improve it.[38] The Imperial Society for Agriculture of Southern Russia, founded in 1828, worked tirelessly to promote agriculture on the steppes.[39] In the late nineteenth and early twentieth centuries, the *zemstvos* (local councils established in 1864) encouraged agriculture with extension services to offer agronomical advice, model farms, depots for improved implements, and education.[40]

The Soviet regime that came to power in 1917 promoted an extreme version of the notion that humans, armed with science and technology, could conquer nature. The collectivization of agriculture on the steppes entailed a massive plow-up of virgin land by tractors in vast collective and state farms. Stalin praised the expansion of agriculture: "The question of cultivating neglected land and virgin soil is of tremendous importance." He denigrated scientists who opposed plowing up virgin land. This was part of the wider attitude that nature was to be transformed in the cause of constructing socialism, and not protected.[41] The perceived need to transform nature was further underlined by renewed drought, harvest failure, and famine in 1946–7. This led to the "Great Stalin Plan for the

Transformation of Nature" of 1948. It called for planting nearly 6 million hectares of trees in shelterbelts along the rivers and around fields in the steppe region. The aims were to protect land from the drying and eroding influence of the easterly winds and to reduce the threat of droughts by making the climate wetter. The plan was based on ideas proposed by Dokuchaev in the 1890s, which were conservationist and based on science, but also on Promethean notions of transforming nature promoted by the pseudo-scientist Trofim Lysenko, a favorite of Stalin's.[42] In 1950, a further plan announced the construction of dams on the major rivers of the steppe region to provide water for irrigation and hydroelectric power.[43]

The collectivization and mechanization of steppe farming, planting of shelterbelts of trees, and building of dams were expensive ways of exploiting the fertile soil, but they did not solve the problems of Soviet agricultural production. After Stalin's death in 1953, Nikita Khrushchev sought a quick fix to the urgent problem of food supply in the Soviet Union. His answer was the Virgin Lands Campaign, launched in 1954, which called for the plowing up and cultivation of "virgin or idle land" on the steppes of southeastern European Russia, southern Siberia, and, in particular, northern Kazakhstan. The original plan was for 13 million hectares to be plowed up and sown to grain; by 1960, the total was nearly 33 million. Hundreds of thousands of people, including Young Communists, mechanics, and agronomists, went east to spearhead and implement the campaign. There was also substantial investment in agricultural equipment and seed. Agronomists devised implements and methods of cultivating land in semi-arid conditions. In some years – 1956 and 1958, for example – the virgin lands yielded bumper crops, and helped Khrushchev consolidate his political position among rivals within the Soviet leadership.[44]

Thus, it is possible to trace a progressive narrative of the environmental history of the steppes in a straight line from Catherine the Great to Khrushchev. But Catherine's hopes were not fully realized. The efforts of the tsarist government, agricultural societies, zemstvos, and agronomists did not stop the recurring droughts and crop failures. The expeditions led by Dokuchaev and Zhilinskii were halted in the late 1890s, due to shortage of funds and political decisions.[45] Collectivization was followed by harvest failure and famine in the steppe region. "The Great Stalin Plan," in the words of Stephen Brain, "met almost none of its stated goals," and collapsed after Stalin's death. Less than half the planned area was planted with trees, and around half the seedlings had died by 1954. Shelterbelts around fields were more successful, and some improvements in crop yields in fields protected by trees were reported. The Promethean hopes of changing the climate of the steppe region, however, did not materialize.[46] There were good reasons why the lands Khrushchev ordered to be plowed up were "virgin": the climate was more extreme, and average rainfall lower, than in the European part of the Soviet Union, and the lands were more exposed to the easterly winds and thus more prone to wind erosion. The "idle" land that was plowed up had been left fallow to assist in retaining moisture, controlling weeds, and allowing soil structure to recover. The agricultural equipment recommended by agronomists was not made available in sufficient quantities; the techniques they advised were disregarded; and migrants to the virgin lands used farming methods more appropriate to the wetter lands west of the Ural Mountains they had come from. Even basic precautions such as contour plowing were ignored. The consequences were poor harvests in dry years, such as 1957 and 1963, soil erosion, and dust storms. In Kazakhstan in 1963, dust storms ruined crops on over 3 million hectares.[47] Subjected to closer scrutiny, therefore, this "progressive" narrative of the environmental history of the steppes turns to dust. Chibilev and Levykin used the experiences of collectivization, the Stalin Plan, and Virgin Lands Campaign to support a declensionist interpretation of human activity degrading the steppe environment. They pointed out, moreover, that Khrushchev had ignored the lessons of the "Dust Bowl" on the Great Plains only two decades earlier.[48]

Is there a "middle ground" interpretation for the environmental history of the steppes? To replicate Cunfer's study of land use over time would probably not be possible due to the shortage of reliable data with the same degree of consistency and comparability as he was able to gather (and as GIS work requires). Large-scale plowing up of the steppes began in the mid-eighteenth century, around a century before plowing in the Great Plains, and there have been many changes in administrative boundaries and political regimes since then which severely complicate any data-gathering exercise for the steppes. It is possible, however, to reach broad conclusions about trends in land use. In 1725, around 6 percent of the land in the steppe region as a whole in European Russia had been plowed up. By the end of the eighteenth century, this had increased to around 15 percent, and by 1861 to nearly 18 percent. These averages concealed variations across the region. The big increases came over the next century. In 1887, when approximately 31 percent of the total region had been plowed up, around half the land in the lower Volga region had been converted to arable, and over two-thirds on the steppes to the north of the Black Sea and the Don region. By 1922, the proportion of land designated as arable (but not necessarily sown that year) in the lower Volga region had reached two-thirds.[49] The twentieth century witnessed a considerable expansion in the proportion of steppe plowed up in southern Siberia and Kazakhstan, and further expansion in the steppe region of European Russia and Ukraine. Chibilev and Grosheva estimated that arable land currently comprises around 57 percent of the total area of the entire Eurasian steppes, but that the proportions are far higher, up to 83 percent, in some parts.[50] Thus, in contrast to the underlying persistence in patterns of land use on the Great Plains since the 1920s, the proportion of the steppes under crops has steadily increased, and is far higher than its American counterpart.

Conclusion

This chapter has focused on the Dust Bowl in the 1930s on the Great Plains and on the drought, crop failure, and famine of 1891–2 on the steppes as lenses through which to consider wider issues in the environmental histories of the grasslands. Three interpretations have been analyzed: declensionist, progressive, and the "middle ground." The first stressed the damage inflicted by humans to grassland ecosystems; the second, the ability of people, armed with science and technology, to transform grasslands to meet their goals. The latter interpretation is no longer sustainable. In the 1950s, moreover, Webb implicitly disavowed such a view when criticizing the mismanagement of the land on the plains.[51] Thus, we are left with the declensionist interpretation and the middle ground, which proposes a more balanced view.

Approaching the story from the perspective of geography and ecology, there can be no dispute that plowing up "virgin" grassland, which has formed over several millennia, and replacing the plant communities with a small number of cultivated crops is the "unjustifiable annihilation of the steppe landscape, the destruction of flora and fauna."[52] Plowing destroys, permanently, the structure of the soil. It has further been argued that arable farming on the grasslands can continue only with massive investment and subsidies. The water in the Ogallala Aquifer has already been exhausted around Lubbock in the Texas panhandle. From this perspective, the solution is conservation and restoration. In both continents, areas of unplowed grassland have been preserved for scientific research, such as the Konza Biological Research Station in Kansas and the Askaniia Nova reserve in southern Ukraine. New protected areas were established in the Orenburg and Rostov regions of Russia in 1989 and 1996. Scientists have also devised techniques to restore, or recreate, grassland environments. Frank and Deborah Popper have promoted the rather extreme idea of taking land on the Great Plains out of production and sowing wild grasses on former arable and pasture land to create a "buffalo commons." Wes Jackson's Land Institute in Kansas has developed ways of farming with hybrid, perennial crops that mimic the ecosystem of the wild grasslands. In Orenburg in the steppe region, Chibilev has argued that steppe agriculture must become ecologically safe and economically profitable as a basis for sustainable land use.[53]

Geoff Cunfer's "middle ground" is based on the premise that people – including farmers who have settled on grasslands – are part of the natural world. James Sherow, following Robert O'Neill, has argued that humans be considered a "keystone species" on the grasslands. The concept that humans are part of the environment opens up new ways of interpreting environmental histories. Moreover, scientists have overturned the view that ecosystems reach stable equilibria, or climaxes, until disrupted by humans.[54] On both the Great Plains and the steppes, for example, there were dust storms before the wholesale plowing up of the grasslands.[55] The notion that the ways of life of indigenous peoples, including Plains Indians and steppe nomads, were in balance with nature has also been challenged.[56] To some extent, the grasslands the agricultural settlers encountered had been shaped by their existing inhabitants. Over the second half of the nineteenth century, Ivan Palimpsestov – an Orthodox clergyman and agricultural specialist – suspected that there had been more forests in the steppe region in the past, and speculated that the nomads had destroyed them by fire.[57] Later research analyzing fossil pollen to some extent confirmed his suspicions.[58]

In 1899, Charles E. Bessey, professor of botany at the University of Nebraska, suggested that the area of forest in Nebraska had been greater a few centuries earlier. He argued that forests had been destroyed by fires and grazing animals, but that they were now returning. "No one who has seen and studied the forest areas in eastern Nebraska," he argued, "will be able to doubt that they are spreading where they are given a fair opportunity by man or his domestic animals."[59] By this time, the populations of Plains Indians and buffalo in the state had fallen sharply. Experiments with controlled burning and grazing regimes on the Konza research station have confirmed their importance in maintaining grassland environments by retarding tree growth.[60] Thus, fire (accidental and human set) and grazing by wild and herded animals played a large role in the formation and maintenance of the grasslands. When the plow broke the steppes and the plains it was only the latest stage in the environmental history of humans on the grasslands, already several millennia long, in which humans have a played an increasingly prominent part.

Notes

1   The similarities were recognized by Russian and American scientists from the late nineteenth century. See A. N. Krasnov, "Travianye stepi Severnogo polushariia," *Izvestiia Obshchestva liubitelei estestvoznaniia, antropologii i etnografii* 81/7, 1894, pp. 1–294; V. V. Dokuchaev, "K uchenie o zonakh prirody," in *Sochineniia*, 9 vols., Moscow and Leningrad, 1949–61

[1899], vol. 6, pp. 398–414; M. A. Carleton, "Russian Cereals adapted for cultivation in the United States," *USDA Division of Botany Bulletin* 23, Washington, DC, GPO, 1900; C. F. Marbut, "Soils of the Great Plains," *Annals of the Association of American Geographers* 13/2, 1923, pp. 41–66.

2   In A. A. Chibilev and O. A. Grosheva, *Ocherki po istorii stepevedeniia*, Ekaterinburg, UrO RAN, 2004, pp. 122–30.

3   Internet Archive, *The Plow that Broke the Plains*, accessed from http://www.archive.org/details/PlowThatBrokethePlains1 on June 10, 2010.

4   US Great Plains Committee, *The Future of the Great Plains*, Washington, DC, The House of Representatives, 1937, pp. 1–2, 49–50, 63; G. F. White, "The Future of the Great Plains Re-Visited," *Great Plains Quarterly* 6/2, 1986, pp. 84–93.

5   D. Worster, *Dust Bowl: The Southern Plains in the 1930s*, 25th anniversary ed., New York, Oxford University Press, 2004, pp. 4–5, 106–7, 140–1, 151.

6   G. Cunfer, *On the Great Plains: Agriculture and Environment*, College Station, TX, Texas A&M University Press, 2005, pp. 9–10, 234–5.

7   W. P. Webb, *The Great Plains*, Lincoln, NE, University of Nebraska Press, 1931, pp. 9, 167–79, 295–318, 333–74.

8   US Great Plains Committee, *The Future of the Great Plains*, pp. 71–89; D. Helms, "Conserving the Plains: The Soil Conservation Service in the Great Plains," *Agricultural History* 64/2, 1990, pp. 58–73.

9   W. H. Droze, *Trees, Prairies, and People: Tree Planting in the Plains States*, Denton, TX, Texas Woman's University, 1977.

10 P. O. Rudolf and S. R. Gevorkiantz, "Shelterbelt Experience in Other Lands," in *Possibilities of Shelterbelt Planting in the Plains Region*, Washington, DC, GPO, 1935, pp. 59–76.

11 Worster, *Dust Bowl*, pp. 210–30.

12 Cunfer, *On the Great Plains*, p. 148.

13 Cunfer, *On the Great Plains*, pp. 3–13, 150–63, 232–40.

14 S. D. Schubert, M. J. Suarez, P. J. Pegion, et al., "On the Cause of the 1930s Dust Bowl," *Science* 303/5665, 2004, pp. 1855–9.

15 Cunfer, *On the Great Plains*, pp. 5–6, 16–36, 200, 236.

16 J. C. Malin, "The Adaptation of the Agricultural System to Sub-Humid Environment," *Agricultural History* 10/3, 1936, pp. 118–41; J. C. Malin, *The Grasslands of North America*, Lawrence, KS, University Press of Kansas, 1947.

17 B. Wills, booklet accompanying 4-CD set, *Take Me Back to Tulsa: Bob Wills & His Texas Playboys*, Beckenham, Proper Records Ltd, 2001, pp. 28, 46.

18 W. Guthrie, *Dust Bowl Ballads*, RCA Victor, 1940.

19 On the ecology and human history of the steppes, see E. M. Lavrenko, Z. V. Karamysheva, and R. I. Nikulina, *Stepi Evrazii*, Leningrad, Nauka, 1991; W. H. McNeill, *Europe's Steppe Frontier, 1500–1800*, Chicago, University of Chicago Press, 1964; W. Sunderland, *Taming the Wild Field: Colonization and Empire on the Russian Steppe*, Ithaca, NY, Cornell University Press, 2004.

20 J. Steinbeck, *A Russian Journal*, London, Minerva, 1994 [1948], pp. 52, 82, 86, 107, 118–41.

21 See W. Guthrie, *Bound for Glory*, New York, Dutton, 1943.

22 For differing interpretations, see R. Conquest, *The Harvest of Sorrow: Soviet Collectivization and the Terror-Famine*, London, Hutchinson, 1986; R. W. Davies and S. G. Wheatcroft, *The Years of Hunger: Soviet Agriculture, 1931–1933*, Basingstoke, PalgraveMacmillan, 2004.

23 E. Lyons, *Assignment in Utopia*, London, Harrap, 1937, pp. 572–3.

24 M. Sholokhov, "Sholokhov i Stalin: Perepiska nachala 30-kh godov," *Voprosy istorii* 3, 1994, pp. 3–25; M. Sholokhov, *Virgin Soil Upturned*, London, Putnam, 1935; M. Sholokhov, *Harvest on the Don*, London, Putnam, 1960.

25 N. E. Saul, *Concord and Conflict: The United States and Russia, 1867–1914*, Lawrence, KS, University Press of Kansas, 1996, pp. 335–64; N. E. Saul, *Friends or Foes? The United States and Soviet Russia, 1921–1941*, Lawrence, KS, University Press of Kansas, 2006, pp. 44–97.

26 D. Moon, "The Environmental History of the Russian Steppes: Vasilii Dokuchaev and the Harvest Failure of 1891," *Transactions of the Royal Historical Society*, 6th series, 15, 2005, pp. 149–74.

27 State Archive of Samara region, Collection 3, Inventory 233, File 1060b, folios 3–5.

28 A. S. Ermolov, *Neurozhai i narodnoe bedstvie*, St Petersburg, 1892, pp. 3–34.

29 N. Adamov, "Meteorologicheskie nabliudeniia 1892–1894 godov," *Trudy Ekspeditsii, snariazhennoi Lesnym Departmentom, pod rukovodstvom professora Dokuchaeva, Nauchnii otdel* 3/1, 1894, pp. 235–42.

30 N. Kravtsov, "Po povodu neurozhaev v 1891 i 1892 godakh," *Sel'skoe khoziaistvo i lesovodstvo* 172, 1893, pp. 317–35; M. N. Raevskii, "Neurozhai 1891 goda...," *Izvestiia Russkogo Geograficheskogo Obshchestva* 28/1, 1892, pp. 1–29; "Besedy ... po voprosu o prichinakh neurozhaia 1891 goda...," *Trudy Vol'nogo Ekonomicheskogo Obshchestva* 1, 1892, pp. 67–144.

31 D. Moon, "The Debate over Climate Change in the Steppe Region in Nineteenth-Century Russia," *Russian Review* 69, 2010, pp. 251–75.

32 V. V. Dokuchaev, "Nashi Stepi prezhde i teper," *Sochineniia* 6, 1951 [1892], pp. 57–89 (pp. 57–61, 87–9).

33 V. S. Solov'ev, "Vrag s Vostoka," *Severnyi vestnik* 7, 1892, pp. 253–64.

34 Moon, "The Environmental History of the Russian Steppes."

35 I. I. Zhilinkskii, *Ocherk rabot ekspeditsii po orosheniiu na iuge Rossii i Kavkaze*, St Petersburg, Bezobrazov, 1892; Russian State Historical Archive, Collection 426, Inventory 1, 1894, Files 29, 30, 55.

36 D. Moon, "The Russian Academy of Sciences Expeditions to the Steppes in the Late Eighteenth Century," *Slavonic and East European Review* 88, 2010, pp. 204–36.

37 L. Wolff, *Inventing Eastern Europe*, Stanford, CA, Stanford University Press, 1994.

38 "Russkie khoziaistvennye periodichestkie izdaniia," *Zhurnal Ministerstva Gosudarstvennykh Imushchestv* 3, 1841, pp. 218–34; K. S. Veselovskii, *Prostranstvo i stepen' naselennosti Evropeiskoi Rossii: Sbornik Statisticheskikh Svedenii o Rossii*, vol. 1, St Petersburg, 1851, pp. 28–9.

39 M. P. Borovskii, *Istoricheskii obzor piatidesiatiletnei deiatel'nosti Imperatorskogo Obshchestva Sel'skogo Khoziaistva Iuzhnoi Rossii*, Odessa, Frantsov, 1878; A. Bychikhin, *K 80-letiiu izdaniia "zapisok" Imperatorskogo Obshchestva sel'skogo khoziaistva iuzhnoi Rossii (1830–1910)*, Odessa, 1911.

40 N. G. Koroleva (ed.), *Zemskoe samoupravlenie v Rossii, 1864–1918*, 2 vols., Moscow, Nauka, 2005.

41 D. R. Weiner, *Models of Nature: Ecology, Conservation and Cultural Revolution in Soviet Russia*, 2nd ed., Pittsburgh, University of Pittsburgh Press, 2000, pp. 122–3, 131.

42 S. C. Brain, "The Great Stalin Plan for the Transformation of Nature," *Environmental History* 15, 2010, pp. 670–700. See also Brain's chapter in this volume.

43 P. R. Josephson, *Industrialized Nature: Brute Force Technology and the Transformation of the Natural World*, Washington, DC, Island Press, 2002, pp. 27–40.

44 M. McCauley, *Khrushchev and the Development of Soviet Agriculture: The Virgin Land Programme, 1953–1964*, London, Macmillan, 1976.

45 V. S. Diakin, *Den'gi dlia sel'skogo khoziaistva, 1892–1914*, St Petersburg, Sankt-Peterburgskii Universitet, 1997, pp. 39–40.

46 Brain, "The Great Stalin Plan," pp. 692–4.

47 McCauley, *Khrushchev and the Development of Soviet Agriculture*, pp. 91, 95, 156–67, 170–4, 181.

48 In Chibilev and Grosheva, *Ocherki po istorii stepevedeniia*, p. 122.

49 Figures calculated from data in M. A. Tsvetkov, *Izmenenie lesistosti evropeiskoi Rossii: s kontsa XVII stoletiia po 1914 god*, Moscow, AN SSSR, 1957, pp. 111–17; *Sbornik statisticheskikh svedenii po Soyuzu S.S.R., 1918–1923*, Moscow, 1924.

50 Chibilev and Grosheva, *Ocherki po istorii stepevedeniia*, pp. 34–5.

51 J. E. Sherow, *The Grasslands of the United States: An Environmental History*, Santa Barbara, CA, ABC-CLIO, 2007, p. 126.

52 A. A. Chibilev, *Priroda znaet luchshe*, Ekaterinburg, UrO RAN, 1999, p. 173.

53 Chibilev and Grosheva, *Ocherki po istorii stepevedeniia*, pp. 107–21, 131–2.

54 Sherow, *The Grasslands of the United States*, pp. xiv–xv, 4–5, 123–55, 280–2; D. R. Weiner, *A Little Corner of Freedom: Russian Nature Protection from Stalin to Gorbachev*, Berkeley, University of California Press, 1999, pp. 374–401.

55 J. C. Malin, "Dust Storms: Part One, 1850–1860," *Kansas Historical Quarterly* 14/2, 1946, pp. 129–44; S. Pallas, "Travels into Siberia and Tatary, Provinces of the Russian Empire," in J. Trusler (ed.), *The Habitable World Described*, vol. 4, London, 1789, p. 315.

56 S. Krech, *The Ecological Indian: Myth and History*, New York, Norton, 1999.

57 I. Palimpsestov, *Stepi iuga Rossii byli-li iskoni vekov stepami i vozmozhno-li oblesit'ikh?*, Odessa, Nitche, 1890.

58 C. V. Kremenetski, "Human Impact on the Holocene Vegetation of the South Russian Plain," in J. Chapman and P. Dolukhanov (eds.), *Landscapes in Flux: Central and Eastern Europe in Antiquity*, Oxford, Oxbow, 1997, pp. 272–87.

59 C. E. Bessey, "Report of the Botanist," *Annual Report of Nebraska State Board of Agriculture*, 1899, pp. 17–23.

60 Konza Prairie Biological Station, accessed from http://kpbs.konza.ksu.edu/currentresearch.html on March 7, 2012.

References

Adamov, N., "Meteorologicheskie nabliudeniia 1892–1894 godov," *Trudy Ekspeditsii, snariazhennoi Lesnym Departmentom, pod rukovodstvom professora Dokuchaeva, Nauchnii otdel* 3/1, 1894, pp. 235–42.

"Besedy ... po voprosu o prichinakh neurozhaia 1891 goda...," *Trudy Vol'nogo Ekonomicheskogo Obshchestva* 1, 1892, pp. 67–144.

Bessey, C. E., "Report of the Botanist," *Annual Report of Nebraska State Board of Agriculture*, 1899, pp. 17–23.

Borovskii, M. P., *Istoricheskii obzor piatidesiatiletnei deiatel'nosti Imperatorskogo Obshchestva Sel'skogo Khoziaistva Iuzhnoi Rossii*, Odessa, Frantsov, 1878.

Brain, S. C., "The Great Stalin Plan for the Transformation of Nature," *Environmental History* 15, 2010, pp. 670–700.

Bychikhin, A., *K 80-letiiu izdaniia "zapisok" Imperatorskogo Obshchestva sel'skogo khoziaistva iuzhnoi Rossii (1830–1910)*, Odessa, 1911.

Carleton, M. A., "Russian Cereals adapted for cultivation in the United States," *USDA Division of Botany Bulletin* 23, Washington, DC, GPO, 1900.

Chibilev, A. A., *Priroda znaet luchshe*, Ekaterinburg, UrO RAN, 1999.

Chibilev, A. A., and Grosheva, O. A., *Ocherki po istorii stepevedeniia*, Ekaterinburg, UrO RAN, 2004.

Conquest, R., *The Harvest of Sorrow: Soviet Collectivization and the Terror-Famine*, London, Hutchinson, 1986.

Cunfer, G., *On the Great Plains: Agriculture and Environment*, College Station, TX, Texas A&M University Press, 2005.

Davies, R. W., and Wheatcroft, S. G., *The Years of Hunger: Soviet Agriculture, 1931–1933*, Basingstoke, PalgraveMacmillan, 2004.

Diakin, V. S., *Den'gi dlia sel'skogo khoziaistva, 1892–1914*, St Petersburg, Sankt-Peterburgskii Universitet, 1997.

Dokuchaev, V. V., "K uchenie o zonakh prirody," in *Sochineniia*, 9 vols., Moscow and Leningrad, 1949–61 [1899], vol. 6, pp. 398–414.

Dokuchaev, V. V., "Nashi Stepi prezhde i teper," *Sochineniia* 6, 1951 [1892], pp. 57–89.

Droze, W. H., *Trees, Prairies, and People: Tree Planting in the Plains States*, Denton, TX, Texas Woman's University, 1977.

Ermolov, A. S., *Neurozhai i narodnoe bedstvie*, St Petersburg, 1892.

Guthrie, W., *Bound for Glory*, New York, Dutton, 1943.

Guthrie, W., *Dust Bowl Ballads*, RCA Victor, 1940.

Helms, D., "Conserving the Plains: The Soil Conservation Service in the Great Plains," *Agricultural History* 64/2, 1990, pp. 58–73.

Internet Archive, *The Plow that Broke the Plains*, accessed from http://www.archive.org/details/PlowThatBrokethePlains1 on June 10, 2010.

Josephson, P. R., *Industrialized Nature: Brute Force Technology and the Transformation of the Natural World*, Washington, DC, Island Press, 2002.

Konza Prairie Biological Station, accessed from http://kpbs.konza.ksu.edu/currentresearch.html on March 7, 2012.

Koroleva, N. G. (ed.), *Zemskoe samoupravlenie v Rossii, 1864–1918*, 2 vols., Moscow, Nauka, 2005.

Krasnov, A. N., "Travianye stepi Severnogo polushariia," *Izvestiia Obshchestva liubitelei estestvoznaniia, antropologii i etnografii* 81/7, 1894, pp. 1–294.

Kravtsov, N., "Po povodu neurozhaev v 1891 i 1892 godakh," *Sel'skoe khoziaistvo i lesovodstvo* 172, 1893, pp. 317–35.

Krech, S., *The Ecological Indian: Myth and History*, New York, Norton, 1999.

Kremenetski, C. V., "Human Impact on the Holocene Vegetation of the South Russian Plain," in J. Chapman and P. Dolukhanov (eds.), *Landscapes in Flux: Central and Eastern Europe in Antiquity*, Oxford, Oxbow, 1997, pp. 272–87.

Lavrenko, E. M., Karamysheva, Z. V., and Nikulina, R. I., *Stepi Evrazii*, Leningrad, Nauka, 1991.

Lyons, E., *Assignment in Utopia*, London, Harrap, 1937.

Malin, J. C., "The Adaptation of the Agricultural System to Sub-Humid Environment," *Agricultural History* 10/3, 1936, pp. 118–41.

Malin, J. C., "Dust Storms: Part One, 1850–1860," *Kansas Historical Quarterly* 14/2, 1946, pp. 129–44.

Malin, J. C., *The Grasslands of North America*, Lawrence, KS, University Press of Kansas, 1947.

Marbut, C. F., "Soils of the Great Plains," *Annals of the Association of American Geographers* 13/2, 1923, pp. 41–66.

McCauley, M., *Khrushchev and the Development of Soviet Agriculture: The Virgin Land Programme, 1953–1964*, London, Macmillan, 1976.

McNeill, W. H., *Europe's Steppe Frontier, 1500–1800*, Chicago, University of Chicago Press, 1964.

Moon, D., "The Debate over Climate Change in the Steppe Region in Nineteenth-Century Russia," *Russian Review* 69, 2010, pp. 251–75.

Moon, D., "The Environmental History of the Russian Steppes: Vasilii Dokuchaev and the Harvest Failure of 1891," *Transactions of the Royal Historical Society*, 6th series, 15, 2005, pp. 149–74.

Moon, D., "The Russian Academy of Sciences Expeditions to the Steppes in the Late Eighteenth Century," *Slavonic and East European Review* 88, 2010, pp. 204–36.

Palimpsestov, I., *Stepi iuga Rossii byli-li iskoni vekov stepami i vozmozhno-li oblesit'ikh?*, Odessa, Nitche, 1890.

Pallas, S., "Travels into Siberia and Tatary, Provinces of the Russian Empire," in J. Trusler (ed.), *The Habitable World Described*, vol. 4, London, 1789.

Raevskii, M. N., "Neurozhai 1891 goda...," *Izvestiia Russkogo Geograficheskogo Obshchestva* 28/1, 1892, pp. 1–29.

Rudolf, P. O., and Gevorkiantz, S. R., "Shelterbelt Experience in Other Lands," in *Possibilities of Shelterbelt Planting in the Plains Region*, Washington, DC, GPO, 1935, pp. 59–76.

"Russkie khoziaistvennye periodichestkie izdaniia," *Zhurnal Ministerstva Gosudarstvennykh Imushchestv* 3, 1841, pp. 218–34.

Saul, N. E., *Concord and Conflict: The United States and Russia, 1867–1914*, Lawrence, KS, University Press of Kansas, 1996.

Saul, N. E., *Friends or Foes? The United States and Soviet Russia, 1921–1941*, Lawrence, KS, University Press of Kansas, 2006.

Sbornik statisticheskikh svedenii po Soyuzu S.S.R., 1918–1923, Moscow, 1924.

Schubert, S. D., Suarez, M. J., Pegion, P. J., et al., "On the Cause of the 1930s Dust Bowl," *Science* 303/5665, 2004, pp. 1855–9.

Sherow, J. E., *The Grasslands of the United States: An Environmental History*, Santa Barbara, CA, ABC-CLIO, 2007.

Sholokhov, M., *Harvest on the Don*, London, Putnam, 1960.

Sholokhov, M., "Sholokhov i Stalin: Perepiska nachala 30-kh godov," *Voprosy istorii* 3, 1994, pp. 3–25.

Sholokhov, M., *Virgin Soil Upturned*, London, Putnam, 1960.

Solov'ev, V. S., "Vrag s Vostoka," *Severnyi vestnik* 7, 1892, pp. 253–64.

Steinbeck, J., *A Russian Journal*, London, Minerva, 1994 [1948].

Sunderland, W., *Taming the Wild Field: Colonization and Empire on the Russian Steppe*, Ithaca, NY, Cornell University Press, 2004.

Tsvetkov, M. A., *Izmenenie lesistosti evropeiskoi Rossii: s kontsa XVII stoletiia po 1914 god*, Moscow, AN SSSR, 1957.

US Great Plains Committee, *The Future of the Great Plains*, Washington, DC, The House of Representatives, 1937.

Veselovskii, K. S., *Prostranstvo i stepen' naselennosti Evropeiskoi Rossii: Sbornik Statisticheskikh Svedenii o Rossii*, vol. 1, St Petersburg, 1851.

Webb, W. P., *The Great Plains*, Lincoln, NE, University of Nebraska Press, 1931.

Weiner, D. R., *A Little Corner of Freedom: Russian Nature Protection from Stalin to Gorbachev*, Berkeley, University of California Press, 1999.

Weiner, D. R., *Models of Nature: Ecology, Conservation and Cultural Revolution in Soviet Russia*, 2nd ed., Pittsburgh, University of Pittsburgh Press, 2000.

White, G. F., "The Future of the Great Plains Re-Visited," *Great Plains Quarterly* 6/2, 1986, pp. 84–93.

Wills, B., *booklet accompanying 4-CD set, Take Me Back to Tulsa: Bob Wills & His Texas Playboys*, Beckenham, Proper Records Ltd, 2001.

Wolff, L., *Inventing Eastern Europe*, Stanford, CA, Stanford University Press, 1994.

Worster, D., *Dust Bowl: The Southern Plains in the 1930s*, 25th anniversary ed., New York, Oxford University Press, 2004.

Zhilinskii, I. I., *Ocherk rabot ekspeditsii po orosheniiu na iuge Rossii i Kavkaze*, St Petersburg, Bezobrazov, 1892.

CHAPTER FIFTEEN

Global Forests

NANCY LANGSTON

Forests across the world are changing rapidly and, by most measures, they are in a great deal of trouble. Healthy forests are central to healthy human communities, and degraded forests are often accompanied by poverty. As an Earth Day 2006 opinion piece in *The New York Times* put it:

> Our forests are the heart of our environmental support system. And yet, in the 36 years that have passed since the first Earth Day, on April 22, 1970, we have lost more than one billion acres of forest, with no end in sight. The people most vulnerable to the disappearance of forests are the poor: nearly three-quarters of the 1.2 billion people defined as extremely poor live in rural areas, where they rely most directly on forests for food, fuel, fiber and building materials … Everywhere, forests prevent erosion, filter and regulate the flow of fresh water, protect coral reefs and fisheries and harbor animals that pollinate, control pests and buffer disease. That is why the single most important action we can take to protect lives and livelihoods worldwide is to protect forests.[1]

If forests are so critical for human communities, why are they so often degraded? It is not an exaggeration to state that forests made human evolution possible, nor is it an exaggeration to claim that the loss of forests would undermine the future of humans on earth. Yet forests and their histories are nevertheless invisible to many people. Many people see forests as little more than attractive backdrops to the real stuff of human history, but our human stories are intimately interconnected with forests. This chapter will explore some of the often invisible links between forest and human histories.

The air that we breathe, the fuel that powers our industrial development, the marine and terrestrial ecosystems that feed us – these are all dependent on forests, not just those from the present, but also those from the past. The carbon in the coal that threatens our future comes from ancient fossilized forests, their photosynthetic energy trapped 300 million years ago.[2] The fecundity of near-shore oceanic habitats – nurseries for the

A Companion to Global Environmental History, First Edition. Edited by J.R. McNeill and Erin Stewart Mauldin.
© 2012 John Wiley & Sons, Ltd. Published 2015 by John Wiley & Sons, Ltd.

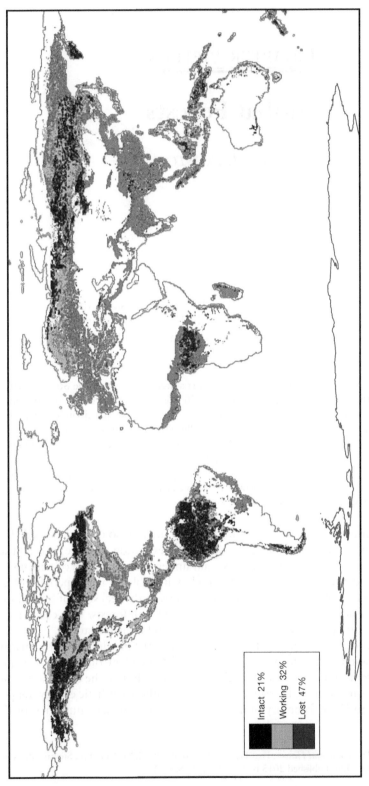

Map 15.1 Changes in global forest cover over 8,000 years. Based on *State of the World's Forests*, 2009, World Resources Institute, http://www.wri. org/image/view/10640/_original. This map is part of an initiative by Global Forest Watch, a group that uses global mapping as a tool to protect forests. "Intact forests" have not yet experienced industrial logging. They are typically large enough to maintain viable populations of wide-ranging species. "Working forests" have been significantly modified by logging and other extractive industries. "Lost forests" are forest land where there has been a period of complete clearance by humans with conversion to another land use. Data on intact forests were gathered in 1997. For more details see http://www.wri.org/publication/last-frontier-forests

Intact 21%

Working 32%

Lost 47%

marine life that feeds us – depends on the carcasses of ancient forests that have made their way to the sea. The soil that supports our grain production builds on the detritus of crumbled forests. We participate each day in an intimate exchange with the forests of the past.

Deep Forest History

As Michael Williams notes in *Deforesting the Earth*, deforestation is as old as the human occupation of the earth. Half of the forest that has vanished from the earth was gone before 1950 (see Map 15.1). But the footprint of humans on a landscape is not always that of a logger's boot leaving destruction in its wake – sometimes, forests spring up in human footsteps, particularly when people suppress fire or build soil for agriculture, and then abandon plots. Particularly during times of war or cultural tumult, forests expand, taking advantage of surprising opportunities.

Forests are dynamic ecosystems marked by disturbance and change. Cut a tree down and it usually grows back, given half a chance. Cut an entire forest down and it often returns – a changed forest, but a forest all the same. The key question in forest history should not be just what kills trees, but what threatens forest resiliency, preventing them from growing back after a disturbance. Resiliency is defined as the potential for a forest to regenerate after disturbances, whether those are anthropogenic disturbances such as logging, ecological disturbances such as herbivory and fire, or biophysical disturbances such as glaciation, erosion, and climate change.

The biophysical template of climate, soils, and glaciation constrains the possibilities for forest resiliency and recovery. You rarely see a red pine forest growing in the middle of the ocean, for example, and mangroves do not thrive on top of snowy peaks. Forests, however, are not merely responding passively to biophysical conditions. They lack the visible agency that marks animals, but plants, even more than animals, are agents of change on earth.[3]

Four and a half billion years ago, before plants evolved, the atmosphere contained very little oxygen, far too little to sustain animal life. The extraordinary innovation of photosynthesis changed this by fixing carbon from the air and releasing oxygen. By about 2.4 billion years ago, in the ancient world of the Devonian, photosynthetic bacteria were producing so much oxygen that the gas created an ozone layer, which in turn absorbed a significant amount of ultraviolet radiation, allowing cells to leave the ocean surface and colonize land. An evolutionary explosion of protoforests soon covered much of the planet. Plants, in other words, through photosynthesis, helped to create the climate and atmosphere that sustains them and ourselves.[4]

Ancient forests pulled carbon dioxide out of the sky and into the pores of the soil, "making the interface between the atmosphere and the earth viable for an explosion of terrestrial life."[5] Forest canopies created the first shade; forest roots bound dirt together into the first true soils; forest cellulose provided fuel for flames; and forest communities created new habitats where terrestrial animals could thrive. The result was a series of evolving feedback relationships that transformed the earth into an interactive system tying together atmosphere, oceans, rocks, soil, bacteria, plants, and terrestrial animals.

This system was dynamic rather than stable. When dinosaurs walked the earth some 200 million years ago, the carbon dioxide level was more than three times our current level (about 360 parts per million in 2010). Dense tropical forests responded to this increased carbon dioxide by spreading across much of the globe, creating a warmer, wetter atmosphere. These forests thrived for millennia, until India eventually collided into Asia. The grinding of tectonic plates pushed up silicate rocks from within the earth's crust, exposing them to the weathering forces of wind and rain. Weathering used up much of the carbon dioxide in the atmosphere, leading to cooling and drying of the earth. Fire frequencies increased in the new climate, and burning fostered the spread of grasses, which in turn increased fires even more – a positive feedback loop that resulted in a steep decline of forests across the globe.[6]

Mammals that had relied on forests disappeared early in the transition from forest to grassland vegetation, while others adapted to the new savannas and grasslands. Morton writes that "among those forced down new evolutionary pathways by the change in the landscape were the African apes; as the woodlands they inhabited shrank and fragmented ... the apes faced a series of challenges that would lead to the evolution of the omnivorous hunting species from which modern humans are evolved."[7] People, in other words, are around because of an evolutionary radiation that resulted from forest change.

About 2 million years ago, the modern ice age began, with continental glaciers periodically expanding over much of the northern temperate zone. Each glacial cycle lasted for about 100,000 years of ice growth, followed by 20,000 years of warming, with massive repercussions for forests in Eurasia and North America that lay in the path of the ice. Forests were scraped away, and when the ice retreated, new forest communities had to start all over again, regaining their hold on formerly glaciated landscapes. Yet forests did not merely follow the retreating ice sheets north. They actually chased the ice north, as recovering forests changed local microclimates, melting the ice before them. When the most recent ice age ended some 11,500 years ago, forests spread back into the temperate zones and carbon dioxide levels rose once again. Wild grains became more fruitful with the increased carbon dioxide, a change which made them far more attractive for human harvesting, ushering in a world of possibilities for settled agriculture.

These glacial histories mean that forest communities that now exist across North America and Europe are relatively young, in evolutionary terms. In effect, because people followed the melting ice north into the grasslands and new forests, human disturbances have been part of those forests for as long as they have existed. It is difficult, if not impossible, to speak of natural or pristine states for these communities.[8]

In contrast, tropical forests were never leveled by glaciation, so they just continued developing as communities, becoming more complex and diverse over the millennia. Yet this greater age means that tropical forests often exist on older soils, where time and weathering have leached nutrients away. Thus, while they tend to have greater ecological diversity than temperate forests because of the greater time for coevolution and development of complex communities, they also tend to have less resilience to human disturbance.

Precolonial Forests

When the most recent ice age ended some 11,500 years ago, forests spread back into the temperate zones. As the ice receded and human populations increased with agricultural expansion, people became significant forces shaping forest-disturbance processes and plant communities. Yet groups of people did not evenly distribute themselves across the landscape, and the intensity of human disturbances varied significantly from one forest stand to another. Fishing, hunting, gathering and management of rice and other native plants, gardening, the location of village sites, and trade networks all shaped forests, but not in homogenous ways.

In most of the northern forests across North America, Russia, and Europe, clearing for farming was restricted in extent, and hunting pressures and fire manipulation were probably the major effects of human populations on forests. Fire, even more than agriculture, was the major process shaping the movement of forests across space and time, determining the location of the ecotone between forest and grassland. In what are now the Great Plains of North America, for example, Indian manipulations of fire likely restricted the extent of forests across the center of the continent.[9]

In Amazonia, where soils are ancient and nutrient-poor, archeologists are increasingly arguing that human presence – particularly farming – may have helped increase the diversity and abundance of some forests, rather than destroying them. Tropical forests typically face high levels of rainfall and, over hundreds of thousands of years, water moving through soils can remove most nutrients, leaving only a clay that cannot hold on to nutrients. About 2,500 years ago, small groups of farmers began creating a particular soil in the Amazon now called "terra preta," or black earth. The writer Charles Mann describes terra preta formation as "a process reminiscent of dropping microorganism-rich starter into plain dough to create sourdough bread." Terra preta, which appears to cover about 10 percent of Amazonia, is generated by a special suite of microorganisms that resists depletion, so intense tropical rains do not leach nutrients from them. By mulching organic material, including fish and aquatic plants, ancient farming communities may have produced a meter of soil in just decades, a startlingly rapid rate. Anna C. Roosevelt, curator of archeology at the Field Museum of Natural History, argues that rather than destroying Amazonian forests, ancient farming communities, with their dense populations and intensive farm plots, may have improved these forests, for "the most luxuriant and diverse" forest growth now occupies mound sites where people once settled.[10]

Asia

In Southeast Asia, archeological evidence similarly shows that people have had a long and complex relationship with tropical forests. Records from 10,000 to 40,000 years ago describe human communities that thrived along the coast of Vietnam, the Malay Peninsula, and Sumatra, hunting a broad range of forest-dwelling species. Semi-sedentary cultivation of wild plant species such as yam was integrated with agroforestry by at least 6,000 years ago, yet human populations remained small. People living in small settlements established swidden fields of small plants such as yams in small forest clearings, enriching local soils. Farmers also domesticated indigenous forest trees such as durian, breadfruit, banana, and coconut, transplanting them from forest interiors to sites closer to their houses and swidden fields.[11]

Agriculture intensified in Thailand about 5,000 years ago. As fields became larger and yams gave way to cereal production, forests were less integral to food systems. Trees became a hindrance to farms, rather than part of the ecological system of food production. As farmers focused on growing cereal grains, they pushed back the forest, both physically and conceptually. Some anthropologists have argued that, in Southeast Asia, people who began engaging in permanent field agriculture essentially "locked themselves out of the forest" conceptually. The forest became fearful and dangerous in agricultural people's cultural mythologies, even though forest-dependent communities continued to view the forest as a source of resources and cultural protection. The anthropologist Poffenberger writes: "Even today, forest dwelling communities like the Semang of the Malay Peninsula seek out the forest because it is 'cool' and therefore 'healthy,' while neighboring Melayu and Temair people regard it as disease-ridden and 'too cold'."[12]

Expanding trade networks in the first millennium led to new pressures on forests across the globe. In Asia, trade and associated rain-forest clearance began in the first century CE, when an extensive trade in ceramics began, linking ports throughout Southeast Asia with the interiors of rain forests. The trade in forest products helped create new governance structures and relations between peoples. The upland rain forests were beyond the administrative control of the royal courts, which had to establish new exchange relationships with the forest villages in the interior. Expanding sea trade in the modern world intensified the demands on forests across the globe and helped to link European and Asian centers of power. As European nations depleted their forests of suitable ship timber, they turned to Asia, particularly for lightweight woods from Asian rain forests. As early as the seventeenth century, the Dutch began negotiating contracts with Javanese rulers for access to teak forests, and commercial timber extraction became widespread in the nineteenth century. From the 1850s on, Burma, Thailand, and much of the lowland Philippines were intensely harvested.[13]

Even with new trade relations, forest peoples were often able to retain their distinctive identities and cultural practices. The power of the precolonial state was limited in forests, and local communities maintained substantial autonomy on what Mahesh Rangarajan calls "the fringes of the cultivated arable."[14] In particular, precolonial forests were usually explicitly gendered spaces, and forests were of fundamental economic and cultural importance to the lives of women. Forests provided the foods that sustained families with protein, minerals, and vitamins lacking in grains, and women collected those forest foods. Trees provided the fodder that sustained the small livestock that women usually tended, and fuel for cooking and heating. Understories provided habitat for the medicinal plants that women collected, developing intricate cultural practices in the process.

Europe

Within European and Mediterranean forests, similar transformations followed the expansion of agriculture and trade. During the beginning of the Neolithic period (c. 11,000 years BP), the climate became more arid, and a slow shift from hunting and gathering to agriculture began. Early farmers cleared many lowland forests for fields, and by 3,000 years ago, the lowland plains of northern Greece had lost significant forest coverage. As human populations increased, the search for arable land led people up the hillsides, where they cleared forests on steeper hills more prone to erosion. Cutting trees for fields, however, probably affected forest regeneration less than introducing goats. Repeated browsing by goats seems to have overpowered the resiliency of many forests, leaving them unable to regenerate.[15]

New technologies in shipbuilding made it safer for ships to sail the winds of the Mediterranean, which stimulated trade and economic growth from 600 BCE on. Ships required wood, so forests made their way into the sea, first sailing atop the waves and eventually sinking to the bottom of the sea, where the ruins of abandoned fleets became nurseries for new generations of marine life. Miners added to pressures on Mediterranean and European forests, cutting down great swathes of timber to provide the heat needed for smelting. Silt increasingly filled in river deltas and harbors along the Mediterranean coasts, ruining ports that trade depended upon. With these combined pressures on forests, many Greek writers between 500 BCE and 25 CE commented on the rapid replacement of forests with pastures and fields.[16]

Material conditions and sociopolitical considerations such as the state, property rights, regulation, or economics do not entirely explain human transformations of forests.[17] We also need to pay attention to culture: the intellectual, spiritual, and religious networks of beliefs that affect peoples' relationships with the natural world, as the literary scholar Robert Pogue Harrison argues in *Forests: The Shadow of Civilization*.[18]

According to Harrison, western civilization has defined itself in opposition to the forest. Harrison argues that binary distinctions – between right and wrong, male and female, order and chaos, light and dark, history and the future – lie at the basis of civilization and patriarchy. Yet it is these distinctions and dualisms which the wild forest continually confuses. Forests have a way of destabilizing and reversing simple matters of right and wrong, natural and cultural. Outside the boundaries of society lies the forest, a place of refuge for outcasts, the mad, lovers, hermits and saints, and lepers. While doctrinal Christian attitudes toward forests were essentially hostile, saints' legends, for example, tell a different story: one devout soul after another took to the forest, and in its refuge they lived in the intimate presence of their gods. Laws go astray in the forests. Conventional distinctions collapse; the profane becomes sacred; the outlaw becomes the guardian of higher justice; the virtuous knight turns into a wild man; the wild man turns into a virtuous man; and the straight way becomes a circuitous path. Forests unsettle; they overturn stability, they confuse clear distinctions – but in that confusion, profound learning can occur. Forests are not just places of fear and evil. They are places of transformation: places where the human and wild meet and get entangled in a web of myth, ritual, stories, worship, and fear.

Harrison maintains that the Scientific Revolution and the Enlightenment attempted to empty the forests of this confusion. They tried to make the forests places purely of reason and production, where all that mattered about a forest was what it could produce for human needs. With the rationalization of forest management in the eighteenth century, forests became the object of a new science, which tried to reduce the messy, fertile complexity of myth and undergrowth to timber. An entire science of measuring the wood in a forest sprung up, the definition of a forest became little more than its timber. Many foresters trained in Europe found employment in colonial outposts, confronting the bewildering diversity of forests far from home.

Colonialism led to key shifts in forests and human communities, as over the course of several centuries, with the help of the new scientific foresters, ecological communities were transformed into collections of resources exported to feed the demands of distant markets. Many forest historians have noted that ecological simplification has been a problematic by-product of colonial forestry; James C. Scott's *Seeing Like a State* (1998) is unique in connecting this ecological simplification to a larger state project. Scott's work asks: what are the connections between ecological simplification and government power? Why do bureaucrats and governments so often try to simplify ecosystems? How does land tenure affect forest change? Scott argues that from the rise of the modern state in the eighteenth century, those in authority have tried to organize society and ecosystem through centralized, top-down plans that simplify human and ecological connections, to further the state functions of taxation, conscription, and the maximization of the state's resources. He shows how centralized planning and "high modernism" have often led to radical ecological simplification and human disaster.

Property rights define the relationships between people, states, and forests, yet property rights are not static systems. The meanings of property are dynamic, and as institutions change within a state, so too do beliefs about property and access to forests. Property rights create relationships that are fluid yet often naturalized in such a way that their historical contingency becomes invisible to people who wield these rights as tools to control the behavior of others.[19]

Across the world, it is a reasonable generalization to state that most forests were traditionally some form of common property regimes. No single person owned all the rights to a forest; unlike agricultural lands, which did tend to be individual private property in many cultures, forests had broader – but not unrestricted – access. Customary tenure systems traditionally regulated access to common property resources within a forest such as fuel wood, grazing, and what foresters now awkwardly term "non-timber forest resources" such as berries and game. Customary tenure systems were based not on the authority of a centralized state, but rather on

> the values of a particular social group, and it is these values which confer legitimacy on local decision-making. Since patterns of interest within a particular social group continually evolve, due to changing conditions, such as population pressure and the value of resources, so the customs and practice in relation to how resources are managed will also evolve.[20]

Customary tenure systems were not perfect, of course, nor were they necessarily equitable. Powerful groups within a community could control preferential access to certain resources, and often socially marginal groups (such as women) were excluded from resource use.

With the growing power of the state, statutory tenure codes were drawn up by centralized governments, reflecting the values and interests of the state. One of the major forces behind the compulsory enclosures, Scott argues, was the tax collectors, who wanted a more detailed and accurate map of who owned and owed what. Customary tenure systems favored local knowledge, while statutory systems favored professional knowledge. Power shifted from those who knew the land to those who knew the law: "State simplifications such as maps, censuses, cadastral lists, and standard units of measurement represent techniques for grasping a large and complex reality; in order for officials to be able to comprehend aspects of the ensemble, that complex reality must be reduced to schematic categories."[21]

One of Scott's key examples is that of colonial forestry. For tax purposes, colonial foresters had to figure out a way to measure standing timber, and the most efficient and accurate way to do so was to legislate that the only legal forest was the measurable, regulated forest. The state simplified forests and land-tenure systems to make forests easier to tax, regulate, and ultimately control.

The links between people and forests were often invisible to the cadre of professional foresters who followed colonial powers around the world. By the nineteenth century, with the deterioration of customary tenure systems under colonial regimes of taxation and land allocation, forests lost many of their traditional protections from overuse. Professional foresters saw the forests being depleted and, being almost completely ignorant of the complex tenure systems that had traditionally regulated access to the forest, they drew the erroneous conclusion that the problem was the customary tenure systems – not the breakdown of these tenure systems. In an effort to slow the deterioration of forests that resulted from the deterioration of tenure systems, colonial powers called on a new generation of technically trained foresters who attempted to use forest science, quantification, and conservation laws to slow forest destruction. Ironically, these attempts at forest conservation ignored the root causes of depletion, and so the result was often increased exploitation, accompanied by a centralization of decision-making that often led to ecological simplification and a failure to protect forest resources or communities dependent on forests, particularly women.

In colonial contexts, women lost many of their customary rights to forest access, and the diversity of forest life that sustained these uses also diminished. As colonial foresters transformed ecologically complex communities into professionally managed, sustained-yield forests, the diversity of women's work vanished in the forests, just as did much of the diversity of plant and animals that had supported their work – and the traditional tenure rights that gave women access to forests. Likewise, the very idea that women had anything important to do with forests vanished. With the increasing growth of sustained-yield forestry, "working forests" had room only for timber – not for medicinal plants, mushrooms, firewood, fish, insects, and herbs. "Working forests" became the province of "working men": of loggers, professional foresters, and corporate accountants. In the process, women lost many of their customary rights to forests, and women were increasingly defined as peripheral to the concerns of forestry.[22]

Yet even as colonialism led to intensified deforestation, it also shaped the roots of modern environmental concern, by providing "a context in which those on the periphery could witness and think critically about such change."[23] As historian Richard Grove notes in *Green Imperialism*, colonial powers in the eighteenth century expressed grave concerns about colonial deforestation – particularly in Caribbean islands – and its potential to lead to climate change and drought.[24] Desiccation concerns led to the first forest conservation policies of many of Britain's colonial states. For all the social, ecological, and political chaos sparked by nineteenth-century state interventions in forests, colonial forestry was not just "a set of simplifying practices exported from Europe and applied in the European colonies."[25] Colonial foresters encountered new ecologies, cultures, and politics, and those encounters transformed scientific practices, cultural perceptions, and developing environmental concern.

North America

Colonialism transformed forest communities, but it did not always mean a decrease in forests, at least initially. North America provides a useful case study. European exploration, trade, and wars from the 1600s to the 1800s altered American Indian groups' relationships with forests. Those groups who first entered into trade relations with the Europeans, particularly with French fur traders, often increased local intensity of hunting and trapping. Seasonal movements of people changed as well; as Indian contact with the French increased, they became located in larger, higher-density, and more persistent settlements around trading points, and this had significant effects on forests.[26]

Disease, famine, and wars, however, quickly devastated American Indian populations and, in the eighteenth century, American forests rebounded as human numbers dropped. The Iroquois Wars in the Eastern Great Lakes led to ripple effects across the Great Lakes region, as the Iroquois were the first to acquire guns and pushed westward, causing massive migrations, leading to new social relations, population pressures, and effects on local and regional forests.[27] By 1750, Michael Williams argues, after disease decimated Indian cultures, the forest was "probably thicker and more extensive than … at any time for the previous thousand years."[28]

European sea trade put an end to the eighteenth-century expansion of the forests. Before the advent of iron ships, European trade depended on wood because massive timbers were needed for shipbuilding. An intricate interplay developed between British military and commercial power, and Baltic and North American sources of timber and naval stores such as pitch and tar. The British navy required masts for ships, and conifers from Baltic forests provided the perfect combination of straightness, strength, and durability. The Baltic nations, however, were reluctant to supply the British navy with the timber it needed for masts, because, as Williams notes, "each Baltic supplying nation was at pains not to alienate influential customers and lose essential revenue." The British turned to its North American colonies as a more reliable supply of trees suitable for masts. The tall, straight white pines that dotted northern forests proved to be perfect substitutes. Yet colonists in North America wanted those trees for local building supplies and for their own export trade. In 1691 the Massachusetts Bay Charter reserved "All trees of the diameter of 24 inches and upward at 12 inches from the ground" for the Royal Navy, embittering colonists. By the eighteenth century, the best mast trees had been harvested, leaving very few old-growth white pines in the remaining forests.[29]

European colonists who came to North America were awed and often overwhelmed by forests that appeared inexhaustible. Immigrants hacked and sawed their way through forests that soon proved to be anything but limitless. Williams reports that the British artist and writer Basil Hall in the 1820s described fields with "numerous ugly stumps of old trees; others allowed to lie in the grass guarded, as it were, by a set of gigantic black monsters, the girdled, scorched and withered remains of the ancient woods."[30] By 1900, half of the original forest cover in the US had been eliminated, much to many people's surprise.

White pine was the foundation of American lumber industry for more than two and a half centuries. White pine rarely grew in pure stands, but it was a key component of mixed forests from west of the Great Lakes to the Atlantic coast. Commercial logging of white pine at first focused on Maine, which had large stands of pine, plus swift rivers and good lake connections. With excellent sources of water power for sawmills, and good water access to ocean ports, Maine was the first region in North America to develop a profitable large-scale export industry.

The demands of new technologies and new markets led to new stresses on American forests in the early nineteenth century. The combination of two things – large capital investment into steam-powered sawmills and the constant threat of fires – meant Maine lumber barons, to stay in business, needed a rapid return on investments, since neither the forest nor the mill might be around for long. The demand for quick returns meant that efficient logging operations needed to run night and day, with lands logged in the fastest, cheapest way. Little thought was given to future regeneration.

As Maine's pine forests declined, lumbermen and loggers headed west to the Lakes States' pineries. The first step in the development of the Midwest logging industry was the removal of Indian title to the land. Treaties negotiated in the 1830s and 1840s began the process of dispossessing Ho Chunk, Chippewas, and Sioux from tribal forest lands. Speculation in timberlands quickly followed, marked by the migration of the Maine timber baron Isaac Stephenson to Wisconsin in 1845. Stephenson found amazing stands of white pine growing on sandy soils deposited by the glaciers. One Wisconsin acre could yield as much as 222 cubic meters (94,000 board feet) of white pine, while in other places, such as Maine, 23.6 cubic meters (10,000 board feet) had been considered a nice stand.

Before 1845, logging in the Great Lakes States had been driven by farming, not by an export market for timber. Wood had been cut largely for local farm consumption, and logging had provided farmers with cash to buy farming supplies. But by the 1860s, the rate of forest clearing for farming sharply declined as farmers pushed onto the fertile prairies. Those farmers needed lumber, as did a growing industrial economy, and the rate of cutting for industrial logging skyrocketed. By the beginning of the Civil War, 153 million acres of forests were cleared for agriculture – and over 12 times that amount had been cleared for industrial logging. As William Cronon argues in *Nature's Metropolis*, Chicago became the center of industrial transformations of the northern forests.[31] In 1847, the Illinois–Michigan canal was completed, linking the Great Lakes with the Illinois and Mississippi rivers. With the shorter, cheaper, route for lumber, the price soon halved, and a substantial deterrent to settlement on the prairies was removed.

Lumber production grew from 11.8 million cubic meters (5 billion board feet) in 1850, to 30.7 million cubic meters (13 billion board feet) in 1870 – a rate of production that the great pineries of the Lakes States could not sustain for long. Sawmills in the state processed 141.6 million cubic meters (60 billion board feet) of lumber between 1873 and 1897, and, by 1898, the federal forester Filbert Roth estimated only 13 percent of the white pine was still standing. The ecological and human effects of this deforestation were devastating, particularly for the Great Lakes Indian peoples.[32] Most American settlers had thought forests were so abundant that loggers could never reach the end of them. When the white pines that seemed like they would be around for ever were gone in less than four decades, deforestation and its effects became a galvanizing issue for the nascent American environmental movement.[33]

In response, Congress created the federal forest reserve system in 1891, withdrawing millions of acres of federal land from settlement and placing it under the control of the Department of Interior's General Land Office. The legislation, however, provided no means for administration or management of those areas, nor did the law state what kind of use could take place inside the reserves. It was increasingly unclear whether the reserves were intended for use – which most people interpreted as grazing and logging – or for protection. In 1897 Congress passed the Organic Act, which clarified the purposes of the reserves: to protect water flow and to insure a continuous supply of timber. The Organic Act gave the government the authority to use the forests and made it clearer that the forests were not preserves. But different federal agencies struggled for control of the forests, and corruption was rampant.[34]

Finally, in 1905, Gifford Pinchot, the most charismatic forester of his generation, won control of the forest reserves within the Department of Agriculture. Gifford Pinchot created a Forest Service that, he believed, would put an end to wasteful exploitation of resources by the rich for private gain. He set out to protect the forests, not for eternal preservation, but for fair, conservative, sustainable use. He believed that the government had an obligation to put an end to wasteful exploitation of resources by the rich for private gain, and scientific forestry was the tool the federal agencies would use to do this.[35]

When Gifford Pinchot lobbied to create the Forest Service, Americans feared a timber famine that would undermine the basis of economic prosperity. Wood was the building block of civilization, or so Pinchot thought: it supplied the fuel to drive machines; the ties to build the railroad networks; the timber to build housing for a growing population.[36] Few of Pinchot's peers recognized that industrialization was about to have profound effects on forests – effects that were not always negative.

The shift from wood fuels to coal increased carbon-dioxide emissions into the atmosphere, but it also slowed deforestation. Without the shift to fossil fuels, fewer of the world's forests would have survived. With new sources of energy, canals gave way to railroads, steam engines gave way to steam turbines, and eventually they all gave way to the internal combustion engines of the car, truck, and aircraft. These technological innovations unleashed stored energy into the atmosphere, releasing the buried carbon of 100 million years in just a few centuries, changing the earth's climate cycles in ways that scientists are only beginning to comprehend.[37]

Twentieth-Century Forests

Before the twentieth century, logging had its most dramatic effects on the temperate forests of North America, Europe, and Russia. But since 1900, in historian Brian Donahue's words, "the temperate forests have largely stabilized in area (and are now even increasing in volume), while the great onslaught has fallen upon the tropics."[38] Ricardo Carrere and Larry Lohmann's book, *Pulping the South*, shows how net deforestation in the temperate developed world has dropped close to zero, with increasing protections on forests in the northern latitudes. Yet this has been accompanied not by a decrease in wood consumption, but by a shift toward the south – specifically, to the world's moist tropical forests.

Tropical rain forests cover less than 6 percent of the globe, but they contain at least two-thirds of all plant and animal species, making them critical hotspots for biodiversity. Yet while logging in temperate forests has leveled off or even decreased, tropical moist forests have been cleared at very high rates in recent decades. Tropical deforestation occurs not just because of industrial logging, but because poverty and government policies encourage forest clearance. Subsistence and commercial agriculture, industrial ranching, and mining all pressure tropical moist forests as much as commercial logging.

Globally, since the 1950s, tropical rain forest has been reduced by over 60 percent. In some regions, the loss has been even greater. Between 1960 and 1990, Brazil destroyed as much of its Atlantic rain forest as had been lost during the previous three centuries. Less than 7 percent of its original 120 million hectares (463,300 square miles) remains, and much of that exists in fragmented patches rather than in contiguous forest.[39]

In some regions, clearing of tropical moist forests has decreased. Brazil's deforestation rate has dropped from 2,800,000 hectares in 2004 to 750,000 hectares in 2009. Yet these efforts could be severely undermined by climate change. Increasing temperatures and aridity lead to increasing forest fires, which, combined with deforestation, can trigger positive feedback cycles of forest loss in the drier southern and southeastern portions of the Amazon. Eighteen percent of the Amazon is currently cleared, and the loss of only 2 percent more could trigger significant dieback. A global temperature increase of 3.5 degrees Celsius (6.3 degrees Fahrenheit) could lead to the loss of half of the Amazon, according to World Bank reports.[40] With the industry shift toward the tropics has come consolidation, loss of jobs, increase in chipping and pulping, intensive capital investments, and the ecological changes brought about by widespread resort to eucalyptus and pine plantations in the name of efficient forest science.

To many people, tree-planting at first seems entirely a good thing, motivated by Arbor Day impulses. But as Carrere and Lohmann argue, "planting a tree, whether native or exotic, is in itself neither a positive nor a negative process. It is the social and geographical structures within which that tree is planted which make it one or the other." *Pulping the South* shows that reforestation has a complex history, growing from agroforestry projects composed largely of fruit-bearing species such as olives, palms, coffee, cocoa, and apples. Teak and eucalyptus began to be planted in the nineteenth century as a response to depletion of oak in Europe. Nevertheless, extensive industrial tree plantations are a twentieth-century invention, established as a result of overexploitation of native forests for wood. Their justification was the discourse of environmentalism. Yet they developed out of what Carrere and Lohmann term "forestry imperialism," not in response to local needs. Carrere and Lohmann argue that

> the problems modern forestry science sets and solves in short are those thrown up by a politics of centralized control of land aimed at extracting a very few types of raw material in industrial quantities. Working exclusively within mainstream forestry science means not asking questions about, and thus tacitly supporting, that politics. Forestry science is thus not a "neutral tool" which can be detached from its social surrounding and adapted to any political purposes.[41]

Tree plantations, therefore, are a way of responding to problems brought about by the prevailing economic model without addressing their underlying causes: rising demand, decreasing access, and changing climate.

Research in Africa on forest loss and recovery illustrates how politically complex forest science has become since World War II. The rising power of postcolonial, transnational organizations such as the Food and Agriculture Organization of the United Nations (FAO) and environmental nongovernmental organizations has transformed forestry discourses and practices. Conservationists have long argued that deforestation has run rampant throughout Africa, with grave impacts for biodiversity. The political ecologists Melissa Leach and James Fairhead argue that West African forests are not nearly as degraded as often assumed, while James McCann suggests that much less of the Ethiopian highlands was forested in historical eras than conservationists and colonial officials believed.[42] Global nongovernmental and governmental organizations may have systematically exaggerated forest loss, with profound effects on African peoples, blaming them for deforestation which they have not caused (and which may not exist). Assumptions about forest history drive forest policy throughout the world.

If we abandon the myth of pristine forests untouched by people and reject the assumption that people always harm forests, what other concepts can guide forest protection? Resiliency remains a useful concept for a world struggling to protect forests. Some forests have been so simplified that they lack resiliency in the face of change. Industrial tree farms fragment ecological interrelationships to the point that they cannot function without extensive inputs of petrochemicals. Other forests still have enough complexity and diversity to sustain themselves, even as the climate shifts. The critical difference between resilient forests and degraded forests is not the presence of humans, but rather the presence of interconnected communities that allow for functioning ecological and evolutionary processes. In a rapidly changing world, sustaining resilient forests may well become one of the key challenges facing communities.

Notes

1   D. Melnick and M. Pearl, "Not Out of the Woods Just Yet," *The New York Times*, April 20, 2006, accessed from http://www.nytimes.com/2006/04/20/opinion/20melnick.html on March 8, 2012.

2   J. R. Fleming, V. Jankovic, and D. R. Coen (eds.), *Intimate Universality: Local and Global Themes in the History of Weather and Climate*, Sagamore Beach, MA, Science History Publications, 2006, pp. ix–x.

3   O. Morton, *Eating the Sun: How Plants Power the Planet*, New York, HarperCollins, 2008.

4   N. Langston, "Air: Climate Change and Environmental History," in D. Sackman (ed.), *A Companion to American Environmental History*, New York, Wiley, 2010, pp. 33–50.

5   Morton, *Eating the Sun*, p. 232.

6   S. Pyne, *World Fire: The Culture of Fire on Earth*, New York, Holt, 1997, and S. Pyne, *Fire: A Brief History*, Seattle, University of Washington Press, 2001.

7   Morton, *Eating the Sun*, p. 288.

8   M. Williams, *Deforesting the Earth: From Prehistory to Global Crisis*, Chicago, University of Chicago Press, 2003.

9   Williams, *Deforesting the Earth*.

10 C. Mann, "1491," *Atlantic Monthly*, March 2002, accessed from http://www.theatlantic.com/past/docs/issues/2002/03/mann.htm on January 21, 2011.

11 M. Poffenberger, *Communities and Forest Management in Southeast Asia*, Gland, IUCN, 2000.

12 Poffenberger, *Communities and Forest Management in Southeast Asia*, p. 14.

13 Poffenberger, *Communities and Forest Management in Southeast Asia*.

14 M. Rangarajan, "Environmental Histories of South Asia: A Review Essay," *Environment and History* 2, 1996, pp. 129–43.

15 Williams, *Deforesting the Earth*.

16 D. Hughes, *The Mediterranean: An Environmental History*, Santa Barbara, CA, ABC-CLIO, 2005; Williams, *Deforesting the Earth*.

17 S. Schama, *Landscape and Memory*, New York, Knopf, 1995.

18 R. P. Harrison, *Forests: The Shadow of Civilization*, Chicago, University of Chicago Press, 1992.

19 N. Langston, "Teaching World Forest History," *Environmental History* 10, 2005, pp. 20–9; L. Heasley and R. P. Guries, "Forest Tenure and Cultural Landscapes: Environmental Histories in the Kickapoo Valley," in H. M. Jacobs (ed.), *Who Owns America? Social Conflict over Property Rights*, Madison, WI, University of Wisconsin Press, 1998, pp. 182–207.

20 International Institute for Environment and Development, *Land Tenure and Resource Access in West Africa*, London, IIED, 1999, p. 2.

21 J. C. Scott, *Seeing like a State: How Certain Schemes to Improve the Human Condition Have Failed*, New Haven, CT, Yale University Press, 1998, pp. 76–7.

22 B. Agarwal, "A Challenge for Ecofeminism: Gender, Greening, and Community Forestry in India," *Women & Environments International Magazine* 52–3, 2001, pp. 12–16.

23 P. Sutter, "Reflections: What Can U.S. Environmental Historians Learn from non-U.S. Environmental Historiography?," *Environmental History* 8, 2003 pp. 109–29.

24 R. H. Grove, *Green Imperialism: Colonial Expansion, Tropical Island Edens and the Origins of Environmentalism, 1600–1860*, New York, Cambridge University Press, 1995.

25 P. Vandergeest and N. L. Peluso, "Empires of Forestry: Professional Forestry and State Power in Southeast Asia, Part 1," *Environment and History* 12, 2006, pp. 31–64.

26 Williams, *Deforesting the Earth*.

27 The Iroquois Wars (or Beaver Wars) were a series of conflicts in the mid-seventeenth century between the Iroquois Confederation and largely Algonquian-speaking tribes of the Great Lakes and Ohio Valley regions, who were supported by the French.

28 Williams, *Deforesting the Earth*, p. 71.

29   Williams, *Deforesting the Earth*, p. 176.
30   Williams, *Deforesting the Earth*, p. 209.
31   W. Cronon, *Nature's Metropolis: Chicago and the Great West*, New York, W. W. Norton, 1992.
32   S. Flader (ed.), *The Great Lakes Forest: An Environmental and Social History*, Minneapolis, University of Minnesota Press, 1983.
33   M. Williams, *Americans and their Forests: A Historical Geography*, Cambridge, Cambridge University Press, 1989.
34   N. Langston, *Forest Dreams, Forest Nightmares: The Paradox of Old Growth in the Inland West*, Seattle, University of Washington Press, 1995.
35   C. Miller (ed.), *American Forests: Nature, Culture, and Politics*, Lawrence, KS, University Press of Kansas, 1997.
36   Williams, *Americans and their Forests*.
37   Morton, *Eating the Sun*.
38   B. Donahue, "People and Trees: A Review of Deforesting the Earth," *Science* 9, 2003, pp. 907–8.
39   W. Dean, *With Broadax and Firebrand: The Destruction of the Brazilian Atlantic Forest*, Berkeley, University of California Press, 1997.
40   J. Astill, "Seeing the Wood," *The Economist*, September 23, 2010, accessed from http://www.economist.com/node/17062713 on March 8, 2012.
41   R. Carrere and L. Lohmann, *Pulping the South: Industrial Tree Plantations and the World Paper Economy*, London, Zed, 1996, p. 11.
42   J. Fairhead and M. Leach, *Misreading the African Landscape: Society and Ecology in a Forest-Savanna Mosaic*, New York, Cambridge University Press, 1996; J. Fairhead and M. Leach, *Reframing Deforestation: Global Analysis and Local Studies in West Africa*, London, Routledge, 1998; J. McCann, "The Plow and the Forest: Narratives of Deforestation in Ethiopia, 1840–1992," *Environmental History* 2, 1997, pp. 138–59.

References

Agarwal, B., "A Challenge for Ecofeminism: Gender, Greening, and Community Forestry in India," *Women & Environments International Magazine* 52–3, 2001, pp. 12–16.

Astill, J., "Seeing the Wood," *The Economist*, September 23, 2010, accessed from http://www.economist.com/node/17062713 on March 8, 2012.

Carrere, R., and Lohmann, L., *Pulping the South: Industrial Tree Plantations and the World Paper Economy*, London, Zed, 1996.

Cronon, W., *Nature's Metropolis: Chicago and the Great West*, New York, W. W. Norton, 1992.

Dean, W., *With Broadax and Firebrand: The Destruction of the Brazilian Atlantic Forest*, Berkeley, University of California Press, 1997.

Donahue, B., "People and Trees: A Review of Deforesting the Earth," *Science* 9, 2003, pp. 907–8.

Fairhead, J., and Leach, M., *Misreading the African Landscape: Society and Ecology in a Forest-Savanna Mosaic*, New York, Cambridge University Press, 1996.

Fairhead, J., and Leach, M., *Reframing Deforestation: Global Analysis and Local Studies in West Africa*, London, Routledge, 1998.

Flader, S. (ed.), *The Great Lakes Forest: An Environmental and Social History*, Minneapolis, University of Minnesota Press, 1983.

Fleming, J. R., Jankovic, V., and Coen, D. R. (eds.), *Intimate Universality: Local and Global Themes in the History of Weather and Climate*, Sagamore Beach, MA, Science History Publications, 2006.

Grove, R. H., *Green Imperialism: Colonial Expansion, Tropical Island Edens and the Origins of Environmentalism, 1600–1860*, New York, Cambridge University Press, 1995.

Harrison, R. P., *Forests: The Shadow of Civilization*, Chicago, University of Chicago Press, 1992.

Heasley, L., and Guries, R. P., "Forest Tenure and Cultural Landscapes: Environmental Histories in the Kickapoo Valley," in H. M. Jacobs (ed.), *Who Owns America? Social Conflict over Property Rights*, Madison, WI, University of Wisconsin Press, 1998, pp. 182–207.

Hughes, D., *The Mediterranean: An Environmental History*, Santa Barbara, CA, ABC-CLIO, 2005.

International Institute for Environment and Development, *Land Tenure and Resource Access in West Africa*, London, IIED, 1999.

Langston, N., "Air: Climate Change and Environmental History," in D. Sackman (ed.), *A Companion to American Environmental History*, New York, Wiley, 2010, pp. 33–50.

Langston, N., *Forest Dreams, Forest Nightmares: The Paradox of Old Growth in the Inland West*, Seattle, University of Washington Press, 1995.

Langston, N., "Teaching World Forest History," *Environmental History* 10, 2005, pp. 20–9.

Mann, C., "1491," *Atlantic Monthly*, March 2002, accessed from http://www.theatlantic.com/past/docs/issues/2002/03/mann.htm on January 21, 2011.

McCann, J., "The Plow and the Forest: Narratives of Deforestation in Ethiopia, 1840–1992," *Environmental History* 2, 1997, pp. 138–59.

Melnick, D., and Pearl, M., "Not Out of the Woods Just Yet," *The New York Times*, April 20, 2006, accessed from http://www.nytimes.com/2006/04/20/opinion/20melnick.html on March 8, 2012.

Miller, C. (ed.), *American Forests: Nature, Culture, and Politics*, Lawrence, KS, University Press of Kansas, 1997.

Morton, O., *Eating the Sun: How Plants Power the Planet*, New York, HarperCollins, 2008.

Poffenberger, M., *Communities and Forest Management in Southeast Asia*, Gland, IUCN, 2000.

Pyne, S., *Fire: A Brief History*, Seattle, University of Washington Press, 2001.

Pyne, S., *World Fire: The Culture of Fire on Earth*, New York, Holt, 1997.

Rangarajan, M., "Environmental Histories of South Asia: A Review Essay," *Environment and History* 2, 1996, pp. 129–43.

Schama, S., *Landscape and Memory*, New York, Knopf, 1995.

Scott, J. C., *Seeing like a State: How Certain Schemes to Improve the Human Condition Have Failed*, New Haven, CT, Yale University Press, 1998.

Sutter, P., "Reflections: What Can U.S. Environmental Historians Learn from non-U.S. Environmental Historiography?," *Environmental History* 8, 2003 pp. 109–29.

Vandergeest, P., and Peluso, N. L., "Empires of Forestry: Professional Forestry and State Power in Southeast Asia, Part 1," *Environment and History* 12, 2006, pp. 31–64.

Williams, M., *Americans and their Forests: A Historical Geography*, Cambridge, Cambridge University Press, 1989.

Williams, M., *Deforesting the Earth: From Prehistory to Global Crisis*, Chicago, University of Chicago Press, 2003.

Further Reading

Guha, R., "The Prehistory of Community Forestry in India," *Environmental History* 6/2, 2001, pp. 213–38.

Guha, R., *The Unquiet Woods: Ecological Change and Peasant Resistance in the Himalaya*, Berkeley, University of California Press, 2000.

Langston, N., "American Forest History," in K. Brosnan (ed.), *Encyclopedia of American Environmental History*, vol. 2, New York, Facts on File, 2010, pp. 570–5.Shiva, V., *Ecology and the Politics of Survival: Conflicts over Natural Resources in India*, New Dehli and New York, Sage, 1991.

Tucker, R., *Insatiable Appetite: The United States and the Ecological Degradation of the Tropical World*, Berkeley, University of California Press, 2000.

CHAPTER SIXTEEN

Fishing and Whaling

MICAH S. MUSCOLINO

Since the prehistoric era, humans have fished and hunted in the world's lakes, rivers, and inshore waters. From the time they first turned to fishing and whaling for their subsistence, people have altered the structure and function of aquatic and marine ecosystems. The magnitude of human impacts and rates of environmental change have accelerated in tandem with population growth and economic integration. The geographical scale of impacts caused by fishing and whaling has thus tended to expand over time.

The earliest subsistence-based fishing targeted inland and near-shore ecosystems with relatively simple boats and extractive technologies. An extensive trade in maritime products flourished in ancient times and greatly expanded in later periods. After the increase in commercial exploitation of coastal and marine fisheries after 1000 CE, marine-resource frontiers were integrated into developing market economies that fed growing human populations. Since the end of the nineteenth century, modern industrial fisheries made possible intensive and geographically pervasive exploitation of marine ecosystems to supply consumers all over the world.

In medieval and early modern times, fleets turned to new species after exhausting established stocks, developed new technologies that enabled them to exploit previously inaccessible environments, and searched for new resources to maintain their harvests. Even before the dawn of modern industrial fishing and whaling, the earth's marine environments were far from pristine. Yet the transformations that emerged in the twentieth century due to intensified extraction and globalized demand far exceeded everything that came before. By the end of the twentieth century, fishing fleets moved into the last marine-resource frontiers and global catches declined. Despite their vastness, it is clear that the world's oceans are not as inexhaustible as they once appeared.[1] Human predation has severely depleted the sea's bounty, encouraging a recent shift toward aquaculture production. Humans have restructured the seas in profound and increasingly unpredictable ways. In coming years, protecting and rebuilding depleted marine life will

A Companion to Global Environmental History, First Edition. Edited by J.R. McNeill and Erin Stewart Mauldin.
© 2012 John Wiley & Sons, Ltd. Published 2015 by John Wiley & Sons, Ltd.

be of vital importance for ensuring healthy functioning of ecosystems and feeding the 1.2 billion people around the world who depend on seafood as their main source of protein.

Prehistoric and Ancient Fisheries

In all likelihood, our hominid ancestors exploited aquatic and marine ecosystems to some extent for millions of years. But archeological and anthropological data suggest that intensity, diversity, and technological sophistication of aquatic resource use increased significantly with the appearance of anatomically "modern" humans about 150,000 years ago. Since the Pleistocene, humans have exploited aquatic life through gathering, fishing, and hunting. Archeological sites in coastal southern Africa contain evidence of exploitation of shellfish, marine mammals, and fish dating to between 125,000 and 40,000 years ago. Barbed bone spears unearthed in Congo indicate that freshwater-fishing technologies existed 90,000–80,000 years ago.

Fishing and sea-hunting appear to have played a central role in the geographical spread of humans throughout the globe. The Pleistocene peopling of Australia and western Melanesia due to maritime migrations through Island Southeast Asia approximately 60,000–35,000 years ago owed much to the use of marine resources like fish and shellfish. Evidence of early boat use in Japan 35,000–25,000 years ago points to existence of refined methods for utilizing marine resources in the Northwest Pacific. Colonization of California's Channel Island 13,000–12,000 years ago and coastal shell middens on the Andean Coast dating to over 11,000 years ago are further proof of aquatic- and marine-resource exploitation. By 8,000 years ago Brazil's coastal residents gathered shellfish in tidal estuaries, leaving behind massive shell middens known as *sambaquis*. The earliest evidence of whaling comes from a Neolithic site in South Korea from 6000–1000 BCE. Late Pleistocene expansion of maritime peoples into recently deglaciated coastal regions of Scandinavia and early Holocene settlement of some Caribbean islands also suggest diversified coastal economies based on aquatic resources.

Following the end of the last ice age, during the period 6000–2000 BCE, global sea-surface temperatures were approximately 2 degrees Celsius (3.6 degrees Fahrenheit) warmer than present levels. As a result, sustained fishing activities became possible throughout the year in many regions of the world. Archeological evidence from kitchen middens (piles of day-to-day waste) indicates that early coastal populations in northern Europe subsisted primarily on fish, shellfish, seaweed, and other marine life. Human settlement concentrated along the world's coastlines, where abundant seafood gave marine communities a relative advantage over communities that depended solely on hunting and gathering. Fishing and collecting shellfish undergirded Southeast Asian settlements during the sixth millennium BCE, as well as the Jōmon culture in Japan around 10,000 BCE. Studies of fish bones from middens show that by 2000 BCE Japanese fishermen had already sailed far into the Pacific. The Polynesian migrations beginning around 3,500 years ago and Māori colonization of New Zealand around 1,000 years ago were made possible by exploitation of marine life.[2]

In some places and times, early fishing activities had a significant impact upon aquatic, estuarine, and freshwater species and ecosystems. Archeological evidence from coastal sites on several continents strongly suggests that human exploitation could alter the size and abundance of aquatic species, including shellfish, fish, and some marine mammals.[3] Changes in harvesting that came with intensification of human settlement along the South African coast during the middle and late stone age led to decreases in the average size of shellfish like black mussels and limpets.[4] Hunting by Aleutian peoples of the Northeast Pacific diminished sea-otter populations several thousand years ago. As sea-otter numbers declined, the size of sea urchins they preyed upon increased, subjecting kelp forests that the urchins fed upon to depletion.[5] When humans arrived in the Solomon Islands around 900 BCE, mollusks, fish, and turtles appear to have suffered sharp declines. On the other hand, taboos and other social restrictions on resource use probably moderated pressure that Pacific Islanders placed on many lagoons, reefs, and seas.[6] Upper levels of *sambaquis* in Brazil contain immature specimens and a greater species variety than lower ones, suggesting that people abandoned coastal sites and relocated as intensive exploitation made easily harvested or tasty mollusks grow scarce.[7] Shell middens from the Antilles show that with the human settlement of the Caribbean from 400 BCE to 700 CE crab populations declined and the size of fish decreased over time.[8] Yet in these early stages human impacts mainly affected near-shore ecosystems and were constrained by technological, economic, and cultural limits.

In ancient Near Eastern and Mediterranean cultures, fish was of greater importance than meat as a source of protein in human diets. In the sixth and seventh centuries BCE, population growth in the Aegean region led to the establishment of Greek colonies on the Black Sea coast. To finance imports of wine and oil from the Aegean, Black Sea cities began sea-fishing on a commercial scale to supply Athens and other Greek cities. Around 50 BCE, fixed nets and traps for tuna and other migrating fish came into use, involving 60–70 fishers working together. For the next century, tuna-fishing became one of the mainstays of the Black Sea economy. Greek merchants conducted a trade in fish that stretched from the Black Sea and the rivers of Russia to Greek and later Roman markets.[9] Roman demand for marine products also supported large commercial fisheries. The ruins of Roman fish-processing enterprises, which distributed marine products throughout the Mediterranean and beyond, still dot the southern Iberian Peninsula.[10]

In addition to capturing fish, ancient peoples manipulated aquatic organisms and ecosystems through aquaculture. Archeological excavations of Roman villas have revealed ponds stocked with fish for consumption by elites who served elaborate and expensive seafood banquets. The beginnings of fish-farming in East Asia long predate its beginnings in ancient Europe. China developed fishponds for raising herbivorous common carp at least 2,300 years ago. Fish-farming reached high levels of refinement under the Tang (619–908 CE) and Song (960–1270 CE) dynasties. In rural Southeast China, aquaculture was connected to commercial silk production. The "mulberry tree and fishpond" system, which many view as an ecologically sustainable agricultural system, used mud from carp ponds to fertilize mulberry trees. Mulberry leaves fed silkworms, while fish fed on organic matter from trees and silkworm droppings.[11]

Medieval and Early Modern Fisheries

In medieval Europe, expanding human population, production, and exchange acted to intensify human impacts upon fisheries. During the period from 800 to 1500 CE, demographic growth, urbanization, and markets transformed Europe's aquatic and marine ecosystems. Since medieval Christian prohibitions forbade consumption of mammals and birds up to 180 days each year, Europeans had a healthy appetite for fish and seafood. In the early Middle Ages, people exploited inland and estuarine fisheries, largely ignoring coastal fish stocks. Prior to the eleventh century, Europeans mostly consumed freshwater fishes like perch, pike, and trout obtained in local waters, as well as anadromous species like salmon and sturgeon that migrate between freshwaters and the sea at different stages in their lifecycles.

During the high and later Middle Ages, overfishing and degradation of Europe's waterways due to damming and siltation put pressure on fish populations in some lakes, streams, and rivers. Documentary and archeological evidence shows that the fish species that people favored as food were coming under stress. During the thirteenth and fourteenth centuries, sturgeon and salmon markedly declined. By contrast, species that flourished in aquatic environments transformed by human action grew in importance. Eel thrived in turbid, shallow freshwater lakes created by diversion of rivers and other human alterations of Europe's natural landscape. At the same time, people farmed and transported carp throughout Europe for elite consumption.

As freshwater ecosystems deteriorated and demand outstripped local supply, marine fisheries expanded. Early medieval coastal dwellers from Sussex to Sweden had fished for herring along local shorelines for local consumption. But large-scale sea-fishing – especially for cod and herring – did not get started in earnest until after 1000 CE, with coastal residents supplying saltwater fish to inland consumers. After this time, a dramatic shift took place in Europe from eating freshwater species to saltwater fish. Osteological analysis of fish bones has revealed early medieval sites dominated by freshwater and migratory species like eel and salmon, while later settlements indicate widespread consumption of herring, cod, and other marine species. During the eleventh century, a bustling trade in cod – known as stockfish – existed in Europe's urban areas. Commerce in fish and the rise of towns went hand in hand, with fish markets often forming the earliest focal point of medieval towns. The fish trade initially supplied towns, later on spreading to inland rural areas. Substantial commercial fisheries emerged over the eleventh century, and in the twelfth and thirteenth centuries Europeans came to eat larger shares of preserved marine fish marketed over greater distances.[12]

By the thirteenth century, a massive trade in herring put heightened pressures on Europe's marine fishing grounds, aided by improved sea-going vessels and preservation techniques. Along with climatic fluctuations, in the late thirteenth and early fourteenth century this intensified exploitation destroyed herring fisheries in the western Baltic and southern North Sea. Taking advantage of improved curing techniques, the Dutch came to dominate northern European herring fisheries during the sixteenth century, harvesting rich shoals off Scotland. In the 1680s, the herring fishery employed half a million people – an astounding one-fifth of the Dutch population. By the time dominance passed to British fishers in the eighteenth century, early modern harvesting activities had removed huge quantities of biomass from the sea. Dutch, English, Scottish, and Norwegian fleets netted over 100,000 tons of herring as of 1700 and about 200,000 tons in the late 1700s. Catches totaled 300,000 tons in 1870, the equivalent of the total allowable catch for North Sea herring in 2007. By the early nineteenth century, herring stocks had fallen to 50–60 percent of the levels of the seventeenth century.[13]

Even in early modern times, fisheries subject to commercial exploitation followed a familiar pattern. Faced with environmental changes largely of their own making, fishers extended their spatial scope and increased fishing effort. European fleets searched for fish farther afield, venturing to the prolific New World fishing grounds after John Cabot discovered abundant cod stocks near Newfoundland in the 1490s.

In addition to human action, climatic changes may have stimulated geographic expansion. Seawater temperatures below a certain level may inhibit the ability of cod populations to reproduce. With the onset of cooler temperatures during the Little Ice Age in the 1400s, formerly productive cod fishing grounds off of Norway failed. European fishers responded by sailing west (rather than north) to fish for cod off Newfoundland, Nova Scotia, southern Labrador, and eventually Maine. Invention of new fishing boats and equipment in the 1500s made it possible for European fishers to move into these distant waters. New World fisheries soon emerged as a thriving industry that utilized European capital and entrepreneurial energy to tap into resources hardly touched by North America's indigenous inhabitants.[14]

Boats from England, France, and Spain crossed the Atlantic annually to fish Newfoundland's Grand Banks. Intricate distribution networks and far-flung markets transformed fish caught in New World waters into commodities. Each year during the seventeenth century, fleets caught on average 200,000 tons of cod to sell on European markets. From 1625 to 1725, when boats based in New England joined the British and French in fishing these stocks, catch levels rose even higher. Fishing pressure decreased the average size of cod that were caught, but total output continued to grow. By the 1700s, cod catches had doubled to an average of nearly 400,000 tons per year. In addition to providing an accessible and affordable source of protein for expanding European markets, codfish fed colonial settlers in North America and African slave laborers on Caribbean sugar plantations. Newfoundland's inshore stocks came under full exploitation during the mid-eighteenth century, and expansion was made possible by moving into waters off Labrador. As their prey diminished in the 1850s, New England schooners undertook longer voyages to the Gulf of Saint Lawrence and the Grand Banks, where stocks were perceived to be larger.[15]

The history of whaling in medieval and early modern Europe followed a similar path. From the eleventh century, or possibly earlier, Basques hunted whales in the Bay of Biscay for their meat, oil, and bone. Contemporaries perceived falling whale populations in these grounds after the fourteenth century, and whalers ventured into more distant waters to pursue abundant stocks. During the 1500s, European whalers set sail in enlarged ships for previously uncharted western arctic and subarctic waters to find their prey. Between 1530 and 1620, Basque whalers killed tens of thousands of right whales and bowheads in straits near Newfoundland and Labrador. From 1660 to 1701 Dutch and Basques killed 35,000–40,000 whales in the western Arctic, depleting populations considerably and altering whale-migration patterns. Intensified shore-based whaling in northern Europe and off the coast of North America added to the catch. Shore-whaling got under way in Massachusetts in the 1650s and New England's peak shore-whaling years occurred from 1690 to 1725. Colonists killed roughly 3,000 right whales between 1696 and 1734, in addition to numerous pilot whales and other species. By the 1740s hunting had wiped out near-shore whaling grounds and it was no longer a viable trade. In the 1760s, Cape Cod vessels had to go offshore to Labrador and Newfoundland to find whales. Not until the early nineteenth century, after depleting whale populations in the Atlantic, did American whalers bring the hunt to the Pacific and Indian Ocean.[16]

The impact of early modern European commercialized fishing and whaling reached far beyond local ecosystems. Europeans appropriated huge amounts of biomass from the seas of the New World, making the northwest Atlantic an extension of Europe's diminished marine environments. Yet these patterns were not unique to early modern Europe. In East Asia, commercialized societies with growing populations experienced analogous ecological trends in their fisheries.

Marco Polo, who traveled to China and visited the Song dynasty capital of Hangzhou in the 1270s–1290s, described the scale of urban markets for freshwater and marine fish. Merchants transported large quantities of saltwater fish to the capital from coastal areas, while fishermen landed abundant catches in freshwater lakes. Marveling at the scale of the fish trade in Southeast China, Marco Polo noted that

> Anyone who should see the supply of fish in the market would suppose it impossible that such a quantity could ever be sold; and yet in a few hours the whole shall be cleared away; so great is the number of inhabitants who are accustomed to delicate living. Indeed they eat fish and flesh at the same meal.[17]

In China during the Ming (1368–1644) and Qing (1644–1911) dynasties, fishing and aquaculture entered a new phase of development in tandem with deepening commercialization and population growth. In the eighteenth and nineteenth centuries especially, population pressure and the pull of commercial profits led many of China's coastal residents to "use the sea as their fields," stimulating growth in marine fisheries and cultivation of clams, oysters, and seaweed. Freshwater fisheries also flourished, but by the nineteenth century widespread land reclamation impinged upon lakes and rivers, causing serious declines in many areas. Fishers responded by turning to marine fishing grounds in greater numbers. Marketing systems integrated coastal fisheries with urban consumption centers, giving small-scale producers the ability to reap greater profits from fish stocks, while also placing heightened pressure on marine resources.[18] Commercial pressures reached far beyond China's borders, stimulating rapacious extraction of sea cucumbers, shark fins, and other marine products in Southeast Asia and the Pacific Islands to sell as commodities on Chinese markets.[19]

To the north of Japan during the nineteenth century, the expansion of herring fisheries in the island frontier of Hokkaido destroyed near-shore stocks, making it necessary for fishers to move into deeper waters to boost their catches. Following the depletion of herring populations close to their home villages, small-scale fishing households had no choice but to engage in wage labor for large-scale fishing enterprises.[20]

Throughout the medieval and early modern world, commercialized societies with growing populations thus placed greater demands on aquatic and marine ecosystems. Preindustrial fisheries had the capacity to extirpate certain types of marine life. Yet the magnitude of impact paled in comparison with what resulted from modern industrial fishing and whaling.

Modern Fisheries

A major departure in the history of the fisheries came in the mid-nineteenth century with the harnessing of fossil fuels to power transport and exploitation of marine life. Before the 1800s, fresh fish often reached consumers in decayed or damaged condition. Most production was localized, with only a few cities enjoying fresh fish supplied from inshore, riverine, and estuarine fishing grounds. The fresh-fish industry benefited from the growth of coal-fueled railways and steamship navigation in Britain in the 1840s, which made it possible to transport fresh catches to inland marketing centers. During the 1860s, Britain's railways carried 100,000 tons of fish. Markets for fresh fish expanded and prices increased, stimulating expansion of trawling fisheries. Trawling involves dragging nets across the seabed, catching fish indiscriminately and inadvertently destroying habitats on the sea bottom, making it a particularly damaging fishing technique. In the 1840s, England had a trawling fleet of 130; by 1860, the fleet grew to 800. The trawling boom after the 1840s intensified pressures on North Sea fishing grounds and made it necessary for fleets to search out new ones.[21]

Even more significantly, the world's first steam-powered trawler, the *Pioneer*, was registered in Great Britain in 1878. Between 1881 and 1902, the British built 1,573 steam trawlers. It did not take long for steam-powered vessels to deplete Europe's already strained fishing grounds. Some coastal locales imposed prohibitions on trawling due to its impact on herring, cod, and haddock fisheries. Because Britain's steam-trawler fleet focused exclusively on flatfish and other bottom fish, it soon had to seek out new stocks. By the 1890s, fisheries in the North Sea showed signs of overexploitation. Trawlers gradually expanded through the northeastern Atlantic, exploring fishing grounds off the Faeroes, Iceland, and in the Barents Sea.

During the late nineteenth and early twentieth centuries, European countries followed Britain in constructing trawl-fishing fleets. Mechanized high-seas fisheries symbolized modernity and economic development, as well as assertion of national geopolitical power. Germany, for example, subsidized construction of a steam-trawling fleet to help feed its growing population and foster a maritime labor force for its emerging navy. First introduced across the Atlantic prior to World War I, by the 1920s steam-trawling was also well established in North America. Benefiting from new fast-freezing technology, American trawlers in the Georges Banks harvested huge amounts of haddock, which were filleted and transported to inland markets. The haddock fishery boomed in the 1920s, with landings peaking in 1929 at 120,000 tons before crashing to 28,000 tons in 1934 and settling around 50,000 tons from the 1930s to 1960.[22]

The biggest growth in mechanized fishing occurred in Japan. At the end of the nineteenth century, Japan launched a successful modernization program to break into the ranks of the world's imperial powers. Strengthening the nation's sea power by securing a greater share of the world's marine resources motivated the growth of Japan's fisheries. The rise of Japan's modern fishing fleet went hand in hand with its industrialization at home and imperial expansion abroad. In the eyes of Japanese officials and fishery specialists, national survival required taking control of pelagic fishing grounds and using modern technologies to exploit them as efficiently as possible. By 1910 Japanese landings equaled Britain's; in 1938 Japan took over as the world's foremost fishing nation with landings of 3 million tons (three times those of Britain). By the 1930s, Japan had an extensive deep-sea fleet, with crab-canning factory ships in the Bering Sea, tuna fleets in the central and south Pacific, and trawlers fishing the Yellow, East China, and South China seas. Crabbers in Hokkaido suffered declining harvests in the 1930s, and Japan's tuna fleet had to perpetually expand its range of operations as overfishing depleted previously productive fishing grounds.[23]

After exhausting highly valued fish species such as sea bream in the Yellow and East China seas in the 1920s, Japan's trawling fleet set its sights on several varieties of yellow croaker in waters off China's southeast coast, leading to international disputes with Chinese fishing boats that had ventured offshore to pursue productive fishing grounds. Chinese as well as Japanese participants in these disputes understood control of the sea and its resources as a vital means of fortifying the wealth and power of the nation-state. Chinese and Japanese fleets aggressively pursued yellow croaker to avoid giving the fish up to their foreign competitors, but superior technology and diplomatic privilege gave Japanese trawlers a clear upper hand. By the late-1930s, international pressures led to the decline of yellow croaker stocks.[24]

As in the fishery sector, the Industrial Revolution greatly intensified whale hunts. Throughout the nineteenth century, these marine mammals yielded a wealth of products for human use. Sperm-whale oil functioned as a high-quality lubricant for industrial machinery. Oil from various types of whales found wide application as a source of fuel until supplanted in the twentieth century by the rise of the petroleum industry. Manufacturers used baleen or whalebone – a hornlike substance that forms plates in the upper jaw of whales that strain plankton from seawater for food – to make corsets, umbrellas, and other goods. Whalers from the eastern US rounded the cape to hunt whales in the Pacific during the 1820s, and in a matter of decades they exhausted sperm and right whales. By 1890, Americans also hunted out the last bowhead whales in the Bering Sea. In the late nineteenth century, half the world's whaling grounds were abandoned because whales had been hunted to the point of commercial extinction.

Abundant whale populations still remained in the southern oceans of the Antarctic, which supported hundreds of thousands of large baleen whales (rorquals) like blue and fin whales. Because of the size and speed of rorquals, whalers could not catch them in small boats armed with harpoons. Norwegian whalers overcame this challenge in the 1860s by inventing the harpoon cannon, which fired explosive grenades into their prey. Mounted on steam-powered catcher boats that traveled at over 15 knots, the harpoon cannon opened the hunt for rorquals. At first, the Antarctic whale hunt focused on easily caught humpbacks, which declined after 1911. From then on, rorquals were the main targets. Whaling grew even more efficient after 1925, when another Norwegian invented the whaling factory ship. These floating slaughterhouses featured a stern slipway for dragging whales on board and equipment for flensing (stripping blubber) and processing them into oil. Factory ships could capture a 100-ton blue whale and render it into oil and bonemeal in one hour. In addition, Norwegian whalers took to injecting compressed air into dead rorquals so they would float until hauled aboard. These innovations started a new era of profitable whaling. Serial depletion of whale species proceeded apace. British whalers rivaled Norwegians from the 1920s, and in the 1930s Argentine, American, Danish, German, and Japanese whalers joined the hunt. Profits still came primarily from whale oil, which was processed into margarine, soap, and nitroglycerine for explosives. Blue whales, the largest animals on earth, yielded the greatest profits. When blue whales grew scarce between 1913 and 1938, whalers switched to fin whales and other smaller, less valuable species.[25]

During the 1930s, human demands placed a heavy strain upon marine resources in the North Atlantic and North Pacific. Catch per unit effort (CPUE) in the world's fisheries declined steadily between 1906 and 1937, despite a brief boom after World War I. Fleets were working harder and employing improved technologies to catch the same number of fish. World War II reduced deep-sea fishing and whaling activities even more than World War I, giving many previously depleted stocks the chance to undergo a recovery. Yet the wartime hiatus proved short-lived. Larger and more sophisticated fleets resumed harvesting traditional fishing and whaling grounds during the 1950s, while also opening up new ones. The Japanese fleet, left in shambles at the end of World War II, was revived with government assistance and by the 1950s and 1960s it again dominated the whale fisheries of the Pacific.[26]

In contrast to the prewar slaughter, the post-World War II era was marked by gradual regulation of whaling on the high seas. Concerns about dwindling whale stocks had existed almost from the beginnings of modern whaling, and they grew stronger in the twentieth century. The prewar decline of blue-whale stocks had led to regulations administered by the League of Nations in 1935, but they had little effect. The International Whaling Commission (IWC), founded in 1946 to protect whale-oil prices by allotting quotas among major whaling nations, looked after the whaling industry rather than the whales. Catches grew until 1964, when whalers took 66,000 whales. That year, the IWC shifted its focus to conservation via protections on humpbacks and blue whales. Despite this move toward conservation, between 1945 and 1970 at least half a million whales were killed in the North Pacific and the Bering Sea.

During the 1960s, conservation movements and dwindling whale populations drove most whaling fleets out of business, although several nations went on hunting whales in defiance of international regulations. The Soviet Union's fleet killed 90,000 whales from 1949 to 1980, including protected humpback and blue whales. Japan's whaling fleet still operated out of bases in Canada, South America, and Newfoundland, as well as in Japanese waters. In 1982 the IWC placed a moratorium on all commercial whaling. But whalers from Japan, Norway, and Iceland circumvented these restrictions. Japanese whalers catch thousands of minke whales for scientific purposes every year, with much of their meat ending up on markets in Japan.

The whaling industry contracted after the late 1980s, but not before decades of commercial exploitation depleted whale populations species by species. Hunting pushed blue, fin, and humpback whales to the edge of extinction, with sei and Bryde's whales dwindling shortly thereafter. Fin whales declined from a pre-whaling population of as many as 360 million to a present-day total of 56,000. Humpback-whale numbers have fallen from 240 million to a mere 9,000–12,000. Only minke whales exist in numbers anywhere comparable to those that existed prior to the advent of large-scale commercial hunting.

Nevertheless, the moratorium against commercial whaling has given populations a chance to recover. Since the 1990s, populations of most whale species have been on the upswing. Rather than a commodity for human use, whales have come to symbolize environmental protection. In the eyes of many people, whales are animals that deserve protection under all circumstances. Whale-watching is growing in popularity as a form of recreation and ecotourism, most notably in whaling countries like Japan and Norway. However, environmentalists struggle to reconcile whales' status as conservation icons with the aspirations of Native American groups, who see whale-hunting as part of their cultural heritage. In the 1970s and 1980s the Inuit of Alaska won a legal battle with the IWC, environmentalists, and the US government to gain the right to continue hunting bowhead whales. In the 1990s, the Makah tribe of Washington won a similar victory in their effort to catch one grey whale per year. Since Japanese and Norwegian whaling interests make many of the same claims regarding whaling's cultural significance, victories won by indigenous peoples have cast some doubt upon the conservationist argument that Japan and Norway should cease all whale-hunting activities. These difficulties aside, whales have avoided extinction in no small part because of their unusual appeal to conservationists.[27]

If whale populations generally fared better after the 1950s, the same cannot be said of fish and other marine life. The decades after World War II were characterized by worldwide fishing operations and ruthless exploitation of marine resources. In 1950, fishing concentrated on waters in the northern hemisphere off Japan and industrialized countries in Europe and North America. Increased fishery output in the 1950s and

1960s derived largely from migration of fishing fleets into the South Atlantic and South Pacific. Vessels were equipped with diesel engines and refrigeration, which enabled longer hauls and bigger catches. The fishing industry also benefited from technological advances made during World War II like radar, sonar, and nylon for nets. Beginning in the 1960s, the Soviet Union and other Eastern European countries pioneered deepwater trawling on the open oceans with huge factory ships that processed catches. Technological innovations facilitated expansion of fishing grounds and greatly increased the impact of a given amount of fishing effort. Not surprisingly, the result was widespread overfishing.[28]

Fish live underwater and are always on the move, which makes them notoriously difficult to count. Landings can be measured, but they fluctuate unpredictably due to climatic changes and oceanographic events that occur largely independent of human action. Hence, it is extremely hard to determine if changes in abundance result from naturally occurring trends or from fishing activities. Despite these complications, there is no doubt that fishery expansion after 1950 had an unprecedented impact on the marine environment. Global catches increased quickly from the late 1940s to 1973, followed by slower growth thereafter. By the 1980s, fishing fleets caught as much in two years as their ancestors did in the entire nineteenth century. The oceans yielded more fish in the twentieth century than they had in all previous centuries combined. But high catch levels were only maintained through use of larger fleets with greater fishing power.[29]

Increased overall output in the latter half of the twentieth century concealed the collapse of many of the world's most valuable fisheries. Traditional fishing grounds disappeared one after another and fleets switched to previously untargeted species. Localized fishery collapses had a long history. But after World War II collapses grew larger and more frequent. With the application of modern fishing methods, exploitative pressure combined with environmental fluctuations to bring about crashes. The Japanese pilchard fishery collapsed in 1946–9, recovered during the 1970s, and crashed again in 1994. Falling temperatures off Greenland after 1960 and in 1982–4 meant fewer cod at a time of intense fishing. Cod catches collapsed and the industry shifted to shrimp and halibut. Natural fluctuations and overfishing also combined to bring about the decline of the California sardine fishery after 1945, and its total collapse after 1968. These sardine and cod populations have yet to make a comeback.[30]

As prime fish stocks declined, the world's fishing fleets moved on to others, and previously undesirable species grew in value. During the 1980s and 1990s, catches came to include larger shares of previously undesirable "trash fish," harvested because previously desirable species like cod, herring, and tuna had grown scarce. Fishers first targeted large, high-value, typically predatory species. After these resources declined, fleets either moved to other fishing grounds or switched to harvesting other types of fish. Increased fishing has thus resulted in a tendency for the industry to "fish down the food web," shifting to lower trophic levels as higher ones become depleted.[31]

Even the world's most productive fishery fell victim to this pattern of boom and bust. In the mid-1950s, fishermen and boats from the failing California sardine fishery moved south to Peru to fish abundant anchoveta and Chilean jack mackerel stocks in the cold, nutrient-rich Humboldt Current. Agricultural demand for cheap protein to feed pigs, chickens, and other livestock drove the large-scale exploitation of these low-value fish. Production went from 59,000 tons in 1955 to 3 million tons in 1960. By 1962 Peru landed more fish than any country. Peru's landings peaked from 1967 to 1971 with 10–12 million tons, or 20 percent of the world total. Exports of anchoveta meal and oil earned one-third of Peru's foreign exchange. The bust came after 1972, when production fell to 4.7 million tons and stayed around 2–4 million tons for the next 15 years. The fishery's failure coincided with the 1972 El Niño, a recurrent fluctuation in Pacific currents that brings warm, nutrient-poor water to Peru's coast. Production hit rock bottom during the even stronger El Niño of 1982–3. The fishery enjoyed some good years in the early 1990s, but collapsed again in 1997 with another El Niño episode.[32]

After the 1970s, a general crisis existed in virtually all the world's fisheries. The growth of fishery output since 1945 derived from opening up new fishing grounds and intensifying fishing effort in established ones. Most of these new fisheries were in the southern hemisphere. The pilchard fishery in the southeastern Pacific, which thrived in the 1970s and 1980s, was the last. Thereafter, none of the world's fish stocks were left unexploited.

Maintaining fishery output required ever more intensive effort, made possible by government subsidies and improved technologies. Fishing fleets took advantage of satellite imagery to find fish, using bigger trawlers and drift nets to chase after them. As fleets grew more technologically sophisticated, they also got more expensive. Fixed capital sunk in fishing fleets could not stand idle, so they had to fish throughout the year, depleting stocks to the point that the sale of catches could not cover operating costs.[33] Under these circumstances, only hefty state subsidies sufficed to keep fleets in business. Fishery subsidies worldwide, approximately $50 billion a year by 1995, doubled global fishing-fleet tonnage between 1970 and 1995. In 1989, total world revenues from fishing amounted to $70 billion, but total operating costs were $92 billion. Including capital costs, the deficit was $54 billion. State subsidies made up the deficit, distorting fleet size and encouraging overcapacity. With too many boats chasing too few fish, the world's fishing fleet is currently large enough to catch its total annual output four times over.[34]

Intensifying fishing effort yielded higher overall catches, but masked a serious deterioration of commercially harvested fish stocks. In the past, after one fishery declined new ones were found. Especially from 1950 to 1971, this process facilitated huge increases in output. After the 1970s, however, there were no more marine-resource frontiers to exploit. By 2000, more than three-quarters of the world's fisheries were fully exploited or overfished. Every year, fishermen lost 9 million tons to overfishing. The ocean has lost more than 90 percent of large predatory fishes, such as marlin, sharks, and rays. Bottom trawls have destroyed complex seabed ecosystems, leaving nothing but plains of mud in their wake.[35]

The growth of distant-water fishing after World War II inspired efforts to limit access to productive fishing grounds. As early as 1945 president Harry S. Truman declared that the US would protect coastal-fishery resources, not least against competition from Japan's rebuilt fleet. Other nations rushed to make similar declarations. Iceland, in particular, asserted territorial claims over its offshore fisheries to keep out rival European distant-water fleets. Despite fierce opposition from the UK during the "Cod Wars" of the 1950s and 1970s, by 1976 Iceland had expanded its exclusive fishing zone to 200 nautical miles. In the late 1970s, almost all the world's coastal fishing nations claimed 200-mile territorial waters, known as exclusive economic zones (EEZs). These zones enclosed highly productive waters over the earth's continental shelves – which make up only 8 percent of the sea's area but account for over 70 percent of fishery output – within the sovereignty territory of nation-states.[36]

The assertion of claims to EEZs deprived distant-water fleets of access to fishing grounds in the coastal waters of foreign countries. Many traditional fishing powers like

Britain, Germany, and Japan had to scale back their distant-water fleets or shift to fishing entirely on the high seas. At the same time, EEZs gave coastal countries in South America, East Asia, and Southeast Asia the chance to assume leading positions in the world's fishing industry. North American fisheries grew as well, benefiting from the exclusion of European and Japanese fleets and growing domestic demand.[37]

Yet overfishing went on unabated within most EEZs, as governments made heavy investments in fleets to exploit their own territorial waters. Canada declared its EEZ in 1977, taking control of the cod fishing grounds of the Grand Banks, which European fishers had first harvested in the sixteenth century. To exploit the resource more effectively, Canada subsidized an expansion and technological upgrade of its fishing fleet. Output boomed for a time, but the cod of the Grand Banks were soon fished out of existence. Cod catches, nearly a million tons per year in the 1960s, collapsed after 1990. Canada's government declared a moratorium on cod fishing in 1992, which was extended indefinitely. The unpopular decision left as many as 40,000 Newfoundlanders unemployed. Even after nearly two decades without fishing, the cod have yet to recover and the species is now part of Canada's endangered species list. Compared to the 2 billion breeding individuals that existed in the 1960s, Atlantic cod populations have fallen by almost 90 percent.[38]

Even as the world's fishery resources dwindle, demand for marine products has continued to grow. In recent decades, more and more fish has been sold on global markets thanks to rapid and flexible transport. Airplanes carry flash-frozen bluefin tuna caught in the North Atlantic and Mediterranean to supply markets in Japan, which then ship the fish to chefs in New York and Hong Kong. Airfreight has increased sales of desirable species like lobster, shrimp, and salmon to consumers all over the world. In fish markets in Tokyo, New York, and other metropolitan centers, octopus from Senegal, shrimp from China, crab from Russia, salmon from Canada, and abalone from California are sold alongside one another.[39]

Vertical integration has rapidly taken place in the fishing industry, leaving only a few companies large enough to cater to global markets. Even as EEZs consolidated national sovereignty over fishing grounds, multinational corporations secured access agreements to waters off developing countries to harvest fish for European, North American, and Japanese markets. Usually these fishery agreements limit numbers of boats, but not the size of catches. Massive super-trawlers financed by multinational corporations severely deplete coastal fishing grounds off West Africa, damaging the livelihoods of African fishing communities.[40]

Unreliable data make precise determination of current global catches difficult. Regional officials in China have systematically overreported fishery output, inflating data on international catch rates. The unintentionally caught "bycatch" – estimated at 18–40 million tons each year – is simply tossed overboard and goes unreported. After controlling for distortions, fishery researchers have discerned a general trajectory. Catches increased almost fivefold in the late twentieth century, growing from 19 million tons in 1950 to 87 million tons in 2005. Global catches grew rapidly throughout the 1960s and 1970s, although the rate of increase slowly declined. After the late 1980s, global catches no longer increased, peaking in 1988 at around 90 millions tons. A slow decrease of about half a million tons per year ensued and it shows no indication of reversing. As globalization stimulates demand, overall fishery production has stagnated and declined.[41]

Decreasing catches set in after the 1980s because the rate at which fleets opened up new fish grounds no longer compensated for depletion of traditional stocks.

Exploitation of new pools of resources facilitated growth in marine fisheries during the last half of the twentieth century. Newly exploited areas have declined since the 1990s, which more than anything accounts for the decline in global catches. At millennium's end, low-productivity areas on the high seas and inaccessible Arctic and Antarctic waters were the only remaining resource frontiers. If exploitation persists at its current rate, researchers predict that all commercially exploited fish and seafood species will collapse by 2048. For marine-capture fisheries, the boom years have come to an end.[42]

As wild fish populations dwindle, species farmed through aquaculture have flourished. World aquaculture production grew from 5 million tons in 1980 to 25 million tons in 1996. In 1950 aquaculture made up 1 percent of global fishery output, but by the end of the 1990s it exceeded 25 percent. The most rapid growth took place in Asia, where aquaculture enterprises farm commercially valuable fish to satisfy global demand and supply domestic markets. In 1994 China accounted for over 60 percent of the world's aquaculture production, with India, Japan, South Korea, and the Philippines providing another 20 percent. Thanks in large part to its aquaculture output, China is now the world's largest fishery producer. Inland freshwater ponds stocked with carp make up about two-thirds of production, supplemented by saltwater aquaculture centered on shrimp, prawns, and salmon.

Unfortunately, aquaculture has done little to relieve overexploited fishery resources. Since most aquaculture operations grow carnivorous fish, catches of wild fish are used to feed farmed ones. Raising 1 kilogram (2.2 pounds) of farmed salmon consumes 5 kilograms (11 pounds) of fishmeal, usually made from anchovies or menhaden. Aquaculture has grown more efficient in recent years, largely due to the growth of Chinese enterprises that raise herbivorous fish like carp and tilapia. Still, 1 kilogram of farmed fish on average consumes 1.36 kilograms (3 pounds) of wild-caught fish. With aquaculture production on the rise, the total catch used for fish food has increased from 10 million tons in 1997 to 12 million tons in 2001. As aquaculture booms, it takes its toll on species like sardines and herring. A future collapse of wild fish populations would seriously limit aquaculture's potential for growth.

What is more, aquaculture enterprises generate a great deal of pollution in the form of nutrient, antibiotic, and pesticide runoff. In Norway, salmon farms generate as much nitrogen as 4 million people. In the US, aquaculture is a net consumer of energy rather than a net producer. Shrimp-farming is also implicated in destruction of the world's mangrove environments, which has made coastal areas more vulnerable to storms and reduced breeding grounds for fish and other marine life. Aquaculture alone is thus not likely to bring about the recovery of the earth's fisheries.[43]

The reality is that the only way to prevent overfishing will be to fish less. Many marine ecologists advocate establishment of a comprehensive system of marine protected areas (MPAs) in which all exploitation of marine life is banned. This system of reserves, it is held, would not only allow the eventual recovery of the ocean's natural diversity and abundance of life, but also create restocking zones for adjacent areas. Less than 1 percent of ocean area currently enjoys this protected status, which makes it extremely difficult to predict the rate at which recovery will take place. Recovery efforts usually do not start until undeniable evidence of fishery overexploitation has appeared.[44] Unless positive steps are taken to reduce capacity, control the amount of fishing, eliminate waste, and limit damaging fishing methods, the future does not look bright for the world's marine ecosystems.

Notes

1   W. J. Bolster, "Opportunities in Marine Environmental History," *Environmental History* 11, 2006, pp. 567–97.

2   W. Dean, *With Broadax and Firebrand: The Destruction of the Brazilian Atlantic Forest*, Berkeley, University of California Press, 1995, p. 24; J. Erlandson and T. Rick, "Archaeology, Marine Ecology, and Human Impacts on Marine Environments," in T. Rick and J. Erlandson (eds.), *Human Impacts on Ancient Marine Ecosystems: A Global Perspective*, Berkeley, University of California Press, 2008, pp. 1–20 (pp. 4–5); P. Holm, "Fishing," in S. Krech, J. R. McNeill, and C. Merchant (eds.), *Encyclopedia of World Environmental History*, New York, Routledge, 2004, pp. 529–35 (pp. 529–30); C. Roberts, *The Unnatural History of the Sea*, Washington, DC, Island Press, 2007, p. 85.

3   Erlandson and Rick, "Archaeology, Marine Ecology, and Human Impacts on Marine Environments," pp. 5–7.

4   A. Jerardino, G. M. Branch, and R. Navarro, "Human Impact on Precolonial West Coast Marine Environments of South Africa," in T. Rick and J. Erlandson (eds.), *Human Impacts on Ancient Marine Ecosystems: A Global Perspective*, Berkeley, University of California Press, 2008, pp. 279–96.

5   D. Corbett, D. Causey, M. Clementz, et al., "Aleut Hunters, Sea Otters, and Sea Cows: Three Thousand Years of Interactions in the Western Aleutian Islands, Alaska," in T. Rick and J. Erlandson (eds.), *Human Impacts on Ancient Marine Ecosystems: A Global Perspective*, Berkeley, University of California Press, 2008, pp. 43–76; J. Jackson, M. X. Kirby, W. H. Berger, et al., "Historical Overfishing and the Recent Collapse of Coastal Ecosystems," *Science* 293, 2001, pp. 630–8 (pp. 630–1).

6   J. R. McNeill, "Of Rats and Men: A Synoptic Environmental History of the Island Pacific," *Journal of World History* 5, 1994, pp. 299–350 (pp. 305–8).

7   Dean, *With Broadax and Firebrand*, p. 24.

8   S. Fitzpatrick, W. F. Keegan, and K. S. Sealey, "Human Impacts on Marine Environments in the West Indies during the Middle to Late Holocene," in T. Rick and J. Erlandson (eds.), *Human Impacts on Ancient Marine Ecosystems: A Global Perspective*, Berkeley, University of California Press, 2008, pp. 147–64.

9   T. Bekker-Nielsen (ed.), *Ancient Fishing and Fish Processing in the Black Sea Region*, Aarhus, Aarhus University Press, 2004.

10 Holm, "Fishing," pp. 529–30; A. Morales-Muñiz and E. Roselló-Izquierdo, "Twenty Thousand Years of Fishing in the Strait: Archaeological Fish and Shellfish Assemblages from Southern Iberia," in T. Rick and J. Erlandson (eds.), *Human Impacts on Ancient Marine Ecosystems: A Global Perspective*, Berkeley, University of California Press, 2008, pp. 243–78.

11 R. Marks, *Tigers, Rice, Silk, and Silt: Environment and Economy in Late Imperial South China*, Cambridge, Cambridge University Press, 1998, pp. 119–21; C. Nash, *The History of Aquaculture*, New York, Wiley-Blackwell, 2011, pp. 11–24; G. F. Zhong, "The Mulberry Dike–Fish Pond Complex: A Chinese Ecosystem of Land–Water Interaction on the Pearl River Delta," *Human Ecology* 10, 1982, pp. 191–202.

12 R. Hoffmann, "Economic Development and Aquatic Ecosystems in Medieval Europe," *The American Historical Review* 101, 1996, pp. 631–69; R. Hoffmann, "Frontier Foods for Late Medieval Consumers: Culture, Economy, and Ecology," *Environment and History* 7, 2001, pp. 131–67; R. Hoffmann, "A Brief History of Aquatic Resource Use in Medieval Europe," *Helgoland Marine Resource* 59, 2005, pp. 22–30; Roberts, *The Unnatural History of the Sea*, chapter 2.

13 Holm, "Fishing," p. 530; B. Poulsen, *Dutch Herring: An Environmental History*, Amsterdam, Aksant Publishing, 2008.

14 W. J. Bolster, "Putting the Ocean in Atlantic History: Maritime Communities and Marine Ecology in the Northwest Atlantic, 1500–1800," *The American Historical Review* 113, 2008, pp. 19–47 (pp. 27–8); Holm, "Fishing," p. 531; Roberts, *The Unnatural History of the Sea*, chapter 3.

15 Bolster, "Putting the Ocean in Atlantic History," pp. 30–1, 41–2; J. Richards, *Unending Frontier: An Environmental History of the Early Modern World*, Berkeley, University of California Press, 2003, chapter 15; A. A. Rosenberg, W. J. Bolster, K. E. Alexander, et al., "The History of Ocean Resources: Modeling Cod Biomass Using Historical Records," *Frontiers in Ecology and the Environment* 3, 2005, pp. 84–90.

16 Bolster, "Putting the Ocean in Atlantic History," pp. 31–6; Richards, *Unending Frontier*, chapter 16; Roberts, *The Unnatural History of the Sea*, pp. 83–91.

17 J. Gernet, *Daily Life in China on the Eve of the Mongol Invasion, 1250–1276*, Stanford, Stanford University Press, 1962, p. 51.

18 M. Muscolino, *Fishing Wars and Environmental Change in Late Imperial and Modern China*, Cambridge, MA, Harvard University Asia Center and Harvard University Press, 2009, chapter 1.

19 P. Boomgaard, "Resources and People of the Sea in and around the Indonesian Archipelago, 900–1900," in P. Boomgaard and D. Henley (eds.), *Muddied Waters: Historical and Contemporary Perspectives on Management of Forest and Fisheries in Island Southeast Asia*, Leiden, KITLV, 2005, pp. 98–119; McNeill, "Of Rats and Men," pp. 319–26; K. Schwerdtner Máñez and S. Ferse, "The History of Makassan Trepang Fishing and Trade," *PLoS ONE* 5/6, 2010, pp. 1–8.

20 D. Howell, *Capitalism from Within: Economy, Society, and the State in a Japanese Fishery*, Berkeley, University of California Press, 1995, pp. 106–18.

21 Holm, "Fishing," p. 531; Roberts, *The Unnatural History of the Sea*, chapter 10.

22 Holm, "Fishing," p. 531; Roberts, *The Unnatural History of the Sea*, chapter 11; R. Thurstan and C. Roberts, "Ecological Meltdown in the Firth of Clyde, Scotland: Two Centuries of Change in a Coastal Marine Ecosystem," *PlosOne* 5/7, 2010, pp. 1–14.

23 Holm, "Fishing," p. 532; W. Tsutsui, "Landscapes in the Dark Valley: Toward an Environmental History of Wartime Japan," in R. Tucker and E. Russell (eds.), *Natural Enemy, Natural Ally: Toward an Environmental History of Warfare*, Corvallis, OR, Oregon State University Press, 2004, pp. 195–216 (pp. 206–7); W. Tsutsui, "The Pelagic Empire: Reconsidering Japanese Expansionism, 1895–1945," in I. Miller, J. Thomas, and B. Walker (eds.), *Japan at Nature's Horizon*, Honolulu, University of Hawai'i Press, forthcoming.

24 M. Muscolino, "The Yellow Croaker War: Fishery Disputes between China and Japan, 1925–1935," *Environmental History* 13, 2008, pp. 306–24; Muscolino, *Fishing Wars and Environmental Change*, chapter 4.

25 K. Dorsey, "Whale," in S. Krech, J. R. McNeill, and C. Merchant (eds.), *Encyclopedia of World Environmental History*, New York, Routledge, 2004, pp. 1324–7 (p. 1324); J. R. McNeill, *Something New Under the Sun: An Environmental History of the Twentieth-Century World*, New York, W. W. Norton, 2000, pp. 238–41; Roberts, *The Unnatural History of the Sea*, pp. 91–8.

26 McNeill, *Something New Under the Sun*, p. 246.

27 Dorsey, "Whale," pp. 1324–7; K. Dorsey, "International Whaling Commission," in S. Krech, J. R. McNeill, and C. Merchant (eds.), *Encyclopedia of World Environmental History*, New York, Routledge, 2004, pp. 701–2; McNeill, *Something New Under the Sun*, pp. 242–3.

28 Holm, "Fishing," p. 532; Roberts, *The Unnatural History of the Sea*, pp. 188–98, 287–302, 305–16; W. Swartz, E. Sala, S. Tracey, et al., "The Spatial Expansion and Ecological Footprint of Fisheries (1950 to Present)," *PLoS ONE* 5/12, 2010, pp. 1–6.

29 McNeill, *Something New Under the Sun*, pp. 246–7.

30 A. McEvoy, *The Fisherman's Problem: Ecology and Law in the California Fisheries, 1850–1980*, Cambridge, Cambridge University Press, 1986, chapter 6; McNeill, *Something New Under the Sun*, p. 248.

31 C. Clover, *The End of the Line: How Overfishing Is Changing the World and What We Eat*, Berkeley, University of California Press, 2006, pp. 65–6; McNeill, *Something New Under the Sun*, p. 248; D. Pauly, V. Christensen, J. Dalsgaard, et al., "Fishing Down Marine Food Webs," *Science* 279, 1998, pp. 860–3.

32 McEvoy, *The Fisherman's Problem*, p. 155; Holm, "Fishing," p. 532; McNeill, *Something New Under the Sun*, pp. 248–9; Roberts, *The Unnatural History of the Sea*, pp. 192, 323; Swartz, Sala, Tracey, et al. "The Spatial Expansion and Ecological Footprint of Fisheries."

33 McNeill, *Something New Under the Sun*, pp. 249–50.

34 Clover, *The End of the Line*, pp. 136–40.

35 Clover, *The End of the Line*, pp. 37, 66–8; McNeill, *Something New Under the Sun*, pp. 249–50; Roberts, *The Unnatural History of the Sea*, pp. 317–21.

36 Roberts, *The Unnatural History of the Sea*, pp. 189–90, 285.

37 Holm, "Fishing," pp. 532–3; McNeill, *Something New Under the Sun*, pp. 250–1.

38 Clover, *The End of the Line*, pp. 111–15, 123–6; McNeill, *Something New Under the Sun*, p. 251; Roberts, *The Unnatural History of the Sea*, pp. 199–213.

39 T. Bestor, "How Sushi Went Global," *Foreign Policy* 121, 2000, pp. 54–63; Clover, *The End of the Line*, pp. 198–213; Holm, "Fishing," p. 534; Roberts, *The Unnatural History of the Sea*, pp. 279–82.

40 Clover, *The End of the Line*, pp. 41–53; D. Pauly, V. Christensen, S. Guénette, et al., "Towards Sustainability in World Fisheries," *Nature* 418, 2002, pp. 689–95; Roberts, *The Unnatural History of the Sea*, pp. 328–9.

41 Clover, *The End of the Line*, pp. 21–3; Roberts, *The Unnatural History of the Sea*, pp. 317–18, 338–9; R. Watson and D. Pauly, "Systematic Distortions in World Fisheries Catch Trends," *Nature* 414, 2001, pp. 534–6.

42 Roberts, *The Unnatural History of the Sea*, pp. 329–30; Swartz, Sala, Tracey, et al. "The Spatial Expansion and Ecological Footprint of Fisheries"; B. Worm, E. B. Barbier, N. Beaumont, et al., "Impact of Biodiversity Loss on Ocean Ecosystem Services," *Science* 314, 2006, pp. 787–90.

43 P. Aldhous, "Fish Farms Still Ravage the Sea," *Nature News*, February 17, 2004, accessed from http://www.nature.com/news/2004/040217/full/news040216-10.html on March 9, 2012; Clover, *The End of the Line*, pp. 252–69; Holm, "Fishing," pp. 533–5; McNeill, *Something New Under the Sun*, pp. 251–2.

44 Clover, *The End of the Line*, pp. 297–314; Roberts, *The Unnatural History of the Sea*, pp. 349–77; B. Worm, R. Hilborn, J. K. Baum, et al., "Rebuilding Global Fisheries," *Science* 325, 2009, pp. 578–85.

References

Aldhous, P., "Fish Farms Still Ravage the Sea," *Nature News*, February 17, 2004, accessed from http://www.nature.com/news/2004/040217/full/news040216-10.html on March 9, 2012.

Bekker-Nielsen, T. (ed.), *Ancient Fishing and Fish Processing in the Black Sea Region*, Aarhus, Aarhus University Press, 2004.

Bestor, T., "How Sushi Went Global," *Foreign Policy* 121, 2000, pp. 54–63.

Bolster, W. J., "Opportunities in Marine Environmental History," *Environmental History* 11, 2006, pp. 567–97.

Bolster, W. J., "Putting the Ocean in Atlantic History: Maritime Communities and Marine Ecology in the Northwest Atlantic, 1500–1800," *The American Historical Review* 113, 2008, pp. 19–47.

Boomgaard, P., "Resources and People of the Sea in and around the Indonesian Archipelago, 900–1900," in P. Boomgaard and D. Henley (eds.), *Muddied Waters: Historical and*

Contemporary Perspectives on Management of Forest and Fisheries in Island Southeast Asia, Leiden, KITLV, 2005, pp. 98–119.

Clover, C., *The End of the Line: How Overfishing Is Changing the World and What We Eat*, Berkeley, University of California Press, 2006.

Corbett, D., Causey, D., Clementz, M., et al., "Aleut Hunters, Sea Otters, and Sea Cows: Three Thousand Years of Interactions in the Western Aleutian Islands, Alaska," in T. Rick and J. Erlandson (eds.), *Human Impacts on Ancient Marine Ecosystems: A Global Perspective*, Berkeley, University of California Press, 2008, pp. 43–76.

Dean, W., *With Broadax and Firebrand: The Destruction of the Brazilian Atlantic Forest*, Berkeley, University of California Press, 1995.

Dorsey, K., "International Whaling Commission," in S. Krech, J. R. McNeill, and C. Merchant (eds.), *Encyclopedia of World Environmental History*, New York, Routledge, 2004a, pp. 701–2.

Dorsey, K., "Whale," in S. Krech, J. R. McNeill, and C. Merchant (eds.), *Encyclopedia of World Environmental History*, New York, Routledge, 2004b, pp. 1324–7.

Erlandson, J., and Rick, T., "Archaeology, Marine Ecology, and Human Impacts on Marine Environments," in T. Rick and J. Erlandson (eds.), *Human Impacts on Ancient Marine Ecosystems: A Global Perspective*, Berkeley, University of California Press, 2008, pp. 1–20.

Fitzpatrick, S., Keegan, W. F., and Sealey, K. S., "Human Impacts on Marine Environments in the West Indies during the Middle to Late Holocene," in T. Rick and J. Erlandson (eds.), *Human Impacts on Ancient Marine Ecosystems: A Global Perspective*, Berkeley, University of California Press, 2008, pp. 147–64.

Gernet, J., *Daily Life in China on the Eve of the Mongol Invasion, 1250–1276*, Stanford, Stanford University Press, 1962.

Hoffmann, R., "A Brief History of Aquatic Resource Use in Medieval Europe," *Helgoland Marine Resource* 59, 2005, pp. 22–30.

Hoffmann, R., "Economic Development and Aquatic Ecosystems in Medieval Europe," *The American Historical Review* 101, 1996, pp. 631–69.

Hoffmann, R., "Frontier Foods for Late Medieval Consumers: Culture, Economy, and Ecology," *Environment and History* 7, 2001, pp. 131–67.

Holm, P., "Fishing," in S. Krech, J. R. McNeill, and C. Merchant (eds.), *Encyclopedia of World Environmental History*, New York, Routledge, 2004, pp. 529–35.

Howell, D., *Capitalism from Within: Economy, Society, and the State in a Japanese Fishery*, Berkeley, University of California Press, 1995.

Jackson, J., Kirby, M. X., Berger, W. H., et al., "Historical Overfishing and the Recent Collapse of Coastal Ecosystems," *Science* 293, 2001, pp. 630–8.

Jerardino, A., Branch, G. M., and Navarro, R., "Human Impact on Precolonial West Coast Marine Environments of South Africa," in T. Rick and J. Erlandson (eds.), *Human Impacts on Ancient Marine Ecosystems: A Global Perspective*, Berkeley, University of California Press, 2008, pp. 279–96.

Marks, R., *Tigers, Rice, Silk, and Silt: Environment and Economy in Late Imperial South China*, Cambridge, Cambridge University Press, 1998.

McEvoy, A., *The Fisherman's Problem: Ecology and Law in the California Fisheries, 1850–1980*, Cambridge, Cambridge University Press, 1986.

McNeill, J. R., "Of Rats and Men: A Synoptic Environmental History of the Island Pacific," *Journal of World History* 5, 1994, pp. 299–350.

McNeill, J. R., *Something New Under the Sun: An Environmental History of the Twentieth-Century World*, New York, W. W Norton, 2000.

Morales-Muñiz, A., and Roselló-Izquierdo, E., "Twenty Thousand Years of Fishing in the Strait: Archaeological Fish and Shellfish Assemblages from Southern Iberia," in T. Rick and J. Erlandson (eds.), *Human Impacts on Ancient Marine Ecosystems: A Global Perspective*, Berkeley, University of California Press, 2008, pp. 243–78.

Muscolino, M., *Fishing Wars and Environmental Change in Late Imperial and Modern China*, Cambridge, MA, Harvard University Asia Center and Harvard University Press, 2009.

Muscolino, M., "The Yellow Croaker War: Fishery Disputes between China and Japan, 1925–1935," *Environmental History* 13, 2008, pp. 306–24.

Nash, C., *The History of Aquaculture*, New York, Wiley-Blackwell, 2011.

Pauly, D., Christensen, V., Dalsgaard, J., et al., "Fishing Down Marine Food Webs," *Science* 279, 1998, pp. 860–3.

Pauly, D., Christensen, V., Guénette, S., et al., "Towards Sustainability in World Fisheries," *Nature* 418, 2002, pp. 689–95.

Poulsen, B., *Dutch Herring: An Environmental History*, Amsterdam, Aksant Publishing, 2008.

Richards, J., *Unending Frontier: An Environmental History of the Early Modern World*, Berkeley, University of California Press, 2003.

Roberts, C., *The Unnatural History of the Sea*, Washington, DC, Island Press, 2007.

Rosenberg, A. A., Bolster, W. J., Alexander, K. E., et al., "The History of Ocean Resources: Modeling Cod Biomass Using Historical Records," *Frontiers in Ecology and the Environment* 3, 2005, pp. 84–90.

Schwerdtner Máñez, K., and Ferse, S., "The History of Makassan Trepang Fishing and Trade," *PLoS ONE* 5/6, 2010, pp. 1–8.

Swartz, W., Sala, E., Tracey, S., et al., "The Spatial Expansion and Ecological Footprint of Fisheries (1950 to Present)," *PLoS ONE* 5/12, 2010, pp. 1–6.

Thurstan, R., and Roberts, C., "Ecological Meltdown in the Firth of Clyde, Scotland: Two Centuries of Change in a Coastal Marine Ecosystem," *PlosOne* 5/7, 2010, pp. 1–14.

Tsutsui, W., "Landscapes in the Dark Valley: Toward an Environmental History of Wartime Japan," in R. Tucker and E. Russell (eds.), *Natural Enemy, Natural Ally: Toward an Environmental History of Warfare*, Corvallis, OR, Oregon State University Press, 2004, pp. 195–216.

Tsutsui, W., "The Pelagic Empire: Reconsidering Japanese Expansionism, 1895–1945," in I. Miller, J. Thomas, and B. Walker (eds.), *Japan at Nature's Horizon*, Honolulu, University of Hawai'i Press, forthcoming.

Watson, R., and Pauly, D., "Systematic Distortions in World Fisheries Catch Trends," *Nature* 414, 2001, pp. 534–6.

Worm, B., Barbier, E. B., Beaumont, N., et al., "Impact of Biodiversity Loss on Ocean Ecosystem Services," *Science* 314, 2006, pp. 787–90.

Worm, B., Hilborn, R., Baum, J. K., et al., "Rebuilding Global Fisheries," *Science* 325, 2009, pp. 578–85.

Zhong, G. F., "The Mulberry Dike–Fish Pond Complex: A Chinese Ecosystem of Land–Water Interaction on the Pearl River Delta," *Human Ecology* 10, 1982, pp. 191–202.

CHAPTER SEVENTEEN

Riverine Environments

ALAN ROE

From the Nile to the Amazon, from the Euphrates to the Colorado, rivers have long captivated the human imagination. While early civilizations associated life-giving and frequently uncontrollable rivers with deities, "enlightened" statesmen of the eighteenth and nineteenth centuries, as well as the political leaders and technocrats of the twentieth century, constructed myths that rivers could be completely "tamed," "subdued," and even "conquered." Whether they have provided the water for irrigation, the power to turn turbines, or the mystery around which humankind has developed religious beliefs and traditions, rivers have been central to human existence since the scattered settlements in several river valleys formed the embryos of civilization millennia ago. Despite the obvious importance of rivers to human societies, historians have usually relegated them to static backdrops. Over the past 30 years, however, several historians have expanded our view of the complex interplay between human history and rivers, usually specific rivers.[1] Historians must continue to zoom in on individual rivers. Navigating the wider currents of the historical relationship between humans and rivers is also an important task that historians would be remiss to forget and that this chapter will explore.

Civilization developed next to rivers. Between 8000 and 4000 BCE, agriculture emerged in the Tigris–Euphrates, Wei, Indus, and Nile valleys (see Map 17.1). Toward the latter part of the fourth millennium and in the first few centuries of the third millennium BCE, city-states and other organized polities with more centralized sources of authority, specialization, elaborate belief systems, and social stratification formed in these river valleys. Although historians have argued over the degree to which the expansion and management of irrigation was the pivotal factor in state formation, the formation of centralized states undoubtedly allowed humans to expand irrigation.[2] Rivers also provided ancient civilizations with regular reminders of humanity's dependence on nature's rhythms and, at times, helplessness before her capriciousness, often in the form of floods or draught. Rivers' oppositional life-giving and destructive powers perhaps

A Companion to Global Environmental History, First Edition. Edited by J.R. McNeill and Erin Stewart Mauldin.
© 2012 John Wiley & Sons, Ltd. Published 2015 by John Wiley & Sons, Ltd.

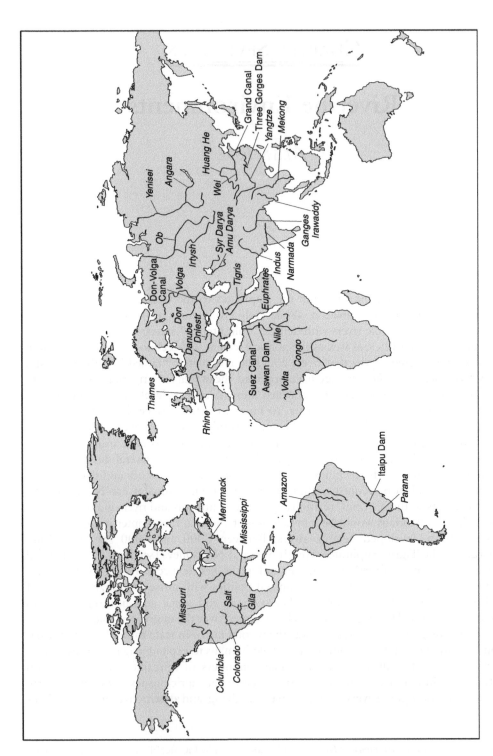

Map 17.1 Rivers featured in Chapter 17

explain why they became so important to early religious traditions. The Egyptians believed that the tears of the goddess Isis caused the Nile's seasonal floods, and European pagan traditions held that the springs where rivers form brought forth children. From the Sumerians to the Jews, civilizations of the Fertile Crescent told, retold, and wrote down different versions of a myth that gods or a single God caused a flood that destroyed much of humanity.

Yet the destructive powers of rivers were frequently unleashed by humanity's failure to understand the ecological consequences of intensive irrigation. Mesopotamia provides an example. Throughout the third millennium BCE, the Sumerians expanded and intensified irrigation throughout the Tigris and Euphrates river valleys, resulting in a population explosion that necessitated the intensification and further expansion of the irrigation regime. This positive feedback loop continually brought more water onto the flat plain, which had little natural drainage and nearly impermeable alluvial soils.[3] Having neither sophisticated drainage techniques nor the knowledge of the benefits of leaving land fallow, the region's inhabitants allowed water to pool up in the fields year after year. The evaporating water from these rivers left layers of salt on the fertile ground, depositing around 150 centimeters (59 inches) of salt on the land's surface over the course of one millennium. Written records reveal that from about 2400 to about 1700 BCE crop yields fell by nearly two-thirds as the region's wheat was frequently replaced with less nutritious barley. In turn, the population of southern Mesopotamia fell from 630,000 in 1900 BCE to 270,000 in 1600 BCE.[4] With declining yields, southern Mesopotamia civilizations could not maintain large armies. In turn, a series of different peoples invaded the area, established societies, and then fell to new invaders during the late third and early second millennium BCE.[5]

More than 2.5 millennia later, the Hohokom peoples, located in the present-day American Southwest, began developing what would become the most extensive irrigation network in the Americas prior to the arrival of Europeans. Beginning around 600 CE, the Hohokom "canal builders" started digging simple diversion canals off the Gila and Salt rivers. By 1200 CE they had dug 560 kilometers (350 miles) of main canals and over 1,600 kilometers (1,000 miles) of distribution canals that went well beyond the flood plains of these rivers. By that time, the Hohokom population reached nearly 300,000 people dispersed over 16 settlements.[6] But over time, the unlined canals became laden with silt, weed infested, and unevenly graded. As the bottom of the canals became soft, settlements along the canals likely endured catastrophic floods.[7] Moreover, their fields became uncultivable after years of irrigating them with saline water. When the Spanish arrived in the sixteenth century, they commented on the "sterile plains" that "appear[ed] as if they had been strewn with salt" amid the abandoned settlements. Having long since forgotten how to forage for food, the Hohokom likely latched onto different nomadic groups, as cultivating crops had become nearly impossible.[8]

The experiences of Mesopotamia and the Hohokom peoples have been repeated on semi-arid lands around the world, whereas in more humid landscapes societies could utilize their rivers – mainly for drinking water, waste disposal, and navigation – more sustainably. The Yangzi, the Danube, the Congo, the Mississippi, the Amazon, and thousands of smaller rivers served human purposes sustainably for centuries. Since the early nineteenth century, however, societies have acquired both incentives and capacity to change rivers more fully.

The emergence of rail, the steamship, and refrigeration made agricultural and farm products even easier to ship to distant markets in the nineteenth and twentieth centuries.

Agriculture developed from a local to a national activity; before long, the importance of agricultural products in international trade provided perhaps the major impetus for digging irrigation canals and damming rivers over the past two centuries. With the emergence of the global marketplace aided by technological advancements, irrigation regimes grew to an extent that dwarfed the work of earlier civilizations.

From 1800 to 1900, the area of irrigated land in the world expanded nearly fivefold from roughly 8 to 40 million hectares. The Indus and Ganges basins, Egypt, the western US, and Australia were the areas in which irrigation expanded the most. The British development of irrigation in the Indus and Ganges river valleys was arguably the most dramatic of these regional transformations.[9] When Alexander the Great ventured through the Indus River valley in the fourth century BCE, he commented on the lush forest environment, replete with Asian lions, elephants, cheetahs, rhinos, gazelles, wild ass, and sheep. Some of this biological diversity was still intact upon the arrival of the British 2,000 years later. But making these vast, "uninhabited" expanses bloom, as the British would say, required chopping down and burning the region's dense forests, reclaiming marshland, and then constructing a vast network of irrigation canals. The British would remake the Indus River into the most extensive irrigation system in the world, a transformation the Pakistani government would continue in the late twentieth century. Irrigation undoubtedly allowed for dramatic increases in food production and hence population growth. For instance, the Ganges Canal, which measured 3 meters (10 feet) deep, 50 meters (164 feet) wide, and nearly 1,500 kilometers (930 miles) long, was responsible for irrigating 600,000 hectares, and perhaps as many as 2.5 million people a year were fed by the resulting expanded agricultural output by the late nineteenth century. This expansion, combined with the British advanced flood-control techniques, strongly informed the imperial narrative of the benefits of British guidance. This narrative ignored the coercion, the social dislocation, and the environmental impacts, such as salinization, water-logging, and the eradication of much of the region's wildlife, caused by these projects.

Control of rivers also proved central to the British imperial strategy in Egypt. Of course, alterations to the Nile long predated British arrival. Beginning in the Old Kingdom, pharaohs forced peasants to dig irrigation canals running away from the river to expand the cultivable area. In their time, the Ottomans sought to expand Egypt's canal network, most notably to increase Alexandria's potable-water supply. As the British began exercising their expanded authority over Egypt in the 1880s, colonial officials argued that controlling the Nile was the most important factor in establishing dominance in the region. Lord Cromer, the governor of Egypt, emphatically stated that "no water should flow wasted into the Mediterranean."[10] Within two decades, improvements to the irrigation system along the Nile had doubled Egypt's cotton production, undoubtedly gaining the British their share of supporters. But the British, like the Ottomans, expanded irrigation largely on the backs of Egyptian peasants through coercion, earning the empire a large number of enemies.

The British and French used irrigation to achieve similar ends in the Mekong and Irawaddy deltas of Southeast Asia. Both nations sought to transform these "gloomy wastelands" of mangrove swamps and rain forests, where large cats, elephants, reptiles, and birds wandered freely, into a "lush garden" in which rice could flourish. Swamps were drained, land reclaimed, irrigation canals dug, and the populations of large mammals largely eradicated as the British and French ordered inhabitants in other parts of the region to be sent to these valleys. In Burma, the British succeeded in expanding rice production by more than a factor of ten; the French expanded it in Indochina nearly sevenfold. But in addition to the aforementioned landscape changes, waste from rice mills went directly into the rivers, creating conditions that contributed to frequent outbreaks of cholera and typhus.[11]

Central Asia's rivers could not produce rice, but with sufficient incentive and technical skill could be harnessed for cotton production. While small-scale agriculture had taken place along the Amu and Syr Darya as early as 3000 BCE, the Russian Empire was the first power to orient the local economy to producing for distant markets, first domestic and eventually international. Tsar Alexander II asserted control over the region in the 1860s in order to develop domestic cultivation of "white gold" as the Civil War in the US threatened the Empire's sources of it.[12] The Soviet Union made a far more totalizing effort to transform the region that brought extensive ecological consequences. Referring to Central Asia, Lenin stated:

> Irrigation is especially important to "lift" agriculture and to ensure pastoralism does not continue and to immediately try to improve the condition of peasants and to start large-scale projects of electrification and irrigation. Irrigation most of all is needed to rebuild the hinterland [*krai*], renew it, bury the past, and strengthen the transition to socialism.[13]

From the late 1920s to the early 1980s, the Soviet Union built dams, drained swamps, uprooted orchards and vineyards, plowed under pasture lands and dug canals from the Amu and Syr Darya beyond the deltaic and littoral zones. By 1985, cotton grew on over 85 percent of the cultivated land in these river basins.[14] Uzbekistan alone was producing more cotton than the entire US and the Soviet Union had become the world's second leading cotton producer behind China.

Consequently, by the mid-1980s, the Amu and Syr Darya rivers brought a mere trickle of water to the Aral Sea. As the Sea contracted, the adjacent regions suffered violent dust storms that originated on the exposed, windswept, seabed. Moreover, over half of the irrigated land had become highly salinized. Soon, over a million hectares had been taken out of production. At the same time, some soils, lakes, swamps, and lush meadows in the lower parts of the river basins dried up altogether, making irrigation nearly impossible in dry years near the Aral itself. The desiccation of subsoils, in turn, reduced vegetation by 20 percent to 40 percent throughout the lower stretches of the river valleys, pushing many mammal and bird species to extinction.[15]

Soviet engineers drew up plans to correct "nature's mistake" by redirecting northerly flowing Siberian rivers south in order to bring more water to the thirsty Central Asian plain. Abandoned in the 1980s, this plan stands as one of the foremost examples of the Soviet Union's ideological commitment to transform nature.[16] While the Soviet plan is striking in the scope of its ambition, it is also indicative of the global obsession with maximizing agriculture production in the twentieth century. Joseph Stalin, Winston Churchill, Jawaharal Nehru, Fidel Castro, and many other political leaders made similar statements expressing the "instrumental reason" that rivers were little more than a resource to be exploited. Increased man power, more sophisticated machinery and engineering as well as the unwavering dedication of political leaders to development (often quite unreflective) led to the construction of large dams, which enabled humanity to expand irrigation by 420 billion hectares from 1950 to 2000.[17] A total of 20 to 30 million hectares of these lands, which were poorly suited to intensive agriculture, suffered severe erosion or salinization by the 1990s while over

80 million hectares showed moderate effects. In 1987, the United Nations Environment Programme (UNEP) estimated that the loss of agricultural land was 5–7 million hectares per year. While India had lost over 11 percent of its agriculturally productive land, China had lost over 23 percent. Turkmenistan and Uzbekistan lost 48 percent and 24 percent respectively.

But salinization and erosion are not always irreversible processes, even in the short term. Exhausted lands were reclaimed several times throughout the twentieth century. In Egypt, for instance, where the arid climate made irrigation essential for farming, inefficient distribution and limited natural drainage in the Nile Valley caused more than 60 percent of the land to become waterlogged or highly saline by 1970. By 1990, better drainage had been provided for 1.4 million hectares, salinity had been reduced by as much as a factor of five, and land productivity, in turn, increased. The World Bank helped the Pakistanis bring back an average of 80,000 hectares into cultivation each year. It also helped with several projects in Central Asia that reclaimed heavily saline lands while encouraging better conservation techniques, such as lining irrigation canals.[18] The expansion of irrigation, even in arid lands, has not always been a declensionist story of irreversible despoliation.

Transportation Canals

Transportation canals have figured at least as prominently as irrigation canals in the ambitions of rulers. Pharaohs and Roman emperors dreamed of a canal at Suez in ancient times. As early as the ninth century, Charlemagne expressed his desire to link the Rhine and Danube river systems through a canal 30 meters (98.5 feet) wide. The Ottoman Sultan Mehmet II began touting a similarly ambitious project – a canal connecting the Volga and Don Rivers – in the late fifteenth century. Later sultans tried to fulfill this ambition but without success. Peter the Great rehatched this idea in the last years of the eighteenth century as he plotted an attack of the Ottomans on the Sea of Azov. Although he sent over 35,000 peasants to dig the canal, he cancelled the project after they had worked only a few weeks. Never short of grand ambitions, Peter conceived the even more grandiose plan of connecting the Amu Darya to the Caspian Sea in order to provide a system of waterways that could take Russian merchants closer to India.[19]

While neither sultans nor tsars carried out such ambitious canal-building projects until the twentieth century, Tang-dynasty emperors in China completed an even more ambitious project in the eighth century. China had built impressive canals as early as the sixth century BCE.[20] Most of these canals were in the northern part of the empire. But, when the Sui dynasty came to power in the sixth century, emperors did not want to relocate the capital even though the empire's economic center of gravity was shifting to the more fertile south. As a solution, they devised the Grand Canal, which carried grain north to feed the large urban populations of the northern plain. For centuries the world's most heavily trafficked transportation thoroughfare, the Grand Canal allowed for the annual transport of 350,000 tons of grain to Beijing on average during the fifteenth century. While the canal enabled several dynasties to keep China's empire intact, it also transformed the ecology of some of the lands in its vicinity, creating a chain of lakes and marshlands to the west. Exceptionally vulnerable to floods, these regions became a haven for outlaws and bandits, largely because they were viewed as unfit for settlement by most Chinese.[21]

The Grand Canal had always required large governmental expenditures for its upkeep.[22] The canal's maintenance costs had become crippling by 1821, accounting for

20 percent of the government's budget. When the Qing rulers cut taxes to placate the population following the Taiping Rebellion (1850–64), maintaining the canal became virtually impossible. Meanwhile, the state began to direct its resources toward servicing the national debt, development of the coasts, and military modernization. Consequently, floods wrought more damage than ever before in the areas near the canal.[23] Confucian tradition emphasized that control of water was one of a leader's important responsibilities. The inability of the late Qing emperors to carry this out likely contributed to the weakening of their rule.

Medieval European states lacked the strong central authority, necessary man power, and technological sophistication to undertake any project on a par with China's Grand Canal. But as the end of the first millennium approached, monasteries, lords, and kings began digging canals to serve various ends. Some local navigation canals were probably being dug by monks in the Low Countries in the tenth and eleventh centuries for moving boats, powering mills, and harvesting fish. From the eleventh to the mid-thirteenth century, English landowners and monks dug canals that wrought extensive environmental changes to the English landscape. Excessively high milldams were responsible for flooding, which provoked many lawsuits. Bypass channels were dug on major rivers in order to allow migratory fish populations to get around the weirs and the traps. Siltation proved a regular problem. By the early thirteenth century, many canals had degraded into disrepair.[24] Without any strong central authority capable of maintaining them, most of them became unnavigable.

The Russian Empire never achieved Peter the Great's grandiose visions of building a network of waterways from Saint Petersburg to Central Asia, nor the later plans for a system that would stretch all the way to Lake Baikal.[25] But by the early nineteenth century, canals connected many of the major rivers in European Russia and thus facilitated trade from the Urals to Saint Petersburg. Iron and bread shipments to Saint Petersburg increased by a factor of four and six respectively.[26] Even by the time railroads, roads, and aviation reduced their value, the Soviet Union continued to value canals for their symbolic power demonstrating mankind's ability to conquer the natural world and transform itself in the process.[27] In the 1930s, Stalin deployed Soviet workers to dig the Baltic–White Sea and Volga–Moscow canals with the most primitive equipment and under deplorable living conditions. Under Nikita Khrushchev's rule (1956–64), Peter's dream of connecting the Caspian Sea and Amu Darya was achieved with the completion of the Karakum Canal.

Pollution

People have long used rivers for disposing of waste. Small quantities posed no hazard. Over the past two centuries, however, the dramatic increase in industrial wastes has frequently destroyed rivers' self-purifying capabilities. So rivers throughout the world have often became unsanitary and putrid, largely unfit for drinking.

Industrial pollution on the Rhine River was among the worst over the past two centuries. Throughout the early to mid-nineteenth century, prior to the unification of Germany, the princes of Westphalia commissioned engineers, most importantly Johann Tulla, to make the Rhine flow in one channel. The Rhine came to resemble a canal more than a river, and its ease of navigation made it Europe's most important thoroughfare. The discovery of coal deposits along the Middle Rhine watershed made it a logical hub for industry as Germany, France, and the Low Countries industrialized in the late nineteenth century. In 1913, over 100 million tons of coal were mined in the region.

As mine shafts reached near-ground water levels, miners had to pump more and more water out and find somewhere to put it. In 1908 alone, mines and factories together discharged over 56 million cubic meters (almost 2 billion cubic feet) of wastewater into the Ruhr – a tributary of the Rhine. One biologist described the Ruhr as a "brown black brew, reeking of prussic acid, containing no trace of oxygen, and absolutely dead."[28] The neighboring Emscher River was even dirtier. In the early 1900s, 1.5 million people, 150 mines, and 100 factories poured their effluents into the Emscher, doubling the flow of the slow-moving stream. One solution to the problem was to channel the effluents directly into the Rhine. Representatives of the chemical industry claimed that the Rhine was too big for the waste to have any effect on it. In the words of one such representative: "You could pour Germany's entire production of sulfuric acid into the Rhine at Cologne for an entire year and you would not find a trace of it downstream at Mulheim."[29]

Despite its size, "Europe's romantic sewer" did not in fact have unlimited self-cleansing potential.[30] By 1975, most of the Upper and Middle Rhine was moderately or strongly polluted while the Lower Rhine was strongly or completely polluted. Only 22 of the 45 native fish species survived and many of these were maintained only through hatcheries. The three most commercially valuable fish – shad, sturgeon, and salmon – had all vanished. By the 1980s, Rhine fish often carried 400 times the concentrations of PCBs – compounds associated with neurotoxicity and endocrine disruption – deemed safe by European standards.[31]

Some English rivers suffered from industrial growth as much as the Rhine. Pollution in the Thames was not new to the industrial era. During the fourteenth century, Edward III had complained of the "dung and other filth [that] had accumulated in diverse places on the banks of the river and ... [of] fumes and other abominable stenches arising there from."[32] In the late eighteenth century, slaughterhouses and tanneries polluted the Thames' waters by dumping almost all their wastes into it. But the situation drastically deteriorated in the nineteenth century as England's industries dramatically expanded. Throughout the century, industries could dump wastewater into the Thames without fear of consequences. The situation was worse on some other rivers. In 1866, a royal commission stated that the water of the Calder River in West Yorkshire made good ink. In industrial Manchester, nearly 70,000 tons of coal ashes were dumped in the city waterways of Mersey Irwell River basin each year. Robert Rawlinson, an engineer, described the waterways of Manchester: "a semi-liquid compound is formed, an accurate idea of which no written description can convey. A thick scum coats the surface (of the rivers), upon which and over which birds can walk."[33] Not surprisingly, waterborne diseases, including typhoid and cholera, were a huge problem throughout Great Britain during the nineteenth century. Across the Atlantic, rivers and the people that drank from them in highly industrialized cities in the US fared little better. Until an engineering project reversed the flow of the Chicago River in 1900, the waste that many Chicagoans and Chicago factories dumped directly into the river went into Lake Michigan – the city's source of drinking water. Waterborne diseases, especially typhoid, were a problem in the Windy City throughout the latter part of the nineteenth century, killing 90,000 people in 1885–6 alone.[34]

The emergence of popular environmentalism in the 1960s provided citizens of Germany, Great Britain, and the US with a platform on which to express their grievances about the horrendous state of their rivers. Eventually, environmental legislation mandating clean water was passed in these countries and throughout much of Europe. Citizens of the USSR, on the other hand, had limited means of pressuring the government to take measures that would help clean up the nation's rivers. While Soviet rivers were as polluted as those of the US and Western Europe during the late 1960s, their condition would only grow worse over the next two decades as the West adopted more stringent environmental regulation. Although the Soviet Union's headlong march toward becoming a highly urbanized and industrialized nation necessitated developing large supplies of municipal water, the Communist Party gave sewage treatment for these areas much lower priority. The USSR tried to make Moscow's sewage system a showpiece in the 1960s, yet over 500,000 cubic meters (17.6 million cubic feet) of untreated industrial sewage were still dumped into the Moscow River in 1970.[35] The river was completely devoid of oxygen in many areas. Some large metropolitan areas, such as Minsk, did not even have sewage-treatment plants.

The rivers of the USSR became even more polluted over the next two decades. Throughout the 1970s and 1980s, chemical factories along the Volga River dumped much of their untreated wastewater directly into it. In the late 1980s, the Dniester River had six times the prescribed limits of pesticide loads due to agricultural runoff. Untreated wastewater in this river increased from 13,000 metric tons to 21,000 metric tons from 1986 to 1991. Consequently, whereas cholera epidemics had been eliminated from much of the industrial world, the citizens of Odessa suffered from several epidemics of this disease throughout the 1980s.[36]

Rivers in areas with relatively undeveloped industry also experienced horrific water pollution during the twentieth century. As the population of the Indian subcontinent burgeoned, millions of Hindus preserved their age-old ritual of bathing in the Ganges and pouring the ashes of family members into it. By 1990, over 450 million people lived in the Ganges River basin. With several million tons of the ashes of funerary urns being dumped into the Ganges and over 70 million people discharging their untreated wastewater directly into it, the Ganges had become a sewer of human refuse and remains by the 1980s. The Indian government attempted to pass some measures to improve the situation, such as the Ganga Action Plan. But, 20 years later, environmental groups lamented that these initiatives seemed to no avail as pollution indicators had not changed or had even grown worse, despite the construction of new sewage-treatment plants and extensive efforts to educate the region's inhabitants about pollution issues.[37]

Dams

Humans have long attempted to manipulate the flow of rivers by constructing dams. The ancient Egyptians built one about 32 kilometers (20 miles) south of Cairo. It probably functioned only for a very short period of time between 2950 and 2750 BCE. While there are scattered accounts of early Sumerian dams, many laws found in Hammurabi's code indicate that dam-building was fairly widespread in the Babylonian era. Just north of Baghdad, the Babylonians constructed an earth and wood dam dedicated to their primary god Marduk in the mid-second millennium BCE. Until it failed around 1200 CE, this dam altered the course of the Tigris. By the seventh century BCE, the Assyrians were building larger dams that directed water to Ninevah's hanging gardens. The Romans built dams throughout North Africa for flood control, water retention, and soil conservation. Of the few dams they built in Italy, Subiaco, located on the Aniene River, towered over all others at 40 meters (130 feet) high – a true architectural phenomenon for its time. Dam-building largely came to a halt in Europe following the collapse of the Roman Empire until the Moors began constructing dams on the Iberian Peninsula in the late first millennium. In the Americas, long before Columbus, the Chimu Empire of the Andean highlands in present-day Peru developed numerous dams on the Chicama and Moche Rivers, some up to a kilometer (0.6 miles) long. The Mayans built several dams on the rivers of the north of the Yucatán Peninsula in order to form reservoirs.[38]

Our knowledge of the ecological transformations due to dams before the nineteenth century is limited. Perhaps the size and relative scarcity of dams made their ecological impact minimal. But the scope and intensity of the environmental impacts caused by dams increased dramatically during the Industrial Revolution. Growing urban areas demanded expanded municipal water supplies. Consequently, cities tried to outdo one another by building bigger dams, not simply for increasing the water supply, but also to demonstrate their respective power to "tame" rivers.[39] While Germany was building dams along the Rhine in the late nineteenth century, New England industrialists built dams along many rivers, but most extensively on the Merrimack. As river water was commodified, industrial associations wrested control of it from petty producers and farmers for the exclusive benefit of textile manufacturers. While industrial production soared in communities like Lowell, Massachusetts, the salmon population in northeastern rivers plummeted.[40]

Nineteenth-century dam-building paled before the economic, social, and ecological impact and symbolic value that dams would have in the twentieth century. Several factors combined to distinguish twentieth-century dam-building from earlier eras. First, increased technological sophistication and man-power resources allowed nations to construct more dams of unprecedented size and capacity.[41] Second, while dams had previously been built predominantly by and for local interests – monasteries, city-states, and provinces – nation-states or entities created by them usually carried out dam-building during the twentieth century.[42] Finally, dam-building and, toward the end of the century, the opposition to dams, became internationalized. While the engineers from powerful nation-states aided developing nations in constructing hydroelectric installations, and organizations, such as the World Bank, funded the construction of many dams, the anti-dam movement of the late twentieth century became an international phenomenon influencing all nations that did not systematically choke off dissent.

While the US and the Soviet Union viewed dams as both a key component to Cold War economic competition and symbolic monuments to their respective industrial might and control over rivers, Third World leaders frequently saw dams as a powerful means of self-assertion that would help break with the colonial past. The rhetoric of dam-building took on different hues as it served different ideological messages. But for the first 80 years of the century dam-building was an almost contagious obsession, best understood by the nearly universally accepted ideology of "developmentalism," or "high-modernist" utopianism, both of which hold industrial and economic modernization as a necessity while overlooking its sometimes devastating environmental and human consequences. Dams have endangered freshwater fish populations, inundated stands of valuable forest, and necessitated the removal of somewhere between 30 and 60 million people from their homes.[43] The proliferation of large dams has made rivers instruments of power rather than mysterious forces that have inherent value in and of themselves.

Few regions of the world have been as transformed socially and ecologically by dam-building as the western US, and environmental historians have spilled more ink on trying to understand hydrodevelopment in the US than on any other region of the world.[44]

Much of the west remained sparsely settled at the turn of the twentieth century. Harnessing the region's rivers, it had long been recognized, was critical to its development. In 1902, Congress passed the Newlands Act, which placed the federal government in charge of developing western irrigation. While the act ostensibly was designed to open more cultivable land to small farmers, the federal government repeatedly wielded control of irrigation and hydroelectric development at the behest of industrial and business interests, which led to a system with absentee landowners with hundreds of wage laborers. While the development of large dams on the Colorado has prevented it from emptying into the Pacific Ocean, prompting some historians to refer to it as "a river no more," vibrant urban centers, such as Las Vegas and Los Angeles, bloomed because of the extensive hydrodevelopment along the Colorado.[45]

Certainly, on some level, most engineers, politicians, and the throngs of migrants to the American West's urban areas accepted dam-building as central to the progress of the nation, or, if one prefers, to building an "empire."[46] But individuals in this history were usually oriented toward the practical short-term impacts their decisions had on the lives of others. Only through closer examinations of individual river histories can we fully respect this fact. The damming of the Tuolumne River in the Hetch Hetchy Valley is one such case. This episode has been recounted innumerable times as a tale of undefiled nature against the callous locomotive of progress. The actual history of this episode was more complex from both a moral and a practical perspective. By 1906, San Francisco officials realized that their city needed more water to continue to grow. While one could question the growth imperative and the choice of location for the dam, the city undoubtedly needed more water when hydrants produced little more than a trickle as fires raged throughout San Francisco following the famous earthquake of 1906. The battle that ensued over damming the Tuolumne became one of the most famous fights in the annals of American nature preservation with John Muir frequently presented as the heroic preservationist fighting futilely to save his beloved Hetch Hetchy. It is important to remember, however, that the San Francisco mayor Robert Phelan aggressively promoted the construction of the Hetch Hetchy Dam out of a deep sense of commitment to the people of San Francisco whose lives he actively sought to improve. Such facts are frequently lost in the Manichean lens through which environmental conflicts are often remembered.[47]

The Hoover, Grand Coulee, Glen Canyon, Bonneville, and many other dams have fueled the development of the west, often with unfortunate environmental consequences. Majestic landscapes were inundated, agricultural land was lost, rivers often dried up before reaching the sea, and in the Northwest salmon populations were devastated. The damming of the Columbia can, indeed, be seen as a tragic tale where a nation's ravenous appetite for industrial and economic development blinded it to everything else. The Bonneville and Grand Coulee dams were touted by populists as projects that would serve the average inhabitants of the Columbia River basin. But the hydroelectric power produced by dams on the Columbia outstripped energy demands in the region. Vast excesses of hydroelectricity capacity created an opportunity for the booming aluminum and plutonium industries to receive subsidized electricity following World War II. Thus, underpriced energy of the Bonneville Power Authority ultimately benefited big industry, despite being touted as a project in service of the common man. At the same time, the salmon population of the Columbia was devastated by dams. Prior to the Columbia dams, between 10 and 16 million salmon traveled from the Pacific Ocean to their spawning grounds in the upper reaches of the river. Today, fewer than 1.5 million salmon, many of which are farm raised, swim in the Columbia's waters. The fish catch dropped by 95 percent from 1930 to 1993. Several Native American tribes who depended on the salmon found their livelihoods destroyed.[48]

Perhaps no country undertook dam-building with comparable ideological gusto to the Soviet Union. Shortly after its formation, the Soviet Union also recognized the transformative potential of harnessing its rivers. When Vladimir Lenin uttered the oft-quoted words, "Communism equals Soviet power plus electrification of the entire country," the Soviet Union ranked 15th in the world in hydropower generation.[49] Upon accession to power, Stalin began enacting ambitious plans for developing hydroelectric power. The propaganda blitz for the first major project, the Dneprostroi Dam on the Dnieper River, established an oft-repeated Soviet trope that dam construction was tantamount to an epic struggle against rivers, necessary to build socialism. At the ceremony celebrating the beginning of the operation for what was then the largest hydroelectric installation in Europe, one official said: "Today the homeland of soviets celebrates the victory over the Dnepr. Here a structure has been created, never before seen in history."[50] Around the same time, Stalin's favorite intellectual, Maxim Gorky, stated that the Soviet Union was obliged to "make mad rivers sane."[51]

The heyday of Soviet hydroelectric development began following World War II. Cold War competition enhanced the leadership's sense of urgency to industrialize rapidly. In a very typically militaristic formulation, two of the state's most prolific propagandists on the transformation of nature wrote: "The offensive on the rivers, the building of large hydroelectric stations – this is the first part of the battle for the transformation of Siberia."[52] According to Soviet authors, the men constructing the dams fully embraced the trope, frequently chanting "We will conquer you Ob" as they built the Krasnoyarsk Hydroelectric Station.[53]

Throughout the 1950s, the Soviet Union built behemoth dams on the Ob, Enisei, Angara, and Irtysh rivers while sending engineers to help build others throughout Africa, Latin America, and Asia. There is little doubt that these hydroelectric giants transformed parts of Siberia into industrial and urban centers, which enabled the Soviet Union to utilize the region's vast untapped natural resources. But Soviet planners never seemed to give much consideration to the social and environmental costs of development. All told, the size of the forest stands lost in the creation of large reservoirs amounted to 28,000 square kilometers (10,800 square miles) – an area almost the size of Belgium. Villages were inundated and people removed to newly built cities that appealed to the Soviet government's technocratic sensibilities, but probably not to those who had to move to these cities. Moreover, Siberia's intensely cold climate demanded the expenditure of tremendous amounts of energy just to make these towns habitable.[54]

The Soviet and US models of large-scale hydroelectric development were emulated throughout the world. Developing nations often looked at dams as highways to modernization. Engineers from both countries offered their expertise in a sort of hydroelectric diplomacy geared toward winning the hearts and minds of the Third World. There were only 10 major dams throughout the world in 1950. By 1995, 305 such dams had been erected.[55] Dams became symbols of national rejuvenation and technological mastery. They demonstrated how ingrained the developmentalist mindset had become among political leaders throughout the world. In the late 1950s, President Nasser of Egypt championed the Aswan High Dam to help modernize the nation and turn the page from its colonial past. The new dam increased water storage from 1 to 5 million cubic meters (35.3 to 176.6 million cubic feet), provided flood protection, and allowed

Egypt to expand its rice, sugarcane, and cotton production dramatically. In the mid-1960s, Brazil and Paraguay drowned millions of hectares of forests while dramatically increasing their respective energy capacities by building what was at the time the largest dam in the world – Itaipu. On Ghana's Volta River, the World Bank and the governments of the US and Great Britain provided much of the funds for the Aksombo Dam, which inundated over 5 percent of the nation's land surface area while becoming an icon of nationalism. With the help of foreign engineers, Pakistan built the Tarbela Dam on the Indus.[56]

China and India undertook the most extensive dam-building efforts in the late twentieth century. The trajectory of Chinese top-down dam-building took root in the prerevolutionary years when Sun Yat-Sen argued that dams should serve broad national interests such as power generation and national defense rather than local interests, which were primarily concerned with flood control, irrigation, and transport. The center dictated to the periphery much more under communist rule after 1949, which enabled the government to undertake a sweeping effort to dam China's rivers. While China had fewer than 1,000 dams before the 1949 revolution, it had over 19,000 by 1990. As large dams increased the state's ability to generate power, they also inundated the homes of over 10 million people.[57]

While India had almost no large dams in 1930s, it had more than 1,500 by the 1980s. Jawaharal Nehru spoke effusively of hydroelectric development, equating dams to temples. Speaking of the Bhakra–Nangal dam in the 1950s, he stated:

> What a stupendous, magnificent work – a work which only that nation can take up which has faith and boldness! … it has become of a nation's will to march forward with strength, determination, and courage … As I walked around the dam site I thought that these days the biggest temple and mosque and gurwara is the place where man works for the good of mankind. Which place can be greater than this, this Bhakra–Nangal, where thousands and hundreds of thousands of men have worked, have shed their blood and sweat and laid down their lives as well? Where can be a greater and holier place than this, which we can regard as higher?[58]

While opponents of dams in India had more opportunity to express misgivings than their counterparts in China, social and environmental costs of dam-building rarely tempered the enthusiasm of Indian national leaders until the 1980s.[59] Dams provided urban areas with a large water supply, allowed India to expand its irrigation system, and produced energy that growing Indian industries required.

Yet the environmental and social costs were immense. Nearly 13,000 square kilometers (5,000 square miles) of forest land were inundated and over half the newly irrigated land was waterlogged or saline by the late twentieth century.[60] Most tragically, as many as 14 million people were displaced from their homes, sometimes by force.

As the social and environmental consequences of large dams were becoming clearer in the late 1970s and early 1980s, many governments encountered more difficulty in carrying out massive projects. A series of dams on the Narmada River, which would have become India's largest river-basin scheme, had long been planned by the Indian government. By the mid-1980s, international NGOs and activists had begun campaigns against these dams, which would submerge tens of thousands of villages and displace millions.[61] The 140-meter-(460-foot-)tall Sardar Sarovar project had received funding from many foreign donors and the World Bank, which championed its modernizing potential. But by late 1992, the World Bank pulled its support for the project because the issues regarding resettlement and the environment had not been properly addressed.

The anti-dam movement was becoming a global phenomenon. In the 1990s, opposition mounted to the Katun Hydroelectric Station in the Altai Mountains of the newly formed Russian Federation. After years of outcry by the literary and growing dissident community in the late Soviet Union, a governmental commission concluded that the project should not move forward due to the "ecological hazard and the threat of loss of historical, cultural, and spiritual values, and erosion of the Altai ethnos."[62] Meanwhile, in the US, groups called for the disassembly of several US dams and in Brazil an organization called the Movement of People Affected by Dams (MAB) publicly deplored the effect of large dams on indigenous peoples. Having become more oriented toward conservation in the wake of the Rio de Janeiro Earth Summit (1992) and the broader upsurge in environmental cooperation in the 1990s, in 1997 the World Bank sponsored the World Commission on Dams, which conducted a thorough review of the costs and benefits of dams in the developing world. While the World Bank had been one of the most sanguine boosters throughout the second half of the twentieth century, the commission concluded that dam-building had proceeded too frequently without the planning or foresight necessary to mitigate environmental and social problems.[63]

As transnational organizations made dam-building much more difficult throughout much of the world, China seemed defiantly to assert its prerogative to modernize unhampered by international opinion. Shortly after the Tienanmen Square massacre in 1989, the Chinese government jumpstarted the long dormant plan, originally conceived by Sun Yat-Sen in 1924, of building a dam in the three gorges of the Yangzi River. Subject to much scrutiny by the foreign community as well as some dissenting Chinese engineers, the Chinese government proceeded headlong in building this dam, which would become the largest hydroelectric project in the world.[64] Opposition to the dam, whether it came from engineers, ecologists, intellectuals, or a variety of experts in different fields, was treated as disloyalty with "counter-revolutionary intent."[65] The Chinese government systematically isolated dissent about environmental impacts of the Three Gorges Dam. While the dam has increased shipping capacity on the Yangzi, produces large amounts of electricity, and provides for flood control, it has resulted in the displacement of 1.2 million people, flooded important archeological sites, and dramatically increased the risk of mudslides in many areas adjacent to the newly formed reservoir.[66]

Conclusion

While large numbers of people undoubtedly continued to engage in rituals that sought to affirm their mystery and spiritual powers throughout the twentieth century, people in the highest positions of influence and power came to view rivers as instruments to exploit for economic and industrial growth. This transformation of perspective, which had slowly evolved over the centuries, accelerated in the twentieth century. But occasional dam failures in the twentieth century have provided reminders that rivers are not, in fact, as tractable as engineers might wish. In the early 1960s torrential rains swelled the Varmont Reservoir in the Italian Alps. Over 350 million cubic meters (12.3 billion cubic feet) of rock then tumbled into the reservoir, causing a huge wave that toppled the towering 260-meter (853-feet) dam, causing a flood that killed 2,600 people. In 1975, over

230,000 people died in China's Henan province when a typhoon collapsed as many as 62 dams on tributaries of the Yangzi. In late 2009, as the hydroelectric cult was again gaining steam in the Russian Federation, the Enisei River breached part of a dam causing a flood that killed 73 people.[67] As sediments pile up at the base of many of the twentieth-century hydroelectric behemoths, similar incidents will undoubtedly occur in the decades to come.

The prudent use of rivers will be critical for the survival of large portions of humanity in the twenty-first century. Dams will become defunct and new ones will be built. Preservationists will clamor against them. There will be no easy solutions. And while history will never provide a compass for leading us into the future, it can provide some guideposts to help us reflect upon the course that we choose. The one thing that history has shown is that the growth of ambition to manipulate rivers has been commensurate with the technological capacity to do so. Yet while the growth of countervailing ideologies, such as environmentalism, has in many cases added breadth to the discourse about the use of rivers and a brake on the speed of development, our collective wisdom to use rivers intelligently still seems to lag behind our desire to transform them.

Notes

1   See D. Worster, *Rivers of Empire: Water, Aridity, and the Growth of the American West*, New York, Oxford University Press, 1985; R. White, *The Organic Machine: The Transformation of the Columbia River*, New York, Hill and Wang, 1994; and M. Cioc, *The Rhine: An Eco-Biography*, Seattle, University of Washington Press, 2000.

2   R. McC. Adams, *Land behind Baghdad: A History of Settlement on the Diyala Plains*, Chicago, University of Chicago Press, 1965, p. 33; P. Bogucki, *The Origins of Human Society*, Oxford, Blackwell, 1999, pp. 167, 355, 359; C. W. Cowan and P. J. Watson, *The Origins of Agriculture: An International Perspective*, Tuscaloosa, AL, University of Alabama Press, 2006, p. 13; V. Scarborough, *The Flow of Power: Ancient Water Systems and Landscapes*, Santa Fe, NM, SAR Press, 2003, p. 9.

3   T. Tvedt, and E. Jacobsson (eds.), *A History of Water, vol. 3, The World of Water*, New York, I. B. Tauris, 2006, p. 111.

4   N. Brown, "Wittfogel and Hydraulic Despotism," in T. Tvedt, E. Jacobsson, R. Coopey, and T. Oestigaard (eds.), *A History of Water, vol. 2, The Political Economy of Water*, New York, I. B. Tauris, 2006, pp. 103–16 (p. 109).

5   C. Ponting, *A Green History of the World*, New York, Penguin Books, 2009; Adams, *Land behind Baghdad*, pp. 50–3.

6   Bogucki, *The Origins of Human Society*, p. 251.

7   S. Krech, *The Ecological Indian*, New York, W. W. Norton, 1999, p. 55.

8   D. R. Abbott (ed.), *Centuries of Decline during the Hohokam Classic Period at Pueblo Grande*, Tuscon, University of Arizona Press, 2003, pp. 53, 59, 227.

9   R. D'Souza, *Drowned and Dammed: Colonial Capitalism and Flood Control in Eastern India*, Oxford, Oxford University Press, 2006, p. 3; P. McCully, *Silenced Rivers: The Ecology and Politics of Large Dams*, London, Zed, 2001, p. 165; I. Stone, *Canal Irrigation in British India*, Cambridge, Cambridge University Press, 2002; Tvedt and Jacobssen, *Water History in World History*, introduction.

10 J. Waterbury, *Hydropolitics of the Nile Valley*, Syracuse, NY, Syracuse University Press, 1979, p. 12; A. Mikhail, *Nature and Empire in Ottoman Egypt: An Environmental History*, Cambridge, Cambridge University Press, 2011, pp. 242–81; D. Worster, "Water in the Age of Imperialism and Beyond," in Tvedt and Jacobsson (eds.), *A History of Water*, vol. 3, *The World of Water*, pp. 5–18.

11 M. Adas, "Continuity and Transformation: Colonial Rice Frontiers and Their Environmental Impact on the Great River Deltas of Mainland Southeast Asia," in E. Burke III and K. Pomeranz (eds.), *The Environment and World History*, Berkeley, University of California Press, 2009, pp. 191–210 (pp. 192–9).

12 I. Matley, "The Golodnaya Steppe: A Russian Irrigation Venture in Central Asia," *Geographical Review* 60/3, 1970, pp. 328–46 (p. 333). The cotton was of a very low quality as compared to the types that would later be imported from the US. By rotating crops, constructing low earthen walls, and planting mulberry trees, they prevented water-logging and salinization. These techniques were, in turn, adopted by the region's first kingdoms, but large-scale irrigation did not take place until after the Russian Empire took control of the region in the mid-nineteenth century.

13 FAN, *Vodohozyaystvyenniye Sistyemi Sryednyey Azii*, Tashkent, Izdatel'stvo FAN, 1971, p. 6; A. Khalid, "Backwardness and the Quest for Civilization: Early Soviet Central Asia in Comparative Perspective," *Slavic Review* 65/2, 2006, pp. 231–51.

14 F. Pearce, *When the Rivers Run Dry: Water – the Defining Crisis of the Twenty-First Century*, Boston, Beacon Press, 2006, p. 203.

15 A. Bekenov, A. Kovshar, and R. Jashenko, *The Problems of Animal World Conservation in the Northern and the Eastern Aral Sea Region: Diagnostic Study for the Development of an Action Plan for the Conservation of the Aral Sea*, Nairobi, UNEP, 1993, p. 51. From 1960 to 1990, the population of nesting birds diminished by 50 percent with the red-billed diving duck, river ducks, ibis, great cormorants, and white and dalmatian pelicans disappearing from the region.

16 D. R. Weiner, *A Little Corner of Freedom: Russian Nature Protection from Stalin to Gorbachev*, Berkeley, University of California Press, 1999, pp. 88–93; D. F. Duke, "Unnatural Union: Soviet Environmental Policies, 1950–1991," PhD thesis, University of Alberta, 1999.

17 McCully, *Silenced Rivers*, pp. 165, 237.

18 D. L. Umali, "Irrigation-Induced Salinity: A Growing Problem for Development and the Environment," World Bank Technical Paper Number 215, Washington, DC, 2009, pp. 54–5; McCully, *Silenced Rivers*, p. 206; World Bank, Policy Paper No. 135, 1993.

19 J. Blair and H. Hamerow, *Waterways and Canal Building in Medieval England*, Oxford, Oxford University Press, 2007, p. 3; G. Kerner, *The Urge to the Sea: The Course of Russian History, The Role of Portages, Ostrogs, Monasteries, and Furs*, Berkeley, University of California Press, 1946, p. 65; L. B. Kurat, "The Volga–Don Canal," *Imago Mundi* 10, 1953, pp. 97–8 (p. 98); E. B. Anisimov, *The Reforms of Peter the Great: Progress through Coercion in Russia*, London, M. E. Sharpe, 1993, p. 48; S. L. Zhitkov, *Proekti Soedineniya Vodnikh Putei Rossii*, Saint Petersburg, Sankt Peterburgskaya Tipographiya, 1908, pp. 42–4. Designed to curb Russian southerly advances into the North Caucasus, the Volga–Don Canal was largely forgotten until Suleyman revived the project in the mid-sixteenth century. He planned it to be a staging point for an attack on Astrakhan. His successor, Selim II, sent over 20,000 men up the Don River to begin the project in 1569, but he cancelled it only a few weeks after work commenced.

20 Worster, "Water in the Age of Imperialism and Beyond," p. 5; L. Van Slyke, *Yangtze: Nature, History, and the River*, Stanford, CA, Stanford Alumni Association, 1988. Historical records mention the construction of a 418-kilometer (260-mile) link between the Yellow River and the tributaries of the Huai River system by the name of the Wild Goose canal as early as the sixth century BCE. Archeologists, however, have yet to confirm to this day exactly where this canal lay.

21 Van Slyke, *Yangtze*, pp. 66, 69, 76, 78.

22 M. Elvin, *The Retreat of the Elephants: An Environmental History of China*, New Haven, CT, Yale University Press, 2004, pp. 131–40.

23 Burke and Pomeranz, *The Environment and World History*, p. 132.

24 Blair and Hamerow, *Waterways and Canal Building*, pp. 4, 8–10.

25 A. S. Nikolaev, *Kratkii Istoricheski Ocherk: Razvitiya Vodyanikh i Sukhoputnikh Soobsheniya Torgovikh Portov v Rossii*, Saint Petersburg, Minister Putei Soobscheniya, 1900, p. 12.

26 E. G. Istomina, *Vodni Transport Rossii: V Doreformenni Period*, Moscow, Nauka, 1991, p. 21.

27 Weiner, A Little Corner of Freedom.

28 Cioc, *The Rhine*, pp. 83, 101.

29 C. E. Closmann, "Holding the Line: Pollution, Power, and Rivers in Yorkshire and the Ruhr, 1850–1990," in C. Mauch and T. Zeller (eds.), *Rivers in History: Perspectives on Waterways in Europe and North America*, Pittsburgh, University of Pittsburgh Press, 2008, pp. 89–109 (p. 101).

30 Closmann, "Holding the Line," p. 3.

31 Cioc, *The Rhine*, p. 150; J. R. McNeill, *Something New under the Sun: An Environmental History of the Twentieth-Century World*, New York, W. W. Norton, 2000. For more on Rhine pollution, see P. H. Nienhuis, *Environmental History of the Rhine–Meuse Delta*, New York, Springer, 2008, pp. 329–52.

32 J. Schneer, *The Thames*, New Haven, CT, Yale University Press, 2005, p. 144.

33 Schneer, *The Thames*, p. 146; McNeill, *Something New under the Sun*, p. 131; H. Platt, *Shock Cities: The Environmental Transformation and Reform of Manchester and Chicago*, Chicago, University of Chicago Press, 2005, pp. 207–9.

34 Schneer, *The Thames*, p. 146; M. Melosi, *The Sanitary City: Urban Infrastructure in America from Colonial Times to the Present*, Baltimore, The Johns Hopkins University Press, 2000, p. 38.

35 M. I. Goldman, *Environmental Pollution in the Soviet Union: The Spoils of Progress*, Boston, MIT, 1972, pp. 99, 102.

36 M. Feshbach and A. Friendly, *Ecocide in the USSR: Health and Nature under Siege*, New York, Basic Books, 1992, pp. 17, 125.

37 McNeill, *Something New under the Sun*, p. 129; "Ganga Action Plan Bears no Fruit," *The Hindu*, August 28, 2004, accessed from http://www.hindu.com/2004/08/28/stories/2004082807430400.htm on March 9, 2012.

38 N. A. Smith, *A History of Dams*, London, Peter Davies, 1971, pp. 1, 8–9, 26, 131.

39 H. Ritvo, *The Dawn of Green: Manchester, Thirlmere, and Modern Environmentalism*, Chicago, University of Chicago Press, 2009; McCully, *Silenced Rivers*, p. 14.

40 T. Steinberg, *Nature Incorporated: Industrialization and the Waters of New England*, New York, Cambridge University Press, 1991, p. 164.

41 For more on hydroenergy as a panacea and the utopian visions of its champions, see R. Talbert, *FDR's Utopian: Arthur Morgan of the TVA*, Jackson, MS, University Press of Mississippi, 1987.

42 K. Pomeranz, "The Transformation of China's Environment, 1500–2000," in Burke and Pomeranz (eds.), *The Environment and World History*, pp. 118–64 (p. 134).

43 McCully, *Silenced Rivers*, p. 8. For more on developmentalism, see Burke and Pomeranz, *The Environment and World History*; on high-modernist utopianism, see J. C. Scott, *Seeing Like a State: How Certain Schemes to Improve the Human Condition Have Failed*, New Haven, CT, Yale University Press, 1999.

44 For general treatments on water and the West see Worster, *Rivers of Empire*; M. Reisner, *Cadillac Desert: The American West and Its Disappearing Water*, New York, Penguin Books, 1993; N. Hundley, *The Great Thirst: Californians and Their Water, a History*, Berkeley, University of California Press, 2001.

45 D. Worster, *A River Running West: The Life of John Wesley Powell*, New York, Oxford University Press, 2002, p. 179; P. L. Fradkin, *A River No More: The Colorado River and the West*, Berkeley, University of California Press, 1996. For a counterpoint to the dominance of the federal government in dam-building, see K. Brooks, *Private Power, Public Dams*, Seattle, University of Washington Press, 2009.

46 Empire is the concept which Worster uses in *Rivers of Empire* to explain the development of hydropower in the West.

47 Worster, *Rivers of Empire*; Hundley, *The Great Thirst*; R. W. Righter, *The Battle over Hetch Hetchy: America's Most Controversial Dam and the Birth of Modern Environmentalism*, Oxford, Oxford University Press, 2006, pp. 46, 59.

48 White, *The Organic Machine*, pp. 59, 97.

49 D. M. Yurinov (ed.), *Water Power and Construction of Complex Hydraulic Works during Fifty Years of Soviet Rule*, Washington, DC, United States Department of the Interior, 1978, p. 574.

50 A. D. Rassweiler, *The Generation of Power: The History of Dneprostroi*, Oxford, Oxford University Press, 1988, p. 3.

51 B. Komarov, *The Destruction of Nature in the Soviet Union*, Pluto Press, 1978, p. 61.

52 M. Davidov and T. Tsunts, *Ot Volkhova do Amura*, Moscow, Sovetskaya Rossiya, 1958, p. 325.

53 I. M. Tsunts, *Velikie stroiki na rekakh Sibiri*, Moscow, Gosizdat polit lit, 1956, p. 43.

54 Hill and Gaddy argue that the resources dedicated to making Siberian cities habitable ultimately crippled the Soviet state; F. Hill and C. C. Gaddy, *The Siberian Curse: How Soviet Planners Left Russia Out in the Cold*, Washington, DC, Brookings Institution, 2003.

55 Worster, "Water in the Age of Imperialism and Beyond," p. 11; McCully, *Silenced Rivers*, p. 6.

56 S. Robinson, K. Strzepek, M. El-Said, and H. Lofgren, "The High Dam at Aswan," in R. Bhatia, R. Cesti, M. Scatasta, and R. P. S. Malik, *Indirect Economic Impacts of Dams: Case Studies from India, Egypt and Brazil*, Washington, DC, World Bank, 2008, pp. 227–74; G. Kohlehepp, *Itaipu: Basic Geopolitical and Energy Situation – Socio-Economic and Ecological Consequences of the Itaipu Dam and Reservoir on the Rio Parana*, Wiesbaden, Vieweg und Sohn, 1987; McCully, *Silenced Rivers*, pp. 12, 240.

57 McCully, *Silenced Rivers*, pp. 4, 66, 134.

58 McCully, *Silenced Rivers*, p. 2.

59 E. G. Thukral (ed.), *Big Dams, Displaced People: Rivers of Sorrow, Rivers of Change*, New Delhi, Sage, 1992; M. S. Dreze and S. Singh, *The Dam and the Nation: Displacement and Resettlement in the Narmada Valley*, Delhi, Oxford University Press, 1997, p. 6.

60 M. Mennon, *Large Dams for Hydropower in Northeast India*, Kalpavriksh, South Asia Network on Dams, Rivers and People, 2005.

61 S. Khagram, *Dams and Development: Transnational Struggles for Water and Power*, Ithaca, NY, Cornell University Press, 2004, p. 2.

62 V. V. Kakakin, A. V. Mulin, and I. L. Dmitrieva, "Lessons from an Examination by Experts of the Katun Hydroelectric Station Project," *Gidrotekhnicheskoe Stroitel'stvo* 19, 1993, p. 571.

63 McCully, *Silenced Rivers*, p. xxii; World Commission on Dams, *Dams and Development: A New Framework for Decision-Making*, London, Earthscan, 2000.

64 For more on dissenting Chinese engineers see Z. Jianhong, "The Three Gorges Project: An Enormous Environmental Disaster," in P. Adams and J. Thibodeau (eds.), *Yangtze, Yangtze*, London, Earthscan, 1994, pp. 189–209.

65 L. R. Sullivan, "Introduction: The Three Gorges Dam and the Chinese Polity," in Adams and Thibodeau (eds.), *Yangtze, Yangtze*, pp. xiv–xx (p. xiv).

66 Y. Tan, *Resettlement in the Three Gorges Project*, Hong Kong, Hong Kong University Press, 2008, p. 2.

67 McCully, *Silenced Rivers*, pp. 114, 117; D. Verkhoturov, "Rusgidro net dostoino pomoshi," *Agenstvo Politicheskoi Novosti*, 2009, accessed from http://www.apn.ru/publications/comments21901.htm on March 9, 2012.

References

Abbott, D. R. (ed.), *Centuries of Decline during the Hohokam Classic Period at Pueblo Grande*, Tuscon, University of Arizona Press, 2003.

Adams, R. McC., *Land behind Baghdad: A History of Settlement on the Diyala Plains*, Chicago, University of Chicago Press, 1965.

Adas, M., "Continuity and Transformation: Colonial Rice Frontiers and Their Environmental Impact on the Great River Deltas of Mainland Southeast Asia," in E. Burke III and K. Pomeranz (eds.), *The Environment and World History*, Berkeley, University of California Press, 2009, pp. 191–210.

Anisimov, E. B., *The Reforms of Peter the Great: Progress through Coercion in Russia*, London, M. E. Sharpe, 1993.

Bekenov, A., Kovshar, A., and Jashenko, R., *The Problems of Animal World Conservation in the Northern and the Eastern Aral Sea Region: Diagnostic Study for the Development of an Action Plan for the Conservation of the Aral Sea*, UNEP, Nairobi, 1993.

Blair, J., and Hamerow, H., *Waterways and Canal Building in Medieval England*, Oxford University Press, Oxford, 2007.

Bogucki, P., *The Origins of Human Society*, Blackwell, Oxford, 1999.

Brooks, K., *Private Power, Public Dams*, Seattle, University of Washington Press, 2009.

Brown, N., "Wittfogel and Hydraulic Despotism," in T. Tvedt, E. Jacobsson, R. Coopey, and T. Oestigaard (eds.), *A History of Water*, vol. 2, *The Political Economy of Water*, New York, I. B. Tauris, 2006, pp. 103–16.

Burke, E., III, and Pomeranz, K. (eds.), *The Environment and World History*, University of California Press, Berkeley, 2009.

Cioc, M., *The Rhine: An Eco-Biography*, Seattle, University of Washington Press, 2000.

Closmann, C. E., "Holding the Line: Pollution, Power, and Rivers in Yorkshire and the Ruhr, 1850–1990," in C. Mauch and T. Zeller (eds.), *Rivers in History: Perspectives on Waterways in Europe and North America*, Pittsburgh, University of Pittsburgh Press, 2008, pp. 89–109.

Cowan, C. W., and Watson, P. J., *The Origins of Agriculture: An International Perspective*, Tuscaloosa, AL, University of Alabama Press, 2006.

Davidov, M., and Tsunts, T., *Ot Volkhova do Amura*, Moscow, Sovetskaya Rossiya, 1958.

Dreze, M. S., and Singh, S., *The Dam and the Nation: Displacement and Resettlement in the Narmada Valley*, Delhi, Oxford University Press, 1997.

D'Souza, R., *Drowned and Dammed: Colonial Capitalism and Flood Control in Eastern India*, Oxford, Oxford University Press, 2006.

Duke, D. F., "Unnatural Union: Soviet Environmental Policies, 1950–1991," PhD thesis, University of Alberta, 1999.

Elvin, M., *The Retreat of the Elephants: An Environmental History of China*, New Haven, CT, Yale University Press, 2004.

FAN, *Vodohozyaystvyenniye Sistyemi Sryednyey Azii*, Tashkent, Izdatel'stvo FAN, 1971.

Feshbach, M., and Friendly, A., *Ecocide in the USSR: Health and Nature under Siege*, New York, Basic Books, 1992.

Fradkin, P. L., *A River No More: The Colorado River and the West*, Berkeley, University of California Press, 1996.

"Ganga Action Plan Bears no Fruit," *The Hindu*, August 28, 2004, accessed from http://www.hindu.com/2004/08/28/stories/2004082807430400.htm on March 9, 2012.

Goldman, M. I., *Environmental Pollution in the Soviet Union: The Spoils of Progress*, Boston, MIT, 1972.

Hill, F., and Gaddy, C. C., *The Siberian Curse: How Soviet Planners Left Russia Out in the Cold*, Washington, DC, Brookings Institution, 2003.

Hundley, N., *The Great Thirst: Californians and Their Water, a History*, Berkeley, University of California Press, 2001.

Istomina, E. G., *Vodni Transport Rossii: V Doreformenni Period*, Moscow, Nauka, 1991.

Jianhong, Z., "The Three Gorges Project: An Enormous Environmental Disaster," in P. Adams and J. Thibodeau (eds.), *Yangtze, Yangtze*, London, Earthscan, 1994, pp. 189–209.

Kakakin, V. V., Mulin, A. V., and Dmitrieva, I. L., "Lessons from an Examination by Experts of the Katun Hydroelectric Station Project," *Gidrotekhnicheskoe Stroitel'stvo* 19, 1993, p. 571.

Kerner, G., *The Urge to the Sea: The Course of Russian History, The Role of Portages, Ostrogs, Monasteries, and Furs*, Berkeley, University of California Press, 1946.

Khagram, S., *Dams and Development: Transnational Struggles for Water and Power*, Ithaca, NY, Cornell University Press, 2004.

Khalid, A., "Backwardness and the Quest for Civilization: Early Soviet Central Asia in Comparative Perspective," *Slavic Review* 65/2, 2006, pp. 231–51.

Kohlehepp, G., *Itaipu: Basic Geopolitical and Energy Situation – Socio-Economic and Ecological Consequences of the Itaipu Dam and Reservoir on the Rio Parana*, Wiesbaden, Vieweg und Sohn, 1987.

Komarov, B., *The Destruction of Nature in the Soviet Union*, Pluto Press, 1978.

Krech, S., *The Ecological Indian*, New York, W. W. Norton, 1999.

Kurat, L. B., "The Volga–Don Canal," *Imago Mundi* 10, 1953, pp. 97–8.

Matley, I., "The Golodnaya Steppe: A Russian Irrigation Venture in Central Asia," *Geographical Review* 60/3, 1970, pp. 328–46.

McCully, P., *Silenced Rivers: The Ecology and Politics of Large Dams*, London, Zed, 2001.

McNeill, J. R., *Something New under the Sun: An Environmental History of the Twentieth-Century World*, New York, W. W. Norton, 2000.

Melosi, M., *The Sanitary City: Urban Infrastructure in America from Colonial Times to the Present*, Baltimore, The Johns Hopkins University Press, 2000.

Mennon, M., *Large Dams for Hydropower in Northeast India*, Kalpavriksh, South Asia Network on Dams, Rivers and People, 2005.

Mikhail, A., *Nature and Empire in Ottoman Egypt: An Environmental History*, Cambridge, Cambridge University Press, 2011.

Nienhuis, P. H., *Environmental History of the Rhine–Meuse Delta*, New York, Springer, 2008.

Nikolaev, A. S., *Kratkii Istoricheski Ocherk: Razvitiya Vodyanikh i Sukhoputnikh Soobsheniya Torgovikh Portov v Rossii*, Saint Petersburg, Minister Putei Soobscheniya, 1900.

Pearce, F., *When the Rivers Run Dry: Water – the Defining Crisis of the Twenty-First Century*, Boston, Beacon Press, 2006.

Platt, H., *Shock Cities: The Environmental Transformation and Reform of Manchester and Chicago*, Chicago, University of Chicago Press, 2005.

Pomeranz, K., "The Transformation of China's Environment, 1500–2000," in E. Burke III and K. Pomeranz (eds.), *The Environment and World History*, Berkeley, University of California Press, 2009, pp. 118–64.

Ponting, C., *A Green History of the World*, New York, Penguin Books, 2009.

Rassweiler, A. D., *The Generation of Power: The History of Dneprostroi*, Oxford, Oxford University Press, 1988.

Reisner, M., *Cadillac Desert: The American West and Its Disappearing Water*, New York, Penguin Books, 1993.

Righter, R. W., *The Battle over Hetch Hetchy: America's Most Controversial Dam and the Birth of Modern Environmentalism*, Oxford, Oxford University Press, 2006.

Ritvo, H., *The Dawn of Green: Manchester, Thirlmere, and Modern Environmentalism*, Chicago, University of Chicago Press, 2009.

Robinson, S., Strzepek, K., El-Said, M., and Lofgren, H., "The High Dam at Aswan," in R. Bhatia, R. Cestti, M. Scatasta, and R. P. S. Malik, *Indirect Economic Impacts of Dams: Case Studies from India, Egypt and Brazil*, Washington, DC, World Bank, 2008, pp. 227–74.

Scarborough, V., *The Flow of Power: Ancient Water Systems and Landscapes*, Santa Fe, NM, SAR Press, 2003.

Schneer, J., *The Thames*, New Haven, CT, Yale University Press, 2005.

Scott, J. C., *Seeing Like a State: How Certain Schemes to Improve the Human Condition Have Failed*, New Haven, CT, Yale University Press, 1999.

Smith, N. A., *A History of Dams*, London, Peter Davies, 1971.

Steinberg, T., *Nature Incorporated: Industrialization and the Waters of New England*, New York, Cambridge University Press, 1991.

Stone, I., *Canal Irrigation in British India*, Cambridge, Cambridge University Press, 2002.

Sullivan, L. R., "Introduction: The Three Gorges Dam and the Chinese Polity," in P. Adams and J. Thibodeau (eds.), *Yangtze, Yangtze*, London, Earthscan, 1994, pp. xiv–xx.

Talbert, R., *FDR's Utopian: Arthur Morgan of the TVA*, Jackson, MS, University Press of Mississippi, 1987.

Tan, Y., *Resettlement in the Three Gorges Project*, Hong Kong, Hong Kong University Press, 2008.

Thukral, E. G. (ed.), *Big Dams, Displaced People: Rivers of Sorrow, Rivers of Change*, New Delhi, Sage, 1992.

Tsunts, I. M., *Velikie stroiki na rekakh Sibiri*, Moscow, Gosizdat polit lit, 1956.

Tvedt, T., and Jacobsson, E. (eds.), *A History of Water, vol. 3, The World of Water*, New York, I. B. Tauris, 2006.

Umali, D. L., "Irrigation-Induced Salinity: A Growing Problem for Development and the Environment," World Bank Technical Paper Number 215, Washington, DC, 2009.

Van Slyke, L., *Yangtze: Nature, History, and the River*, Stanford, CA, Stanford Alumni Association, 1988.

Verkhoturov, D., "Rusgidro net dostoino pomoshi," *Agenstvo Politicheskoi Novosti*, 2009, accessed from http://www.apn.ru/publications/comments21901.htm on March 9, 2012.

Waterbury, J., *Hydropolitics of the Nile Valley*, Syracuse, NY, Syracuse University Press, 1979.

Weiner, D. R., *A Little Corner of Freedom: Russian Nature Protection from Stalin to Gorbachev*, Berkeley, University of California Press, 1999.

White, R., *The Organic Machine: The Transformation of the Columbia River*, New York, Hill and Wang, 1994.

World Bank, Policy Paper No. 135, 1993.

World Commission on Dams, *Dams and Development: A New Framework for Decision-Making*, London, Earthscan, 2000.

Worster, D., *A River Running West: The Life of John Wesley Powell*, New York, Oxford University Press, 2002.

Worster, D., *Rivers of Empire: Water, Aridity, and the Growth of the American West*, New York, Oxford University Press, 1985.

Worster, D., "Water in the Age of Imperialism and Beyond," in T. Tvedt and E. Jacobsson (eds.), *A History of Water, vol.3, The World of Water*, New York, I. B. Tauris, 2006, pp. 5–18.

Yurinov, D. M. (ed.), *Water Power and Construction of Complex Hydraulic Works during Fifty Years of Soviet Rule*, Washington, DC, United States Department of the Interior, 1978.

Zhitkov, S. L., *Proekti Soedineniya Vodnikh Putei Rossii*, Saint Petersburg, Sankt Peterburgskaya Tipographiya, 1908.

Further Reading

Adams, P., and Thibodeau, J. (eds.), *Yangtze, Yangtze*, London, Earthscan, 1994.

Berkman, R., and Viscusi, W. K. (eds.), *Damming the West*, New York, Grossman Publishers, 1973.

Bhatia, R., Cestti, R., Scatasta, M., and Malik, R. P. S., *Indirect Economic Impacts of Dams: Case Studies from India, Egypt and Brazil*, Washington, DC, World Bank, 2008.

Billington, D. P., and Jackson, D. C., *Big Dams and the New Deal Era: A Confluence of Engineering and Politics*, Norman, OK, University of Oklahoma, 2006.

Blackbourn, D., *The Conquest of Nature: Water, Landscape, and the Making of Modern Germany*, New York, W. W. Norton, 2006.

Collins, R. O., *The Nile*, New Haven, CT, Yale University Press, 2002.

Kelman, A., *A River and Its City: The Nature and Landscape in New Orleans*, Berkeley, University of California Press, 2003.

Kodiyanplakkal, J. J., "River Cult and Water Management Practices in Ancient India," in T. Tvedt and E. Jacobsson (eds.), *A History of Water, vol. 3, The World of Water*, New York, I. B. Tauris, 2006, pp. 385–408.

Leslie, J., *Deep Water: The Epic Struggle over Dams, Displaced People, and the Environment*, New York, Fararar, Straus, and Giroux, 2005.

Malitsev, V. F., *Bratskaia GES: Sbornik Dokumentov i materialov*, Irkutsk, Vostochnoe Sibirskoe Knizhnoe Izdatel'stvo, 1964.

Ramendra, N. N., *Ideology and Environment: Situating the Origin of Vedic Culture*, Delhi, Aakar Books, 2009.

Tvedt, T., *The River Nile in the Age of the British: Political Ecology and the Quest for Economic Power*, London, I. B. Tauris, 2004.

Wohl, E., *Virtual Rivers: Lessons from the Colorado Front Range*, New Haven, CT, Yale University Press, 2001.

Chapter Eighteen

War and the Environment

Richard P. Tucker

The environmental history of warfare and mass violence has emerged as a dimension of environmental historiography in the past decade. Although it is a new synthesis, it draws on many familiar subjects, from the history of state formations, social structures, and economics to military, demographic, and disease history, as well as historical geography. Military historians have routinely written about the significance of terrain and weather for the planning and management of campaigns. Moreover, they have frequently traced military planners' concern for manipulation of natural resources for their strategic purposes, and even the use of natural processes such as fire as weapons. But their interest lies almost exclusively with the human drama; they see nature as context, but not as consequence, of mass violence.

In contrast, environmental historians have often discussed elements of the history of warfare, such as the sites of military installations. But until recently the organizing focus of their work was rarely on the dynamics of mass violence or the structures of military operations in relation to state, society, economy, and ecology. Most recent studies of the environmental history of war have focused on the industrial era, beginning with the American Civil War, and thus have addressed the leading industrial countries. Centering at first on the global devastations caused by two world wars, studies have broadened to consider the structures and consequences of massive permanent military establishments, especially during the Cold War. Themes include the global reach of the major economies for control of strategic resources, and the impacts on economies and ecologies. Recognizing that most mass violence in recent decades has taken the form of civil wars and insurgencies, rather than wars between states, historians (like their counterparts in public and strategic policy) have begun to join ecologists and policy specialists in analyzing the processes and legacies of "resource wars" and (dis)organized violence.

Studies of premodern regions and long historical themes have also begun to appear, as building blocks toward a full global history. An outline is emerging, yet a full

A Companion to Global Environmental History, First Edition. Edited by J.R. McNeill and Erin Stewart Mauldin.
© 2012 John Wiley & Sons, Ltd. Published 2015 by John Wiley & Sons, Ltd.

perspective on the worldwide history of war's ecological consequences is still to develop in depth. This chapter is an overview of the embryonic synthesis, describing warfare's environmental consequences through global history.

Hunter-Gatherer and Sedentary Farming Cultures

Collective violence between structured social groups is as old as human societies themselves, and resulting ecological change has just as ancient a history.[1] What has been called "tribal warfare" has occurred throughout history, wherever indigenous cultures have not been displaced by modern regimes. In tribal wars the scale and duration of violence was highly localized, limited by the destructive power of hand-tools and the small scale of social organization. Border areas between two communities were perennially contested for control of food and other resources, and shifts in territorial control were frequent.

Attackers typically raided their enemy's fields and food supplies, both to seize resources and to cripple their foes. This often resulted in high mortality rates and depopulation of conquered areas, for raiders made little distinction between fighting men and the families that supported them. Humans suffered, and so did their crops and water sources. As historical anthropologist Lawrence Keeley expresses it, "Except in geographical scale, tribal warfare could be and often was total war in every modern sense. Like states and empires, smaller societies can make a desolation and call it peace."[2] However, as ecologist Jeffrey McNeely adds, "The greatest diversity of terrestrial species today is found in forested areas inhabited by tribal and other indigenous peoples, where relatively large areas of 'unoccupied' territory served as a sort of buffer zone between communities that may be embroiled ... in virtually constant warfare."[3]

Warfare between small communities had inadequate destructive power to cause lasting ecological change, for preindustrial weapons caused little collateral damage. The major exception to this was fire, throughout history the most destructive of tools. Its early ecological impact occurred mainly in semi-arid regions, such as much of Australia, where fire transformed landscapes into stable but species-reduced ecosystems.[4] In more humid climes, where fires did not easily spread, they could rarely provoke much ecological change, whether set in war or in peace. But in tribal settings it is almost arbitrary to distinguish the ecological consequences of collective violence from those of broader social processes, since violent and peaceful times were so closely interwoven. The emergence of state systems changed the scale of impact.

Urban Civilizations with State Systems

Here we meet more complex social and political/military organizations, with more sophisticated and powerful technologies that enabled a larger-scale and more protracted duration of warfare, and concomitantly greater impacts on the natural world. In these "command states," as William McNeill has called them, military-political elites organized entire societies around the constant threat of hostile neighbors.[5] Settled agriculture emerged some 10,000 years ago in the semi-arid lands of the Near East, and the first cities not long thereafter. In the Fertile Crescent of the Tigris and Euphrates river valleys, fortified cities required wood (or charcoal) for forging metal and building chariots, fortifications, battering rams, and siege machines, permanently reducing forest cover in many places. The organization of large-scale human labor made possible perennial agriculture through the construction of an elaborate system of irrigation canals that needed constant maintenance. Lowland areas suffered gradual ecological deterioration from siltation and waterlogging.[6] But irrigation systems could also be deliberately damaged. Repeated conflicts between city-states over water rights and territorial boundaries, including quarrels between upstream and downstream users, put irrigation systems at risk.[7]

In the lands farther west, in the northeast of the Mediterranean basin, ecological changes linked to warfare emerged more than 2,500 years ago, with the appearance of the early Greek city-states. The Mediterranean borderlands are naturally fragile, featuring long hot summers and short wet winters; their topography is mostly mountainous, with soils that are light and easily eroded once natural vegetation is removed. The early Greek city-states observed no overall leader, no imperial authority, and fought frequently among themselves. The leaders of the city-states and adjacent lands raised armies and navies, commanding the human and natural resources they required in order to confront their enemies.

Armies burned their enemies' forests and pillaged their farmlands. Rural people fled to safety in the hill forests or fortified towns ahead of advancing military columns. In the Peloponnesian War (431–404 BCE) that ended the golden age of Athens, the Spartan army repeatedly ravaged the farmlands of Attica, Athens' agricultural base, destroying crops in the effort to starve the city into submission. Attacks such as these were grim precursors of modern "total war," involving civilian targets as much as military. Ever since then writers have described the "devastation" wrought by those campaigns. But what constituted the devastation? Was it disruption of rural communities, destruction of their homes and other buildings, or the long-term productivity of their lands? The short-term impacts were obvious to everyone involved; the longer-term environmental results are more difficult to measure.[8] Many wars since then have posed the same conundrum.

In the same centuries naval warfare also caused ecological damage onshore, as the impacts of militarization spread into peacetime. At the mouths of rivers shipyards appeared, building ships of various designs for peaceable trade, and also the great triremes for naval warfare. Woodsmen gradually cleared forested watersheds upriver to meet the shipyards' needs. As local timber supplies ran low, strategic needs demanded control of more distant stands of timber. In the Peloponnesian War great naval battles between the Spartan alliance and the Athenian Empire destroyed hundreds of triremes; the prime trees of whole forests were lost as the ships burned and sank. This was a bitter irony, since one of Athens' purposes in the struggle for Sicily was to capture forests for shipbuilding.[9]

An entirely greater scale of ecological change resulted from the history of the Roman Republic and Empire. In the third century BCE the rising Roman Republic began to confront the Carthaginian Empire, based on the adjacent North African coastal lowlands, which was then the dominant maritime power in the western Mediterranean. Responding to defeats in the first Punic War (264–241 BCE), Roman commanders launched an intensive campaign of ship-building in numerous coastal cities around Italy and beyond. The environmental result was forest depletion in many local hill regions and watersheds, with consequent soil erosion and coastal siltation. By the time Carthaginian general Hannibal initiated the Second Punic War (219–201 BCE), the Roman fleet had become dominant. Thirteen years of annual campaigning in southern Italy impoverished the land, as both armies attempted to deprive each other of provisions. A half century later Roman armies, by then the largest and most professional (to that date) in the history of the western world, finally conquered and obliterated Carthage, consolidating Rome's control of North Africa. They damaged croplands and irrigation systems, but not severely, since Rome needed grain from today's northern Tunisia and Morocco to feed the capital city of 1 million people or more.[10]

In the course of these years Rome's command structure was geared to imperial wars, leading to inexorable expansion, looking northward from Italy on land. Rome confronted many enemies in Europe and beyond, mobilizing heavy infantry units to complement its cavalry. As the Roman armies moved northward in the conquest of Gaul and then southern Germany and Britain, their engineers built a system of all-weather roads so superbly engineered that some are still in use today. On the northern frontiers of the empire, as far as the Rhine and beyond, a string of fortified military cantonments sustained garrisons of troops. These military installations were the nuclei of the domestication of entire landscapes, as the roads and fortifications enabled peasants to clear hundreds of patches of forest for settled agriculture, even in the midst of chronic skirmishes between the Romans and their Germanic adversaries. Villages appeared on cleared forest land, where peasants worked to provide food and fiber for the garrisons.[11]

By the fifth century the Roman Empire in the north had fully collapsed in the face of its enemies. Population declined, trade networks and urban centers shriveled, and many cleared lands reverted to second-growth forest, often for centuries afterward. Indeed the same was true in much of the Italian peninsula as well. But both the human structures and the landscape changes survived the collapse of Roman authority into the early Middle Ages.

In contrast, in the arid Middle East, the Roman Empire survived in its Greek-speaking form, known as the Byzantine Empire. However, it soon faced formidable threats to its survival. Arab clans operating in border zones between arable and desert lands posed a perennial threat of pillaging raids on more highly developed civilizations. Starting with the Prophet Muhammad (570–632 CE), Islam introduced greater unity, discipline, and hierarchy into these Arab tribes. By the time the first Arab Muslim armies penetrated into the Fertile Crescent in the late 630s, social disruptions on irrigated lands resulting from frequent battles had resulted in long-term decline of the region's irrigation systems and agricultural productivity, and some reversion to pasture. The Muslim overlords of the Umayyad and then Abbasid dynasties encouraged the revival of rural productivity, partly to enhance state revenues, which were then used to further military campaigns. But long-term ecological decline could only be partially reversed.[12]

A blow of permanent proportions came from the northeast, when nomadic cavalry warfare erupted from the Eurasian steppes. In the long history of confrontations between pastoral and sedentary peoples, there has never been more appalling destruction than in the Mongols' conquest of other civilizations. But the ecological consequences were highly variable over the long run. In the vast Iranian plateau, cities on the Silk Road and other trade routes were located on rivers where reliable water supplies from rivers or adjacent mountains could be used to irrigate natural grasslands. Mongol cavalry conquered one city after another in the late 1200s. As one historian writes, the immediate impact

amounted to a holocaust. The populations of many cities and towns were systematically exterminated. Whole regions were depopulated by invading armies and by the influx of Turkish and Mongol nomads who drove the peasants from the land. The conquerors plundered their subjects, made them serfs, and taxed them ruinously. The result was a catastrophic fall in population, income, and state revenue ... Substantial territories were turned from agriculture into pasturage.[13]

The long-term consequences for irrigation systems and agriculture were highly variable, for some post-conquest rulers followed plunder with reconstruction. Yet, although some Mongol rulers rebuilt cities, repaired irrigation and farming systems, and fostered trade, the decline in productivity (both agricultural and biotic) of these managed landscapes was to some degree permanent.

In Iraq the drama had far more lasting results for both people and the land. There the Mongol invaders raced through the Abbasid Empire's heartland and captured its capital, Baghdad, in 1259, massacring the entire population of the city. "Baghdad and Iraq never again recovered their central position in the Islamic world. The immediate effects of the invasion were the breakdown of civil government and the consequent collapse of the elaborate irrigation works on which the country depended for its prosperity, even for its life."[14] Throughout the turbulent history of the Middle East since then, irrigation systems have been vulnerable targets for armies.[15]

The example of the Mongol conquests in southwest Asia illustrates one of the ambiguities in the environmental history of warfare. From the point of view of pastoralists such as the Mongols, the ideal biome was grassland suitable for raising livestock. The destruction of irrigation works, orchards, palm groves, and vineyards, if it led to more grass, horses, and sheep, amounted to environmental improvement. Over a generation or two, Mongol rulers sometimes changed their outlook, preferring the higher revenues that irrigated agriculture could provide over the expansion of pastureland and herds.

Han China experienced a contrasting range of long-term struggles with non-Chinese cultures in the northwestern steppes and the southern mountains.[16] On China's northwest frontier, facing perennial threats from nomadic warriors of the central Asian grasslands, emperors built the Great Wall and other fortifications and cleared forests on some adjacent lands for security. They also protected forest zones, especially in Manchuria, to guard against invading cavalry.[17] In contrast, they pursued a policy of imperial conquest in the southern frontier region, where the rugged Guizhou Plateau was home to a wide range of tribal cultures, especially the Miao, who resisted Han civilization for centuries. Like the Romans, the Han armies built roads and garrison settlements to move military columns and pacify the region, opening it to agricultural settlement and forest reduction by immigrants from the north. The long-term expansion of Chinese culture and of the Chinese state, frequently a matter of military conquest, also involved the eventual transformation of forest, wetland, and grassland into farmers' fields.[18]

South of the great Himalayan mountain barrier, monsoon India presented a very different climatic cycle. In most of the Indian subcontinent the three to four months of heavy summer monsoon rains alternate annually with a long dry season. Hindu kingdoms, especially in the Ganges basin, developed the capacity to mount extended wars for control of river basins and forested hilly hinterlands, employing weapons essentially similar to those of the Chinese in their destructive power. The increasing scale of military forces gradually became a key element in north Indian warfare. On long campaigns, huge royal armies led by elephant corps devoured food and fodder resources wherever they went. In the upper Ganges basin, from 997 CE onward, Muslim conquest states from farther northwest – the Delhi Sultanate and then the greatest of them all, the Mughal Empire – mounted sustained campaigns that had widespread environmental impacts. The Mughal imperial army at its apogee, in its late-seventeenth-century campaigns in central India, was a mobile city of nearly 1 million fighters, camp followers, and suppliers, who stripped wide areas of the land as they moved. Cavalrymen swept the countryside, depopulating villages, while their horses and the royal elephant corps required massive amounts of fodder.[19] Rural society and its biological base could take years or decades to recover from the disruption. That process has not been studied in detail; it seems likely that the locally complex and varied impacts of these dislocations on farm and forest were rarely permanent. The exception to this, in the Indian subcontinent as elsewhere, was areas of irrigated agriculture, both in semi-arid lowlands and on terraced mountain terrain, since both are vulnerable to siltation and erosion, which are difficult to reverse.

Throughout the premodern world, many conflicts took the form of frontier wars, fought between two non-state societies, two states, or as wars of conquest pursued by ambitious powers. Often protracted and intermittent, these wars were similar in many ways to modern guerrilla warfare and counterinsurgency, though they did not produce the devastation that is caused by today's counterinsurgency weapons. They were characterized by seasonal skirmishes and raids, fortified outposts, capture of loot including movable natural resources, and, probably most significant, the dislocation of rural populations. Many were fought in mountainous or hilly areas, on forested slopes with easily eroded soils.

To take just one of many instances, in the central Himalayas in the eighteenth century, the armies of the rising Gurkha state based in the Kathmandu valley of Nepal conquered peripheral hill areas of western Nepal and Kumaon. The conquerors destroyed crop terraces and irrigation systems, and cut hardwood timber as booty. In the monsoon climate, this resulted in soil erosion, permanently reducing the agricultural productivity of the land and density of forest cover, even after the British colonial army displaced the Gurkha legions and brought peace after 1815.[20]

Medieval and Early Modern Europe

In the wake of the fourth- and fifth-century invasions from the north that crippled Rome's empire, urban centers and trade networks shrank drastically. As populations fell, agricultural lands reverted to secondary woodlands until an era of population growth and forest clearances ushered in the high Middle Ages beginning as early as the tenth century.[21]

Warfare was endemic in medieval Europe just as elsewhere. Warriors' weapons were no more powerful than in earlier imperial days, so the major environmental impacts of organized violence revolved around fortifications and siege warfare. Here the importance of social hierarchies in shaping warfare's environmental impact is clear. Lords on manorial estates and the serfs who worked their lands were both warriors whenever military campaigning demanded. The norms and styles of chivalry evolved, expressed in dominance of war by mounted knights on heavy horses.

Lords of the land built massive fortifications surrounded by earthen ramparts with wooden palisades.[22] Sieges of these fortresses and fortified towns often lasted for entire summer seasons, when invading armies could be mobilized and maintained. Many campaigns were renewed each summer for years. Peasant armies, levied by their lords, foraged for food and fodder and sought to deny susbistence to their adversaries. They often destroyed the crops and woods they could not use themselves.[23] In the twilight zone between mass violence and peaceful times, including after campaigns were over and temporary troops disbanded, brigandage (often hardly distinguishable from regular soldiering) festered. Peasant populations were terrorized by raids on food and livestock. Lands deserted when rural people became refugees reverted toward natural woodlands and wetlands, with concomitantly increasing species diversity. The short-term damage to partially domesticated landscapes was evident to anyone with eyes.

Perhaps the greatest example of destruction was the Hundred Years War (1337–1453) in France. English armies crossed to northwestern France in repeated campaigns to conquer fortified cities and occupy their agricultural hinterlands.[24] Not yet professional armies, they were raised for the duration of specific campaigns, with impressed peasants fighting under their feudal lords. Alternatively, private armies of mercenaries were organized by adventurers, either paid by kings or more often given license to plunder. For provisions they lived off the land as they moved; their rewards were booty; officers were also granted lands to settle. This form of warfare had direct environmental consequences. Sieges of Calais and other fortified cities caused severe damage to the surrounding countryside, often from skirmishes between two sides. Just as in southern Italy in Hannibal's time, pillaging the land could be either random or systematic terror inflicted on the rural populace.

Warfare often coincided with disease epidemics; in tandem they reduced population. The greatest example in Eurasian history was the 1348–51 bubonic plague, which killed something like one-half of Europe's people in the midst of the Hundred Years War. The mortality was likely intensified for both military and civilian populations in the disrupted conditions of war zones.[25] In the postwar and post-plague stillness once-tilled farms were deserted, reverting to pasture or secondary woodlands where wildlife flourished and local biodiversity increased. In the longer run these changes were usually reversed, for farmers sooner or later renewed agricultural landscapes with the return of peace and security. In general terms, it took about 150 years for European population to recover to the levels before the plague, a process slowed by the intermittent ravages of war.

The destructive power of weapons began to accelerate when gunpowder was introduced into Europe from China in the 1300s, and was followed by the development of steadily more powerful cannons. In response, fortifications became far more elaborate by the 1500s. The Military Revolution was in full swing, accelerating arms races on both land and sea.[26] In the Thirty Years War (1618–48) northern and central Europe degenerated into chaos, as anarchic military bands repeatedly pillaged the land until the region reached a point of general exhaustion.[27]

In the aftermath much of Europe saw the emergence of centralizing states with ever-expanding professional armies, supported by vastly expanded government revenues and fiscal administration.[28] Disciplined armies with better organized supply lines meant reduced environmental damage in the lands of neutral populations. Though there had always been close relations between rulers and civilian suppliers, this era showed the clear emergence of a "military-industrial complex," in which governments coordinated closely with their suppliers. Taxation became more regular, as military economies became more systematized and provided support for accelerating lethality.[29] Bankers and merchants could follow the temptations of profiteering on a previously unknown scale – a driving force behind warfare, though not always visible. In all, Europe's expanding imperial states would lead toward both global conquest and ever-greater scale of destructive power in the industrial era.

Global Empires in the Early Modern Era

Except for the impacts of regional empires such as Rome, Han China, and Mughal India, the ecological consequences of wars were largely local until Europe began to extend its hegemony globally in the sixteenth century. Since then the world has changed fundamentally, as the era of the imperial nation-state and large-scale capital and industry accelerated the technological impacts associated with global trade and transport.[30]

The frontier wars of European conquest from around 1500 onward were the cutting edge. Over a half millennium, in a process completed in the twentieth century, European empires, later joined by the US, dismantled non-state societies in temperate forests, savanna lands, and tropical rain forests. Their arms, and their military-bureaucratic capacity to sustain warfare over long periods, had many long-range environmental consequences.[31]

Early ecological damage outside Europe reflected the navies' needs for construction timber and naval stores. By the 1700s, European navies began cutting the hardwood and white-pine stands of northeastern North America, the coastal hardwoods of Brazil, the mahogany and cedars of Cuba, and later the teak forests of monsoon Asia, to find substitutes for the depleted English oak and Scandinavian conifers.[32] These environmental costs of naval warfare were confined to the land; the seas themselves suffered little pollution or biological reduction from naval wars until the great wars of the twentieth century.

The most fundamental ecological impacts of Europe's global conquests occurred in the Americas, where Europeans brought with them epidemic diseases that inaugurated a holocaust for Native Americans. Up to 90 percent of the indigenous American population died by the late sixteenth century.[33] This depopulation led to widespread abandonment of cultivated lands and reversion to secondary forest, often for long periods. In Latin America even in the 1500s the impacts of conquest registered on lowland coastal zones and riverine forests, the highlands of Mexico and the Andes, where sheep and goats came to rule degraded pasturelands, and also the wide natural grasslands where cattle soon prevailed.[34]

In an ironic case of warfare and epidemic disease, by the 1700s Iberian-Americans became relatively resistant to malaria and yellow fever by virtue of growing up in lands where these infections were prevalent. The dreaded twin diseases were their allies in defending their colonial empire against armies from northern Europe, made up of men whose childhoods did not include experience with tropical disease, and who therefore remained fully susceptible as adults. Malaria and yellow fever played this partisan role in warfare until the collapse of the Spanish empire in the Americas during the Napoleonic Wars.[35]

In North American woodland ecosystems the impact of endemic frontier warfare was somewhat different. There Europeans were able to follow up their conquests by settling on the land and clearing temperate forests far more readily than they could anchor themselves in tropical rain-forest zones.[36] In contrast to Latin America, where populations did not recover to their pre-1492 levels until around 1800, the native populations of North America were fully replaced by North European immigrants in much shorter order. By 1800 they – and the slaves they owned – had cleared most of the eastern forests from Georgia to Quebec.

Wars of the Industrial Era

The great escalation of modern warfare and its environmental impacts began in Europe in the 1790s, when revolutionary France and Napoleon expanded both the intensity of warfare and its continent-wide reach.[37] Mass mobilization created huge semi-trained armies that were inadequately supplied. From 1793 onward French armies moved into Belgium and beyond, ravaging rural lands to the north as they moved. The era of patriotic armies had begun, though the disciplined logistics of the industrial era were only beginning to emerge.

The Napoleonic wars also disrupted intercontinental transport of food supplies, in one case resulting in a major long-term change in cropping patterns. The British naval blockade after 1805 cut off supplies of cane sugar from the Caribbean to French ports. In response, new techniques of extracting sugar from beets led to an explosion of sugar-beet farming in the heavy soils and cool climate of northern Europe. Meanwhile the former slaves of Haiti turned their labor from half-deserted cane plantations in the fertile lowlands to subsistence cropping in the erosive hill woodlands, and Haiti gradually became one of the most degraded landscapes in the Americas. In this way Europe's revolutionary wars had unintended ecological consequences across the ocean.[38]

From the mid-nineteenth century onward Western European and American industry produced a quantum expansion of destructive capacity, through revolutionary innovations in mass production. By the late 1800s highly accurate breech-loading Enfield, Mauser, and Springfield rifles and Maxim machine guns transformed the battlefield, and more powerful explosives began to ravage both urban and rural targets. Moreover, railroads and steamships gave industrialized nations far greater mobility and international reach. In addition to their civilian uses, they moved troops and materiel rapidly, inexpensively, and far, making possible the conquest of the rest of the world.[39]

Nineteenth-century Africa experienced the culmination of Europe's conquests.[40] Conquest and colonization began slowly, in a loose sense as early as the 1440s. But it gathered pace, especially in southernmost Africa after 1815, and extended to the entire continent by 1910. In southeastern Africa, for example, the Zulu wars of the early 1800s led to British control of the coastal lowlands and interior hills, and the Zulu people were gradually forced to settle on the semi-arid high plains of the interior.[41] Among the colonies that Germany claimed after 1885 was Tanganyika, now combined with Zanzibar to form Tanzania. The forest resources of Tanzania came under the management of officials trained in the authoritarian German forestry tradition, sharply restricting the rights of access and trade for the local people. With resource access among their grievances, many Tanzanians revolted in the Maji Maji rebellion of 1905–6, an early war of resistance against European colonial rule.[42] The flora and fauna resources of the colonies would see many contestations. But these first studies of the environmental impacts of Europe's conquest wars in sub-Saharan Africa give only fragmentary hints at the overall picture. Wars of colonial expansion in Africa and Asia inevitably involved ecological dimensions which future historians will perhaps illuminate.

At century's end in South Africa's Boer War, which pitted two white populations against each other, both the British colonial army and the Boers were armed with modern weapons. Savagery on both sides lasted far longer than anyone had anticipated. Under political pressure to bring the war to a close, the British commander, Lord Kitchener, resorted to environmental warfare, intended to undermine the ability of his enemies to feed themselves. As one historian reflects, "to shatter Boer resistance once and for all, Kitchener began systematically to destroy Boer farms ... Between 1900 and 1902 the British burned thousands of Boer farms and herded their wretched inhabitants into 'concentration camps'."[43] Sixty percent of the entire Boer population, some 120,000 Afrikaners, were moved into those camps, while their farms withered. A careful analysis of the environmental aftermath on those wide lands would be in order.

The US Civil War had already given a grim demonstration of the ecological dangers of the new industrial warfare. When it began in 1861, no one expected the war to grind on for over four years, but its glacial momentum toward exhaustion of the South produced widespread destruction of croplands and fodder resources by Northern armies, extending in its last two years to deliberate scorched-earth campaigns such as General Sherman's famous March to the Sea in its last two years. As so often when wars drag on, destruction of the enemy's fields and livestock seemed an acceptable way to end the conflict, even if it visited starvation on civilians.[44] That experience trained Northern soldiers to destroy the food supplies of Native Americans in the West, including their herds of bison, as a strategy in the conquest of that frontier.[45]

Although environmental historians have yet to tackle them, two other gigantic civil wars at the same time featured abundant environmental destruction as well. In China the Taiping Rebellion (1850–64), which killed about 20 million people – roughly 40 times as many as the US Civil War – involved armies rampaging through peasant landscapes in some of the most productive parts of the country. Simultaneously in India, the so-called Sepoy Rebellion or Munity of 1857–8, brought destruction to the irrigated landscapes of the Ganges valley. In both China and India, the prominence of irrigation infrastructure made it easy for an army to undermine the food-production capacity of its enemies.

The US Civil War also produced environmental impacts half a world away from the scenes of battle. The northern navy's blockade of southern ports interdicted the South's raw-cotton exports to Europe, forcing English cloth mills to look elsewhere for their supplies as long as the war lasted. On the rich black soils of the Malwa Plateau in central India, small farmers responded to a spike in cotton prices in Bombay in early 1862, switching from subsistence food crops to cotton for export to England.[46] Diverse lands from Egypt to Polynesia were sown to cotton in these years, and some of them remained in production for a century or more.

In Europe in the same decade of the 1860s, Germany harnessed the Industrial Revolution to accelerate military mobilization. Rapid victories over Austro-Hungary (1866) and then France (1870–1) resulted from skillful movement of the German armies over the new railway networks, with communications provided by the new telegraph, while more powerful artillery damaged woodlands and cities. Great Britain, faced with the new challenge from Germany, strove to maintain its control of the seas by producing rapid innovations in naval technology, which required that military planners and industrialists work closely together.[47] The mid-nineteenth-century wars and the concomitant arms race were merely overtures to the two world wars that followed, when the environmental impacts of warfare became truly global.

World War I

Contemporaries called this the Great War (1914–18), in which the military-industrial complex fatefully matured. As the first campaigns of the war bogged down in a four-year stalemate along hundreds of miles of trenches in Flanders and northeastern France, munitions were consumed to the fullest extent that each country's factories could produce them. Millions of bomb and shell craters left puddles, ponds, and mud where crops and woods had stood before. On both sides of the war, improved long-distance food transport enabled mass armies to be sustained year-round, and battles to be fought almost endlessly. On occasion, armies deliberately deprived both enemy units and civilians of food, fiber, and fodder, by ravaging land and destroying stored crops. In early 1917, as the German armies withdrew from the Somme battlefields, they systematically destroyed nearly every building, fence, well, bridge, and tree over an area 105 by 30 kilometers (65 by 20 miles), to deprive the advancing enemy of sustenance and cover.[48] Industrial-scale combat presumably proved ecologically destructive in other theaters besides the western front as well. Heavy fighting occurred in Iraq and Palestine, in Macedonia and at Gallipoli in Turkey, in northeastern Italy, and in the western reaches of what was then the Russian Empire. And on a small scale, combat took place in parts of Africa as well. The environmental dimensions of these conflicts deserve more attention than they have so far received.[49]

The war also saw the first large-scale use of chemical warfare: first the chlorine gas that German chemists had synthesized, and later mustard gas and other experimental chemicals. By the war's end chemical attacks produced 1.3 million casualties, including 90,000 deaths; mustard gas and other chemical agents temporarily poisoned lands on and near the battlefields. It is difficult to assess the immediate environmental impact because no one tried to record or measure it. But its carryover effect was extensive. Chemical warfare increased the scale of chemical industries, demonstrated the value of scientific research to governments, and inspired the postwar development of agricultural pesticides. And military aircraft became the backbone of postwar crop-dusting, increasing the scale on which pest control was economical.[50]

Throughout Europe and even overseas, forests came under unprecedented wartime pressures. Seemingly endless bombardments in battle zones shattered forests that had been carefully managed for centuries. Behind the lines for hundreds of miles, massive emergency fellings of timber were carried out. Only the great forest zone of Russia escaped heavy exploitation, since imperial Russia's transport system was still rudimentary. From beyond Europe the British, Canadians, and Americans organized large timber shipments from North America and even India's monsoon forests.[51] Perhaps equally important in the longer run, government forestry agencies in many countries took greater control over forest resources during the war. The immediate postwar recovery period saw reforestation programs in both Europe and North America, in which single-species tree plantations replaced the greater variety of species in the former natural forests.

World War II

Between the two world wars further acceleration of military industry enabled militarized states to mobilize far greater resources from around the world than a quarter-century before, and impose new levels of destruction. When Hitler's armies invaded Poland in late 1939, he did not envision a war in which 70 million people would die. His own country ultimately suffered some of the most total devastation, particularly at the hands of the Allied air forces. Incendiary bombs produced by the rapidly maturing chemical industry leveled 130 German cities, killing some 600,000 civilians. The postwar reconstruction, physical as well as social, would be daunting.[52]

In combat zones the forests of Europe were once again badly damaged by fighting. Behind the lines of combat, timber was cut at the most urgent rates that the limited available workforce could achieve, and great forests of Norway and Poland were looted of their timber wealth. This time, even more than in the previous war, the battle zones of Europe, North Africa, and the Middle East could call upon timber resources from other continents. Harvesting machinery and transport networks from forest roads to harbor facilities to oceanic ships were more highly developed than in the previous war. (Russia's vast forest resources east of the Urals were still largely inaccessible.)[53]

World War II marked another watershed in the history of warfare: for the first time in a major war more soldiers died in battle than of disease. In the Pacific theater, until 1943 malaria caused nearly ten times as many casualties as battles for the American forces. The tide turned with the new insecticide called DDT, which killed the mosquitoes that transmitted malaria and the lice that transmitted typhus. DDT almost totally controlled the disease among the troops before the war's end.[54] The surge in manufacturing of DDT, along with its miraculous reputation, created a massive postwar market. No one at the time foresaw the massive environmental damage which DDT would produce following peacetime.[55]

The war in the Pacific had impacts on both island systems and the aquatic environment that had no previous parallel in that ocean's web of life. Small islands support limited varieties of plant and animal species. Coral atolls have thin or fragile soils; they are exceptionally vulnerable to the impacts of human conflict. On both steep volcanic islands and coral atolls throughout the Pacific, the fighting produced fundamental ecological degradation of forests, watersheds, coastal swamplands, and coral reefs.

Diseases, of both humans and livestock, had spread into the Pacific with traumatic impacts ever since the 1770s, but the Pacific War accelerated the process in some instances. Allied disease-control teams succeeded in containing the spread of malaria from the islands where it was already widespread, but the cattle tick reached New Caledonia, infecting livestock and damaging agricultural systems there. In this war, as in others, the spread of epidemic diseases could accelerate longer trends in human and animal populations.

The war at sea had paradoxical effects on marine resources. Commercial fisheries and whaling fleets were largely destroyed, docked, or transformed into military uses until 1945, leaving fish stocks and marine-mammal populations to recover somewhat, though submarine warfare killed some whales, and any increase in their numbers would prove to be very temporary.[56]

In Japan itself the war had tragic ecological as well as human impacts. Before Pearl Harbor (1941), the Japanese war machine had attempted to command the mineral and forest resources of Southeast Asia, to compensate for its limited resource base at home, but that effort was stopped by Allied counterattacks. The war's impact on Japan's domestic forest resources was severe: the loss of timber imports (including from the northwest coast of North America) meant intensive forest fellings, even in ancient stands that had been preserved for centuries. The direct result was loss of soil and damage to water regimes. In Japan (just as in many countries) food production expanded urgently, especially on marginal lands.[57]

American incendiary bombing almost totally destroyed Japan's urban areas, which had been built largely of wood. And the ultimate environmental disaster, the impact of nuclear bombs, was also Japan's fate, when Hiroshima and Nagasaki were leveled on August 6 and 9, 1945. The two cities were rapidly rebuilt after the war, and the local flora made a surprisingly rapid recovery from radioactive pollution, yet the human costs of the two bombs are still being counted.

In the course of the war the US suffered relatively little long-term damage to its domestic resources and ecosystems or to its additional source areas in Latin America. In sharp contrast, its allies, Britain and the USSR, suffered extensive economic and environmental damage from combat. American military industry had grown exponentially, and military-industrial coordination had reached high levels. Hence World War II sowed the seeds of later troubles, which began to be evident as the Cold War deepened after 1948.

The Late Twentieth Century

The global arms race after 1945 produced incalculable acceleration of every tool of destruction. One of the smallest weapons, though multiplied almost countless times, has been the land mine. Some 100 million unexploded anti-personnel mines remain around the planet now, littering rural Vietnam, Cambodia, Afghanistan, Angola, and many other war-torn countries, and grievously retarding the restoration of postwar farms, pastures, forests, and water regimes. These and a Pandora's box of other weapons have spread through many unstable regions of the postcolonial world, in Africa and elsewhere. Grim contributions to wars both civil and transboundary, they have also extracted a wide-spread ecological toll on forests, savannas, and farmlands.[58]

Equally widespread by the time the Cold War ended in 1991, long-term pollution effects of military industry left many locations severely poisoned. Weapons-production sites and testing grounds in the US required massively expensive cleanups of a broad spectrum of toxic wastes. Even more appalling, large areas of Soviet and Eastern European land and air had become virtual wastelands, and even the Arctic Ocean north of Russia was severely polluted.[59]

Chemical warfare reached a new level of destruction in the Second Vietnam War (1961–75), as the US Air Force applied Agent Orange and other defoliants to the forests of Indochina. In addition to 14 million tons of bombs and shells, American planes sprayed 44 million liters of Agent Orange and 28 million liters of other defoliants over Vietnam. The result was serious damage to 1.7 million hectares of upland forest and mangrove marshes, widespread soil poisoning or loss of soil, and destruction of wildlife and fish habitat.[60] Most ominous of all, nuclear technology designed for both bombs and power plants became the greatest environmental threat in history, though its primary impact resulted from the peacetime armament race rather than from actual warfare. Until international nuclear-testing freeze conventions came into effect, locations such as Soviet sites in Central Asia and Britain's testing grounds in central Australia became uninhabitable for almost all forms of life. And in the southern Pacific Ocean, islands and their coastal reefs, their civilian populations removed, became unfit for life as a result of American and French nuclear-weapons testing.[61] Beyond that, in the nuclear industrial complex, many weapons-production and storage sites became highly radioactive. In the US, nuclear facilities in Washington state, Colorado, and elsewhere became radioactive sewers. Soviet nuclear-weapons sites were even more devastated.[62]

Finally, twentieth-century warfare has made a major contribution to warming of the global atmosphere. The Persian Gulf War of 1991 was the most notorious case of atmospheric pollution in wartime, as the plumes of burning oil wells darkened skies for months far downwind. It now seems that the fires caused less regional and global air pollution than was feared in their immediate aftermath, though they dropped heavy pollution on nearby desert, farmland, and the Gulf's waters.[63] More pervasively important, military establishments consume great amounts of fossil fuels, contributing directly to global warming.

Conclusion

In the long perspective of global environmental history, warfare and the preparation for war stand out as a central dimension of how societies, states, and economies have been organized. In the industrial era the accelerating scale of technology and social institutions has brought many benefits to humanity. But the price has been high, and it is accelerating. Accelerating damage to the biosphere is the ultimate testimony to that price.

Notes

1 For the origins of warfare, see L. H. Keeley, *War before Civilization*, Oxford, Oxford University Press, 1996; R. C. Kelly, *Warless Societies and the Origin of War*, Ann Arbor, MI, University of Michigan Press, 2000; and K. F. Otterbein, *The Evolution of War: A Cross-Cultural Study*, New Haven, CT, HRAF, 1970. For the early history of species extinctions see P. S. Martin and R. G. Kein (eds.), *Quaternary Extinctions: A Prehistoric Revolution*, Tucson, AZ, University of Arizona Press, 1984.

2 Keeley, *War before Civilization*, p. 108.

3 J. A. McNeely, "Biodiversity, War, and Tropical Forests," *Journal of Sustainable Forestry* 16/3, 2003, pp. 1–20 (p. 2).

4 T. Flannery, *The Future Eaters: An Ecological History of the Australian Lands and People*, New York, George Braziller, 1994; S. J. Pyne, *World Fire: The Culture of Fire on Earth*, New York, Holt, 1995.

5 W. H. McNeill, *The Pursuit of Power: Technology, Armed Force, and Society since AD 1000*, Chicago, University of Chicago Press, 1982, chapter 1.

6 For the general process, see T. Jacobsen and R. M. Adams, "Salt and Silt in Ancient Mesopotamian Agriculture," *Science* 128, 1958, pp. 1251–8.

7 P. H. Gleick, "Water, War and Peace in the Middle East," *Environment* 36/3, 1994, pp. 6–15, 35–42 (pp. 10–11); also H. Hatami and P. H. Gleick, *Chronology of Conflict over Water in the Legends, Myths, and History of the Ancient Middle East*, Oakland, CA, Pacific Institute, 1993.

8 J. D. Hughes, *Pan's Travail: Environmental Problems of the Ancient Greeks and Romans*, Baltimore, The Johns Hopkins University Press, 1994; J. V. Thirgood, *Man and the Mediterranean Forest: A History of Resource Depletion*, London, Academic Press, 1981; V. D. Hanson, *Warfare and Agriculture in Classical Greece*, Berkeley, University of California Press, 1998.

9 R. Meiggs, *Trees and Timber in the Ancient Mediterranean World*, Oxford, Clarendon Press, 1982.

10 A. Goldsworthy, *Roman Warfare*, London, Cassell, 2000.

11 A. Goldsworthy, *How Rome Fell*, New Haven, CT, Yale University Press, 2009; C. R. Whittaker, *Frontiers of the Roman Empire: A Social and Economic Study*, Baltimore, The Johns Hopkins University Press, 1994.

12 J. M. Wagstaff, *The Evolution of Middle Eastern Landscapes: An Outline to AD 1840*, Totowa, NJ, Barnes & Noble, 1985, pp. 152–83.

13 I. Lapidus, *A History of Islamic Societies*, Cambridge, Cambridge University Press, 1988, pp. 278–82. See also I. P. Petrushevsky, "The Socio-Economic Condition of Iran under the Il-Khans," in J. A. Boyle (ed.), *The Cambridge History of Iran*, vol. 5, Cambridge, Cambridge University Press, 1968, pp. 483–517; P. Christensen, *The Decline of Iranshahr: Irrigation and Environments in the History of the Middle East, 500 BC to AD 1500*, Copenhagen, Museum Tusculanum Press, 1993.

14 B. Lewis, *The Middle East: A Brief History of the Last 2,000 Years*, New York, Simon and Schuster, 1995, p. 99.

15 For the threat to water-management systems in more recent Middle Eastern wars, see P. H. Gleick, "Water and Conflict: Fresh Water Resources and International Security," *International Security* 18/1, 1993, pp. 79–112; Gleick, "Water and Conflict."

16 For long perspectives on China as a whole, see M. Elvin, "Three Thousand Years of Unsustainable Growth: China's Environment from Archaic Times to the Present," *East Asian History* 6, 1993, pp. 7–47; J. R. McNeill, "China's Environmental History in World Perspective," in M. Elvin and L. Ts'ui-jung (eds.), *Sediments of Time: Environment and Society in Chinese History*, Cambridge, Cambridge University Press, 1998, pp. 36–47 (pp. 36–8, 46–7).

17 N. K. Menzies, *Forest and Land Management in Imperial China*, London, St. Martin's Press, 1994, pp. 22, 59–61; P. Perdue, *China Marches West*, Cambridge, MA, Harvard University Press, 2005.

18 M. Elvin, *The Retreat of the Elephants: An Environmental History of China*, New Haven, CT, Yale University Press, 2004, chapter 8.

19 J. Gommens, *Mughal Warfare*, London, Routledge, 2002, chapter 4; S. Digby, *Warhorse and Elephant in the Delhi Sultanate*, Oxford, Orient Monographs, 1971; R. Tucker and E. Russell (eds.), *Natural Enemy, Natural Ally: Toward an Environmental History of War*, Corvallis, OR, Oregon State University Press, 2004, pp. 42–64.

20 R. P. Tucker, "The British Empire and India's Forest Resources: The Timberlands of Assam and Kumaon, 1914–1950," in J. F. Richards and R. Tucker (eds.), *World Deforestation in the Twentieth Century*, Durham, NC, Duke University Press, 1988, pp. 91–111.

21 C. Darby, "The Medieval Clearances," in W. L. Thomas (ed.), *Man's Role in the Changing the Face of the Earth*, Chicago, University of Chicago Press, 1956, pp. 183–216; N. J. G. Pounds, *An Historical Geography of Europe, 450 B.C. – A.D. 1330*, Cambridge, Cambridge University Press, 1973, chapter 3.

22 J. Brauer and H. van Tuyll, *Castles, Battles and Bombs: How Economics Explains Military History*, Chicago, University of Chicago Press, 2008, chapter 2.

23 M. Keen, *Medieval Warfare: A History*, Oxford, Oxford University Press, 1999; R. Beaumont, *War, Chaos and History*, Westport, CT, Praeger, 1994, chapter 2.

24 For a vivid account, see B. W. Tuchman, *A Distant Mirror: The Calamitous 15th Century*, New York, Ballantine, 1978.

25 W. H. McNeill, *Plagues and Peoples*, Garden City, NY, Anchor Books, 1976; K. Kiple (ed.), *The Cambridge History of Human Diseases*, Cambridge, Cambridge University Press, 1993.

26 C. J. Rogers (ed.), *The Military Revolution Debate*, Boulder, CO, Westview Press, 1995. For the crucial Dutch role at sea, see L. Sicking, *Neptune and the Netherlands: State, Economy and War at Sea in the Renaissance*, Leiden, Brill, 2004.

27 G. Parker (ed.), *The Thirty Years War*, London, Routledge, 1997. For southern Germany, see P. Warde, *Ecology, Economy and State Formation in Early Modern Germany*, Cambridge, Cambridge University Press, 2006; for the Swedish role, see J. Myrdal, "Food, War and Crisis: The Seventeenth Century Swedish Empire," in A. Hornborg, J. R. McNeill, and J. Martinez-Alier (eds.), *Rethinking Environmental History: World-System History and Global Environmental Change*, Lanham, MD, AltaMira Press, 2007, pp. 89–99.

28 M. van Creveld, *Supplying War*, Cambridge, Cambridge University Press, 1977.

29 See the view of the militarized state as predatory in C. Tilly, "War Making and State Making as Organized Crime," in P. B. Evans, D. Rueschemeyer, and T. Skocpol (eds.), *Bringing the State Back In*, Cambridge, Cambridge University Press, 1985, pp. 169–91.

30 C. I. Archer, J. R. Ferris, H. H. Herwig, and T. H. E. Travers, *World History of Warfare*, Lincoln, NE, University of Nebraska Press, 2002, chapter 11.

31 For a lively debate among historical anthropologists over the changes in indigenous societies in what some call the "tribal zone" in response to European penetration, see R. B. Ferguson and N. L. Whitehead (eds.), *War in the Tribal Zone: Expanding States and Indigenous Warfare*, Santa Fe, SAR Press, 1992.

32   R. G. Albion, *Forests and Sea Power: The Timber Problem of the Royal Navy, 1652–1862*, Cambridge, Cambridge University Press, 1926; P. W. Bamford, *Forests and French Sea Power, 1660–1789*, Toronto, University of Toronto Press, 1956; S. W. Miller, *Fruitless Trees: Portuguese Conservation and Brazil's Colonial Timber*, Stanford, CA, Stanford University Press, 2000; G. Wynn, *Timber Colony: A Historical Geography of Early Nineteenth Century New Brunswick*, Toronto, University of Toronto Press, 1981.

33   A. W. Crosby, *The Columbian Exchange: Biological and Cultural Consequences of 1492*, Westport, CT, Greenwood Press, 1972, and his *Ecological Imperialism: The Biological Expansion of Europe, 900–1900*, Cambridge, Cambridge University Press, 1986; Kiple, *The Cambridge History of Human Diseases*, pp. 305–33; K. Kiple and S. V. Beck (eds.), *Biological Consequences of the European Expansion, 1450–1800*, Aldershot, Ashgate, 1997; N. D. Cook, *Born to Die: Disease and the New World Conquest, 1492–1650*, Cambridge, Cambridge University Press, 1998; M. Livi Bacci, *Conquest: The Destruction of the American Indios*, Cambridge, Polity, 2008.

34   There are indicative references to these trends in many works, though no one has yet studied the subject systematically. See C. Sauer, *The Early Spanish Main*, Berkeley, University of California Press, 1966; E. Melville, *A Plague of Sheep*, Cambridge, Cambridge University Press, 1994; M. J. MacLeod, *Spanish Central America: A Socioeconomic History, 1520–1720*, Berkeley, University of California Press, 1973; W. Dean, *With Broadax and Firebrand: The Destruction of the Brazilian Atlantic Forest*, Berkeley, University of California Press, 1995.

35   J. R. McNeill, *Mosquito Empires: Ecology and War in the Greater Caribbean, 1620–1914*, New York, Cambridge University Press, 2010.

36   M. Williams, *Americans and Their Forests*, Cambridge, Cambridge University Press, 1989.

37   For the transformations of industrial warfare initiated in those campaigns, see van Creveld, *Supplying War*, chapters 2–3.

38   R. P. Tucker, *Insatiable Appetite: The United States and the Ecological Degradation of the Tropical World*, Berkeley, University of California Press, 2000, pp. 25–36.

39   D. R. Headrick, *The Tools of Empire: Technology and European Imperialism in the Nineteenth Century*, Oxford, Oxford University Press, 1981; Tucker and Russell, *Natural Enemy, Natural Ally*, pp. 65–92; T. Sunseri, "Forests, Social Rebellion, and Social Control in the Rufiji Region, German East Africa," *Environmental History*, July 2003, pp. 430–51.

40   D. R. Headrick, *Power over People: Technology, Environments, and Western Imperialism, 1400 to the Present*, Princeton, NJ, Princeton University Press, 2010.

41   Tucker and Russell, *Natural Enemy, Natural Ally*, pp. 65–92.

42   T. Sunseri, *Wielding the Ax: State Forestry and Social Conflict in Tanzania, 1820–2000*, Athens, OH, Ohio University Press, 2009.

43   G. Wawro, *Warfare and Society in Europe, 1792–1914*, London and New York, Routledge, 2000, p. 144.

44   See Tucker and Russell, *Natural Enemy, Natural Ally*, pp. 93–109; L. M. Brady, "The Wilderness of War: Nature and Strategy in the American Civil War," *Environmental History* 10, July 2005, pp. 421–47; G. Linderman, *Embattled Courage: The Experience of Combat in the American Civil War*, New York, Free Press, 1987, chapter 10, "A Warfare of Terror."

45   M. E. Neely Jr., *The Civil War and the Limits of Destruction*, Cambridge, MA, Harvard University Press, 2007, chapter 5.

46   J. F. Richards and M. McAlpin, "Cotton Cultivating and Land Clearing in the Bombay Deccan and Karnatak: 1818–1920," in R. P. Tucker and J. F. Richards (eds.), *Global Deforestation and the Nineteenth-Century World Economy*, Durham, NC, Duke University Press, 1983, pp. 68–104.

47   McNeill, *The Pursuit of Power*, chapter 8; P. A. C. Koistinen, "The 'Industrial-Military Complex' in Historical Perspective: World War I," *Business History Review* 41/4, 1967, pp. 379–403.

48   Beaumont, *War, Chaos and History*, p. 140.

49 W. K. Storey, *The First World War: A Concise Global History*, Boulder, CO, Rowman & Littlefield, 2009, is a start.

50 E. Russell, *War and Nature: Fighting Humans and Insects with Chemicals from World War I to Silent Spring*, Cambridge, Cambridge University Press, 2001.

51 Tucker and Russell, *Natural Enemy, Natural Ally*, pp. 110–21.

52 J. M. Diefendorf, "Wartime Destruction and the Postwar Cityscape," in C. E. Closmann (ed.), *War and the Environment: Military Destruction in the Modern Age*, College Station, TX, Texas A&M University Press, 2009, pp. 171–92.

53 Tucker and Russell, *Natural Enemy, Natural Ally*, pp. 121–38; B. M. Barr, "Perspectives on Deforestation in the U.S.S.R.," in Richards and Tucker (eds.), *World Deforestation in the Twentieth Century*, pp. 230–61.

54 E. Russell, "The Strange Career of DDT: Experts, Federal Capacity, and 'Environmentalism' in World War II," *Technology and Culture* 40, 1999, pp. 770–96.

55 J. Bennett, *Natives and Exotics: World War II and Environment in the Southern Pacific*, Honolulu, University of Hawai'i Press, 2009.

56 Tucker and Russell, *Natural Enemy, Natural Ally*, pp. 252–69. There was a parallel, and equally temporary, resurgence of the cod population of the North Atlantic during the war, when German submarines prevented the Allies' fishing fleets from operating. See M. Kurlansky, *Cod: A Biography of the Fish that Changed the World*, New York, Walker & Co., 1997.

57 Tucker and Russell, *Natural Enemy, Natural Ally*, pp. 195–216.

58 McNeely, "Biodiversity, War, and Tropical Forests."

59 P. Josephson, "War on Nature as Part of the Cold War: The Strategic and Ideological Roots of Environmental Degradation in the Soviet Union," in J. R. McNeill and C. R. Unger (eds.), *Environmental Histories of the Cold War*, Cambridge, Cambridge University Press, 2010, pp. 21–50.

60 A. K. Biswas, "Scientific Assessment of the Long-Term Environmental Consequences of War," in J. E. Austin and C. E. Bruch (eds.), *The Environmental Consequences of War: Legal, Economic, and Scientific Perspectives*, Cambridge, Cambridge University Press, 2000, pp. 303–15 (p. 307). See also A. H. Westing's earlier works, including *Ecological Consequences of the Second Indochina War*, Stockholm, Almqvist and Wiksell International, 1976, and *Herbicides in War: The Long-Term Ecological and Human Consequences*, London and Philadelphia, Taylor & Francis, 1984.

61 S. Firth, *Nuclear Playground*, Honolulu, University of Hawai'i Press, 1987; B. and M.-T. Danielsson, *Poisoned Reign: French Nuclear Colonialism in the Pacific*, Harmondsworth, Penguin Books, 1986; M. D. Merlin and R. M. Gonzales, "Environmental Impacts of Nuclear Testing in Remote Oceania, 1946–1996," in McNeill and Unger (eds.), *Environmental Histories*, pp. 167–202.

62 M. Feshbach and A. Friendly, *Ecocide in the USSR*, New York, Basic Books, 1992; Z. Wolfson [Boris Komarov], *The Geography of Survival: Ecology in the Post-Soviet Era*, Armonk, M. E. Sharpe, 1994.

63 S. A. S. Omar, E. Briskey, R. Misak, and A. A. S. O. Asem, "The Gulf War Impact on the Terrestrial Environment of Kuwait: An Overview," in Austin and Bruch (eds.), *Environmental Consequences*, pp. 316–37.

References

Albion, R. G., *Forests and Sea Power: The Timber Problem of the Royal Navy, 1652–1862*, Cambridge, Cambridge University Press, 1926.

Archer, C. I., Ferris, J. R., Herwig, H. H., and Travers, T. H. E., *World History of Warfare*, Lincoln, NE, University of Nebraska Press, 2002.

Bamford, P. W., *Forests and French Sea Power, 1660–1789*, Toronto, University of Toronto Press, 1956.

Barr, B. M., "Perspectives on Deforestation in the U.S.S.R.," in J. F. Richards and R. Tucker (eds.), *World Deforestation in the Twentieth Century*, Durham, NC, Duke University Press, 1988, pp. 230–61.

Beaumont, R., *War, Chaos and History*, Westport, CT, Praeger, 1994.

Bennett, J., *Natives and Exotics: World War II and Environment in the Southern Pacific*, Honolulu, University of Hawai'i Press, 2009.

Biswas, A. K., "Scientific Assessment of the Long-Term Environmental Consequences of War," in J. E. Austin and C. E. Bruch (eds.), *The Environmental Consequences of War: Legal, Economic, and Scientific Perspectives*, Cambridge, Cambridge University Press, 2000, pp. 303–15.

Brady, L. M., "The Wilderness of War: Nature and Strategy in the American Civil War," *Environmental History* 10, July 2005, pp. 421–47.

Brauer, J., and van Tuyll, H., *Castles, Battles and Bombs: How Economics Explains Military History*, Chicago, University of Chicago Press, 2008.

Christensen, P., *The Decline of Iranshahr: Irrigation and Environments in the History of the Middle East, 500 BC to AD 1500*, Copenhagen, Museum Tusculanum Press, 1993.

Cook, N. D., *Born to Die: Disease and the New World Conquest, 1492–1650*, Cambridge, Cambridge University Press, 1998.

Crosby, A. W., *The Columbian Exchange: Biological and Cultural Consequences of 1492*, Westport, CT, Greenwood Press, 1972.

Crosby, A. W., *Ecological Imperialism: The Biological Expansion of Europe, 900–1900*, Cambridge, Cambridge University Press, 1986.

Danielsson, B., and Danielsson, M.-T., *Poisoned Reign: French Nuclear Colonialism in the Pacific*, Harmondsworth, Penguin Books, 1986.

Darby, C., "The Medieval Clearances," in W. L. Thomas (ed.), *Man's Role in the Changing the Face of the Earth*, Chicago, University of Chicago Press, 1956, pp. 183–216.

Dean, W., *With Broadax and Firebrand: The Destruction of the Brazilian Atlantic Forest*, Berkeley, University of California Press, 1995.

Diefendorf, J. M., "Wartime Destruction and the Postwar Cityscape," in C. E. Closmann (ed.), *War and the Environment: Military Destruction in the Modern Age*, College Station, TX, Texas A&M University Press, 2009, pp. 171–92.

Digby, S., *Warhorse and Elephant in the Delhi Sultanate*, Oxford, Orient Monographs, 1971.

Elvin, M., *The Retreat of the Elephants: An Environmental History of China*, New Haven, CT, Yale University Press, 2004.

Elvin, M., "Three Thousand Years of Unsustainable Growth: China's Environment from Archaic Times to the Present," *East Asian History* 6, 1993, pp. 7–47.

Ferguson, R. B., and Whitehead, N. L. (eds.), *War in the Tribal Zone: Expanding States and Indigenous Warfare*, Santa Fe, SAR Press, 1992.

Feshbach, M., and Friendly A., *Ecocide in the USSR*, New York, Basic Books, 1992.

Firth, S., *Nuclear Playground*, Honolulu, University of Hawai'i Press, 1987.

Flannery, T., *The Future Eaters: An Ecological History of the Australian Lands and People*, New York, George Braziller, 1994.

Gleick, P. H., "Water and Conflict: Fresh Water Resources and International Security," *International Security* 18/1, 1993, pp. 79–112.

Gleick, P. H., "Water, War and Peace in the Middle East," *Environment* 36/3, 1994, pp. 6–15, 35–42.

Goldsworthy, A., *How Rome Fell*, New Haven, CT, Yale University Press, 2009.

Goldsworthy, A., *Roman Warfare*, London, Cassell, 2000.

Gommens, J., *Mughal Warfare*, London, Routledge, 2002.

Hanson, V. D., *Warfare and Agriculture in Classical Greece*, Berkeley, University of California Press, 1998.

Hatami, H., and Gleick, P. H., *Chronology of Conflict over Water in the Legends, Myths, and History of the Ancient Middle East*, Oakland, CA, Pacific Institute, 1993.

Headrick, D. R., *Power over People: Technology, Environments, and Western Imperialism, 1400 to the Present*, Princeton, NJ, Princeton University Press, 2010.

Headrick, D. R., *The Tools of Empire: Technology and European Imperialism in the Nineteenth Century*, Oxford, Oxford University Press, 1981.

Hughes, J. D., *Pan's Travail: Environmental Problems of the Ancient Greeks and Romans*, Baltimore, The Johns Hopkins University Press, 1994.

Jacobsen, T., and Adams, R. M., "Salt and Silt in Ancient Mesopotamian Agriculture," *Science* 128, 1958, pp. 1251–8.

Josephson, P., "War on Nature as Part of the Cold War: The Strategic and Ideological Roots of Environmental Degradation in the Soviet Union," in J. R. McNeill and C. R. Unger (eds.), *Environmental Histories of the Cold War*, Cambridge, Cambridge University Press, 2010, pp. 21–50.

Keeley, L. H., *War before Civilization*, Oxford, Oxford University Press, 1996.

Keen, M., *Medieval Warfare: A History*, Oxford, Oxford University Press, 1999.

Kelly, R. C., *Warless Societies and the Origin of War*, Ann Arbor, MI, University of Michigan Press, 2000.

Kiple, K. (ed.), *The Cambridge History of Human Diseases*, Cambridge, Cambridge University Press, 1993.

Kiple, K., and Beck, S. V. (eds.), *Biological Consequences of the European Expansion, 1450–1800*, Aldershot, Ashgate, 1997.

Koistinen, P. A. C., "The 'Industrial-Military Complex' in Historical Perspective: World War I," *Business History Review* 41/4, 1967, pp. 379–403.

Kurlansky, M., *Cod: A Biography of the Fish that Changed the World*, New York, Walker & Co., 1997.

Lapidus, I., *A History of Islamic Societies*, Cambridge, Cambridge University Press, 1988.

Lewis, B., *The Middle East: A Brief History of the Last 2,000 Years*, New York, Simon and Schuster, 1995.

Linderman, G., *Embattled Courage: The Experience of Combat in the American Civil War*, New York, Free Press, 1987.

Livi Bacci, M., *Conquest: The Destruction of the American Indios*, Cambridge, Polity, 2008.

MacLeod, M. J., *Spanish Central America: A Socioeconomic History, 1520–1720*, Berkeley, University of California Press, 1973.

Martin, P. S., and Kein, R. G. (eds.), *Quaternary Extinctions: A Prehistoric Revolution*, Tucson, AZ, University of Arizona Press, 1984.

McNeely, J. A., "Biodiversity, War, and Tropical Forests," *Journal of Sustainable Forestry* 16/3, 2003, pp. 1–20.

McNeill, J. R., "China's Environmental History in World Perspective," in M. Elvin and L. Ts'ui-jung (eds.), *Sediments of Time: Environment and Society in Chinese History*, Cambridge, Cambridge University Press, 1998, pp. 36–47.

McNeill, J. R., *Mosquito Empires: Ecology and War in the Greater Caribbean, 1620–1914*, New York, Cambridge University Press, 2010.

McNeill, W. H., *Plagues and Peoples*, Garden City, NY, Anchor Books, 1976.

McNeill, W. H., *The Pursuit of Power: Technology, Armed Force, and Society since AD 1000*, Chicago, University of Chicago Press, 1982.

Meiggs, R., *Trees and Timber in the Ancient Mediterranean World*, Oxford, Clarendon Press, 1982.

Melville, E., *A Plague of Sheep*, Cambridge, Cambridge University Press, 1994.

Menzies, N. K., *Forest and Land Management in Imperial China*, London, St. Martin's Press, 1994.

Merlin, M. D., and Gonzales, R. M., "Environmental Impacts of Nuclear Testing in Remote Oceania, 1946–1996," in J. R. McNeill and C. R. Unger (eds.), *Environmental Histories of the Cold War*, Cambridge, Cambridge University Press, 2010, pp. 167–202.

Miller, S. W., *Fruitless Trees: Portuguese Conservation and Brazil's Colonial Timber*, Stanford, CA, Stanford University Press, 2000.

Myrdal, J., "Food, War and Crisis: The Seventeenth Century Swedish Empire," in A. Hornborg, J. R. McNeill, and J. Martinez-Alier (eds.), *Rethinking Environmental History: World-System History and Global Environmental Change*, Lanham, MD, AltaMira Press, 2007, pp. 89–99.

Neely, M. E., Jr., *The Civil War and the Limits of Destruction*, Cambridge, MA, Harvard University Press, 2007.

Omar, S. A. S., Briskey, E., Misak, R., and Asem, A. A. S. O., "The Gulf War Impact on the Terrestrial Environment of Kuwait: An Overview," in J. E. Austin and C. E. Bruch (eds.), *The Environmental Consequences of War: Legal, Economic, and Scientific Perspectives*, Cambridge, Cambridge University Press, 2000, pp. 316–37.

Otterbein, K. F., *The Evolution of War: A Cross-Cultural Study*, New Haven, CT, HRAF, 1970.

Parker, G. (ed.), *The Thirty Years War*, London, Routledge, 1997.

Perdue, P., *China Marches West*, Cambridge, MA, Harvard University Press, 2005.

Petrushevsky, I. P., "The Socio-Economic Condition of Iran under the Il-Khans," in J. A. Boyle (ed.), *The Cambridge History of Iran*, vol. 5, Cambridge, Cambridge University Press, 1968, pp. 483–517.

Pounds, N. J. G., *An Historical Geography of Europe, 450 B.C. – A.D. 1330*, Cambridge, Cambridge University Press, 1973.

Pyne, S. J., *World Fire: The Culture of Fire on Earth*, New York, Holt, 1995.

Richards, J. F., and McAlpin, M., "Cotton Cultivating and Land Clearing in the Bombay Deccan and Karnatak: 1818–1920," in R. P. Tucker and J. F. Richards (eds.), *Global Deforestation and the Nineteenth-Century World Economy*, Durham, NC, Duke University Press, 1983, pp. 68–104.

Rogers, C. J. (ed.), *The Military Revolution Debate*, Boulder, CO, Westview Press, 1995.

Russell, E., "The Strange Career of DDT: Experts, Federal Capacity, and 'Environmentalism' in World War II," *Technology and Culture* 40, 1999, pp. 770–96.

Russell, E., *War and Nature: Fighting Humans and Insects with Chemicals from World War I to Silent Spring*, Cambridge, Cambridge University Press, 2001.

Sauer, C., *The Early Spanish Main*, Berkeley, University of California Press, 1966.

Sicking, L., *Neptune and the Netherlands: State, Economy and War at Sea in the Renaissance*, Leiden, Brill, 2004.

Storey, W. K., *The First World War: A Concise Global History*, Boulder, CO, Rowman & Littlefield, 2009.

Sunseri, T., "Forests, Social Rebellion, and Social Control in the Rufiji Region, German East Africa," *Environmental History*, July 2003, pp. 430–51.

Sunseri, T., *Wielding the Ax: State Forestry and Social Conflict in Tanzania, 1820–2000*, Athens, OH, Ohio University Press, 2009.

Thirgood, J. V., *Man and the Mediterranean Forest: A History of Resource Depletion*, London, Academic Press, 1981.

Tilly, C., "War Making and State Making as Organized Crime," in P. B. Evans, D. Rueschemeyer, and T. Skocpol (eds.), *Bringing the State Back In*, Cambridge, Cambridge University Press, 1985, pp. 169–91.

Tuchman, B. W., *A Distant Mirror: The Calamitous 15th Century*, New York, Ballantine, 1978.

Tucker, R. P., "The British Empire and India's Forest Resources: The Timberlands of Assam and Kumaon, 1914–1950," in J. F. Richards and R. Tucker (eds.), *World Deforestation in the Twentieth Century*, Durham, NC, Duke University Press, 1988, pp. 91–111.

Tucker, R. P., *Insatiable Appetite: The United States and the Ecological Degradation of the Tropical World*, Berkeley, University of California Press, 2000.

Tucker, R., and Russell, E. (eds.), *Natural Enemy, Natural Ally: Toward an Environmental History of War*, Corvallis, OR, Oregon State University Press, 2004.

van Creveld, M., *Supplying War*, Cambridge, Cambridge University Press, 1977.

Wagstaff, J. M., *The Evolution of Middle Eastern Landscapes: An Outline to AD 1840*, Totowa, NJ, Barnes & Noble, 1985.

Warde, P., *Ecology, Economy and State Formation in Early Modern Germany*, Cambridge, Cambridge University Press, 2006.

Wawro, G., *Warfare and Society in Europe, 1792–1914*, London and New York, Routledge, 2000.

Westing, A. H., *Ecological Consequences of the Second Indochina War*, Stockholm, Almqvist and Wiksell International, 1976.

Westing, A. H., *Herbicides in War: The Long-Term Ecological and Human Consequences*, London and Philadelphia, Taylor & Francis, 1984.

Whittaker, C. R., *Frontiers of the Roman Empire: A Social and Economic Study*, Baltimore, The Johns Hopkins University Press, 1994.

Williams, M., *Americans and Their Forests*, Cambridge, Cambridge University Press, 1989.

Wolfson, Z. *[Boris Komarov]*, *The Geography of Survival: Ecology in the Post-Soviet Era*, Armonk, M. E. Sharpe, 1994.

Wynn, G., *Timber Colony: A Historical Geography of Early Nineteenth Century New Brunswick*, Toronto, University of Toronto Press, 1981.

CHAPTER NINETEEN

Technology and the Environment

PAUL JOSEPHSON

Humans in all societies and cultures have interacted with and attempted to control the natural environment with increasingly sophisticated and powerful technologies – instruments, tools, machinery, and other devices and various techniques. While the relationship between nature and technology has been sufficiently pronounced to chart it since the time of the first agricultural societies some 10,000 years ago, its presence became more ubiquitous and obvious in the Middle Ages, and especially from the Industrial Revolution – about 1750 onward. Belief in the power of humans to improve upon nature through technology, and faith in progress, grew stronger during the European Enlightenment (c. 1680–1800). In some respects this direct relationship reflects the effort of entrepreneurs, craftsmen and women, scientists, and engineers to bring industrial understandings and efficiencies to nature. By studying river flow and topology, for example, engineers could straighten the rivers, reclaim wetlands and put that land into agricultural production, and dam the rivers at various places to hold water, even during dry seasons, for use in irrigation and power generation. The river itself thereby became a kind of technology. In the twentieth century, technology and the environment became even more interconnected as large-scale, expensive, and government-supported projects became routine.

Human–nature interaction through technology has been manifested in a number of positive and negative ways: flood control, yet destruction of ecosystems; production of metals, yet extensive despoliation of the land through rapacious mining activities, air and water pollution; vast increases in the variety and quantity of crops, yet overuse of biocides, and of monocultures including more recently genetically modified organisms whose full environmental impact remains uncertain.

This chapter examines the relationships between technologies and environments beginning with the origins of agriculture and with farming technologies such as plows and tractors. But its heart concerns the industrial age, and technologies associated with

A Companion to Global Environmental History, First Edition. Edited by J.R. McNeill and Erin Stewart Mauldin.
© 2012 John Wiley & Sons, Ltd. Published 2015 by John Wiley & Sons, Ltd.

industrial and urban life. It highlights the impacts of technologies such as the private auto-mobile and the large-scale systems embraced by modern states in the twentieth century. Wherever one looks in environmental history, there are technologies at work, and techno-logical change is bound up with environmental change. In the last 200 years, however, the power of technologies to affect the environment has attained unprecedented levels.

Agriculture, Nature, and Technology

For the past 10,000 years humans have utilized the environment in pursuit of domes-ticated – and eventually hybridized – plants and animals. In the process, they pushed back forests, cut and mowed meadows, plowed prairies, created impoundments for water, reclaimed swamps, irrigated fields, and employed slash-and-burn (among other) agricultural methods. Slash-and-burn methods to clear fields had short-term fertilizing benefits for the soil, thanks to nutrients contained in the ash of burned biomass. But the soils lost that fertility quickly, and so population pressures often led to more forest falling to fire and axe. Only higher productivity in farming could relieve the need to clear more land wherever populations were rising. Technological innovations that increased productivity and lowered labor inputs were relatively gradual until the Industrial Revolution and the attendant changes in power generation (steam, electric-ity, internal combustion engine). In the twentieth century, the application of chemical fertilizers and the rise of large-scale agribusinesses founded on these capital inputs accelerated environmental change.

There is significant archeological evidence that agriculture developed in the Fertile Crescent of Southwest Asia, in East Asia, and in South America at the end of the last ice age some 10,000 years ago. Domesticated plants, such as grains, spread quickly. Corn or maize was the most prevalent crop grown in the Americas, where agriculture first appeared between 8,000 and 9,000 years ago; other crops were beans, potatoes, squash, peanuts, and cotton. Turkeys were also domesticated, as were sunflowers, tobacco, and squash. Agriculturalists took the first steps of irrigation, reclamation of land from swamps, terracing of hillsides, and so on, for example among the Aztecs whose agriculture sup-ported an expanding empire. In northern Africa, the Indian subcontinent, and China agriculture also blossomed. In the latter, rice production developed, and spread to Korea and Japan with new methods of weeding, the adoption of new irrigation techniques, and cultivation of different strains of rice.[1]

In classical antiquity, both the Greeks and the Romans emphasized cultivation of crops for trade and export. Grain, wine, and olive oil became items of long-distance trade and large-scale production. The Roman world collapsed in the fourth and fifth centuries CE, but its large estates laid the groundwork for the feudalism in which peasants toiled for a manor lord. By the ninth and tenth centuries, improvements to feudal agriculture contributed to a more reliable source of food and to a more sedentary lifestyle for many people. This in turn gave rise to villages, towns, and cities, themselves revealing of the relationship between technology and the environment (see below).

Everywhere that agriculture supported sedentary societies, technological innova-tions accelerated in a variety of areas with environmental impact because of the sources of materials and nature of processes: pottery for household and decorative use; clay bricks for building structures; spinning, weaving, and other textile crafts; tanning and leatherworking; metallurgy; hunting tools, such as traps, nets, slings, and bows; and scythes and plows.[2]

The Plow

The plow, one of the oldest agricultural technologies, has clear environmental impacts – cutting and turning the soil, contributing to erosion with the help of wind and rain, flattening land, and accelerating the use of nutrients – that must be balanced against its crucial contribution to increased food production. The first plows, made of wood, were developed around 8,000 BCE in Mesopotamia. They scratched furrows in the soil, preparing it for planting. But they did not turn the soil, bringing deeper nutrients up where roots could easily get at them. Eventually iron plowshares were added, allowing farmers to cultivate tougher soils. But farmers normally followed the contours of the land, plowing only where soils were light and slopes modest.

In the European Middle Ages, through the application of stronger materials and better designs, and the addition of teams of animals – oxen for example – the plow gained the ability to turn heavier wetter soils more easily and over larger areas. Sometime in the sixth century, improved harnesses enabled tandem and multiple-ox teams to extend the amount of land under cultivation. The invention of the collar enabled horses, which were faster and more maneuverable than oxen, to work larger areas more quickly. Also the harrow was employed to crumble clods after plowing, the scythe to cut hay to feed the oxen, and the pitchfork to handle hay. The rigid padded collar turned the horse into a draft animal and enabled the rich, heavy soils of northwest Europe to be plowed, while the forest, moor, and swamp were "attacked with ax and spade." In the Middle Ages the Dutch began their extensive reclamation of land from the sea to increase cultivated land.[3] Manuring – both "natural" and through application – helped the productivity of soils.

The necessity of plowing harder ground to expand cultivation in the face of population pressures and exhaustion of older fields led to the development of plows with blades of sturdier materials. This eventually led to plows of steel, which enabled plowing in a straight line for more efficient planting and harvesting. In penetrating more deeply, heavier plows allowed greater aeration and nutrient redistribution, and the use of draft animals allowed farmers to plow long rows even in difficult or rocky terrain, which accelerated forest clearing.[4]

Once fields had been significantly tamed by centuries of plowing, it was possible to use plows with multiple shares that created more uniform furrows, although this procedure likely required more energy to pull the plow. Ultimately, the tractor solved the problems of energy, but also led farmers to believe there was no limit to the amount of land under cultivation. (The American Dust Bowl was the result of prolonged drought and the loosening of topsoil from plowing.) It is important to remember that soil, particularly the uppermost layers such as the topsoil and humus, is involved in a delicate balance of nutrients and life, of water and air.[5]

The Industrial Revolution and Agriculture

During the Industrial Revolution (eighteenth and nineteenth centuries), inventors turned to a variety of advances from other fields of economic activity – power generation, new materials, and so on – to increase agricultural production. The new devices, powered by animals, steam, and eventually the internal combustion engine rather than by human muscles – gave the appearance of control over nature, and encouraged agronomists to pursue large-scale changes in the landscape. Many of the early technological innovations originated in Britain, where the Industrial Revolution took off, and the US, where high wages encouraged the substitution of machinery for labor. Such devices – reapers, rakes, milking mowers, horse-drawn mowers, hay loaders, seed drills, and so on – had a cyclical effect on supply and demand, calling forth the industrialization of agriculture. For example, the cotton gin quickly separated seeds from cotton, enabling more planting. Milking machines enabled the industrial production of milk, lowered costs, and provoked increased demand, which led to further improvements in the machines, the establishment of larger dairy herds, the pushing back of forests to enable grazing of more milk cattle – and so on.[6] The grain elevator (1843, Buffalo, New York) is another example of the industrial impetus to production. It enabled more rapid handling of grain from the Midwest to eastern markets, encouraged increases in production per unit of land, and stimulated the cultivation of more land. Prior to the grain elevator, teams of workers moved grain in bags rather than in bulk. Because many of the machines were expensive, not all farmers could afford them, and this triggered the still ongoing process of the consolidation of farming into larger farms, with many people leaving the countryside for positions in factories.

The Tractor

The internal combustion engine found its place in agriculture in the early twentieth century in the tractor, and shortly in a host of other machines that enabled farmers to reshape the landscape faster than ever before. No more would land contours prevent reshaping of the farm, nor would forests, stumps, rocks, boulders, streams, and rivers create insurmountable obstacles to agriculture. The increasingly powerful and nimble tractor appeared in large numbers, in some senses not only enabling farms to extend to the horizon, but almost requiring the planting of monocultures of corn, wheat, soybeans, and other cash crops beyond the limits of demand. The tractor, according to this "technologically determinist" argument, pushed agricultural production beyond the needs of demand to production for the sake of production: the bulldozer (for clearing land), the tractor and attachments (for planting crops), and the harvester and other machines (for harvesting crops) could operate seemingly without limits. By the second half of the twentieth century the result was megafarms, overproduction of crops, cheap food, government price supports, costly inputs of fertilizers, biocides and other chemicals, many of which spoiled the land, then entered waterways and eventually bays and oceans, causing extensive damage such as algal blooms.

Gas tractors appeared in the early 1900s. They were cumbersome and heavy, but met a growing demand particularly in wheat-growing regions. In the 1910s such companies as Ford, International Harvester, John Deere, and others began manufacturing lighter, maneuverable, and inexpensive models that could be used for plowing, mowing, and reaping wheat, corn, or cotton, raising livestock, and producing dairy products. They quickly replaced steam engines and draft animals. Other innovations (low-pressure tube tires that did less damage to fields and the power take-off that permitted various attachment tools to be driven by the tractor) essentially led to the replacement of the horse by the end of the 1920s in the US.[7] On the eve of the Great Depression there were 100 different companies that marketed some 250 models and types of machines. There were 920,000 tractors in the US in 1930, 1.6 million in 1940, and 2.4 million in 1945.[8] Tractors altered not only the landscape, but the social structure of the environment as small farms, small towns, and small businesses dwindled and were replaced by agribusinesses, and rural life gave way to industrial agriculture.

The tractor also gained importance in the second half of the twentieth century in Europe and the USSR. In the USSR political leaders and economic managers believed that the most efficient way to modernize agriculture was to apply modern technology – the tractor – to massive collective farms that they coercively forced on the peasantry in the 1930s. Although the numbers of tractors produced lagged significantly behind those in the US with perhaps only 80,000 rudimentary models produced, the Soviets employed them in megafarms that rapidly changed the landscape – and the peasant way of life.

Concentrated Animal Feed Operations (CAFOs)

By the end of the twentieth century, modern agriculture in North America, Europe, and Japan and many other places besides had become capital intensive, costly, and large scale, with significant environmental costs that belatedly generated attention. One of the major forms of this agriculture is the CAFO. The CAFO arose in the 1970s and especially 1980s in part from the increasingly scientifically based nature of agriculture, which enabled the development of crops and animals raised to move as quickly as possible to market. Interestingly, the science grew out of a public–private interface through which national government agriculture departments provided research funds and programs to develop the crops, often hand in hand with multinational corporations. For example, the research budget of the US Department of Agriculture grew to over $1 billion annually in the 2000s and supported such products as corn monocultures and GMOs in concert with multinational corporations such as Monsanto, and even special turf for athletic fields of service to professional sports.[9] The crops and their attendant chemical and other inputs were not only costly, but appropriate only for large-scale agriculture, not the smaller farmer whose production almost by definition has a significantly smaller environmental impact.

CAFOs rely on standard, "mass produced" animals. Their environmental costs may be difficult to remediate. A large CAFO includes 1,000 cattle, 2,500 hogs over 25 kilograms (55 pounds), or 125,000 broiler (not laying) chickens. A medium CAFO has 300 to 999 cattle, 750 to 2,499 hogs of 25 kilograms or more, and 37,500 to 124,999 chickens. CAFOs confine large numbers of animals in close quarters in which feed, manure, sick animals, and disease share space – often metal buildings and pens that restrict the behavior and movement of animals who rarely have access to sunlight or fresh air. Many animals are mutilated to adapt them to factory-farm conditions (e.g., "de-beaking" chickens and turkeys, "docking" [amputating] the tails of cows and pigs). Workers add antibiotics to the feed to fight the diseases and accelerate fattening.

An obvious environmental cost of CAFOs is a liquid manure system in which the animals' urine and feces are mixed with water, and stored, often in huge open-air lagoons. Little provision has been made for removing the manure and offal of the animals that contain a variety of pathogens and whose millions of gallons of liquid contaminants can leach into groundwater. The manure can be sprayed on crops as fertilizer, but then runs off into surface waters. There is growing evidence that CAFOs increase local healthcare and pollution abatement costs, and lead to property depreciation. Since CAFOs are operations of large corporations, they typically do not contribute significantly to the local economy, but provide low-paying, dangerous, and unskilled jobs. However, owing to economies of scale, CAFOs have low "unit" (animal) costs, provided they do not take into consideration social, public health, and environmental costs.[10]

Genetically Modified Organisms (GMOs)

GMOs deserve mention for their potential environmental impact. GMOs emerged over the course of the twentieth century on the foundation of advances in genetics. The first field tests with crops were conducted in the 1980s, and the first GMOs were approved for commercial use in the 1990s. Governments and corporations have spent billions of dollars on research, commercial development, and regulation, with the European Union taking a lead in regulating GMOs. Given the initially unexpected yet far-reaching environmental impacts of domestication of plants and animals, the plow, the tractor, and CAFOs, GMOs will likely have unanticipated social and environmental impacts, especially in the US with its relatively weak regulatory framework. Many aspects of GMO regulation are voluntary, and depend on companies to consult on safety issues or share data with the government. To date, the US Food and Drug Administration has an ad hoc procedure for judging safety and efficacy of GMOs, while more than 40 GMOs have reached the market. These are mostly products with herbicide tolerance and insect resistance (soybeans, corn) produced by Monsanto, DuPont/Pioneer, Syngenta, and Dow/Mycogen.

The goal of these technologies is to lessen the application of biocides, but some biocides must still be used. In addition, a fear is that GMOs may transfer genes to wild plants, might become weeds, might have toxicities that express themselves later, and may have other hard to predict negative impacts.[11] As Eric Chivian and Aaron Bernstein argue, any detriment to biodiversity such as that which will almost inevitably be produced with the introduction of GMOs must be treated with complete circumspection.[12]

Fisheries and Aquaculture

Fisheries have been under threat since human interest in damming and harnessing rivers and streams found expression in impoundments, mills, and factories. While dams and millponds obstructed the migration of anadromous fish and destroyed habitat, pollution also had an impact on fish populations. Pollution became a greater problem throughout the twentieth century both because of industry and because of the increasing use of chemical fertilizers and biocides in agriculture that flowed as run-off into rivers, streams, lakes, and other bodies of water. The increased use of nitrogen as a fertilizer in the 1960s and 1970s led to harmful algal blooms in coastal waters. In the late 1990s, even after efforts at remediation and regulation, fisheries suffered because of agricultural pollution especially in bays, gulfs, and inland seas. Among the most affected bodies of water were the Gulf of Mexico, Chesapeake Bay, the Baltic Sea, the Yellow Sea, the Black Sea, and the Adriatic Sea.

While pollution took its toll on aquatic ecosystems, significant depletion of fisheries also resulted from overfishing, itself in part the result of modern technology. In the postwar years, on the wave of larger ships for trawling, new plastics for nets, sonar for locating fish, industrial processing facilities, and industrial refrigerators and freezers to hold fish, such fishing nations as Canada, Portugal, the USSR, and the US pursued rapacious harvests; now many fish such as cod are under threat.[13]

Aquaculture has a long history, especially in China. In modern times, it took off late in the twentieth century when yields in marine-capture fisheries started to stagnate. In the twenty-first century, aquaculture became the mass-production technology of monocultures of fish specially selected for rapid growth and harvest, and may soon surpass wild fisheries in production. The main forms are pond fisheries for carp and catfish, offshore facilities for salmon, tilapia and other fish, and shrimp. Aquaculture has its own very significant environmental problems. While aquaculture seems more efficient than nature at manufacturing fish products, the costs of this capital-intensive technology include overuse of antibiotics and the risk that farmed fish will escape and outcompete the remaining wild ones. Aquaculture (of seaweed) contributed to the largest-ever algal bloom in the Yellow Sea in 2009.[14] In addition, aquaculture has a big carbon footprint, with products – salmon, shrimp, eels, and so on – shipped great distances to wealthy urban populations.

Take, for instance, shrimp aquaculture, much of which is based in mangroves along the Chinese, Indonesian, Thai, and other South Asian coasts. Approximately 5 million metric tons of shrimp are produced annually. Shrimp aquaculture, which increased nine-fold during the 1990s, accounts for one-third of the shrimp produced globally. The majority of farmed shrimp is exported to the US, European Union, and Japan. Yet shrimp aquaculture often organized in mangroves has raised significant environment concerns. According to several studies, mangroves are "an ecosystem in danger." They have long been used as wasteland for dumping. On top of this, the authorities use her-bicides in mangroves, which seem to be 5 to 10 times more sensitive than other forest species.[15]

Urbanization, Industrialization, and the Environment

Industrialization has had a fundamental environmental impact. In Europe and North America, it led to rapid deforestation in pursuit of lumber for construction, firewood, and more farmland, rapacious quarrying, mining, and smelting operations, the presence of tanning and other increasingly polluting chemical operations, and the dumping of wastes willy-nilly into public waterways, lakes, and ponds. Seeking power for their mills, industrialists dammed rivers and streams, and extracted and burned coal and petroleum with nary a thought to environmental consequences. By the end of the nineteenth century, most New England streams had lost their salmon populations. Attempts to restore fish runs in major rivers lost to the power of industrialists who resisted regulation and avoided the expense of fish ladders, which in fact rarely work.[16]

The significant environmental problems connected with industrialization have been well known for two centuries. In *Hard Times* (1854) and other novels, Charles Dickens captured the human costs of industrialization in fiction. In 1897, John Glaister, presi-dent of the Sanitary and Social Economic Section of the Philosophical Society of Glasgow, lamented the serious pollution of Scottish rivers caused by industry and a growing popu-lation that dumped sewage indiscriminately. He worried that not "a single river … in Scotland … is in a state of even comparative purity." The fish had disappeared from the now "dark, grimy and slimy" water with "silted beds and heaped up banks, composed of all manner of foreign material, always unpleasant to see, and sometimes unpleasantly odorous." Glaister welcomed the appointments of royal commissions to investigate means of preventing pollution, but only by the end of the twentieth century had regula-tions to define and control pollution, and to punish polluters, begun to have a significant positive impact.[17]

For centuries wastes from industries and municipalities were dumped haphazardly into waterways or, in the best cases, carted elsewhere for near-surface burial. Simultaneously, as the dangers of modern pollution grew clear, engineers became ever more skilled at tapping natural resources, generally ignoring environmental issues. Using modern science and large-scale technologies such as factories and power plants, they set out to manufacture goods on a scale unimaginable a few decades earlier. Workers mined ores deep in the earth at great risk to themselves or removed overburden with manpower, machines, and explosives, leaving great scars and polluted waterways behind. The result has been the thorough transformation of nature and a legacy of hazardous waste that at the end of the twentieth century seemed intractable.

Air Pollution and Its Historical Roots

Significant air pollution has also accompanied industrialization and urbanization. Firewood, peat, coal, oil, and gas consumed in heating, melting, welding, smelting, and power generation (on automobiles, see below) produce carbon monoxide, carbon dioxide, sulfur and nitrogen oxides, and traces of metals and radioactive substances. Over the years since 1800, coal has been the most important source of air pollution. Yet one early-twentieth-century observer – correctly – called coal "fundamental to the civilization of the present era."[18] It has been difficult to wean manufacturers and power producers from coal, largely because of its relatively low price (which does not reflect its environmental costs). India and China in the late twentieth century and other rapidly industrializing countries consumed vast quantities of the stuff, normally without pollution control.

In the mid- to late nineteenth century, several governments began to take some steps to regulate air pollution, but these efforts were largely ineffective until after World War II. Many ordinances exempted existing manufacturing plants, permitted higher densities of smoke than they ought, or were deliberately vague. The agencies created and the laws promulgated were too weak, the fines too small, and the belief that economic production trumped pollution control too deep-seated. Pittsburgh grew infamous for cloudy darkness at midday. Worse still, a killer smog blanketed London, England, from December 5 to December 9, 1952, killing at least 4,000 people. These and other serious smog incidents indicated that real action was required, although it would be another 30 years in most cases before effective regulation. Technological advances would only slightly solve the problem of smoke through more efficient combustion or filtering.

In the industrial age, air pollution eventually gave rise to problems on a global scale. The emissions of carbon dioxide in particular, mainly from the combustion of coal and oil, gradually made the atmosphere more efficient at trapping heat from the sun's rays. By the 1980s, the lower atmosphere and oceans had begun to warm up at rates unprecedented, perhaps for millions of years, in the earth's history. Late in the twentieth century scientists identified the direct connection between fossil-fuel emissions and climate change, although some individuals continue to deny the link. International efforts to check emissions and contain climate change to date have borne minimal results. It may well require radically new energy technologies to prevent runaway climate change – time will tell.[19]

Industrial Pollution and Hazardous Waste

In *A Civil Action* (1996) Johnathan Harr discusses industrial pollution and its great human and environmental costs in Woburn, Massachusetts. Dozens of tanneries and piggeries were established in the 1700s along the Aberjona River. By 1884, 26 Woburn tanneries employed 1,500 people. Extensive pollution was visible in the discoloration of the water, extensive fish kills, and cattle refusing to drink the water.[20] Toxic chemicals used in tanning led to cancers, birth defects, and other human health problems. Many tanneries disposed of their refuse in open pits that leached into the soil and ground water; tanneries refused to build individual water-treatment works. In response to the obvious pollution, the Massachusetts General Court passed legislation at the beginning of the twentieth century that prohibited discharge of any substance that might be injurious to public health or create a public nuisance, and established a fine of $500, but the pollution continued.

The tanning industry spurred a related industry to supply the chemicals needed for tanning when it turned to chromium salts rather than vegetable tannins. Chromium and other metals gained broad application, for example in the manufacture of alloys, chemicals, pigments, electroplating, fungicides, wallpaper, film, and ink. Toxic chromium enters the environment in natural ways, but more through human mishandling – through leakage, poor storage, and improper disposal. Chromium contamination persists in Long Island, New York, Corvallis, Oregon, at the United Chrome Products site, in the infamous Woburn, Massachusetts, W. R. Grace Company site, and in dozens of sites throughout the world.[21]

Other metals, such as lead, had similar long-term impacts on the environment. People have long understood that lead contributes to neurological problems, but, because of its wide applications in pipes, paints, ceramic glazes, batteries, and elsewhere, and because of the seeming obeisance of officials to industry, lead maintained a significantly unhealthful presence in many countries into the late twentieth century.

One of the interesting aspects of the relationship between the environment and technology has been the role of governments not only in gathering information but in regulating industrial and other business activities in the interest of environmental and public health. Many governments have stepped into this role grudgingly, fearful of interfering in market mechanisms, or hampering economic growth. For example, the US Public Health Service (PHS) determined to allow tetraethyl lead back into gasoline as an anti-knocking agent in 1926 after it had been removed voluntarily from the market, based on an incomplete study of gas-station workers that found them to be healthy – when no blood-lead tests were available. The PHS recommended further study, but the industry took the report as a "clean bill of health." Subsequent studies of the health impacts of lead between the 1930s and the 1960s were "funded by, and heavily biased toward, the industry" and were intended to prove that high levels of lead were "both normal and harmless"; many studies ultimately were deliberately deceptive.[22]

Some specialists estimate that from the 1920s lead poisoning contributed to the deaths of thousands of children in the US; perhaps 200 children died annually as a result of lead poisoning. The government did not prohibit residential use until the 1970s, after the creation of the Environmental Protection Agency (EPA). (Lead-based paint was first identified as the source of deadly childhood poisoning in Australia in 1904.) In 1971 the US Congress passed the Lead-Based Poisoning Prevention Act yet delayed implementation of its official ban until 1977,[23] and, while the problem of exposure to lead has diminished, some groups, "mainly minority and poor children living in the inner city, suffer from high rates of lead poisoning," since lead residues remain after decades of use in paints.[24] Such countries as China and Mexico still permit leaded gasoline and continue to have significant environmental problems associated with this metal.[25]

In sum, industrial technologies have been and remain a significant source of pollution. According to one estimate, the US in 1935 alone generated 5.7 million tons of hazardous waste. Today, US industry produces hundreds of millions of tons of waste annually, excluding radioactive waste (which represents 85 percent of the world's total hazardous waste), and there is no inventory to tell us how much or what kinds. Between 2,000 to 10,000 waste sites that require remediation are underfunded. According to the 2000 Toxics Release Inventory of the EPA, over 2.95 million metric tons of toxic chemicals from about 2,000 industrial facilities are annually released into the environment, including nearly 45,360 metric tons (100 million pounds) of recognized carcinogens.[26] We know that other countries produce significant quantities of toxic chemicals and hazardous wastes, that there is international transport of these wastes and chemicals, and that many of them are improperly used and haphazardly stored. The figures from the US suggest that the problem is massive and on a global scale.

Cities as Technologies with Environmental Impacts

Cities are technologies of their own sort comprised of a wide range of systems: housing, industrial production, distribution, transportation, sewage, water treatment, power generation, illumination, heating and cooling, culture (libraries, museums), and so on.[27] These technologies represent both the pinnacle of human ingenuity and the hubris of believing that humans fully understand or can predict the social and environmental consequences of their activities.

To examine just one of these technologies from the point of view of environmental issues, consider water pollution, waste disposal, and disease and illness in connection with the environment. People have long recognized the relationship between water and illness, but only in the late nineteenth century, with the rise of the germ theory of disease, did they comprehend the direct relationship between filth, sickness, and water. Still, cities belatedly built sewage systems that did little more than carry waste water further downstream. The philosophy was to dilute pollution and move it elsewhere, not to treat it. This led to the challenges both of providing cities with clean water and of treating waste. Since town officials had considerably more power than rural residents, in many instances they were able to lay claim to entire watersheds well outside metropolitan areas, establish reservoirs there, and build pumping stations, canals, and pipelines to bring clean water to urban residents. This was true of New York City, which tapped water from the Adirondack Mountains 100 years ago; of Boston and eastern Massachusetts, which built the Quabbin Reservoir in the western part of the state in the 1930s after having evicted hundreds of residents of four towns, razed their homes, and moved their cemeteries (except for those of Native Americans);[28] and of Las Vegas, Nevada, which is draining the aquifer under the northern 90 percent of the state at great cost to fragile arid environments and to the lives and activities of people who live in them.[29]

While water-treatment facilities were widely installed in the US, Europe, and elsewhere in the first half of the twentieth century, many of the facilities were soon overburdened by growing populations and little understanding of the nature of public health. For instance, storm drains were at first hooked into treatment facilities that became overloaded during heavy storms. Waste treatment often came to towns and villages much later than it did to cities, In 1950s Eastern Europe and the USSR, for example, many hospitals and schools remained without running water. In the US, in spite of some progress, significant challenges remained. In 1939, F. W. Mohlman, research director of the Sanitary District of Chicago, lamented the fact that industries discharged untreated wastes that polluted streams and overloaded sewage-treatment facilities. He criticized meat packers, paper industries and others for ignoring the problem of waste and following a policy of passive resistance to pollution control and remediation.[30]

The Rise of Preservation and Conservation Movements

During the eighteenth and nineteenth centuries the governments of Europe and North America were actively involved in the promotion of industrialization and the management of natural resources. This management entailed an effort to catalog and exploit the great mineral and natural wealth available to the nations and their citizens. The goals were to support military, economic, public-health, and other programs. Government activities included support of scientific expeditions to the interior, for example that of Meriwether Lewis and William Clark from the Missouri River Basin to the Pacific Coast in 1804–6, or those of Siberian explorers supported by the Russian tsar, the charting of coasts and rivers, soundings of harbors, surveying of lands and the timbers remaining on them, and meteorological studies. Governments provided funds to growing networks of public and private universities for agricultural, silvicultural, and other research activities important to economic growth.[31]

Several states also began to regulate resource exploitation in modest ways – through closure of forests, establishment of users' fees for water or access to grasslands, and various taxes – to ensure conservation of resources. In the US the Coastal Survey joined with the Army Corps of Engineers to bring about improvements on the nation's waterways and harbors. In Prussia foresters in service of the state advanced *Forstwissenschaft* to bring order and regular growth to the stands of trees (only belatedly recognizing how their pruning and clearing of the forest floor deprived the trees of essential nutrients for growth). In England, parliament continued a long tradition of protecting the forests established under the crown to ensure access to materials needed for the navy. In Norway the government created a Fisheries Inspectorate in the mid-nineteenth century to assist coastal fishing communities in maintaining their livelihoods under the pressure of technological change.

A number of philosophers, authors, artists, and others worried publicly that these efforts were not sufficient given the impact of human action on the environment. They included John James Audubon, John Muir, Henry David Thoreau, and George Perkins Marsh, to name only the American examples. They and others like them saw in the effects of industrialization upon land, air, and water a threat to human well-being as well as to nature itself. They inaugurated, or at least solidified, a cultural critique of industrialization and modern technology that lives on in the twenty-first century.

The profligate use of resources and pollution led at the end of the nineteenth century to broader conservation movements in North America, in Europe, and in colonial lands. The movements were often connected to professional associations of geologists, botanists, and biologists. During the Progressive era (1890–1910) in the US, scientists and engineers from a variety of specialties sought to improve the quality of American life through the application of scientific management. They believed that applied science could determine the availability of natural and mineral resources across the greatest number of uses for the greatest number of citizens for centuries to come. They believed that technicians and engineers, not legislators, should deal with these issues. Among their achievements were the establishment of a series of bureaucracies and the passage of laws to reclaim and irrigate land, improve inland waterways for navigation, save forest stock, and so on. Gifford Pinchot, an engineer of forestry and first head of the US

National Forest Service, proudly proclaimed the birth of scientific forestry under the banner of conservation. Another achievement of the Progressive era was the establishment of parks, national monuments, and wilderness areas.[32]

The Automobile

The automobile has had significant environmental impact, not only from its engine, but from the creation of roads and the extensive pollution from the manufacturing process. Initially, many enthusiasts touted the cleanliness of the automobile as compared to horses whose urine and manure were a great problem in cities. But by the early twenty-first century there were 247 million registered motor vehicles in the US alone, and a total of over 800 million in the world. Transportation has become the largest single source of air pollution in the US. It produced over half the carbon monoxide, a third of the carbon dioxide, over a third of the nitrogen oxides and volatile organic compounds, and almost a quarter of the hydrocarbons in the atmosphere in 2006. This pollution contributes to smog, to severe respiratory problems, and to untold deaths. Significant water pollution results from runoff of oil, gasoline, and other automotive fluids, brake dust, vehicle exhaust, and so on. The hundreds of millions of engines used in snowmobiles, personal watercraft, ATVs, gardening equipment, chain saws, and so on also pollute extensively. On top of this, as many as 1 million animals are struck on roads daily.

Vehicle pollution in the twentieth century contributed mightily to urban air pollution and human health problems. Cities where abundant sunshine mixed with tailpipe exhaust suffered from photochemical smog. Perhaps the worst case was Mexico City, where the topography prevented winds from dispersing pollution quickly, and 4 million vehicles clogged the streets by 1990. Respiratory problems from Mexico City's smog typically killed 6,000 to 12,000 people annually in the late twentieth century. Tehran, Los Angeles, Athens, Bangkok, and dozens of other sunny cities suffered somewhat less.

Highways, roads, and ubiquitous parking lots create another set of problems. A four-lane highway requires a minimum of 3.6 meters (12 feet) per lane, plus shoulders and median, so that total right-of-way width may exceed 30 meters (100 feet). Entrance and exit ramps also require significant takings of land. Early highways and roads were often built in or near floodplains since the land was relatively flat and required less grading. But covering floodplains with tar or concrete prevented porous soils from absorbing rain, and the construction of bridge abutments over rivers also interrupted natural processes of absorption and contributed to floods: the water cannot seep through the roads and it backs up at bridge abutments. Road-construction projects have destroyed the solitude and quality of life in many neighborhoods, and roads impinge significantly – and intentionally – on poor people's neighborhoods.[33]

The impact of highways has been nearly universal. The German highways (Autobahnen) were important from the points of view of public-works projects to put people back to work, and as symbols of Nazi power. But construction engineers who pushed for sensitivity to environmental issues were marginalized. Adolf Hitler recognized the importance of highways to move troops and for their propaganda value, so they had to be built quickly.[34] The Autobahnen were 2,128 kilometers (1,322 miles) in length at the end of World War II. Extensive construction commenced anew in the 1950s. By 1984 they were over 8,000 kilometers (4,971 miles) in length. More construction commenced after the unification of East and West Germany, making the Autobahn network the third largest superhighway system in the world, after those in the US and China.[35]

Highways contribute to environmental degradation not only through clearing of vegetation and removal of topsoil, both of which accelerate erosion, but also by facilitating increased traffic and migration. In the 1970s, Brazil planned a 3,220-kilometer (2,000-mile) highway that cut through the Amazon, opening the rain forest to settlement and accelerating development of timber and mineral resources. Yet engineers failed to understand the unstable nature of Amazonian soils and the roads' contribution to flooding. The roads accelerated in-migration of peasants, whose efforts at farming and cattle raising triggered rapid deforestation. Harvest yields were dismal, since the nutrient-poor soils were quickly exhausted, and new forest had to be cleared annually. Rampant erosion followed clear-cutting and forest burning. Since the highways went in, roughly 15 percent of the Brazilian Amazon has been deforested. The gaping gashes in the forest, which follow the pattern of the highway layout, are visible from space.[36]

Large-Scale Technological Systems and the State

In the twentieth century a number of states pursued such large-scale technological systems as highways and railways, irrigation networks, hydropower stations, nuclear power stations, and so on to improve navigation and commerce, raise agricultural production, and increase electricity production. They did so through various state-funded and state-run agencies across all economic systems and levels of economic development, from nineteenth-century European powers to twentieth-century African nations, and from the capitalist US to the socialist USSR. State power combined with engineering hubris. Government bureaucracies increasingly took the lead in new projects. A major result was the social and ecological transformation of society.[37]

The Soviet development model, with its emphasis on rapid industrial growth and collectivization of agriculture, was significantly costly from an environmental point of view. Worker safety and pollution-control equipment were an afterthought. Factories, mines, smelters, and the like were usually located near bodies of water, ensuring that solid and liquid pollution more quickly spread through the environment. Smoke-belching factories, fossil-fuel boilers that used low-grade fuels without scrubbers, heavy-metal refineries – all polluted substantially. Workers' housing was located near factories. No site, no matter how unique an ecosystem, seemed immune to planners' designs, including even Lake Baikal, home to 1,500 endemic species. Since the 1960s, Lake Baikal has suffered from extensive pollution from pulp and paper mills built on its shorelines, erosion due to deforestation, and railway construction in the surrounding region. It remains under threat from pollution in the twenty-first century.[38]

A contributing factor to the extensive alteration of ecosystems in the USSR was the momentum that design and construction organizations acquired. As a construction organization grew, it often became the seed of another organization that departed the region in search of its own projects. Metrostroi in Moscow (which built the Moscow subway) contributed thousands of employees to Kuibyshevgestroi, which built hydroelectric power stations on the Volga River. Kuibyshevgestroi then shed 5,000 employees toward Angarastroi, which built massive and environmentally costly power stations in Siberia. These projects inundated vast quantities of fertile farmlands, towns, and cities. The most hubristic project of them all, a plan to divert up to 10 percent of the water of some Siberian rivers to Central Asia through transfer canals, some of them up to 1,500 kilometers (932 miles) in length, built in part through excavation by nuclear explosions, was abandoned after years of planning only because of the breakup of the USSR.

In Brazil, state electrification agencies illuminated literally and figuratively the development of the Amazonian interior, and accelerated the destruction of the rain forest. By the end of the twentieth century massive hydroelectric projects had contributed to significant loss of forest, both through felling and through inundation.[39] Under Nehru and afterward, India embarked on a sprawling program to build hydropower stations and eventually nuclear reactors that Nehru considered "temples" of India's future economic growth and military prowess.[40] In the US, under the Bureau of Reclamation, the Army Corps of Engineers, the Tennessee Valley Authority, and the Bonneville Power Administration, the nation undertook a widespread building program beginning in the 1930s to tame rivers, reclaim land, and produce electricity. In China, from the 1950s onward, Communist leaders embarked on a ruthless, state-sponsored program to increase agricultural and industrial productivity with devastating impacts on families, society, and the environment.[41] Many of these efforts support military installations.

Military Technology

The military presents a special case, and an extremely costly one from the point of view of environmental degradation. Since the creation of gunpowder, militaries have experimented with ordnance and explosives that have scarred the landscape, and in their manufacture have produced hazardous waste. Unexploded shells from wars decades earlier continue to turn up. In the testing of weaponry, militaries have taken larger and larger holdings of land, in the twentieth century to test jets and rockets whose fuels are highly toxic. Significant quantities of haphazardly stored toxic chemical, biological, and radioactive wastes have leaked into the environment (or are stored precariously), spoiling hundreds of square kilometers of land and polluting waterways. Paradoxically, when military units leave bases and test sites closed, the ecosystems often recover rapidly and reveal surprising biodiversity.

Testing of weapons of mass destruction, especially by the US and the USSR, but also by France, the UK, and China, has had a devastating environmental impact. These include the toxic processes to mine and enrich ores and to process nuclear fuels. The US has conducted over 1,050 tests, over 300 of them above ground; the USSR 715 tests, of which 219 were above ground; France 210 tests, and so on. These tests have left radioactivity in unsafe amounts in Nevada, Semipalatinsk, Novaia Zemlia, Lop Nor, Polynesia, Bikini Atoll, and elsewhere. The nuclear industries enterprise has left behind millions of gallons of low- and high-level radioactive waste, hundreds of thousands of tons of solid waste and spent nuclear fuel, much of it initially poorly handled and stored, and to this day awaiting a final, safe storage.[42] Long secret, one of the most notorious nuclear accidents before Chernobyl was the Kyshtym waste-dump explosion in September 1957 that released a huge amount of radioactivity into the atmosphere, which spread over 800 square kilometers (300 square miles) and led to the evacuation of at least 100,000 people – a week later – who had no idea what had happened. Eighty-two percent of the land has been returned to use for forestry and agricultural purposes. The CIA kept the disaster secret in the US because of fear that the American public would turn against nuclear power.[43]

Technological Failure and the Environment

The names of several environmental disasters connected with industry signify the extent to which technological failure and environmental disaster have become linked: Bhopal, India, Chernobyl, Ukraine, and Love Canal near Buffalo, New York. On December 2, 1984, a Union Carbide factory in Bhopal, India, leaked methyl isocyanate gas from failed tanks, killing 2,259 people immediately, and 3,787 people overall. On April 26, 1986, a nuclear reactor at the Chernobyl nuclear power station exploded, releasing vast quantities of radio-activity into the surrounding area and the atmosphere, killing 33 people in the initial explosion. A few firefighters who arrived on the scene also died at Chernobyl, and at least 5,000 and perhaps as many as 50,000 excess cancer deaths resulted. Beginning in 1942, the Hooker Chemicals and Plastics Company began dumping various chemical wastes, over 21,000 tons in all, including halogenated organics, pesticides, chlorobenzenes, and dioxin. Dumping ceased in 1952, and in 1953, the landfill was covered and deeded to the Niagara Falls Board of Education. Subsequently, developers turned the area into "Love Canal," a town with homes and schools. Over the years, as toxic waste leaked from rusted, buried drums, birth defects and blood diseases became rampant, and the US federal government was forced to step in and pay for the relocation of more than 800 families. These three examples stand for hundreds of industrial accidents that have occurred over the past 200 years with acute environmental, and human health, consequences. There will be more.

Conclusion

From the Paleolithic to the present technologies have helped humankind make its imprint on the global environment. Lately, some technologies (e.g., catalytic converters in automobile engines) have helped to check that impact as well. Through most of human history, it was agriculture and agricultural technologies that mattered most, mainly because most people were farmers. The ax, firestick, and plow were probably the most environmentally transformative technologies.

Since the onset of industrialization and the widespread adoption of fossil fuels, the technologies wielded by human society have grown far more powerful in every respect, including their power to affect the natural environment. Today no corner of the earth, not the glaciers of Antarctica nor the vast waters of the Pacific Ocean, remains unaffected by the impacts of industrial technologies. The cumulative impact of modern technologies has proven revolutionary, economically and socially, but environmentally as well. That impact today threatens the stability of several crucial biogeophysical systems, especially climate. Ironically, every imaginable exit strategy from the conundrum of climate change involves technology. In the century to come we might develop energy technologies that radically reduce green-house-gas emissions. One way or another, for better and for worse, technology and the environment will remain tightly linked in the future, at least as much so as in the past.

Notes

1   L. Jinhui, *Rice Culture of China*, 2002, accessed from http://www.china.org.cn/english/2002/ Oct/44854.htm on March 13, 2012.
2   T. Schultz, *Two Effects of the Development of Agriculture on Early Societies*, 2010, accessed from http://www.brighthub.com/environment/science-environmental/articles/93829.aspx on September 27, 2011.

3   F. Gies and J. Gies, *Cathedral, Forge and Waterwheel: Technology and Invention in the Middle Ages*, New York, HarperCollins, 1994, pp. 44–9.

4   Gies and Gies, *Cathedral, Forge and Waterwheel*.

5   F. L. Pryor, "The Invention of the Plow," *Comparative Studies in Society and History* 27/4, 1985, pp. 727–43; M. Williams, *Deforesting the Earth: From Prehistory to Global Crisis*, Chicago, University of Chicago Press, 2003, pp. 102–42.

6   Illinois State Museum, *Early Mechanization of Farming*, 2011, accessed from http://www.museum.state.il.us/exhibits/agriculture/htmls/technology/horse-drawn/tech_horse-drawn_early.html on March 13, 2012.

7   W. J. White, *Economic History of Tractors in the United States*, 2001, accessed from http://eh.net/encyclopedia/article/white.tractors.history.us on March 13, 2012.

8   F. Jones, *Farm Gas Engines and Tractors*, New York, McGraw-Hill, 1932, pp. 253–6; J. Ellickson and J. Brewster, "Technological Advance and the Structure of American Agriculture," *Journal of Farm Economics* 29/4, 1947, pp. 827–47; R. T. McMillan, "Effects of Mechanization on American Agriculture," *The Scientific Monthly* 69/1, 1949, pp. 23–8.

9   J. Hightower, *Hard Times and Hard Tomatoes*, Rochester, VT, Schenkman Books, 1978; R. Wolf, "Industrializing Agriculture," *North American Review* 285/1, 2000, pp. 43–8; L. A. Jackson and M. Villinski, "Reaping What We Sow: Emerging Issues and Policy Implications of Agricultural Biotechnology," *Review of Agricultural Economics* 24/1, 2002, pp. 3–14.

10 FDA, *Statement on Foodborne E. coli O157:H7 Outbreak in Spinach*, 2006, accessed from http://www.fda.gov/NewsEvents/Newsroom/PressAnnouncements/2006/ucm108761.htm on March 13, 2012; C. Stofferahn, *Industrial Farming and Its Relationship to Community Well-Being: An Update of a 2000 Report by Linda Lobao*, 2007, University of South Dakota, Department of Sociology, accessed from http://www.und.edu/org/ndrural/Lobao%20&%20Stofferahn.pdf on March 13, 2012; Sustainable Table, *Factory Farming: The Issues*, 2010, accessed from http://www.sustainabletable.org/issues/factoryfarming on March 13, 2012.

11 M. Mellon and J. Rissler, *Environmental Effects of Genetically Modified Food Crops – Recent Experiences*, 2010, accessed from http://www.ucsusa.org/food_and_agriculture/science_and_impacts/impacts_genetic_engineering/environmental-effects-of.html on March 13, 2012.

12 E. Chivian and A. Bernstein, "Embedded in Nature: Human Health and Biodiversity," *Environmental Health Perspectives* 112/1, 2004, pp. A12–A13.

13 R. Smith, *Scaling Fisheries*, Cambridge, Cambridge University Press, 1994; see also Muscolino (Chapter 16) in this volume.

14 D. Liu, J. K. Keesing, Z. Dong, et al., "Recurrence of the World's Largest Green-Tide in 2009 in Yellow Sea, China," *Marine Pollution Bulletin* 60, 2010, pp. 1423–34.

15 O. Linden and A. Jernelov, "The Mangrove Swamp: An Ecosytem in Danger," *Ambio* 9/2, 1980, pp. 81–8; L. Lebel, N. H. Tri, A. Saengnoree, et al., "Industrial Transformation and Shrimp Aquaculture in Thailand and Vietnam: Pathways to Ecological, Social and Economic Sustainability?," *Ambio* 31/4, 2002, pp. 311–23; World Wildlife Fund, *Shrimp Aquaculture*, 2011, accessed from http://www.worldwildlife.org/what/globalmarkets/aquaculture/dialogues-shrimp.html on March 13, 2012.

16 R. Judd, *Common Lands, Common People: The Origins of Conservation in Northern New England*, Cambridge, MA, Harvard University Press, 1997.

17 J. Glaister, "Pollution of Scottish Rivers," *Proceedings of the Philosophical Society of Glasgow*, 1897, pp. 1–5.

18 E. C. Jeffrey, "The Origin and Organization of Coal," *American Academy of Arts and Sciences* 15/1, 1924, pp. 5–6.

19 S. Weart, *The Discovery of Global Warming*, Cambridge, MA, Harvard University Press, 2008.

20 EPA, *Wells G and H Site Remedial Investigation Report*, 1986, accessed from http://ce547.groups.et.byu.net/woburn/nus/index.php on March 13, 2012.

21   C. Palmer and P. Wittbrodt, "Affecting the Remediation of Chromium-Contaminated Sites,"
     Environmental Health Perspectives 92, 1991, pp. 25–40; E. S. Blair, K. Svitana, C. Manduca,
     et al., *Leather Tanning in Woburn*, 2009, accessed from http://serc.carleton.edu/woburn/
     issues/tanning.html on March 13, 2012.

22   W. Graebner, "Hegemony through Science: Information Engineering and Lead Toxicology,
     1925–1965," in D. Rosner and G. Markowitz (eds.), *Dying for Work: Workers' Safety and
     Health in Twentieth-Century America*, Bloomington, IN, Indiana University Press, 1989,
     pp. 140–59 (p. 140).

23   Pollution Issues, *Lead*, 2010, accessed from http://www.pollutionissues.com/Ho-Li/Lead.
     html on March 13, 2012.

24   H. W. Mielke, "Lead in the Inner Cities," *American Scientist* 87/1, 1999, pp. 62–73 (p. 62).

25   H. Ren, J. Wang, and X. Zhang, "Health Risk Assessment of Lead Pollution in Inner-City
     Environment in Shenyang, China," *Chinese Journal of Geochemistry* 25, 2006, Supplement 1,
     p. 56.

26   Pollution Issues, *Industry*, 2010, accessed from http://www.pollutionissues.com/Ho-Li/
     Industry.html on March 13, 2012.

27   M. V. Melosi, "Urban Pollution: Historical Perspective Needed," *Environmental Review*
     3/3, 1979, pp. 37–45; M. V. Melosi, "The Place of the City in Environmental History,"
     Environmental History Review 17/1, 1993, pp. 1–23.

28   J. Naglack, "Quabbin Reservoir: The Elimination of Four Small New England Towns,"
     Historical Journal of Western Massachusetts 3/2, 1974, pp. 51–9; D. Howe, *Quabbin: The Lost
     Valley*, Ware, The Quabbin Book House, 1951. See also MDCR, *Quabbin Reservoir*, 2011,
     accessed from http://www.mass.gov/dcr/parks/central/quabbin.htm on March 13, 2012.

29   J. E. Deacon, A. E. Williams, C. Deacon Williams, and J. E. Williams, "*Fueling Population
     Growth in Las Vegas: How Large-Scale Groundwater Withdrawal Could Burn Regional
     Biodiversity*," *Bioscience* 57/8, 2007, pp. 688–98.

30   F. W. Mohlman, "The Disposal of Industrial Wastes," *Sewage Works Journal* 11/4, 1939,
     pp. 646–56.

31   J. Scott, *Seeing Like a State*, New Haven, CT, Yale University Press, 1998. On the history of
     conservation, see S. Hays, *Conservation and the Gospel of Efficiency: The Progressive
     Conservation Movement, 1890–1920*, Pittsburgh, University of Pittsburgh Press, 1959;
     R. Nash, *Wilderness and the American Mind*, 4th ed., New Haven, CT, Yale University Press,
     2001; C. Miller, *Gifford Pinchot and the Making of Modern Environmentalism*, Washington,
     DC, Island Press, 2001.

32   Hays, *Conservation and the Gospel of Efficiency*; Miller, *Gifford Pinchot*.

33   On the impact of roads on the quality of life in New York City, and on various neighborhoods
     in particular, see R. Caro, *The Power Broker*, New York, Knopf, 1974.

34   T. Zeller, *Driving Germany: The Landscape of the German Autobahn, 1930–1970*, New York,
     Berghahn Books, 2007.

35   *Autobahn History*, 2010, accessed from http://www.german-autobahn.eu/index.asp?page=history
     on March 13, 2012.

36   P. M. Fearnside, "Brazil's Cuiabá- Santarém (BR-163) Highway: The Environmental Cost of
     Paving a Soybean Corridor through the Amazon," *Environmental Management* 39/5, 2007,
     pp. 601–14.

37   D. Worster, *Rivers of Empire: Water, Aridity and the Growth of the American West*, New York,
     Oxford, 1985.

38   M. Feshbach and A. Friendly, *Ecocide in the USSR: Health and Nature under Siege*, New
     York, Basic Books, 1992.

39   W. Dean, *With Broadax and Firebrand: The Destruction of the Brazilian Atlantic Forest*,
     Berkeley, University of California Press, 1997.

40   A. Parthasarathi, "Science and Technology in India's Search for a Sustainable and Equitable
     Future," *World Development* 18/2, 1990, pp. 1693–6.

41 J. Shapiro, *Mao's War against Nature*, Cambridge, Cambridge University Press, 2001.

42 I. Kudrik, C. Digges, A. Nikitin, et al., "The Russian Nuclear Industry: The Need for Reform," *Bellona Foundation Report* 4, 2004.

43 Z. Medvedev, *Nuclear Disaster in the Urals*, New York, Norton, 1979.

References

Autobahn History, 2010, accessed from http://www.german-autobahn.eu/index.asp?page=history on March 13, 2012.

Blair, E. S., Svitana, K., Manduca, C., et al., *Leather Tanning in Woburn*, 2009, accessed from http://serc.carleton.edu/woburn/issues/tanning.html on March 13, 2012.

Caro, R., *The Power Broker*, New York, Knopf, 1974.

Chivian, E., and Bernstein, A., "Embedded in Nature: Human Health and Biodiversity," *Environmental Health Perspectives* 112/1, 2004, pp. A12–A13.

Deacon, J. E., Williams, A. E., Deacon Williams, C., and Williams, J. E., "Fueling Population Growth in Las Vegas: How Large-Scale Groundwater Withdrawal Could Burn Regional Biodiversity," *Bioscience* 57/8, 2007, pp. 688–98.

Dean, W., *With Broadax and Firebrand: The Destruction of the Brazilian Atlantic Forest*, Berkeley, University of California Press, 1997.

Ellickson, J., and Brewster, J., "Technological Advance and the Structure of American Agriculture," *Journal of Farm Economics* 29/4, 1947, pp. 827–47.

EPA, *Wells G and H Site Remedial Investigation Report*, 1986, accessed from http://ce547.groups.et.byu.net/woburn/nus/index.php on March 13, 2012.

FDA, *Statement on Foodborne E. coli O157:H7 Outbreak in Spinach*, 2006, accessed from http://www.fda.gov/NewsEvents/Newsroom/PressAnnouncements/2006/ucm108761.htm on March 13, 2012.

Fearnside, P. M., "Brazil's Cuiabá- Santarém (BR-163) Highway: The Environmental Cost of Paving a Soybean Corridor through the Amazon," *Environmental Management* 39/5, 2007, pp. 601–14.

Feshbach, M., and Friendly, A., *Ecocide in the USSR: Health and Nature under Siege*, New York, Basic Books, 1992.

Gies, F., and Gies, J., *Cathedral, Forge and Waterwheel: Technology and Invention in the Middle Ages*, New York, HarperCollins, 1994.

Glaister, J., "Pollution of Scottish Rivers," *Proceedings of the Philosophical Society of Glasgow*, 1897, pp. 1–5.

Graebner, W., "Hegemony through Science: Information Engineering and Lead Toxicology, 1925–1965," in D. Rosner and G. Markowitz (eds.), *Dying for Work: Workers' Safety and Health in Twentieth-Century America*, Bloomington, IN, Indiana University Press, 1989, pp. 140–59.

Hays, S., *Conservation and the Gospel of Efficiency: The Progressive Conservation Movement, 1890–1920*, Pittsburgh, University of Pittsburgh Press, 1959.

Hightower, J., *Hard Times and Hard Tomatoes*, Rochester, VT, Schenkman Books, 1978.

Howe, D., *Quabbin: The Lost Valley*, Ware, The Quabbin Book House, 1951.

Illinois State Museum, *Early Mechanization of Farming*, 2011, accessed from http://www.museum.state.il.us/exhibits/agriculture/htmls/technology/horse-drawn/tech_horse-drawn_early.html on March 13, 2012.

Jackson, L. A., and Villinski, M., "Reaping What We Sow: Emerging Issues and Policy Implications of Agricultural Biotechnology," *Review of Agricultural Economics* 24/1, 2002, pp. 3–14.

Jeffrey, E. C., "The Origin and Organization of Coal," *American Academy of Arts and Sciences* 15/1, 1924, pp. 5–6.

Jinhui, L., *Rice Culture of China*, 2002, accessed from http://www.china.org.cn/english/2002/Oct/44854.htm on March 13, 2012.

Jones, F., *Farm Gas Engines and Tractors*, New York, McGraw-Hill, 1932.

Judd, R., *Common Lands, Common People: The Origins of Conservation in Northern New England*, Cambridge, MA, Harvard University Press, 1997.

Kudrik, I., Digges, C., Nikitin, A., et al., "The Russian Nuclear Industry: The Need for Reform," *Bellona Foundation Report* 4, 2004.

Lebel, L., Tri, N. H., Saengnoree, A., et al., "Industrial Transformation and Shrimp Aquaculture in Thailand and Vietnam: Pathways to Ecological, Social and Economic Sustainability?," *Ambio* 31/4, 2002, pp. 311–23.

Linden, O., and Jernelov, A., "The Mangrove Swamp: An Ecosytem in Danger," *Ambio* 9/2, 1980, pp. 81–8.

Liu, D., Keesing, J. K., Dong, Z., et al., "Recurrence of the World's Largest Green-Tide in 2009 in Yellow Sea, China," *Marine Pollution Bulletin* 60, 2010, pp. 1423–34.

McMillan, R. T., "Effects of Mechanization on American Agriculture," *The Scientific Monthly* 69/1, 1949, pp. 23–8.

MDCR, *Quabbin Reservoir*, 2011, accessed from http://www.mass.gov/dcr/parks/central/quabbin.htm on March 13, 2012.

Medvedev, Z., *Nuclear Disaster in the Urals*, New York, Norton, 1979.

Mellon, M., and Rissler, J., *Environmental Effects of Genetically Modified Food Crops – Recent Experiences*, 2010, accessed from http://www.ucsusa.org/food_and_agriculture/science_and_impacts/impacts_genetic_engineering/environmental-effects-of.html on March 13, 2012.

Melosi, M. V., "The Place of the City in Environmental History," *Environmental History Review* 17/1, 1993, pp. 1–23.

Melosi, M. V., "Urban Pollution: Historical Perspective Needed," *Environmental Review* 3/3, 1979, pp. 37–45.

Mielke, H. W., "Lead in the Inner Cities," *American Scientist* 87/1, 1999, pp. 62–73.

Miller, C., *Gifford Pinchot and the Making of Modern Environmentalism*, Washington, DC, Island Press, 2001.

Mohlman, F. W., "The Disposal of Industrial Wastes," *Sewage Works Journal* 11/4, 1939, pp. 646–56.

Naglack, J., "Quabbin Reservoir: The Elimination of Four Small New England Towns," *Historical Journal of Western Massachusetts* 4/1, 1975, pp. 51–9.

Nash, R., *Wilderness and the American Mind*, 4th ed., New Haven, CT, Yale University Press, 2001.

Palmer, C., and Wittbrodt, P., "Affecting the Remediation of Chromium-Contaminated Sites," *Environmental Health Perspectives* 92, 1991, pp. 25–40.

Parthasarathi, A., "Science and Technology in India's Search for a Sustainable and Equitable Future," *World Development* 18/2, 1990, pp. 1693–6.

Pollution Issues, *Industry*, 2010, accessed from http://www.pollutionissues.com/Ho-Li/Industry.html on March 13, 2012.

Pollution Issues, *Lead*, 2010, accessed from http://www.pollutionissues.com/Ho-Li/Lead.html on March 13, 2012.

Pryor, F. L., "The Invention of the Plow," *Comparative Studies in Society and History* 27/4, 1985, pp. 727–43.

Ren, H., Wang, J., and Zhang, X., "Health Risk Assessment of Lead Pollution in Inner-City Environment in Shenyang, China," *Chinese Journal of Geochemistry* 25, 2006, Supplement 1, p. 56.

Schultz, T., *Two Effects of the Development of Agriculture on Early Societies*, 2010, accessed from http://www.brighthub.com/environment/science-environmental/articles/93829.aspx on September 27, 2011.

Scott, J., *Seeing Like a State*, New Haven, CT, Yale University Press, 1998.

Shapiro, J., *Mao's War against Nature*, Cambridge, Cambridge University Press, 2001.

Smith, R., *Scaling Fisheries*, Cambridge, Cambridge University Press, 1994.

Stofferahn, C., *Industrial Farming and Its Relationship to Community Well-Being: An Update of a 2000 Report by Linda Lobao*, 2007, University of South Dakota, Department of Sociology, accessed from http://www.und.edu/org/ndrural/Lobao%20&%20Stofferahn.pdf on March 13, 2012.

Sustainable Table, *Factory Farming: The Issues*, 2010, accessed from http://www.sustainabletable.org/issues/factoryfarming on March 13, 2012.

Weart, S., *The Discovery of Global Warming*, Cambridge, MA, Harvard University Press, 2008.

White, W. J., *Economic History of Tractors in the United States*, 2001, accessed from http://eh.net/encyclopedia/article/white.tractors.history.us on March 13, 2012.

Williams, M., *Deforesting the Earth: From Prehistory to Global Crisis*, Chicago, University of Chicago Press, 2003.

Wolf, R., "Industrializing Agriculture," *North American Review* 285/1, 2000, pp. 43–8.

World Wildlife Fund, *Shrimp Aquaculture*, 2011, accessed from http://www.worldwildlife.org/what/globalmarkets/aquaculture/dialogues-shrimp.html on March 13, 2012.

Worster, D., *Rivers of Empire: Water, Aridity and the Growth of the American West*, New York, Oxford, 1985.

Zeller, T., *Driving Germany: The Landscape of the German Autobahn, 1930–1970*, New York, Berghahn Books, 2007.

Further Reading

Callan, S., and Thomas, J., *Environmental Economics and Management: Theory, Policy, and Applications*, 3rd ed., Mason, OH, Thompson, 2004.

Rome, A., *The Bulldozer in the Countryside: Suburban Sprawl and the Rise of American Environmentalism*, New York, Cambridge University Press, 2001.

Weiner, D., *A Little Corner of Freedom: Russian Nature Protection from Stalin to Gorbachev*, Berkeley, University of California Press, 1999.

CHAPTER TWENTY

Cities and the Environment

JORDAN BAUER AND MARTIN V. MELOSI

Origins of the Field

Although relatively new, urban environmental history as a formally recognized field emerged primarily in the US in the late 1970s and early 1980s from the work of scholars conducting research in environmental history, urban history, the history of technology, and public-health history. Of course, books and articles relevant to the subject predate those years. Why the field developed when it did is subject to speculation, but attention to urban environments makes sense if only because the level of urbanization globally was striking by the early twentieth century: 80 percent in Great Britain, more than 60 percent in the Netherlands and Germany, 50 percent in the US, and 45 percent in France. By 2000, every part of the world was at least 40 percent urbanized.

The founding of the Public Works Historical Society brought together a few urban historians, historians of technology, and public-works practitioners in 1975. The organization commemorated the nation's "public-works heritage" in the bicentennial year through the publication of *History of Public Works in the United States, 1776–1976* and by promoting the study of public-works history.[1] While not focusing directly on environmental issues, several of the members incorporated environmental themes in their studies of infrastructure and public works. Most conspicuously the founding of the American Society for Environmental History (also in 1976) provided an intellectual haven for aspiring urban environmental historians, as did the older Society for the History of Technology (1958).[2]

From a geographic standpoint, the field has shown its greatest vitality in the US and Europe, but is becoming more global.[3] Historians on both sides of the Atlantic, at least in the early years, tended to concentrate on the late nineteenth and early twentieth centuries, exploring the impacts of industrialization and political and social reform. Much of the research output has been case-study- or single-city-driven. Case studies allow schol-

A Companion to Global Environmental History, First Edition. Edited by J.R. McNeill and Erin Stewart Mauldin.
© 2012 John Wiley & Sons, Ltd. Published 2015 by John Wiley & Sons, Ltd.

ars to examine the relationship of a specific city to its environment over time. Since the early 1980s, an international network of urban environmental scholars has produced a workable intellectual exchange of theories and topics. A number of international history conferences and workshops also led to the creation of new scholarly associations devoted to the study of urban environmental history – including the Urban Environmental Roundtable held biennially in Western Europe – as well as collaborations on influential edited volumes that approach the environmental history of cities from a variety of perspectives.[4] Professional groups also have formed internationally in recognition of the importance of environmental history, including the well-established European Society for Environmental History (1999), the Society for Latin American and Caribbean Environmental History, the Association of East Asian Environmental History, and the Network in Canadian History & Environment. The International Water History Association or IWHA (2001), although founded in Europe, cast a wide net in focusing on water issues throughout the world – especially through its journal *Water History*. The IWHA boasts a membership spanning almost every continent.

Some have argued that European scholars took their lead from American colleagues in studying the urban environment. Others, such as Geneviève Massard-Guilbaud and Peter Thorsheim, suggest that, while European urban environmental history remains "relatively frail," its traditions "at least partly" grew out of the study of urban technical networks and rich scholarship on the history of cities in general.[5] A good example was scholarly output from a 1983 conference held in Paris, and hosted by French geographer Gabriel Dupuy and American historian Joel Tarr, which brought together mostly French and American academics to discuss their work on urban technical networks and systems, and resulted in the award-winning *Technology and the Rise of the Networked City in Europe and America*.[6] The organizers stressed in the resulting book:

> Technology is critical to the city-building process and the operation of cities, but historians have not paid serious attention to its vital role in shaping the urban environment until the last decade or so. Although technology and cities have always been interdependent, only since the advent of industrialism in the nineteenth century have urban technological networks evolved.[7]

This assertion guided much of the work in the field both in the US and in Europe, and introduced American scholars to the possibilities of transatlantic and comparative research. The role played by technology has been central to urban environmental history, in contrast to the general field of environmental history. Changes in infrastructure and related technologies have drawn substantial attention, as have city services of all kinds.[8] "Without using the term environmental history," Massard-Guilbaud and Thorsheim stated, "the [1983] conference ... explored topics that would soon be considered major themes in urban environmental history, such as the impact of city building and technical choices on the environment."[9]

Dieter Schott added that, "European environmental history had overall never been dominated by an 'agro-ecological perspective,'" in societies where the study of urban growth and development had been longstanding.[10] Others in Europe have drawn additional lines between urban environmental history practiced there and in the US. Peter Brimblecombe and Christian Pfister stated that, "Unlike their colleagues in North America, historians in Europe have long been reluctant to explore the dimensions of past relations between societies and their natural habitat."[11] Brimblecombe, Pfister, and others were intent on creating "an area of dialogue" between the social and the natural sciences to improve understanding of ecological problems.

The use of "reluctant" to express the lag in the role of historians to embrace environmental history may have been an unfortunate word choice. Like in the US, the roots of European environmental history could be found in the past of several countries. In reality, there has not been a "European" environmental history per se, but a collection of histories emerging from individual countries. Some would suggest that environmental history in Europe remains fragmented. In several cases, scholars in Europe operated in isolation from each other due to national and language barriers or rifts emerging during the Cold War. The lack of institutional recognition and support had also constrained the growth of the field there.[12]

To be sure, the discipline of history has had a longer and different evolution in Europe than in the US. Formalizing the field of urban environmental history in Europe also appears to have a geographical orientation. To the north, Germany, the UK, and Sweden produced some early, important work in environmental history, drawing upon studies in geography, landscape history, and forest history. In France, Spain, and Italy, the field developed somewhat more slowly. The Annales School in France, however, proved to be a model for environmental history beyond French borders over several decades. In Italy agricultural history, nature conservation, art history, and historical geography have been important. Agrarian ecology – as well as pollution studies – intrigued scholars in Spain. Finland has more recently blossomed as a center of study for environmental history across a wide spectrum, while interest in the Czech Republic and other countries in Eastern Europe is slowly emerging.[13] More so than in the US, the development of urban environmental history in Europe seemed to develop hand-in-hand with environmental history in general.

The lag in the development of the field in other parts of the world may be attributed to a variety of causes. For example, since the works discussed in this chapter exist only in English, there is a gap in truly understanding the breadth of urban environmental history not only in Europe but elsewhere. The tendency to concentrate on books misses some of the intellectual vitality in this burgeoning subject area represented in articles or through presentations at scholarly meetings. In addition, historical traditions in some parts of the world have placed priorities on different lines of inquiry than in North America and in Europe. Paul Sutter has made a compelling argument that "the most notable aspect of non-U.S. environmental historiography lies in its focus on colonialism and imperialism as environmental processes. Indeed, studies of the environmental implications of colonial and imperial encounters have largely fueled the rapid growth of non-U.S. environmental historiography."[14]

It is not that cities rarely get discussed in the environmental literature of many non-US and non-European countries, but – so far – they have not been studied within the constructs currently dominating the field of urban environmental history. In addition – and this is true for any geographic region – as research agendas in the field of environmental history itself have changed, the lines between urban environmental history and other inquiries have blurred, been ignored, or have been absorbed into broader narratives.

Major Debates

Lewis Mumford once wrote that "The city is a fact in nature, like a cave, a run of mackerel or an ant-heap." He also wrote that the city was an expression of human civilization, a physical site for communal living, "man's greatest work of art."[15] Mumford appears to argue, on the one hand, that the city is in nature, a part of nature, but on the other, that "Nature, except in a surviving landscape park is scarcely to be found near the metropolis."[16] It is this polarity – city in nature versus city as outside of nature – that has been a central point of debate among scholars who study environmental history in general and cities in particular.

Martin Melosi has argued that historians continue "to create a barrier between the natural world and the city." That is, "they treat cities – the built environment – as an artificial creation of humans, who are themselves regarded as outside of nature."[17] While urban environmental historians have successfully placed cities within the field of environmental history, they still tend to draw a distinction between nature and the built environment in their works. They challenge, however, the idea that cities should be excluded from environmental history because cities are inherently artificial. The scholarship also takes into account that perceptions of nature have guided the ways in which cities were built and how the changing cultural, social, political, technological, and economic conditions shape the urban environment.[18]

Broadly speaking, urban environmental history encompasses the study of the natural history of the city, the history of city-building, the place of the city within a larger environmental context, and their possible intersections. Networked systems, city services, urban technology, public health, pollution, and spatial transformation are and have been important topics for urban environmental history. Scholarship has examined the causes of environmental change: rapid population growth, urbanization, changes in technology, energy use, economic growth, politics, and so on. The decisions to establish cities in particular locations have historically relied heavily on humans' understanding of the environment. Settlement selections were influenced by proximity to waterways and abundant food supplies to ensure resources, transportation, commerce, and defense. The Chicago School of sociology, founded at the University of Chicago during World War I, is credited with first stimulating interest in many of these issues. Those members of the school committed to an ecological perspective concentrated on factors determining urban spatial patterns and their social impacts.[19]

Urban environmental historians – as well as other environmental historians – have tended to focus on what might be called "the declensionist narrative," or the argument that human agency through the building of cities and other actions (e.g., industrialization) contributed mightily to harming, modifying, or destroying nature. Some early work focused on "the environmental crisis of the city," emphasizing the destructive impact of the Industrial Revolution on urban development through crowding, air and water pollution, noise, and waste accumulation.[20] This view contributed to studying the urban environment from an internal – how cities worked – as opposed to an external – how cities influenced their surroundings – perspective. As John McNeill has suggested, "Urban environmental history originally focused mainly on pollution and sanitation, but diversified so as to encompass the development of technical systems generally, the provisioning and metabolism of cities."[21]

Such a focus often led to developing metaphors to describe urban development. Among the most popular was some version of "the organic city," which in its most literal form related the structure and operation of the city to that of the human body. The theory has obvious flaws and has been used in the past to support the interests of elites, but nevertheless elicited striking images of community interdependency and the rational functioning of many city services. Few scholars embraced organic theory without reservation, opting for the idea of cities as "open systems" that were neither self-contained nor functioning outside a larger world. More persuasive is the notion that cities could be viewed as a special kind of ecosystem that was dynamic rather than static.[22]

To rise above the idea that cities were simply a series of technical networks, some historians such as Joel Tarr, Sabine Barles, and Verena Winiwarter have utilized the concept of the "urban metabolism" to explain how cities function. Cities, according to these urban environmental historians, have metabolisms; they use water, food, and oxygen, and produce garbage, sewage, and pollution. Winiwarter viewed urban metabolism in a broad social and cultural context in which all inputs and outputs had to be taken into account, be they biological, physical, social, or technical. This included utilizing a variety of resources, producing waste, or impacting the relationship among people.[23]

Particularly useful have been efforts to link city development to larger patterns of urbanization and to non-urban environments. Rapid global urbanization, for example, has brought about changes beyond city boundaries. William Cronon was among the first to point the way for placing the city in a wider environmental and regional context. His seminal work, *Nature's Metropolis*, demonstrates how the city and its hinterland are inextricably linked through the environment and the market economy. He shows not only the physical linkage between city and hinterland, but also how cities actually transform raw materials from the countryside – turning forests into lumber – into products to be bought and sold. This process of commoditization distinguishes the function of cities, at least in an economic sense, from the broader environment.[24]

A new generation of urban environmental historians, including Andrew Isenberg, Ellen Stroud, Matthew Klingle, Karl Appuhn, Ari Kelman, Emmanuel Kreike, Peter Thorsheim, and others, has attempted to complement, revise, and in some cases challenge the approaches of a first generation of urban environment historians such as Joel Tarr, Martin Melosi, Bill Luckin, and William Cronon. *The Nature of Cities*, edited by Isenberg, makes a conscious effort to move beyond the existing literature on the urban environment, but also to distance itself from the stark dichotomy between nature and the built environment. Isenberg argues against the notion of an organic city, but concludes that it still remains popular among some historians. The claim that the essays in the volume "mark a new direction in urban environmental history" is based upon the pointed focus on cultural history, including race and class conflict, consumerism, and the cultural construction of place. Isenberg notes that a common feature of the volume is "a concern with power: political power, imperial power, and the power of historians, archeologists, geographers, and others to shape our understanding of past and present landscapes." A "pernicious problem," he argues, in the use of "the undiluted form of the organism model is its tendency to obscure the power relations that are central to urban history."[25]

A very good example of the "new" urban environmental history is Matthew Klingle's *Emerald City*. In his study, Klingle is skeptical about how the field of environmental history is sometimes trapped "in an analytical device of its own making." "By juxtaposing nature and culture as pure categories," he argues, "environmental historians have demonstrated the independence of nature as well as the consequences of human actions." This often "yields stereotypical stories of decline." He would rather see "humans and nature as tangled together" and to avoid a dualism that asks if elements in cities are natural or unnatural, or if cities themselves are one or the other. While Klingle tries to redefine the merger of the natural and the built into something new (nature "co-evolving with humans"), his major preoccupation is to demonstrate that "History is inseparable from place." In this sense he is interested in process, rather than in looking only at the internal workings of, in this case, Seattle. Nature and cities may not follow diverse paths – or even parallel paths – but touch at key points in history without being one and the same.[26]

Themes and Topics

While historians question existing interpretations of the urban environment, much of the scholarship, as stated above, continues to follow a case-study approach as opposed to major overviews, concentrates on single cities, and focuses on the nineteenth and twentieth centuries. In addition, the content and major themes of European urban environmental history and that of other areas are similar to much of American scholarship. Studies tend to revolve around questions of urban growth, infrastructure, health, and pollution. Interest in the impacts of race, class, and gender in the urban environment – which are preoccupations of American historical scholarship in general – has come into its own in the US more than elsewhere, but that is changing.

An important early theme cutting across local and national lines has been – and in many cases remains – the development of and changes in urban technical networks and construction. As urban populations increased, the need for a reliable water supply and waste management took on new importance. Whereas city dwellers had once been on their own to address issues of sanitation, the growth in size and scale of cities required that populations respond collectively. Historians found that decision-making and implementation of these systems were not only influenced by political, economic, and technical considerations, but also informed by broad cultural issues, public health, and scientific knowledge of the time.[27] A fairly recent work, geographer Matthew Gandy's *Concrete and Clay*, explores the interrelationship between the physical and cultural in his analysis of the building of New York City, emphasizing the process of urbanization more than studying a fixed place. In so doing, he hoped "to build a conception of urban nature that is sensitive to the social and historical contexts that produce the built environment and imbue places with cultural meaning." Michael Rawson follows a comparable path for Boston in *Eden on the Charles*. One of the very few attempts to bring environmental history to the suburbs is Adam Rome's influential *The Bulldozer in the Countryside* that discusses the fundamental transformative impact of development on what was the urban hinterland.[28]

There is, however, no overarching international history of urban development which incorporates an environmental framework. For the twentieth century, J. R. McNeill's *Something New under the Sun* is a good starting point. The environmental history of individual cities, however, is a promising recent trend that began with Andrew Hurley's *Common Fields*. The volume of essays covers a wide array of topics on St. Louis rather than sustaining a single narrative, which has opened possibilities for future research into a number of themes and issues. *Common Fields* inspired several more volumes on specific cities, mainly in the US, each with a different emphasis. There have been only a few attempts to complete similar books on cities outside the US – most recently Montreal and, earlier, Sydney – but in the next several years more books on international cities are likely to appear.[29]

Advances in knowledge and theories about public health during the mid-nineteenth century, such as bacteriological theories of disease, guided human action in response to environmental concerns in the city. For instance, until the late nineteenth century, ideas about public health revolved around the miasma or filth theory of disease that identified decaying material and stenches as the primary causes of sickness.

This non-contagionist theory dictated what city services were given priority. As a result, developing potable water supplies and ridding the streets of trash and filth took precedence. Environmental sanitation, however, removed potentially risky matter from human senses; although it was a useful tool, it did not eliminate disease. In addition, urban fires were a common problem and therefore the construction of early waterworks systems reflected a need to address conflagrations with the same new water systems. Nineteenth-century sanitarians, most notably Englishman Edwin Chadwick, influenced the ways city leaders and engineers thought about solving health problems by emphasizing the physical conditions of the city and their impact on the population. With the advent of the germ – or bacteriological – theory of disease after 1880, primitive methods of dealing with health risks were replaced with more effective approaches based on chemical testing and immunization.

The field of public-health history has a long tradition of its own with many fine contributors. Among more recent scholars dealing with the urban environment, the work of Christopher Hamlin, Nancy Tomes, Werner Troesken, and Suellen Hoy stands out.[30]

Technology has played an important part in the ways in which cities develop and change. Along with rapid population growth from migration and immigration, changes in technology stimulated the physical expansion of the city. Particularly during industrialization, city development greatly relied on more elaborate forms of technology, such as the horse-drawn streetcar and the steam-powered railway, which transformed urban space. Transportation has been a key topic for urban environmental historians to explore, ranging from public transit to the automobile, with a lengthy number of studies. One novel approach, pioneered by Clay McShane and Joel Tarr, was to look back before intense mechanization to the central role of the horse in urban transport.[31]

The physical growth of the city required the expansion of existing and the creation of new public services and infrastructure. Technology transformed the physical environment of cities, made them more livable, and alleviated many public-health problems. The literature on urban infrastructure and city services is substantial, ranging from water supplies to sewerage systems, from electrical power to solid-waste disposal. A sampling of research in this area will lead to more recent work by a wide array of scholars.[32]

Historians have shown that, while technology solved some urban problems, it created new obstacles, and that the solution to one environmental dilemma often simply produced another problem. The technologies that were built to reduce domestic and industrial waste simply moved it to another sink – be it air, water, or land – or to a site where pollution was supposedly safe from the urban public. For example, sanitary engineers and urban planners designed and built sewer systems that moved effluent from domestic residences and industries to nearby rivers and streams, and although this eliminated the problem for much of the city, those downstream had to contend with the waste pollution. No one understands that issue better than Joel Tarr, whose seminal book, *The Search for the Ultimate Sink*, is an excellent starting point for reflecting on this issue and urban pollution in general.[33]

The decision-making process about sanitation and waste in particular has been a long-standing issue of concern for urban environmental historians. What choices did cities possess? What drove those choices? Who was responsible for making decisions? What were the consequences? Martin Melosi's *Garbage in the Cities* reflects on these questions in the area of solid-waste management.[34] His new book, *Precious Commodity*, attempts to do the same thing for water supply.[35] The decision-making process was informed by conceptions of the environment and public health, and by contemporary cultural values.

Choices about the design of a system, for instance, were shaped by factors such as changes in technology, immediacy of and proximity to a problem, and political, social, and cultural conditions. These choices in turn had consequences for urban residents and city leaders. In many cases, if a problem did not directly affect a particular population, it might not be addressed.

Beyond the impact of new urban technologies, economic forces also affected urban pollution. In the earliest scholarship on the urban environment the correlation between urbanization and industrialization usually always was center stage. The argument asserted that industrialization accelerated the rate of urbanization as factories were built in cities, populations migrated for urban jobs, and capital and commodities flowed into and out of the city. While smoke and other pollutants in a somewhat perverse way symbolized economic progress and prosperity, they were extraordinarily destructive in the long run. Reform responses thus challenged the status quo and questioned why governments and the people themselves ignored the environmental risks.

Such a historical scenario resulted in much new scholarship but has proved to be too narrow in assessing urban pollution, its origins, and its impacts. The most obvious retort is that industrialization need not be the key factor in generating pollution; there are multiple potential sources. One only has to look at urban growth in developing countries to observe pollution generation growing out of stimuli other than industrialization. More recent scholarship focusing on several forms of pollution introduce these complexities and have moved beyond the simple correlation between industrialization and urbanization, or at least have refined it. Common topics for research on urban pollution have focused on water-quality issues, sewerage, solid waste, industrial pollution, smoke, and even odors and noise. Smoke pollution, on an international scale at least, has attracted the most scholarly attention. Air pollution more generally has not.[36]

The role of humans in exacerbating disaster is a topic of urban environmental history that has caught attention in the last two decades or so. This scholarship argues that city-building worsens natural phenomena such as wildfires, flooding, hurricanes, and earthquakes because structures built on fault lines, paved-over floodplains and rivers, bulldozed wildlife terrains, and coastal wetlands alter ecosystems in key ways, and place the built environment in harm's way. Constructing cities in vulnerable locations has made them more susceptible to environmental disasters. Cities' ecological disasters have actually become "unnatural" in the face of "natural" disasters because of certain land-use decisions.[37]

Since the 1990s, the issues of race, class, and gender have become fertile ground for studying the US urban environment. Key pioneering work on the urban environment and gender has been written by Susan Strasser, and more recently by Elizabeth Blum.[38] Andrew Hurley's *Environmental Inequalities* is an important early effort to correlate race, class, and urban development. This book connected urban and environmental history with cultural history in a powerful way. Consequently, some historians are now examining who has power in the urban environment and are also concerned with what they call the "social allocation" of environmental hazards. These scholars question who is and is not benefiting from human action in city development and they argue that human and nonhuman worlds coexist and coevolve. Scholarship also has used the categories of race and class to analyze how decision-making about environmental conditions transformed cities across the world into landscapes of privilege and exclusion. For instance, in places such as Gary, Indiana, Chicago, Illinois, and Manchester, England, the middle class and the wealthy were able to escape industrial pollution by moving to the urban fringe and fighting industrial expansion into suburbs.[39] The more affluent thus had little initiative in maintaining a healthy environment in the city from which they were far removed. Working-class and lower-income residents on the other hand depended on industries because they worked in the factories that polluted the urban environment. Real-estate and land-use practices meanwhile confined these populations – and pollution – to certain neighborhoods. Some scholars have argued that urban pollution was a social construction as well because it became a problem only when city residents deemed it to be. If anything, class has probably played a larger role in European environmental history than it has in the US. Yet, other elements that would broaden and deepen an understanding of the social and cultural setting and how they impact the environment and how it impacts them – especially race and gender – are less well developed in Europe.

A feminist perspective on the environment and the emergence of the environmental justice movement recently have become central issues for several American environmental historians. While this trend was slow to develop elsewhere, in the last few years it has become an international occurrence. "Environmental justice" – a term that originated among African-American movement leaders – and its implications were less well understood in Europe until the last few years. Following the lead of scholars such as sociologist Robert Bullard, "environmental justice" is more frequently incorporated into new studies throughout the world, albeit with sometimes less of the political baggage than the term carries in the US, and with more attention to the broader concept of "social justice" that is not so dependent exclusively on race or racial issues.

Some of the first environmental justice studies tied race to the intentional siting of facilities that emit pollution, to problems of solid-waste collection, and to other unhealthy conditions that affect the neighborhoods of people of color. The concern here is that race determines the disproportionate amount of environmental risk that people confront. While environmental justice is meant to give a positive spin to equal protection from environmental risk, several studies emphasize who has social power in terms of fighting environmental inequalities and discriminatory municipal land-use practices, and who does not. Some scholars, however, question the definition of environmental justice as too narrow, not giving sufficient attention to class alongside race.[40]

Future Research

Today the world's urban populations continue to confront many of the same concerns of previous generations. Resource use, urban pollution, the need for clean water, waste disposal, erosion, flooding, drought, and the call for sustainable development remain vital issues for cities across the world. Urbanization of the world's environment, according to J. R. McNeill, has "revolutionized the human condition," by transforming people's quality of life and by bringing about uniquely urban problems.[41] As cities grow and change, the needs stay the same: to make urban environments adaptable to change. Today we live in an increasingly urbanized world; a fact that makes understanding the modern city in which we live and our relationship to nature of great importance.

Historical scholarship on the urban environment is a useful tool in helping to understand the current status of cities in the world through a deep appreciation of important issues embedded in the past. While the young field of urban environmental history has made a good start, there is much left to be done. Taking new and different vantage points might be most useful. Especially necessary is greater attention to comparative work, especially that which crosses national and continental borders. In a review of

Michele Dagenais, Irene Maver, and Pierre-Yves Saunier's *Municipal Services and Employees in the Modern City* (2003), Stefan Couperus spoke about urban government studies, suggesting that "Municipal history has too often been confined to national contexts, lurking in the shadow of national history."[42] Couperus' comment is applicable to urban environmental history because of its generally local orientation and its connection to a national identity in the grossest sense. A major exception may be thematic studies that explore a particular technology (sewerage) or a particular form of pollution (smoke), but many of these studies are also nation-bound.

Few historians have ventured into comparative study of the urban environment in order to move away from a strictly nation-bound approach to urban history. One important exception is Harold Platt's *Shock Cities*. Platt, an early pioneer in the field of urban environmental history, studied both Chicago and Manchester in depth, finding common and disparate themes in the histories of their growth and development. A few others also have looked to city comparisons, but it is most uncommon, as is the further development of histories focusing on the connection between city and hinterland pioneered by William Cronon. Connections of all types are necessary – city and suburb, the urban and the nonurban, and regionalization of cities. Char Miller's *Cities and Nature in the American West* attempts to find common themes in regional urban development, but it too is just a beginning.[43]

There certainly are a variety of topics yet to be explored, most especially in the areas of urban energy use and urban communication. It is not that these topics have been totally ignored, but they remain clearly understudied. In the truest sense, almost anything associated with cities and urban development can be and should be considered worthy of investigation under the auspices of urban environment. While urban environmental history on a global stage still seems rather embryonic, a number of special issues of journals, conferences and symposia, and cyberspace discussions are pointing more and more toward expanding the field.[44]

Rather than provide a laundry list of topics yet to be pursued, suffice it to say that urban environmental history would be well served by reinterpreting and reinventing ways to understand the relationship between humans and the cities they build, live in, despoil, and expand – and by taking vantage points on as many issues as can conceivably be considered "urban." It is the myriad physical and cultural relationship between humans and cities globally that deserves further pondering.

Notes

1 Public Works Historical Society, *History of Public Works in the United States, 1776–1976*, Chicago, American Public Works Association, 1976.

2 J. A. Tarr, "Urban History and Environmental History in the United States: Complementary and Overlapping Fields," in C. Bernhardt (ed.), *Environmental Problems in European Cities in the 19th and 20th Century*, Munster, Waxmann, 2001, pp. 25–40; J. K. Stine and J. A. Tarr, *At the Intersection of Histories: Technology and the Environment*, 2001, accessed from http://www.h-net. org/~environ/historiography/ustechnology.htm on March 13, 2012; M. V. Melosi, "The Place of the City in Environmental History," *Environmental History Review*, 17/1, 1993, pp. 1–23.

3 Tarr, "Urban History and Environmental History in the United States," pp. 25–6; M. V. Melosi, "Humans, Cities, and Nature: How Do Cities Fit in the Material World?," *Journal of Urban History* 36/1, 2010, pp. 3–21 (pp. 4–5); S. Sorlin and P. Warde, "The Problem of the Problem of Environmental History: A Re-Reading of the Field," *Environmental History* 12/1, 2007, pp. 107–30 (pp. 108–10).

4 D. Schott, B. Luckin, and G. Massard-Guilbaud (eds.), *Resources of the City: Contributions to an Environmental History of Modern Europe*, Aldershot, Ashgate, 2005.

5 G. Massard-Guilbaud and P. Thorsheim, "Cities, Environments, and European History," *Journal of Urban History* 33/5, 2007, pp. 691–701 (pp. 691–2).

6 J. A. Tarr and G. Dupuy (eds.), *Technology and the Rise of the Networked City in Europe and America*, Philadelphia, Temple University Press, 1988.

7 Tarr and Dupuy, *Technology and the Rise of the Networked City*, p. xiii.

8 Stine and Tarr, *At the Intersection of Histories*.

9 Massard-Guilbaud and Thorsheim, "Cities, Environments, and European History," p. 692.

10 D. Schott, "Urban Environmental History: What Lessons Are there to Be Learnt?," *Boreal Environment Research* 9, December 2004, pp. 519–28 (p. 520).

11 P. Brimblecombe and C. Pfister (eds.), *The Silent Countdown: Essays in Environmental History*, Berlin, Springer, 1990.

12 K. J. W. Oosthoek, *What Is Environmental History*, n.d., accessed from http://www.eh-resources.org/environmental_history.html on March 13, 2012; T. Myllyntaus, *Writing about the Past with Green Ink: The Emergence of Finnish Environmental History*, n.d., accessed from http://www.h-net.org/~environ/historiography/finland.htm on March 13, 2012.

13 J. R. McNeill, "Observations on the Nature and Culture of Environmental History," *History and Theory* 42, December 2003, pp. 5–43; M. Cioc, B. Linner, and M. Osborn, "Environmental History Writing in Northern Europe," *Environmental History* 5, July 2000, pp. 396–406; M. Bess, M. Cioc, and J. Sievert, "Environmental History Writing in Southern Europe," *Environmental History* 5, October 2000, pp. 545–56; Myllyntaus, *Writing about the Past with Green Ink*.

14 P. Sutter, "What Can U.S. Environmental Historians Learn from Non-U.S. Environmental Historiography?," *Environmental History* 8/1, 2003, pp. 109–29 (p. 110).

15 L. Mumford, *The Culture of Cities*, New York, Harcourt Brace Publishing, 1938, p. 5.

16 Mumford, *The Culture of Cities*, p. 252.

17 Melosi, "Humans, Cities, and Nature," p. 4.

18 Melosi, "Humans, Cities, and Nature," p. 4.

19 M. V. Melosi, *The Sanitary City: Urban Infrastructure in America from Colonial Times to the Present*, Baltimore, The Johns Hopkins University Press, 2000.

20 M. V. Melosi (ed.), *Pollution and Reform in American Cities, 1870–1930*, Austin, TX, University of Texas Press, 1980.

21 J. R. McNeill, *Something New under the Sun: An Environmental History of the Twentieth-Century World*, New York, W. W. Norton, 2000, pp. 6–7.

22 Melosi, "Humans, Cities, and Nature," pp. 3–4.

23 Bernhardt *Environmental Problems in European Cities*, pp. 105–19; Schott, "Urban Environmental History," pp. 522–3.

24 W. Cronon, *Nature's Metropolis: Chicago and the Great West*, New York, W. W. Norton, 1991. See also K. Brosnan, *Uniting Mountain and Plain: Cities, Law, and Environmental Change along the Front Range*, Albuquerque, NM, University of New Mexico Press, 2002; S. P. Hays, "Toward Integration in Environmental History," *Pacific Historical Review* 70/1, 2001, pp. 59–67.

25 A. C. Isenberg (ed.), *The Nature of Cities: Culture, Landscape, and Urban Space*, Rochester, VT, University of Rochester Press, 2006, pp. xiii–xiv.

26 M. Klingle, *Emerald City: An Environmental History of Seattle*, New Haven, CT, Yale University Press, 2007, pp. xii–xiii, 4, 9.

27 See C. Bernhardt and G. Massard-Guilbaud (eds.), *The Modern Demon: Pollution in Urban and Industrial European Societies*, Clermont-Ferrand, Presses Universitaires Blaise Pascal, 2002; J. W. Konvitz, *The Urban Millennium: The City-Building Process from the Early Middle Ages to the Present*, Carbondale, IL, Southern Illinois University Press, 1985; Tarr and Dupuy, *Technology and the Rise of the Networked City*; Melosi, *The Sanitary City*; M. Hard

and T. Misa (eds.), *Urban Machinery: Inside Modern European Cities*, Cambridge, MA, MIT, 2010; J. Benidickson, *The Culture of Flushing: A Social and Legal History of Sewage*, Vancouver, UBC Press, 2007.

28 M. Gandy, *Concrete and Clay: Reworking Nature in New York City*, Cambridge, MA, MIT, 2002, pp. 6–10; M. Rawson, *Eden on the Charles: The Making of Boston*, Cambridge, MA, Harvard University Press, 2010; A. Rome, *The Bulldozer in the Countryside: Suburban Sprawl and the Rise of American Environmentalism*, New York, Cambridge University Press, 2001.

29 McNeill, "Observations on the Nature and Culture of Environmental History"; A. Hurley (ed.), *Common Fields: An Environmental History of St. Louis*, St. Louis, MI, Missouri Historical Society Press, 1997; A. N. Penna and C. E. Wright (eds.), *Remaking Boston: An Environmental History of the City and Its Surroundings*, Pittsburgh, University of Pittsburgh Press, 2009; W. Deverell and G. Hise, *Land of Sunshine: An Environmental History of Metropolitan Los Angeles*, Pittsburgh, University of Pittsburgh Press, 2006; C. Miller, *On the Border: An Environmental History of San Antonio*, Pittsburgh, University of Pittsburgh Press, 2001; J. A. Tarr (ed.), *Devastation and Renewal: An Environmental History of Pittsburgh and Its Region*, Pittsburgh, University of Pittsburgh Press, 2005; M. V. Melosi and J. A. Pratt (eds.), *Energy Metropolis: An Environmental History of Houston and the Gulf Coast*, Pittsburgh, University of Pittsburgh Press, 2007; C. E. Colten, *Transforming New Orleans and Its Environs: Centuries of Change*, Pittsburgh, University of Pittsburgh Press, 2001; S. Castonguay and M. Dagenais, *Metropolitan Natures: Environmental Histories of Montreal*, Pittsburgh, University of Pittsburgh Press, 2011; D. H. Coward, *Out of Sight: Sydney's Environmental History, 1851–1981*, Canberra, Australian National University Press, 1988.

30 C. Hamlin, *Public Health and Social Justice in the Age of Chadwick: Britain, 1800–1854*, Cambridge, Cambridge University Press, 1998; N. Tomes, *The Gospel of Germs: Men, Women, and the Microbe in American Life*, Cambridge, MA, Harvard University Press, 1998; W. Troesken, *Water, Race and Disease*, Cambridge, MA, MIT, 2004; and S. Hoy, *Chasing Dirt: The American Pursuit of Cleanliness*, New York, Oxford University Press, 1995.

31 C. McShane and J. A. Tarr, *The Horse in the City: Living Machines in the Nineteenth Century*, Baltimore, The Johns Hopkins University Press, 2007. See also M. Rose, B. Seely, and P. Barrett, *The Best Transportation System in the World: Railroads, Trucks, Airlines, and American Public Policy in the Twentieth Century*, Columbus, OH, Ohio State University Press, 2006.

32 See A. Guillerme, *The Age of Water: The Urban Environment in the North of France, A.D.300–1800*, College Station, TX, Texas A&M University Press, 1988; G. Cooper, *Air-Conditioning America: Engineers and the Controlled Environment, 1900–1960*, Baltimore, The Johns Hopkins University Press, 1998; S. Elkind, *Bay Cities and Water Politics*, Lawrence, KS, University Press of Kansas, 1998; M. Engler, *Designing America's Waste Landscapes*, Baltimore, The Johns Hopkins University Press, 2003; J. Hassan, *A History of Water in Modern England and Wales*, Manchester, Manchester University Press, 1998; J. Peterson, *The Birth of City Planning in the United States, 1840–1917*, Baltimore, The Johns Hopkins University Press, 2003; M. Tebeau, *Eating Smoke: Fire in Urban America, 1800–1950*, Baltimore, The Johns Hopkins University Press, 2003; C. Zimring, *Cash for Your Trash: Scrap Recycling in America*, New Brunswick, NJ, Rutgers University Press, 2005; D. Reid, *Paris Sewers and Sewermen: Realities and Representation*, Cambridge, MA, Harvard University Press, 1993; M. Hietala, *Services and Urbanization at the Turn of the Century: The Diffusion of Innovations*, Helsinki, Finnish Historical Society, 1987; and S. Halliday, *The Great Stink of London: Sir Joseph Bazalgette and the Cleansing of the Victorian Metropolis*, Stroud, Sutton Publishing, 2001.

33 J. A. Tarr, *The Search for the Ultimate Sink: Urban Pollution in Historical Perspective*, Akron, OH, University of Akron Press, 1996.

34 M. V. Melosi, *Garbage in the Cities: Refuse, Reform, and the Environment*, Pittsburgh, University of Pittsburgh Press, 2005.

35 M. V. Melosi, *Precious Commodity: Providing Water for America's Cities*, Pittsburgh, University of Pittsburgh Press, 2011.

36 See S. Moseley, *The Chimney of the World: A History of Smoke Pollution in Victorian and Edwardian Manchester*, Manchester, White Horse Press, 2001; D. Stradling, *Smokestacks and Progressives: Environmentalists, Engineers, and Air Quality in America, 1881–1951*, Baltimore, The Johns Hopkins University Press, 1999; P. Thorsheim, *Inventing Pollution: Coal, Smoke, and Culture in Britain since 1800*, Athens, OH, Ohio University Press, 2006; P. Brimblecombe, *The Big Smoke: A History of Air Pollution in London since Medieval Times*, London, Methuen, 1987; F. Uekoetter, *The Age of Smoke: Environmental Policy in Germany and the United States, 1880–1970*, Pittsburgh, University of Pittsburgh Press, 2009; W. H. TeBrake, "Air Pollution and Fuel Crises in Pre-Industrial London, 1250–1650," *Technology and Culture* 16, 1975, pp. 337–59; D. Davis, *When Smoke Ran like Water: Tales of Environmental Deception and the Battle against Pollution*, New York, Basic Books, 2003; M. Anderson, "The Conquest of Smoke: Legislation and Pollution in Colonial Calcutta," in D. Arnold and R. Guha (eds.), *Nature, Culture, Imperialism*, Delhi, Oxford University Press, 1995, pp. 293–335. See also S. Dewey, *Don't Breathe the Air: Air Pollution and U.S. Environmental Politics, 1945–1970*, College Station, TX, Texas A&M University Press, 2000; S. Fang and H. Chen, "Air Quality and Pollution Control in Taiwan," *Atmospheric Environment* 30, 1996, pp. 735–41; M. Hashimoto, "History of Air Pollution Control in Japan," in H. Nishimura (ed.), *How to Conquer Air Pollution: A Japanese Experience*, Amsterdam, Elsevier, 1989, pp. 1–94.

37 See T. Steinberg, *Acts of God: The Unnatural History of Natural Disaster in America*, New York, Oxford University Press, 2000; M. Davis, *Ecology of Fear: Los Angeles and the Imagination of Disaster*, New York, Vintage Books, 1999; G. Clancey, *Earthquake Nation: The Cultural Politics of Japanese Seismicity, 1868–1930*, Berkeley, University of California Press, 2006; E. Klineberg, *Heat Wave: A Social Autopsy of Disaster in Chicago*, Chicago, University of Chicago Press, 2002.

38 S. Strasser, *Waste and Want: A Social History of Trash*, New York, Holt, 1999; E. Blum, *Love Canal Revisited: Race, Class, and Gender in Environmental Activism*, Lawrence, KS, University Press of Kansas, 2008.

39 A. Hurley, *Environmental Inequalities: Class, Race, and Industrial Pollution in Gary, Indiana, 1945–1980*, Chapel Hill, NC, University of North Carolina Press, 1995; H. L. Platt, *Shock Cities: The Environmental Transformation and Reform of Manchester and Chicago*, Chicago, University of Chicago Press, 2005.

40 See J. Sze, *Noxious New York: The Racial Politics of Urban Health and Environmental Justice*, Cambridge, MA, MIT, 2006; S. Washington, *Packing Them In: An Archaeology of Environmental Racism in Chicago, 1865–1954*, Lanham, MD, Lexington Books, 2005; E. McGurty, *Transforming Environmentalism: Warren County, PCBs, and the Origins of Environmental Justice*, New Brunswick, NJ, Rutgers University Press, 2007; B. L. Allen, *Uneasy Alchemy: Citizens and Experts in Louisiana's Chemical Corridor Disputes*, Cambridge, MA, MIT, 2003; J. E. Castro, *Water, Power and Citizenship: Social Struggle in the Basin of Mexico*, London, PalgraveMacmillan, 2006; S. Lerner, *Diamond: A Struggle for Environmental Justice in Louisiana's Chemical Corridor*, Cambridge, MA, MIT, 2005.

41 McNeill, "Observations on the Nature and Culture of Environmental History," p. 35.

42 S. Couperus, *Review of Michele Dagenais, Irene Maver, and Pierre-Yves Saunier's Municipal Services and Employees in the Modern City*, 2004, accessed from http://h-net.msu.edu/cgi-bin/logbrowse.pl?trx=vx&list=h-urban&month=0404&week=c&msg=w22nNEQBRFgkSDswCmwEmA&user=&pw= on March 27, 2012.

43 C. Miller (ed.), *Cities and Nature in the American West*, Reno, NV, University of Nevada Press, 2010.

44 A good place to observe the growing range of interest in environmental history in general, and urban environmental history in particular, is Environmental History Resources, online at http://www.eh-resources.org/index.html.

References

Allen, B. L., *Uneasy Alchemy: Citizens and Experts in Louisiana's Chemical Corridor Disputes*, Cambridge, MA, MIT, 2003.

Anderson, M., "The Conquest of Smoke: Legislation and Pollution in Colonial Calcutta," in D. Arnold and R. Guha (eds.), *Nature, Culture, Imperialism*, Delhi, Oxford University Press, 1995, pp. 293–335.

Benidickson, J., *The Culture of Flushing: A Social and Legal History of Sewage*, Vancouver, UBC Press, 2007.

Bernhardt, C. (ed.), *Environmental Problems in European Cities in the 19th and 20th Century*, Munster, Waxmann, 2001.

Bernhardt, C., and Massard-Guilbaud, G. (eds.), *The Modern Demon: Pollution in Urban and Industrial European Societies*, Clermont-Ferrand, Presses Universitaires Blaise Pascal, 2002.

Bess, M., Cioc, M., and Sievert, J., "Environmental History Writing in Southern Europe," *Environmental History* 5, October 2000, pp. 545–56.

Blum, E., *Love Canal Revisited: Race, Class, and Gender in Environmental Activism*, Lawrence, KS, University Press of Kansas, 2008.

Brimblecombe, P., *The Big Smoke: A History of Air Pollution in London since Medieval Times*, London, Methuen, 1987.

Brimblecombe, P., and Pfister, C. (eds.), *The Silent Countdown: Essays in Environmental History*, Berlin, Springer, 1990.

Brosnan, K., *Uniting Mountain and Plain: Cities, Law, and Environmental Change along the Front Range*, Albuquerque, NM, University of New Mexico Press, 2002.

Castonguay, S., and Dagenais, M., *Metropolitan Natures: Environmental Histories of Montreal*, Pittsburgh, University of Pittsburgh Press, 2011.

Castro, J. E., *Water, Power and Citizenship: Social Struggle in the Basin of Mexico*, London, PalgraveMacmillan, 2006.

Cioc, M., Linner, B., and Osborn, M., "Environmental History Writing in Northern Europe," *Environmental History* 5, July 2000, pp. 396–406.

Clancey, G., *Earthquake Nation: The Cultural Politics of Japanese Seismicity, 1868–1930*, Berkeley, University of California Press, 2006.

Colten, C. E., *Transforming New Orleans and Its Environs: Centuries of Change*, Pittsburgh, University of Pittsburgh Press, 2001.

Cooper, G., *Air-Conditioning America: Engineers and the Controlled Environment, 1900–1960*, Baltimore, The Johns Hopkins University Press, 1998.

Couperus, S., *Review of Michele Dagenais, Irene Maver, and Pierre-Yves Saunier's Municipal Services and Employees in the Modern City*, 2004, accessed from http://h-net.msu.edu/cgi-bin/logbrowse.pl?trx=vx&list=h-urban&month=0404&week=c&msg=w22nNEQBRFgkSDswCmwEmA&user=&pw= on March 27, 2012.

Coward, D. H., *Out of Sight: Sydney's Environmental History, 1851–1981*, Canberra, Australian National University Press, 1988.

Cronon, W., *Nature's Metropolis: Chicago and the Great West*, New York, W. W. Norton, 1991.

Davis, D., *When Smoke Ran like Water: Tales of Environmental Deception and the Battle against Pollution*, New York, Basic Books, 2003.

Davis, M., *Ecology of Fear: Los Angeles and the Imagination of Disaster*, New York, Vintage Books, 1999.

Deverell, W., and Hise, G., *Land of Sunshine: An Environmental History of Metropolitan Los Angeles*, Pittsburgh, University of Pittsburgh Press, 2006.

Dewey, S., *Don't Breathe the Air: Air Pollution and U.S. Environmental Politics, 1945–1970*, College Station, TX, Texas A&M University Press, 2000.

Elkind, S., *Bay Cities and Water Politics*, Lawrence, KS, University Press of Kansas, 1998.

Engler, M., *Designing America's Waste Landscapes*, Baltimore, The Johns Hopkins University Press, 2003.

Fang, S., and Chen, H., "Air Quality and Pollution Control in Taiwan," *Atmospheric Environment* 30, 1996, pp. 735–41.

Gandy, M., *Concrete and Clay: Reworking Nature in New York City*, Cambridge, MA, MIT, 2002.

Guillerme, A., *The Age of Water: The Urban Environment in the North of France, A.D.300–1800*, College Station, TX, Texas A&M University Press, 1988.

Halliday, S., *The Great Stink of London: Sir Joseph Bazalgette and the Cleansing of the Victorian Metropolis*, Stroud, Sutton Publishing, 2001.

Hamlin, C., *Public Health and Social Justice in the Age of Chadwick: Britain, 1800–1854*, Cambridge, Cambridge University Press, 1998.

Hard, M., and Misa, T. (eds.), *Urban Machinery: Inside Modern European Cities*, Cambridge, MA, MIT, 2010.

Hashimoto, M., "History of Air Pollution Control in Japan," in H. Nishimura (ed.), *How to Conquer Air Pollution: A Japanese Experience*, Amsterdam, Elsevier, 1989, pp. 1–94.

Hassan, J., *A History of Water in Modern England and Wales*, Manchester, Manchester University Press, 1998.

Hays, S. P., "Toward Integration in Environmental History," *Pacific Historical Review* 70/1, 2001, pp. 59–67.

Hietala, M., *Services and Urbanization at the Turn of the Century: The Diffusion of Innovations*, Helsinki, Finnish Historical Society, 1987.

Hoy, S., *Chasing Dirt: The American Pursuit of Cleanliness*, New York, Oxford University Press, 1995.

Hurley, A. (ed.), *Common Fields: An Environmental History of St. Louis*, St. Louis, MI, Missouri Historical Society Press, 1997.

Hurley, A., *Environmental Inequalities: Class, Race, and Industrial Pollution in Gary, Indiana, 1945–1980*, Chapel Hill, NC, University of North Carolina Press, 1995.

Isenberg, A. C. (ed.), *The Nature of Cities: Culture, Landscape, and Urban Space*, Rochester, VT, University of Rochester Press, 2006.

Klineberg, E., *Heat Wave: A Social Autopsy of Disaster in Chicago*, Chicago, University of Chicago Press, 2002.

Klingle, M., *Emerald City: An Environmental History of Seattle*, New Haven, CT, Yale University Press, 2007.

Konvitz, J. W., *The Urban Millennium: The City-Building Process from the Early Middle Ages to the Present*, Carbondale, IL, Southern Illinois University Press, 1985.

Lerner, S., *Diamond: A Struggle for Environmental Justice in Louisiana's Chemical Corridor*, Cambridge, MA, MIT, 2005.

Massard-Guilbaud, G., and Thorsheim, P., "Cities, Environments, and European History," *Journal of Urban History* 33/5, 2007, pp. 691–701.

McGurty, E., *Transforming Environmentalism: Warren County, PCBs, and the Origins of Environmental Justice*, New Brunswick, NJ, Rutgers University Press, 2007.

McNeill, J. R., "Observations on the Nature and Culture of Environmental History," *History and Theory* 42, December 2003, pp. 5–43.

McNeill, J. R., *Something New under the Sun: An Environmental History of the Twentieth-Century World*, New York, W. W. Norton, 2000.

McShane, C., and Tarr, J. A., *The Horse in the City: Living Machines in the Nineteenth Century*, Baltimore, The Johns Hopkins University Press, 2007.

Melosi, M. V., *Garbage in the Cities: Refuse, Reform, and the Environment*, Pittsburgh, University of Pittsburgh Press, 2005.

Melosi, M. V., "Humans, Cities, and Nature: How Do Cities Fit in the Material World?," *Journal of Urban History* 36/1, 2010a, pp. 3–21.

Melosi, M. V., "The Place of the City in Environmental History," *Environmental History Review*, 17/1, 1993, pp. 1–23.

Melosi, M. V. (ed.), *Pollution and Reform in American Cities, 1870–1930*, Austin, TX, University of Texas Press, 1980.

Melosi, M. V., *Precious Commodity: Providing Water for America's Cities*, Pittsburgh, University of Pittsburgh Press, 2011.

Melosi, M. V., *The Sanitary City: Urban Infrastructure in America from Colonial Times to the Present*, Baltimore, The Johns Hopkins University Press, 2000.

Melosi, M. V., and Pratt, J. A. (eds.), *Energy Metropolis: An Environmental History of Houston and the Gulf Coast*, Pittsburgh, University of Pittsburgh Press, 2007.

Miller, C. (ed.), *Cities and Nature in the American West*, Reno, NV, University of Nevada Press, 2010.

Miller, C., *On the Border: An Environmental History of San Antonio*, Pittsburgh, University of Pittsburgh Press, 2001.

Moseley, S., *The Chimney of the World: A History of Smoke Pollution in Victorian and Edwardian Manchester*, Manchester, White Horse Press, 2001.

Mumford, L., *The Culture of Cities*, New York, Harcourt Brace Publishing, 1938.

Myllyntaus, T., *Writing about the Past with Green Ink: The Emergence of Finnish Environmental History*, n.d., accessed from http://www.h-net.org/~environ/historiography/finland.htm on March 13, 2012.

Oosthoek, K. J. W., *What Is Environmental History*, n.d., accessed from http://www.eh-resources.org/environmental_history.html on March 13, 2012.

Penna, A. N., and Wright, C. E. (eds.), *Remaking Boston: An Environmental History of the City and Its Surroundings*, Pittsburgh, University of Pittsburgh Press, 2009.

Peterson, J., *The Birth of City Planning in the United States, 1840–1917*, Baltimore, The Johns Hopkins University Press, 2003.

Platt, H. L., *Shock Cities: The Environmental Transformation and Reform of Manchester and Chicago*, Chicago, University of Chicago Press, 2005.

Public Works Historical Society, *History of Public Works in the United States, 1776–1976*, Chicago, American Public Works Association, 1976.

Rawson, M., *Eden on the Charles: The Making of Boston*, Cambridge, MA, Harvard University Press, 2010.

Reid, D., *Paris Sewers and Sewermen: Realities and Representation*, Cambridge, MA, Harvard University Press, 1993.

Rome, A., *The Bulldozer in the Countryside: Suburban Sprawl and the Rise of American Environmentalism*, New York, Cambridge University Press, 2001.

Rose, M., Seely, B., and Barrett, P., *The Best Transportation System in the World: Railroads, Trucks, Airlines, and American Public Policy in the Twentieth Century*, Columbus, OH, Ohio State University Press, 2006.

Schott, D., "Urban Environmental History: What Lessons Are there to Be Learnt?," *Boreal Environment Research* 9, December 2004, pp. 519–28.

Schott, D., Luckin, B., and Massard-Guilbaud, G. (eds.), *Resources of the City: Contributions to an Environmental History of Modern Europe*, Aldershot, Ashgate, 2005.

Sorlin, S., and Warde, P., "The Problem of the Problem of Environmental History: A Re-Reading of the Field," *Environmental History* 12/1, 2007, pp. 107–30.

Steinberg, T., *Acts of God: The Unnatural History of Natural Disaster in America*, New York, Oxford University Press, 2000.

Stine, J. K., and Tarr, J. A., *At the Intersection of Histories: Technology and the Environment*, 2001, accessed from http://www.h-net.org/~environ/historiography/ustechnology.htm on March 13, 2012.

Stradling, D., *Smokestacks and Progressives: Environmentalists, Engineers, and Air Quality in America, 1881–1951*, Baltimore, The Johns Hopkins University Press, 1999.

Strasser, S., *Waste and Want: A Social History of Trash*, New York, Holt, 1999.

Sutter, P., "What Can U.S. Environmental Historians Learn from Non-U.S. Environmental Historiography?," *Environmental History* 8/1, 2003, pp. 109–29.

Sze, J., *Noxious New York: The Racial Politics of Urban Health and Environmental Justice*, Cambridge, MA, MIT, 2006.

Tarr, J. A. (ed.), *Devastation and Renewal: An Environmental History of Pittsburgh and Its Region*, Pittsburgh, University of Pittsburgh Press, 2005.

Tarr, J. A., *The Search for the Ultimate Sink: Urban Pollution in Historical Perspective*, Akron, OH, University of Akron Press, 1996.

Tarr, J. A., "Urban History and Environmental History in the United States: Complementary and Overlapping Fields," in C. Bernhardt (ed.), *Environmental Problems in European Cities in the 19th and 20th Century*, Munster, Waxmann, 2001, pp. 25–40.

Tarr, J. A., and Dupuy, G. (eds.), *Technology and the Rise of the Networked City in Europe and America*, Philadelphia, Temple University Press, 1988.

Tebeau, M., *Eating Smoke: Fire in Urban America, 1800–1950*, Baltimore, The Johns Hopkins University Press, 2003.

TeBrake, W. H., "Air Pollution and Fuel Crises in Pre-Industrial London, 1250–1650," *Technology and Culture* 16, 1975, pp. 337–59.

Thorsheim, P., *Inventing Pollution: Coal, Smoke, and Culture in Britain since 1800*, Athens, OH, Ohio University Press, 2006.

Tomes, N., *The Gospel of Germs: Men, Women, and the Microbe in American Life*, Cambridge, MA, Harvard University Press, 1998.

Troesken, W., *Water, Race and Disease*, Cambridge, MA, MIT, 2004.

Uekoetter, F., *The Age of Smoke: Environmental Policy in Germany and the United States, 1880–1970*, Pittsburgh, University of Pittsburgh Press, 2009.

Washington, S., *Packing Them In: An Archaeology of Environmental Racism in Chicago, 1865–1954*, Lanham, MD, Lexington Books, 2005.

Zimring, C., *Cash for Your Trash: Scrap Recycling in America*, New Brunswick, NJ, Rutgers University Press, 2005.

Further Reading

Flanagan, M., *Seeing with Their Hearts: Chicago Women and the Vision of the Good City, 1871–1933*, Princeton, NJ, Princeton University Press, 2002.

Juuti, P., Katko, T., and Vurinen, H. (eds.), *Environmental History of Water: Global Views on Community Water Supply and Sanitation*, London, IWA Publishing, 2007.

Parker, D. S., "Civilizing the City of Kings: Hygiene and Housing in Lima," in R. F. Pineo and J. A. Baer (eds.), *Cities of Hope: People, Protests, and Progress in Urbanizing Latin America, 1870–1930*, Boulder, CO, Westview Press, 1998, pp. 153–78.

Rosen, C. M., and Tarr. J. A., "The Importance of an Urban Perspective in Environmental History," *Journal of Urban History* 20, 1994, pp. 299–310.

CHAPTER TWENTY-ONE

Evolution and the Environment

EDMUND RUSSELL

Science provides wonderful tools to help historians understand people and environments of the past. The workhorse sciences for environmental historians traditionally have been ecology and public health. Recently, however, we have seen a growing interest among historians in evolutionary biology. Of particular interest has been the way in which human beings have shaped the evolution of populations of nonhuman species, and how such evolution has circled back to shape human experience. This chapter explains the basics of evolutionary theory, identifies human activities that have led to evolution in nonhuman species, and illustrates the ability of an evolutionary approach to revise our understanding of familiar episodes by analyzing the role of anthropogenic evolution in the Industrial Revolution.[1]

The idea that human beings can affect evolution is unfamiliar to many of us. A common view is that evolution is speciation, that it took place in the distant past, that "nature" carried it out, and that it occurred via natural selection.[2] These are all elements of evolution, but by no means represent all of it. Evolutionary biologists define evolution as changes in inherited traits (or the genetic makeup) of populations of organisms over generations.[3] Evolution may produce changes in a population radical enough to spawn a new species, but changes that fall short of speciation also are evolution. Evolution did happen in the distant past, but it also happens today and will continue to occur in the future.[4] "Nature" (usually taken to mean everything other than people) certainly affects evolution, but human beings do too. Natural selection is one important mechanism, but other processes – such as sexual selection, chance, and human behavior – also affect evolution.[5]

A good example illustrating this broader conception of evolution comes from African elephants. Historically, almost all African elephants (male and female) have grown tusks. Tusks helped elephants forage for food, mark territory, and defend themselves. In some parts of Africa today, however, a third or more of the elephants do not grow tusks.

A Companion to Global Environmental History, First Edition. Edited by J.R. McNeill and Erin Stewart Mauldin.
© 2012 John Wiley & Sons, Ltd. Published 2015 by John Wiley & Sons, Ltd.

Elephants can lose tusks through breakage, or people might cut them off, but the elephants of interest here never grow tusks at all. Tusklessness in these elephants is an inherited genetic trait. Inherited tusklessness was rare until the twentieth century, and then it became common (reaching over 50 percent) in some populations.[6] At first blush, this development seems to counter evolutionary theory. Why would a beneficial trait (tusks) become less common in a population?

The proximate answer is human hunting. The ivory trade created a huge demand for elephant tusks. Hunters, legally and illegally, have killed large numbers of tusked elephants (often taking only the tusks and leaving the rest of the carcass to rot). Ivory hunters had no motive to shoot tuskless elephants, so these elephants survived. In evolutionary terms, hunters had reversed selective pressures. Before heavy human hunting, selection favored tusked elephants, which had a higher chance of surviving and reproducing than tuskless elephants. Now selection favored tuskless elephants, which came to enjoy a higher chance of surviving and reproducing than tusked elephants. The proportion of tusked elephants in populations declined, and the proportion of tuskless elephants grew. Since only the survivors reproduced, and since parents passed on their traits to their offspring, the proportion of tuskless elephants grew over generations.[7]

Notice how this example fulfills the definition of evolution. The frequency of an inherited trait (tusklessness) changed (increased) over generations in a population of organisms (elephants in parts of Africa). This evolution took place recently, not in the distant past. It produced measurable changes in the population but did not create a new species. It happened as a result of human actions, not solely under the hand of "nature" or natural selection. It is an example of anthropogenic evolution.

Tuskless African elephant populations also illustrate how chance can affect evolution. In one South African population, the rate of tusklessness is unusually high even for heavily hunted populations, reaching 50 percent overall and 98 percent among females. Hunters were responsible for a big increase in tusklessness in the population, and for reducing the number of elephants that founded the protected herd to a small size (11 individuals). Then a chance effect called *drift* appears to have pushed the frequency of tusklessness higher. Drift is the random change in frequency of traits in a population. Small populations are especially, but not uniquely, susceptible to drift because a difference of just a few individuals' reproducing can have a big effect on the next generation. In this herd, the founder population consisted of eight females, four or five of them tuskless. If just two of the three or four tusked female ancestors happened not to produce offspring, and if that same thing happened several years running, the proportion of tuskless elephants would quickly grow. So here we have an illustration of how a process other than natural selection has led to evolution, and we have seen that evolution can occur as a combination of human actions (hunting and founding a herd with few individuals) and "natural" processes (drift).[8]

Hunting has shaped the evolution of populations of other animals, too. In Canada, the size of bighorn sheep on Ram Mountain declined over the twentieth century. Hunters prized large males, especially those with big horns that looked great mounted on the wall, and selectively harvested these rams. As a result, the proportion of smaller rams in the population rose. Body size is partly an inherited trait. So in this case, hunting shaped the evolution of a population of wild bighorn sheep by selecting against large animals.[9]

Fishing has also reduced the size of wild animals. By selectively harvesting larger fish, fishers gave smaller fish an advantage in the game of survival. Over time this process reduced the average size of adult fish in some populations. In North American lakes, the average size of whitefish dropped by half (from 2 kilograms to 1 kilogram) between the 1940s and the 1970s. Removal of older fish probably played a role, but so did genetic change. Small fish could slip through the holes in the mesh of fishing nets, while large fish could not, giving small adults a selective advantage. Heavy fishing also appears to be responsible for reducing the average size of Atlantic cod and Pacific salmon.[10]

Eradication efforts join hunting and fishing on the list of human activities with evolutionary impact. Here we use eradication to refer to efforts to eliminate a certain organism from a certain area, ranging in scale from individual bodies to the entire globe. Farmers around the world have spent billions of dollars on pesticides designed to kill insects, weeds, and pathogens that threaten crops and animals. Public-health officials have sprayed vast areas with insecticides to kill insects that transmit disease. Doctors have prescribed billions of dollars of antibiotics to stifle infections. These practices have benefited human beings by boosting food supply and reducing disease. None of these efforts has provided a permanent solution because target populations have evolved resistance to the poisons. The global effort to eradicate malaria after World War II provides an excellent example. This project relied on insecticides (to kill malaria-carrying mosquitoes) and drugs (to kill malaria plasmodia inside human bodies), and it achieved remarkable success. Some have estimated that it saved 15–25 million lives. Unfortunately for subsequent humans, mosquitoes and plasmodia evolved resistance to the chemicals that once killed them. By chance, a few plasmodia carried genes that enabled them to survive medications, and a few mosquitoes carried genes that enabled them to survive insecticides. Genes for resistance were rare in plasmodium and mosquito populations until spraying killed off the susceptible members of populations and left resistant individuals behind to reproduce. Resistant parents passed genes for resistance to their offspring, and the proportion of resistant individuals increased with every spraying until specific drugs and insecticides became useless. The World Health Organization gave up on the project in the early 1970s, and malaria came roaring back. By 2000, an estimated 2 million people were dying each year of malaria.[11] When first-choice medicines no longer kill pathogens, doctors turn to second-choice medicines that often are more expensive. In the US, antibiotic resistance has raised the cost of medicine by an estimated 30 billion dollars per year.[12]

Human beings set out to inflict mortality on the populations described so far; they have also shaped the evolution of populations with which they had no intention of interacting at all. The major mechanism in "invisible" evolution has been environmental modification. By modifying a habitat, people have created selective pressures on potentially all the organisms that live in the habitat. The most numerous organisms in any habitat are also the smallest, so they easily pass without notice. Usually we know what has happened to these organisms only because biologists recorded data, and they can document only the tiniest fraction of populations affected in this way.

Two examples from Spain illustrate possible selection by environmental modification on very different geographic scales. Let us begin on the small end of the scale with a winery near Cordoba. Winemakers consciously interacted with grapes and yeast to make wine, but their impact extended beyond those two species. Fruit flies were attracted to the winery because of the abundance of fruit. Biologists gathered flies from three places: flies walking on the yeasty surface of wine in barrels in a cellar, flies 2–5 meters away from the barrels near the entrance to the cellar, and flies 500 meters away in fields. They found that many members of the wine-walking population carried a gene for making an enzyme called alcohol dehydrogenase, which detoxifies alcohol. The frequency of the trait was lower in the population near the door of the cellar and lowest in the population farthest from the cellar. The researchers hypothesized that the wine-barrel environment selected for individuals making alcohol dehydrogenase, and the outdoor environment selected for individuals with a different version (allele) of the gene. The genetic makeup of the population near the door may have reflected the mixing of individuals from the two selective regimes (barrels and outdoors). So, without realizing it, winemakers may have been affecting the evolution of fruit-fly populations by modifying the environment in which the flies lived.[13]

The example at the other end of the scale is global, and again we look at insects. Fruit flies live on O Pedroso Mountain in northwest Spain. The genetic makeup of the population varied along a gradient, with one version of a chromosome more common in the south and another version more common in the north. As average temperatures on the mountain rose between 1976 and 1991, the southern chromosome version became more common at the northern end of the range. This pattern is consistent with evolution in the population to adapt to warmer temperatures. Populations of other species, such as pitcher-plant mosquitoes in North America, have also changed genetically as temperatures have risen. These populations are probably bellwethers of evolution yet to come in the face of global climate change.[14]

The examples so far have focused on ways in which people have affected evolution accidentally, but we have also done so intentionally. Usually people have not thought about their actions in this way. They might have said that they were breeding plants or animals, or that they were genetically engineering organisms, rather than saying that they were shaping evolution. But they have been evolutionary actors nonetheless. By altering the inherited traits and genetic makeup of populations of organisms over generations, they have influenced evolution.

Charles Darwin knew this. He owed much of his theory of evolution by natural selection to the plant and animal breeders of his day. He joined not one but two pigeon-fancier clubs to gain first-hand knowledge of breeding, and he surveyed large numbers of plant and animal breeders to learn about their techniques. The data he gathered on domestic populations provided the bulk of the evidence in *On the Origin of Species*, and he published an important but less famous work titled *Variation of Animals and Plants under Domestication*.[15]

Darwin chose the term *selection* with the explicit aim of highlighting the connection between the actions of human beings and the actions of Nature (he capitalized the word). In his day, *selection* referred to what today we call *breeding*. He attached the adjective *natural* to *selection* to expand the meaning beyond human actions to include nonhuman forces. Darwin knew *selection* was not ideal because it seemed to imply intentionality on the part of Nature, which was not part of his theory, but he decided that the risk of confusion was outweighed by the advantage of linking selection in the barnyard with selection in the wild. He wrote, "The term [*natural selection*] is so far a good one as it brings into connection the production of domestic races by man's power of selection, and the natural preservation of varieties and species in a state of nature."[16]

Darwin suggested that animal owners practiced two kinds of selection. In *methodical selection*, breeders selected specific males and females to mate with the goal of influencing the traits of the next generation. Methodical selection (also known as selective mating) was intentional selection. In *unconscious selection*, owners kept the individual animals that performed their job the best and discarded the rest. Unconscious selectors did not aim to shape the traits of future generations; they simply kept animals that did their jobs well. Unconscious selectors let animals choose their own mates, but the only options on a given farm were individuals with traits the owner wanted. The effect was much the same as under methodical selection – individuals with desirable traits mated with each other and passed along their traits to the next generation – but the effect was accidental rather than intentional. Natural selection, Darwin argued, acted like unconscious selection. Both shaped the traits of future generations by influencing survivorship.[17]

Darwin took pains to highlight *methodical selection* and *unconscious selection* in his books, but his successors have largely overlooked these terms in favor of *artificial selection*. Biologists cite Darwin as a source for *artificial selection*, and he did use the term, but it played a minor role in his work – it appears twice in *On the Origin of Species* and once in *Variation of Animals and Plants under Domestication*. The latter book is a long study of the impact of breeding on domestic plants and animals, so the near-omission of *artificial selection* is noteworthy. Neither *Origin* nor *Variation* indexed *artificial selection*. On the other hand, both books indexed *unconscious selection* and *methodical selection*. *Unconscious selection* made seven appearances in *Origin* and 43 in *Variation*. *Methodical selection* showed up seven times in *Origin* and 23 times in *Variation*.

Dropping Darwin's terms has had unfortunate effects. One is that biologists largely lost track of unconscious selection and sometimes restricted the definition of *artificial selection* to selective mating.[18] Another problem was that this revision converted the advantage of *selection* into a liability. Darwin used *selection* to highlight commonalities between people and "nature." Juxtaposing *artificial selection* against *natural selection* created a false, non-Darwinian dichotomy between people and "nature" instead. It has led people to interpret *artificial* as meaning *false* or *feigned*, which has led in turn to seeing artificial selection as producing false or feigned evolution (in contrast to what people think of as real evolution produced by natural selection). This in turn has led a number of scholars to believe that people cannot affect evolution.

Biologists have used *artificial selection* in ways that are mutually contradictory. Sometimes they have made conscious intent to change populations essential to the definition, and in other cases they have included accidental effects.[19] One feature unites usages: *artificial* identifies something people did. The central purpose of *artificial*, then, is to reinforce the idea that people and nature are fundamentally different and distinct, and therefore the effects of their actions are fundamentally different and distinct. It seems better to avoid *artificial selection* in favor of Darwin's *methodical selection* and *unconscious selection*.

Human beings have been modifying domestic plants and animals since at least the agricultural revolution of about 12,000 years ago. Most likely the domestication resulted from unconscious selection. Berries provide a good example. Given variation in the traits of berries in a patch, hunter-gatherers probably ate the berries that tasted best and grew to the largest size. They would have defecated seeds of these berries near their camps, and the plants that grew from them would have produced berries something like their parents. So people were practicing unconscious selection for berries most suited for human food. All else being equal, people would have harvested nearby plants before plants that were far away, and they probably would have continued to select for the best-tasting and biggest berries over time, and this steady selection would have made plants more suitable for living with and serving human ends. Other traits that typically came with domestication of many plants were simultaneous ripening and, in grains, seeds that clung tightly to stalks rather than seeds that fell off easily. Unconscious selection can explain these traits, too, for people would have harvested from plants that ripened at the same time, and tight-clinging grains would have been harvested more often than grains that easily fell off at the touch of a hand or breeze.[20]

It is not clear when people started practicing methodical selection, and adoption of the practice varied with time, space, and type of organism, but a surge of interest arose in the eighteenth century in England among animal breeders. At a time when most farmers let animals choose their own mates, breeders paired individual males and females with desirable traits. At first they focused mainly on sires as the means for affecting a herd's traits, though later they gave closer consideration to the traits of females as well. Breeders wanted to produce predictable results in the next generation, and they found that mating close relatives with each other produced the most consistent results. Called *inbreeding*, this practice influenced evolution not only by raising the frequency of certain traits in a population, but also by reducing overall variation.[21]

Today human beings have developed the ability to widen the genetic variation within populations in a way unavailable to earlier breeders. Genetic engineering enables scientists to move genes across widely different taxonomic groups. Some bacteria and rice plants now carry human genes with instructions for producing useful compounds. Genetic engineering also can narrow genetic variation. If genetic engineering endows a plant or animal with an unusually beneficial trait, it provides an incentive for farmers to adopt that strain of plant or animal and abandon other types, and genetically identical strains of corn or cotton now grow over large areas.[22]

So far we have focused primarily on proximate, biological dimensions of anthropogenic evolution. Now we widen our focus to include the historical dimensions of these processes. Historians bring special expertise for studying social causes and effects of anthropogenic evolution. Social forces are evolutionary forces, and evolutionary changes are social forces. By synthesizing the strengths of history and evolutionary biology, we can develop a deeper and broader understanding of the past than either discipline could provide on its own.

We can illustrate the value of linking history and biology by returning to earlier examples of anthropogenic evolution. Biology has explained how people selected for certain traits and encouraged drift through heavy hunting of game animals. The value of history lies in explaining why and how people came to hunt animals with those traits in the first place. There was nothing inevitable about the hunting of elephants for ivory. If there were, Africans long ago would have selected for tusklessness. And there was nothing inevitable about killing large bighorn sheep. Hunters could have killed small males as well as large ones.

History is good at situating these sorts of practices in larger historical periods and patterns. Heavy ivory hunting in Africa derived from trade, artistic tastes, leisure activities, and musical technology. Trade enabled consumption in one part of the globe to shape evolution in another. Art created demand for ivory to carve into sculptures. The game of billiards created demand for ivory balls. The popularity of the piano created demand for ivory keys. Markets for these products, largely outside Africa, made it profitable to kill huge numbers of African elephants for their tusks, which in turn selected for tusklessness.[23]

Political history also helps us understand evolution in populations of large mammals. Political historians use the term *state capacity* to refer to the ability of governments to accomplish their goals. Both low and high state capacities have affected evolution. Low state capacity has contributed to evolution in populations of elephants. Ivory hunting has taken place in places where it was illegal. If nations enforced anti-poaching laws, selection for tusklessness would likely have plummeted. But low state capacity has interfered with law enforcement. Cash-strapped governments have little money to spend on game wardens, and bribery has encouraged officials to look the other way while someone hunts. Conversely, high state capacity has influenced the evolution of bighorn sheep. Canada enforces its game laws, which prohibit hunters from taking animals below a specified size. By forcing hunters to harvest only large animals, government policy selected against large body size and for small body size in populations of bighorn sheep.

These examples show the potential of evolutionary history to widen the ambit of historical fields, including those that seem unlikely candidates for such an analysis. How many historians of art, music, leisure, and politics have thought to ask whether the processes they study have affected the evolution of populations of elephants? The first step is to consider the role of nonhuman species in history at all, which has been the central project of environmental historians. The second step is to include evolution in a checklist of potential human impacts on populations of other species.[24]

Along with studying evolution as a consequence of human actions, we should study it as a cause. We can see both in the history of the global Malaria Eradication Project. Human beings carried out certain actions (using insecticides and drugs), which prompted evolution in nonhuman populations (resistance in mosquitoes and plasmodia). This evolution encouraged a new human action (abandoning the project). Suggesting that evolution can influence human actions does not mean it determines human actions. Like any other historical process, it creates a condition, and people choose, from among a range of options, how they will respond to the condition. In fact, people did not give up on eradication the first time they encountered resistance. They substituted new insecticides and drugs for the useless older ones. Mosquitoes and plasmodia evolved resistance to the new poisons, which led to more substitutions, and so on. Some have likened the introduction of a stream of new technologies to counter continual evolution in pests and pathogens to the Red Queen in *Alice in Wonderland*, who had to run as fast as she could to stay in place.[25]

The process we have just described is an example of *coevolution*. Biologists introduced the concept of coevolution to explain how the traits of plants and insect pollinators worked so well together. They suggested that insect and plant populations had continually evolved in response to each other – change in one led to change in the other, which led to change in the first, which led to another change in the other, and so on in an endless spiral. Since then the concept has been expanded from genetic traits to include cultural traits. Here *culture* is taken to mean ideas about how to do things. Genes did not determine that people would spray pesticides or take drugs; these behaviors grew out of culture. In human beings, the pace of cultural evolution has far surpassed the rate of genetic evolution. Cultural evolution is a large and complex topic that we cannot explore in depth here, but other writers have done so admirably.[26]

One of the virtues of new fields is their ability to shed new light on familiar topics. To demonstrate the ability of evolutionary history to play this role, we turn to an example: the Industrial Revolution.[27] It is an important episode in history, and it is one that, at first blush, would seem to have little to do with evolution. One common narrative suggests that the Industrial Revolution succeeded because it replaced biology with technology. No longer did nature limit human potential, runs the argument, because people had burst its bounds. Technology housed in factories enabled human beings to produce useful goods faster, more cheaply, and in larger quantities than hand labor could provide. Mineral energy sources, such as coal and oil, powered the machines that turned out these products, releasing human beings from the bounds set by the energy that plants could provide to human muscles. In place of natural products, synthetic materials surged in popularity.[28]

Most historians believe the Industrial Revolution began in England in the late eighteenth century. To explain the origins of the revolution, historians have identified several types of Englishmen who played important roles, including inventors of new machines, entrepreneurs who set up factories housing the new machines, companies that expanded transportation systems, workers who became more industrious, and politicians who encouraged investment by passing laws defending property rights. While differing in emphasis, many of these accounts share a focus on human beings (Englishmen) working within one country (England) developing new technology. Revisionists have challenged the anthropocentric focus of these explanations by arguing for the importance of England's coal deposits and other materials to fuel industrialization. They have also challenged the geographic limit of these explanations by arguing for the importance of the slave trade and colonies as a source of resources, such as capital, raw materials, and markets.[29]

Cotton textiles made up one of the most important components of the Industrial Revolution, and we will use cotton as a case study of industrialization. Cotton was one of the first sectors to mechanize, it played a key role in developing the factory system, and it supplied a high percentage of England's exports. Historians have described a series of inventions that developed in a stimulus-response fashion and revolutionized textile-making. They have highlighted the invention of the fly shuttle (1733), the spinning jenny (1764), the water frame (1769), and the mule (1779).[30] These machines had a major quantitative impact on the rate and quantity of textile production.[31]

This sequence makes a wonderful example of the transformative power of technological innovation, which is one reason it figures so prominently in the work of historians of the Industrial Revolution. The key features are consistent with the story of the Industrial Revolution in general. The innovation was technological. The innovators were Englishmen. The site of innovation was England. The period of innovation was the mid- to late eighteenth century. Weaving these elements together makes a beautiful tapestry of explanation. Some loose ends hang at the edge of this tapestry, however, and pulling them leads us to a new interpretation of cause and effect that I will state as propositions.

One loose thread is the availability of spinnable cotton. Many historians have described cotton as inherently suited to machine spinning, often explaining that cotton was relatively uniform. This inherent suitability, they have suggested, explains why cotton industrialized before wool or linen.[32] But we should not take cotton's traits for granted any more than we take the invention of new machines for granted. People developed cotton suited to machine spinning as well as the machines to do the spinning.

My first proposition is that anthropogenic evolution set the stage for the Industrial Revolution by developing cotton fiber that was suitable for spinning by machine. Wild cottons grew short hairs useless for spinning. People domesticated cotton in both the Old World and the New World around 5,000 years ago, and over time the length of fiber in domesticated populations increased. We do not know the mechanism, but I suspect that unconscious selection by human beings was the primary force. Selective mating may also have been used. Long spinnable fibers were not a gift of nature, but the product of people and plants interacting in a way that developed an important trait. These human selectors did not intend to create fibers suited to machine spinning. To the extent they had a use in mind, it would have been hand spinning. But suitability for machines was the effect. Evolution under domestication increased the quantity as well as the quality of fiber as populations grew larger bolls than in the past and adapted to a variety of growing conditions.[33]

A second loose thread is crediting the ability of the English to make all-cotton cloth to new machines. The English began making all-cotton cloth in the eighteenth century. Before that, some historians have claimed, the closest they came was to make fustian, a cloth with cotton weft and linen warp. They relied on linen warp because their cotton thread broke too easily when used as warp. Some historians have credited the manufacture of all-cotton cloth in England to new spinning machines, suggesting that the mule or water frame created the ability of England to produce warp-strength cotton thread for the first time.[34] Unfortunately for this theory, the English made all-cotton cloth by 1726, and large amounts of it in the 1750s, long before the appearance of the water frame (1769) and the mule (1779).[35]

My second proposition is that the introduction of New World cotton probably created the ability of the English to make cotton warp and all-cotton cloth. England relied on Old World cotton imports (mainly from the Levant) in the seventeenth century.[36] Cotton from the West Indies flooded into England in the eighteenth century, around the time cotton warp appeared. New World cotton grew longer fibers than Old World cotton. Longer fibers make stronger thread, which might explain the newfound ability to spin cotton warp.[37]

My third proposition is that New World cotton made the mechanization of spinning possible. Inventors introduced successful machines for spinning after England began making warp-strength cotton thread. The added strength that long fibers conferred was critical for thread to withstand the rigors of machine spinning (and later machine weaving). In the late eighteenth and early nineteenth centuries, hard-nosed factory owners bought high-priced New World cotton almost exclusively even though they could have purchased low-priced Old World cotton. Spinners eventually learned to use Old World cotton in machine spinning, but they almost always mixed it with New World cotton to add strength. The "cotton famine" created by the American Civil War in the mid-nineteenth century hit England especially hard because the Old World produced little long-fibered cotton. An exception, Egypt, was able to do so because it grew a New World species.[38]

The essential role of long-fibered cotton sheds new light on the old debate about the slave trade and industrialization. The debate has focused most famously on whether profits from the slave trade underwrote capital expansion in England. It has broadened to consider the role of the slave trade in supplying raw materials and in creating markets for English goods. These debates have been largely about quantities of money and goods. I suggest that the slave trade played a key role because of its impact on quality of cotton in England. The triangular trade between England, Africa, and the New World brought New World cottons to England. Because England's main slave-trade port was Liverpool, which served the Lancashire textile industry, the slave trade concentrated New World cottons in Lancashire around the time when Lancashire inventors developed the new spinning machines.[39]

A genetic difference between Old World and New World cottons probably contributed to differences in fiber length. Domestic New World cotton species had twice as many chromosomes and genes as domestic Old World cotton species. Doubling the number of genes would have roughly doubled the chance of mutations, and mutations were the raw material of evolution because they generated new traits that selection could encourage. So England's machine spinners probably owed their success to New World genetics as well as to Amerindians who had selected for valuable traits.[40]

Now we can sum up an account of the Industrial Revolution grounded in evolutionary history. Some wild cotton species in the New World developed genomes twice the size of some Old World cotton species, which created the opportunity for twice as many mutations and new traits. People in the New World and Old World developed a domestic relationship with these cottons. Fibers lengthened more in domestic New World cotton populations than in Old World populations. The slave trade delivered long New World fiber to Liverpool, endowing England with the evolutionary inheritance of the New World. Long New World fiber made stronger thread than short Old World fiber. When New World fiber flowed from Liverpool into surrounding Lancashire, it created an opportunity for inventors to capitalize on increased thread strength. They responded by inventing new machines to spin thread. Notice how this account revises the standard story of industrialization. Biological innovation (longer fiber) made technological innovation (new machines) possible. Innovation by Amerindians in the New World made innovation by Englishmen in England possible. Innovation essential for industrialization did not occur only in the eighteenth century; it began 5,000 years earlier. Technology did not change around static "nature"; change in traits of organisms made change in technology possible. Industrialization did not free human beings from dependence on "nature"; industrialization depended on "nature" to succeed.

Evolutionary history, as a field of history, does a service in calling attention to the effects of human beings on populations of other species. It complements the long-term focus in environmental history on the impact of human beings on ecology (that is, our impact on the distribution and abundance of organisms). Evolutionary history shows that, in addition to moving species around the earth, and in addition to making populations of organisms larger and smaller, people have changed the traits of populations. We have fiddled, sometimes intentionally and often accidentally, with the nature of organisms themselves. We have altered their genetic makeup in important ways. Historians have long stressed the importance of locating ideas and institutions in time and space, and of seeing how particular circumstances shaped the world in particular ways. Evolutionary history helps us situate the genetic composition of organisms in time and space as well. The traits of many organisms are historical products, and they are historical forces.[41]

Notes

1   For a fuller explanation of the ideas in this chapter, see E. Russell, *Evolutionary History: Uniting History and Biology to Understand Life on Earth*, New York, Cambridge University Press, 2011.
2   Scholars have voiced these beliefs in conversations and in questions after formal presentations.
3   I thank Michael Grant for this definition. Similar definitions are in D. Futuyma, *Evolutionary Biology*, 3rd ed., Sunderland, MA, Sinauer Associates, 1998; and B. Charlesworth and D. Charlesworth, *Evolution: A Very Short Introduction*, Oxford, Oxford University Press, 2003, p. 5.
4   J. Weiner, *The Beak of the Finch: A Story of Evolution in Our Time*, New York, Knopf, 1994, pp. 70–82; P. Grant, *Ecology and Evolution of Darwin's Finches*, Princeton, NJ, Princeton University Press, 1999.

5   Futuyma, *Evolutionary Biology*, pp. 297, 349, 586; S. R. Palumbi, *Evolution Explosion: How Humans Cause Rapid Evolutionary Change*, New York, W. W. Norton, 2001; Russell, *Evolutionary History*, chapter 2.

6   H. Jachmann, P. S. M. Berry, and H. Imae, "Tusklessness in African Elephants: A Future Trend," *African Journal of Ecology* 33, 1995, pp. 230–5; E. Abe, "Tusklessness amongst the Queen Elizabeth National Park Elephants, Uganda," *Pachyderm* 22, 1996, pp. 46–7.

7   Jachmann, Berry, and Imae, "Tusklessness in African Elephants."

8   A. M. Whitehouse, "Tusklessness in Elephant Population of the Addo Elephant National Park, South Africa," *Journal of the Zoological Society of London* 257, 2002, pp. 249–54.

9   D. W. Coltman, P. O'Donoghue, J. T. Jorgenson, et al., "Undesirable Evolutionary Consequences of Trophy Hunting," *Nature* 426, 2003, pp. 655–8, accessed from http://www.nature.com/nature/journal/v426/n6967/abs/nature02177.html on March 15, 2012.

10 P. Handford, G. Bell, and T. Reimchen, "A Gillnet Fishery Considered as an Experiment in Artificial Selection," *Journal of Fisheries Research Board of Canada* 34, 1977, pp. 954–61; D. P. Swain, A. F. Sinclair, and J. M. Hanson, "Evolutionary Response to Size-Selective Mortality in an Exploited Fish Population," *Proceedings of the Royal Society, Series B* 274, 2007, pp. 1015–22; E. M. Olsen, M. Heino, G. R. Lilly, et al., "Maturation Trends Indicative of Rapid Evolution Preceded the Collapse of Northern Cod," *Nature* 428, 2004, pp. 932–5; W. E. Ricker, "Changes in the Average Size and Average Age of Pacific Salmon," *Canadian Journal of Fisheries and Aquatic Sciences* 38, 1981, pp. 1636–56; Russell, *Evolutionary History*, chapter 3.

11 D. Brown, "Wonder Drugs' Losing Healing Aura," *Washington Post*, June 26, 1995, p. A1; R. S. Phillips, "Current Status of Malaria and Potential for Control," *Clinical Microbiology Reviews* 14/1, 2001, pp. 208–26; J. F. Trape, "The Public Health Impact of Chloroquine Resistance in Africa," *American Journal of Tropical Medicine and Hygiene* 64/1–2, 2001, pp. 12–17; J. A. Najera, "Malaria Control: Achievements, Problems, and Strategies," *Parassitologia* 43/1–2, 2001, pp. 1–89; Palumbi, *Evolution Explosion*, pp. 137–8.

12 D. Brown, "TB Resistance Stands at 11% of Cases," *Washington Post*, March 24, 2000, p. A14; S. B. Levy, *The Antibiotic Paradox: How Miracle Drugs Are Destroying the Miracle*, New York, Plenum Press, 1992, p. 279; Palumbi, *Evolution Explosion*, p. 85.

13 A. Alonso-Moraga, A. Muñoz-Serrano, J. M. Serradilla, and J. R. David, "Microspatial Differentiation of *Drosophila melanogaster* Populations In and Around a Wine Cellar in Southern Spain," *Genetics Selection Evolution* 20/3, 1988, pp. 307–14; Russell, *Evolutionary History*, chapter 4.

14 F. Rodriguez-Trelles and M. A. Rodriguez, "Rapid Micro-Evolution and Loss of Chromosomal Diversity in *Drosophila* in Response to Climate Warming," *Evolutionary Ecology* 12, 1998, pp. 829–38; W. Bradshaw, and C. M. Holzapfel, "Genetic Shift in Photoperiodic Response Correlated with Global Warming," *Proceedings of the National Academy of Sciences of the United States of America* 98, 2001, pp. 14509–11; S. C. Thomas and J. G. Kingsolver, "Natural Selection: Responses to Current (Anthropogenic) Environmental Changes," in *Encyclopedia of Life Sciences*, London, Macmillan, 2002, pp. 659–64; Russell, *Evolutionary History*, chapter 5.

15 C. Darwin, *On the Origin of Species by Means of Natural Selection, or the Preservation of Favoured Races in the Struggle for Life*, London, Odhams Press, 1872 [1859]: see p. 45 on pigeon clubs; C. Darwin, *Variation of Animals and Plants under Domestication*, Baltimore, The Johns Hopkins University Press, 1998 [1868]: see p. 2 on the importance of evidence in *Variation* for the argument in *Origin*.

16 Darwin, *Variation of Animals and Plants under Domestication*, p. 6.

17 Darwin, *On the Origin of Species*, pp. 52–63.

18 Futuyma, *Evolutionary Biology*, p. 765; H. Curtis, *Biology*, New York, Worth, 1983, p. 1088.

19 Futuyma, *Evolutionary Biology*; Curtis, *Biology*; S. M. Carlson, E. Edeline, L. Asbjørn Vøllestad, et al., "Four Decades of Opposing Natural and Human-Induced Artificial Selection Acting on Windermere Pike (*Esox lucius*)," *Ecology Letters* 10, 2007, pp. 512–21.

20　J. Diamond, *Guns, Germs, and Steel: The Fates of Human Societies*, New York, W. W. Norton, 1999, pp. 114–30; Russell, *Evolutionary History*, chapter 6.

21　H. Ritvo, *The Animal Estate: The English and Other Creatures in the Victorian Age*, Cambridge, MA, Harvard University Press, 1987.

22　Palumbi, *Evolution Explosion*; Russell, *Evolutionary History*, chapter 7.

23　E. A. Alpers, *Ivory and Slaves: Changing Pattern of International Trade in East Central Africa to the Later Nineteenth Century*, Berkeley, University of California Press, 1975; A. Sheriff, *Slaves, Spices, & Ivory in Zanzibar: Integration of an East African Commercial Empire into the World Economy, 1770–1873*, London, J. Currey, 1987; E. Closson and R. Golding, *History of the Piano*, 2nd ed., London, Elek, 1976 [1947], p. 154; B. C. Eastham and W. C. Hu, *Chinese Art Ivory*, Ann Arbor, MI, Ars Ceramica, 1976, p. 86; Walters Art Gallery, *Ivory, the Sumptuous Art: Highlights from the Collection*, Baltimore, MD, The Gallery, 1983, p. 40.

24　For reviews of evolutionary history in environmental history and the history of technology, see Russell, *Evolutionary History*, chapters 10–11; S. R. Schrepfer and P. Scranton, (ed.), *Industrializing Organisms: Introducing Evolutionary History*, New York, Routledge, 2004; E. Russell, "Evolutionary History: Prospectus for a New Field," *Environmental History* 8, 2003, pp. 204–28.

25　L. V. Valen, "A New Evolutionary Law," *Evolutionary Theory* 1, 1973, pp. 1–30.

26　P. R. Ehrlich and P. H. Raven, "Butterflies and Plants: A Study in Coevolution," *Evolution* 18, 1964, pp. 586–608; C. J. Lumsden and E. O. Wilson, *Genes, Mind, and Culture: The Coevolutionary Process*, Cambridge, MA, Harvard University Press, 1981; D. Futuyma and M. Slatkin (eds.), *Coevolution*, Sunderland, MA, Sinauer Associates, 1983; W. H. Durham, *Coevolution: Genes, Culture, and Human Diversity*, Stanford, CA, Stanford University Press, 1991; J. M. Gowdy, *Coevolutionary Economics: The Economy, Society, and the Environment*, Boston, Kluwer Academic Publishers, 1994; F. Fenner and P. J. Kerr, "Evolution of the Poxvirus, including the Coevolution of Virus and Host in Myxomatosis," in S. S. Morse (ed.), *The Evolutionary Biology of Viruses*, New York, Raven Press, 1994, pp. 273–92; M. D. Rausher, "Coevolution and Plant Resistance to Natural Enemies," *Nature* 411/6839, 2001, pp. 857–64.

27　Russell, *Evolutionary History*, chapter 9.

28　D. S. Landes, *The Unbound Prometheus: Technological Change and Industrial Development in Western Europe from 1750 to the Present*, Cambridge, Cambridge University Press, 2003, pp. 1–3; J. Mokyr, *The Lever of Riches: Technological Creativity and Economic Progress*, New York, Oxford University Press, 1990, p. 82; F. Braudel, *A History of Civilizations*, New York, A. Lane, 1993, p. 377; W. H. McNeill, *A World History*, Oxford, Oxford University Press, 1999, pp. 420–3; K. Marx, *The Poverty of Philosophy*, Chicago, Charles H. Kerr, 1910, p. 119; W. H. Shaw, "The Handmill Gives You the Feudal Lord: Marx's Technological Determinism," *History and Theory* 18, 1979, pp. 155–76; F. Engels, "Socialism: Utopian and Scientific," in R. C. Tucker (eds.), *The Marx–Engels Reader*, 2nd ed., New York, W. W. Norton, 1978, pp. 683–717 (p. 690).

29　P. Mantoux, *The Industrial Revolution in the Eighteenth Century: An Outline of the Beginnings of the Modern Factory System in England*, London, Jonathan Cape, 1928, p. 25; P. O'Brien, T. Griffiths, and P. Hunt, "Political Components of the Industrial Revolution: Parliament and the English Cotton Textile Industry, 1660–1774," *Economic History Review* 44/3, 1991, pp. 395–423; D. A. Farnie and D. J. Jeremy (ed.), *The Fibre That Changed the World: The Cotton Industry in International Perspective, 1600–1990s*, Oxford, Oxford University Press, 2004; P. Deane, *The First Industrial Revolution*, Cambridge, Cambridge University Press, 1979; P. Deane and W. A. Cole, *British Economic Growth 1688–1959: Trends and Structure*, Cambridge, Cambridge University Press, 1967; R. M. Hartwell, *The Industrial Revolution and Economic Growth*, London, Methuen, 1971; O. L. May and K. E. Lege, "Development of the World Cotton Industry," in C. W. Smith and J. T. Cothren (eds.), *Cotton: Origin, History, Technology, and Production*, New York, John Wiley, 1999, pp. 67–76;

J. de Vries, *The Industrious Revolution: Consumer Behavior and the Household Economy, 1650 to the Present*, New York, Cambridge University Press, 2008; K. Pomeranz, *The Great Divergence: Europe, China, and the Making of the Modern World Economy*, Princeton, NJ, Princeton University Press, 2000; A. Hornborg, "Footprints in the Cotton Fields: The Industrial Revolution as Time–Space Appropriation and Environmental Load Displacement," in A. Hornborg, J. McNeill, and J. Martinez-Alier (eds.), *Rethinking Environmental History: World-System History and Global Environmental Change*, New York, AltaMira Press, 2007, pp. 259–72; E. Williams, *Capitalism and Slavery*, Chapel Hill, NC, University of North Carolina Press, 1944; K. Morgan, *Slavery and the British Empire: From Africa to America*, Oxford, Oxford University Press, 2007; K. Morgan, *Slavery, Atlantic Trade and the British Economy, 1660–1800*, Cambridge, Cambridge University Press, 2000.

30 Mokyr, *The Lever of Riches*, pp. 96–8; Landes, *The Unbound Prometheus*, pp. 84–5; C. K. Harley, "Cotton Textile Prices and the Industrial Revolution," *The Economic History Review* 51/1, 1998, pp. 49–83 (p. 50).

31 Braudel, *A History of Civilizations*; J. Mokyr (ed.), *The British Industrial Revolution: An Economic Perspective*, San Francisco, Westview Press, 1993; Landes, *The Unbound Prometheus*, pp. 41–5; McNeill, *A World History*, p. 420; A. P. Wadsworth and J. Mann, *The Cotton Trade and Industrial Lancashire, 1600–1780*, New York, Augustus M. Kelley, 1968.

32 M. Mehta, *The Ahmedabad Cotton Textile Industry: Genesis and Growth*, Ahmedabad, New Order, 1982, p. 38; O'Brien, Griffiths, and Hunt, "Political Components of the Industrial Revolution," p. 415; Landes, *The Unbound Prometheus*, p. 83; Mokyr, *The British Industrial Revolution*, p. 100.

33 C. L. Brubaker, F. M. Bourland, and J. F. Wendel, "Origin and Domestication of Cotton," in Smith and Cothren (eds.), *Cotton*, pp. 3–31; K. M. Butterworth, D. C. Adams, H. T. Horner, and J. F. Wendel, "Initiation and Early Development of Fiber in Wild and Cultivated Cotton," *International Journal of Plant Sciences* 170, 2009, pp. 561–74; J. F. Wendel, C. L. Brubaker, and T. Seelanan, "The Origin and Evolution of *Gossypium*," in J. M. Stewart, D. M. Oosterhuis, J. J. Heitholt, and J. R. Mauney (eds.), *Physiology of Cotton*, Dordrecht, Springer, 2010, pp. 1–18.

34 Mokyr, *The British Industrial Revolution*, p. 98; May and Lege, "Development of the World Cotton Industry"; T. Ellison, *The Cotton Trade of Great Britain*, London, Frank Cass, 1968; E. Baines, *History of the Cotton Manufacture in Great Britain*, New York, Augustus M. Kelley, 1966, p. 183.

35 Wadsworth and Mann, *The Cotton Trade and Industrial Lancashire*, pp. 175–6, 275.

36 Wadsworth and Mann, *The Cotton Trade and Industrial Lancashire*, pp. 15–17; Brubaker, Bourland, and Wendel, "Origin and Domestication of Cotton"; Ellison, *The Cotton Trade of Great Britain*, p. 16.

37 Wadsworth and Mann, *The Cotton Trade and Industrial Lancashire*, pp. 520–1.

38 C. R. Benedict, R. J. Kohel, and H. L. Lewis, "Cotton Fiber Quality," in C. W. Smith and J. T. Cothren (eds.), *Cotton: Origin, History, Technology, and Production*, New York, John Wiley, 1999, pp. 269–88 (p. 283).

39 Williams, *Capitalism and Slavery*; Morgan, *Slavery and the British Empire*; Morgan, *Slavery, Atlantic Trade and the British Economy*.

40 A. E. Percival, J. E. Wendel, and J. M. Stewart, "Taxonomy and Germplasm Resources," in C. W. Smith and J. T. Cothren (eds.), *Cotton: Origin, History, Technology, and Production*, New York, John Wiley, 1999, pp. 33–63; R. Hovav, B. Chaudhary, J. A. Udall, et al., "Parallel Domestication, Convergent Evolution and Duplicated Gene Recruitment in Allopolyploid Cotton," *Genetics* 179/3, 2008, pp. 1725–33; J. J. Doyle, L. E. Flagel, A. H. Paterson, et al., "Evolutionary Genetics of Genome Merger and Doubling in Plants," *Annual Review of Genetics* 42, 2008, pp. 443–61; L. E. Flagel and J. Wendel, "Gene Duplication and Evolutionary Novelty in Plants," *New Phytologist* 183, 2009, pp. 557–64.

41 This chapter largely summarizes material in Russell, *Evolutionary History*. In addition to the people thanked in the book, I am grateful to John McNeill and Erin Stewart Mauldin for

their assistance with this chapter. A grant from the National Science Foundation (number SES-0220764) supported research for this chapter. Any opinions, findings, and conclusions or recommendations expressed in this material are those of the author and do not necessarily reflect the views of the National Science Foundation.

References

Abe, E., "Tusklessness amongst the Queen Elizabeth National Park Elephants, Uganda," *Pachyderm* 22, 1996, pp. 46–7.

Alonso-Moraga, A., Muñoz-Serrano, A., Serradilla, J. M., and David, J. R., "Microspatial Differentiation of *Drosophila melanogaster* Populations In and Around a Wine Cellar in Southern Spain," *Genetics Selection Evolution* 20/3, 1988, pp. 307–14.

Alpers, E. A., *Ivory and Slaves: Changing Pattern of International Trade in East Central Africa to the Later Nineteenth Century*, Berkeley, University of California Press, 1975.

Baines, E., *History of the Cotton Manufacture in Great Britain*, New York, Augustus M. Kelley, 1966.

Benedict, C. R., Kohel, R. J., and Lewis, H. L., "Cotton Fiber Quality," in C. W. Smith and J. T. Cothren (eds.), *Cotton: Origin, History, Technology, and Production*, New York, John Wiley, 1999, pp. 269–88.

Bradshaw, W., and Holzapfel, C. M., "Genetic Shift in Photoperiodic Response Correlated with Global Warming," *Proceedings of the National Academy of Sciences of the United States of America* 98, 2001, pp. 14509–11.

Braudel, F., *A History of Civilizations*, New York, A. Lane, 1993.

Brown, D., "TB Resistance Stands at 11% of Cases," *Washington Post*, March 24, 2000, p. A14.

Brown, D., "Wonder Drugs' Losing Healing Aura," *Washington Post*, June 26, 1995, p. A1.

Brubaker, C. L., Bourland, F. M., and Wendel, J. F., "Origin and Domestication of Cotton," in C. W. Smith and J. T. Cothren (eds.), *Cotton: Origin, History, Technology, and Production*, New York, John Wiley, 1999, pp. 3–31.

Butterworth, K. M., Adams, D. C., Horner, H. T., and Wendel, J. F., "Initiation and Early Development of Fiber in Wild and Cultivated Cotton," *International Journal of Plant Sciences* 170, 2009, pp. 561–74.

Carlson, S. M., Edeline, E., Asbjørn Vøllestad, L., et al., "Four Decades of Opposing Natural and Human-Induced Artificial Selection Acting on Windermere Pike (*Esox lucius*)," *Ecology Letters* 10, 2007, pp. 512–21.

Charlesworth, B., and Charlesworth, D., *Evolution: A Very Short Introduction*, Oxford, Oxford University Press, 2003.

Closson, E., and Golding, R., *History of the Piano*, 2nd ed., London, Elek, 1976 [1947].

Coltman, D. W., O'Donoghue, P., Jorgenson, J. T., et al., "Undesirable Evolutionary Consequences of Trophy Hunting," *Nature* 426, 2003, pp. 655–8, accessed from http://www.nature.com/nature/journal/v426/n6967/abs/nature02177.html on March 15, 2012.

Curtis, H., *Biology*, New York, Worth, 1983.

Darwin, C., *On the Origin of Species by Means of Natural Selection, or the Preservation of Favoured Races in the Struggle for Life*, London, Odhams Press, 1872 [1859].

Darwin, C., *Variation of Animals and Plants under Domestication*, Baltimore, The Johns Hopkins University Press, 1998 [1868].

de Vries, J., *The Industrious Revolution: Consumer Behavior and the Household Economy, 1650 to the Present*, New York, Cambridge University Press, 2008.

Deane, P., *The First Industrial Revolution*, Cambridge, Cambridge University Press, 1979.

Deane, P., and Cole, W. A., *British Economic Growth 1688–1959: Trends and Structure*, Cambridge, Cambridge University Press, 1967.

Diamond, J., *Guns, Germs, and Steel: The Fates of Human Societies*, New York, W. W. Norton, 1999.

Doyle, J. J., Flagel, L. E., Paterson, A. H., et al., "Evolutionary Genetics of Genome Merger and Doubling in Plants," *Annual Review of Genetics* 42, 2008, pp. 443–61.

Durham, W. H., *Coevolution: Genes, Culture, and Human Diversity*, Stanford, CA, Stanford University Press, 1991.

Eastham, B. C., and Hu, W. C., *Chinese Art Ivory*, Ann Arbor, MI, Ars Ceramica, 1976.

Ehrlich, P. R., and Raven, P. H., "Butterflies and Plants: A Study in Coevolution," *Evolution* 18, 1964, pp. 586–608.

Ellison, T., *The Cotton Trade of Great Britain*, London, Frank Cass, 1968.

Engels, F., "Socialism: Utopian and Scientific," in R. C. Tucker (eds.), *The Marx–Engels Reader*, 2nd ed., New York, W. W. Norton, 1978, pp. 683–717.

Farnie, D. A., and Jeremy, D. J. (ed.), *The Fibre That Changed the World: The Cotton Industry in International Perspective, 1600–1990s*, Oxford, Oxford University Press, 2004.

Fenner, F., and Kerr, P. J., "Evolution of the Poxvirus, including the Coevolution of Virus and Host in Myxomatosis," in S. S. Morse (ed.), *The Evolutionary Biology of Viruses*, New York, Raven Press, 1994, pp. 273–92.

Flagel, L. E., and Wendel, J., "Gene Duplication and Evolutionary Novelty in Plants," *New Phytologist* 183, 2009, pp. 557–64.

Futuyma, D., *Evolutionary Biology*, 3rd ed., Sunderland, MA, Sinauer Associates, 1998.

Futuyma, D., and Slatkin, M. (eds.), *Coevolution*, Sunderland, MA, Sinauer Associates, 1983.

Gowdy, J. M., *Coevolutionary Economics: The Economy, Society, and the Environment*, Boston, Kluwer Academic Publishers, 1994.

Grant, P., *Ecology and Evolution of Darwin's Finches*, Princeton, NJ, Princeton University Press, 1999.

Handford, P., Bell, G., and Reimchen, T., "A Gillnet Fishery Considered as an Experiment in Artificial Selection," *Journal of Fisheries Research Board of Canada* 34, 1977, pp. 954–61.

Harley, C. K., "Cotton Textile Prices and the Industrial Revolution," *The Economic History Review* 51/1, 1998, pp. 49–83.

Hartwell, R. M., *The Industrial Revolution and Economic Growth*, London, Methuen, 1971.

Hornborg, A., "Footprints in the Cotton Fields: The Industrial Revolution as Time–Space Appropriation and Environmental Load Displacement," in A. Hornborg, J. McNeill, and J. Martinez-Alier (eds.), *Rethinking Environmental History: World-System History and Global Environmental Change*, New York, AltaMira Press, 2007, pp. 259–72.

Hovav, R., Chaudhary, B., Udall, J. A., et al., "Parallel Domestication, Convergent Evolution and Duplicated Gene Recruitment in Allopolyploid Cotton," *Genetics* 179/3, 2008, pp. 1725–33.

Jachmann, H., Berry, P. S. M., and Imae, H., "Tusklessness in African Elephants: A Future Trend," *African Journal of Ecology* 33, 1995, pp. 230–5.

Landes, D. S., *The Unbound Prometheus: Technological Change and Industrial Development in Western Europe from 1750 to the Present*, Cambridge, Cambridge University Press, 2003.

Levy, S. B., *The Antibiotic Paradox: How Miracle Drugs Are Destroying the Miracle*, New York, Plenum Press, 1992.

Lumsden, C. J., and Wilson, E. O., *Genes, Mind, and Culture: The Coevolutionary Process*, Cambridge, MA, Harvard University Press, 1981.

Mantoux, P., *The Industrial Revolution in the Eighteenth Century: An Outline of the Beginnings of the Modern Factory System in England*, London, Jonathan Cape, 1928.

Marx, K., *The Poverty of Philosophy*, Chicago, Charles H. Kerr, 1910.

May, O. L., and Lege, K. E., "Development of the World Cotton Industry," in C. W. Smith and J. T. Cothren (eds.), *Cotton: Origin, History, Technology, and Production*, New York, John Wiley, 1999, pp. 67–76.

McNeill, W. H., *A World History*, Oxford, Oxford University Press, 1999.

Mehta, M., *The Ahmedabad Cotton Textile Industry: Genesis and Growth*, Ahmedabad, New Order, 1982.

Mokyr, J. (ed.), *The British Industrial Revolution: An Economic Perspective*, San Francisco, Westview Press, 1993.

Mokyr, J., *The Lever of Riches: Technological Creativity and Economic Progress*, New York, Oxford University Press, 1990.

Morgan, K., *Slavery and the British Empire: From Africa to America*, Oxford, Oxford University Press, 2007.

Morgan, K., *Slavery, Atlantic Trade and the British Economy, 1660–1800*, Cambridge, Cambridge University Press, 2000.

Najera, J. A., "Malaria Control: Achievements, Problems, and Strategies," *Parassitologia* 43/1–2, 2001, pp. 1–89.

O'Brien, P., Griffiths, T., and Hunt, P., "Political Components of the Industrial Revolution: Parliament and the English Cotton Textile Industry, 1660–1774," *Economic History Review* 44/3, 1991, pp. 395–423.

Olsen, E. M., Heino, M., Lilly, G. R., et al., "Maturation Trends Indicative of Rapid Evolution Preceded the Collapse of Northern Cod," *Nature* 428, 2004, pp. 932–5.

Palumbi, S. R., *Evolution Explosion: How Humans Cause Rapid Evolutionary Change*, New York, W. W. Norton, 2001.

Percival, A. E., Wendel, J. E., and Stewart, J. M., "Taxonomy and Germplasm Resources," in C. W. Smith and J. T. Cothren (eds.), *Cotton: Origin, History, Technology, and Production*, New York, John Wiley, 1999, pp. 33–63.

Phillips, R. S., "Current Status of Malaria and Potential for Control," *Clinical Microbiology Reviews* 14, 2001/1, pp. 208–26.

Pomeranz, K., *The Great Divergence: Europe, China, and the Making of the Modern World Economy*, Princeton, NJ, Princeton University Press, 2000.

Rausher, M. D., "Coevolution and Plant Resistance to Natural Enemies," *Nature* 411/6839, 2001, pp. 857–64.

Ricker, W. E., "Changes in the Average Size and Average Age of Pacific Salmon," *Canadian Journal of Fisheries and Aquatic Sciences* 38, 1981, pp. 1636–56.

Ritvo, H., *The Animal Estate: The English and Other Creatures in the Victorian Age*, Cambridge, MA, Harvard University Press, 1987.

Rodriguez-Trelles, F., and Rodriguez, M. A., "Rapid Micro-Evolution and Loss of Chromosomal Diversity in *Drosophila* in Response to Climate Warming," *Evolutionary Ecology* 12, 1998, pp. 829–38.

Russell, E., "Evolutionary History: Prospectus for a New Field," *Environmental History* 8, 2003, pp. 204–28.

Russell, E., *Evolutionary History: Uniting History and Biology to Understand Life on Earth*, New York, Cambridge University Press, 2011.

Schrepfer, S. R., and Scranton, P. (ed.), *Industrializing Organisms: Introducing Evolutionary History*, New York, Routledge, 2004.

Shaw, W. H., "The Handmill Gives You the Feudal Lord: Marx's Technological Determinism," *History and Theory* 18, 1979, pp. 155–76.

Sheriff, A., *Slaves, Spices, & Ivory in Zanzibar: Integration of an East African Commercial Empire into the World Economy, 1770–1873*, London, J. Currey, 1987.

Swain, D. P., Sinclair, A. F., and Hanson, J. M., "Evolutionary Response to Size-Selective Mortality in an Exploited Fish Population," *Proceedings of the Royal Society, Series B* 274, 2007, pp. 1015–22.

Thomas, S. C., and Kingsolver, J. G., "Natural Selection: Responses to Current (Anthropogenic) Environmental Changes," in *Encyclopedia of Life Sciences*, London, Macmillan, 2002, pp. 659–64.

Trape, J. F., "The Public Health Impact of Chloroquine Resistance in Africa," *American Journal of Tropical Medicine and Hygiene* 64/1–2, 2001, pp. 12–17.

Valen, L. V., "A New Evolutionary Law," *Evolutionary Theory* 1, 1973, pp. 1–30.

Wadsworth, A. P., and Mann, J., *The Cotton Trade and Industrial Lancashire, 1600–1780*, New York, Augustus M. Kelley, 1968.

Walters Art Gallery, *Ivory, the Sumptuous Art: Highlights from the Collection*, Baltimore, MD, The Gallery, 1983.

Weiner, J., *The Beak of the Finch: A Story of Evolution in Our Time*, New York, Knopf, 1994.

Wendel, J. F., Brubaker, C. L., and Seelanan, T., "The Origin and Evolution of *Gossypium*," in J. M. Stewart, D. M. Oosterhuis, J. J. Heitholt, and J. R. Mauney (eds.), *Physiology of Cotton*, Dordrecht, Springer, 2010, pp. 1–18.

Whitehouse, A. M., "Tusklessness in Elephant Population of the Addo Elephant National Park, South Africa," *Journal of the Zoological Society of London* 257, 2002, pp. 249–54.

Williams, E., *Capitalism and Slavery*, Chapel Hill, NC, University of North Carolina Press, 1944.

CHAPTER TWENTY-TWO

Climate Change in Global Environmental History

SAM WHITE

The study of past climate change has recently emerged both as a critical topic in global environmental history and as its own interdisciplinary subject of historical research. As global warming has raised awareness of current and future climate issues, it has also raised questions about earlier climate fluctuations, their impacts, and how societies have coped with change and variability. Drawing on new methods and evidence, researchers around the world have begun to reconstruct past climates and their role in human history. This growing field has shed new light on important historical developments, and in time may offer new perspectives and lessons for the challenge of global warming.

Background, Methods, Concepts

The study of climate in human affairs reaches back to the earliest works of history. Ancient authors such as Herodotus described hot and cold "climes" and pondered their impact on the character and constitution of societies. Such ideas persisted through the Middle Ages and into works of Enlightenment thinkers such as Montesquieu and Voltaire. In the late nineteenth and early twentieth centuries, the study of climate sometimes merged with notions of racism and racial hierarchy, producing simplistic climatic explanations of history, as in the work of Elsworth Huntington. Such "climate determinism" produced a backlash among subsequent scholars, many of whom have resisted climatic explanations ever since.[1]

Nevertheless, with recent advances in climatology and especially the discovery of global warming, climate has resurfaced as a serious subject for archeologists and historians. Scholars as far back as Edward Gibbon (1737–94) may have noted differences between past and present temperatures, but only in the mid-twentieth century did researchers assemble firm evidence of significant historical climate change. Studies of the Little Ice Age (see below) by Fernand Braudel, Gustav Utterström, and Emmanuel Le

A Companion to Global Environmental History, First Edition. Edited by J.R. McNeill and Erin Stewart Mauldin.
© 2012 John Wiley & Sons, Ltd. Published 2015 by John Wiley & Sons, Ltd.

Roy Ladurie in the 1950s opened the first serious debates about climate and history in early modern Europe.[2] By 1958, measurements of rising atmospheric carbon dioxide began to support theories of a man-made greenhouse-gas effect, an idea first proposed in the 1890s. These discoveries spurred the development of increasingly detailed and sophisticated scientific measurements of past and present climate, which have continued ever since.[3] At the same time, a handful of scholars began the painstaking work of compiling and then verifying and quantifying written observations of past weather. By gathering and analyzing this data, researchers since the 1980s have forged a new interdisciplinary field known as "historical climatology," which seeks to reconstruct past climates and their role in human history.[4]

In general, historical climatologists have used three types of evidence: instrumental measurements, climate proxies, and written records. Weather instruments such as the thermometer offer the most accurate information, but these only date back to the seventeenth century and only in a few parts of the world. Therefore, most climate reconstructions rely on so-called proxies, which are physical records of past climates. For instance, dendroclimatologists drill cores from old trees to measure their annual growth rings. If the size of these growth rings lines up well with some weather variable – such as spring temperature or summer rainfall – then climatologists can calibrate the correlation with modern instrumental data and extrapolate back in time, using older growth rings as a "proxy" for preinstrumental measurements. Apart from tree rings, researchers now employ a wide variety of proxies, including ice cores, cave deposits, lake sediment layers, and buried pollen. In some instances historical climatologists can also make use of written sources describing the weather, which range from diaries to chronicles to official archives.

These different types of historical evidence present different strengths and weaknesses. For instance, tree-ring measurements are often very precise, but they usually only date back a few hundred years. Pollen samples and bore holes can offer thousands of years of data but only at a low resolution –the average temperature for each century but not each season. Ice cores, on the other hand, can provide thousands of years of data at high resolution, but they are found in very cold places where few people have lived. Written records can provide very specific information about weather events, but the descriptions are often subjective and hard to quantify. To overcome these limitations, historical climatologists must combine and compare a range of sources to form a complete and accurate picture.

While most climatologists focus on large patterns of global warming, researchers in historical climatology usually deal with particular climate events and weather systems that have had the greatest impact on human society. For instance, northern European countries were traditionally most affected by extreme winters or cold, wet summers that could ruin their grain crops. Inhabitants of Mediterranean and Middle Eastern lands, on the other hand, have been more vulnerable to spring droughts. Almost half the world's population, in South and East Asia, has traditionally depended on the success or failure of annual monsoon rains to grow crops like rice.

To understand changes and fluctuations in these weather patterns, researchers look for external factors (or "forcings") that can alter the earth's climate. Over very long timescales, for instance, changes in the earth's orbit known as Milankovitch cycles have played a role in the beginning and end of ice ages. Over shorter timescales, volcanic eruptions, slight variations in solar radiation, and now the emission of greenhouse gasses have all affected basic weather patterns. At the same time, natural fluctuations in important climate cycles such as the North Atlantic Oscillation, the El Niño–Southern Oscillation (ENSO) over the Pacific, and the migration of the Intertropical Convergence Zone (ITCZ) of rain and low pressure over the equator have all contributed to climate variability from year to year.

Over the past few decades, historical climatologists have made significant progress in reconstructing and understanding both long-term shifts and also shorter fluctuations and extremes in past climate. In the years ahead, the greatest challenge for the field will be to better integrate those discoveries into our understanding of global history. Presently the level of historical analysis often falls short of the level of climatological analysis, and popular historians and some climatologists have made hasty and dramatic conclusions about climate and the collapse of civilizations. Nevertheless, a growing number of archeologists and historians have now begun to rethink their subject in light of new information from historical climatology. Their work has continued to uncover the more subtle, contingent, and sometimes unexpected ways that climate has shaped the human past. In the following sections, this chapter outlines some major findings of historical climatology and current theories regarding the role of climate in human history.

Climate in Prehistory

Humans first evolved in an ice-age world far colder and drier than the present. For more than 90 percent of our species' existence, this unfavorable climate made farming and permanent settlements impossible, dictating our ancestral nomadic hunter-gatherer lifestyle. Extreme cold during a "glacial maximum" roughly 73,000 years ago – probably the result of a tremendous eruption of Mt. Toba – may even have produced the population and genetic bottleneck that led to emergence of modern *Homo sapiens* from our archaic predecessors.[5] It was these modern humans, sometimes known as Cro-Magnon, who emigrated from Africa and colonized the still glacial earth between c. 50,000 and 14,000 years ago.

Ice-age climate fluctuations probably played a major role in one of our early ancestors' greatest environmental impacts. As humans occupied a mostly steppe and tundra world, populations developed a culture of large-game hunting. Mounting archeological evidence suggests that some combination of human predation and climate fluctuations led to the extinction of many large genera of mammals and birds, including most of the largest species in Australia and the Americas, such as mammoths and giant ground sloths.[6]

The North American extinctions in particular corresponded with dramatic fluctuations in global temperature as the last ice age gave way to the current interglacial, known as the Holocene. This transition took place when slight changes in the earth's orbit warmed the northern hemisphere, causing glaciers to retreat in a positive feedback loop, as newly exposed land and sea released more CO_2 and absorbed more solar radiation, warming the earth more and causing yet more ice to melt. However, when the ice sheets collapsed too quickly, the rapid outflow of freshwater into the North Atlantic could shut off the warm Gulf Stream, which is driven by sinking salty water. These episodes, known as Heinrich Events, produced periods of sudden cooling called stadials. The largest stadial occurred around 18,000 to 14,700 years ago, followed by the warmer Bølling-Allerød interstadial, which was followed in turn by another cold spell called the Younger Dryas about 12,700 to 11,900 years ago.[7]

During warm episodes, rising global temperatures and precipitation fostered denser human populations and richer, more complex material cultures. Sometimes referred to as the Mesolithic, the period at the end of the Pleistocene and start of the Holocene witnessed the emergence of societies with more diverse resource use (such as fishing), more complex technologies (such as pottery), and eventually permanent settlements in place of nomadic hunting and gathering.[8] The relationship between climate change and the origins of agriculture, which began in the Fertile Crescent sometime after 10,000 BCE, remains more complex and contested. On the one hand, agriculture would have been all but impossible during the ice age. On the other hand, archeologists remain divided on whether agriculture was a natural outcome of more favorable climate and denser settled populations in the Holocene, or a specific social response to the onset of more difficult conditions during the Younger Dryas, which may have forced some populations to concentrate on gathering grains. Implicit in this debate is the issue of whether climate fluctuations are inherently harmful or whether such environmental challenges can spur creative, productive human responses.[9]

The first few millennia after the Younger Dryas marked a so-called climate "optimum" of warmth and strong monsoon rains, which reached even into the Sahara and Middle East. However, climatologists have discerned a significant cold and drought event around 8,200 years ago, and then a widespread deterioration of climate around 6,000 years ago. This Mid-Holocene Transition from roughly 4000 to 3000 BCE witnessed a shift to drier conditions in much of the world, including North Africa, the Middle East, the Indus region, and possibly northern China and Peru. At this time more arid conditions may have forced populations to migrate to fertile river valleys, perhaps driving the emergence of the world's first city-states and empires, such as Sumer and Harappa. This theory proves especially persuasive for the rise of civilization in Egypt, where climate refugees from the Sahara would have contributed to social stratification and agricultural specialization in the crowded Nile Valley.[10]

Climate and Crisis in the Ancient World

By 3000 BCE, global climate had settled more or less into modern patterns. Nevertheless, climate fluctuations on a smaller scale continued to play a major role in human history, especially in the first city-states and empires of the Middle East. As archeologists have observed, more complex political and economic systems, though usually more adaptable to small annual variations, risk greater catastrophe during serious climate fluctuations – much as settlers along a river can build levees to survive small floods but then risk getting trapped if the water overtakes them in a major deluge.[11]

The strongest evidence for such a climate disaster comes from around 2200 BCE. This episode, perhaps the result of a volcanic event, witnessed major droughts from North Africa through North India and possibly into China. Although some archeologists including Karl Butzer have cautioned against broad historical inferences, others such as Harvey Weiss have made a compelling case that serious climate deterioration drove economic and political crises in Egypt, Syria, Mesopotamia, and possibly parts of China and the Indus civilization of present-day Pakistan.[12] Other scholars have proposed a second major climate disaster in the late bronze age (c. 1200–1100 BCE) associated with mass migrations in both the eastern Mediterranean and northern China. However, this theory has not yet received as much evidence or support.[13]

By the late first millennium BCE, Europe and the Mediterranean had entered into another climate optimum. Evidence from European tree rings points to warm summers and more reliable spring rains from around 300 BCE to 250 CE, during the expansion and peak of the Roman Empire. Starting in the late third century CE, tree ring and speleothem (cave deposit) studies indicate that the climate became cooler, drier, and more erratic. Although there is little direct historical evidence, this climate deterioration may have played a role in Celtic and Germanic invasions and the crisis and collapse of the western Roman Empire over the following two centuries.[14] China, too, may have suffered from a period of colder climate during the collapse of the Han dynasty (206 BCE–220 CE) and the centuries of political fragmentation that followed.[15]

Other historical climatologists have discerned a worldwide climatic disaster in the 530s CE. A combination of ice-core data and Byzantine eyewitness accounts indicates that a major volcanic eruption launched a veil of dust and sulfates into the upper atmosphere and created unusual "dry fogs" in the Mediterranean region. This atmospheric anomaly may have been the cause of serious harvest failures in the Byzantine Empire. The weakened famine refugees may in turn have spread the so-called Plague of Justinian (541–2), which killed a large part of the Byzantine population. While the Byzantines were the worst affected, the impact of the eruption reached around the world: The same event has also been associated with serious droughts in the rising Maya civilization in the Yucatán, perhaps the cause of the so-called "Maya hiatus" of population loss and settlement abandonment c. 530–630 CE.[16]

Medieval Warm, Medieval Cold

Between the disasters of the 530s and the onset of the Little Ice Age 1,000 years later, climate shifts continued to play a significant role in human history. Whereas historians and climatologists once wrote of a "medieval warm" period, most have now adopted the term "medieval climate anomaly" to reflect the considerable complexities and variations of this era. While certain regions in certain periods did enjoy unusually warm and favorable climate, others proved less fortunate.[17]

The early Middle Ages witnessed episodes of severe cold in much of the northern hemisphere. Climatic deterioration and poor harvests likely played a role in the decline of population and agriculture in Europe's so-called dark age of the fifth and sixth centuries CE; and observers during the Carolingian period also recorded frequent storms and flooding.[18] In China, volcanic weather in the 630s and 930s brought severe cold to the north, causing famines in the Tang kingdom (618–907) and in neighboring Turkic empires.[19] Other volcanic eruptions may have contributed to a series of major droughts in Yucatán over the late ninth century, at the peak of the classic Maya civilization.[20] While much debated by scholars, mounting evidence suggests that these droughts, combined with pervasive food shortages and environmental pressures, precipitated widespread population loss and the abandonment of most Maya urban centers.[21]

The high Middle Ages, from around 1000 to 1300 CE, brought a period of unusually warm and favorable climate to northern Europe. While poor seasons occurred from time to time, generally dry summers and mild winters contributed to more reliable harvests, underpinning the population expansion and urban growth of the age. While not comparable to present global warming, European temperatures were on the whole distinctly higher than in the centuries immediately before or following. China, too, may have enjoyed a period of relative warmth throughout the later Song dynasty (960–1279).

Yet in other parts of the world, this period proved less favorable. In Central Asia, for instance, much of the era was marked by a strong Siberian high-pressure cell, creating unusually severe winters.[22] Historian Richard Bulliet has argued that this cooling drove the Seljuk Turks to invade Iran and Anatolia in the eleventh century;[23] and Chinese historical climatologists have found correlations between these periods of unusual cold and drought and nomad migrations on China's northern and western frontiers.[24] Furthermore, this period of general warming witnessed frequent strong "La Niña" conditions, which brought more reliable rains to much of South and East Asia, but also more frequent and severe droughts to the Americas.[25] In the Andes, ice-core and lake sediment data indicate that the climate turned more arid in the late ninth century. While the evidence is much debated, severe droughts and falling lake levels in the twelfth century may have brought about the collapse of the Tiwanaku civilization, whose agriculture relied on irrigation from Lake Titicaca.[26] (In subsequent centuries, the warming trend may have contributed to the rise of other Andean civilizations, such as the Inca, who relied on high-altitude terraced fields for food production.[27]) In the southwestern US, tree-ring studies reveal intense recurring droughts in the twelfth and thirteenth centuries, coming at the peak of the Hohokam and Anasazi civilizations. These droughts probably undermined their delicate irrigation systems for growing maize, which forced them to abandon their major population centers.[28]

Following this period of generally higher temperatures, much of the world underwent another phase of cooling in the early to mid-1300s. The change came most abruptly to northern Europe, which suffered a succession of extremely cold, wet springs and summers in the 1310s. The harsh weather brought a succession of bad harvests followed by widespread pestilence in cattle and sheep. In countries already facing strong population pressure from the previous centuries of growth, these disasters unleashed the Great Famine, with widespread mortality.[29] The ecological pressures, poor climate, and chronic malnutrition of the early fourteenth century have also been implicated in the rapid spread of the Black Death of the 1340s, in which a third of Europe's population may have died. More recently, historian Timothy Brook has identified a similar phase of climatic deterioration, harvest failures, and famines leading up to the Black Death and the collapse of the Yuan dynasty (1271–1368) in China.[30]

Perhaps the most closely studied episode of this period has been the crisis of Viking settlements in the North Atlantic. Taking advantage of the unusual warmth during the tenth and eleventh centuries, Scandinavian Vikings had sent expeditions across northern Europe and beyond, reaching Iceland and then Greenland and even Newfoundland. Wherever they landed, the colonists brought their European livestock and pastoral practices with them. The Vikings (and their sheep, pigs, and cattle) who settled in these remote outposts then encountered overwhelming hardships during cold spells of the 1300s and 1400s. First the Western and then the Eastern Greenland settlements either migrated or succumbed to starvation once their animals had died in the long winters. In Iceland, meanwhile, clearance for fuel and pasture had caused severe deforestation and erosion during the first period of settlement. In the cold fourteenth and fifteenth centuries, the Icelanders suffered serious famine and population loss, forcing them to diversify from pastoralism into fishing.[31] Therefore, some scholars have looked to the histories of Viking Greenland and Iceland as parables of the dangers of cultural conservatism and the importance of flexibility and adaptation to environmental changes. Others, however, have pointed instead to the Vikings' relative success in preserving these difficult colonies for more than three centuries.[32]

Along with the colder temperatures of the fourteenth and fifteenth centuries came a return to more persistent El Niño conditions and consequently monsoon failures in the

Pacific region. In Cambodia, the Angkor civilization with its vast temple complexes had relied on elaborate hydraulic works to support its rice agriculture. During the fourteenth and early fifteenth centuries, severe droughts may have undermined Angkor's irrigation and food production, precipitating foreign invasions and then the abandonment of the kingdom.[33] In the South Pacific, meanwhile, these El Niños may have helped Polynesian sailors colonize the last remote islands by slowing or even reversing the normal easterly trade winds. However, subsequent El Niño-related droughts may also have contributed to the crises that soon overtook many new island settlements once they had depleted their easiest available natural resources.[34]

The Little Ice Age

As the last great global climate anomaly before modern global warming, the Little Ice Age of the late sixteenth to early eighteenth centuries has received the most detailed research by both climatologists and historians. Over the past decades, a range of studies have confirmed the existence of a general cooling phase, lowering temperatures on average by perhaps 1 to 2 degrees Celsius (1.8 to 3.6 degrees Fahrenheit). The precise cause of the Little Ice Age remains uncertain, but a combination of solar forcing and volcanic activity could probably explain most or all of the climatic anomaly.[35] The Little Ice Age was not just cold but often highly variable: historical climatologists have now identified distinct phases and regional variations in climate throughout this period. While Europeanists have led the field, research on the Little Ice Age has now spread to other parts of the globe, opening new insights into the role of climate in the modern-era world.

During the late fifteenth and sixteenth centuries, an interval of relatively warm and stable climate had helped promote the growth of agriculture and population across Eurasia. By the late 1500s, populations in Europe, China, and the Middle East had more than recovered from the Black Death and were once again facing rising prices and shortages of land, leaving them vulnerable to the onset of the Little Ice Age. In northern Europe in particular, historians have used more detailed records of weather, prices, and vital statistics to demonstrate strong links among the cold wet springs and summers of the later sixteenth century and frequent crop failures, inflation, vagrancy, and high mortality.[36] Even relatively advanced European economies, such as that of England, suffered real famines during the 1580s and 1590s following poor grain harvests. Only the highly urbanized and commercialized Dutch population seems to have weathered the Little Ice Age without substantial economic or demographic losses. Detailed written sources from the Ming dynasty (1368–1644) also indicate an unfavorable climate shift during this period, bringing more droughts in the north and floods along the major river valleys of central and southern China.[37]

At the turn of the seventeenth century, following a major eruption of Mt. Huaynaputina in 1600, the Little Ice Age entered into one of its coldest phases.[38] In Russia, Chester Dunning has argued that a period of extreme winters, harvest failures, and famine in 1601–3 helped turned a political succession crisis into the widespread outbreak of vagrancy, violence, and civil war over the following decade known as the Time of Troubles.[39] The severe cold also reached North America, likely contributing to the high mortality of some of the first European settlements in present-day Canada and the US. English colonies at Roanoke (1585) and Jamestown (1607) also had the misfortune to begin during one of the deepest droughts in Virginia for the past millennium.[40]

Perhaps the worst climate-related crisis of this period occurred in the Ottoman Empire, which then ruled most of the Middle East, Balkans, and North Africa. The population of Ottoman lands had roughly doubled since the late 1400s, creating serious population pressure and inflation in some regions by the later sixteenth century. From the late 1560s to 1580s, the onset of the Little Ice Age brought a succession of severe winters and spring droughts, creating several waves of harvest failures and shortages. In the 1590s, the eastern Mediterranean underwent its longest drought in the past 600 years, causing widespread famine. At the same time, a disease of livestock wiped out most of the sheep and cattle in Anatolia, the Balkans, and the Crimea. Locked in a difficult war with the Habsburg Empire, the Ottoman state imposed high taxes and requisitions on the starving peasantry, fueling a major uprising in Anatolia called the Celali Rebellion (1596–1610). Recurring Little Ice Age drought and cold contributed to the widespread violence, flight, and famine that followed, which left much of the Ottoman countryside depopulated by the early 1600s.[41]

These events foreshadowed a wider outbreak of disasters in the mid-seventeenth century sometimes called the "general crisis." Recent research implicates the role of Little Ice Age climate anomalies in this contemporary wave of famines, wars, and rebellions across much of the globe.[42] China suffered decades of unprecedented cold, drought, and famine, which hastened the fall of the Ming dynasty and the conquest of the Manchu Qing in the 1640s. A third or more of the country's population may have died in the disasters.[43] In the West African Sahel, at the edge of the Sahara, serious recurring droughts disrupted agriculture and commerce. The range of tsetse flies, which are fatal to livestock, retreated south with the drier climate, depriving the region's farmers of their natural protection against pastoral invaders and desert raiders on horseback.[44] In western Europe, ongoing Little Ice Age weather events and harvest failures contributed to the high mortality of the Thirty Years War (1618–48) and possibly the outbreak of political disorders including the French Fronde and the English Civil War.[45]

The last major phase of the Little Ice Age from around 1680 to 1710 is often identified with the "Maunder Minimum" of low sunspot activity. Europe experienced some of its coldest winters of the past millennium, contributing to another wave of harvest failures and high prices. Scotland and Finland suffered severe cold and famines in the 1690s in which a tenth and fifth of their respective populations may have perished.[46] In the Ottoman Empire, a return of freezing winters and erratic precipitation brought more famine and unrest, derailing a potential recovery from the disasters of the late sixteenth and early seventeenth centuries.

The Little Ice Age also witnessed a high occurrence of strong El Niños, some bringing serious monsoon failures to South and Southeast Asia.[47] Mughal India lost millions in famines of the 1630s and 1680s. In Indonesia, a succession of droughts and epidemics aggravated a demographic and economic crisis of the mid-seventeenth century, as the Dutch East India Company seized control of the region's trade.[48] In parts of Spanish America, particularly Mexico and the American Southwest, the late 1500s and 1600s also brought a number of significant droughts, some leading to serious food shortages and popular unrest.[49]

Beyond these economic and political crises, historians have only just begun to explore the cultural dimensions of climate change in the early modern world. Some early modern European art, for instance, reveals the impression of harsh Little Ice Age winter landscapes, including Pieter Brueghel's famous *Hunters in the Snow*. Likewise, the second act of Shakespeare's *A Midsummer Night's Dream* describes contemporary volcanic weather anomalies, including hazy skies, warm winters, cold summers, harvest failures, and murrains, belying the play's lighthearted comedy. Recently, German historical climatologist Wolfgang Behringer has looked for further evidence of a Little Ice Age mentality in everything from changes in clothing and architectural styles to the rise of witchcraft trials and the spread of severe religious doctrines in the late 1500s.[50]

From the Little Ice Age to Global Warming

Over the 1700s and 1800s, the unusual cold of the Little Ice Age gave way to the more moderate temperatures of the early twentieth century, against which climatologists now measure the onset of global warming. However, volcanic events and strong El Niños produced several more episodes of severe erratic weather. For instance, the 1783 eruption of Laki in Iceland not only brought famine to that island, but created volcanic weather around the northern hemisphere, including unusual cold and harvest failures in Europe, especially Ireland, and in the northern US and Canada.[51] Following the eruption, an extreme El Niño event of the late 1780s and 1790s brought serious droughts to India, Japan, Mexico, Peru, and France, among other countries. Millions died of famine around the world; and these natural disasters likely contributed to popular uprisings including the French Revolution.[52] El Niño droughts of the 1790s also brought hardship to the first British settlers of Australia – a country that has struggled to practice agriculture in its unpredictable ENSO-influenced climate ever since.[53] During the 1810s, another wave of volcanic activity brought a brief return of Little Ice Age weather, culminating in the 1815 Tambora eruption and famous "year without a summer" in 1816. Parts of Europe and America witnessed frosts well into June and July, creating what one historian has described as "the last great subsistence crisis in the Western world."[54] Other major climate-induced disasters of the modern age include a series of major El Niños, tropical monsoon failures, and famines over the late 1800s that Mike Davis has termed "Late Victorian Holocausts,"[55] and the intense droughts of the 1930s that contributed to the American Dust Bowl.[56]

During the past century, however, human activities have overtaken natural variability as the leading cause of climate change. Although some climatologists such as William Ruddiman would trace human impacts on the environment all the way back to prehistoric deforestation and agriculture,[57] the most notable man-made effects have come from fossil fuels since the Industrial Revolution. By the 1890s, Swedish physicist Svante Arrhennius predicted that burning coal would create an atmospheric "greenhouse effect," and by the 1930s, British meteorologist Guy Stewart Callendar found data to argue that global warming had already begun. Since 1957, when measurements first began, the level of heat-trapping carbon dioxide in the atmosphere has risen from roughly 315 to 390 parts per million; and global temperatures have begun to rise swiftly beyond levels seen in the last millennium, a trend most visible since the 1990s. Projections for the twenty-first century point to a further rise of 2 to 4 degrees Celsius (3.6 to 7.2 degrees Fahrenheit), depending on further greenhouse-gas emissions, with consequences that could range from severe droughts to stronger storms to more frequent heat waves.[58]

Beset by scientific complexities and political controversies, global warming has thus far proven a difficult subject for historians. Nevertheless, some notable studies of twentieth-century climate change already point to some of the challenges to be faced and possible patterns for future climate crises. For instance, Michael Glantz and collaborators have emphasized the persistent problems of population pressure and short-term thinking driving unsustainable agriculture in semi-arid lands, now faced with the prospects of more severe and frequent drought.[59] In his history of melting glaciers and glacial lake outbursts in Peru, Mark Carey has stressed not only the immediate loss of lives and property in the disasters but the way in which these disasters have been used to promote outside political and economic agendas, without always addressing local risks and vulnerabilities.[60] As the record of climate change and impacts grows, the historiography of global warming will certainly expand as well, opening a new field of study for environmental historians.

Conclusion

Like other environmental factors from disease to natural disasters, climate has proven a powerful if often overlooked force in the human past. From prehistory to present times, climate changes have played a part in human evolution and migration, the rise and fall of civilizations, the success and failure of colonies, and the stability and crisis of states and empires. At times, climate changes have had a swift and decisive historical impact, especially in vulnerable populations and marginal environments. However, as this survey suggests, the role of climate in history has usually been more complex and contingent. The consequences of climate change and variability have depended on local environmental, economic, and political conditions, and, in many cases, the links between climate and history remain uncertain or unproven.

Even more so than other topics in environmental history, the study of climate in history has been and will be propelled by its contemporary relevance in a world of rising temperatures. Already, historical examples point to certain themes in the human experience of climate changes that may prove relevant for the future: the critical role of a few key weather patterns, the dangers of difficult environments, the social dimensions of climate disasters, and the importance of cultural and political flexibility in the face of change. In time, with more detailed and comparative studies of past climate, crisis, and adaptation, historians may be in a position to offer, if not specific policies, then at least parallels and parables for the challenges of global warming.

Notes

1   C. Glacken, *Traces on the Rhodian Shore: Nature and Culture in Western Thought from Ancient Times to the End of the Eighteenth Century*, Berkeley, University of California Press, 1967; J. Fleming, *Historical Perspectives on Climate Change*, New York, Oxford University Press, 1998.

2   G. Utterström, "Climatic Fluctuations and Population Problems in Early Modern History," *The Scandinavian Economic History Review* 3, 1955, pp. 3–47; E. Le Roy Ladurie, *Times of Feast, Times of Famine: A History of Climate since the Year 1000*, New York, Noonday Press, 1971.

3   On the history of global warming science, see S. Weart, *The Discovery of Global Warming*, Cambridge, MA, Harvard University Press, 2008.

4   For an overview of studies and methods, see e.g., R. Brázdil, C. Pfister, H. Wanner, et al., "Historical Climatology in Europe: The State of the Art," *Climatic Change* 70, 2005, pp. 363–430.

5   S. H. Ambrose, "Late Pleistocene Human Population Bottlenecks, Volcanic Winter, and Differentiations of Modern Humans," *Journal of Human Evolution* 34, 1998, pp. 623–51.

6   For a summary of evidence, see e.g., P. Koch and A. Barnosky, "Late Quaternary Extinctions: State of the Debate," *Annual Review of Ecology, Evolution, and Systematics* 37, 2006, pp. 215–50.

7   B. Fagan provides an overview for a popular audience in *The Long Summer: How Climate Changed Civilization*, New York, Basic Books, 2004. For recent theories of glacial cycles, see G. Denton, "The Last Glacial Termination," *Science* 328, 2010, pp. 1652–6.

8   For an excellent popular overview of Mesolithic climatology and archeology, see S. Mithen, *After the Ice: A Global Human History 20,000–5000* BC, Cambridge, MA, Harvard University Press, 2003.

9   For an overview of theories see A. M. Rosen, *Civilizing Climate: Social Responses to Climate Change in the Ancient Near East*, Lanham, MD, AltaMira Press, 2007.

10   N. Brooks, "Cultural Responses to Aridity in the Middle Holocene and Increased Social Complexity," *Quaternary International* 151, 2006, pp. 29–49; D. Anderson, K. Maasch, and D. Sandweiss, *Climate Change and Cultural Dynamics: A Global Perspective on Mid-Holocene Transitions*, London, Elsevier, 2007.

11   Fagan, *The Long Summer*.

12   H. Dalfes, G. Kukla, and H. Weiss (eds.), Third Millennium B.C. Climate Change and Old World Collapse, Berlin, Springer-Verlag, 1997. On China, see also C. An, "Climate Change and Cultural Response around 4000 cal. yr. bp in the Western Part of Chinese Loess Plateau," *Quaternary Research* 63, 2005, pp. 347–52.

13   For the Mediterranean see, e.g., R. Bryson, H. Lamb, and D. Donley, "Drought and the Decline of Mycenae," *Antiquity* 48, 1974, pp. 46–50; and J. Neumann, "Climatic Changes in Europe and the Near East in the Second Millenium BC," *Climatic Change* 23, 1993, pp. 231–45. For a further overview of climate shifts in the ancient Middle East, see A. S. Issar, and M. Z. Zohar, *Climate Change: Environment and Civilization in the Middle East*, Berlin, Springer, 2004. For China, see C. C. Huang, "Climatic Aridity and the Relocations of the Zhou Culture in the Southern Loess Plateau of China," *Climatic Change* 61, 2003, pp. 361–78.

14   U. Büntgen, W. Tegel, K. Nicolussi, et al., "2500 Years of European Climate Variability and Human Susceptibility," *Science* 331, 2011, pp. 578–82; and I. Orland, M. Barmatthews, N. Kita, et al., "Climate Deterioration in the Eastern Mediterranean as Revealed by Ion Microprobe Analysis of a Speleothem that Grew from 2.2 to 0.9 ka in Soreq Cave, Israel," *Quaternary Research* 71, 2009, pp. 27–35.

15   E.g., B. Yang, A. Braeuning, K. R. Johnson, and S. Yafeng, "General Characteristics of Temperature Variation in China during the Last Two Millennia," *Geophysical Research Letters* 29, 2002, pp. 1324–7.

16   For an overview of historical climatology in the 530s, see J. Gunn, *The Years without Summer: Tracing A.D. 536 and Its Aftermath*, Oxford, Archaeopress, 2000.

17   On the debate over medieval temperatures, see, e.g., R. S. Bradley, M. K. Hughes, and H. F. Diaz, "Climate in Medieval Time," *Science* 302, 2003, pp. 404–5.

18   Büntgen, Tegel, Nicolussi, et al. "2500 Years of European Climate Variability and Human Susceptibility"; F. Cheyette, "The Disappearance of the Ancient Landscape and the Climatic Anomaly of the Early Middle Ages: A Question to be Pursued," *Early Medieval Europe* 16, 2008, pp. 127–65; P. Dutton, "Thunder and Hail over the Carolingian Countryside," in D. Sweeney (ed.), *Agriculture in the Middle Ages: Technology, Practice, and Representation*, Philadelphia, University of Pennsylvania Press, 1995, pp. 111–37.

19   J. Fei, J. Zhou, and Y. Hou, "Circa ad 626 Volcanic Eruption, Climatic Cooling, and the Collapse of the Eastern Turkic Empire," *Climatic Change* 81, 2007, pp. 469–75; J. Fei and J. Zhou, "The Possible Climatic Impact in China of Iceland's Eldgjá Eruption Inferred from Historical Sources," *Climatic Change* 76, 2006, pp. 443–57.

20   For recent reconstructions of the Maya droughts see, e.g., D. A. Hodell, M. Brenner, and J. H. Curtis, "Climate and Cultural History of the Northeastern Yucatan Peninsula, Quintana Roo, Mexico," *Climatic Change* 83, 2007, pp. 215–40.

21 There is an extensive literature on the Maya collapse. R. Gill makes the case for drought in *The Great Maya Droughts: Water, Life, and Death*, Santa Fe, NM, University of New Mexico Press, 2000; D. Webster favors a multicausal environmental explanation in *The Fall of the Ancient Maya*, London, Thames and Hudson, 2002. For a recent synthesis, see J. Yaeger and D. A. Hodell, "The Collapse of Maya Civilization: Assessing the Interaction of Culture, Climate, and Environment," in D. H. Sandweiss and J. Quilter (eds.), *El Niño, Catastrophism, and Cultural Change in Ancient America*, Cambridge, MA, Harvard University Press, 2008, pp. 187–242.

22 R. D'Arrigo, et al., "1738 Years of Mongolian Temperature Variability Inferred from a Tree-Ring Width Chronology of Siberian Pine," *Geophysical Research Letters* 28, 2001, pp. 543–6.

23 R. Bulliet, *Cotton, Climate, and Camels in Early Islamic Iran: A Moment in World History*, New York, Columbia University Press, 2009.

24 E.g., J. Fang and G. Liu, "Relationship between Climatic Change and the Nomadic Southward Migrations in Eastern Asia during Historical Times," *Climatic Change* 22, 1992, pp. 151–69.

25 See, e.g., B. Rein, A. Lückge, and F. Sirocko, "A Major Holocene ENSO Anomaly during the Medieval Period," *Geophysical Research Letters* 31, 2004, L17211.

26 E.g., C. R. Ortloff and A. L. Kolata, "Climate and Collapse: Agro-Ecological Perspectives on the Decline of the Tiwanaku State," *Journal of Archaeological Science* 20, 1993, pp. 195–221; and M. Binford, "Climate Variation and the Rise and Fall of an Andean Civilization," *Quaternary Research* 47, 1997, pp. 235–48. For opposing views, see, e.g., C. L. Erickson, "Neo-Environmental Determinism and Agrarian 'Collapse' in Andean Prehistory," *Antiquity* 73, 1999, pp. 634–42; and P. R. Williams, "Rethinking Disaster-Induced Collapse in the Demise of the Andean Highland States: Wari and Tiwanaku," *World Archaeology* 33, 2002, pp. 361–74.

27 A. Chepstow-Lusty, "Putting the Rise of the Inca Empire within a Climatic and Land Management Context," *Climate of the Past* 5, 2009, pp. 375–88.

28 E.g., L. Benson, K. Petersen, and J. Stein, "Anasazi (Pre-Columbian Native-American) Migrations during the Middle-12th and Late-13th Centuries – Were They Drought Induced?," *Climatic Change* 83, 2007, pp. 187–213.

29 For a narrative, see W. Jordan, *The Great Famine*, Princeton, NJ, Princeton University Press, 1996.

30 T. Brook, *The Troubled Empire: China in the Yuan and Ming Dynasties*, Cambridge, MA, Belknap Press, 2010, chapter 3.

31 For an overview, see, e.g., the articles in W. Fitzhugh and E. Ward (eds.), *Vikings: The North Atlantic Saga*, Washington, DC, Smithsonian Institution Press, 2000.

32 Cf. J. Diamond, *Collapse: How Societies Choose to Fail or Succeed*, New York, Viking, 2005, and J. Berglund, "Did the Medieval Norse Society in Greenland Really Fail?," in P. A. McAnany and N. Yoffee (eds.), *Questioning Collapse: Human Resilience, Ecological Vulnerability, and the Aftermath of Empire*, New York, Cambridge University Press, 2010, pp. 45–70.

33 B. M. Buckley, K. J. Anchukaitis, D. Penny, et al., "Climate as a Contributing Factor in the Demise of Angkor, Cambodia," *Proceedings of the National Academy of Sciences* 107, 2010, pp. 6748–52.

34 P. D. Nunn, "Environmental Catastrophe in the Pacific Islands around A.D. 1300," *Geoarchaeology* 16, 2000, pp. 715–40, and P. D. Nunn, and J. M. Britton, "Human–Environment Relationships in the Pacific Islands around A.D. 1300," *Environment and History* 7, 2001, pp. 3–22.

35 E.g., T. Crowley, "Causes of Climate Change over the Past 1000 Years," *Science* 289, 2000, pp. 270–7, and M. Mann, "Global Signatures and Dynamical Origins of the Little Ice Age and Medieval Climate Anomaly," *Science* 326, 2009, pp. 1256–60.

36 For an overview of such research, see C. Pfister and R. Brázdil, "Climatic Variability in Sixteenth-Century Europe and Its Social Dimension: A Synthesis," *Climatic Change* 43, 1999, pp. 5–53.

37 Brook, *The Troubled Empire*.

38 S. L. De Silva and G. A. Zielinski, "Global Influence of the AD 1600 Eruption of Huaynaputina, Peru," *Nature* 393, 1998, pp. 455–8; A. Schimmelmann, M. Zhao, C. C. Harvey, and C. B. Lange, " A Large California Flood and Correlative Global Climatic Events 400 Years Ago," *Quaternary Research* 49, 1998, pp. 51–61.

39 C. Dunning, *Russia's First Civil War: The Time of Troubles and the Founding of the Romanov Dynasty*, University Park, PA, Pennsylvania State University Press, 2001.

40 D. W. Stahle, M. K. Cleaveland, D. B. Blanton, et al., "The Lost Colony and the Jamestown Droughts," *Science* 280, 1998, pp. 564–7; D. Blanton, "Drought as a Factor in the Jamestown Colony, 1607–1612," *Historical Archaeology* 34, 2000, pp. 74–81.

41 S. White, *The Climate of Rebellion in the Early Modern Ottoman Empire*, New York, Cambridge University Press, 2011.

42 For an overview, see G. Parker, "Crisis and Catastrophe: The Global Crisis of the Seventeenth Century Reconsidered," *American Historical Review* 113, 2008, pp. 1053–79.

43 Brook, *The Troubled Empire*. For regional studies, see also P. Perdue, *Exhausting the Earth: State and Peasant in Hunan, 1500–1850 AD*, Cambridge, MA, Harvard University Press, 1987; and R. B. Marks, *Tigers, Rice, Silk, and Silt: Environment and Economy in Late Imperial South China*, New York, Cambridge University Press, 1998. For physical records of climate change and correlation with dynastic cycles, see D. D. Zhang, C. Y. Jim, G. C.-S. Lin, et al., "Climatic Change, Wars and Dynastic Cycles in China Over the Last Millennium," *Climatic Change* 76, 2006, pp. 459–77.

44 G. E. Brooks, *Landlords and Strangers: Ecology, Society, and Trade in Western Africa, 1000–1630*, Boulder, CO, Westview Press, 1993; J. Webb, *Desert Frontier: Ecological and Economic Change along the Western Sahel, 1600–1850*, Madison, University of Wisconsin Press, 1995. For more recent reconstructions of West African precipitation, see T. Shanahan, "Atlantic Forcing of Persistent Drought in West Africa," *Science* 324, 2009, pp. 377–80.

45 See, e.g., E. Le Roy Ladurie, *Histoire humaine et comparée du climat, vol. 1, Canicules et glaciers (XIIIᵉ-XVIIIᵉ siècles)*, Paris, Fayard, 2004, chapter 8.

46 On the climatology of the late Maunder Minimum, see, e.g., J. Luterbacher, R. Rickli, E. Xoplaki, et al., "The Late Maunder Minimum: A Key Period for Studying Decadal Climate Change in Europe," *Climatic Change* 49, 2001, pp. 441–62. On the Scottish and Finnish famines see Le Roy Ladurie, *Canicules et glaciers*, chapter 9.

47 R. Grove and J. Chappell, "El Niño Chronology and the History of Global Crises during the Little Ice Age," in R. Grove and J. Chappell (eds.), *El Niño: History and Crisis*, Cambridge, White Horse Press, 2000, pp. 5–34.

48 A. Reid, "The Seventeenth-Century Crisis in Southeast Asia," *Modern Asian Studies* 24, 1990, pp. 639–59; P. Boomgaard, "Crisis Mortality in Seventeenth-Century Indonesia," in T. Liu, J. Lee, D. S. Reher, et al. (eds.), *Asian Population History*, New York, Oxford University Press, 2001, pp. 191–220.

49 See especially G. Endfield, *Climate and Society in Colonial Mexico*, London, Wiley-Blackwell, 2008.

50 W. Behringer, *A Cultural History of Climate*, Cambridge, Polity, 2010.

51 See, e.g., R. B. Stothers, "The Great Dry Fog of 1783," *Climatic Change* 32, 1996, pp. 79–89.

52 See R. Grove, "Global Impact of the 1789–93 El Niño," *Nature* 393, 1998, pp. 318–19.

53 J. Gergis, D. Garden, and C. Fenby, "The Influence of Climate on the First European Settlement of Australia: A Comparison of Weather Journals, Documentary Data and Palaeoclimate Records, 1788–1793," *Environmental History* 15, 2010, pp. 485–507; T. Sherratt, T. Griffiths, and L. Robin (eds.), *A Change in the Weather: Climate and Culture in Australia*, Canberra, National Museum of Australia Press, 2005.

54 J. D. Post, *The Last Great Subsistence Crisis in the Western World*, Baltimore, The Johns Hopkins University Press, 1977.

55   M. Davis, *Late Victorian Holocausts: El Niño Famines and the Making of the Third World*, New York, Verso, 2001.
56   S. D. Schubert, M. J. Suarez, P. J. Pegion, et al., "On the Cause of the 1930s Dust Bowl," *Science* 303, 2004, pp. 1855–9.
57   W. F. Ruddiman, *Plows, Plagues, and Petroleum: How Humans Took Control of Climate*, Princeton, NJ, Princeton University Press, 2005.
58   Intergovernmental Panel on Climate Change, *Fourth Assessment Report*, New York, Cambridge University Press, 2007.
59   M. H. Glantz (ed.), *Drought Follows the Plow: Cultivating Marginal Areas*, New York, Cambridge University Press, 1994.
60   M. Carey, *In the Shadow of Melting Glaciers: Climate Change and Andean Society*, New York, Oxford University Press, 2010.

References

Ambrose, S. H., "Late Pleistocene Human Population Bottlenecks, Volcanic Winter, and Differentiations of Modern Humans," *Journal of Human Evolution* 34, 1998, pp. 623–51.

An, C., "Climate Change and Cultural Response around 4000 cal. yr. BP in the Western Part of Chinese Loess Plateau," *Quaternary Research* 63, 2005, pp. 347–52.

Anderson, D., Maasch, K., and Sandweiss, D., *Climate Change and Cultural Dynamics: A Global Perspective on Mid-Holocene Transitions*, London, Elsevier, 2007.

Behringer, W., *A Cultural History of Climate*, Cambridge, Polity, 2010.

Benson, L., Petersen, K., and Stein, J., "Anasazi (Pre-Columbian Native-American) Migrations during the Middle-12th and Late-13th Centuries – Were They Drought Induced?," *Climatic Change* 83, 2007, pp. 187–213.

Berglund, J., "Did the Medieval Norse Society in Greenland Really Fail?," in P. A. McAnany and N. Yoffee (eds.), *Questioning Collapse: Human Resilience, Ecological Vulnerability, and the Aftermath of Empire*, New York, Cambridge University Press, 2010, pp. 45–70.

Binford, M., "Climate Variation and the Rise and Fall of an Andean Civilization," *Quaternary Research* 47, 1997, pp. 235–48.

Blanton, D., "Drought as a Factor in the Jamestown Colony, 1607–1612," *Historical Archaeology* 34, 2000, pp. 74–81.

Boomgaard, P., "Crisis Mortality in Seventeenth-Century Indonesia," in T. Liu, J. Lee, D. S. Reher, et al. (eds.), *Asian Population History*, New York, Oxford University Press, 2001, pp. 191–220.

Bradley, R. S., Hughes, M. K., and Diaz, H. F., "Climate in Medieval Time," *Science* 302, 2003, pp. 404–5.

Brázdil, R., Pfister, C., Wanner, H., et al., "Historical Climatology in Europe: The State of the Art," *Climatic Change* 70, 2005, pp. 363–430.

Brook, T., *The Troubled Empire: China in the Yuan and Ming Dynasties*, Cambridge, MA, Belknap Press, 2010.

Brooks, G. E., *Landlords and Strangers: Ecology, Society, and Trade in Western Africa, 1000–1630*, Boulder, CO, Westview Press, 1993.

Brooks, N., "Cultural Responses to Aridity in the Middle Holocene and Increased Social Complexity," *Quaternary International* 151, 2006, pp. 29–49.

Bryson, R., Lamb, H., and Donley, D., "Drought and the Decline of Mycenae," *Antiquity* 48, 1974, pp. 46–50.

Buckley, B. M., Anchukaitis, K. J., Penny, D., et al., "Climate as a Contributing Factor in the Demise of Angkor, Cambodia," *Proceedings of the National Academy of Sciences* 107, 2010, pp. 6748–52.

Bulliet, R., *Cotton, Climate, and Camels in Early Islamic Iran: A Moment in World History*, New York, Columbia University Press, 2009.

Büntgen, U., Tegel, W., Nicolussi, K., et al., "2500 Years of European Climate Variability and Human Susceptibility," *Science* 331, 2011, pp. 578–82.

Carey, M., *In the Shadow of Melting Glaciers: Climate Change and Andean Society*, New York, Oxford University Press, 2010.

Chepstow-Lusty, A., "Putting the Rise of the Inca Empire within a Climatic and Land Management Context," *Climate of the Past* 5, 2009, pp. 375–88.

Cheyette, F., "The Disappearance of the Ancient Landscape and the Climatic Anomaly of the Early Middle Ages: A Question to be Pursued," *Early Medieval Europe* 16, 2008, pp. 127–65.

Crowley, T., "Causes of Climate Change over the Past 1000 Years," *Science* 289, 2000, pp. 270–7.

Dalfes, H., Kukla, G., and Weiss, H. (eds.), *Third Millennium B.C. Climate Change and Old World Collapse*, Berlin, Springer-Verlag, 1997.

D'Arrigo, R., et al., "1738 Years of Mongolian Temperature Variability Inferred from a Tree-Ring Width Chronology of Siberian Pine," *Geophysical Research Letters* 28, 2001, pp. 543–6.

Davis, M., *Late Victorian Holocausts: El Niño Famines and the Making of the Third World*, New York, Verso, 2001.

De Silva, S. L., and Zielinski, G. A., "Global Influence of the AD 1600 Eruption of Huaynaputina, Peru," *Nature* 393, 1998, pp. 455–8.

Denton, G., "The Last Glacial Termination," *Science* 328, 2010, pp. 1652–6.

Diamond, J., *Collapse: How Societies Choose to Fail or Succeed*, New York, Viking, 2005.

Dunning, C., *Russia's First Civil War: The Time of Troubles and the Founding of the Romanov Dynasty*, University Park, PA, Pennsylvania State University Press, 2001.

Dutton, P., "Thunder and Hail over the Carolingian Countryside," in D. Sweeney (ed.), *Agriculture in the Middle Ages: Technology, Practice, and Representation*, Philadelphia, University of Pennsylvania Press, 1995, pp. 111–37.

Endfield, G., *Climate and Society in Colonial Mexico*, London, Wiley-Blackwell, 2008.

Erickson, C. L., "Neo-Environmental Determinism and Agrarian 'Collapse' in Andean Prehistory," *Antiquity* 73, 1999, pp. 634–42.

Fagan, B., *The Long Summer: How Climate Changed Civilization*, New York, Basic Books, 2004.

Fang, J., and Liu, G., "Relationship between Climatic Change and the Nomadic Southward Migrations in Eastern Asia during Historical Times," *Climatic Change* 22, 1992, pp. 151–69.

Fei, J., and Zhou, J., "The Possible Climatic Impact in China of Iceland's Eldgjá Eruption Inferred from Historical Sources," *Climatic Change* 76, 2006, pp. 443–57.

Fei, J., Zhou, J., and Hou, Y., "Circa AD 626 Volcanic Eruption, Climatic Cooling, and the Collapse of the Eastern Turkic Empire," *Climatic Change* 81, 2007, pp. 469–75.

Fitzhugh, W., and Ward, E. (eds.), *Vikings: The North Atlantic Saga*, Washington, DC, Smithsonian Institution Press, 2000.

Fleming, J., *Historical Perspectives on Climate Change*, New York, Oxford University Press, 1998.

Gergis, J., Garden, D., and Fenby, C., "The Influence of Climate on the First European Settlement of Australia: A Comparison of Weather Journals, Documentary Data and Palaeoclimate Records, 1788–1793," *Environmental History* 15, 2010, pp. 485–507.

Gill, R., *The Great Maya Droughts: Water, Life, and Death*, Santa Fe, NM, University of New Mexico Press, 2000.

Glacken, C., *Traces on the Rhodian Shore: Nature and Culture in Western Thought from Ancient Times to the End of the Eighteenth Century*, Berkeley, University of California Press, 1967.

Glantz, M. H. (ed.), *Drought Follows the Plow: Cultivating Marginal Areas*, New York, Cambridge University Press, 1994.

Grove, R., "Global Impact of the 1789–93 El Niño," *Nature* 393, 1998, pp. 318–19.

Grove, R., and Chappell, J., "El Niño Chronology and the History of Global Crises during the Little Ice Age," in R. Grove and J. Chappell (eds.), *El Niño: History and Crisis*, Cambridge, White Horse Press, 2000, pp. 5–34.

Gunn, J., *The Years without Summer: Tracing A.D. 536 and Its Aftermath*, Oxford, Archaeopress, 2000.

Hodell, D. A., Brenner, M., and Curtis, J. H., "Climate and Cultural History of the Northeastern Yucatan Peninsula, Quintana Roo, Mexico," *Climatic Change* 83, 2007, pp. 215–40.

Huang, C. C., "Climatic Aridity and the Relocations of the Zhou Culture in the Southern Loess Plateau of China," *Climatic Change* 61, 2003, pp. 361–78.

Intergovernmental Panel on Climate Change, *Fourth Assessment Report*, New York, Cambridge University Press, 2007.

Issar, A. S., and Zohar, M. Z., *Climate Change: Environment and Civilization in the Middle East*, Berlin, Springer, 2004.

Jordan, W., *The Great Famine*, Princeton, NJ, Princeton University Press, 1996.

Koch, P., and Barnosky, A., "Late Quaternary Extinctions: State of the Debate," *Annual Review of Ecology, Evolution, and Systematics* 37, 2006, pp. 215–50.

Le Roy Ladurie, E., *Histoire humaine et comparée du climat, vol. 1, Canicules et glaciers (XIIIᵉ-XVIIIᵉ siècles)*, Paris, Fayard, 2004.

Le Roy Ladurie, E., *Times of Feast, Times of Famine: A History of Climate since the Year 1000*, New York, Noonday Press, 1971.

Luterbacher, J., Rickli, R., Xoplaki, E., et al., "The Late Maunder Minimum: A Key Period for Studying Decadal Climate Change in Europe," *Climatic Change* 49, 2001, pp. 441–62.

Mann, M., "Global Signatures and Dynamical Origins of the Little Ice Age and Medieval Climate Anomaly," *Science* 326, 2009, pp. 1256–60.

Marks, R. B., *Tigers, Rice, Silk, and Silt: Environment and Economy in Late Imperial South China*, New York, Cambridge University Press, 1998.

Mithen, S., *After the Ice: A Global Human History 20,000–5000* BC, Cambridge, MA, Harvard University Press, 2003.

Neumann, J., "Climatic Changes in Europe and the Near East in the Second Millenium BC," *Climatic Change* 23, 1993, pp. 231–45.

Nunn, P. D., "Environmental Catastrophe in the Pacific Islands around A.D. 1300," *Geoarchaeology* 16, 2000, pp. 715–40.

Nunn, P. D., and Britton, J. M., "Human–Environment Relationships in the Pacific Islands around A.D. 1300," *Environment and History* 7, 2001, pp. 3–22.

Orland, I., Barmatthews, M., Kita, N., et al., "Climate Deterioration in the Eastern Mediterranean as Revealed by Ion Microprobe Analysis of a Speleotherm that Grew from 2.2 to 0.9 ka in Soreq Cave, Israel," *Quaternary Research* 71, 2009, pp. 27–35.

Ortloff, C. R., and Kolata, A. L., "Climate and Collapse: Agro-Ecological Perspectives on the Decline of the Tiwanaku State," *Journal of Archaeological Science* 20, 1993, pp. 195–221.

Parker, G., "Crisis and Catastrophe: The Global Crisis of the Seventeenth Century Reconsidered," *American Historical Review* 113, 2008, pp. 1053–79.

Perdue, P., *Exhausting the Earth: State and Peasant in Hunan, 1500–1850 AD*, Cambridge, MA, Harvard University Press, 1987.

Pfister, C., and Brázdil, R., "Climatic Variability in Sixteenth-Century Europe and Its Social Dimension: A Synthesis," *Climatic Change* 43, 1999, pp. 5–53.

Post, J. D., *The Last Great Subsistence Crisis in the Western World*, Baltimore, The Johns Hopkins University Press, 1977.

Reid, A., "The Seventeenth-Century Crisis in Southeast Asia," *Modern Asian Studies* 24, 1990, pp. 639–59.

Rein, B., Lückge, A., and Sirocko, F., "A Major Holocene ENSO Anomaly during the Medieval Period," *Geophysical Research Letters* 31, 2004, L17211.

Rosen, A. M., *Civilizing Climate: Social Responses to Climate Change in the Ancient Near East*, Lanham, MD, AltaMira Press, 2007.

Ruddiman, W. F., *Plows, Plagues, and Petroleum: How Humans Took Control of Climate*, Princeton, NJ, Princeton University Press, 2005.

Schimmelmann, A., Zhao, M., Harvey, C. C., and Lange, C. B., " A Large California Flood and Correlative Global Climatic Events 400 Years Ago," *Quaternary Research* 49, 1998, pp. 51–61.

Schubert, S. D., Suarez, M. J., Pegion, P. J., et al., "On the Cause of the 1930 Dust Bowl," *Science* 303, 2004, pp. 1855–9.

Shanahan, T., "Atlantic Forcing of Persistent Drought in West Africa," *Science* 324, 2009, pp. 377–80.

Sherratt, T., Griffiths, T., and Robin, L. (eds.), *A Change in the Weather: Climate and Culture in Australia*, Canberra, National Museum of Australia Press, 2005.

Stahle, D. W., Cleaveland, M. K., Blanton, D. B., et al., "The Lost Colony and the Jamestown Droughts," *Science* 280, 1998, pp. 564–7.

Stothers, R. B., "The Great Dry Fog of 1783," *Climatic Change* 32, 1996, pp. 79–89.

Utterström, G., "Climatic Fluctuations and Population Problems in Early Modern History," *The Scandinavian Economic History Review* 3, 1955, pp. 3–47.

Weart, S., *The Discovery of Global Warming*, Cambridge, MA, Harvard University Press, 2008.

Webb, J., *Desert Frontier: Ecological and Economic Change along the Western Sahel, 1600–1850*, Madison, University of Wisconsin Press, 1995.

Webster, D., *The Fall of the Ancient Maya*, London, Thames and Hudson, 2002.

White, S., *The Climate of Rebellion in the Early Modern Ottoman Empire*, New York, Cambridge University Press, 2011.

Williams, P. R., "Rethinking Disaster-Induced Collapse in the Demise of the Andean Highland States: Wari and Tiwanaku," *World Archaeology* 33, 2002, pp. 361–74.

Yaeger, J., and Hodell, D. A., "The Collapse of Maya Civilization: Assessing the Interaction of Culture, Climate, and Environment," in D. H. Sandweiss and J. Quilter (eds.), *El Niño, Catastrophism, and Cultural Change in Ancient America*, Cambridge, MA, Harvard University Press, 2008, pp. 187–242.

Yang, B., Braeuning, A., Johnson, K. R., and Yafeng, S., "General Characteristics of Temperature Variation in China during the Last Two Millennia," *Geophysical Research Letters* 29, 2002, pp. 1324–7.

Zhang, D. D., Jim, C. Y., Lin, G. C.-S., et al., "Climatic Change, Wars and Dynastic Cycles in China Over the Last Millennium," *Climatic Change* 76, 2006, pp. 459–77.

Further Reading

Historical climate change is a highly diverse multidisciplinary field with now thousands of publications scattered among many periodicals and presses. Key articles on past climate regularly appear in major scientific journals such as *Nature* and *Science* and in specialist periodicals such as *Climate Change* and *Climate of the Past*. The references for this chapter provide a small representative sample. Unfortunately, there is still no standard introductory text for students. The pathbreaking work of H. Lamb, *Climate, History, and the Modern World* (London, Routledge, 1995), is now somewhat dated. W. Behringer's *A Cultural History of Climate* (Cambridge, Polity, 2010) covers mostly European climate history, with a focus on the cultural history of the Little Ice Age. Brian Fagan has written several short histories (of uneven quality) summarizing research on different periods of climate change for a popular audience. Another popular history – J. Diamond's *Collapse* (New York, Viking, 2005) – argues for the role of climate in the downfall of Viking Greenland, the Maya, and the Anasazi. Several of the best European histories remain untranslated: R. Glaser's *Klimageschichte Mitteleuropas* (Darmstadt, Primus Verlag, 2001) covers Germany; C. Pfister's *Wetternachhersage: 500 Jahre Klimavariationen und Naturkatastrophen* (Bern, Paul Haupt, 1999) examines Switzerland; and E. Le Roy Ladurie *Histoire humaine et comparée*, 3 vols. (Paris, Fayard, 2004–9), offers a broad survey of climate and history in Europe over the last millennium. The extensive research of environmental historian Richard Grove and various collaborators has explored the role of El Niños in history, especially in South and Southeast Asia. Among other environmental and political histories emphasizing the role of climate, see (in the reference list) Marks (1998) and Brook (2010) on China, Brooks (1993) and Webb (1995) on Africa, and Bulliet (2009) and White (2011) on the Middle East.

CHAPTER TWENTY-THREE

Industrial Agriculture

MEREDITH MCKITTRICK

Introduction

Over the course of the past century, humanity's relationship to its food supply – once the most visible aspect of our relationship to our environment – has radically changed. From 1900 to 2000, the world's human population grew from 1.7 billion to 6 billion. Yet, by the end of the twentieth century, the world had so much food that the number of overnourished people surpassed, for the first time in history, the number of undernourished people. This was accompanied by a major shift in human diets toward animal products and foods that are highly processed and high in sugars and saturated fats – a shift that required vast amounts of energy inputs relative to the calories ultimately made available for human digestion. Food consumers spent a smaller proportion of their incomes on food than ever before, and food producers were fewer in number than ever before. A world that had been overwhelmingly rural in 1900 had, by 2000, nearly half of its human population residing in urban areas.

This ocean of calories was produced on a sea of fossil fuels. Indeed, it was the availability of cheap fossil fuels, and the technology to utilize them, that paved the way for all other changes: the mechanization of planting, cultivating, and harvesting; the manufacture of synthetic fertilizers and pesticides; a sixfold increase in land under irrigation; the removal of vast numbers of people from agricultural work; the refrigerated storage of food and its transportation over long distances; the spread of new high-yield crops; and the processing of what farmers harvest into products ranging from canned tomato sauce and cheese to chicken nuggets and Coca-Cola. For most of its history agriculture's energy base was the solar energy available for photosynthesis in a given growing season. But agriculture now relies on the stored sunlight of past eons as well as the sunlight we see today.

Our numbers on the planet and where those numbers dwell, how we extract sustenance from our environment and the impact of such extraction on the world's ecosystems – all

A Companion to Global Environmental History, First Edition. Edited by J.R. McNeill and Erin Stewart Mauldin.
© 2012 John Wiley & Sons, Ltd. Published 2015 by John Wiley & Sons, Ltd.

of these have been transformed by changes in food production over the past century. While the first global agricultural revolution, the one of 8,000–5,000 BCE, also altered ecosystems and human population maps, the impacts of the more recent revolution differ in kind as well as in degree. Biology has been subordinated to the logic of industrial production, and a rapidly declining number among us know how to produce our own sustenance.

The industrialization of agriculture required more than just technological change. It was also driven by a changing mindset that began to view farming as another industry, akin to making cars or textiles. Agriculture was brought under the same ethos of simplification, bureaucratization, and legibility that was a hallmark of the twentieth century.[1] Agriculture has been largely marginalized in studies of industrialization except when it is viewed as a contributing factor.[2] This is in part because agriculture seems less important in industrial nations when measured as a percentage of economic output. But the environmental changes that are considered direct outgrowths of the Industrial Revolution are all rooted first and foremost in our ability to feed ever more non-farmers with the output of ever fewer farmers, and to do so cheaply and reliably.[3]

Industrial agriculture developed within a specifically American context of land abundance combined with robust industrial development and a government that actively shaped the direction of both agriculture and scientific research. Until the 1940s, those circumstances kept industrial agriculture centered in the US and they ultimately foreclosed other possible routes toward increasing the global harvest. In China, for example, farmers in the nineteenth and early twentieth centuries applied waste products to restore soil fertility. In many parts of the world, farmers used polycultural systems – growing more than one crop in a given plot – to deter pests, minimize loss of soil nutrients, and obtain per-acre yields far beyond those achieved by American farmers.[4] But such systems required the one input most lacking in the US: labor. That modern agricultural systems rely on monocultures of staple crops and high levels of chemical inputs ultimately, if indirectly, derives from this fact.

The twentieth-century transformation of world agriculture has had plenty of critics. Yet those nations that have not industrialized their agricultural sector are by and large those that enjoy the least food security today. The question is whether our current food supply's reliance on fossil fuels and the simplification of ecological processes will ultimately leave all of humankind more food insecure over the longer term.

Origins

The components of this new system of food production arose primarily in Europe. In seventeenth- and eighteenth-century northern Europe, changes in crop-rotation practices and increased use of legumes as fodder crops enhanced the food supply, allowing the feeding of more non-farmers, freeing labor to work in factories, and ultimately contributing to the industrialization of manufacturing.[5] Gregor Mendel's forays into plant genetics and Justus von Liebig's identification of the chemical basis of plant nutrition in the nineteenth century led to modern plant breeding and the manufacture of synthetic nitrogen and other fertilizers in the twentieth century.[6] In the meantime, guano from Peru and sodium nitrate mined in Chile offered American and European farmers who could afford them a way to augment nitrogen in their soils. The feeding of oilseed cakes, an industrial by-product, to livestock allowed farmers to keep more animals and thereby increased the availability of manure for fertilizer. Such agricultural intensification made sense in places where land was scarce and prices for farm products were high. Between 1830 and 1880, British farmers increased their purchases of feed tenfold and their purchase of inorganic fertilizers by a factor of 30.[7]

Across the Atlantic, farmers in the nineteenth-century US were moving into the seemingly endless fertile lands of the West. With abundant soils and scarce labor, these farmers had little use for expensive artificial fertilizers or systems of recycling organic waste: ensuring the long-term fertility of the prairies they were plowing was one of the last things on their minds. Instead, they experimented with the other major thrust of agricultural change in the nineteenth century – mechanization – in order to increase yields per unit of labor. Between 1830 and 1860, horse-drawn reapers and mowers – and later threshers and binders – replaced scythes, cradles, and massive quantities of human labor in the harvesting of wheat and hay; horse-drawn seed drills planted wheat and corn; and horse-drawn plows were adapted to turning over prairie soils. These were followed by steam-powered machinery, adopted in the flat, semi-arid lands of California and the Great Plains.[8]

The impact of mechanization was limited until the internal-combustion engine was applied to farm machinery. But mechanization was what Deborah Fitzgerald has called the "opening wedge" of industrial agriculture.[9] Machines were expensive, and the only way to justify their expense was to use them to their fullest extent. Agricultural machinery was also specialized to particular crops and intolerant of uneven terrain. Machinery therefore dictated massive monocultures on level terrain, growing a crop that could be consumed even when harvested in immense quantities all at once. Wheat fit the bill: a staple that did well on the flat, semi-arid lands of the West, wheat was singularly amenable to mechanical harvesting and processing, and it was storable and transportable.

If steam power had a limited impact on mechanization, it revolutionized transportation. The expanding infrastructure of railroads and steamships allowed this expanding American harvest to travel the globe and resulted in an increasingly commercialized agriculture. American farmers who sold 30 percent of what they harvested in 1820 were selling 60 percent of their harvest by 1860.[10] And the trend was just beginning: the years after 1860 saw vast expansion in American agriculture, as hundreds of millions of acres were put under the plow. At the same time, the revolution in transportation allowed other wheat farmers to enter the global market: Australia, Canada, Argentina, and Russia joined the US in exporting wheat and, with the exception of Russia, did so by embracing the new machines.

These processes were more than the sum total of the individual decisions of countless farmers. US farm policy was grounded in deeper ideas about national identity and destiny. Until 1890, it focused on distributing land as cheaply as possible to as many people as possible. This attachment to the notion of a nation of small farmers, an agrarian idealism traceable to Thomas Jefferson, not only motivated the 1862 Homestead Act and related policies that gave settlers access to public land at low or no cost; it also has been reflected in continued and consistent public support for government assistance to farmers even as small farms have given way to agribusiness. But support for agriculture also emerged, particularly after 1890, in the context of a society in which farm productivity and incomes were declining relative to those in the industrial sector. The 1862 establishment of the US Department of Agriculture and the land-grant system of agricultural colleges was followed by federal support for state agricultural experimental stations in the 1870s. In the early twentieth century, the government created the Cooperative Extension Service to educate and train farmers, it exempted farm cooperatives from anti-trust legislation, and it paid for the construction of farm-to-market roads and rural electrification programs. Beginning in 1933 (and continuing to the present), it guaranteed farm incomes in the face of rising surpluses and global competition.[11]

Governments in Europe and Japan also established agricultural research stations and educational services for farmers in the late nineteenth and early twentieth centuries, a new kind of state intervention in food production. Rather than simply taxing the harvest, as states had done for millennia, national governments in the late nineteenth and early twentieth centuries began to insert themselves into the process of producing food, seeking to shape how farmers farmed. These new institutions promoted a view of farming as properly the domain of "experts" – scientists, engineers, bureaucrats – and a new kind of farmer, who was to be a manager and bookkeeper as well as a cultivator or a shepherd. This was a vision of what Deborah Fitzgerald calls "the industrial ideal" in agriculture.[12]

This vision was not widely embraced on the farm. When Tom Campbell, who farmed wheat on 100,000 acres of Montana using a veritable fleet of tractors, proclaimed in 1919, "Modern farming is 90% engineering and 10% agriculture," few farmers anywhere in the world would have agreed with him.[13] New ideas about what farming should look like were formed off the farm by non-farmers working in factories, new government agencies, and land-grant universities. American farmers in the early 1900s enjoyed historically high prices for their harvests and possessed little incentive to change the way they farmed. Many feared that the practices advocated by "experts" would increase their dependency on institutions beyond their control. This was a "resisted revolution."[14] And so, from southern cotton regions where labor remained cheap and mechanization ill-suited to the demands of the dominant crop, to diversified farms in the Midwest and in the rocky, hilly Northeast where machines designed for monoculture and flat terrain were impractical, most farmers changed little about how they farmed.[15] Even after the application of the combustion engine to farm machinery around 1900, it would be years before corn benefited from the technologies that had reshaped wheat cultivation, decades before cotton and rice did, and still more decades before tomatoes or beef cattle could be brought into a system of mechanized, capital-intensive, highly managed production.[16]

In 1920 and 1921, as Europe's agricultural sector recovered from war, the short-lived golden age of American farming collapsed and farmers who had borrowed money to buy land and livestock during the good times found themselves insolvent.[17] The spate of farm failures that followed – akin to the farm crisis that hit the US in the 1980s – left banks holding vast tracts of virtually worthless agricultural lands. The notion that agriculture was the last major economic sector where chaos and inefficiency reigned and that ignorant and unskilled farmers were to blame took root in the consciousness of economists and businessmen, urban residents and scientists. In their quest to "fix" farming, they looked to the factory. In agriculture that would mean scaling up production, standardizing procedures and products, treating farmers as managers rather than artisans, mechanizing as much of the process as possible, and measuring success in terms of efficiency of labor and other inputs.[18]

Marked improvements in farm equipment powered by combustion engines sped this process. American farms had 10,000 tractors in 1910; in 1925, 158,000 tractors were sold.[19] Widespread rural electrification also played a transformative role: "grain and fruit dryers, electric incubators, milking machines, and electrical lighting in farm buildings and yards enabled countless farmers to automate, speed up, or standardize farm tasks, which in turn required farmers to take on more obligations."[20] Individual farmers increased output to offset low prices, but this further drove down prices. It was a logic that would continue (eventually with the backing of taxpayer money) and that would ultimately dictate much about how industrial agricultural practices and the changing human diet were linked. In the short term, it placed farmers who did not purchase new equipment at a distinct disadvantage, while sending those who did into debt and forcing them to cultivate more land to pay for such investments. Farm failures and farm consolidations continued, and the average size of the American farm began to grow. By 1930, the trajectory of agriculture's industrialization had been established.[21]

Reshaping Nature

Mechanization opened vast tracts of land to farming in the US, Canada, Australia, Argentina, and, as late as the 1950s, in the Soviet Union. With fewer farmers needed to cultivate more acreage, rural populations and social institutions declined. From 1920 to 1980, the US population more than doubled, but the number of farmers fell by 24 million. The average farm size rose from 139 acres in 1910 to 444 acres in 2005.[22] This pattern of fewer farmers on larger farms was eventually replicated in virtually every modernizing agricultural economy except Japan.[23]

But mechanization did not in itself fundamentally change the biological processes that underlay agriculture. That required chemical inputs, genetic changes in the plants and animals farmers produced, and new methods of managing those plants and animals. The industrial inputs went far beyond the fossil fuels that powered tractors and transportation systems, and the industrial technologies were ultimately as much organizational as material.

The major breakthrough in chemical use came in the 1910s with the discovery of nitrogen fixation via ammonia synthesis – the so-called Haber–Bosch process, named after the German chemists who invented it and made it commercially viable.[24] It required only a cheap supply of electricity – and thus a cheap supply of fossil fuels. Since the advent of agriculture, nitrogen had been the primary nutrient that limited what a piece of land could yield, and there was no easy way to increase it. Manures and other organic wastes contained only small quantities of nitrogen, and the sodium nitrate mined in Chile in the nineteenth century was expensive and finite. Nitrogen makes up most of the Earth's atmosphere, but only with the Haber–Bosch process could that atmospheric abundance be converted into a form plants could use. The results revolutionized food production: Kenneth Pomeranz has argued that chemical fertilizers, particularly nitrogen, "may be the 20th century's most important invention" while Vaclav Smil, in his study of nitrogen's role in food production, argues that 40 percent of the world's population is now dependent for its very existence on the synthetic nitrogen that boosts crop yields while seemingly reducing the need for fallowing land.[25]

Nitrogenous fertilizer could not accomplish these things by itself, however; it worked alongside other technologies. Among these were pesticides, herbicides, insecticides, and fungicides. During World War II, DDT was used to control malaria and typhus among Allied troops, while 2,4-D, a broadleaf herbicide, was used to increase wartime crop yields in Britain. These were joined by a proliferation of new chemicals through the 1950s and beyond. Fertilizer use in the US rose from 800,000 metric tons in 1946 to 17 million metric tons in 1947. Between 1960 and 1995, these trends went global: nitrogen fertilizer use rose by a factor of 7, and phosphorous by a factor of 3.5; the usage of each is likely to triple again by 2050.[26] Pesticide use doubled globally over roughly the same three decades, and tripled in the US. Agrochemicals have been key to growing world food supplies not just because they raise yields per acre but because they have largely allowed farmers to end crop rotations without apparent adverse effects in the short term, although this will likely not be true of the longer term.[27]

A third component of industrialization in agriculture emerged within the biological forms themselves. From beets, whose "seeds" are actually seedpods that germinate as clusters of plants requiring extensive thinning, to cotton, whose bolls grew low to the ground and collected dirt and weeds that clogged early mechanical harvesters, plants had to be remade for machinery. Scientists hybridized corn to increase yields (with adequate applications of nitrogen fertilizer), resist disease, and withstand the rigors of mechanical harvesting. Breeders created sugar beets that produced single seeds rather than pods to simplify the process of thinning. They bred cotton plants that produced bolls at heights accessible to machines and tomato plants whose fruits ripened simultaneously to enable mechanical harvesting.[28]

All these successes were also projects of simplification: for plants to come under a regimen of mass production, variety needed to be eliminated. An individual farm laborer could assess an individual corn stalk or tomato plant and choose which ears or fruits to harvest; a machine treated every plant in the field in the same way and thus every plant in the field needed to behave the same way. The thousands of plant varieties that represented the product of millennia of farmers selecting for characteristics specific to particular places and preferences were gradually abandoned in favor of a mere handful of varieties that met the needs of an industrial system; 90 percent of the world's food supply today comes from just 30 varieties of plants.[29]

Performing similar feats of standardization for animals was more difficult, requiring changes to the ways in which they grew and reproduced. The goal, ultimately, was an animal that could satisfy consumer demand for a steady supply of inexpensive meat with particular characteristics. Breeding and management focused on maximizing weight gain, minimizing feed, and standardizing body structures and growth rates. Artificial technologies of reproduction meant that the seasonality of meat availability could give way to year-round abundance.

This process again began in the US, and it began with chickens. Prior to the 1920s chickens had been bred – when they were consciously bred at all – for egg, not meat, production. Chicken was a delicacy in the American diet, more expensive than steak or lobster. Americans consumed far less chicken than beef or pork; it was only in 1990 that American poultry consumption passed beef consumption. A "dressed" chicken from the 1920s would look unfamiliar to most Americans today – not least because Americans don't generally see poultry meat in its whole form; only 13 percent of chicken is now sold as whole birds.[30] The 1920s-era chicken would have been scrawny – less than 1.4 kilograms (3 pounds) while it was still alive – narrow-breasted, and tough to a modern palate. A combination of publicly funded science and private investment resulted by the 1950s in a new kind of broiler chicken. A single breed now dominates the commercial market: 65 percent larger at slaughter than its 1920 counterparts, it reaches slaughter weight in 7 weeks rather than 16. To cater to American tastes for white meat, the percentage of muscle tissue that is breast meat doubled.[31]

The rapid growth and the accumulation of muscle tissue around the breast came at the expense of internal organ and joint development: modern broiler chickens are susceptible to a host of musculoskeletal, metabolic, and immunodeficiency problems. These are birds unsuited to a barnyard. Industrializing meat production thus involved far more than altering the genetics of the animal. The management of production was itself transformed to meet the requirements of a more fragile animal but also the requirements of efficiency that large-scale production and narrow profit margins demanded.[32] Chickens were moved indoors to temperature-controlled sheds. Intensive confinement meant protection from the vagaries of nature; it also meant less energy expended on movement for animals and farmers alike, lowering animal feed costs and human labor costs. It increased the number of animals a given farmer could raise, as long as feed could be trucked in, and it greatly increased the external energy requirements of poultry production.

The patterns in chicken production were soon mirrored in beef and pork production. Intensified breeding and changes in animal management resulted in faster-maturing cattle and pigs whose meat better suited consumer preferences. By the 1950s, cattle raised to about 320 kilograms (700 pounds) on grass were being brought to feedlots to be fattened on grain for slaughter. Feedlots were near sources of pasture to minimize cattle weight loss in transit, and they were therefore far from areas of grain production. Rather than hauling massive quantities of grain across the Midwest to these feedlots, farmers increasingly began tapping aquifers to irrigate nearby lands for grain production. In Kansas between 1950 and the 1970s, the quantity of irrigated land rose tenfold.[33] Pork production remained centered in the Midwest, but it too came to rely upon confinement systems as pigs were put "on concrete" and in specially designed pens to minimize both high piglet mortality and the labor investment farmers had to make in moving pigs and water supplies between pastures. As with cattle feedlots and large-scale poultry houses, moving pigs indoors allowed the model of the assembly line to be applied to livestock production. "Work simplification specialists" produced systems for saving farm labor, and the hogs themselves in effect became workers whose time was to be used to maximum advantage. By the 1960s pigs had largely disappeared from pastures in the US.[34]

The commercial livestock industry therefore came to rely not just on genetic changes but also on new technologies of managing animals, new knowledge about animal growth and nutrition, and government policies that created incentives for specific kinds of agricultural practices. Electricity controlled the climate, lighting, and automatic feeders in vast new confinement systems; it ran milking machines on dairy farms, and it ran the immense hatcheries that provided a year-round supply of broiler chicks to farmers. Research into animal nutrition, much of it government-funded, led to the creation of commercial animal feeds from grain made abundant and cheap by government price supports. Livestock producers mixed antibiotics into feed to reduce disease susceptibility among crowded animals, but most importantly to promote faster growth – a practice that was nearly universal for all US livestock by the 1970s. In 1954, 222,000 kilograms (490,000 pounds) of antibiotics were added to livestock feed; by the late 1990s, half of the 22.3 million kilograms (50 million pounds) of antibiotics produced in the US each year were used in the livestock industry, mostly to promote rapid growth.[35]

These emerging models of industrial livestock production were capital-intensive; like the industrialization of crop production, they tended to lead to larger farm operations and consolidation across sectors. The 181,400 hog farms in Iowa in 1940 had dwindled to 10,500 by 2000.[36] And there was vertical integration in the entire system of livestock production: hog and poultry farmers increasingly worked as contractors for – and under the supervision of – large agro-industrial firms that also processed and marketed the meat. One result was an abundance of cheap meat on the American table. Chicken was by the 1990s often cheaper than the potatoes it was served with.[37] Ultimately, this system of meat production – and the increasing prevalence of meat in human diets – would spread to other parts of the world alongside other aspects of industrial agriculture.

The success of each innovation rested on the others. Mechanization required monocultures and standardized plants and animals. Monocultures decreased biodiversity on farms, making crops more susceptible to diseases and pests, and reduced soil fertility that had once been replenished through polyculture, fallow, and crop rotations. New varieties of plants and animals, bred for a single characteristic such as high yield or rapid maturation, were often deficient in other respects. Chemical inputs in the form of inorganic fertilizers, pesticides, or antibiotics compensated for these deficiencies.[38] These inputs could begin to operate as a sort of treadmill. For reasons that are still debated, ever-larger quantities of fertilizers have to be applied to maintain high yields. Meanwhile, insecticides eliminated not only damaging species but also those species that preyed on the pests; the resulting collapse in biological controls tied farmers further to the use of insecticides. The result was that farmers spent increasing quantities of their income on capital inputs – even the hybrid seed had to be purchased anew each year from the growing agribusiness sector – and increased outputs by cultivating still larger farms to pay for them. In 1900, US farmers spent 45 percent of their farm income on external inputs – a high percentage relative to most of the rest of the world; by 1990, the figure exceeded 80 percent.[39] Globally, pesticide use has perhaps doubled since the 1980s, while nitrogen fertilizer use rose by a factor of 7 between 1960 and 1995.[40] Resistance – of weeds and insects alike – to agrochemicals resulted in higher levels of application. American farmers applied 22.3 million kilograms (50 million pounds) of insecticide to their crops in 1948 and lost 7 percent of their crop to insects; in 2001, they applied 453.6 million kilograms (1 billion pounds) of insecticide and lost 13 percent of their crop to insects.[41]

This massive increase in agrochemicals is responsible for the most pernicious environmental consequences of industrial agriculture. The catastrophic impact of DDT on wild bird populations, made infamous by Rachel Carson's *Silent Spring* in 1962, ultimately led to the banning of DDT in the US. (It is still used and manufactured in other countries, including China and India, mainly for mosquito control.) But all pesticides and synthetic fertilizers have significant ecological impacts. The most widely used herbicide in the world remains 2,4-D; among its unintended consequences is increased infestation of farmers' fields by weedy grasses, necessitating the development of other, often more toxic and more persistent, herbicides. It has also been implicated in hermaphroditism among frogs and remains a suspected carcinogen. Glyphosate (Roundup) is now the most popular herbicide in the US, and corn has been bred for glyphosate resistance. But – another unintended consequence – weeds, too, are now developing resistance, leading to another escalation of the battle between industrial agriculture and Mother Nature. The same process is occurring with insecticides.

Meanwhile, agrochemicals have major impacts on water systems, polluting groundwater, rivers, and oceans alike. Less than half the nitrogen and phosphorous applied to fields contributes to plant growth; the rest washes into waterways. Agriculture is the main source of groundwater pollution in the US. The combination of agrochemical runoff and soil erosion that suspends solids in water has created dead zones in marine systems around the world and is the major threat to the world's most famous coral reefs.[42]

Industrial Agriculture Goes Global

The full scale of industrial agriculture's environmental impact became apparent only after its impressive gains in yield per acre and per labor hour had drawn attention from other parts of the world. The successes and failures of efforts to replicate these gains elsewhere demonstrate how the component parts of industrial agriculture are connected, as well as what is required to adapt industrial agriculture to distinctive biological and sociopolitical environments.

The revolutionary governments that swept to power in Russia in 1917 and China in 1949 vowed to modernize their nations' economies in a fraction of the time it had taken the industrialized West. While the focus in each was on industry, both also borrowed ideas about modernizing agriculture but did so selectively and without the participation of farmers – with catastrophic results.

The Soviet Union came into existence – and faced massive famine – as ideas and practices in temperate-zone agriculture were changing fast. After a decade of tumult in the countryside, Stalin launched the collectivization campaign of 1929–36, which staked its success on rapidly mechanizing the farming sector.[43] Soviet leaders found eager supporters in the US: proponents of industrial agriculture like the Montana wheat farmer Tom Campbell, charged with helping coordinate the management and mechanization of the massive collective farms, joined the American experts who flocked to the Soviet Union.[44]

As in the US, however, mechanization was by itself insufficient to transform global food production. The Americans noted the absence of an "industrial psychology" in the Soviet Union, bemoaning for example the lack of interest on the part of top bureaucrats and peasants alike in management techniques that would guarantee the availability of machines on farms at the appropriate times. Most American visitors believed such obstacles would inevitably be overcome and were oblivious to the catastrophic famine that swept parts of the Soviet Union between 1932 and 1934, when the government confiscated the wheat crop, which had become the sole focus of agricultural output.[45]

Until the 1970s, Soviet agricultural productivity was plagued by a top-down emphasis on grand, transformative schemes, by a lack of resources devoted to agriculture relative to industry, and by hostile climatic conditions. In the 1950s, Soviets raised harvests as the US had in an earlier century, by colonizing new lands for agriculture. What became known as the Virgin Lands Campaign was agricultural expansion built on mechanization – 50,000 tractors were shipped to the new farms – and nutrient reserves in formerly uncultivated soil. By the decade's end, wind-induced soil erosion and nutrient-depleted soil sent yields plunging. Ultimately one-quarter of the land colonized under the Virgin Lands Campaign was abandoned.[46] While much of the rest of the world's agricultural output soared between 1954 and the 1970s, Soviet gains, for all the focus on radical change, were far more modest.[47]

Chinese agriculture fared far worse. In the 1930s, with virtually no use of modern agricultural technology, China's farmers were feeding more people per hectare than the world averages today by using such highly efficient techniques as recycling wastes and complex intercropping. But like other parts of the world, they barely kept up with demand, producing a per capita caloric intake that was barely sufficient and sometimes lacking.

Mao's Great Leap Forward, meant to modernize China's economy in a few short years, was particularly disastrous for the country's agriculture. The Chinese government ordered peasants to cultivate less acreage and devote their labors to the futile task of smelting backyard steel. It also appropriated food from the countryside based on falsified harvest figures. This combination of actions sent food supplies plummeting between 1958 and 1961. China averaged 2,053 calories per person per day in 1958 and 1,453 in 1960. Perhaps 45 million people died in the worst recorded famine in world history, and China's food supply did not return to 1958 levels until 1974.[48] Domestic turmoil kept the agricultural sector virtually stagnant, or worse, at a time when China's neighbors doubled their rice and wheat harvests.

China demonstrates, even more starkly than the Soviet Union, the interdependence of the components of the emerging industrial agriculture – mechanization, chemical inputs, improved seed, scientific understandings of biological processes and, in particular, the bureaucratic notions of efficiency and rationality that underlay the application of these components. It was ultimately impossible to pull large quantities of labor into the industrial economy without either devastating agriculture or applying industrial technology to it. In China, those technologies were absent or misapplied. Mechanization required access to fossil fuels, and China had no internal oil supply, and virtually no mechanization, until the late 1960s. High yields required synthetic nitrogen, and China used very little, and produced none, until the 1970s. But the dramatic gains in harvests also required management practices not possible in Mao's China. Transportation and storage were regulated by the state, and badly so, so that food that might have fed millions sat by the roadside, rotted, or was infested by vermin. State-sponsored science and farmer education existed, but they were based primarily on ideology, not experimental results, and were run by people who knew little about agriculture. The government bureaucracy was intensely involved in food production, but its accounting and distribution networks served political, not economic, ends.[49]

This is not to argue that industrial agriculture can function only in a capitalist economy of private farmers. China and the Soviet Union ultimately improved agricultural yields under command economies, and China, clearly the greater agricultural success of the two, is both the world's largest consumer of nitrogen fertilizer and a food exporter. But both regimes accomplished this under bureaucrats who looked a lot like bureaucrats in the US, using technologies and practices that were built on oddly similar notions of the place of agriculture and the state in a modern economy. When Jenny Leigh Smith writes of the Soviet Union that "industrial farms increased the dependency of farmers on the state by limiting producer self-sufficiency" and "rural workers came to rely upon the same state-sponsored food networks as urban residents," she could be referring to virtually any part of the world practicing modern agriculture.[50]

While China and the Soviet Union embarked on their experiments in breakneck agricultural transformation, industrial agriculture modeled on the US system began spreading to other parts of the world. The most dramatic manifestation of this is what became known, retrospectively, as the Green Revolution – a program begun after World War II, at the initiative of the US government and private foundations, to create new high-yielding varieties (HYVs) of rice and wheat in order to avert what was predicted to be a global food crisis and thus a threat to global stability.[51]

Current views of the Green Revolution hail it as the miracle that defused the population bomb and a way for developing countries to compete with food-secure economic powers – or condemn it as the process by which rural inequalities were entrenched in Asia and Latin America and the critical moment at which small farmers became dependent on the power of multinational corporations. It is regarded as an environmental good, sparing the necessity of converting tens of millions of acres of wild land to agriculture, and an environmental menace, prompting unprecedented levels of soil degradation and water pollution. But fundamentally the Green Revolution spread the twentieth-century agricultural revolution – with its reliance on fossil fuels, chemistry, technology, and industrial capital – to much of the rest of the world. The Green Revolution raised the total world harvest, but it neither erased inequalities between farmers in rich and poor nations nor ensured an equitable distribution of the world food supply. It also exported the input treadmill. And, as with all agricultural change over the past century, it brought with it a number of environmental and socioeconomic consequences that remain today.

The initiative, funding, and vision of agriculture that made the Green Revolution possible all originated in the US. In 1943, the Rockefeller Foundation created a research center in Mexico aimed at developing wheat varieties that were high yielding, disease resistant, and suited to a variety of growing conditions. By the 1950s, Rockefeller scientists had introduced dwarf wheat with short, stout stalks that could hold large seed heads produced by generous inputs of irrigation water and synthetic nitrogen without lodging, or falling over. By 1962 Mexico was exporting wheat, and in 1963 Norman Borlaug, the scientist who led the breeding effort, sent seeds to India, where they were up to three times as productive as local varieties.[52] In 1953 the foundation established another center, this one in the Philippines, aimed at replicating these results for rice. As in the case of wheat, the goal was dwarf rice that could grow in a variety of tropical environments, independent of the amount of daylight (photoperiod) or the length of growing season. Their success was even more dramatic than it had been for wheat. Researchers released their first improved seed in 1965 and over the next two years received reports of rice harvests that had doubled and tripled per acre. The push to create new HYVs continued in the 1970s, as did adoption of the new seeds. In 1971, the Consultative Group on International Agricultural Research (CGIAR) was founded, aimed at coordinating a growing network of research centers around the developing world. The scope of breeding spread to new crops, including sorghum, millet, barley, and cassava – over 8,000 new varieties of plants for 11 different major food crops. But the success stories were wheat and rice: In 1970, 10 to 15 percent of wheat and rice crops in the developing world were planted in new varieties; by 1991, three-quarters were. In many countries, growth in food production outpaced population growth, sometimes significantly so.[53]

HYVs were bred to respond to a technological package of synthetic fertilizers, pesticides, mechanization, and irrigation – indeed, without these inputs they did no better, and sometimes performed worse, than traditional varieties.[54] From the beginning, the focus was on raising yields by substituting capital inputs for land and labor. Industrial technologies of food production and a hefty reliance on fossil fuels were thus woven into this vision of global agriculture. But underlying the entire program of increasing grain yields was an assumption that national security was linked to an abundant national food supply – an assumption shared by the US and the most enthusiastic adopters of Green Revolution technology, such as India and Mexico (and one that still informs much agricultural policy around the world today).

That assumed link between national security and national food supply meant that the socioeconomic impacts of agricultural change within a country were of little concern to either the US or the elites in developing countries who embraced the focus on high yields. The new HYVs came to countries that already had profound inequalities in access to the very inputs that were essential to the success of these seeds. Irrigation, land, and access to credit to purchase fertilizer and pesticides and seeds – these were not evenly distributed anywhere and they were particularly inequitable in some of the regions targeted by the Green Revolution, including most of Latin America and parts of Asia.[55] That these inequalities were exacerbated by the Green Revolution was not something that concerned its supporters.[56]

The technological package that came with the Green Revolution carried a large environmental footprint. The amount of farmland under irrigation exploded; today, agriculture accounts for 70 percent of the fresh water humans use – and levels of human water use are proving unsustainable around the world. Large-scale use of pesticides, artificial fertilizers, and monocropping increased pollution in waterways, reduced soil fertility and increased soil salinization, and reduced biodiversity in fields. Virtually all environmental damage now caused by rice cultivation is attributed to Green Revolution technology. The crop that accounts for 25 percent of the world's caloric intake produces perhaps 10–15 percent of global methane emissions because it is now grown in constantly flooded paddy fields. Under traditional methods of cultivation, paddies had astonishing biodiversity – up to 700 species of animals in a single hectare of Philippine rice paddy. But the addition of high levels of agrochemicals eliminated the algae that formed the base of this food chain, thereby reducing other species and eliminating important local food sources in the form of frogs, fish, snakes, and snails. The low rates of fertilizer uptake by plants resulted in heavy runoff that polluted waterways, ultimately causing red tides along stretches of Asian coastline; and the persistence of many pesticides meant a significant buildup of those substances in soils, water, and the food chain over several decades.[57] These undesirable impacts were often exacerbated by hefty government subsidies for fertilizers and pesticides that encouraged their overuse. On the other hand, without the Green Revolution, crop yields would have been almost 25 percent lower in developing countries, grain prices would be higher, and far more land would have been placed under cultivation.[58]

Preexisting agrarian social relations also shaped the Green Revolution's environmental and social impacts. In Latin America, with its pronounced rural inequalities, the introduction of HYVs raised the value of land, providing powerful landowners with an incentive to expand their holdings at the expense of smallholders. Mechanization reduced the need for labor, prompting those same wealthy landowners to eject tenants from their farms. Landless and unemployed peasants sometimes moved to cities, but they also began cultivating marginal lands unsuited to agriculture, including steep, highly erodible hillsides.[59] And everywhere, success with the new varieties depended on access to the required inputs. Whether a society granted poor farmers on small plots access to credit, irrigation, and transportation networks shaped who could obtain – and market – the yields obtained at research stations. And continued access to credit mattered a great deal, since new seed, fertilizer, and pesticides had to be purchased every year.

The Green Revolution had a further unintended consequence. In its focus on maximizing available calories through grain monoculture, it created an insidious form of malnutrition in the form of micronutrient deficiencies. Farmers focused on grain production at the expense of legumes, pulses, fruits, and vegetables. Prices for these latter foods rose while prices for grain fell, and the diets of the poor became accordingly less diverse. Further, these new varieties of wheat and rice were relatively poor at pulling micronutrients such as iron, zinc, and selenium out of the soil and thus delivered less nutrition than older varieties. One group of researchers argues that up to 40 percent of the world's population now faces debilitating diseases due to micronutrient deficiencies.[60]

The Green Revolution raised the supply of basic food staples – sometimes dramatically – at a time when world population was growing rapidly. It did not radically transform the map of global food production, however. The industrial agricultural revolution continued to unfold in the West and Japan, and these wealthier countries saw the labor efficiency of agriculture rise at far faster rates than it did in the developing world. The Green Revolution ultimately brought only a few countries self-sufficiency in staple foods, and the developing world as a whole remained a net importer of food.[61]

There are two reasons for this persistent global imbalance. The first is that the Green Revolution completely bypassed Africa. The crop varieties that were the focus of the most intensive research are not well adapted to African growing conditions, and the most successful have never been important grain crops on most of the continent. Few African farmers could access the fertilizers and pesticides necessary for high yields. And the centrality of irrigation to the success of Green Revolution crops meant they were unsuited to a continent where 4 percent of farmland is irrigated and where 40 percent of rural residents live in arid and semi-arid environments. There has been much talk of a new Green Revolution for Africa, but it is one that will have to be built on a different foundation.[62]

The second reason for the continuing imbalance in global agricultural output is the role of states in shaping who grows what where. The governments of wealthy states protected their farmers by propping up farm incomes and creating trade barriers to lock out food imports from countries with lower costs of production (especially lower labor costs). These policies kept global prices for many crops artificially low. Meanwhile, the governments of poorer states protected urban consumers by ensuring a supply of affordable food – usually by underpaying farmers in the countryside so that they had little incentive to raise productivity, particularly in the face of cheap food flooding in from the West. Further, in both rich and poor countries, food security is often translated into the politically popular concept of self-sufficiency in food, which leads to policies promoting agriculture in places where it has neither a competitive nor an ecological advantage.[63]

Who feeds whom therefore has much to do with non-environmental factors: the US is endowed with good soils, but it is also endowed with a government that rewards the overproduction of grains. Switzerland may be generally lacking in good soils and level terrain, but its subsidies to agriculture, among the highest in the world, ensure abundant pastoral activity amid the Alps. The same is true in crowded Japan and frigid Norway.[64] The propensity of wealthy, urban societies to protect their farmers, combined with abundant land and industrial models of production, has meant that the US continues to export more food than any nation in the history of the world. US agriculture is not as efficient in terms of land or energy usage as that of other nations, but at least one-third of the massive American harvest is sent abroad each year – 20 to 25 percent of the corn, 33 percent of the soybeans, and 40 to 50 percent of the wheat and rice feed the rest of the world.[65] Even countries that have been largely bypassed by the twentieth-century agricultural revolution, such as most of Africa, have had their food systems reshaped by it. Africans buy imported food because it is cheaper than what their own farmers grow.

Nutritional Transitions

But perhaps more important from an environmental perspective is the fact that up to 70 percent of the massive cereal and legume harvest of the US feeds the livestock that form the basis of the high-protein diets of the affluent world. This compounds the environmental impacts of industrial grain agriculture by an order of magnitude, because animals require many pounds of feed to add a pound of (edible) muscle tissue.[66] In addition, the industrial production of meat carries with it unique environmental consequences. The Delmarva Peninsula in the mid-Atlantic US produces 600 million of the roughly 8 billion chickens Americans now consume each year – and generates more waste than a city of 4 million people. As industrialized poultry production shifted toward the southern US, such waste was initially a boon that restored fertility to surrounding exhausted soils. But no farmland can absorb such quantities of manure over a significant period of time, and all centers of industrial livestock production, whether cattle feedlots, hog farms, or poultry producers, now have enormous problems of waste disposal. The waste ultimately contaminates waterways, creating anaerobic conditions and "dead zones."[67]

Increased meat consumption in the US was the front line of a larger global nutritional transition that has intensified since the 1970s, in which human diets have incorporated a higher proportion of fats, saturated fats, and sugars. These dietary changes result from several factors. Urbanization, even without an increase in income, prompts people to shift to diets higher in processed and convenience foods – but economic growth in developing countries has accelerated this nutritional transition. Another major factor is the decline in price and increase in availability of animal products and processed and convenience foods, including cooking oils derived from vegetables and oilseeds.[68]

The impact of these dietary changes on human health – in the form of a global obesity epidemic and sharp increases in the prevalence of diseases like Type 2 diabetes – has attracted a great deal of attention. But the impact of these changes on the environment is similarly dramatic. The additional processing of food requires the same inputs as any manufacturing process – fossil fuels, electricity, and the like. As more people shifted to this kind of diet, the environmental footprint of agriculture grew, in the form of both the greater quantity of land required to produce animal products and greater energy expenditures on the processing and retailing end of food production. Already in the US, the energy expended on food processing and retailing is nearly double that expended in agriculture itself (including the production of fertilizer). This is a pattern that is becoming more common worldwide.

The so-called Livestock Revolution has often been overlooked amid the attention given to the Green Revolution. But people have redirected their food dollars more profoundly as a result of the nutrition transition than they did when HYVs were introduced: the market value of the increased meat and milk consumption globally is more than twice the market value of the increased cereal consumption that accompanied the Green Revolution. This Livestock Revolution has likely just begun, given that people in the developing world still consume only about one-quarter as many animal products as people in wealthy nations.[69] There are still many people who would jump at the chance to diversify what are fairly rudimentary diets.

China has seen particularly dramatic changes in its diet, a counterpart to the broader economic changes of the past several decades. Between 1983 and 1997, meat and milk consumption per capita nearly tripled in China. (Since 1997 China, generally considered a lactose-intolerant population, has seen urban consumption of fresh milk products rise 25 percent annually.[70]) Given China's share of world population, such dietary changes have a huge impact on global agriculture and its environmental footprint. But growth in meat and milk consumption is projected to grow at similar rates across Asia and sub-Saharan Africa, and at only slightly slower rates in Latin America. As diets shift, industrialized livestock production will become increasingly global. Add to this the shift toward biofuels, and it is obvious that the pressure to continue raising yields, and the demand for industrial inputs that make those yields possible, will continue unabated.[71]

Conclusion: Industrial Agriculture and the Global Environment

The industrialization of agriculture was an uneven and halting process that emerged in a piecemeal fashion, a direct consequence not only of the broader sociopolitical dynamics that also shaped the formation of industries but additionally of agriculture's rootedness in biological processes. Wheat and pork, the American Corn Belt and Indonesia's rice lands, are not interchangeable. Technologies of industrial production were developed and applied within specific biological, political, and social contexts. And yet, paradoxically, the globalized nature of industrial agriculture largely divorced the process of growing food from its geographic specificity; nations and communities now eat what they do not grow and grow what they do not eat.

The environmental impacts of modern agriculture are similarly distant from the consumers of food – and even, to an extent, from the producers. In the US, subsidized grain that could be transported on highways built with public money ultimately allowed for geographical separation between regions of grain and poultry production: the poultry industry increasingly moved away from the Corn Belt and into the US South, a region deficient in grain but well supplied with cheap labor that raised and slaughtered the billions of chickens that fed Americans by the 1990s.[72] On a transnational scale, African agriculture remained land-intensive and low yielding, which meant that the continent's growing population was fed largely through expanded food imports. Those tripled between 1963 and 1974 and doubled again during the next few years.[73] Water pollution from fertilizer and pesticide runoff, like other environmental costs of industrial agriculture, has not been a notable problem in sub-Saharan Africa – but that pollution somewhere else is nonetheless partly a result of the need to feed people in Africa.

This complex web of connections grounded in food and fiber make quantifying the environmental impacts of industrial agriculture difficult if not impossible. In the passionate and partisan literature on modern agriculture, critics offer sweeping, and often justifiable, indictments. The current system's reliance on fossil fuels is complete. US farming in 1940 burned less than one calorie of energy for every calorie of food it produced; by 1990 it required 2.3 calories to produce a calorie of food. Forty percent of those energy requirements came from the manufacture of chemical fertilizers and pesticides. But this far understates the energy required for food production, most of which actually is directed at what happens after food leaves the farm: the transportation, processing, packaging, and retailing of food. In terms of its energy requirements, agriculture has become much less efficient, not just in the US but around the globe, and most of these extra energy requirements are met by fossil fuels.[74] Smil argues that there has been an eightyfold increase in external energy inputs and that these have come mostly from fossil fuels.[75]

The component parts of industrial agriculture evolved together, and they are not easily disentangled. Remove one, and the system ceases to function. Given the finite supply of fossil fuels, this is a major vulnerability in industrial agriculture. But fossil fuels are not the only essential input. If subtherapeutic antibiotics in animal feed ultimately proved to have highly undesirable effects on the health of humans or the environment – a very real possibility – the entire industrial system of animal production, and thus the supply of relatively inexpensive meat, would be in jeopardy. If aquifers and other sources of irrigation water run dry – an increasingly likely scenario in some regions – the current method of crop production becomes unsustainable in many places.[76]

At the same time, contemporary high-yield agriculture rests on an increasingly small number of plant and animal varieties. Many of these are less robust than older varieties and are cocooned by an array of industrial inputs.[77] If a lethal disease or pest resistant to our current array of chemicals struck a major crop like wheat, corn, or rice, we would have two choices: shift to a less susceptible variety or rapidly develop a means of containing the outbreak. The former means that yields would plummet; the latter assumes that we can stay one step ahead of nature's process of evolution. The successful biological control of the mealybug that threatened cassava harvests in the late 1970s and early 1980s is reason for optimism; the much-cited statistic that US farmers increased pesticide use tenfold in the 45 years after World War II while simultaneously losing twice as much of their crop to pests is reason for considerably less optimism.[78]

Defenders of the current agricultural system respond to critics by arguing – correctly – that it would be impossible to feed the world's billions with the farming methods of a century ago. They also argue that marginal, low-input production causes the bulk of environmental damage in poor countries.[79] The reliance on fossil fuels, a narrow range of carefully bred species, and manufactured inputs is a necessary component of this success. Further, intensifying agricultural yield per acre has arguably prevented millions of hectares of land from being converted to agriculture in the first place. There is some truth to these claims. Mechanization alone dramatically reduced the amount of land needed to grow food: in 1910, 27 percent of harvested acres in the US went to feed horses and mules; in 1960 the figure was 5 percent.[80] It has been argued that, without the Green Revolution, India would have needed to double its land in agriculture.[81] Similar arguments are now made in defense of the shift toward genetically modified (GM) crops. Since human population is expected to peak at slightly over 9 billion around 2050, global food supply will have to expand not only to feed people shifting to a more resource-intensive diet, but also to offer even basic nutrition to another 2 billion people.

While critics claim that the brittleness of the current system invites disaster, particularly when global climate change is considered, defenders ask what the alternative would have been. This is a difficult question to answer. Because food and statecraft have long been linked, it is no surprise that the world's current agricultural system is as much a product of politics as it is of biological processes and scientific innovation. Beginning with the US, decisions made by governments around the world have helped shape the form of agriculture that emerged. Technical advances in agriculture first reflected the advantages and constraints of the US, with its abundant land, access to industrial technology, relative shortage of labor, and strong state support for both farmers and science. Something as seemingly trivial as the fact that the US presidential primaries begin in Iowa affected the global supply of corn by offering incentives to American farmers to maximize production and by supporting technologies to make this possible. That glut of corn in turn largely shaped the industrial livestock system for poultry, pork, and beef. The same pattern plays out now in support for corn-based biofuels.[82]

But industrial agriculture has adapted to many sociopolitical environments and served many political agendas. In Indonesia in the early 1980s, the government subsidized 85 percent of the cost of pesticides – a politically popular move that encouraged overuse, resulting in insect resistance and crop losses.[83] In Namibia, sub-Saharan Africa's driest country, a politically popular program of food self-sufficiency has resulted in high-input (and intensely irrigated) agricultural schemes – and wheat yields that are more than double the global average.[84] Across the world, Japanese policies promoted –and attained – self-sufficiency in rice at the cost of heavy reliance on imported inputs derived from petroleum products – again justified in terms of food security but also environmental protection and rural development.[85]

In the end, however, this template for agricultural modernization, bound as it was to industrial notions of efficiency and productivity, and the Western world's commitment to transforming the developing world in its own image, foreclosed the exploration of possible alternatives in much of the twentieth century. There were roads not traveled as the modern system of food production was being developed, because one choice inevitably foreclosed others. For example, polyculture systems can be highly productive, but they are not easily mechanized and thus not amenable to large-scale production, and they require significant labor. Once mechanization became the foundation of modern agriculture and farm consolidation became accepted as synonymous with progress, research monies were unlikely to be directed toward increasing the yields of such systems. As concerns about the cost and future availability of inputs continue to grow and as the limits to the current system become more apparent, notions of what is possible are beginning to shift. In wealthy countries, consumers are demanding food that is produced differently. More significantly, perhaps, agricultural research in the developing world is beginning to take seriously notions of agro-ecology – working with, rather than against, ecological systems to increase food production – arguing that any emphasis on increasing yields needs to combine older technologies and new understandings of ecological systems with judicious, even sparing, use of industrial inputs to guarantee long-term success and sustainability. Whether industrial agriculture will remain an ideal – or even a necessity – in the next century remains to be seen.

Notes

1   K. Brown, "Gridded Lives: Why Montana and Kazakhstan Are Nearly the Same Place," *American Historical Review* 106/1, 2001, pp. 17–48; J. Scott, *Seeing Like a State: How Certain Schemes to Improve the Human Condition Have Failed*, New Haven, CT, Yale University Press, 1998.

2   B. Page, "Restructuring Pork Production, Remaking Rural Iowa," in D. Goodman and M. Watts (eds.), *Globalising Food: Agrarian Questions and Global Restructuring*, London, Routledge, 1997, pp. 97–114; D. Grigg, *The Transformation of Agriculture in the West*, Oxford, Blackwell, 1992, p. 10. P. Lains and V. Pinilla summarize the economic history debates over the contributions of increased agricultural productivity to industrialization in *Agriculture and Economic Development in Europe since 1870*, London, Routledge, 2009, pp. 6–9.

3   J. R. McNeill details the changes of the past century in *Something New under the Sun: An Environmental History of the Twentieth-Century World*, New York, W. W. Norton, 2000.

4   J. Clay, *World Agriculture and the Environment*, Washington, Island Press, 2004, pp. 396–400.

5   R. C. Allen, "The Nitrogen Hypothesis and the English Agricultural Revolution: A Biological Analysis," *Journal of Economic History* 68, 2008, pp. 182–210; Grigg, *The Transformation of Agriculture in the West*.

6   D. Fitzgerald, *The Business of Breeding: Hybrid Corn in Illinois, 1890–1940*, Ithaca, NY, Cornell University Press, 1990, pp. 26–8; V. Smil, *Enriching the Earth: Fritz Haber, Carl Bosch, and the Transformation of World Food Production*, Cambridge, MA, MIT, 2001, pp. 5–8.

7   M. Knibbe, "Feed, Fertilizer, and Agricultural Productivity in the Netherlands, 1880–1930," *Agricultural History* 74, 2000, pp. 39–57; Grigg, *The Transformation of Agriculture in the West*; D. Goodman, B. Sorj, and J. Wilkinson, *From Farming to Biotechnology: A Theory of Agro-Industrial Development*, Oxford, Blackwell, 1987, pp. 29–30; also Smil, *Enriching the Earth*. Not all innovation emerged in Europe: Japan's experiments in the late nineteenth and early twentieth centuries with breeding dwarf wheat and rice would later make a major contribution to the Green Revolution, providing important breeding stock for scientists in the 1950s. But these innovations also built on Mendelian genetics. See McNeill, *Something New under the Sun*, pp. 219–20; T. Ogura, *Japanese Economic History, 1930–1960*, vol. 6, *Agricultural Developments in Modern Japan*, London, Routledge, 2000.

8   H. Drache, *History of U.S. Agriculture and Its Relevance to Today*, Danville, IL, Interstate Publishers, 1996, pp. 103–15; D. Fitzgerald, *Every Farm a Factory: The Industrial Ideal in American Agriculture*, New Haven, CT, Yale University Press, 2003, pp. 93–5.

9   Fitzgerald, *Every Farm a Factory*, p. 6.

10 Drache, *History of U.S. Agriculture*, pp. 115–16.

11 A. Effland, "U.S. Farm Policy: The First 200 Years," *Agricultural Outlook*, 2000, pp. 21–5.

12 Brown, "Gridded Lives"; Scott, *Seeing Like a State*.

13 Quoted in Fitzgerald, *Every Farm a Factory*, p. 129.

14 D. Danbom, *The Resisted Revolution: Urban America and the Industrialization of Agriculture, 1900–1930*, Ames, IA, Iowa State University Press, 1979.

15 Danbom, *The Resisted Revolution*; Fitzgerald, *Every Farm a Factory*.

16 G. Federico, *Feeding the World: An Economic History of Agriculture, 1800–2000*, Princeton, Princeton University Press, 2005, p. 91.

17 Danbom, *The Resisted Revolution*, p. 121.

18 Fitzgerald, *Every Farm a Factory*, pp. 20–3.

19 Goodman, Sorj, and Wilkinson, *From Farming to Biotechnology*, p. 22.

20 Fitzgerald, *Every Farm a Factory*, pp. 115–16.

21 Danbom, *The Resisted Revolution*, p. 137.

22 Goodman, Sorj, and Wilkinson, *From Farming to Biotechnology*, pp. 24–5; E. W. F. Peterson, *A Billion Dollars a Day: The Economics and Politics of Agricultural Subsidies*, Malden, MA, Wiley-Blackwell, 2009, pp. 124–5.

23 Ogura discusses the peculiarities of Japanese agricultural modernization in *Japanese Economic History*.

24 Smil, *Enriching the Earth*.

25 K. Pomeranz, "Advanced Agriculture," in J. Bentley (ed.), *The Oxford Handbook of World History*, Oxford, Oxford University Press, 2011, pp. 246–66 (pp. 254–5); Smil, *Enriching the Earth*, pp. 160–1. Seemingly because constant cultivation exhausts land in other ways that are only coming to be understood, which is one reason why ever-increasing quantities of fertilizer are needed.

26 Clay, *World Agriculture and the Environment*, p. 50.

27 Clay, *World Agriculture and the Environment*, p. 53.

28 Goodman, Sorj, and Wilkinson, *From Farming to Biotechnology*, p. 35; Fitzgerald, *Every Farm a Factory*, p. 14; P. Conklin, *A Revolution Down on the Farm: The Transformation of American Agriculture since 1929*, Lexington, KY, University Press of Kentucky, 2008, pp. 119–20.

29 Clay, *World Agriculture and the Environment*, p. 49.

30 M. Ollinger, J. MacDonald, and M. Madison, "Poultry Plants Lowering Cost and Increasing Variety," *Food Review* 23/2, 2000, pp. 2–7 (p. 3).

31 W. Boyd and M. Watts, "Agro-Industrial Just-in-Time: The Chicken Industry and Postwar American Capitalism," in D. Goodman and M. Watts (eds.), *Globalising Food: Agrarian Questions and Global Restructuring*, London, Routledge, 1997, pp. 139–65 (pp. 143–4); W. Boyd, "Making Meat: Science, Technology, and American Poultry Production," *Technology and Culture* 42, 2001, pp. 631–64 (pp. 637–8).

32 Boyd, "Making Meat," p. 633.
33 R. Horowitz, *Putting Meat on the American Table: Taste, Technology, Transformation*, Baltimore, The Johns Hopkins University Press, 2006, pp. 135–6.
34 M. Finlay, "Hogs, Antibiotics, and the Industrial Environments of Postwar Agriculture," in S. Schrepfer and P. Scranton (eds.), *Industrializing Organisms: Introducing Evolutionary History*, New York, Routledge, 2004, pp. 237–60.
35 Boyd, "Making Meat."
36 Finlay, "Hogs, Antibiotics, and the Industrial Environments of Postwar Agriculture."
37 A point made by Boyd and Watts in "Agro-Industrial Just-in-Time," p. 193.
38 McNeill summarizes these consequences nicely in *Something New under the Sun*, pp. 223–4.
39 McNeill, *Something New under the Sun*, p. 216.
40 Clay, *World Agriculture and the Environment*, pp. 50–2.
41 Clay, *World Agriculture and the Environment*, p. 53.
42 Clay, *World Agriculture and the Environment*, pp. 49–50.
43 Z. Medvedev, *Soviet Agriculture*, New York, W. W. Norton, 1987, p. 38; G. Ioffe and T. Nefedova, *Continuity and Change in Rural Russia: A Geographical Perspective*, Boulder, CO, Westview Press, 1997, p. 38.
44 Fitzgerald, *Every Farm a Factory*, chapter 6.
45 Fitzgerald, *Every Farm a Factory*, pp. 177–9.
46 Medvedev, *Soviet Agriculture*, chapter 6; M. McCauley, *Khrushchev and the Development of Soviet Agriculture: The Virgin Land Programme 1953–1964*, London, Macmillan, 1976; McNeill, *Something New under the Sun*, p. 215.
47 Medvedev, *Soviet Agriculture*, p. 395.
48 V. Smil, *China's Past, China's Future: Energy, Food, Environment*, New York, Routledge, 2004, pp. 74–6; Dikötter (2010), pp. 127–39.
49 Smil, *China's Past, China's Future*; F. Dikötter, *Mao's Great Famine: The History of China's Most Devastating Catastrophe, 1958–1962*, New York, Walker & Co., 2010.
50 J. L. Smith, "The Soviet Farm Complex: Industrial Agriculture in a Socialist Context, 1945–65," PhD thesis, MIT, 2006, pp. 5–6.
51 N. Cullather, *The Hungry World: America's Cold War Battle against Poverty in Asia*, Cambridge, MA, Harvard University Press, 2010; M. Latham, *The Right Kind of Revolution: Modernization, Development, and U.S. Foreign Policy from the Cold War to the Present*, Ithaca, NY, Cornell University Press, 2011.
52 McNeill, *Something New under the Sun*, p. 221.
53 Latham, *The Right Kind of Revolution*, pp. 109–16; R. Paarlberg, *Food Politics: What Everyone Needs to Know*, Oxford, Oxford University Press, 2010, pp. 56–7.
54 Federico, *Feeding the World*, p. 12. This was in part because pest resistance was often lost when breeding focused solely on yields.
55 Paarlberg argues in *Food Politics* that Asian farmers had these things and Latin American farmers did not; Griffin and others see the entrenchment or augmentation of rural inequality in Asia as well; see K. Griffin, *The Political Economy of Agrarian Change: An Essay on the Green Revolution*, 2nd ed., London, Macmillan, 1979.
56 J. H. Perkins, *Geopolitics and the Green Revolution: Wheat, Genes, and the Cold War*, New York, Oxford University Press, 1997.
57 Clay, *World Agriculture and the Environment*, pp. 396–9.
58 C. Juma, *The New Harvest: Agricultural Innovation in Africa*, Oxford, Oxford University Press, 2011, pp. 5–6.
59 Paarlberg, *Food Politics*, pp. 59–63.
60 R. Welch, G. Combs Jr., and J. Duxbury, "Toward a 'Greener' Revolution," *Issues in Science and Technology Online*, 1997, accessed from http://www.issues.org/14.1/welch.htm on March 15, 2012. These trends do not just affect the rural poor, however; the diets of wealthier and urbanized societies, which are also lower in plant-based foods, also tend to be deficient in micronutrients.

61 McNeill, *Something New under the Sun*, pp. 225–6.
62 Paarlberg, *Food Politics*, pp. 64–5; Juma, *The New Harvest*, p. 8. M. Smale and T. Jayne discuss the reasons high-yielding maize did not transform African agriculture in "'Seeds of Success' in Retrospect: Hybrid Maize in Eastern and Southern Africa," in S. Haggblade and P. Hazell (eds.), *Successes in African Agriculture: Lessons for the Future*, Baltimore, The Johns Hopkins University Press, 2010, pp. 71–112. Whether that foundation will be a model of small farmers utilizing sustainable and affordable inputs, as advocated by agro-ecologists, or whether it will be the massive farms now being created by foreign companies leasing immense tracts of African land at virtually no cost remains to be seen.
63 Peterson, *A Billion Dollars a Day*.
64 Peterson, *A Billion Dollars a Day*.
65 Smil, *Enriching the Earth*, p. 164.
66 The efficiency of feed conversion to protein is 10–15 percent for pork and 5–8 percent for beef; it is markedly higher for modern broiler chickens, who are about three times more efficient than older breeds. See Smil, *Enriching the Earth*, pp. 164–5, and G. B. Havenstein, P. R. Ferket, and M. A. Qureshi, "Growth, Livability, and Feed Conversion of 1957 versus 2001 Broilers When Fed Representative 1957 and 2001 Broiler Diets," *Poultry Science* 82, 2003, pp. 1500–8.
67 Boyd, "Making Meat," pp. 643–4.
68 A. Drewnowski and B. Popkin, "The Nutrition Transition: New Trends in the Global Diet," *Nutrition Reviews* 55/2, 1997, pp. 31–43 (p. 34). Although in all cases of nutrition transition the proportion of fats and sweeteners has risen, it is also important to note that the first impact of changing diets is greater diversity and, in general, the addition of nutrients and proteins that were in short supply. Drewnowsky and Popkin, "The Nutrition Transition," p. 39.
69 C. Delgado, "Rising Consumption of Meat and Milk in Developing Countries Has Created a New Food Revolution," *Journal of Nutrition* 133, 2003, pp. 3907s–3910s (pp. 3907s–3908s).
70 F. Fuller and J. C. Beghin, *China's Growing Market for Dairy Products*, accessed from http://www.card.iastate.edu/iowa_ag_review/summer_04/article5.aspx on March 15, 2012.
71 Delgado, "Rising Consumption of Meat and Milk," p. 3909s.
72 Boyd and Watts, "Agro-Industrial Just-in-Time," pp. 201–6.
73 D. Grigg, *The World Food Problem, 1950–1980*, Oxford, Blackwell, 1985, pp. 134–5.
74 Center for Sustainable Systems, "U.S. Food System Factsheet," pub. no. CSS01-06, University of Michigan, 2010; McNeill, *Something New under the Sun*, p. 224.
75 Cited in Clay, *World Agriculture and the Environment*, p. 5.
76 S. Postel, *Pillar of Sand: Can the Irrigation Miracle Last?*, New York, W. W. Norton, 1999.
77 Clay, *World Agriculture and the Environment*, p. 376.
78 Center for Sustainable Systems, "U.S. Food System Factsheet."
79 Clay, *World Agriculture and the Environment*, p. 8.
80 But bear in mind that land in pasture lacks many of the environmental liabilities of land in grain monoculture.
81 Paarlberg, *Food Politics*, p. 63.
82 Paarlberg, *Food Politics*, pp. 103–4.
83 Paarlberg, *Food Politics*, p. 63.
84 Clay, *World Agriculture and the Environment*, p. 370; Namibian Agronomic Board, *Home page*, accessed from http://www.nab.com.na/ on March 15, 2012.
85 Peterson, *A Billion Dollars a Day*, pp. 184–91.

References

Allen, R. C., "The Nitrogen Hypothesis and the English Agricultural Revolution: A Biological Analysis," *Journal of Economic History* 68, 2008, pp. 182–210.
Boyd, W., "Making Meat: Science, Technology, and American Poultry Production," *Technology and Culture* 42, 2001, pp. 631–64.

Boyd, W., and Watts, M., "Agro-Industrial Just-in-Time: The Chicken Industry and Postwar American Capitalism," in D. Goodman and M. Watts (eds.), *Globalising Food: Agrarian Questions and Global Restructuring*, London, Routledge, 1997, pp. 139–65.

Brown, K., "Gridded Lives: Why Montana and Kazakhstan Are Nearly the Same Place," *American Historical Review* 106/1, 2001, pp. 17–48.

Center for Sustainable Systems, "U.S. Food System Factsheet," pub. no. CSS01-06, University of Michigan, 2010.

Clay, J., *World Agriculture and the Environment*, Washington, Island Press, 2004.

Conklin, P., *A Revolution Down on the Farm: The Transformation of American Agriculture since 1929*, Lexington, KY, University Press of Kentucky, 2008.

Cullather, N., *The Hungry World: America's Cold War Battle against Poverty in Asia*, Cambridge, MA, Harvard University Press, 2010.

Danbom, D., *The Resisted Revolution: Urban America and the Industrialization of Agriculture, 1900–1930*, Ames, IA, Iowa State University Press, 1979.

Delgado, C., "Rising Consumption of Meat and Milk in Developing Countries Has Created a New Food Revolution," *Journal of Nutrition* 133, 2003, pp. 3907S–3910S.

Dikötter, F., *Mao's Great Famine: The History of China's Most Devastating Catastrophe, 1958–1962*, New York, Walker & Co., 2010.

Drache, H., *History of U.S. Agriculture and Its Relevance to Today*, Danville, IL, Interstate Publishers, 1996.

Drewnowski, A., and Popkin, B., "The Nutrition Transition: New Trends in the Global Diet," *Nutrition Reviews* 55/2, 1997, pp. 31–43.

Effland, A., "U.S. Farm Policy: The First 200 Years," *Agricultural Outlook*, 2000, pp. 21–5.

Federico, G., *Feeding the World: An Economic History of Agriculture, 1800–2000*, Princeton, Princeton University Press, 2005.

Finlay, M., "Hogs, Antibiotics, and the Industrial Environments of Postwar Agriculture," in S. Schrepfer and P. Scranton (eds.), *Industrializing Organisms: Introducing Evolutionary History*, New York, Routledge, 2004, pp. 237–60.

Fitzgerald, D., *The Business of Breeding: Hybrid Corn in Illinois, 1890–1940*, Ithaca, NY, Cornell University Press, 1990.

Fitzgerald, D., *Every Farm a Factory: The Industrial Ideal in American Agriculture*, New Haven, CT, Yale University Press, 2003.

Fuller, F., and Beghin, J. C., *China's Growing Market for Dairy Products*, accessed from http://www.card.iastate.edu/iowa_ag_review/summer_04/article5.aspx on March 15, 2012.

Goodman, D., Sorj, B., and Wilkinson, J., *From Farming to Biotechnology: A Theory of Agro-Industrial Development*, Oxford, Blackwell, 1987.

Griffin, K., *The Political Economy of Agrarian Change: An Essay on the Green Revolution*, 2nd ed., London, Macmillan, 1979.

Grigg, D., *The Transformation of Agriculture in the West*, Oxford, Blackwell, 1992.

Grigg, D., *The World Food Problem, 1950–1980*, Oxford, Blackwell, 1985.

Havenstein, G. B., Ferket, P. R., and Qureshi, M. A., "Growth, Livability, and Feed Conversion of 1957 versus 2001 Broilers When Fed Representative 1957 and 2001 Broiler Diets," *Poultry Science* 82, 2003, pp. 1500–8.

Horowitz, R., *Putting Meat on the American Table: Taste, Technology, Transformation*, Baltimore, The Johns Hopkins University Press, 2006.

Ioffe, G., and Nefedova, T., *Continuity and Change in Rural Russia: A Geographical Perspective*, Boulder, CO, Westview Press, 1997.

Juma, C., *The New Harvest: Agricultural Innovation in Africa*, Oxford, Oxford University Press, 2011.

Knibbe, M., "Feed, Fertilizer, and Agricultural Productivity in the Netherlands, 1880–1930," *Agricultural History* 74, 2000, pp. 39–57.

Lains, P., and Pinilla, V., *Agriculture and Economic Development in Europe since 1870*, London, Routledge, 2009.

Latham, M., *The Right Kind of Revolution: Modernization, Development, and U.S. Foreign Policy from the Cold War to the Present*, Ithaca, NY, Cornell University Press, 2011.

McCauley, M., *Khrushchev and the Development of Soviet Agriculture: The Virgin Land Programme 1953–1964*, London, Macmillan, 1976.

McNeill, J. R., *Something New under the Sun: An Environmental History of the Twentieth-Century World*, New York, W. W. Norton, 2000.

Medvedev, Z., *Soviet Agriculture*, New York, W. W. Norton, 1987.

Namibian Agronomic Board, *Home page*, accessed from http://www.nab.com.na/ on March 15, 2012.

Ogura, T., *Japanese Economic History, 1930–1960*, vol. 6, *Agricultural Developments in Modern Japan*, London, Routledge, 2000.

Ollinger, M., MacDonald, J., and Madison, M., "Poultry Plants Lowering Cost and Increasing Variety," *Food Review* 23/2, 2000, pp. 2–7.

Paarlberg, R., *Food Politics: What Everyone Needs to Know*, Oxford, Oxford University Press, 2010.

Page, B., "Restructuring Pork Production, Remaking Rural Iowa," in D. Goodman and M. Watts (eds.), *Globalising Food: Agrarian Questions and Global Restructuring*, London, Routledge, 1997, pp. 97–114.

Perkins, J. H., *Geopolitics and the Green Revolution: Wheat, Genes, and the Cold War*, New York, Oxford University Press, 1997.

Peterson, E. W. F., *A Billion Dollars a Day: The Economics and Politics of Agricultural Subsidies*, Malden, MA, Wiley-Blackwell, 2009.

Pomeranz, K., "Advanced Agriculture," in J. Bentley (ed.), *The Oxford Handbook of World History*, Oxford, Oxford University Press, 2011, pp. 246–66.

Postel, S., *Pillar of Sand: Can the Irrigation Miracle Last?*, New York, W. W. Norton, 1999.

Scott, J., *Seeing Like a State: How Certain Schemes to Improve the Human Condition Have Failed*, New Haven, CT, Yale University Press, 1998.

Smale, M., and Jayne, T., "'Seeds of Success' in Retrospect: Hybrid Maize in Eastern and Southern Africa," in S. Haggblade and P. Hazell (eds.), *Successes in African Agriculture: Lessons for the Future*, Baltimore, The Johns Hopkins University Press, 2010, pp. 71–112.

Smil, V., *China's Past, China's Future: Energy, Food, Environment*, New York, Routledge, 2004.

Smil, V., *Enriching the Earth: Fritz Haber, Carl Bosch, and the Transformation of World Food Production*, Cambridge, MA, MIT, 2001.

Smith, J. L., "The Soviet Farm Complex: Industrial Agriculture in a Socialist Context, 1945–65," PhD thesis, MIT, 2006.

Welch, R., Combs, G., Jr., and Duxbury, J., "Toward a 'Greener' Revolution," *Issues in Science and Technology Online*, 1997, accessed from http://www.issues.org/14.1/welch.htm on March 15, 2012.

Further Reading

Moyazer, M., and Roudart, L., *The History of World Agriculture from the Neolithic Age to the Current Crisis*, New York, Monthly Review Press, 2006.

Stoll, S., *The Fruits of Natural Advantage: Making the Industrial Countryside in California*, Berkeley, University of California Press, 1998.

Tauger, M. B., *Agriculture in World History*, New York, Routledge, 2011.

CHAPTER TWENTY-FOUR

Biological Exchange in Global Environmental History

J. R. McNeill

Introduction

Biological exchange can refer to any number of things. Here it refers above all else to long-distance transfers of crops, domesticated animals, and disease-causing microbes, or pathogens. This choice is intended to emphasize biological exchanges that carried the greatest and most direct historical significance. Thus more attention will be paid to sugarcane and horses than to dandelions and squirrels.

But it is in some respects an arbitrary choice. For one thing, it is anthropocentric, ignoring all manner of biological exchanges that (so far) have had negligible effects on human affairs. While this may be a dubious procedure from an ecological standpoint, it is legitimate within the discipline of history in which human concerns must remain front and center. For another, the choice focuses attention on the movement of biological species rather than the movement of genes among species. In the not-so-distant future, it may be that the intentional and accidental transfers of genes among species will prove more important to human affairs than the movement of species around the globe. That day has not yet come. But the genetic exchanges and manipulation that make high-yield wheat and rice possible have already influenced history in important ways, so omitting them, as done here, ignores part of the story.[1] The choice also privileges sudden, long-distance, intercontinental movements, such as the arrival of maize in Africa, over more local and gradual ones such as Roman efforts to spread grape vines throughout Europe or Chinese endeavors to extend rice cultivation. It also underplays the transfers of medicinal plants, weeds, pests, fungi, insects, birds, and countless other creatures that have had some impact on human history. The chapter aims to explore the role of the most important biological exchanges for human history.

Biological exchange was carried out sometimes intentionally and sometimes accidentally. People carried animals and crops from one place to another by careful and

A Companion to Global Environmental History, First Edition. Edited by J.R. McNeill and Erin Stewart Mauldin.
© 2012 John Wiley & Sons, Ltd. Published 2015 by John Wiley & Sons, Ltd.

conscious design, but normally brought seeds, pests, weeds, and microbes inadvertently. Even intentional introductions often brought unexpected consequences, as in the famous case of rabbits in Australia. A few nineteenth-century immigrant rabbits, brought out from Britain to augment food supplies, multiplied into tens of millions of nibbling pests, destroying the grasslands on which Australian sheep and cattle grazed. An intentional introduction led to vast unintended consequences. Episodes such as this illustrate an important principle in biological exchange, or more strictly speaking in ecological invasions, one called ecological release: exotic immigrant species often leave their predators and parasites behind, and can flourish in their new homes far more prolifically than in their old niches, generating great disruptions in the ecology of the lands that receive them. This, often termed biological invasion or bio-invasion, happens in about one out of every hundred exotic introductions. In terms of total biomass, by far the most success-ful ecological invaders in recent millennia have been lowly earthworms. Despite their numbers, and their hungry rampage through the leaf litter of North American forests, they are far from the most consequential.[2]

Successful biological invasion has happened often enough that today governments make great efforts to prevent most forms of biological exchange. For most of human history, people experimented eagerly with it. Now fear of biological invasions and consequent crop damage or the marginalization or extinction of indigenous species has restricted experimentation. A biological protectionism has grown up. It has dubious intellectual underpinnings, since it is often unclear what is indigenous and what is not, and occasionally carries darkly chauvinist overtones.[3]

Biological exchange still proceeds, indeed in quantitative terms faster than ever.[4] But now it is more furtive and more accidental than at any time in recent millennia. It is at present less historically important than at certain times in the past, although in theory that could change at any moment.

Before Agriculture

For most of the four-billion year history of life on earth, most terrestrial species stayed put. Natural barriers inhibited species' migrations and divided the earth into separate biogeographical provinces. Their frontiers changed slowly. Only birds, bats, flying insects, and creatures that were good swimmers bucked the trend and leapt from prov-ince to province. A few earthbound species did so occasionally, thanks to sea-level changes and land bridges that temporarily united continents. The earliest hominins and humans exemplified this partitioning of the planet. They confined themselves to east and southeast Africa for many tens of thousands of years. They probably helped to rearrange the biota of their preferred range, at least from the time they domesticated fire, perhaps half a million years ago. A more intensive fire regime favors fire-compatible species, often grasses. Early humans may even have purposely used fire to change ecology, if they understood that more fire and more grass led to more big and tasty herbivores.

When humans first left Africa for new horizons, perhaps around 100,000 years ago (see Map 24.1), they brought fire with them and presumably had similar ecological effects on the balance between forest and grasslands. They may additionally have acci-dentally brought some African species with them as they trekked into southwestern Asia, maybe fleas, or lice, or other persistent hitchhikers. They probably brought some archaic pathogens with them. They walked eastward to south and southeast Asia and by 40,000 years ago, if not before, made the open-water crossing to Australia. Others hiked

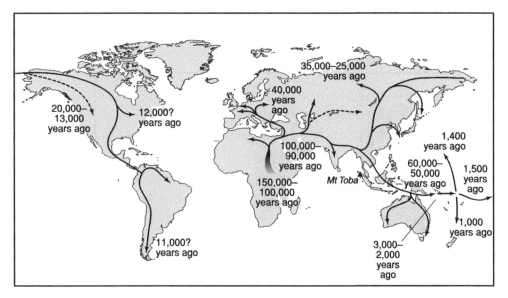

Map 24.1   Human settlement of the globe

northwestward into Europe. By 15,000 years ago, and perhaps before, a small group walked across the land bridge of Beringia to the Americas, probably following herds of caribou. (Abandoning anthropocentrism, one might consider the first American humans as exotic invaders inadvertently brought by caribou.)

When first coming to America, or soon thereafter, migratory humans intentionally brought one species: the dog. Somewhere in western Asia, sometime before 16,000 years ago, some wolves gradually became dogs, partnering with humans in the unconscious interest of mutual survival. The first intentional intercontinental biological introduction was that of dogs to the Americas. There they probably played a role helping humans in the rather sudden extinction of many large mammals (climate change may also have played a part). However that set of extinctions happened, it carried major consequences: the Americas were left with almost no native species suitable for domestication, pushing human history in the Americas onto different paths than it might otherwise have taken. People also brought one dog species, the dingo, to Australia about 3,500 years ago. The dingo, an able hunter, also helped sweep some native Australian species into the dustbin of natural history.[5]

Dogs notwithstanding, for the long span of history before 10,000 BCE – the Paleolithic – biological exchanges were rare and usually inconsequential. There were surely some that have left no trace, especially likely with pathogens. But in the absence of domesticates (other than dogs late in the Paleolithic) there could be no intentional biological introductions – other than of humans themselves.

Agrarian Societies and Overland Biological Exchange to 1400 CE

Toward the end of the last ice age, beginning about 11,000 years ago, people went on a domestication spree. Current opinion now holds that the process took place independently at least seven times in seven places, first in the so-called Fertile Crescent of Southwest Asia, and subsequently in both northern and southern China, New Guinea,

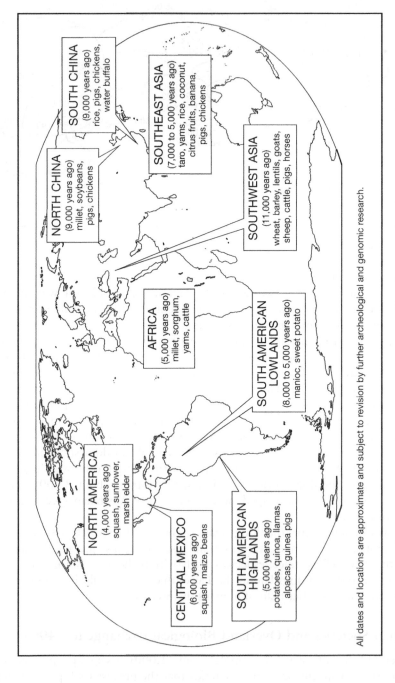

SOUTH CHINA
(9,000 years ago)
rice, pigs, chickens,
water buffalo

SOUTHEAST ASIA
(7,000 to 5,000 years ago)
taro, yams, rice, coconut,
citrus fruits, banana,
pigs, chickens

NORTH CHINA
(9,000 years ago)
millet, soybeans,
pigs, chickens

SOUTHWEST ASIA
(11,000 years ago)
wheat, barley, lentils, goats,
sheep, cattle, pigs, horses

AFRICA
(5,000 years ago)
millet, sorghum,
yams, cattle

**SOUTH AMERICAN
LOWLANDS**
(8,000 to 5,000 years ago)
manioc, sweet potato

NORTH AMERICA
(4,000 years ago)
squash, sunflower,
marsh elder

CENTRAL MEXICO
(6,000 years ago)
squash, maize, beans

**SOUTH AMERICAN
HIGHLANDS**
(5,000 years ago)
potatoes, quinoa, llamas,
alpacas, guinea pigs

All dates and locations are approximate and subject to revision by further archeological and genomic research.

Map 24.2   Transitions to agriculture, 11,000 to 4,000 BCE. From J. R. McNeill and W. H. McNeill, *The Human Web*, W. W. Norton, 2003, p. 27

Sahelian Africa, South America, Mesoamerica and perhaps in North America and Southeast Asia as well. Thereafter biological exchange, both intentional and unintentional, accelerated. Those species most easily domesticated and most useful for human designs were deliberately carried from one place to another. In some cases the transported species brought their own pathogens and parasites with them. Humans often moved less often and less far in their lives as they settled down to agriculture, but they took more with them when did move (see Map 24.2).

By random chance, most of the easily domesticated and especially useful species lived in Eurasia. Those plants and animals sensitive to climate conditions or to day length (several flowering plants take their cues to bloom from day length) traveled fairly well on east–west routes within Eurasia. The plants and animals on which early agriculture and herding rested spread, almost instantaneously by the standards of the past, although in fact it took a few millennia.[6] This process no doubt proved highly disruptive biologically, as alien species invaded new biogeographic provinces. It also proved highly disruptive historically, obliterating peoples who did not adapt to the changing biogeography, the changing disease regimes, and the changed political situations brought on by the spread of farmers, herders, and eventually states. Out of this turmoil of Afro-Eurasian biological exchange emerged the great ancient civilizations from the Hwang He (Yellow River) to the Nile, in place by 3000 BCE in western Asia and Egypt and by 1500 BCE in East Asia. They all based their societies on intersecting but not identical suites of plants and animals, in which wheat, cattle, pigs, goats, sheep, and horses usually figured prominently. These societies and states were in effect the winners in a political shuffle created by the drift toward domestication: they were the ones who took fullest advantage of the novel opportunities for demographic growth, state-building, military specialization, and conquest.

One thing in particular that these "winners" had was easily stored grains as their staple foods. In practice this normally meant wheat and barley in drier climates and rice in rainier ones. Storable foods can be, and were, traded over distances that took weeks to traverse, whereas root crops and other foods that spoil quickly must be consumed soon after harvest. Storable grains can also be, and were, hoarded by the powerful and used to reward followers. Long-distance trade and large states were much more likely to develop where wheat or rice were grown (or, in the Americas, maize) than elsewhere. Since trade and states were an attractive route to wealth and power, ambitious people encouraged the cultivation of storable grains wherever possible: in short, states and trade spread grains, and grain cultivation favored states and trade.

The part of this ecological story that archeological and genetic evidence best (meaning: least incompletely) reveals is the spread of Southwest Asian crops and animals into Europe. Wheat, barley, cattle, pigs, sheep, goats, and horses formed the heart of the Southwest Asian agricultural complex, but peas, lentils, chickpeas, broad beans, and other garden crops were nutritionally important too. Via Anatolia and the Balkans this complex filtered into Europe, averaging about 1 kilometer (0.6 miles) per year, reaching Britain around 4000 BCE. It came to the Baltic later, to Mediterranean shores earlier. Recent genetic evidence implies that it came with human migrants who substantially replaced the previous occupants, but no doubt in many cases European foragers took up herding and farming when presented with the example. They were already in the habit of adjusting to rapid climate change, as the glaciers melted back and cold steppes became forests. All the important domestic animals and most of the crops of Europe as of 2000 BCE were exotics, almost all from Southwest Asia.[7]

As of 2000 BCE, New Guinea, Africa, and the Americas had experienced much more limited biological exchange. The scanty evidence suggests the cultivated zones of highland New Guinea remained highly isolated until modern times. In Africa crops spread mainly along the Sahel, extending from Senegambia to Ethiopia. In the Americas cultivation spread more widely in tropical latitudes but nonetheless interactions of all sorts, biological and otherwise, remained modest in scale compared to Asia, especially the corridor from the Nile to the Indus, the world's most intensively interactive zone around 2000 BCE. One of the constraints on crop movements in the Americas was that the main cultigen of Mesoamerica, an ancestor to today's maize called *teosinte*, had to adjust genetically to different diurnal rhythms before it could prosper at different latitudes. This requirement slowed its northward migration from Mesoamerica, and slowed the diffusion of the concept and practice of agriculture.

Over the next 3,000 or so years, the pace of biological exchange sped up, but in fits and starts. When conditions became favorable for trade, for example when large states imposed peace over broad areas, biological exchange picked up. At the apogee of the Roman and Han empires, roughly 100 BCE to 200 CE, trans-Asian routes, collectively known as the Silk Road, saw heightened traffic, inevitably including plants, animals, and pathogens. China acquired grapes, African sorghum, camels, and donkeys, while Rome received cherries, apricots, peaches, walnuts, and possibly smallpox and measles too. The great plagues of the late second century, which reduced the population of the Roman Empire by perhaps a third, probably included smallpox in a starring role. Garlic, first mentioned in Chinese texts in 510 CE, might also have arrived from western Asia with Silk Road traders. Northern India and Central Asia, of course, took part in these exchanges as well, although the evidence of their involvement remains slender.

In Eurasian history two further moments of heightened biological exchange followed, both thanks to propitious political conditions. The second (the Rome–Han era being the first) and least consequential came during the early Tang dynasty (in China), in the seventh and eighth centuries CE. The Tang rulers (618–907) came from diverse ethnic and cultural backgrounds, and for a century and a half showed keen interest in foreign trade, technology, culture (e.g., Buddhism), as well as plants and animals. The Tang court imported exotica: curious creatures, aromatic plants, ornamental flowers, and so forth. Much of this was inconsequential in social and economic terms, but some of it, such as the cultivation of cotton (imported from India), was not. Cotton textiles soon became a significant part of the Chinese economy, and remained so ever after. The Tang dynasts were culturally receptive to strange plants and animals, but at least as importantly their political power on the western frontier, and the geopolitical situation generally before 750, promoted the trade, travel, and transport that make biological exchange likely. For roughly a century and half (600–750) the numerous polities of Central Asia were frequently consolidated into only a few, simplifying travel and lowering protection costs. A handful of large empires held sway throughout Central Asia, making the connections between China, India, Persia – the latter united under the Sassanid dynasts until 651 – and the Levant safer than usual. Although initially disruptive of peace and trade, the Islamic conquests in southwestern Asia soon temporarily improved conditions for trans-Asian travel. But this geopolitical arrangement fell apart after 751, when Muslims defeated Tang armies, and after 755 when rebellion shook the Tang to its foundations. Thereafter both the stability of the geopolitical situation and the receptivity of the Tang to things foreign changed, waned more often than waxed, and the opportunities for biological exchange diminished.[8]

The third moment came with the Pax Mongolica of the thirteenth and fourteenth centuries. The Mongols established their rule with liberal resort to brutality, but once in power enforced order and encouraged trade. By this time most of the most promising exchanges of plants and animals had already taken place. But the heightened transport across the desert-steppe corridor of Central Asia may have brought carrots and a species of lemon to China, and a form of millet to Persia. Quite possibly, it also allowed the quick diffusion from Central Asia, or perhaps from Yunnan in southwestern China, of the bacillus that causes bubonic plague, provoking the famous Black Death, the worst bout of epidemics in the recorded history of Europe, Southwest Asia, and North Africa. Plague may also have afflicted the heartlands of China in these centuries, and perhaps sub-Saharan Africa too, although the evidence is ambiguous.

Two further biological exchanges within Eurasia deserve special attention for their historical significance on regional scales. The first is the case of so-called champa rice, a strain first cultivated in Vietnam. By the deliberate policy of a Song emperor, champa grains were distributed widely to Chinese farmers beginning in 1012 CE. Champa rice was more resistant to drought than Chinese strains, and much quicker to mature. With champa rice, farmers could double-crop or even triple-crop in a single year, or they could develop crop rotations, alternating rice and soy for example. Within 150 years, champa rice was the commonest variety in many parts of China from the Yangzi southward, and opened up new terrain, especially poorly watered hillslopes, to rice cultivation. The enormous economic expansion of the Song, the emergence of a more commercial economy, and even the development of an iron and coal protoindustrialization in the north all depended on food surpluses that champa rice made possible.[9]

At roughly the same time, the expansion of an Islamic trading world that linked the Indian Ocean shores with those of the Mediterranean brought about a whole range of crop transfers within western Eurasia (and North Africa). Between the tenth and thirteenth centuries, overland and seaborne commercial networks facilitated by the relative peace supervised by the Baghdad-based Abbasid caliphate (750–1258, although from about 900 to about 1120 they often ruled in name only) brought sugar, cotton, rice, and citrus fruits from India to Egypt and the Mediterranean. These plants, and the cultivation techniques that came with them, worked a small agricultural revolution on the hot and often malarial coastlands of North Africa, Anatolia, and southern Europe.[10] Many coastal plains could now be brought under cultivation on a regular basis, often for the first time since the Roman Empire. Sugar and cotton could flourish despite unskilled and unmotivated slave labor; their introduction and spread may have quickened the pace of the slave-raiding that kept Mediterranean and Black Sea populations anxious for centuries. Keeping armies of laborers at work on malarial coasts – in the Levant, Egypt, Cyprus, Crete, Sicily, Tunisia, and Andalusia to mention a few centers of sugar production – required constant resupply from poorly defended peasantries. Sometimes this quest took slave raiders to the Black Sea coasts, but it also took slave merchants across the Sahara and along Atlantic African coasts. Saadi dynasts, originally based in the Sous and Draa river valleys of Morocco, brought sugar and African slaves together in a profitable mix beginning around 1350. They extended their plantation economy and their state, taking over most of Morocco by 1550, by which time their economic formula of sugar and African slaves would be transplanted to Atlantic islands such as the Canaries and Madeira, to Principe and São Tomé, and then to the Americas.

While this process of Eurasian (and North African) biological exchange never truly came to an end, it slowed whenever political conditions weakened interregional contacts.

It waned after 200 CE, with the erosion of the Pax Romana and Pax Sinica which had so encouraged long-distance travel and trade within Eurasia. It slowed especially after the end of the Pax Mongolica around 1350. But by that time, sugarcane had taken root in India and the Mediterranean; wheat had spread widely throughout most of its potential range, as had cattle, pigs, horses, sheep, and goats. In other words, less and less that was worth doing was left to do even when political conditions encouraged biological exchange.

Meanwhile, on other continents, similar if smaller-scale processes of biological exchange and homogenization were in train. In the Americas, maize spread from its Mesoamerican home both north and south, despite the difficulties in adapting to different day lengths at different latitudes. Maize cultivation allowed much denser settlement, and underpinned almost every major concentration of population in North America – salmon-eating peoples in the Pacific Northwest being the likeliest exception. Maize is superb at converting sunshine into edible calories, but as a staff of life it is deficient in important nutrients. So where Amerindians took up maize as their main subsistence crop, they tended to become more numerous, but less healthy, and smaller in stature. By 1000 CE maize had reached the Atlantic seaboard of North America and as far north as Quebec.[11]

In Africa, connections along the Nile corridor brought domesticated cattle south of the Sahara allowing the development of pastoral societies from what is now Sudan to South Africa. By 500 BCE or so, traffic along the Nile corridor introduced malaria to the Mediterranean world, where it remained a scourge for the next 25 centuries. The Bantu migrations of some 2,000 years ago probably diffused several Sahelian crops, mainly sorghums and millets, throughout eastern and southern Africa, and possibly brought infectious diseases that ravaged the indigenous, previously isolated, populations of southern Africa. Bantu immigrants almost surely spread malaria as they went, inadvertently easing their colonization. Sometime around 300 CE some unknown African Columbus inaugurated regular, if at first infrequent, travel across the western half of the Sahara, connecting the Sahel and the Maghreb. The trailblazers needed domesticated camels for trans-Saharan routes (as one needed a stout ship to cross the Atlantic), and good camel saddles, available only by 200 CE or so. Trans-Saharan routes gradually became routine crossings, if always dangerous, and the avenue by which (linguistic evidence suggests) large horses entered West Africa. Large horses allowed cavalry, a new military format for West Africa, and aided in the construction of states and empires far larger than any before, such as those of Mali and Songhai (c. 1230–1591). West Africa acquired an equestrian aristocracy. This aristocracy, and the states they created, were sustained economically in part by slave-raiding, an activity that mounted warriors could pursue much more efficiently than could anyone on foot.[12]

These biological exchanges in the Americas and Africa must have been ecologically and politically tumultuous, although the evidence concerning their impacts is very sparse except in the case of horses in West Africa.

Seaborne Biological Exchange and Biological Invasion before 1400 CE

Intercontinental biological exchange has a long pedigree as the aforementioned example of the first human and canine arrivals in Australia and the Americas shows. Throughout the Southwest Pacific, initial human settlement of unpopulated islands wrought major ecological changes, including numerous extinctions, especially of birds, from about

4,000 years ago until the colonization of New Zealand roughly a millennium ago. All these instances were cases of invasions of "naive" lands – continents and islands that had no prior exposure to humanity and its fellow travelers, or to the intensified fire regime that human presence normally brought. This helps to explain the dramatic effects, particularly the rash of extinctions, that followed upon the first human settlement of Australia, New Zealand, and the Americas.

In one of the enduring mysteries in the history of biological exchanges, the sweet potato, a native of South America, somehow arrived in central Polynesia by 1000 CE and subsequently spread widely throughout Oceania. It is a delicate crop and could not survive a driftwood voyage. No one doubts that people transported it, although no one knows just when, how, or who. It eventually became a staple food in the western Pacific, highland New Guinea, and to a lesser extent the East Asian archipelagos and mainland. The legendary maritime skills of the Polynesians, and the paucity of useful native species on their islands, combined to make Polynesia, proportionally speaking, home to exotic plants and animals more often than almost anywhere on earth.[13]

Once mariners had figured out the annual rhythms of the monsoon winds, the Indian Ocean provided a reliable sailing environment and became another pathway for seaborne biological exchange. Some time before 500 CE, somebody brought bananas, Asian yams, and taro to East Africa. These Southeast Asian crops had much to recommend them, as they do well in moist conditions, whereas the millets and sorghum that Bantu expansion brought into central and southeastern Africa were adapted to drier conditions. Plantains, of which bananas are one variety, had existed in the wild from India to New Guinea. Linguistic and genetic evidence suggests they arrived on the East African coast as early as 3,000 years ago, and reached the forest zone to the west of the great lakes (now Congo) around 2,000 years ago, just about the time of the Bantu migrations. Quite possibly the success of Bantu speakers, often attributed to their use of iron, owed something to their successful adoption of these exotic Asian crops. As relative newcomers to East and southern Africa, Bantu migrants had less invested in prevailing ecological patterns and fewer disincentives to experiment. Bananas, taro, and yams were probably introduced to East Africa more than once, and almost surely were brought again in the Austronesian settlement of Madagascar that took place soon before 500 CE. These Asian crops assisted crucially in the epic (but unrecorded) colonization of Central Africa's moist tropical forests by farmers, as well as in the settlement of Madagascar.[14]

Several other significant intercontinental biological transfers took place before 1400 CE, mainly between Africa and Asia, a route that posed minimal obstacles to sailors. Africa's pearl millet, derived from a West African savanna grass, today is the world's sixth most important cereal. It was brought to South Asia some 3,000 years ago, and today accounts for about 10 percent of India's cereal acreage. East African sorghum entered India at about the same time, and eventually became India's second most important grain after rice. Sorghum stalks were useful as fodder for India's cattle. Finger millet, also from Africa, made it to India only around 1,000 years ago, but became the staple grain in Himalayan foothill communities and in India's far south. The main effect of the transfer of African crops to South Asia was to provide India with drought-resistant dryland crops, opening new areas to settlement and providing a more reliable harvest where water supplies were uncertain. This effect mirrors that of Asian crops in East and southeastern Africa, in which the introduced species encouraged settlement of humid forestlands. These examples suggest a very lively world of crop exchange – and probably weeds, diseases, and animals too – around the Indian Ocean

Rim from around 3,000 to 1,500 years ago. The regular monsoon winds of the Indian Ocean helped make this region of the world precocious in its maritime networks and hence in biological exchange.

Some scholars are convinced that significant biological exchange between the Americas and other continents, conducted via sailing craft, also took place in the millennia before 1400 CE. The evidence – to my mind not genuinely convincing in most cases – ranges from the presence of hookworms in both Asia and the Americas to artwork that seems to show maize in India in the fifth century CE. Some ancient texts can be translated so as to indicate chili peppers or pumpkins, both American crops, existed in China or India long ago. Sixteenth-century Spanish texts claim that mulberry trees (an East Asian species) already existed in Mexico before the conquistadors arrived, and that people there already knew how to make paper from mulberry bark, allegedly showing that the tree had been brought purposely. Scholars have advanced claims for nearly 100 human-induced plant transfers to or from the Americas before Columbus. If half of these were true, then the period that followed would amount only to a continuation of an ongoing genuinely global migration of species. If, as I suspect, most of these claims are untrue, then the period that followed was one of revolutionary biotic change, especially in the Americas.[15]

Biological Globalization after 1400 CE

After 1400, mariners, most of them Atlantic Europeans, linked almost every nook and cranny of the humanly habitable earth into a biologically interactive web. The world's seas and deserts no longer served to isolate biogeographical provinces. The land surface of the Earth became a world without biological borders, as plants, animals, and diseases migrated wherever ecological conditions permitted their spread, although how soon and how thoroughly they did so often depended on transport technologies and skills, and patterns of trade, production, and politics.

Columbus inaugurated regular exchanges across the Atlantic in 1492. On his second voyage he deliberately brought an arkful of species new to the Americas. Over the next few centuries his followers brought still more in an ongoing process now known to historians as the Columbian exchange, after the title of Alfred Crosby's 1972 book on the subject.[16] The most conspicuous result was that Amerindians acquired a large suite of new plants and animals, as well as devastating diseases hitherto unfamiliar to them. Those diseases included smallpox, measles, mumps, whooping cough, and influenza, all of which had become fairly widespread in the sprawling interactive zone from Japan to Senegambia. They were endemic, childhood diseases (sometimes called "crowd diseases" because they require large, interacting populations to stay in circulation) that contributed to high rates of infant mortality. But most adults were survivors, and either partially resistant or fully immune to most or all of these infections. In addition to the crowd diseases, the Columbian exchange brought some lethal vector-borne African diseases to the Americas, such as yellow fever and malaria. In the case of yellow fever, the relevant mosquito vector *Aedes aegypti*, was also an import from Africa.[17] In the Americas all these were new pathogens, so no Amerindians carried any useful resistance. Thus exotic infections ravaged the American hemisphere between 1500 and 1650, lowering populations by 50 to 90 percent in one of the two largest-scale demographic disasters in world history (the other being the Black Death).

In one important respect, this horrific experience was an extension of older patterns. Ever since the rise of dense human populations living cheek by jowl with domesticated herd animals, which means approximately since 3500 BCE, infectious disease has assisted some people at the expense of others. By and large, those who lived in close quarters with many others, and who routinely swapped bacteria and viruses with animals, developed stronger resistance to more infectious diseases than did others (and those in such populations who did not tended to die in infancy). They maintained a quiver full of lethal infectious diseases, circulating among children, which they unwittingly launched against other populations with less extensive disease experience. This differential experience with, and resistance to, infectious disease goes a long way toward explaining why large, dense populations usually expanded territorially at the expense of smaller, more scattered ones.[18] In the case of the pathogen component of the Columbian exchange, this pattern asserted itself on a grand scale.

The Amerindians had little in the way of lethal infectious disease to export to Africa and Eurasia. When the first migrants had arrived in North America 14,000 years previously, they passed through northeastern Siberia and Alaska during an ice age. Brutal cold is not conducive to the survival of most pathogens, so they arrived relatively free from infection. Beyond that, they left Siberia when no animals but dogs had been domesticated, so that the infections derived from cattle, camels, and pigs (e.g., measles, smallpox, influenza) had not yet evolved. Once in the Americas, Amerindians did not domesticate any herd animals other than alpacas and llamas, which seem, by chance, not to have hosted pathogens that evolved into agents of human disease. As regards pathogens, the Columbian exchange was a notably one-sided affair.

The same was true of domesticated animals. The Americas had little in the way of domesticated animals, and what they had did not travel well. American turkeys spread to other continents, but nowhere did they become important. Alpacas and llamas never prospered outside their native Andes, although scattered populations do exist elsewhere. On the other hand, Eurasian species flourished when transported to the Americas. Cattle, goats, sheep, pigs, and horses were the most important large animal immigrants. They all found empty niche space in the Americas, especially cattle and horses on the vast grasslands of the pampas, the llanos, and the prairies. The new animals provided Amerindians with new sources of animal protein, of which they had comparatively little before 1492, and of hides and wool. Horses and oxen offered an important source of traction, making plowing feasible in the Americas for the first time, and improving transportation possibilities through wheeled vehicles and greater variety of pack animals. That extended the potential of commerce and specialization in production, which could be socially divisive but at the same time raised overall economic production. The new animals introduced new frictions to the Americas too. They habitually munched and trampled crops, provoking quarrels between herders and farmers of the sort familiar throughout Africa and Eurasia but hitherto almost unknown in the Americas.[19]

As in West Africa, horses in North America upset the political order. The Amerindians of the prairies acquired horses from Spanish Mexico in the seventeenth century and some of them quickly mastered the equestrian arts. On horseback, they became far more adept as bison hunters, solving any subsistence problems as long as the bison lasted. Moreover, those with horses easily inflicted military defeat on those without, so by 1850 or so peoples such as the Sioux and Comanche had built considerable territorial empires on the basis of mounted warfare.[20]

Horses and humans share an affinity that often affected human history. Horses are by far the easiest animals for people to ride (elephants and camels come far behind), and thus for the millennia between their first domestication and the advent of motorized vehicles, horses represented the apogee in human (terrestrial) mobility. Their speed and power helped make them militarily important until the early twentieth century. Horses and humans are the only species that can sweat profusely and thus sustain strenuous exercise for hours on end. Together they can chase other species – wild cattle, bison, deer, bears, kangaroos – until their quarry overheats and collapses. Together they can fight cavalry actions that last all day, something beyond the capacity of war elephants and camels (the only other species often harnessed for military purposes). In their ability to cool their bodies by profuse sweat, horses and humans were made for each other. They made a formidable tandem wherever they went, often too strong for human populations without horses to withstand.

The Columbian exchange was more reciprocal when it came to crops. The Eurasian staples of wheat, rye, barley, and rice found welcoming niches in the Americas. Sometimes those niches had to be created through human (and animal) labor, as for example with rice, both Asian and African, in Brazil, Surinam, and South Carolina. Some of the new crops could survive in cold and dry landscapes where the indigenous crops fared poorly: Saskatchewan does better growing wheat than maize. Aside from grains, the Americas also acquired citrus fruits, bananas, grapes, and figs from Eurasia, and millets, sorghums, yams, okra, and watermelon from Africa. So the new crops extended the possibilities of American agriculture somewhat, and allowed a more varied diet. But they did not constitute a vast improvement, because the Americas already had maize and potatoes and plenty of fruits and vegetables.

Drug crops changed the Americas at least as profoundly as the food crops. Sugar, originally a grass from New Guinea, but a commercial crop in South Asia, China, and the Mediterranean, came to Brazil and the Caribbean in the sixteenth and seventeenth centuries. Both a mild drug and a food, it became the mainstay of a plantation economy based on African slave labor, an extension of the pattern pioneered by the Saadi monarchs in Morocco (see above). Coffee, from Ethiopia and Arabia, also became a plantation crop in the eighteenth century. Without these imported crops, the plantation complex of the Americas would have been a much smaller matter, and the Atlantic slave trade far smaller as well.

Maize and potatoes, together with cassava, tomatoes, cacao, peanuts, sweet potato, pumpkins, squashes, pineapples, and a handful of others, formed the Americas' contribution to global food crops. South America also gave tobacco to the world. Some of these crops had revolutionary consequences in sizeable regions of Africa and Eurasia. Potatoes, for example, nicely suited soil and climate conditions in northern Europe from Ireland to Russia. Their arrival and spread played a considerable role in allowing the surge of population growth there in the eighteenth and nineteenth centuries, which helped supply both the man power for overseas empires and the labor for the Industrial Revolution. Potatoes stored well, especially in cold climates, and they contain excellent nutrition. In their native Andes, where production and storage had been raised to a high art, potatoes had helped fuel the expansion of the Inka empire in the fifteenth century. A few centuries later they played a broadly similar role in northern Europe. Their career in Europe, especially Ireland, also shows the dangers of over-reliance on a single staple. The Irish potato famine of 1845–52, a result of crop failure caused by a potato disease, cost Ireland a quarter of its population – a million died and million emigrated.[21]

The potato's impact on northern Europe's food supply required another biological migrant: clover. Clover, although not from the Americas, was essential to the impact of the Columbian exchange in Europe. Clover hosts microorganisms that have the rare capacity to fix nitrogen from the air in the soil, putting it where plant roots can absorb it. All plants need nitrogen, and for many it is the nutrient in shortest supply. Potatoes are especially nitrogen-hungry crops, and cannot yield well consistently without nitrogen supplements to the soil. While clover grew wild throughout Europe and southwest Asia, the earliest evidence of clover cultivation comes from Islamic Andalusia in the thirteenth century. Castilians soon learned the value of clover cultivation, and the practice of using clover in rotation with food crops spread northward with Spaniards (in this respect the territorial expansion of the House of Habsburg proved a blessing to all Europeans). Lombardy and the Low Countries became centers of clover cultivation. By 1620, English farmers had begun to plant clover, and by 1850 it was in widespread use throughout northern Europe. Clover's nitrogen-fixing properties allowed the potato to have its nutritional and demographic impact. And because clover makes an excellent fodder crop, it also allowed higher cattle populations and helped raise milk- and meat-production levels throughout northern Europe. Although as yet unsung by historians, clover can claim an impact on European history as great as that of the potato.[22]

Maize had a more diffuse impact than the potato and clover. It did well in conditions as varied as those of southern Europe, China, and large swathes of Africa. Maize allowed farmers to till new lands, because it prospered where grains and tubers would not. It undergirded population growth and famine resistance in China and southern Europe, but nowhere was it more influential than in Africa, where today it is the single most important food. In the two centuries after 1550, maize became a staple in Atlantic Africa, from Angola to Senegambia. Different varieties suited the several different rainfall regimes in Africa, and improved African chances of surviving drought. Maize stores much better than millets, sorghums, or tubers. It thus allowed chiefs and kings to maximize their power by centralizing the storage and distribution of food. In the West African forest zone, maize encouraged the formation of larger states than had existed before. The Asante kingdom embarked on a program of expansion after the 1670s, spearheaded by maize-eating armies which could carry their food with them on distant campaigns. Maize also served well as a portable food for merchant caravans, which contributed to commercialization in Atlantic Africa, including the slave trade. The slave trade could more easily reach well inland if merchants, and their human property, had an easily portable food supply. Storehouses of maize also made it more practical to imprison large numbers of slaves in the infamous barracoons of the West African coast. Just as sugar and coffee heightened the demand in the Americas for African slaves, so maize in Africa increased the practicality of the slave trade.[23]

Cassava, also known as manioc, was the Americas' other great contribution to African agriculture. A native of Brazil, cassava is admirably suited to drought and poor soils and resistant to many crop pests. It too did well in many parts of Africa, and like maize provided a portable food that underlay state formation and expansion in West Africa and Angola. It too made slave caravans easier to supply and thus more profitable to run. Cassava, however, like potatoes but unlike maize, need not be harvested at a particular season; it may be left in the ground for weeks or more without spoiling. It is an ideal crop for people who might need to run away for their own safety, for example people routinely subject to slave-raiding. In this respect it served as a counter to maize and was self-negating as regarded the slave trade: unharvested, it allowed peasantries to flee and survive slave raids, while maize (and harvested cassava) helped slavers to conduct and extend their business.

The Columbian exchange was the largest-scale, fastest, and most important intercontinental biological transfer in world history. But it was not the only one that followed upon the navigational exploits of Columbus' generation. A modest trans-Pacific exchange resulted from Magellan's voyages, at first affecting chiefly the Philippines. A Tagalog folk song which celebrates the abundance of house-gardens mentions 18 plants, none of them native to the Philippines. All 18 came from the Americas, Africa, or East Asia. When in the 1570s the Philippines became the trade hub connecting China and the Americas, the archipelago also became a center for invasive and introduced species. Few places on Earth have quite as global a biota as the Philippines.[24]

Two centuries after the Spanish launched the trans-Pacific trade connecting Mexico to the Philippines, Captain James Cook charted broad expanses of the Pacific, opening a new age of trade involving hundreds of small (and not so small) Pacific islands. The most striking result was sharp depopulation in the wake of repeated epidemics, as had happened before in the Americas and was happening simultaneously in Australia. Pacific islands also lost large numbers of endemic species, birds especially, when introduced species ran wild. They also acquired a broader array of food crops, augmenting the rather narrow traditional roster of cultigens of most islands.[25]

Many American food crops became important in East Asia, such as sweet potatoes, maize, and peanuts in south China. They probably first arrived not across the Pacific in the wake of Magellan, but across the Atlantic and the Indian Oceans via Portuguese and Dutch traders. Biological exchange around the Pacific Rim did not necessarily involve direct voyages across the Pacific.[26]

Taken together the whirlwind of intercontinental biological exchange in the centuries between 1500 and 1800 brought astounding changes around the world. It led to demographic catastrophes in lands unfamiliar with the crowd diseases. It improved the quantity and reliability of food supplies almost everywhere. In cases where horses were new, it reshuffled political relations by providing a new basis for warfare.

All these early modern exchanges carried historical consequences. European imperialism, in the Americas, Australia, and New Zealand, simultaneously promoted and was promoted by the spread of European (or more usually Eurasian) animals, plants, and diseases. Europeans brought a biota that unconsciously worked as a team to favor the spread of European settlers, European power, and Eurasian species, and thereby to create what Alfred Crosby, the foremost historian of these processes, called neo-Europes – including Australia, New Zealand, most of North America, southern Brazil, Uruguay, and Argentina.

Beyond the neo-Europes, something of a neo-Africa emerged in the Americas. More than 10 million Africans arrived in the Americas in slave ships between 1550 and 1850. In those same ships came yellow fever and malaria, which profoundly influenced settlement patterns in the Americas, because they were so lethal to people without prior experience of them. Slave ships also brought West African rice, which became the foundation of the coastal economy in South Carolina and Georgia in the eighteenth century, and important in Surinam as well. Other African crops came too: okra, sesame, and (although not in slave ships) coffee. All this combined to create a neo-Africa in the lowlands from Bahia to the Chesapeake. It was a world of African people and culture, where indigenous peoples almost died out and indigenous culture, including foodstuffs, was partially absorbed into the neo-African matrix. In this it resembled the neo-Europes. But at the same time neo-Africa was profoundly different from the neo-Europes: until the Haitian Revolution (1791–1804), nowhere in the Americas were Africans and people of African descent politically, economically, or socially powerful, whereas Europeans and their descendants soon dominated the neo-Europes.

As the Columbian exchange indicates, sailing ships brought the continents together as never before. But sailing vessels did not prove hospitable conveyances to every form of life. They filtered out a few species that could not for one reason or another survive a long journey. The age of steam, and then of air travel, broke down remaining barriers to biological exchange, accelerating the dispersal of old and new migratory species alike. Although its greatest impacts came in the sixteenth and seventeenth centuries, in a sense the Columbian exchange never ended. American raccoons, grey squirrels, and muskrats for example colonized parts of Europe in the nineteenth and twentieth centuries. European starlings spread throughout North America. Nor did the Magellan exchange come to an end. If anything, it sped up in the nineteenth and twentieth centuries, not least because of deliberate introductions of species such as the kiwi fruit (from China to New Zealand to Chile and California) and eucalyptus trees (from Australia to almost everywhere).

In the eighteenth and nineteenth centuries deliberate introductions became an increasingly institutionalized enterprise. Botanical gardens undertook to spread useful plants far and wide, especially within the confines of European empires. In the most famous example, British plant prospectors took rubber-tree seeds from their native turf in Brazilian Amazonia to Kew Gardens outside London, and from there to British Malaya. A rubber-plantation economy soon blossomed in Malaya, undermining the rubber-tapping business in Brazil by 1912. Dutch authorities managed to get seeds of cinchona trees (native to the eastern slopes of the Peruvian Andes) to Java, where by the 1870s they were producing commercial quantities of quinine, a drug that offered protection against malaria. Cheap quinine made the European empires, and in the early 1940s the Japanese empire too, far more practical in malarial lands than they could otherwise have been. In Australia and New Zealand, settlers organized societies dedicated to the purpose of importing familiar plants and animals from Britain, typically regarded as superior to the native species of the Antipodes. Botanical gardens, plant prospectors, and acclimatization societies all combined with improved transportation technology to sustain biological exchanges in the nineteenth and twentieth centuries.[27]

Inevitably, accidental and unwelcome biological exchange continued as well. Pests such as coffee rust or phylloxera (a menace to grape vines) circulated around the world thanks to improved and intensified transport in the nineteenth century. Cholera escaped from its native haunts around the Bay of Bengal and became a global scourge in the early nineteenth century. At the end of the century rinderpest, an extremely lethal cattle virus, spread to East and southern Africa, wiping out as much as 90 percent of the herds and bringing destitution to pastoral peoples (and opening niche space for wildlife). Around 1880 someone brought the colorful Amazonian flower known as water hyacinth to the Bengal Delta, where it colonized and clogged the waterways by 1910, impeding both navigation and cultivation of rice and jute. The demobilization of millions of World War I soldiers and sailors in 1918, and their quick movements around the world by steamship, spread an influenza virus that killed 30–60 million people, most of them in India. The influenza killed many more people than did the war itself. Faster and more frequent transport opened floodgates anew for invasive species.[28]

Conclusion

Faster and more frequent transport and travel continue to promote biological exchange. Today there are some 44,000 regular air routes, constantly, if inadvertently, moving insects, seeds, and germs around the world.[29] Aquatic species shuttle around the world's harbors in the ballast water carried by ocean-going ships. This was the path taken by the zebra mussel in the 1970s from the Caspian Sea to the lakes and rivers of North America, where its prolific colonization costs billions of dollars annually. Many governments try hard to keep out weeds, pests, and germs, but every success is provisional, valid only until the next arrival of invasive species. And some governments, having other priorities, do not bother. So the long-term process of biological globalization continues, and will inevitably continue.

Like cultural globalization, it will never be full and final. Rubber trees will not colonize Iceland, nor will caribou roam Borneo. But within the limits prescribed by climate, soils, and other ecological conditions – these limits change but they can never disappear – biological exchange will continue to take place and to affect human history. It may now seem that it will never again have the influence that it did in the era of the Columbian exchange 400 and 500 years ago. As regards the exchange of economically useful plants and animals this is probably true: almost all that can usefully happen has already happened, due to determined efforts over the centuries. But in the realm of pests and pathogens it could well be otherwise. In biological history four or five centuries is the merest flash. In the long run strange things will happen, as they so often have before.

Notes

1 On these high-yield crops and the Cold War, for instance, see N. Cullather, *The Hungry World: America's Cold War Battle against Poverty in Asia*, Cambridge, MA, Harvard University Press, 2010.

2 For orientation, see D. Simberloff and M. Rejmank (eds.), *Encyclopedia of Biological Invasions*, Berkeley, University of California Press, 2011. Northern North America lost its earthworms during the last glaciation. In recent centuries invasive earthworm species have colonized the region, devouring leaf litter and thereby changing the ecology of forests and soils.

3 These themes are explored for the US c. 1870–1930 in P. Coates, *American Perceptions of Immigrant and Invasive Species*, Berkeley, University of California Press, 2006.

4 P. Hulme, "Trade, Transport, and Trouble: Managing Invasive Species Pathways in an Era of Globalization," *Journal of Applied Ecology* 46, 2009, pp. 10–18.

5 T. Flannery, *The Future Eaters*, Chatswood, Reed, 1994; T. Flannery, *The Eternal Frontier*, New York, Atlantic Monthly Press, 2001.

6 J. Diamond, *Guns, Germs and Steel*, New York, W. W. Norton, 1997.

7 B. Cunliffe, *Europe between the Oceans*, New Haven, CT, Yale University Press, 2008; L. Cavalli-Sforza, *Genes, Peoples, and Languages*, Berkeley, University of California Press, 2000.

8 E. Schafer, *The Golden Peaches of Samarkand: A Study of T'ang Exotics*, Berkeley, University of California Press, 1963.

9 F. Bray and J. Needham, *Science and Civilization in China*, vol. 6, part 2, *Agriculture*, Cambridge, Cambridge University Press, 1984, pp. 492–5.

10 A. Watson, *Agricultural Innovation in the Early Islamic World: The Diffusion of Crops and Farming Techniques, 700–1100*, Cambridge, Cambridge University Press, 1983. See also a challenge to parts of Watson's position in M. Decker, "Plants and Progress: Rethinking the Islamic Agricultural Revolution," *Journal of World History* 20, 2009, pp. 187–206.

11 J. D. Rice, *Nature and History in the Potomac Country*, Baltimore, The Johns Hopkins University Press, 2009, pp. 26–7.

12 R. Law, *The Horse in West African History*, Oxford, Oxford University Press, 1980.

13 J. R. McNeill, "Of Rats and Men: A Synoptic Environmental History of the Island Pacific," *Journal of World History* 5, 1994, pp. 299–349.

14 J. Iliffe, *Africans: The History of a Continent*, Cambridge, Cambridge University Press, 1995.

15 Enthusiastic support for these views is found in J. L. Sorenson and C. L. Johannessen, "Biological Evidence for Pre-Columbian Transoceanic Voyages," in V. Mair (ed.), *Contact and Exchange in the Ancient World*, Honolulu, University of Hawai'i Press, 2006, pp. 238–97; and J. L. Sorenson and C. L. Johannesson, *World Trade and Biological Exchanges before 1492*, New York, iUniverse, 2009.

16 A. W. Crosby, *The Columbian Exchange: Biological and Cultural Consequences of 1492*, Wesport, CT, Greenwood Press, 1972.

17 J. Webb, *Humanity's Burden: A Global History of Malaria*, New York, Cambridge University Press, 2009; J. R. McNeill, *Mosquito Empires: Ecology and War in the Greater Caribbean, 1620–1914*, New York, Cambridge University Press, 2010; L. P. Lounibos, "Invasions by Insect Vectors of Human Disease," *Annual Review of Entomology* 47, 2002, pp. 233–66.

18 W. H. McNeill, *Plagues and Peoples*, New York, Doubleday, 1976.

19 E. G. K. Melville, *A Plague of Sheep*, New York, Cambridge University Press, 1994.

20 P. Hämäläinen, *The Comanche Empire*, New Haven, CT, Yale University Press, 2008.

21 A recent summary of potato history is J. Reader, *Propitious Esculent*, London, Heinemann, 2008.

22 T. Kjaergaard, "A Plant that Changed the World: The Rise and Fall of Clover, 1000–2000," *Landscape Research* 28, 2003, pp. 41–9.

23 J. McCann is the starting point for the history of maize in Africa: *Maize and Grace: Africa's Encounter with a New World Crop*, Cambridge, MA, Harvard University Press, 2005. On the relevance of introduced crops to the slave trade, see S. Alpern, "The European Introduction of Crops into West Africa in Precolonial Times," *History in Africa* 19, 1992, pp. 13–43.

24 The song is quoted in C. C. Mann, *1493: Uncovering the New World Columbus Created*, New York, Knopf, 2011, pp. 385–6.

25 McNeill, "Of Rats and Men." See also D'Arcy (Chapter 12) in this volume.

26 For a review, S. Mazumdar, "The Impact of New World Food Crops on the Diet and Economy of China and India, 1600–1900," in R. Grew (ed.), *Food in Global History*, Boulder, CO, Westview Press, 2000, pp. 58–78.

27 R. H. Grove, *Green Imperialism: Colonial Expansion, Tropical Island Edens and the Origins of Environmentalism, 1600–1860*, New York, Cambridge University Press, 1995; R. Drayton, *Nature's Government: Science, Imperial Britain, and the "Improvement" of the World*, New Haven, CT, Yale University Press, 2000; W. Dean, *Brazil and the Struggle for Rubber*, New York, Cambridge University Press, 1987.

28 S. McCook, "Global Rust Belt: *Hemileia vastatrix* and the Ecological Integration of World Coffee Production since 1850," *Journal of Global History* 1, 2006, pp. 177–95; C. Hamlin, *Cholera: The Biography*, Oxford, Oxford University Press, 2009; I. Iqbal, "Fighting with a Weed: Water Hyacinth and the State in Colonial Bengal, c. 1910–1947," *Environment and History* 15, 2009, pp. 35–59; J. Taubenberger and D. Morens, " 1918 Influenza: The Mother of All Pandemics," *Revista Biomédica* 17, 2006, pp. 69–79.

29 A. Tatem and S. Hay, "Climatic Similarity and Biological Exchange in the Worldwide Airline Transportation Network," *Proceedings of the Royal Society: Biological Sciences* 274, 2007, pp. 1489–96.

References

Alpern, S., "The European Introduction of Crops into West Africa in Precolonial Times," *History in Africa* 19, 1992, pp. 13–43.

Bray, F., and Needham, J., *Science and Civilization in China*, vol. 6, part 2, *Agriculture*, Cambridge, Cambridge University Press, 1984.

Cavalli-Sforza, L., *Genes, Peoples, and Languages*, Berkeley, University of California Press, 2000.

Coates, P., *American Perceptions of Immigrant and Invasive Species*, Berkeley, University of California Press, 2006.

Crosby, A. W., *The Columbian Exchange: Biological and Cultural Consequences of 1492*, Wesport, CT, Greenwood Press, 1972.

Cullather, N., *The Hungry World: America's Cold War Battle against Poverty in Asia*, Cambridge, MA, Harvard University Press, 2010.

Cunliffe, B., *Europe between the Oceans*, New Haven, CT, Yale University Press, 2008.

Dean, W., *Brazil and the Struggle for Rubber*, New York, Cambridge University Press, 1987.

Decker, M., "Plants and Progress: Rethinking the Islamic Agricultural Revolution," *Journal of World History* 20, 2009, pp. 187–206.

Diamond, J., *Guns, Germs and Steel*, New York, W. W. Norton, 1997.

Drayton, R., *Nature's Government: Science, Imperial Britain, and the "Improvement" of the World*, New Haven, CT, Yale University Press, 2000.

Flannery, T., *The Eternal Frontier*, New York, Atlantic Monthly Press, 2001.

Flannery, T., *The Future Eaters*, Chatswood, Reed, 1994.

Grove, R. H., *Green Imperialism: Colonial Expansion, Tropical Island Edens and the Origins of Environmentalism, 1600–1860*, New York, Cambridge University Press, 1995.

Hämäläinen, P., *The Comanche Empire*, New Haven, CT, Yale University Press, 2008.

Hamlin, C., *Cholera: The Biography*, Oxford, Oxford University Press, 2009.

Hulme, P., "Trade, Transport, and Trouble: Managing Invasive Species Pathways in an Era of Globalization," *Journal of Applied Ecology* 46, 2009, pp. 10–18.

Iliffe, J., *Africans: The History of a Continent*, Cambridge, Cambridge University Press, 1995.

Iqbal, I., "Fighting with a Weed: Water Hyacinth and the State in Colonial Bengal, c. 1910–1947," *Environment and History* 15, 2009, pp. 35–59.

Kjaergaard, T., "A Plant that Changed the World: The Rise and Fall of Clover, 1000–2000," *Landscape Research* 28, 2003, pp. 41–9.

Law, R., *The Horse in West African History*, Oxford, Oxford University Press, 1980.

Lounibos, L. P., "Invasions by Insect Vectors of Human Disease," *Annual Review of Entomology* 47, 2002, pp. 233–66.

Mann, C. C., *1493: Uncovering the New World Columbus Created*, New York, Knopf, 2011.

Mazumdar, S., "The Impact of New World Food Crops on the Diet and Economy of China and India, 1600–1900," in R. Grew (ed.), *Food in Global History*, Boulder, CO, Westview Press, 2000, pp. 58–78.

McCann, J., *Maize and Grace: Africa's Encounter with a New World Crop*, Cambridge, MA, Harvard University Press, 2005.

McCook, S., "Global Rust Belt: *Hemileia vastatrix* and the Ecological Integration of World Coffee Production since 1850," *Journal of Global History* 1, 2006, pp. 177–95.

McNeill, J. R., *Mosquito Empires: Ecology and War in the Greater Caribbean, 1620–1914*, New York, Cambridge University Press, 2010.

McNeill, J. R., "Of Rats and Men: A Synoptic Environmental History of the Island Pacific," *Journal of World History* 5, 1994, pp. 299–349.

McNeill, W. H., *Plagues and Peoples*, New York, Doubleday, 1976.

Melville, E. G. K., *A Plague of Sheep*, New York, Cambridge University Press, 1994.

Reader, J., *Propitious Esculent*, London, Heinemann, 2008.

Rice, J. D., *Nature and History in the Potomac Country*, Baltimore, The Johns Hopkins University Press, 2009.

Schafer, E., *The Golden Peaches of Samarkand: A Study of T'ang Exotics*, Berkeley, University of California Press, 1963.

Simberloff, D., and Rejmank, M. (eds.), *Encyclopedia of Biological Invasions*, Berkeley, University of California Press, 2011.

Sorenson, J. L., and Johannessen, C. L., "Biological Evidence for Pre-Columbian Transoceanic Voyages," in V. Mair (ed.), *Contact and Exchange in the Ancient World*, Honolulu, University of Hawai'i Press, 2006, pp. 238–97.

Sorenson, J. L., and Johannesson, C. L., *World Trade and Biological Exchanges before 1492*, New York, iUniverse, 2009.

Tatem, A., and Hay, S., "Climatic Similarity and Biological Exchange in the Worldwide Airline Transportation Network," *Proceedings of the Royal Society: Biological Sciences* 274, 2007, pp. 1489–96.

Taubenberger, J., and Morens, D., "1918 Influenza: The Mother of All Pandemics," *Revista Biomédica* 17, 2006, pp. 69–79.

Watson, A., *Agricultural Innovation in the Early Islamic World: The Diffusion of Crops and Farming Techniques, 700–1100*, Cambridge, Cambridge University Press, 1983.

Webb, J., *Humanity's Burden: A Global History of Malaria*, New York, Cambridge University Press, 2009.

CHAPTER TWENTY-FIVE

Environmentalism in Brazil: A Historical Perspective

José Augusto Pádua

Introduction

In order to understand the history of environmentalism in Brazil, it must be placed within the broader context of European expansion and the emergence of new countries after the end of the colonial system.[1] The intellectual and social movements related to the environmental issue in Brazil cannot be dissociated from similar movements that took place at several different moments in Western history. The external influence is quite evident, yet the Brazilian case shows the superficiality of schematic perspectives that treat the issue as if it were a one-sided relationship composed exclusively of influence from outside forces; as if environmentalism in Brazil were a mere replication, generally delayed and outdated, of collective concerns and actions that first manifested themselves in Europe and the US.

This type of reading, particularly in the 1970s, caused many of the early critics – from both the right and the left of the political spectrum – of the environmental movement that was then organizing itself in Brazil to argue that environmental concerns were exotic and "misplaced," given the country's social and economic reality. For a developing country, the main priorities needed to be growth and job creation.

Recent historical research has started to reveal a picture that is far more complex. First, by showing the historical breadth of environmental concerns in Brazil: the protoenvironmentalist debate, beginning toward the end of the eighteenth century, with its perception of the problems brought about by deforestation and soil degradation, was no less consistent in Brazil than it was in Europe and the US during that same period.[2] Second, by showing that the formation of environmentalist groups that adhered to contemporary molds – self-aware environmentalism focused on action in the public realm – in essence followed the same timeline in Brazil as it did throughout the West: the first groups began to appear in 1971, at the time Greenpeace was being founded in

A Companion to Global Environmental History, First Edition. Edited by J.R. McNeill and Erin Stewart Mauldin.
© 2012 John Wiley & Sons, Ltd. Published 2015 by John Wiley & Sons, Ltd.

Canada and Friends of the Earth in the US.[3] As such, the study of Brazilian environmentalism can serve as an important component to understanding the birth and evolution of environmentalism on a global scale.

A comparative international approach, however, must not obscure the local characteristics and peculiarities of the Brazilian history. As is often the case with social phenomena, Brazilian environmentalism was forged through a combination of endogenous and exogenous factors.[4]

First, for reasons that warrant a far more extensive analysis, Brazil was not at the forefront of the social and technological transformations brought about by modern capitalism. The burden of slavery endured until the end of the nineteenth century and Brazil reached the mid-twentieth century with an essentially rural population, low levels of education, and a high concentration of wealth in the hands of a relatively small group of elites. Modern urban social contact, which is often the engine of political creativity, was geographically quite limited. Yet, beginning in the second half of the twentieth century, the Brazilian social framework experienced a genuine mutation as the country's economic growth reached some of the highest levels seen anywhere on the planet, despite passing through periods of crisis, stagnation, and inflation.[5] The percentage of the country's urban population rose from 16 percent in 1940 to 81.3 percent in 2000. After two long periods of dictatorial reign, from the 1930s onward, political democracy was consolidated in the 1980s and has come to be associated, during the last decade, with strong policies of wealth redistribution overseen by leftist parties.[6]

The characteristics mentioned above warrant further consideration from an environmental point of view. Until the mid-twentieth century, criticism regarding the destruction of nature in Brazil was focused on archaic rural practices, such as slash-and-burn deforestation. At the end of the twentieth century, despite the continuation of archaic practices in several regions, such as the large-scale slash-and-burn operations in the Amazon rain forest, the criticism began to focus on the ecological effects of a rapid and aggressive process of capitalist modernization.[7]

The intensity of the socioeconomic and geographic shifts resulted in strong movements toward: (1) the expansion and remodeling of urban landscapes, with an increase in pollution and the destruction of traditional buildings and neighborhoods; (2) the expansion of infrastructure, particularly hydroelectric plants and highways; (3) the expansion of industrial areas and the storage of contaminants; (4) the opening of new frontiers for ranching and farming in regions previously covered with tropical forests and other native ecosystems and occupied by traditional (often indigenous) populations with low demographic density; (5) the conversion of areas of long-standing traditional agricultural practices, with established rural populations, into large-scale agribusiness operations based on the use of machines and agrochemicals. Each of these dynamics necessarily fomented different types of environmental conflicts involving the local rural and urban communities. Yet, the fact that these dynamics largely occurred during dictatorial regimes helped to increase the aggressiveness of these conflicts and to complicate, although not fully impede, reactions on the part of society and in the media. This was especially true during the transition from the 1960s to the 1970s, when Brazil was growing at very fast rates under a series of military governments.[8]

This aggressive and rapid urban and industrial growth helps explain the growth of environmental awareness in Brazilian society beginning in the 1970s. However, this specific factor must also be associated with the deeper cultural aspects of the environmental debate in Brazil. Perhaps the most striking element of Brazil to the outside world, and to the nation's own sense of identity, is its territory of over 850 million hectares – the result of complex political engineering that began in the nineteenth century and guaranteed the unification of the mosaic of regions that made up Portuguese America. Brazil is the fifth largest country in the world (see Map 25.1). Almost all of the territory is located in a tropical area, rich in ecological assets, particularly the Amazon rain forest, which has gained increasing planetary relevance in the twenty-first century. The scientific and geopolitical significance that the modern world places on factors such as the concentration of biodiversity, the abundance of fresh water, the incidence of solar radiation, and the capacity of the vegetation to store carbon has guaranteed that the debate over the ecological importance and the ultimate destiny of the Brazilian territory will continue unabated. Every day, the media presents scientific discoveries, social conflicts, and political disputes related to the territory's future. The environmental issue has gained increasing prominence in Brazil's public agenda. The UN Conference on Environment and Development, held in Rio de Janeiro in 1992, firmly established Brazil's role in the international politicization of this issue, helping to intensify its domestic and local politicization, as well.

While the scope of the current debate has expanded considerably, the issues surrounding nature and territory have deep roots in Brazilian political culture. Since the country's independence, in 1822, Brazil's territorial breadth has been considered a major asset that destiny offered to ensure the progress and greatness of the country. In fact, arguments used in favor of independence were based on Brazil's natural bounty. In letters of advice received by Brazilian diplomats who were involved in the 1825 negotiations in London over the recognition of Brazilian independence, one of the central arguments was that "such an extensive empire, provided by nature with some of the world's best ports, with extensive coastline, and a variety of rich natural products, demands to be a separate and independent power."[9]

In the country's subsequent process of political consolidation under a monarchical order, images of nature, and its primitive indigenous inhabitants, were used by the official culture, which was heavily influenced by romanticism, in order to enhance the standing of Brazil in the eyes of Western civilization. The highly idealized self-image being spread at the time was one of a legitimate, solid, and peaceful monarchy that, step by step, was civilizing a splendorous but wild tropical territory.

Even with the end of the monarchy, in 1889, and the profound historical transformations experienced during the twentieth century, the image of a vibrant and diverse natural world as an essential trademark of Brazilian reality remained strong. This widespread sentiment was later appropriated by the new environmentalism, beginning in the 1970s, which helped garner sympathy among the public and in the media toward the fight against the destruction of Brazil's "national heritage."

During the nineteenth century, however, it was possible to observe the apparently paradoxical coexistence of two equally intense movements: a culture of praise for nature and the practice of constant aggression against some of its manifestations. The romantic tradition, upon which the ideology of Brazilian natural grandeur was based, did not concern itself much with the actual destruction of natural spaces. Brazil's fairly rhetorical and superficial romanticism, with many artists and intellectuals depending upon imperial patronage, took a moderated stand, rather uninterested in producing a radical defense of traditional landscapes and peoples against the advance of the modern world – a critique of civilization found in certain sectors of the European and North American romanticism. That said, from the end of the eighteenth century, until the first two decades of the

Map 25.1 Brazil

twentieth century, a powerful critique of the destruction of forests and natural resources was produced by intellectuals of another type, linked to Enlightenment rationalism, and to the notion of an intelligent use of nature in order to avoid waste and guarantee Brazil's progress. This intellectual tradition involved dozens of authors and produced consistent and creative ideas and arguments.[10]

In broad terms, with the risk of being a bit anachronistic, it could be said that this tradition launched "environmentalism" in Brazil. Clearly, the word itself was not used and the intellectual debate at that time differed greatly from that of contemporary environmentalism, which is more centered upon transformative action in the public sphere. Brazil's overall historical context was also quite different. But history is nothing if not an interplay between continuities and discontinuities. The idea of using scientific logic to criticize the irrationality of the economic practices that transformed nature, for example, is a clear connection to modern environmentalism. Many of the issues tackled in the past, such as deforestation and slash-and-burn cultivation, remain at the heart of the Brazilian environmental debate. In addition, the nineteenth-century intellectuals did not restrict themselves to making isolated observations, but rather manifested a general awareness of the importance of society's relationship with the natural world in guaranteeing the survival and continuity of society as a whole.

Generally speaking, the discovery in Brazil and elsewhere of these "environmentalist" – in the sense used by Richard Grove in *Green Imperialism*[11] – intellectual traditions from the seventeenth and eighteenth centuries shows that we must broaden our understanding of the modernity of the environmental question. It is not related solely to the consequences of the twentieth century's large-scale urban industrial transformation, but also to a series of other macrohistorical processes that preceded it. These processes include European colonial expansion and the incorporation of vast regions of the planet into an economic world under its dominion, including ecosystems that had not previously been a part of the Western historical experience. The implementation of predatory productive practices in the colonial world gave rise to intellectual protests that are a part of the formation of the modern world's environmental sensibility.

In the case of Brazilian environmentalism, it must be noted that the strong critiques put forth by intellectuals prior to the twentieth century brought about very little in the way of practical consequences at the time they were made. Furthermore, they did not serve as an inspiration to post-1970s contemporary environmentalism. In fact, the memory that such an intellectual tradition existed had itself been lost. Even prominent authors in the history of Brazilian thought had their environmental concerns overshadowed by their other ideas. The discovery of this established tradition of Brazilian "environmentalism" is the result of rather recent historiographical work. Unlike in the US, where contemporary environmentalism claimed to be a descendent of nineteenth-century thinkers such as John Muir or Henry David Thoreau, the prevailing belief in Brazil until rather recently was that the environmental discussion was something totally new, born in dialogue with ideas and movements that originated in Europe and the US at the end of the twentieth century.

However, nowadays it is possible to consider Brazilian environmentalism within a broader historical perspective, which can be divided into four major periods. These periods do not of course have clearly defined borders. They instead merely indicate the basic lines of differentiation between different historical moments:

1. 1822–1930: A strong "environmentalist" intellectual debate, with few practical or political consequences.
2. 1930–70: The environmental debate is essentially restricted to the community of natural scientists, but with the appearance of the first broader institutional and legal initiatives.
3. 1970–90: The appearance, within the context of a transition from military dictatorship to democracy, of an environmentalism based on confrontation in the public sphere, led, on one side, by middle-class activist groups in dialogue with international movements and, on the other, by movements in poorer communities composed of the rural and urban working classes.
4. 1990–present: The development of a powerful professional environmentalism and the branching out of environmental interests, with a significant role in public opinion and influence on governmental action and the formulation of public policies. This is a moment that could be referred to as the "environmental turning point" in Brazilian society and politics, with very real consequences in terms of the administration of the territory and its resources.

The analyses that follow will explore the fundamental characteristics of each of these periods.

The "Environmentalist" Intellectual Debate: 1822–1930

From the end of the eighteenth century, Brazilian scholars – some of whom later participated in the struggles for independence – began to publish critiques of a series of problems that are now considered part of the environmental agenda: deforestation, slash-and-burn operations, soil erosion, the loss of species, and climate change. The University of Coimbra and the Lisbon Academy of Sciences served as the original loci for these debates, with Enlightenment science, physiocracy, and the economy of nature as their primary theoretical instruments.

Despite occasional romantic influences, the intellectual posture of these critiques was essentially rationalist and anthropocentric. The natural world was praised for its economic potential, and its reckless destruction seen as a sign of ignorance and lack of concern for the future. Environmental devastation was not seen, as it is in today's language, as the "price of progress," but instead as the "price of backwardness," a result of the permanence of rudimentary social and technological practices, a burden of a colonial past.

Within this intellectual framework, formulations of considerable breadth and creativity were introduced, which went far beyond the mere description of specific problems. In 1799, for example, José Gregório de Moraes Navarro, a judge in the interior of the captaincy of Minas Gerais, published in Lisbon a small and fascinating volume entitled *Discurso sobre o Melhoramento da Economia Rústica no Brasil* (*Discourse on the Improvement of Brazil's Rural Economy*). He formulated an integrated vision of an Earth equipped with great resilience, "the common Mother of all living creatures," that even while suffering floods and revolutions on her axis, continued to maintain "the fruitful germ of her fertility." Despite this benevolence, she endured "the ungratefulness of men, who seem to work continuously to destroy and annihilate her natural productions, and to consume and weaken her primitive substance." He noted that in Brazil many ranches and villages were being abandoned due to the destruction of soils, because the colonists, "after reducing all trees to ashes, after depriving the land of its most vigorous substance, left it covered with grasses and ferns." His proposal was to promote lasting social progress through the planting of trees and a reform of productive techniques, so as to "aid the fertility of the earth through means that experience and industry revealed to be the most convenient."[12]

Bearing in mind the differences of historical context, it would be no exaggeration to establish a connection between these ideas and contemporary issues such as the generic relationship between humanity and the planet and the sustainability or unsustainability of societies.

Another author, José Bonifácio de Andrada e Silva, who became one of the primary leaders of the independence movement of 1822, while still in Portugal in 1815, adopted an interesting long-term historical perspective for thinking about the problem of deforestation, approaching what would now be called environmental history: "Everyone who has studied the important influence of forests and trees on the overall natural economy knows that the countries that lose their forests are almost completely sterile or unpopulated. This happened in Syria, Phoenicia, Palestine, Cyprus and other lands." Eight years later, in 1823, as he defended the gradual abolition of slavery before the Constitutional Assembly of a newly independent Brazil, he returned to the issue with even more radicalism, blaming the slave-based economy for the territory's climatic degradation and putting the country's very future at risk:

> Our precious forests are disappearing, victims of fire and the destructive machetes of ignorance and egoism ... and as time goes forward, there will be a decrease in the fertilizing rains that favor vegetation and feed our springs and rivers, without which, our beautiful Brazil, in less than two centuries, will be reduced to the arid deserts of Libya. The day (this terrible and fatal day) will come in which an outraged nature will find itself avenged of the many errors and crimes that have been committed.[13]

There is not enough space here to detail the thoughts of the more than 50 authors who wrote critiques of the destruction of the natural world prior to the end of the monarchical period in 1889.[14] The examples above serve merely to indicate the richness of this intellectual tradition. Throughout the decades, Brazilian authors discussed issues such as: the best way to prevent the extinction of species of Amazon flora and fauna; the relationship between deforestation and the severe droughts in the Northeast; the evils of monoculture farming with respect to the degradation of soils; and the risk of railways turning into instruments of devastation. Slavery was a central theme, in that some authors considered the abundance of forced labor and the lack of small property owners to be incompatible with the establishment of an agriculture that was technically more elaborate and nondestructive to the territory.

At a time where the institutionalization of scientific work and intellectual life was small, the discussions were carried out in generic cultural spaces (such as the National Museum and the Brazilian Institute of History and Geography), professional associations (such as the Imperial Academy of Medicine or the Imperial Institutes of Agriculture), or within sectors of the public administration of a state that was still weak and under construction.

One of the few practical initiatives on the part of the state, with surprisingly important results for the future, was the reforestation project in the Tijuca Mountains of Rio de Janeiro, which helped to reconstitute the forested hills of the country's capital city, which had, in part, been destroyed by the first wave of export-oriented coffee cultivation during the first decades of the nineteenth century. The efforts, which took place between 1862 and 1887, were due more than anything to the initiative of intellectuals and public figures who participated in the international debate on silviculture and urban forestry. In order to guarantee state support for the project, the primary argument put forth was of the necessity of preserving the sources of the streams that provided water to the city. Despite the strong pragmatic appeal involved in the issue, government support over time was fairly limited. Regardless, the migration of the coffee plantations to other regions and the reforestation efforts wound up guaranteeing the continuity of the green mountains that to this day are a part of the city's cultural landscape.[15] For the project's masterminds, however, it was expected to be merely the beginning of a much farther-reaching reforestation policy for Brazil, a policy that was never implemented.

The same destiny awaited the first proposal for the creation of a national parks system in Brazil, put forth in 1876 by the engineer and abolitionist leader André Rebouças. The idea was generous and comprehensive, focusing first on regions of particular beauty, such as the Paraná River's Iguaçu Falls and Seven Falls of Guaíra, and the Araguaia River's Bananal Island. The suggestions made by Rebouças were ignored and the first Brazilian national park would only come into existence in 1937.

The inability of the monarchical government to implement a policy providing greater protection to natural resources must be considered from a historical and sociological perspective. In the nineteenth century, there was almost a complete lack of social and political forces that might have transformed the critiques made by intellectuals into a relevant collective action or public policy. Civil society, in the age of slavery, was rudimentary. The state was weak and relatively poor, especially given that its tax base was limited (for example, land, which was the largest source of wealth at the time, was not taxed). The governments were highly dependent on the support of regional elites who maintained their power through environmentally devastating productive practices.

The situation was essentially the same, or even worse, under the republican regime that came into power in 1889. Political decentralization strengthened the hand of the governors and the regional elites over each province's economy, incentivizing the open exploitation of natural resources. Careless and shortsighted use of the forests was a constant practice, aggravated by the introduction of railways into sparsely populated regions of the territory.

In addition, during the initial decades of the republican period (1889–1930), criticism of environmental destruction cooled to a certain degree, largely due to the growing prestige of positivist and new scientific outlooks that viewed tropical nature as a hostile environment that needed to be transformed through engineering and other forms of intervention. Still, the romantic heritage of the greatness and generosity of Brazilian nature remained strong in the popular imagination.

A small but influential number of intellectuals maintained the tradition of environmental criticism. This is the case of the renowned writer Euclides da Cunha, author of *Os Sertões* (translated into English in 1944 as *Rebellion in the Backlands*), one of the major works of Brazilian literature from that period. In articles published in the press, especially during the year 1901, he warned against what he called the "desert makers." According to the author, Brazilians acted as "a harmful geological agent and a terribly barbarous element, antagonistic to the very nature that surrounds us." The indigenous peoples began the destruction through slash-and-burn techniques. European colonization followed and aggravated the impact. At the beginning of the twentieth century, the devastation continued and followed in the wake of locomotives that, day by day, destroyed "in the smoke of boilers, hectares of our flora."[16]

Years later, in 1915, another important thinker of the early twentieth century, the political philosopher and jurist Alberto Torres, formulated one of the most comprehensive and consistent environmental critiques produced in Brazil up to that point. The critique became even more important due to the significant political influence of its author, which extended into the 1930s and beyond. Torres offered a radical critique of the artificiality and economic and social inefficiency of the old Brazilian republic, where a small group of elites benefited from an institutional formalism while life for the overwhelming majority of the population was marked by backwardness, ignorance, and poverty. To live in a real and dynamic democracy, in his view, it was first necessary to organize the country and modernize its laws, education, infrastructure, economy, and living conditions. A temporary period of authoritarian governments would be beneficial if it were able to advance this program and overcome the inertia of a false republicanism. The proposal made by Torres offered an alternative to communism and fascism and served as an inspiration to the military officers and civilians who carried out the revolution of 1930 and set Brazil on a path of industrial development.

When it came to the environmental question, Torres was fiercely nationalist, criticizing the imperialism of the major powers in their efforts to control the planet's natural resources. Furthermore, he argued that these resources were finite, were being exhausted, and were in need of a policy of restoration. In his view, "man has been a relentless and voracious destroyer of the earth's riches. The entire history of humanity has been marked by the devastation and exhaustion of soils, the burning of treasures and forests, the sacking of minerals from the bosom of the earth and the sterilization of its surface."[17] This behavior, Torres believed, was linked to the impression that the planet's treasures and products were eternal. Upon organizing its economy, therefore, Brazil should seek to halt the sacking and destruction of its resources, promoting a policy of conservation and restoration of that which he called the country's "sources of life," such as forests, rivers, and the health of the human beings themselves.

Despite the prestige enjoyed by some intellectual critics of the destruction of natural resources during the early years of the republic, concrete political actions to confront environmental problems were virtually nonexistent. One clear example can be found in the fact that Brazil's president in 1920, Epitácio Pessoa, expressed regret that Brazil was the only country with large forests that did not have a forestry law. Commissions were created in the Brazilian parliament in order to draw one up, without success. A Forest Service, launched in 1925, was largely decorative, without any significant resources or legal capacities. The inaugural forest code only came into existence in 1934. However, by that point Brazil was living in another reality, which deserves to be explained in a more comprehensive manner.

Natural Scientists and Organized Initiatives: 1930–70

In 1930 a political revolution took place whose aim was to leave behind the economic apathy and elitist formalism of the so-called "República Velha" (Old Republic). During the following decades, which featured authoritarian governments broken up by a few moments of democracy, the growth of the Brazilian economy and population was intense. Despite the fact that the power of large rural landowners remained significant, capable of blocking agrarian reform, industrialization and urbanization continued unabated. By the middle of the century, during the administration of Juscelino Kubitschek (1956–60), the developmentalist approach could be summed up with the motto "50 years of progress in 5," which led to the creation of the new capital (Brasília) and a big boost in industry (especially the automobile industry).

In the context of this strong developmentalist paradigm in national politics, critical thought regarding the destruction of natural resources lost its political strength. Some important thinkers, such as the sociologist Gilberto Freyre and the historian Caio Prado Júnior, occasionally came out against the destruction of forests, the destruction of soils, and the contamination of rivers. But, overall, the prevailing view was that Brazil's massive territory should be occupied by a growing population.

Yet, "environmental" criticism continued in certain academic settings focused on the natural sciences. In addition, these natural scientists launched some initiatives aimed at bringing environmentalism into the realm of organized action in the public sphere. It is also during this time that we can see the beginnings of certain, rather limited, governmental actions aimed at protecting natural resources.

During the first half of the twentieth century, in places such as the National Museum and the São Paulo Museum, a group of naturalists that included Alberto Sampaio,

Candido de Mello Leitão, and Frederico Hoehne criticized the destruction of forests and loss of important flora and fauna. This critique was reinforced by foreign scientists, such as Germany's Hermann von Ihering and Sweden's Alberto Loefgren, who were drawn to work in the country because of the exuberance of its nature.[18]

In 1934, the First Brazilian Conference on the Protection of Nature was organized at Rio de Janeiro's National Museum by scientists such as Alberto Sampaio and other civic entities such as the Society of the Friends of Trees and the Society of the Friends of Alberto Torres. Traditional academic organizations, such as the Brazilian Institute of History and Geography, provided institutional support. Formal sponsorship came from the country's president and the leader of the 1930 revolution, Getúlio Vargas.[19] The 1934 conference included several disciples of Torres.

A more direct legacy of the discussion that took place in the 1930s can be found in the legislation that was passed. The new regime took into account the concerns Torres had with the unregulated and careless use of natural resources, bringing together politicians and scientists in order to create a series of legal codes to better regulate the use of these resources. This led to the introduction, in 1934, of a Forestry Code, a Water Code, a Mining Code, and a Hunting and Fishing Code.

While these codes did include some regulations regarding conservation, they were more related to the implementation of more efficient economic uses of natural resources. But the mere existence of such legislation was not enough to provoke real oversight of the exploration and use of natural resources. The fact that the issue remained essentially absent in the media, combined with the near total lack of actors in civil society willing to push for compliance with the law, resulted in the codes being little more than a legal formality. One of the few concrete measures actually taken was the creation of some national parks, beginning with Itatiaia (in the state of Rio de Janeiro) in 1937. The creation of these parks, however, did not signify any real conservationist engagement at the federal level. It was more a political formality, as it was thought that a modern country should have a few protected areas in order to preserve places of special scenic or biological value.

Another point that is central to understanding the lack of a substantive environmental policy can be found in the almost complete lack, within the state apparatus, of specific institutions created to oversee the use of natural resources. The Brazilian Institute for Forestry Development, an independent agency responsible for implementing a forestry policy in Brazil that included conservation, was only created in 1967, under a military dictatorship that came into power in 1964. Prior to this, there existed only small departments within the Ministry of Agriculture and a few institutes concerned with the promotion of specific products such as pine and yerba mate (the source of a popular tea-like drink). Regardless, the 1967 institute, as the name implies, was more focused on development than forest preservation. The following year, the state of São Paulo, the most developed in the country, became the first and only state to create a specific public entity to look after environmental sanitation, particularly in urban areas.[20]

Some specific non-governmental initiatives were carried out by scientists and specialists interested in the environment. In 1953, a group of professors and students from São Paulo created an Association for the Defense of Flora and Fauna in order to fight, with little success, against the destruction of remaining forests in the southern part of the state. The leader of this group, Paulo Nogueira-Neto, who would later become a professor of ecology at the University of São Paulo, played an important environmental policy role during the period that followed, as we will see. In Rio de Janeiro, continuing in the vein of an environmentalism of natural scientists, the Brazilian Foundation for the Conservation of Nature was launched in 1958, but had little impact on the national political scene.[21]

The fact is that the environmental issue was practically absent from the concerns of both state and society. The imperative of economic development was too strong. An environmentally aware reader feels genuine anguish upon reading the chapters that refer to the decades from the 1930s to the 1960s in the overview provided by Warren Dean (1995) of the history of the destruction of the Atlantic Forest, a region that hosted the vast majority of the Brazilian population and economic activity until the end of the twentieth century. The country was growing economically, and companies openly exploited its natural resources, without any environmental barriers. There was a near-total lack of public entities or civic groups working to denounce environmental abuses. The main issue, for a large majority of intellectuals from both the left and the right, consisted solely of the economic and social dimensions of the debate over Brazil's future. Meanwhile, Brazilian nature suffered intensive deforestation, degradation, and contamination.

Confrontation in the Public Sphere: 1970–90

The situation began to change in the 1970s, a period of substantial creativity and excitement in Brazilian society. Since 1964, Brazil had been living under a military dictatorship that blocked the traditional channels of popular participation. Yet, surprisingly, beginning in the mid-1970s, Brazilian society began to create, from the ground up, new channels of participation, based on participative democracy and independent of the traditional leftist parties, many of which were dogmatic and sectarian. This included the introduction of neighborhood associations; new rural and urban unions; movements in defense of rural laborers, landless workers and agrarian reform; movements in favor of women's rights; and movements in favor of Afro-Brazilians and indigenous peoples, among others. An important social agent at that time was the Catholic Church, which was then deeply influenced by the so-called Liberation Theology. According to this belief, it was a religious duty to participate in the public realm in favor of the poor and oppressed, helping to organize society in a grassroots manner, in order to fight for a better world. The prestige of the Church served as a protective umbrella for grassroots initiatives. The new democratic left, strengthened by social movements, led to the political foundation of the Workers' Party in 1980, which would, in 2002, win the Brazilian presidency.

Simultaneously, a strong countercultural scene emerged: avant-garde artistic movements, alternative communities, Eastern and esoteric spiritual movements, organic foods, and so on. Society as a whole was experiencing a moment of strong growth in the economy, in consumption, and in the liberation of mores and sexuality. In other words, it was a dynamic confluence that challenged the conservatism still dominant in many sectors.

Within the context of this social excitement, Brazilian environmentalism also began to flourish. It was expressed through two parallel movements, which only subsequently began to join forces: (1) a middle-class, urban environmentalism, and (2) an "environmentalism of the poor," mainly rural, but that would also express itself in certain urban contexts.

The first type of environmentalism was manifested through the foundation of many relatively small groups of nonprofessionals who acted at the local level through activities aimed at raising public awareness and entering into conflict with projects that were potentially destructive to the environment. The tendency was to act through networks, forming coalitions of groups and individuals in favor of specific environmental objectives. These groups were in large part created in resistance to the intense processes of urban industrial transformation of the previous decades. The main targets were the destruction of urban green spaces, air and water pollution, real-estate speculation, food contamination, the creation of nuclear plants, and environmental destruction caused by infrastructure projects. In 1977, for example, roughly 70 groups from the state of São Paulo formed a Community Heritage Defense Commission, which carried out a successful campaign against the construction of a new airport in a forest preserve in the city of Caucaia. In the state of Rio de Janeiro, similar groups held demonstrations against the destruction of sparsely populated beaches and small fishing communities by real-estate speculation and by industrial fishing practices. Similar initiatives took place in several states throughout Brazil.[22]

An emblematic struggle, also in 1977, was that of the naturalist Augusto Ruschi against the destruction of a private forest preserve that he had established in the state of Espírito Santo. This struggle represented a clear bridge between the previous period and the new dynamics now taking shape. Ruschi was an established and renowned specialist in orchids and hummingbirds, connected with the National Museum, who had created a private biology museum in his inland city. Since the 1940s, he had been a lone voice against the destruction of forests in his state. When he saw an industrial agricultural project threatening the forest preserve that belonged to his museum, he took a radical step and, armed with a rifle, declared that he would personally kill anyone who destroyed the forest. In the new context that emerged in the 1970s, his decision had a major impact, stimulating acts of solidarity by established groups in several parts of the country. Academic and scientific environmentalism transformed itself into an environmentalism based on confrontation in the public sphere. Ruschi's forest preserve survived.[23]

The influence of exogenous environmental ideas and examples, which the expanding media coverage brought into the country, also inspired the activities of Brazilian groups. In 1979, the political amnesty decreed by the military regime – which had started to promote a transition to democracy – led to the return of leftist activists to Brazil who, during their exile in Europe and the US, had modified their old Marxist ideas and incorporated the ideology of the new green parties. People such as Fernando Gabeira and Alfredo Sirkis introduced themselves as leaders of the new confrontational environmentalism, engaging in the struggle against the construction of nuclear plants, the spread of agrochemicals, and so forth. But the primary force came from the endogenous reactions to the aggressive model of industrial development in place in Brazil.[24]

In any event, the federal government, with its mix of military authoritarianism and a modernizing technocracy, began to awaken to the need to create stronger environmental institutions in Brazil. A small federal Department of the Environment was created in 1973, in large measure as a response to the United Nations Conference on the Human Environment (held in Stockholm in 1972). The department, with its well-intentioned team but slender resources, carried out some legislative and political initiatives in terms of controlling pollution and creating protected areas. Until 1986, its boss, who served through several military governments, was Paulo Nogueira-Neto, the respected ecologist. His most important institutional initiative was the creation, in 1981, of a National Environmental Policy Law, rather advanced for the time, particularly in its creation of a National Council for the Environment (CONAMA) where, as still happens today, the participation of specialists from a variety of ministries is comple-mented with the presence of representatives from civil society.[25]

Meanwhile, in the heart of society, more confrontational voices continued to express themselves, sometimes through clear interactions between the international and the local. An important figure in this vein was José Lutzenberger, an agronomist from a cosmopolitan family of German exiles. After years spent working for agrochemical companies outside Brazil, he resigned in protest against the damage caused by agrochemicals and decided to return to Brazil to promote the spread of ecological values. In 1971, in the southern state of Rio Grande do Sul, he created the Gaucha Association for the Protection of the Natural Environment (AGAPAN), which became an important milestone in the creation of modern Brazilian environmentalism. With a radicalized and well-informed discourse, serving as a critic of the irresponsibility of large businesses in favor of citizens' rights, in 1976 Lutzenberger published the "Brazilian Ecological Manifesto," which outlined an agenda of action for groups that were spreading throughout the country.

At the same time, another branch of Brazilian environmentalism with little contact with the urban middle-class movements and practically no international connections, rose from the ground up. It was an empirical environmentalism of poor communities fighting for their livelihoods in the face of the expansion of industrial capital in their traditional spaces. This phenomenon has emerged in many other countries.[26] Rural com-munities living on local subsistence agriculture, especially in midwestern Brazil, organ-ized themselves against the expansion of mechanized farming operations run by businessmen in search of abundant, cheap land in the Brazilian interior. Communities of small landowners, especially in the South, started to organize in order to defend them-selves against the large hydroelectric projects that displaced thousands of people and flooded their lands. Groups of women who worked extracting babassu nuts in the north-eastern interior fought against the arrival of large farms that aimed to destroy the palm forests. In the eastern Amazon region, in a movement that gained the most international attention, forest communities that extracted rubber and Brazil nuts organized under the leadership of Chico Mendes in order to defend their way of life against the destructive advance of the modern livestock farms. Their famous "human chains" brought together men, women, and children who put themselves in front of the tractors that would have otherwise destroyed the forests.[27] In certain urban settings, unions and the families of workers started to organize in order to fight against the contamination and degradation of their homes and workplaces. This collection of popular struggles, rural and urban, can be compared with what came to be known in the US as the "environmental justice" movement, even though the focus of the Brazilian movements was less on racism and more on the poor and exploited.

During the 1980s, the movements associated with the "environmentalism of the poor" began a process of integration with the middle-class environmentalism, bringing the different struggles together under a broader concept of environmental struggle. There was also more direct involvement with groups acting internationally. This process was facilitated by the improved institutional articulation of popular environmental struggles. In 1985, for example, the myriad struggles against hydroelectric projects united on a national scale to form the Movement of People Affected by Dams. The same happened in the Amazon region in 1985, with the National Council of Rubber Tappers. In 1984, the Movement of Landless Rural Workers was founded, which gave rise to countless struggles for agrarian reform and later came to argue in favor of ecologically based family farming. The tendency of these movements evolved from a simple act of resistance to the proposal of alternatives. After struggling against the destruction of their way of life, they began to suggest innovative policies that would allow for environmentally sustainable improvements to their social conditions. One clear example was the proposal made by rubber tappers of "extractive reserves," a new concept in protected areas, where the traditional populations would continue to live in nature preserves and help in their conservation. At the same time, they would have the right to use to fruits of the natural production of the ecosystems without damaging their structural foundations (nuts, rubber, fish, etc.).

The closing years of the 1980s also bore witness to the violent reaction of local powers, especially in rural settings. The assassination of Chico Mendes, in 1988, shocked the world and, at the same time, became a symbol of the integration of the struggles for social justice and environmental protection. The death of Chico Mendes, in the context of the environmental conflicts that had been playing themselves out in various parts of the country, had a considerable impact on domestic Brazilian politics. The federal government, no longer military after 1985, was forced to take various reformist measures with respect to its environmental policy. In 1989, the Brazilian Institute of the Environment and Renewable Natural Resources was created, which unified all the agencies that separately dealt with environmental issues. An overarching program called "Our Nature" was established, including the creation of the first extractive reserves in the country.

In addition, in October 1988 a new Brazilian Constitution was introduced by a constitutional convention. It advanced principles such as the notion that an "ecologically balanced environment" was among the rights of the people, who had the duty to defend it. This principle provided legal justification for the struggles and stimulated a variety of innovative environmental policies. The new constitution also ordered the demarcation of indigenous lands as federal reserves, which protected vast swathes of the Amazon rain forest (approximately 20 percent in total).

An important sign of this political change was the 1990 invitation to José Lutzenberger to assume control of Brazil's federal Department of the Environment (the Ministry of Environment was created in 1992). Even though he held the job for a relatively short period of time and had few concrete accomplishments – he resigned in 1992, disillusioned with the bureaucracy and political infighting in Brasília – his participation in the government symbolized recognition, by political institutions, of the public prestige enjoyed by the confrontational environmentalism that had emerged in the 1970s.

Professionalization of Environmentalism: 1990–present

At the beginning of the 1990s, environmental issues were gaining prominence throughout Brazil and the wider world. The UN's decision to hold a major conference on the environment and development in 1992 in Rio de Janeiro reinforced this trend. This decision obviously led to a substantial increase in the visibility of the environmental debate in Brazil, which began to receive increasing amounts of attention in the media.

A key occurrence at this moment was the opening throughout the country of local offices of the major international environmental organizations, reinforcing the professionalism and high technical quality of non-governmental environmentalism. Many of these organizations had been developing contacts and specific projects since the 1970s, but the introduction of Brazilian offices happened at roughly the same time.

This was the case, to cite only a few examples, with the WWF (1990), Conservation International (1990), Greenpeace (1991), and The Nature Conservancy (1994). The professionalism of these organizations, which hired excellent local specialists, has allowed them to play an increasingly important role in the public presentation of environmental issues, given that the media seeks out the prestige and reliable information that they provide.

Together with this movement, and influenced by it, came the growing professionalism of certain domestic organizations that emerged in the 1990s or in some cases before. These groups began to enjoy significantly larger budgets, fed by international foundations or local sources. Such was the case with SOS Mata Atlântica (1986), Fundação Biodiversitas (1989), Instituto de Pesquisas Ecológicas (1992), Instituto Socioambiental (1994), as well as groups more concerned with regional issues such as IMAZON (1990) in Pará state and Instituto Centro de Vida (1991) in Mato Grosso state. Following the examples of the international organizations, professional Brazilian environmentalism increasingly began participating in the technical debate and the organization of public action. Coordination between national and international initiatives is now, in fact, commonplace. The small volunteer organizations have not disappeared, but they lost a portion of the national visibility that they once had, instead concentrating on more local struggles. In any event, the National Registry of Environmental Organizations, organized by the Ministry of the Environment, now lists 580 organizations.

Another phenomenon from this period that deserves mention is the tendency to create networks that bring together small organizations – and sometimes large ones as well – in an effort to achieve a more powerful impact. The most important network is the Brazilian Forum of NGOs and Social Movements for the Environment and Development, created in 1990 as part of the nongovernmental preparations for the 1992 conference. As the name implies, the forum's scope is broader than that of the strictly environmental organizations of years past, given that it includes unions, associations focused on regional human development, farmers' movements, and others. The message is that the environment has become an issue of primary interest for different sectors in civil society. Nowadays, the forum includes 608 participating entities. Other networks exist at the regional level. Still others are focused on specific issues of national interest, for example, the Brazilian Network for Environmental Justice, created in 2005 with more than 100 unions and popular organizations, and of the National Agroecology Coalition, created in 2002, which brings together hundreds of producers and organizations focused on supporting sustainable agriculture.

This expansion of environmental action in Brazilian society in recent decades goes beyond the NGOs and the so-called social movements. In the 1980s, the sociologists Eduardo Viola, Hector Leis and other collaborators created the concept of "multisectoral environmentalism," which can also be applied internationally. In essence, the idea states that there was a rise in the quantity and quality of the social actors responsible for environmental discourse and action. Environmentalism ceased to be defended exclusively by the so-called "environmental groups" and was appropriated, with adaptations, by sectors from business, organized labor, the arts, religion, science, education, and so on. It also began to permeate the activities of government employees at the municipal, state, and federal levels, as well as those of diplomats. In short, the environmentalist vision ceased to be the basis of a "new social movement" and instead inspired a "historical movement" expressed through the change in behavior of different social actors.[28]

The growth of this multisectoral environmentalism led to concrete results at the judicial and institutional levels. Without question there has been a considerable rise in environmental consciousness within Brazilian society since the 1990s, which is reflected in stronger laws and institutions and in more effective results from environmental policies. After centuries of deforestation, for example, we can observe an almost frenetic swing toward forest conservation in the last decade. Brazil was responsible for around 74 percent of the protected areas created worldwide after 2003.[29] As a consequence of this move – together with other strong federal policies – deforestation in the Amazon was reduced by more than 70 percent between 2004 and 2010. Amazonian deforestation is declining even in the current context of strong economic growth.

Other environmental policies and actions are being carried out in urban and rural settings, with differing degrees of success. A landmark achievement was the Environmental Crimes Law, which was passed in 1998 and for the first time clearly defined the environmental crimes, punishments, and fines to be considered by Brazil's executive and judicial branches of government.

Despite the advances, enormous challenges remain. Countless problems related to pollution and degradation of the quality of life affect urban, industrial spaces. The uncontrolled use of agrochemicals menaces many rural settings. As regards deforestation, despite the advances observed in the Amazon, there is significant resistance from agribusiness leaders and their legislative representatives, hoping to weaken the forest code and open new areas to deforestation in the name of agricultural exports.

At the same time, it is important to note the major institutional and cultural advances that have occurred in recent decades with respect to the environmental problem in Brazil. They result primarily from the vibrant and multisectoral movement that emerged from society in the 1970s. Brazilian environmentalism has never before been as visible on the public scene, in the media, and in the educational and scientific institutions. Moreover, the state has never before been more prepared, institutionally and legally, to confront Brazil's environmental problems.

An important sign of the political growth of Brazilian multisectoral environmentalism is the trajectory of the woman who led the recent policy of deforestation reduction in the Amazon, the former minister for the environment, Marina Silva. She grew up in a poor family of rubber tappers in the Amazon and was a close partner of Chico Mendes. As minister for the environment in the leftist government of President Lula da Silva, she was able to accomplish the impressive results mentioned above. In 2009, she resigned from the government and from the Workers' Party in order to join the small Brazilian Green Party and propose a new vision of sustainable development for the country as a whole. In the October 2010 presidential elections, Marina Silva received approximately 20 million votes, almost 20 percent of the total. This was not enough to win her the presidency, but it indicates widespread support for new environmental policies.

In any event, the larger problem facing Brazilian society is the transition from confronting problems in isolation to the implementation of a new developmental model that combines human development and environmental protection. Only such a new model would be able to use the ample ecological resources of Brazil to promote a democratic and sustainable society, founded on an economy that must be increasingly based on the territory's high potential for clean and renewable sources of energy. That appears to be the future of Brazilian environmentalism.

Notes

1 J. A. Pádua, "European Colonialism and Tropical Forest Destruction in Brazil: Environment Beyond Economic History," in J. R. McNeill, J. Pádua, and M. Rangarajan (eds.), *Environmental History – As If Nature Existed*, New Delhi, Oxford University Press, 2010, pp. 130–48.

2 J. A. Pádua, "Annihilating Natural Productions: Nature's Economy, Colonial Crisis and the Origins of Brazilian Political Environmentalism (1786–1810)," *Environment & History* 6/3, 2000, pp. 255–87; J. A. Pádua, *Um sopro de destruição: Pensamento político e crítica ambiental no Brasil escravista (1786–1888)*, Rio de Janeiro, Jorge Zahar Editor, 2002.

3 K. Hochstetler and M. Keck, *Greening Brazil: Environmental Activism in State and Society*, Durham, NC, Duke University Press, 2007.

4 J. A. Pádua, "The Birth of Green Politics in Brazil: Exogenous and Endogenous Factors," in W. Rudig (ed.), *Green Politics II*, Edinburgh, Edinburgh University Press, 1992, pp. 134–55.

5 B. Fausto, *A Concise History of Brazil*, Cambridge, Cambridge University Press, 1999; W. Dean, *With Broadax and Firebrand: The Destruction of the Brazilian Atlantic Forest*, Berkeley, University of California Press, 1995.

6 B. McCann, *The Throes of Democracy: Brazil since 1989*, London, Zed, 2008.

7 J. A. Pádua, "Biosphere, History and Conjuncture in the Analysis of the Amazon Problem," in M. Redclift and G. Woodgate (eds.), *The International Handbook of Environmental Sociology*, London, Edward Elgar, 1997, pp. 403–17.

8 A. Rouquié, *Le Brésil au XXIe siècle: Naissance d'un nouveau grand*, Paris, Fayard, 2006.

9 C. P. Valle, *Risonhos lindos campos: Natureza tropical, imagem nacional e identidade Brasileira*, Rio de Janeiro, Editora SENAI, 2005, p. 156.

10 Pádua, *Um sopro de destruição*.

11 R. H. Grove, *Green Imperialism: Colonial Expansion, Tropical Island Edens and the Origins of Environmentalism, 1600–1860*, New York, Cambridge University Press, 1995.

12 Pádua, "Annihilating Natural Productions," pp. 255, 256, and 257.

13 J. A. Pádua, "Nature Conservation and Nation Building in the Thought of a Brazilian Founding Father: José Bonifácio (1763–1838)," *CBS Working Paper 53*, University of Oxford, Center for Brazilian Studies, 2004, pp. 4, 12.

14 Pádua, *Um sopro de destruição*; J. A. Pádua, "Down by Blind Greed: The Historical Origins of Criticism Regarding the Destruction of the Amazon River Natural Resources," in T. Tvedt, G. Chapman, and R. Hagen (eds.), *A History of Water, Series II*, vol. 3, London, I. B. Tauris, 2011, pp. 176–92.

15 J. Drummond, "The Garden in the Machine: An Environmental History of Brazil's Tijuca Forest," *Environmental History* 1/1, 1996, pp. 83–105.

16 E. D. Cunha, *Obra completa*, Rio de Janeiro, Nova Aguilar, 1995, p. 2005.

17 A. Torres, *O problema nacional Brasileiro*, São Paulo, Editora Nacional, 1978 [1914], p. 91.

18 J. L. Franco and J. A. Drummond, "Wilderness and the Brazilian Mind (I): Nation and Nature in Brazil from the 1920s to the 1940s," *Environmental History* 13/4, 2008, pp. 724–50.

19 J. L. Franco and J. A. Drummond, "O cuidado da natureza: A fundação Brasileira para a conservação da natureza e a experiência conservacionista no Brasil (1958–1992)," *Textos de História* 17/1, 2009, pp. 59–84.

20 Dean, *With Broadax and Firebrand*.

21 Franco and Drummond, "O cuidado da natureza."

22 Hochstetler and Keck, *Greening Brazil*.

23 R. Medeiros, *Ruschi, o agitador ecológico*, Rio de Janeiro, Record, 1995.

24 Pádua, "The Birth of Green Politics in Brazil."

25 R. Guimarães, *The Ecopolitics of Development in the Third World: Politics and Environment in Brazil*, Boulder, CO, L. Rienner, 1991.

26 J. Martinez-Alier, *The Environmentalism of the Poor: A Study of Ecological Conflicts and Valuation*, Cheltenham, Edward Elgar, 2003.

27 S. Hecht and A. Cockburn, *The Fate of the Forest: Developers, Destroyers and Defenders of the Amazon*, Chicago, Chicago University Press, 2010.

28 E. Viola and H. Leis, " O ambientalismo multissensorial no Brasil para além da Rio-92: O desafio de uma estratégia globalista viável," in E. J.Viola, H. R. Leis, I. Scherer-Warren, et al. (eds.), *Meio ambiente, desenvolvimento e cidadania: Desafios para as ciências sociais*, 2nd ed., São Paulo, Cortez, 1995, pp. 134–60.

29 N. Jenkins and L. Joppa, "Expansion of the Global Terrestrial Protected Area System," *Biological Conservation* 142/10, 2009, pp. 2166–74.

References

Cunha, E. D., *Obra completa*, Rio de Janeiro, Nova Aguilar, 1995.

Dean, W., *With Broadax and Firebrand: The Destruction of the Brazilian Atlantic Forest*, Berkeley, University of California Press, 1995.

Drummond, J., "The Garden in the Machine: An Environmental History of Brazil's Tijuca Forest," *Environmental History* 1/1, 1996, pp. 83–105.

Fausto, B., *A Concise History of Brazil*, Cambridge, Cambridge University Press, 1999.

Franco, J. L., and Drummond, J. A., "O cuidado da natureza: A fundação Brasileira para a conservação da natureza e a experiência conservacionista no Brasil (1958–1992)," *Textos de História* 17/1, 2009, pp. 59–84.

Franco, J. L., and Drummond, J. A., "Wilderness and the Brazilian Mind (I): Nation and Nature in Brazil from the 1920s to the 1940s," *Environmental History* 13/4, 2008, pp. 724–50.

Grove, R. H., *Green Imperialism: Colonial Expansion, Tropical Island Edens and the Origins of Environmentalism, 1600–1860*, New York, Cambridge University Press, 1995.

Guimarães, R., *The Ecopolitics of Development in the Third World: Politics and Environment in Brazil*, Boulder, CO, L. Rienner, 1991.

Hecht, S., and Cockburn, A., *The Fate of the Forest: Developers, Destroyers and Defenders of the Amazon*, Chicago, Chicago University Press, 2010.

Hochstetler, K., and Keck, M., *Greening Brazil: Environmental Activism in State and Society*, Durham, NC, Duke University Press, 2007.

Jenkins, N., and Joppa, L., "Expansion of the Global Terrestrial Protected Area System," *Biological Conservation* 142/10, 2009, pp. 2166–74.

Martinez-Alier, J., *The Environmentalism of the Poor: A Study of Ecological Conflicts and Valuation*, Cheltenham, Edward Elgar, 2003.

McCann, B., *The Throes of Democracy: Brazil since 1989*, London, Zed, 2008.

Medeiros, R., *Ruschi, o agitador ecológico*, Rio de Janeiro, Record, 1995.

Pádua, J. A., "Annihilating Natural Productions: Nature's Economy, Colonial Crisis and the Origins of Brazilian Political Environmentalism (1786–1810)," *Environment & History* 6/3, 2000, pp. 255–87.

Pádua, J. A., "Biosphere, History and Conjuncture in the Analysis of the Amazon Problem," in M. Redclift and G. Woodgate (eds.), *The International Handbook of Environmental Sociology*, London, Edward Elgar, 1997, pp. 403–17.

Pádua, J. A., "The Birth of Green Politics in Brazil: Exogenous and Endogenous Factors," in W. Rudig (ed.), *Green Politics II*, Edinburgh, Edinburgh University Press, 1992, pp. 134–55.

Pádua, J. A., "Down by Blind Greed: The Historical Origins of Criticism Regarding the Destruction of the Amazon River Natural Resources," in T. Tvedt, G. Chapman, and R. Hagen (eds.), *A History of Water, Series II*, vol. 3, London, I. B. Tauris, 2011, pp. 176–92.

Pádua, J. A., "European Colonialism and Tropical Forest Destruction in Brazil: Environment Beyond Economic History," in J. R. McNeill, J. Pádua, and M. Rangarajan (eds.), *Environmental History – As If Nature Existed*, New Delhi, Oxford University Press, 2010, pp. 130–48.

Pádua, J. A., "Nature Conservation and Nation Building in the Thought of a Brazilian Founding Father: José Bonifácio (1763–1838)," CBS Working Paper 53, University of Oxford, Center for Brazilian Studies, 2004.

Pádua, J. A., *Um sopro de destruição: Pensamento político e crítica ambiental no Brasil escravista (1786–1888)*, Rio de Janeiro, Jorge Zahar Editor, 2002.

Rouquié, A., *Le Brésil au XXIᵉ siècle: Naissance d'un nouveau grand*, Paris, Fayard, 2006.

Torres, A., *O problema nacional Brasileiro*, São Paulo, Editora Nacional, 1978 [1914].

Valle, C. P., *Risonhos lindos campos: Natureza tropical, imagem nacional e identidade Brasileira*, Rio de Janeiro, Editora SENAI, 2005.

Viola, E., and Leis, H., " O ambientalismo multissensorial no Brasil para além da Rio-92: O desafio de uma estratégia globalista viável," in E. J.Viola, H. R. Leis, I. Scherer-Warren, et al. (eds.), *Meio ambiente, desenvolvimento e cidadania: Desafios para as ciências sociais*, 2nd ed., São Paulo, Cortez, 1995, pp. 134–60.

Further Reading

Dean, W., *Brazil and the Struggle for Rubber*, New York, Cambridge University Press, 1987.

Drummond, J., Franco, J. L., and Oliveira, D., "Uma análise sobre a história e a situação das unidades de conservação no Brasil," in R. Ganem (ed.), *Conservação da biodiversidade-legislação e políticas públicas*, Brasília, Edições Câmara dos Deputados, 2011, pp. 341–86.

Franco, J. L., and Drummond, J. A., "Wilderness and the Brazilian Mind (II): The First Brazilian Conference on Nature Protection," *Environmental History* 14/1, 2009, pp. 82–102.

Hochstetler, K., and Keck, M., *Greening Brazil: Environmental Activism in State and Society*, Durham, NC, Duke University Press, 2007.

PART IV

Environmental Thought and Action

CHAPTER TWENTY-SIX

Environmentalism and Environmental Movements in China since 1949

BAO MAOHONG
(translated by Yubin Shen)

There are two types of environmentalism: one is environmentalism as intellectual thought or ideology, the other is environmentalism as a social movement. Chinese environmentalism as intellectual thought or ideology first developed as a set of laws and regulations intended to aid economic development, but actually served as early Chinese environmental protection legislation. After the United Nations Conference on the Human Environment in Stockholm in 1972 – a watershed moment in Chinese environmental history – governmental and public awareness of environmental issues increased significantly. Furthermore, as the country's economy rocketed forward during the 1980s and 1990s, the rapid increase in the seriousness of pollution, industrial waste, and soil loss helped make environmentalist thought a universal ideology, not a capitalist one. Chinese environmentalism as a social movement can be divided into official environmentalism and popular environmentalism. Official environmentalism refers to environmental polices and their implementations by governments at all levels, and popular environmentalism refers to the activities of NGOs in environmental governance. Chinese environmental NGOs, although still relatively few in number, are now playing an important role in the social transformation of Chinese society.

Early Chinese Environmental Thought and Legislation (1949–72)

When the People's Republic of China (PRC) was established in 1949, the new regime had to deal with many severe problems. As a result of many years of warfare, mainland China suffered from frequent and lethal epidemics. From 1900 to 1949, over a million people were infected with plague, and the case mortality rate was 89 percent. Schistosomiasis spread through 12 provinces, covering an area of more than 2 million square kilometers, and infecting more than 11 million people. The average life span in

A Companion to Global Environmental History, First Edition. Edited by J.R. McNeill and Erin Stewart Mauldin.
© 2012 John Wiley & Sons, Ltd. Published 2015 by John Wiley & Sons, Ltd.

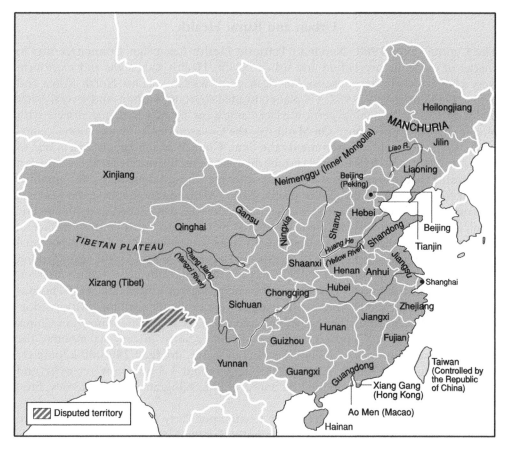

Map 26.1 China: provinces

China was only 35 years. Water shortages and soil erosion were also serious problems. Before 1949, the Yellow River transported 1.4 billion tons of sediments annually to its lower reaches and estuary (see Map 26.1). Soil erosion resulted in soil impoverishment and even desertification in the middle and upper reaches of the Yellow River, while frequent floods menaced its middle and lower reaches, and soil salinization spread along its banks. Deforestation occurred on a large scale, and even in Manchuria (a region with the richest forest reserves in China) the forest area decreased 18 percent, and the forest growing stock decreased by 14.3 percent from 1929 to 1944. The destruction of forests further contributed to water and soil-erosion problems. According to investigations in 1953, the Liao River in northeastern China annually transported more than 40 million cubic meters of sediment to Yingkou Port. Investigations in 1950 indicated that the water depth of Yingkou Port decreased by 0.3 to 0.6 meters each year.[1]

By today's concept and standard of "environment," all those are serious environmental problems. Because there was no such concept of environment in China at that time, however, it is not surprising that there was no specific government agency in charge of environmental issues. As a result, these environmental problems were dealt with by different government agencies.

Urban and Rural Health

The Central (after 1988: National) Patriotic Health Campaign Committee was in charge of epidemic prevention and urban health. Health campaigns and patriotism were connected as a response to American germ warfare against North Korea and China. On February 29, 1952, US planes invaded Andong, Fushun, and several other places in China, launching germ-warfare campaigns by disseminating insects with plague and other infections.[2] On March 14, the Government Administration Council of the Central People's Government (the State Council of the People's Republic of China or PRC from 1949 to 1954) established the Central Patriotic Health Campaign Committee to lead anti-germ warfare and other mass health campaigns. The committee put forward four guidelines for health work: "Medicine should serve the workers, peasants and soldiers; preventive medicine should take precedence over therapeutic medicine; Chinese traditional medicine should be integrated with Western medicine; and health work should be combined with mass movements." A Second National Health Conference was held in Beijing from December 8 to 13, and Mao Zedong wrote a dedication for it: "Get mobilized, pay attention to hygiene, reduce diseases, improve health conditions, and destroy the enemy's germ-warfare attacks." This was the first time epidemic-prevention campaigns were explicitly combined with patriotism, or the "War to Resist US Aggression and Aid Korea" as it was known in China. The Patriotic Health Campaign had considerable results: in only six months, the committee had disposed of 15 million tons of waste, dredged 280,000 kilometers of ditches, built and rebuilt 4.9 million toilets, rebuilt 1.3 million water wells, and eliminated more than 44 million rats and 2 million kilograms of mosquitoes, flies, and lice.

On September 20, 1957, the Chinese Communist Party (CCP) stated in its Third Plenary Session of the Eighth Central Committee that the duties and aims of the Patriotic Health Campaign were: "To eliminate the Four Pests, pay attention to hygiene, eradicate diseases, lift the spirits, change and improve habits, and transform the nation." The slogan "Eliminate the Four Pests" referred to rats (and other noxious creatures), sparrows (and other pest birds), flies, and mosquitoes. Because they harm human hygiene and damage the human food supply, the Patriotic Health Campaign was to eliminate them in seven years:

> According to reports from the Ministry of Food, each year in China, rats consume more than 350 million kilograms of human food. Experiments by the Institute of Zoology, Chinese Academy of Sciences, indicate that a sparrow consumes about 3 kilograms of grain seed, therefore the total amount of food consumed and damaged by sparrows in China is no less than that consumed and damaged by rats.[3]

As mentioned above, the Patriotic Health Campaign had already had dramatic results in this area, but after the push to "Eliminate the Four Pests" was combined with the Great Leap Forward, the Chinese masses were mobilized to help kill these animals and insects. By November of 1958, the Chinese had killed more than 1.88 billion rats, 1.96 billion sparrows and huge amounts of mosquitoes, flies, and maggots. In the meantime, according to official figures, they also disposed of 29.5 billion tons of wastes, collected 61.1 billion tons of manure, dredged 1.65 million kilometers of ditches, and built and rebuilt 85 million toilets.[4]

The approach used to eliminate pests was also applied to eradicating epidemic and endemic diseases. On December 21, 1955, Mao Zedong ordered in his *Circular Requesting Opinions on the Seventeen Articles of Agricultural Work*: "Within seven years, [we will] basically eliminate a number of those diseases most harmful to human beings and livestock, such as schistosomiasis, filariasis, bubonic plague, encephalitis, cattle plague, and hog cholera."[5] It was reported that, with mass mobilizations, Yujiang County, Jiangxi Province, had successfully eliminated schistosomiasis in 1958. Excited to read this news, Mao wrote two poems of "Farewell to the Plague Spirit" to commemorate the campaign.[6]

Although the Patriotic Health Campaign used wartime measures to deal with health work, it should be pointed out that the campaign did incorporate the opinions of the scientific community. During the campaign against sparrows, for instance, many biologists, including some Soviet technological experts who were sent to aid China, were strongly against eliminating sparrows. As a result, the Chinese Academy of Sciences had no choice but to send a report on the beneficial nature of sparrows to Mao Zedong on November 27, 1959. Mao willingly accepted this suggestion, and ordered in *Circular of the Central Committee of the Chinese Communist Party on Health Work* drafted on March 18, 1960, "Do not eliminate sparrows any more, and replace them with bedbugs, and now the slogan is 'eliminate rats, bedbugs, flies, and mosquitoes.'"[7] Later, the State Council announced in a report:

> We have eliminated enough sparrows now, and our food production has also increased year by year, thereby sparrows' damage to food production has been considerably mitigated; in the meantime, the area covered by forest and fruit trees has increased greatly, while sparrows are natural enemies to forest pest insects. Thus, we should not eliminate sparrows any longer, and sparrows should be replaced with bedbugs according to the "Eliminate the Four Pests" announced in this program.[8]

While the Patriotic Health Campaign engaged in epidemic prevention, it also addressed urban and rural environmental sanitation. One of the best examples of an urban sanitation initiative from this period is the improvement of the Long Xu Ditch in Beijing. The Long Xu Ditch was basically used as an open drain in Jinyu Chi Area, Chongwen District. Household waste and wastewater from local small dye factories all flowed into this ditch. It was a breeding ground for flies and mosquitoes and there were always animal carcasses floating in it – the smell could be detected miles away. After 1949, the Beijing Municipal Government began to reconstruct the Long Xu Ditch. The ditch was filled, and the open drain was converted to a covered one. Jinyu Chi Area was rebuilt to be a scenic landscape, and the living standards of local residents were improved. Another project in Beijing was the large-scale reconstruction of the city's river network. The municipal government accomplished this mainly by dredging the rivers, as well as adding new water pools to the preexisting topography. To some extent, this improved sanitation and the health of Beijing's urban environment.

In rural areas, the government implemented a policy called "Two Kinds of Management and Five Kinds of Reconstruction" ("liang guan, wu gai"). Residents were taught how to manage water and night soil, and reconstruct water wells, toilets, animal pens, kitchen stoves, and so forth. These measures altered Chinese peasants' traditional lifestyles and some types of production. To some extent, they not only strengthened peasants' physical constitutions and lengthened their life spans, but also improved rural hygiene and the environment.

Soil and Water Conservation

The Ministry of Water Resources handled soil- and water-conservation projects. Since 1952, the Ministry had been in charge of regulating the development of mountain areas and managing rivers. After the Government Administration Council of the Central People's Government (after 1954, the State Council) sent a circular ordering the expansion of soil- and water-conservation schemes, three National Working Conferences on Soil and Water Conservation were held in 1955, 1957, and 1958. The ministry enhanced soil and water conservation by managing 700,000 square kilometers of land, restoring over 113,000 square kilometers of prime farmland (including 86,670 square kilometers of terraced fields, and 26,670 square kilometers of silt-storage dams and sand dunes reactivated for farmland), planting over 400,000 square kilometers of trees, 40,000 square kilometers of grasses, and 40,000 square kilometers of forest plantations. It also brought about noticeable achievements: total output value was increased by 70 billion RMB yuan, and annually food and fruit production was increased by 17 million tons and 25 million tons respectively. The annual amount of soil erosion was decreased by 1.5 billion tons, and the water-holding capacity was increased by 25 billion cubic meters.[9]

At that time, there were two models in managing water loss and soil erosion and promoting food production: one was the Gaoxigou model, the other was the Dazhai model. Gaoxigou was an administrative village in Mizhi County, Shaanxi Province. Its topography was a huge mountain valley over 10 kilometers long; farming was very difficult in the bottomlands of the valley. From 1952 to 1965, helped by experts from the Yellow River Conservancy Commission, Gaoxigou villagers built terraced fields, water and soil barrages, and silt-trap dams for improving farmland in the bottom of the valley. However, those terraced fields and silt dams could not withstand mountain floods. Villagers in Gaoxigou gradually understood that, in order to manage the valley, they had to manage the slopes. After 1958, Gaoxigou villagers began to plant pea shrubs and pines on the mountain slopes, in addition to building a small water dam. By 1961, Gaoxigou village had converted 800 *mu* (1 *mu* = 667 square meters) from their total 3,000 *mu* farmlands back into forests. Although the area of farmlands decreased, the food production gradually increased year by year: the per *mu* grain yield in 1949 was 40 *jin* (1 *jin* = 0.5 kilograms), in 1956 it was increased to 60 *jin*, in 1958 it was 89 *jin*, and then Gaoxigou transformed from a village with a food shortage to a village with a food surplus. Due to drought, the per *mu* grain yield in 1959 was 80.9 *jin*, and in 1960 with a drought of 150 days, it was 80.6 *jin*. In 1961, the per *mu* grain yield increased to 103.7 *jin*, more than twice the yield of 1957.[10] Gaoxigou also increased forage grasslands by 200 *mu*, and developed animal husbandry to provide fertilizers for farming. Within this positive cycle, Gaoxigou developed a system of thirds, in which agriculture, forestry, and animal husbandry each took up one third of the land. This system brought about a "miracle." As a slogan put it: "Water will not run downhill, and silt will not come out of the valley."

The other agricultural model came from Dazhai, a small mountain village in Xiyang County, Shanxi Province. Its 700 *mu* of farmland were distributed through several mountain valleys. This village frequently suffered droughts, and, when it did rain, it suffered severe flooding. As a result, Dazhai was a village of extreme poverty. Dazhai peasants declared a war against their disastrous environment, and strived to manage mountains and rebuild farmlands. Starting in the winter of 1955 (after several failures), they finally built up mellow-soil fields, which could conserve soil moisture and prevent soil erosion, and could still provide a harvest even if they suffered droughts and floods. Furthermore, they constructed high-yield fields by rational close planting. In total, Dazhai built over 800 *mu* of agrarian lands, 6 kilometers of winding mountain highways, 6 water reservoirs, and 14 kilometers of winding mountain drainage ditches. The per *mu* grain yield increased from 237 *jin* in 1952 to 774 *jin* 10 years later. And even in 1963, when Dazhai suffered several natural disasters, the per *mu* grain yield was still over 700 *jin*. From 1952 to 1963, the village managed to contribute 1,758,000 *jin* of grain to the state, and each household contributed 2,000 *jin* of grain each year.[11]

Both agricultural models attracted the attention of the authorities. On January 18, 1962, the *People's Daily* reported Gaoxigou's success and added an editorial praising it as "a model of conserving water and soil by the mass[es] themselves." On December 28, 1963, the *People's Daily* introduced the successful experience of Dazhai against natural disasters. On February 10, the *People's Daily* published a report entitled "The Dazhai Road." In an additional editorial, "A Good Model of Building Mountain Areas with Revolutionary Spirit," the *People's Daily* pointed out that on the basis of the People's Communes, the Chinese people could triumph over natural disasters by self-reliance. Although the Dazhai model was introduced in national newspapers later than the Gaoxigou model, it aligned better with Mao Zedong's wishes to solve agricultural problems as well as to boost the building of socialism in China. Therefore Mao, on many occasions, said that "Agriculture has to rely on the Dazhai Model," or spoke of "the Dazhai Spirit." In the First Plenary Session of the Third National People's Congress (December 20, 1964 to January 4, 1965), Premier Zhou Enlai officially introduced the Dazhai spirit to the whole nation in his "Report on the Work of the Government," calling upon the Chinese to learn from Dazhai. It was the beginning of the national "Learn from Dazhai in Agriculture" Campaign.[12]

The "Learn from Dazhai in Agriculture" Campaign was intended to uphold the spirit of self-reliance and hard work, focus on farmland capital construction, apply the "Agricultural Constitution of Eight Characters,"[13] and increase agricultural production and efficiency. All Chinese peasants were mobilized to participate in this campaign to manage mountains and rivers, reconstruct land, and build farmland on a considerable scale. In 1965, national grain production reached 194.5 million tons, cotton production reached 2 million tons, and sugar production was 15.4 million tons.[14] In 1978, the area of irrigated farmland in China reached 800 million *mu* – a 60 percent increase since 1965 – and the number of irrigation wells numbered over 200,000 – a 936 percent increase.

The examples taken from the Patriotic Health Campaign and the "Learn from Dazhai in Agriculture" Campaign indicate that, after the establishment of the PRC, Chinese peasants were mobilized en masse, not just during times of war, but also in peacetime to tackle domestic issues. Although these campaigns were designed to protect and improve the Chinese economy and standard of living, rather than the environment, the two campaigns in fact had solved serious environmental issues of the time. They may have had many negative environmental consequences, and sometimes worked in a way unrecognizable as an "environmental movement," but the two campaigns should not be forgotten in the history of world environmentalism.

Changes in Chinese Environmental Thought and Consciousness (1972–2011)

In June 1972, the United Nations held its Conference on the Human Environment in Stockholm. The Chinese government sent a large delegation to the conference with two aims. First, China wanted to take advantage of this international arena to oppose discrimination against and invasions of the Third World by the US. Second, China wished to publicize the PRC's efforts in environmental protection and its achievements in socialist construction, and thus secure a stake in international environmental policy-making.[15] Participation in the conference exposed the Chinese delegation to the environmental issues affecting the global community as well as the mass environmental movements occurring in other countries. This conference, then, was the introduction of modern environmentalism into China, and represents a watershed moment in the development of Chinese environmentalism. It may be considered as the official beginning of its environmental movement.

China realized the importance of environmental issues, but its policy was still very closely tied to state ideology. Chinese officials continued to insist that environmental pollution and destruction were products of capitalism, and the environmental struggle in capitalist states was a manifestation of class struggle in the realm of environment. Zhou Enlai argued that, "Now pollution is becoming a big problem in the world. Waste gas, waste water and waste residues have harmed the US greatly, but the Nixon administration could not solve them. Capitalist states, like the US and Japan, are unable to solve environmental problems."[16] Capitalist states could not solve industrial pollution because of their system of private ownership, the anarchy of their production, and their orientation toward profits. But Zhou Enlai also said:

> We can definitely solve industrial pollution because our socialist planning economy serves the masses. While we are engaged in economic construction, we should pay closer attention to solving this issue and avoid doing something absolute, which may create problems for our descendants. The reason that the Soviet Union has environmental problems is that it has become a revisionist country.[17]

This last statement is significant because it indicates that there was, during the 1970s, a change in Chinese environmental thought: the Chinese authorities recognized that socialism could also bring about environmental problems, but they still insisted that socialism could solve environmental problems far better than capitalism because of its superiority.

Mao died in 1976 and after a brief interlude Deng Xiaoping (1904–97) ascended to power. He launched a series of reform policies aimed at strengthening and modernizing China. After the Reform and Open Door policies were implemented in 1978, the spectacular growth of the Chinese economy made its environmental problems more and more serious.[18] Accordingly, the Chinese understanding of the environment again shifted, and the Chinese authorities began to treat pollution and other issues as the inevitable products of industrialization, rather than simply as a feature of capitalism. An important point to remember is that, while the West industrialized over the course of 200 years, China industrialized in only 30 – this compressed the environmental problems of industrialization into a much smaller time frame. The compression of China's economic development into a few decades made its environmental problems seem compounded, because their scale was so large and their appearance so rapid.

In 1984, the Chinese government announced in the Second National Conference on Environmental Protection that "environmental protection is a strategic task in the process of modernization in China and that it is a basic national policy." However, the implementation of environmental governance was less than satisfactory. When there were contradictions between economic development and environmental protection, or in international environmental cooperation and struggles, the Chinese government had always emphasized its role as a developing country. That implied that China would try its best to avoid the old path of "pollution first, treatment later," but within the prevailing international political and economic order, the Chinese people's right to existence and economic development still should be given priority.

During this period, the term "ecological environment" (*shengtai huanjing*) came into use, which could help understand the complexity of Chinese environmental thought. When the PRC's Constitution was revised in 1982, Huang Bingwei, geographer and member of the National People's Congress (or NPC), argued that the term "ecological balance" (*shengtai pingheng*) in the draft should be changed to "ecological environment." He believed that the "ecological balance" was not always beneficial to human society, thereby the state should not improve and protect the "ecological balance." And since, according to Joseph Stalin, whose writings still carried weight for some Chinese thinkers despite the Sino-Soviet rift, "environment is the natural world around humankind," "balance" here needed to be revised as "environment." His suggestion was accepted and, because of its usage in the Constitution, the term "ecological environment" became popular in China. However, more and more Chinese scholars have questioned the legitimacy of this term, for it cannot easily be translated into other languages, and it also lacks a clear academic definition. In 1998, even Huang himself admitted that the term "ecological environment" was a mistake, and he suggested that the China National Committee for Terms in Sciences and Technologies revise it.[19] In 2005, this committee invited experts in ecology, environmental science, and linguistics to discuss meanings, usage, and translations of "ecological environment" and "build ecological environment."[20] Most of them agreed that this term should be corrected, but quite a few insisted there was no need to revise it, since "ecological environment" had been in popular usage for so long. This term is still used in Chinese environmental culture, and no one has clarified its meaning.[21]

There were some changes in Chinese environmental thought at the turn of this century. The great floods along the Yangzi in 1998 and increasingly serious dust storms coincided with the successful application to host the 2008 Olympics, with a new strategy for development of the western parts of the country, and the process of cutting back internal demand after the Asian financial meltdown of 1997. All these factors together pushed the government to reflect on the purpose of development and revise, again, official Chinese environmental thought. Any changes, however, required theoretical justification. The CCP searched Marxist works and finally found that Friedrich Engels, in his *Dialectics of Nature*, stated that nature would take its revenge if humans overutilized it, and that humans and nature are constantly interacting.[22]

In the 16th National Congress of the Chinese Communist Party (2002), President Hu Jintao advocated a new official socioeconomic ideology, the "Scientific Development Concept," in which one key element is "to balance a harmonious development of Man and Nature." Then in 2005, Hu advocated "realizing an all-around, coordinated and sustainable development," and "building a resource-conserving and environment-friendly society." Premier Wen Jiabao stressed the importance of the "Three Changes" in the 6th National Environmental Protection Conference,[23] held on April 17–18, 2006:

First, to move from a mode of growth that stresses the economy to one which balances the economy and the environment. Second, to move from a situation in which environmental protection holds the economy back, to one where they develop in tandem; from a passive and remedial model of environmental protection to a proactive, protective method. Third, to move from the use of policy and administrative methods to protect the environment to the combined use of legal, economic and technical methods, alongside political intervention when necessary, to adapt to new circumstances and accelerate innovation.

In the "Report to 17th National Congress of the CCP," the Central Committee explicitly advocated "building an ecological civilization." From these types of statements, we can conclude that current Chinese environmental thought emphasizes the harmonious relations between man and nature and between economic development and environmental protection – that the successful management of human–nature interaction is a prerequisite for building a harmonious society and a harmonious world. Compared with previous ideas, this new policy is more comprehensive and reasonable, and better complies with the trends of international environmentalism. Nevertheless, there are still some disappointing gaps between theory and reality, academia and politics, and economic development and environmental protection. A telling example is the different understandings of the term "ecological civilization."

The term "ecological civilization" first officially appeared in the "Report to 17th National Congress of the CCP" (2007). In the part entitled "Building a Moderately Prosperous Society," this report mentions economic, political, cultural, social, and then ecological "civilization." An ecological civilization would be the "forming [of] an energy- and resource-efficient and environment-friendly structure of industries, patterns of growth and modes of consumption." In the rest of the report, however, many themes are outlined and elaborated upon – "Promoting Sound and Rapid Development of the National Economy," "Unswervingly Developing Socialist Democracy," "Promoting Vigorous Development and Prosperity of Socialist Culture," "Accelerating Social Development with the Focus on Improving People's Livelihood," and "Opening Up New Prospects for Modernization of National Defense and the Armed Forces" – but there is no further treatment of any environmental goals. Environmental issues are only briefly mentioned in the fourth point of "Promoting Sound and Rapid Development of the National Economy," where it is indicated that "ecological civilization" is just one of the five steps to "Building a Moderately Prosperous Society." Xia Guang, a scholar in the State Environmental Protection Administration (SEPA), admitted this point.[24]

In Chinese scholarship, there have been a number of different explanations of this political term. Some scholars perceive "ecological civilization" as one important part of society, in parallel with material, spiritual, and political civilizations, and as a civilization coordinating the relationship between man and nature. Others have defined "ecological civilization" as a new type of civilization which follows after agricultural civilization and industrial civilization – a higher level of human civilization succeeding and surpassing industrial civilization. For them, "ecological civilization" includes the entirety of social relations of humankind. Pan Yue even insisted that "ecological civilization" is the sum of all material and spiritual achievements through harmonious developments of humans, nature, and society. The West has lost the opportunity to lead this ecological civilization, he maintains, but socialism has the power to lead the whole world into this new, postindustrial civilization.[25] Some others have a more balanced view, arguing that, vertically speaking, ecological civilization is a new type of civilization following primitive, agricultural, and industrial civilization; horizontally speaking, it parallels material, political, and spiritual civilizations; and comparing different stages of civilization in China and the West, the path China should currently take is "to take up the lessons from industrial civilization, and then follow the path of ecological civilization."[26] As opposed to the disputes surrounding the term "ecological environment," these writings on "ecological civilization" seem to indicate that Chinese scholars have gone beyond the limit of official ideology, and expect China to play a more crucial role in the total transformation of global society. The two cases – "ecological environment" and "ecological civilization" – show that there is a slight disagreement between official ideology and academic thought, and this disagreement has to some extent resulted in a disjuncture between Chinese environmental thought and the reality of environmental policy-making.

Chinese modern environmental consciousness has experienced a noticeable evolution since the mid-twentieth century. The changes in Chinese environmental consciousness may be illustrated by several surveys, taken in different periods. In 1993, Renmin University of China, in Beijing, conducted a sampling survey of 200 people on Chinese environmental consciousness. Sixty-three percent of respondents believed that environmental problems in China had already influenced their living standard; 55.6 percent insisted that local governments should be held responsible for environmental destruction; and when asked how to improve the environment, 55.6 percent chose "to intensify environmental education," and 57.9 percent chose "to strictly implement related laws."[27] In 2008, in a survey by the Chinese Environmental Promotion Association, 76.4 percent of those surveyed recognized that environmental problems in China were "very serious" or "serious." But only 18 percent of respondents said they would take the initiative "to learn about environmental protection." While 24 percent said that they would "advise others to practice green consumption," and 26 percent claimed they "often take environmental protection measures," over 47 percent said they would not report those who violate environmental laws to the government.[28] Before the opening of the NPC and the National Committee of the Chinese People's Political Consultative Conference in 2011, an Internet poll indicated that 93 percent of the 3,207 participants believed "environmental pollution in China is very serious." Asked about the causes of environmental pollution, 75 percent of respondents insisted that "local governments sacrificed environment for economic interest," and 11 percent thought "it is due to supervision failure." Thirty-nine percent of respondents thought it appropriate to raise citizens' consciousness of environmental protection by severe punishment, while 11 percent thought media and public figures should exert more effort in raising citizens' consciousness concerning environmental protection.[29]

From these three surveys, we can conclude that Chinese have gradually recognized the seriousness of environmental problems. They could reach an agreement on the role of local governments in environmental pollution and the importance of severe punishment in solving environmental problems, but their consciousness of solving these problems by themselves is still quite weak. In fact, it reflects the character of governance structures in China, an example of "strong government and weak society" in the realm of the environment.

In sum, Chinese environmental thought is always changing, and has gradually been transformed from a time when the understanding of environmental problems was heavily influenced by state ideology to a time when Chinese environmental thought more closely matched the world's understanding of these problems. However, by comparing the changes of official environmental thought with changes in academic understanding and popular environmental consciousness, it becomes clear that the Chinese people still rely on the government to improve the environment. It is an embarrassing reality in today's China that the government is responsible for both the success and the failure of environmental protection. The Chinese people hate pollution, but they do not want to engage in environmental protection themselves.

Environmental Governance within the Chinese State

China could still be described as a nation with a strong government and a weak society.[30] This description is apt when describing environmental governance in China. Despite increased involvement in policy-making by NGOs since the 1990s, the crafting and implementation of environmental policy has been, and still is, largely handled through the central and local governments.[31] Within the central government, environmental administrations experienced a complicated evolution, operating under many different names and at various levels of organization. Since the 1970s, Chinese environmental protection administrations have been "promoted" once nearly every 10 years. This means the agency which previously dealt with environmental governance is renamed and given an increase in power and responsibility.

In 1974, the first administrative agency to deal with environmental protection was established as the Leading Group of Environmental Protection. In 1982, however, this group was abolished, and a Ministry of Urban and Rural Construction and Environmental Protection was set up. One of the departments of this new ministry was a Bureau of Environmental Protection. In 1988, this Bureau became the State Environmental Protection Bureau, but it was now under the authority of the State Council. In 1998, it was renamed as the State Environmental Protection Administration (SEPA), and finally, in March 2008, it was elevated to the Ministry of Environmental Protection during the 11th NPC. The mission of this new ministry is to "develop and organize the implementation of national policies and plans for environmental protection, draft laws and regulations, and formulate administrative rules and regulations for environmental protection, take charge of the overall coordination, supervision and management of key environmental issues."[32] This latest promotion to a ministry-level organization will help overcome previous difficulties in policy-making. With each new incarnation – from a group to a bureau, an administration, and finally a ministry – the governmental body in charge of environmental protection has passed into a new stage and acquired greater standing within the Chinese state.

The changes in the central government were reflected in changes in local environmental protection agencies. Over time, however, the development of reforms and the growing power of local governments have eroded the power of state-directed environmental policies. So while the Ministry of Environmental Protection has the right to direct local agencies' operations, the personnel and financial issues of local environmental protection agencies are still controlled by local governments. In an attempt to solve this problem, the central government established five regional environmental supervision centers for the Eastern, Southern, Northwestern, Southwestern and Northeastern regions in 2006. The main responsibilities of these centers included: supervising the implementation of environmental regulations, laws, and standards in local regions; processing cases of big and serious environmental pollution and ecological destruction; coordinating trans-regional and trans-river-basin environmental disputes; and supervising the responses to those emergent serious environmental incidents. The five regional centers separate their personnel, finance, and property from local governments, and, without local governments' interference, they could play important roles in governing transriver basin and transregional environmental problems.

China has developed a relatively comprehensive system of environmental policies. From 1973 to 1978, Chinese environmental policy was mainly aimed at cleaning up pollution. In 1979, however, the Environmental Protection Law of the PRC (for trial implementation) was passed, and after 10 years of practice, it became an official law in 1989. In this legislation, the NPC promulgated 11 specific laws, targeting forestry, grasslands, fisheries, mineral resources, water, soil conservation, wildlife protection, land management, marine environments, and the prevention of air and water pollution. At the same time, the central government implemented a new policy that construction projects should result in a unified economic, social, and environmental benefit. The policy emphasized that, "Who develops, should protect; who destroys, should restore; who uses, should compensate."[33] Furthermore, the central government implemented the "polluter pays" principle of the Organization of Economic Cooperation and Development (OECD), charging polluters extra fees for discharging polluted effluents.

After 1998, SEPA worked together with other major ministries to promulgate a new system of environmental economic policies:

1. The green tax: Enterprises that implement environment-friendly behaviors will be rewarded with tax incentives; those enterprises seen as unfriendly to the environment will be responsible for direct pollution taxes based on the amount of their pollution emitted as well as indirect environmental tax on their production.
2. Environmental charges: Raising the level of charges to take full account of environmental factors, pricing, and fees to promote energy-saving.
3. Green capital markets: Environment-friendly enterprises will be provided with "green loans" while those environment-unfriendly enterprises will be limited in access to loans.
4. Ecological compensation: Those who have brought about ecological disasters should be responsible for compensation, and those injured should be compensated financially.
5. Trading of Waste Discharge Permits: Reducing the total cost of pollution control, and mobilizing the enthusiasm of polluters to prevent pollution.
6. Green trade: Collecting an environmental compensation tax from those who export low-value mine products or wild animals, taxing imported cars with high exhaust emissions, and providing green labels to imported scrap iron and other recycled metals.
7. Green insurance: To encourage enterprises to purchase environmental liability insurance; insurance companies will provide compensation to the victims of pollution.

One policy which exemplifies the government's recent efforts at environmental management – and the problems facing policy-makers – is the "Energy Conservation and Emission Reduction" program. In response to the international community's pressure on China to deal with its environmental problems, the Chinese government (in its 11th Five-Year Plan on National Economic and Social Development, 2006–2010) declared that China would reduce the energy intensity of its economy by 20 percent, and reduce the amount of major pollutants by 10 percent.[34] To achieve this aim, many measures were taken, but their effects were not satisfying because of trade monopolies held by state enterprises and local governments' lack of cooperation in implementing the energy-reduction program. In an attempt to speed up the achievement of China's energy aims, SEPA launched four initiatives (or "storms") between 2005 and 2009, specifically targeting state enterprises and local governments.

First, in January 2005, SEPA published the names of 30 large industrial projects and 46 thermal power plants which had not installed desulfurization treatment programs. It also served administrative punishment notices for five large companies, including the China Three Gorges Corporation. Second, in 2006, SEPA halted several large petrochemical and transportation projects along rivers (whose total invested value was about 29 billion RMB). These first two initiatives, or storms, focused on large state enterprises that were keeping China from reaching its energy-reduction goals, while the next two actions were directed at local governments' administrative noncompliance in implementing state policies. In 2007, SEPA publicized 82 projects whose actions violated the National Environmental Assessment Law and the "Three Simultaneities System," and demanded that 23 of those 82 projects modify their actions to fit with these laws. These projects spanned 22 provinces and 12 kinds of industries, and had a total value over 112 billion RMB. After the blue-green algae bloom in Tai Lake, in the Yangzi delta plain, in 2007, SEPA launched its fourth initiative: to temporarily stop all approval assessments on projects in major river systems, and supervise and inspect 38 existing enterprises. While these four "storms" achieved some of their goals and represented the proactive approach the Chinese state environmental administration took toward implementing the "Energy Conservation and Emission Reduction" policy, SEPA's efforts were hampered by the same issue it set out to solve: the lack of implementation of state policies at the local level. The cooperation between local governments and contaminative enterprises is the key to the great gap between policy and result, theory and reality in China's environmental governance.

The reason that local governments ally with polluting enterprises is because they have common interests. In 1980, the state issued a new fiscal policy which said that local governments would be responsible for providing the funding for local development and public service projects. This policy was intended to force local governments to stimulate the economy in their communities. Thus, while China is a centralized state, it relies on the agency of local governments. This means, however, that local governments are in a constant struggle to expand their financial operations, and see private industry as an ally in economic development. Many local governments turn a blind eye to, or even aid in, the environmentally destructive practices of industries or enterprises for the sake of the local economy. Meanwhile, many of these private companies have little incentive to spend money on environmental management: before 1978, they were all state and collective enterprises, and not subject to market competition. After 1978, when many of these companies were privatized, it became necessary to keep costs as low as possible in order to compete with state-owned enterprises. This meant that environmental issues were not given priority. Although Chinese policy since the 1990s has mandated that environmental protection should be integrated into economic development, many local enterprises ignore state policy in order to maximize profits.[35]

Thus, while the Chinese government has established powerful agencies for environmental protection and relatively comprehensive environmental policies, the effects of these policies are rarely satisfying. Qu Gepin, former director of SEPA, lamented that China's economic plans have been "overachieved" every year, but in the last 25 years environmental protection plans have not been achieved once. As a result, SEPA has had to collaborate with environmental non-governmental organizations (ENGOs) to deal with local governments and polluting enterprises.

Environmental NGOs in China

In China, ENGOs refer to those social organizations protecting the environment, but they could not be defined by the conventional theoretical concepts of the West, for their relationship with the state is different.[36] Specifically speaking, they are civic organizations that aim to protect the environment without looking for profit and aim to provide public environmental service without help from organs of administrative power. The first ENGO in China was the China Society for Environmental Sciences, established by the Chinese government in 1978. After its establishment, however, the growth of new ENGOs in China stagnated for over a decade, until the Friends of Nature was founded in Beijing in 1994. Since then, ENGOs in China have developed at a much faster rate. In 2006, there were 2,768 ENGOs in China, which employed 224,000 staff members: 69,000 full-time and 155,000 part-time employees.[37]

There are two types of ENGOs: grassroots ENGOs, such as Friends of Nature, and ENGOs supported by the government, such as the All-China Environment Federation. Regardless of the type of ENGO, in the current Chinese political system no ENGOs can go beyond the actions allowed it by the government: ENGOs may serve as a bridge between the government and the masses and take part in international discussions, but they must practice discretion and not overstep their bounds.

Even so, ENGOs have managed to play an important role in raising the environmental consciousness of the Chinese, using the media to expose pollution and other environmental problems to the public. The use of the media is an effective way to address environmental issues in China, because it results in mass pressure and forces political leaders to pay attention. As a result, ENGOs in China maintain good relations with the media, often expanding their influence and winning support from the public through the media. At the same time, ENGOs utilize mass education techniques, such as publishing books, holding lectures, and organizing training workshops. ENGOs have tried to propel Chinese society and economy in a "green" direction, sometimes with excellent results. For example, in 2004, several ENGOs, including Friends of Nature and Global Village, took action in combating the Yuanmingyuan Lakebed Anti-Seepage Project – a proposal that involved installing an impermeable membrane on the bed of a small lake in Beijing in the hopes of saving water – and pushed SEPA to hold a public hearing on the environmental impact assessment. Finally, they succeeded in stopping the project and restoring the water levels of the lake.

ENGOs have also used their professional knowledge to help shape citizens' environmental legal consciousness and to defend the legal rights of those suffering from environmental injustice. Environmental damages not only corrode relations between man and nature, but also result in many human victims. Some ENGOs have made efforts to assist those engaged in environmental lawsuits. For example, since its establishment in 1999, the Center for Legal Assistance to Pollution Victims at the China University of Political Science and Law (a law school in Beijing) has provided legal assistance to over 10,000 plaintiffs who were suffering the impacts of pollution and helped over 50 people to file lawsuits or pursue administrative redress.

ENGOs have tried their best to influence the formation and implementation of environmental policy. Because ENGOs claim to speak for the public good, they are usually able to muster wide public support for their activities and force the government to pay more attention to their suggestions during policy-making and implementation. One successful case is ENGOs' prevention of several water power station construction projects along the Nu River. (The Nu, also known as the Salween, rises on the Tibetan Plateau and flows south through China and Burma to the Bay of Bengal.) In 2003, the State Reform and Development Committee passed a plan in which two reservoirs and 13 dams were to be constructed in the middle and lower reaches of the Nu River. On hearing this news, several ENGOs in China launched the Great Anti-Dam and Defense of the Nu River Movement. They held lectures and forums through television, print, and Internet media to bring exposure to the Nu River dam plan. Then, in November 2003, Chinese ENGOs submitted a proposal to a World Rivers and Public Anti-Dam Conference held in Thailand, and successfully lobbied ENGOs from over 60 countries to sign a protest letter to put pressure on the PRC government. And in both of the two expert hearings held by SEPA, ENGOs strongly supported those experts who were against this dam plan. These efforts forced Premier Wen Jiabao to delay the developmental project in February 2004. To a certain degree, this case demonstrates that ENGOs in China have the potential to change government environmental policy-making. Yu Xiaogang, the leader of a popular ENGO called "Green Rivers," stated that the significance of ENGOs is in their ability to influence decision-making.[38]

In the way that Chinese environmental governance is currently structured, the central government plays the leading role. But its actions are often strongly resisted by local governments – for a variety of reasons – and businesses, which chafe at regulation and environmental mandates. To deal with opposition to environmental policies, the central government sometimes collaborates with ENGOs when necessary. However, the government will only collaborate with ENGOs when it suits the government. Unlike their counterparts in industrial countries, ENGOs in China cannot put strong pressure on the government, and have not formed a civil society which could balance the power of the government.

ENGOs in China should have more space to contribute to society. This would bring them in line with their foreign counterparts, and allow them to contribute to the "harmonious society." However, in the current situation – rapid economic development without accompanying political reform – there is an alternative trend among ENGOs: those supported by the government have enjoyed rapid growth, while those based on the grassroots are still stagnant. In the long run, this might be considered merely a transition phase during China's social transformation. We could regard it as a stage when grassroots ENGOs are accumulating power. Time will tell.

Environmentalism in China since 1949 has been closely related to the progress of China's political and economic reforms. Although Chinese environmentalism has made great strides over the last half century and enjoyed relative success, there is still a gap between environmental movements in China and those entrenched in other parts of the world. Currently, there are two major shortcomings in the development of Chinese environmentalism. First, ENGOs are still few in number and weak in strength, especially those not under the umbrella of the government. Second, local governments do not lead in movements for environmental protection, but rather stand in opposition to environmentalist movements – allying themselves with businesses in the pursuit of local economic growth. With the continuing development of the Chinese economy and the boosting of political reform, however, it is likely that Chinese environmentalism will experience a structural transformation: the influence from civil society will grow more powerful, and will bring into balance the influence of state, society, and markets. This will allow all kinds of forces to play a role in the development of Chinese environmental governance.

Notes

1 B. Yi and Y. Yiguang, "Qing mo yilai dongbei senling ziyuan ji qi huanjing daijia," *Zhongguo nong shi* 3, 2004, pp. 115–23.

2 Note from the editors: this is a famously controversial point. The US government and many scholars deny all allegations of biological warfare in the Korean War context. Some Western scholars, however, accept the allegations. In any case, the Chinese government acted on the belief, or at the very least on the assertion, that the allegations were true.

3 "Chu si hai," *Renmin ribao*, 12 January 12, 1956, p. 1.

4 D. Li, "Guanyu 1958 nian aiguo weisheng yundong de jiben qingkuang he jin hou de renwu," *Xinhua ban yue kan* 2, 1959, pp. 54–6.

5 Mao Z., vol. 6, *Mao Zedong wen ji*, Beijing, Remin chuban she, 1999, p. 509.

6 Mao Z., *Mao Zedong shi ci ji*, Beijing, Zhongyang wenxian chuban she, 1996, pp. 104–5.

7 Mao Z.,"Zhonggong zhongyang guanyu weisheng gongzuo de zhishi," in Zhonggong zhongyang wenxian yanjiu shi (ed.), *Jian guo yilai zhongyao wenxian xuan bian*, vol. 13, Beijing, Zhonggong zhongyang wenxian chuban she, 1996, n.p.

8 State Council, "Wei tiqian shixian quanguo nongye fazhan gangyao er fendou," report in the Second Session of the Second National People's Congress, April 6, 1960.

9 T. Guo, "Zhongguo shuitu baochi chengjiu yu zhanwang," *Shuili shuidian keji jinzhan* 17/4, 1997, pp. 7–10 and p. 32 (p. 8).

10 "Guanyu san nian lai shuili shuibao gongzuo jiben zongjie yu jin hou yijian," Mizhi County Archive, Record Group No. 0020, 1961.

11 Zhonghua renmin gonghe guo guojia nongye weiyuan hui bangong ting (ed.), *Nongye jiti hua zhongyao wenjian huibian*, Beijing, Zhongyang dangxiao chuban she, 1981, p. 806.

12 J. Li, *Nongye xue dazhai yundong shi*, Beijing, Zhongyang wenxian chuban she, 2011, pp. 74–83.

13 Based on the practice of Chinese peasants and scientific findings, the "Agricultural Constitution of Eight Characters" comprised eight measures summarized by Mao Zedong in 1958 to increase agricultural yield; the eight Chinese characters are *tu, fei, shui, zhong, mi, bao, guan*, and *gong*. *Tu* means soil improvement, survey, and plan of land utilization; *fei* is rational fertilization; *shui* means building irrigations and rational use of water; *zhong* is bleeding and promoting high-quality seeds; *mi* is rational close planting; *bao* is protecting plants against pests; *guan* is field management; and *gong* means renovation of production instruments.

14 Guojia tongji ju guomin jingji zonghe tongji si (ed.), *Xin Zhongguo liushi nian tongji ziliao huibian*, Beijing, Zhongguo tongji shuban she, 2010, p. 37.

15 G. Qu and P. Jinxin (eds.), *Huanjing juexing: renlei huanjing huiyi he Zhongguo di yi ci huanjing baohu huiyi*, Beijing, Zhongguo huanjing kexue chuban she, 2010, pp. 3–7.

16 G. Qu, "Zhou Enlai shi xin Zhongguo huanjing baohu shiye de kaituo zhe he dianji ren," in Mengxiang yu qidai: Zhongguo huanjing baohu de guoqu yu weilai, Beijing, Zhongguo huanjing kexue chuban she, 2004, pp. 34–44.

17 Zhongguo kexue yuan jishu xinxi yanjiu suo (ed.), *Guowai wuran kalian*, Beijing, Renmin chuban she, 1975, p. 88.

18 The Reform and Open-Door policies were economic reforms intended to jumpstart the Chinese economy. These reforms included the decollectivization of agriculture, the introduction of private businesses, and allowing foreign investment in the country's industry. As a result, China has experienced an extremely high rate of growth since the 1980s, averaging a 9.5 percent increase in the economy per annum.

19 See Huang, B., "Dili xue zonghe gongzuo yu kuai xueke yanjiu," in Huang Bingwen yuanshi xueshu sixiang yanjiu hui wenji (ed.), *Ludi xitong kexue yu dili zonghe yanjiu: Huang Bingwen*

yuanshi xueshu sixiang yanjiu hui wenji, Beijing, Kexue chuban she, 1999, pp. 12–13; Huang Bingwei wenji bianji zu (ed.), *Dili xue zonghe yanjiu: Huang Bingwei wenji*, Beijing, Kexue chuban she, 2003, p. xv.

20   See "Re dian ci nan dian ci zongheng tan," *Keji shuyu yanjiu* 2, 2005, pp. 20–38.

21   Y. Hou, "Shengtai huanjing shuyu chansheng de teshu shidai," *Zhongguo lishi dili luncong* 1, 2007, pp. 116–23.

22   F. Engels, "Dialectics of Nature," in Zhonggong zhongyang makesi engesi sidalin zhuzuo bianyi ju (ed.), *Ma en quanji*, vol. 20, Beijing, Renmin chuban she, 1971, p. 519.

23   See http://www.gov.cn/ldhd/2006-04/23/content_261716.htm, accessed on April 16, 2012.

24   G. Xia, "Shengtai wenming gainian you duo yi xing," *Zhonguo huanjing bao*, September 22, 2009, p. 2.

25   Y. Pan, "Shengtai wenming jiang cujing Zhongguo tese shehui zhuyi de jianshe," *Liao wang*, October 22, 2007, pp. 38–9.

26   W. Ye, *lecture*, Ecological Civilization Forum, Beijing University, 2010, accessed from http://pkunews.pku.edu.cn/xwzh/2011-01/03/content_191808.htm on March 19, 2012.

27   D. Liu et al. (eds.), *Huanjing wenti: cong zhong ri bijiao yu hezuo de guandian kan*, Beijing, Zhongguo renmin daxue chuban she, 1995, pp. 88–90.

28   Zhongguo huanjing wenhua cujin hui (ed.), *Zhongguo gongzhong huanbao zhishu: 2008 niandu baogao*, Beijing, Zhongguo huanjing wenhua cujin hui, 2009.

29   X. Chen and C. Bin, "Nian 'Liang Hui' Jianya Fenxi Baogao: Huanjing Wuran Zhuanti," 2011, accessed from http://news.163.com/11/0314/13/6V42PS5100014JB6.html on March 19, 2012.

30   See M. Bao, "The Evolution of Chinese Environmental Policy and Its Efficiency," *Conservation and Society* 4/1, 2006, pp. 36–54.

31   M. Bao, *Chūgoku no kankyō gabanansu to tōhoku ajia no kankyō kyōryoku, trans.* H. Kitagawa, Tokyo, Harushobō, 2009, pp. 6–7.

32   "Guowu yuan bangong ting guanyu yinfa huanjing baohu bu zhuyao zheze nei she jigou he renyuan bianzhi guiding de tongzhi," accessed from http://www.civillaw.com.cn/jszx/law-center/content.asp?no=30560 on April 16, 2012.

33   The policy referred to is known as the "Three Simultaneities System."

34   In 2011 the energy intensity of China's economy, measured by energy use per unit of GDP generated, was about 5 percent higher than that of the US and 10 percent higher than the world average.

35   M. Bao, "Qiye de huanjing zeren yu zhongguo de huanjing zhili," unpublished paper, n.d.

36   For more details, see M. Bao, "Environmental NGOs in Transforming China," *Nature and Culture* 4/1, 2009, pp. 1–16.

37   See Zhonghua huanbao lianhe hui (ed.), "Zhongguo huanjing fei zhengfu zuzhi fazhan lanpi shu," *Zhonghua huanbao lianhe hui huikan* 5, 2006, original text from official website of Zhonghua huanbao lianhe hui.

38   T. Fu, "Zhongguo minjian huanjing zuzhi de fazahn," in L. Congjie (ed.), *2005 nian: Zhongguo de huanjing weiju yu tuwei*, Beijing, Shehui kexue wenxian chuban she, 2006, p. 240.

References

Bao, M., *Chūgoku no kankyō gabanansu to tōhoku ajia no kankyō kyōryoku*, trans. H. Kitagawa, Tokyo, Harushobō, 2009.

Bao, M., "Environmental NGOs in Transforming China," *Nature and Culture* 4/1, 2009, pp. 1–16.

Bao, M., "The Evolution of Chinese Environmental Policy and Its Efficiency," *Conservation and Society* 4/1, 2006, pp. 36–54.

Bao, M., "Qiye de huanjing zeren yu zhongguo de huanjing zhili," unpublished paper, n.d.

Chen, X., and Bin, C., "Nian 'Liang Hui' Jianya Fenxi Baogao: Huanjing Wuran Zhuanti," 2011, accessed from http://news.163.com/11/0314/13/6V42PS5100014JB6.html on March 19, 2012.

"Chu si hai," *Renmin ribao*, 12 January 12, 1956, p. 1.

Engels, F., "Dialectics of Nature," in Zhonggong zhongyang makesi engesi sidalin zhuzuo bianyi ju (ed.), *Ma en quanji*, vol. 20, Beijing, Renmin chuban she, 1971, p. 519.

Fu, T., "Zhongguo minjian huanjing zuzhi de fazahn," in L. Congjie (ed.), *2005 nian: Zhongguo de huanjing weiju yu tuwei*, Beijing, Shehui kexue wenxian chuban she, 2006, p. 240.

"Guanyu san nian lai shuili shuibao gongzuo jiben zongjie yu jin hou yijian," *Mizhi County Archive*, Record Group No. 0020, 1961.

Guo, T., "Zhongguo shuitu baochi chengjiu yu zhanwang," *Shuili shuidian keji jinzhan* 17/4, 1997, pp. 7–10 and p. 32.

Guojia tongji ju guomin jingji zonghe tongji si (ed.), *Xin Zhongguo liushi nian tongji ziliao huibian*, Beijing, Zhongguo tongji shuban she, 2010.

Hou, Y., "Shengtai huanjing shuyu chansheng de teshu shidai," *Zhongguo lishi dili luncong* 1, 2007, pp. 116–23.

Huang, B., "Dili xue zonghe gongzuo yu kuai xueke yanjiu," in Huang Bingwen yuanshi xueshu sixiang yanjiu hui wenji (ed.), *Ludi xitong kexue yu dili zonghe yanjiu: Huang Bingwen yuanshi xueshu sixiang yanjiu hui wenji*, Beijing, Kexue chuban she, 1999, pp. 12–13.

Huang Bingwei wenji bianji zu (ed.), *Dili xue zonghe yanjiu: Huang Bingwei wenji*, Beijing, Kexue chuban she, 2003.

Li, D., "Guanyu 1958 nian aiguo weisheng yundong de jiben qingkuang he jin hou de renwu," *Xinhua ban yue kan* 2, 1959, pp. 54–6.

Li, J., *Nongye xue dazhai yundong shi*, Beijing, Zhongyang wenxian chuban she, 2011.

Liu, D., et al. (eds.), *Huanjing wenti: cong zhong ri bijiao yu hezuo de guandian kan*, Beijing, Zhongguo renmin daxue chuban she, 1995, pp. 88–90.

Mao, Z., *Mao Zedong shi ci ji*, Beijing, Zhongyang wenxian chuban she, 1996.

Mao, Z., *Mao Zedong wen ji*, vol. 6, Beijing, Remin chuban she, 1999.

Mao, Z.,"Zhonggong zhongyang guanyu weisheng gongzuo de zhishi," in Zhonggong zhongyang wenxian yanjiu shi (ed.), *Jian guo yilai zhongyao wenxian xuan bian*, vol. 13, Beijing, Zhonggong zhongyang wenxian chuban she, 1996, n.p.

Pan, Y., "Shengtai wenming jiang cujing Zhongguo tese shehui zhuyi de jianshe," *Liao wang*, October 22, 2007, pp. 38–9.

Qu, G., "Zhou Enlai shi xin Zhongguo huanjing baohu shiye de kaituo zhe he dianji ren," in *Mengxiang yu qidai: Zhongguo huanjing baohu de guoqu yu weilai*, Beijing, Zhongguo huanjing kexue chuban she, 2004, pp. 34–44.

Qu, G., and Jinxin, P. (eds.), *Huanjing juexing: renlei huanjing huiyi he Zhongguo di yi ci huanjing baohu huiyi*, Beijing, Zhongguo huanjing kexue chuban she, 2010.

"Re dian ci nan dian ci zongheng tan," *Keji shuyu yanjiu* 2, 2005, pp. 20–38.

State Council, "Wei tiqian shixian quanguo nongye fazhan gangyao er fendou," report in the Second Session of the Second National People's Congress, April 6, 1960.

Xia, G., "Shengtai wenming gainian you duo yi xing," *Zhonguo huanjing bao*, September 22, 2009, p. 2.

Ye, W., *lecture*, Ecological Civilization Forum, Beijing University, 2010, accessed from http://pkunews.pku.edu.cn/xwzh/2011-01/03/content_191808.htm on March 19, 2012.

Yi, B., and Yiguang, Y., "Qing mo yilai dongbei senling ziyuan ji qi huanjing daijia," *Zhongguo nong shi* 3, 2004, pp. 115–23.

Zhongguo huanjing wenhua cujin hui (ed.), *Zhongguo gongzhong huanbao zhishu: 2008 niandu baogao*, Beijing, Zhongguo huanjing wenhua cujin hui, 2009.

Zhongguo kexue yuan jishu xinxi yanjiu suo (ed.), *Guowai wuran kalian*, Beijing, Renmin chuban she, 1975.

Zhonghua huanbao lianhe hui (ed.), "Zhongguo huanjing fei zhengfu zuzhi fazhan lanpi shu," *Zhonghua huanbao lianhe hui huikan* 5, 2006, original text from official website of Zhonghua huanbao lianhe hui.

Zhonghua renmin gonghe guo guojia nongye weiyuan hui bangong ting (ed.), *Nongye jiti hua zhongyao wenjian huibian*, Beijing, Zhongyang dangxiao chuban she, 1981.

Further Reading

Mertha, A., *China's Water Warriors: Citizen Action and Policy Change*, Ithaca, NY, Cornell University Press, 2008.

Ho, P., and Vermeer, E. (eds.), *China's Limits to Growth: Greening State and Society*, Oxford, WileyBlackwell, 2006.

Naughton, B., *The Chinese Economy: Transitions and Growth*, Cambridge, MA, MIT, 2006, chapter 20.

Managi, S., and Kaneko, S., *Chinese Economic Development and the Environment*, London, Edward Elgar, 2010.

CHAPTER TWENTY-SEVEN

Religion and Environmentalism

JOACHIM RADKAU
(translated by Peter Engelke)

Research Deficits

One of the earliest themes of environmental history is the religious roots of environmentalism. Thus it is surprising that "religion and environmentalism" is among those subjects where research until now has made little headway. The literature leaves much to be desired.[1] This is all the more surprising because recently religion has again found political significance and garnered a measure of attention which it had not received in some time. Hans-Ulrich Wehler, a German social historian for whom religion never was an important theme, writes with perceptible unease in the foreword to the last volume of his *Deutsche Gesellschaftsgeschichte* that "the last one and a half decades" have seemingly "undertaken a breathtaking revaluation of religion."[2] But the renewed scholarly attention to religion still has its gaps.

Religion has not yet received its due within environmental history. For a long time, there has been in the Anglo-American as well as in the German-speaking realm an ever-increasing wealth of literature on the actual or alleged affinities between religion and environmentalism; but in essence this involves for the most part a normative but not a historical-empirical approach. It is a literature of hope, not of sober analysis.[3] Not only is there little progress in the literature; one can even recognize steps backward. In the massive opus by James G. Frazer, *The Golden Bough: A Study in Magic and Religion* (1922),[4] the work of a great scholarly passion that treats animistic cults of tree spirits and corn mothers around the world with singular comprehensiveness, the author used language so suggestive that he subsequently had to defend himself against the charge that he saw the basis of all religious history in the worship of trees. To date, there exists no research of comparable format, although there would be value in rigorous research into the animistic origins of modern environmental awareness. The idea of nature's vengeance for the injuries caused by humankind – the foundational

A Companion to Global Environmental History, First Edition. Edited by J.R. McNeill and Erin Stewart Mauldin.
© 2012 John Wiley & Sons, Ltd. Published 2015 by John Wiley & Sons, Ltd.

idea of modern environmental protest – leads ultimately to the animistic idea that trees, mountains, and rivers are beings with spirits of their own.

The great work of Clarence J. Glacken, *Traces on the Rhodian Shore: Nature and Culture in Western Thought from Ancient Times to the End of the 18th Century* (1967),[5] presented the theological roots of environmental awareness from antiquity to the early modern period with particular thoroughness. But it remains unique in its quality. The magnum opus of Xavier de Planhol, *Les fondements géographiques de l'histoire de l'Islam* (1968),[6] a rich source for the environmental historian and soothing in its sobriety nowadays when the theme of Islam stirs up passions, has no successor. In fact, to date the Islamic world has the largest gaps in the environmental history literature.[7]

The literature about the relationship between religion and environmentalism is a branch of the history of ideas, often far removed from the experience of daily life. As a general rule, it neglects whatever practical effects the ideas about humans and nature may have had. The sociologist Niklas Luhmann, who kept an ironic distance from contemporary environmentalism, nonetheless argued in his *Ecological Communication* (1986) for the concept of a world-historical relationship between ecological ideas and action, wherein religion stands at the beginning of history. The romantic notion of a primordial, magical-religious mediated unity between humankind and nature is hidden in this anti-romantic sociologist's view of history. He asks "how [a] society's processing of environmental information is structured." And he observes: "These societies could visualize spiritual things better than earthly things. Their ecological self-regulation is hence to be sought in mythical-magical ideas, in taboos and in the ritualization of dealing with environmental conditions for survival." These mythical-magical ideas correspond, however, to an entirely pragmatic rationality that actually – so one could argue – needs no religious foundation. The paradigm for Luhmann is the "pig cycle" of the Maring of New Guinea, made famous by Roy A. Rappaport in *Ecology: Meaning, and Religion* (1979): "Whenever the pig population increases too much and they begin to ravage the food supply, strong ritualized justifications come into play to arrange a great feast that reestablishes a balance in the number of pigs and regulates the protein consumption of the tribe."[8] One could counter: are magical rituals essential to explain a festival of slaughter? Luhmann uses indigenous peoples as a contrast to modern industrial society's environmentalism, which he portrays as sociologically unenlightened and hopelessly dilettantish.

Luhmann is part of a long tradition of scholars interested in links between the earliest religions and nature. Some ethnologists believe that at the origin of religious history around the world there was a belief in Mother Earth. A book by Albrecht Dieterich (a religious scholar esteemed by Max Weber), *Mutter Erde: Ein Versuch über Volksreligion* (1905), had a worldwide impact and led to the discovery of more and more mother-earth religions. So much so that later anthropologists had problems slowing this zeal for identifying mother-earth religions and replacing it with a more differentiated view of prehistoric cults. Dieterich invoked the famous pronouncement of the Shawnee chief Tecumseh to General William Henry Harrison: "The earth is my mother and on her bosom I will repose."[9] During the "ecological revolution" of the 1960s and 1970s, the notion of an "ecological Indian," together with his veneration of Mother Earth, experienced a great vogue;[10] but in reaction, revisionists pointed to signs that Native American mother-earth religions had relatively modern origins and may have arisen under European influence.[11] But, whether old or new, can one trust that those people who venerated Mother Earth treated the natural environment carefully and reverently?

That is not certain; mothers are often exploited. On the other hand, the practical effects of the Old Testament's commandment to "subdue the earth" were also ambivalent in the course of history: it could have been understood not only as carte blanche for the exploitation of nature, but also as an invocation for responsibility toward nature. Not without reason did Noah's Ark become an icon of the ecology movement: a symbol of responsibility for the Earth, wherein the environmentalist ascended to the role of a divinely sanctioned patriarch.

The historian must constantly pay close attention when it comes to the practical effects of religious ideas. Even tree cults could justify the felling of trees. In the tree cult, the rationality of modern forestry can be found: trees must be felled as soon as they stop growing. The anthropologist Marvin Harris, in *Cannibals and Kings* (1977), delights in discovering everywhere in religious history – from the Aztecs' human sacrifices through the Indian sanctification of the cow – a tangible pragmatism.[12] Perhaps he goes too far with his penchant for sweeping theses; but surely one can assume that religions – even if their theology is focused on the afterlife – in their daily effects accord with the vital needs of their followers.

Buddhism prohibits the killing of animals. Madhav Gadgil and Ramachandra Guha write, in their *Ecological History of India*, that because of Buddhism, environmentalism has been rooted in Indian culture for over 2,000 years.[13] However, this claim is very difficult to establish empirically. Gadgil and Guha indicate that, in the elite culture of some Indian regions, the hunt was of great significance. As in Europe, it was the ruler's passion for the hunt that prompted the protection of entire forests in ancient India. Religion does not govern daily life independently from cultural and natural conditions. Tibetan nomads who avow Buddhism cannot forego a meat diet, and naturally they do not want to starve – Buddhism or no Buddhism. The physicist Hans Peter Duerr, a leading figure of German environmentalism, justly mocks the fact that "only luxury-loving middle-class citizens" who do not know hunger can imagine that consciousness and not material existence determines social existence.[14]

East of Kathmandu, the capital of Nepal, stands the beautiful forest of Deopatan, deemed holy because the god Shiva, having taken on the form of a gazelle buck, is said to have retired there with his companion Parvati for a 1,000-year coitus. But such marathons of lust are rare even among gods and are insufficient for large-scale forest protection. The Indologist Axel Michaels emphatically denies that the veneration of holy trees leads to forest protection; the real logic is rather exactly opposite: "The sanctity of the tree is its singularity."[15]

The ethnologist Clifford Geertz researched religion's influence on daily life especially intensively. He writes: "no one, not even a saint, constantly lives in a world formulated by religious symbols, and the majority of men live in it only at moments."[16] Geertz devotes special attention to the interplay of religion and ecology.[17] He uses Morocco and Indonesia – the westernmost and easternmost Islamic countries – as examples, because everyday Islam develops very differently under entirely distinct ecological and cultural conditions. Indonesia's Islamic culture is shaped by intensive wet-rice cultivation while the semi-nomadism of the Moroccan mountains brings out an entirely different culture and also a different human nature. In Morocco one finds "the restless, aggressive, extro-verted sheikh husbanding his resources, cultivating his reputation, and awaiting his opportunity"; in Indonesia "the settled, industrious, rather inward plowman of twenty centuries, nursing his terrace, placating his neighbours, and feeding his superiors. In Morocco civilization was built on nerve; in Indonesia, on diligence."[18] Today, Greenpeace finds many supporters in Indonesia, whereas the Near East and North Africa is the nearest thing to a blank spot on the world map of environmentalism.

Geertz's publications are a good antidote against the trend toward sweeping judgments about Islam. Modern religious scholars differentiate between the "great" and "small" traditions within the religions: the foundational ideas of the great theologians and the religion-defined rules of daily life springing up from the agricultural calendar that defined rural routine. When it comes to the practical effects of religion, environmental historians are concerned with the "small" traditions. However, these are often unprepossessing and can only be painstakingly established from the sources. The "great" traditions are more intellectually attractive and offer a more impressive global perspective on history – or at least appear to offer one.

In the beginning of environmental-historical research in the 1960s, the religious-historical approach appeared to be the best path to a global environmental history. Lynn White's address of 1966, "The Historical Roots of Our Ecological Crisis," became a foundational document and a type of holy text of environmental history. White, a prominent medievalist, ultimately reduces the destructive exploitation of natural resources to the Dominium-terrae commandment of the Old Testament God found in the Book of Genesis: "subdue the earth." In contrast, he sees in ancient religions a continuation of prehistoric animism. It is worthwhile to quote him at length because this view of history swayed the perceptions of many environmentalists of the first generation, insofar as they sought a historical orientation:

> The victory of Christianity over paganism was the greatest psychic revolution in the history of our culture ... Our daily habits of action, for example, are dominated by an implicit faith in perpetual progress which was unknown either to Greco-Roman antiquity or to the Orient. It is rooted in, and is indefensible apart from, Judeo-Christian teleology. The fact that Communists share it merely helps to show what can be demonstrated on many other grounds: that Marxism, like Islam, is a Judeo-Christian heresy ... Especially in its Western form, Christianity is the most anthropocentric religion the world has seen ... In Antiquity every tree, every spring, every stream, every hill had its own genius loci, its guardian spirit ... By destroying pagan animism, Christianity made it possible to exploit nature in a mood of indifference to the feelings of natural objects ... For nearly two millennia Christian missionaries have been chopping down sacred groves, which are idolatrous because they assume spirit in nature.[19]

Occasionally, however, White becomes uncertain. What he says holds only for Western Christendom, he admits, and not to the same extent for Orthodox Christianity. Apparently basic religious ideas are not decisive on their own, but rather only in connection with cultural, social, and intellectual conditions. For White, the shining exception to the rule in the history of Western Christianity is Francis of Assisi, whom he proposes as a patron saint for the ecology movement. The Beatniks, the predecessors of the hippies,[20] are St. Francis' American counterpart, in the logic of White's view of history: "The Beatniks, who are the basic revolutionaries of our time, show a sound instinct in their affinity for Zen Buddhism, which conceives of the man–nature relationship as very nearly the mirror image of the Christian view. Zen, however, is as deeply conditioned by Asian history as Christianity is by the experience of the West."[21]

White's association of the Judeo-Christian tradition and anti-ecological behavior proved controversial. The world historian Arnold Toynbee enthusiastically agreed with him.[22] In contrast, René Dubos, the inspirational spirit of the Stockholm international environmental conference of 1972, reacted ambivalently. He also did not contest the religious roots of the problem of environmentally unsound behavior, but he judged the historic role as well as the actual potential of Christianity more positively than did White. One causes no harm to nature through prudent use, thought Dubos; through gardening, one draws nearer to paradise.[23]

At the same time, the Chinese-born geographer Yi-fu Tuan destroyed the myth of an Eastern harmony with nature. Naive outsiders confused the great dreams of philosophical hermits with the thoroughly contrary reality of their everyday world – which showed conspicuous signs of environmental degradation.[24] Mark Elvin went so far as to characterize the entirety of Chinese environmental history as "three thousand years of unsustainable growth."[25] The German sinologist Gudula Linck had already recognized an ecological "vicious circle." "Yin and Yang – forget it!" she cried at a 1986 environmental history conference. As she wrote three years later, the bond with nature proclaimed by Taoists is not an "expression of a lived unity with nature – unless we are speaking of a few outsiders and individualists – but rather of cultural pessimism and civilizational skepticism regarding the harm done to nature by humankind."[26]

This does not settle the question of the significance of Eastern religions for environmentalism. Precisely because it is so difficult to find an answer on broad empirical grounds, the answers are subject to scholarly fashion: in Germany in the 1970s and 1980s, the trend was toward social history, and in the 1990s and afterward it was the new cultural turn.[27] Mark Elvin recognized a dilemma of source analysis, for which there is no definitive solution: "do our sources mainly reflect the dominant tendencies of an age, or are they more often reactions, by far-seeing and sensitive thinkers, against these dominant tendencies?"[28] This question could be directed at the documents of modern environmentalism.

The Protestant Ethic and the Spirit of Environmentalism

The religious taproot of the environmental movement in Protestant countries is even more marked than the historic connection between Protestantism and capitalism – because capitalism also has medieval and Catholic roots. One is tempted to replace the famous Max Weber thesis of the "elective affinity" between the "Protestant ethic and the spirit of capitalism" with a new thesis connecting Protestantism and ecologism. In environmental awareness one recognizes much of the Protestant awareness of sin that can be assuaged by neither confession nor the granting of an indulgence. In the idea of nature's vengeance, one easily recognizes a secularized theology in the form of a stern God who strictly punishes violations of his commandments.

The significance of Protestant traditions for environmental awareness is suggested by differences within Europe as well as within the US. In Europe, environmentalism arose first and foremost in Protestant countries; and within the US, its main wellspring was in the Northeast among cultural elites who were predominantly Protestant and in some cases heir to Puritan traditions. The "hotbed" of the American environmental movement was the Hudson Valley,[29] the glorification of which by the nineteenth-century Hudson River school of painting recalls German Rhine romanticism. The Hudson Valley Riverkeepers became a model for a broader environmentalism.[30] The contrast to the southern states was so prominent, not only in the overall political climate but also in environmental awareness, that it tempts one to formulate an ecological appendage to

Weber's thesis. "Clearly a liberal electorate combined with a morally-shaped political culture is the best guarantee for especially innovative environmental policy," comments Kristine Kern in this regard,[31] especially in view of the comparatively weak environmentalism in the South and Midwest.[32] The genesis of the environmental movement cannot be explained simply by rising ecological duress, the geography of which does not match well with the geography of environmental concern. One must also consider the cultural and historical matrix within which environmentalism might grow or wither.

In purely theological terms, one would expect that religious fundamentalism in the US, which took the Bible's creation myth literally and resisted Darwin, would have been especially receptive to ecological messages. But the opposite was almost entirely the case. The American fundamentalists as a rule so far have been conservative hardliners, associating environmentalism with a culture of protest that they generally condemn. Like the connection between Protestantism and capitalism, it is not easy to determine the character of the connection between Protestantism and environmentalism. In stark contrast to the century-long discussion over the "Weber thesis," so far we lack a prolonged discussion about this connection.

The two most important environmental history publications on the Protestant connection are the books by Mark Stoll, *Protestantism, Capitalism, and Nature in America* (1997), and John Gatta, *Making Nature Sacred: Literature, Religion, and Environment in America from the Puritans to the Present* (2004).[33] Each is rich in ideas and contains a multitude of sources. But John Gatta does not discuss Mark Stoll's thesis, and neither of the two deals with the protracted discussion about the "Weber thesis," which contains potential stimuli for environmental-historical discussions. Both authors invoke Ralph Waldo Emerson and the significance of American transcendentalism for the genesis of environmentalism. But just what kind of connection is this? How does one conceptually grasp and empirically substantiate it? Is it a direct connection, like the one Max Weber saw between the Protestant ethic and the "spirit of capitalism," or rather an indirect connection, mediated by third factors? Is it a dialectic type: environmentalism as a backlash against an unnatural business culture shaped by Puritanism, while sensual forms of religiosity – whether in southern Europe or in Latin America – need environmentalism less? The nature of the links remains unclear.

Environmental awareness of course may take several forms. There is the depressive and guilty type of environmental awareness that parallels Max Weber's "inner-worldly asceticism." In contrast, Roderick Nash depicts an excitement for wildness, almost a revolt against New England's Puritanical tradition.[34] Mark Stoll also ascribes the willful law-breaking of the cult of wilderness championed by Dave Foreman, who called for "ecological sabotage" (ecotage), to the tradition of Protestant-American environmentalism. Stoll cites the founder of Earth First!: "We are involved in the most sacred crusade ever waged on Earth."[35] That is a type of religiosity that invites the release of emotional fury.

The Norwegian mountaineer-philosopher Arne Naess (1912–2009), who hung on ropes over a Norwegian fjord to forestall a local dam project, sketched the tenets of Deep Ecology. Founded by Naess in 1973, the Deep Ecology splinter movement bears the marks of a pantheistic natural religion, a philosophy of joy and guide to well-being in a completely worldly life. He recognized the problem that the selfless love of nature demanded by the ecofundamentalists attracts only sourpusses and misanthropes or hypocrites, because humans are seldom fully selfless. His solution was not to "confuse the limited ego" with one's true self, but rather to create a broad feeling of this self that bursts the boundaries of our bodies, and contains the entirety of nature.[36] This extended sense of self conquers the fear of death and with it the basis of all sadness.

Hubert Weinzierl, a leading personality of German conservation and environmental protection since the beginnings of environmentalism, never denied the emotional and spiritual drivers of conservation within ecology. As he said, the presence of a lynx hallows a forest. He proclaimed that concepts such as "eco" and "bio" should "no longer remain burdened with morality and asceticism, but rather must be associated with joie de vivre ... Therefore, sustainability must become a cult. Then funlovers may also partici-pate: because is there any greater joy than to live passionately in agreement with nature?"[37] This demonstrates the change in the type of person characteristic of the ecoscene, and one of the most significant transformative processes in environmentalism in general. At the inception of modern environmentalism its adherents had neo-puritanical traits and were mocked for them. No longer.

This change was, however, at the same time a return to some of the historic roots of conservation. From the outset, euphoria as well as anxiety informed environmental awareness. The enthusiasm for the flourishing of nature, even for wild nature, could intensify to the level of ecstasy. The romantic landscape painting from the Rhine to the Hudson culminated not in Naturalism, but rather in a supernatural shimmer of transfigured nature. The spiritual component is evident, especially in the cult of wilderness. The protection of wilderness is not justified purely by principles of ecology; many gardens are richer in species than many wildernesses. Where conservation culminates in the cordoning off of a zone from which humans are absolutely excluded, this follows unmistakably in the tradition of the sacred taboo.

One finds spiritual motives above all in early wilderness conservation in the US. "Nature" as an ideal contains secularized religious thought. No figure in American environmentalism occupies such prophet-like status as much as John Muir, the champion of wilderness and the national parks. For him, the fight for the wild forest was "a part of the eternal conflict between right and wrong."[38] Like one of those Old Testament prophets, whom Max Weber called the "titans of the holy curse,"[39] he damned the city of San Francisco for its takeover of Yosemite National Park's Hetch Hetchy Valley, "for no holier temple has ever been consecrated to the heart of man."[40] "Remember Hetch-Hetchy!" was the battle cry of the Sierra Club, founded by Muir in 1892. This ideal was so powerful that the Sierra Club did not protest against nuclear power plants in the 1960s, hoping that the use of nuclear power would prevent the construction of additional hydroelectric power plants.[41] The commitment to wilderness – so wrote David Brower in 1957 – took its power from the "pressure of conscience, of innate knowledge that there are certain things we may not ethically do to the only world we will ever have."[42] As Mark Stoll showed, not only John Muir but also his rival Gifford Pinchot was "something of an evangelist or even prophet." Pinchot "was quite religious, believing in an idiosyncratic mixture of his mother's Puritan heritage and his father's Swedenborgian spiritualism."[43] In the US, both the conservation and the preservation movements, to use the language of American environmental history, had Protestant roots.

The Apocalypse Needs Prophets

One can find an affinity of environmentalism for religion from the inner logic of worry about the environment; one can show it in individual motives; but one can also demonstrate it through a series of leading figures in the environmental movement. Let us begin with the inner logic. Visions of humanity's coming ecological suicide follow in the old traditions of the Christian apocalypse. For the Westphalian cleric Person Gobelinus, who between 1406 and 1418 wrote his *Way of the World* (*Cosmidromius*), it was obvious that world history teetered on a slippery slope and that the end was near; he expected the end of the world in 1426 and died in time to avoid it in 1422. His basic thesis was that "the alpha already contained the omega." The fall from grace in paradise already contained the end of the world. Similarly, one reads again and again in green world histories, from Clive Ponting to Jared Diamond, that the ecological suicide of the Easter Islanders, who in blind cupidity cut down their last tree, represents a prototype for the ecological suicide of humanity.[44] (Probably there was no ecological suicide on Easter Island, rather murder occurred through the actions of colonists and traders[45] – actually a more provocative point for eco-activists!)

The first revelations of the eco-apocalypse appeared on both sides of the Atlantic around 1970.[46] Michael Egan, in a chapter in his biography of Barry Commoner in which he traces the development of his hero into an environmental activist, provides the subtitle "The New Jeremiad." Egan notes Commoner's saying: "If you can see light at the end of the tunnel you are looking the wrong way."[47] At the same time he portrays him as "exceptionally charismatic," but nevertheless a charismatic prophet of calamity who never amassed a large and durable legion of followers. On this point, he cites the sociologist Deborah Lynn Guber, who tracked the "green revolution" in the US through surveys: "By downplaying environmental progress and by using exaggerated doomsday warnings to motivate public awareness and concern, the environmental movement has sacrificed its own credibility by giving in to the politics of Chicken Little."[48]

Egan reminds us, with his analysis of Commoner, of Max Weber, the creator of the modern concept of charisma. Weber's wife Marianne had the feeling during World War I that her husband, when speaking about the prophet Jeremiah, was at the same time speaking of himself.[49] "Charisma" actually means "gift of grace"; but with Max Weber the charismatic person has at the same time the aptitude for gloomy apocalyptic visions that can pull the masses out of their daily routine and set them in motion. He resembles a madman when the spirit comes over him: "Jeremiah becomes like a drunk and trembles in all his appendages."[50] The eco-apocalypse seems made for charismatic persons in Weber's sense.

One thing, however, does not fit into the general equation of religious and ecological prophets: for Max Weber, the hour of the charismatic person occurred in times of general despair. The Israelites' fear of the cruelty and power of the Assyrians put them in a suitably apocalyptic mood to listen to prophets. In contrast, the great hour of the ecology movement occurred around 1970 and then again around 1990, times when the apocalyptic nightmare of nuclear war diminished. Doubts arose about whether or not prophets of doom fit in this historic situation. In the history of the ecological era, again and again we encounter frustrated charismatic persons. Herbert Gruhl, who was the first of many failed aspirants to leadership of the German Greens, recognized that the Greens had a "panicked fear of leaders."[51] The environmental movement loved discussion much more than leadership and allegiance to leaders, and so belongs to the tradition of the Enlightenment more than to that of millennial movements.

Charismatic Animals in Noah's Ark

Animals may have charisma too, at least for some environmentalists. It appears, if one scans the available literature, that the theme of animals is marginal for many modern environmentalists. But themes of this type are tackled relatively easily with a concrete historical approach. The only comprehensive work to date on the environmental history of the Ottoman Empire revolves around the position of certain animals in the Islamic tradition.[52] Just recently, the German-American literary historian Bern Hüppauf ("Hüpper" is the North German word for "frog") published a comprehensive cultural history of the human–frog relationship, which treats the details of the theological and magical traditions of this relationship.[53] Because the frog was among the evil creatures in medieval theology, it is one of the beneficiaries of secularization. The same is true of cats, which were burned as demonic creatures on St. John's Day up until the seventeenth century.[54] Only with the onset of the Enlightenment did the great turn toward the love of cats begin: "When I play with my cat," Montaigne asked himself, "how do I know that she is not playing with me rather than I with her?"[55] People fond of animals often wanted to have them as pets and imagined them anthropomorphically. Hunters, on the other hand, often took into account the connection between animal and habitat much more than animal lovers and thereby presaged ecological thinking.

The philosopher G. W. F. Hegel (1770–1831), in his *Phenomenology of Mind*, differentiated within "natural religion" between the peaceful and selfless "innocence of the flower religion" and the aggressive "animal religion." In modern societies as well, animal protectors are often more aggressive than tree protectors; many conservationists, however, show no clear religious motivation. Be that as it may, certain animals play a prominent role as icons of environmentalism. This role is inexplicable in terms of ecology's scientific rationality. The term "charismatic megafauna" – lions, tigers, elephants, rhinos, pandas, polar bears, recently joined by the long-maligned wolves – has become a standard one in discussions of biological conservation. It is, however, ecologically as well as sociopolitically questionable for wilderness protection to fixate too much on these charismatic creatures by using a term which is oriented toward media impact.[56] This emphasis on charismatic megafauna can affect local peoples, who have to deal with tigers and elephants, and thus can resemble contempt for humankind; or it can distract from the careless treatment of the rest of our environment.

Small creatures also acquired charismatic status in conservation circles. Bats, for example, had previously evoked associations with vampires. But these night creatures acquired a face as a result of the refinements of modern photography. In the 1990s, a pan-European bat project was launched (European Batnight Project). Many priests joined in the cause with great relish, because bats enjoy dwelling in church towers and, through bat protection, the churches obtained an ecological legitimacy. In today's ecclesiastical vocabulary, "ecumenical" often becomes synonymous with "ecological."

Conservationists who took ecology seriously followed the media hype surrounding the charismatic animals with mixed feelings. Henry Makowski commented on the general storm of outrage about seal deaths in the North Sea in 1988, which in reality were probably caused by a virus rather than through pollution: "Never before in its history could conservation more effectively transmit and push through its demands as over the dying seals"[57] with their sad and reproachful gaze. Wolfgang Haber, a grand old man of German ecology, points out that the focus of species conservation on "charismatic" animals in the end amounts to a transformation of the landscape into a giant zoological garden![58] The history of whale conservation after 1945 is telling: the fixation on these charismatic large mammals distracted attention from efforts to achieve comprehensive marine protection policy – and finally could not come to a resolution quickly without a global oceans regime.

In the *Confessions of an Eco-Warrior* by Dave Foreman (1991), one recognizes in particular an identification with the dangerous grizzly bear, the *Ursus arctos horribilis*. "A Grizzly Bear snuffling along Pelican Creek in Yellowstone National Park with her two cubs has just as much right to life as any human has, and is far more important ecologically." Going a step further, the cofounder of the activist environmental group Earth First! wrote: "John Muir said that if it ever came to a war between the races, he would side with the bears. That day has arrived." For a motto, Foreman borrowed a line from another great wilderness prophet, Aldo Leopold: "Relegating grizzlies to Alaska is about like relegating happiness to heaven; one may never get there." And in conclusion he issued the rallying cry that is entirely in the spirit of Deep Ecology: "Run those rivers, climb those mountains, encounter the Griz ... And piss on the developers' graves." And again: "Pissing on what was politically correct. And thereby doing sacred work."[59] Foreman's concept of the saint is connected with disgust for the unnatural puritanical tradition.[60] This wild enthusiasm for wilderness is an entirely different type of religiosity!

When we leave the high plane of intellectual history, the relationship between religion and environmental awareness becomes most vivid and arresting if we concentrate on certain personalities. Religion in environmentalism is not so visible in large institutions and organizations, but rather as individual spiritual desire – as personal experience, as the experience of a great unity of self and nature. It is revealing to trace such experiences in different biographies.

The Ecological Reinvention of Buddhism and Spiritual Vagabonds

In environmentalism the Protestant background predominates. However, Buddhism enjoyed popularity among environmentalists in the era of ecology. At the center of Buddhism stands a peaceful bond with all living things, an insight that long ago became bound to elements of Hinduism and more recently informed the teachings of Gandhi and Albert Schweitzer. Leading the way in this was Fritz (later E. F., for Ernst Fritz) Schumacher (1911–77), an economist from Bonn who emigrated to England in 1936. There he climbed the ladder to lead the statistical division of the National Coal Board and as a coal lobbyist first developed a critical attitude toward nuclear technology. Even while warning of nuclear risks, he remained for life a gentle and cheerful prophet. He did not proclaim the coming apocalypse but rather pointed to signs of hope and made pragmatic suggestions that could actually be implemented here and now, without the need for a new humankind in a new world.

For those who search for a religiously based environmentalism with entirely practical effects, Schumacher offers an impressive example. Already in the early 1950s, he turned to Buddhism, long before it came into fashion in the West and the Dalai Lama rose to become the "Protestants' secret Pope." Stays in Burma were instrumental for Schumacher. So was the acquaintanceship with another German emigrant who was a brilliant and extraordinarily knowledgeable teacher of Buddhism: Edward (originally Eberhard) Conze (1904–79).[61] He advised his students first and foremost to immerse themselves in the "clear serenity of the mind." This was the friendly and warm Buddhism of the Middle Way, far distant from both the cool Japanese Zen Buddhism, which was compatible – as it turned out – with the religion of modern management, and the tantric sexual mysticism of Himalayan Buddhism, which later enraptured the hippies and others. For Schumacher, impressions in Burma and India brought enlightenment,[62] as there he saw craftsmen working as in olden times: destitute according to Western standards but joyful in their skills – which were more refined than those of many Western assembly-line and office workers. These impressions, together with his own experiences with "organic" gardening, formed Schumacher's new worldview: a magical triangle of quality of life, ecological cultivation, and alternative technology.

From the early 1960s, he gave Buddhism's Middle Way an entirely practical and technical turn. His concept of "Appropriate Technology,"[63] – at the outset only a slogan – increasingly acquired concrete form. Although conceived at first for developing countries like India, it also inspired reconsideration of technological alternatives in the West. In his view, the Buddhist ethic agreed in one respect with the Protestant ethic of Max Weber: that human self-realization occurs through work. But this self-realization through work is also spoiled by modern technological trends, which made humans into cogs in a giant industrial wheel or shunted them into unemployment. Middle, mediating technology, as he understood it, embraced the achievements of modern technology but preserved a character of craftsmanship that developed in the worker multifaceted capabilities while not creating mass unemployment. Appropriate technology should be so inexpensive that small enterprises in the Third World could afford it. Schumacher's was a very personal interpretation of Buddhism; many other Buddhist teachers recommended an equanimity that leaves the world as it is and does not concern itself with a vain struggle to change things.

Up through the late 1960s, economists mocked Schumacher's ideas. But in 1973, with his book *Small is Beautiful*, he became a guru who, even in the US (where big was usually better), spoke in front of packed halls. With requests to speak piling up from all over the world, his health rapidly declined. The title, which he did not even come up with himself, was probably the most successful aspect of the book.[64] *Small is Beautiful*, in its genial simplicity, short and crisp like a maxim out of Mao's "Red Book" (which had impressed Schumacher),[65] served perfectly as a rousing rallying cry. As for the content of the book, Schumacher's fundamental ideas had become commonplace enough already that his pages seemed almost a smorgasbord of truisms.

Let us take a great leap from Europe to Japan: the non-Western country that was first to keep pace with the West, not only in terms of industrialization but also with respect to environmentalism. If in Japan one seeks a counterpart to Rachel Carson, one soon finds the name Michiko Ishimure (born 1927). Her book *Kugai Jodo (Paradise in the Sea of Sorrow)* first appeared in 1969 and was, like *Silent Spring*, a combination of non-fiction, polemical pamphlet, and poetic fantasy. The poisoning of Minamata Bay through methyl mercury leaks by the chemical company Chisso was known, to those few who wanted to know it, from 1956. More than 1,000 residents died agonizing deaths from eating poisoned fish over the years. But in official and public discourse, this unprecedented environmental catastrophe was assiduously kept secret. Disclosures began in the press around 1967; but it was Michiko Ishimure who, through her book, brought the extent of human suffering to general public awareness, with gripping language and at the same time extremely symbolic imagery.

As happened with *Silent Spring*, *Kugai Jodo* was compared with *Uncle Tom's Cabin* in its disturbing effect.[66] Precisely because the writer's tone is rather restrained, her portrayal gets under the reader's skin. Michiko Ishimure wrote of her home city and a bay that had once been considered a paradise. The author revealed, however, that her book was not just documentation, but also a projection of fantasies: "The figures in Paradise in the Sea of Sorrows are not only reincarnations of farmers from the Yanaka village, but also projections of ourselves in the past, i.e., incarnations of our ancestors who believed in certain values and a certain philosophy of life that we forfeited long ago. The figures in this book are rooted in that which one could identify as the original topsoil of our thinking."[67] The title hints at "jodo," the "paradise of the pure country" of Amida Buddhism (one of the Japanese variants of this faith).[68] Although there are traces in the book of the anti-capitalist bitterness of 1968 – which was as strong in Japan as elsewhere – the author displays no political activism and declares no radical ideology.

For Ishimure, the Minamata catastrophe signified the fundamentally destructive character of modern civilization. Herewith she joined with the radical anti-modernism and apocalyptic tone of the philosopher-priest Ivan Illich (1926–2002), whom she approached.[69] But this split her off from the mainstream of the environmental movement. Her work has Buddhist but also archaic-animistic features; it arose from the belief in a primordial animated nature whose soul is destroyed by modernity.[70] Western ecoliterature is not a stranger to such thinking, but among the Japanese it possessed a religiously based gravity that one does not easily find among Western nature lovers. She later founded, with kindred spirits, a Buddhist community and led the life of a nun.

Matters of faith and environment were often just as intertwined in Europe as in Asia. The relationship of the German Greens to religion and spirituality is a delicate subject.[71] The German Green Party was formed in 1980 and enjoyed quick electoral and parliamentary success. It argued against nuclear power, nuclear weapons, and pollution in general. The American environmental activist Charlene Spretnak in 1984 published one of the first books about the Greens. She was strongly influenced by conversations with two Green Party leaders, Petra Kelly (1947–92) and Rudolf Bahro (1935–97). Spretnak got the impression – from today's perspective, almost absurd and hardly comprehensible – that a spiritual movement was at the core of the Green Party. Bahro, an East German who had gone to prison for writing *The Alternative: Critique of the Really Existing Socialism* (1977), was permitted to go to West Germany in 1979. There, starting in 1983, he went in search of spiritual experiences and henceforth became addicted to prophetic speech at Green Party meetings. At the Greens' national convention in Hamburg in December 1984, he announced: "The race with the Apocalypse can only be won if these times become a spiritual era, a Pentecost with a living spirit that is poured over everyone as equally as possible." This became too much for the Green leader and ex-clergywoman Antje Vollmer. Quite rightly she countered that Bahro "really forgot something: that one did not make oneself into a prophet, rather one is made into one when people follow and have trust in what one says."[72] In fact, that was the point: a prophet must be seen as credible by his disciples. Bahro, who was narcissistic to the point of autism, lacked this credibility. When, due to his aversion to bureaucratization, he sought to thwart the creation of a more efficient organization out of the chaos of the Greens' office, this was the last straw for his former ally Petra Kelly.[73] As the parliamentary faction protocol at the end of 1983 reads: "In general, Rudolf Bahro is charged with creating an apocalyptic mood."[74]

The Green parliamentary faction offered no audience for a prophet of the apocalypse. For people with religious antennae, Bahro increasingly seemed a hopeless wacko. As a DDR dissident he had become "a great admirer of Luther";[75] he then turned to Meister Eckart, Spinoza, Buddha, Lao-Tse, and other thinkers.[76] Bahro was unsuccessful in his quest to become a green guru; his charisma remained bound to a historic moment. Despite all his eccentricity, his type was common for his time, commoner at the lower rungs of the ecoscene than at the leadership level. In his own view he was "searching," but from a critical outsiders' perspective he indiscriminately took up contradictory spiritual trends, from the cool Japanese Zen meditation to the steamy sweat lodges of Indian shamans.

The most internationally famous of the German Greens was Petra Kelly. She had lived in the US from 1959 to 1970 and graduated from American University. Her mix of spirituality and politics was shaped not only by her Catholic childhood in Bavaria but by her American experiences and contacts. "From the beginning, my engagement with the Greens had very, very much to do with my deep religious-spiritual orientation," Petra Kelly emphasized, "because I consider the authentic green international movement not only a political but also a political-spiritual movement."[77] She knew that "a number of forces within the Greens ... strongly reject such a description." She implied that such Greens were not "authentic." For Kelly, internationalism and spirituality were connected. More courageously than most other politicians in Bonn, she established contacts with civil-rights campaigners in the DDR years before the Berlin Wall fell in 1989. But as the cry rang out, "Germany, one Fatherland," she was as distracted as the majority of the Green leadership. The crossing of national borders held for her something this side of transcendence, typical of the exoticism propagated in intellectual environmentalism. "She was not of this world, she had an unbelievable spiritual dimension," her former lover Lukas Beckmann recalled after her death.[78]

But "spiritual" is a broad concept. Of what type was her spirituality? Journalists already believed to have sensed, many years before her death, a "martyr-like aura" around her.[79] That fit the apocalyptic mood of the time in which she lived.[80] In the hour of the apocalypse, parliamentary politics would no longer have made sense; one could only bear witness like the martyrs, the "witnesses." In the meantime she spoke of the "erotic character of real religiosity" and named as examples not only "tantric yoga" but also "Tao."[81] Her examples turned out to be only an indiscriminate stringing together of spiritual buzzwords. Over time, she herself was less and less capable of love or of meditation. To quote Beckmann once more: "Petra assimilated everything and so to speak embodied it. She was a type of medium."[82]

In the end, her true saint was the Dalai Lama, whom she met for the first time in Bonn in 1987 and for whom she had unrestrained adoration. From this time on, she spoke often of "His Holiness the Dalai Lama," without regard for the feelings of embarrassment of those many Greens who did not share her worshipful attitude. Controversy and confrontation suited her, and in her nature she was as removed from serenity and Buddhism's "Middle Way" as possible. She never really emerged from the good-versus-evil thinking that arose out of the rigid provincial Catholicism in which she grew up. She became completely caught up in the demand for full Tibetan autonomy, which fundamentally amounted to a disentanglement of Tibet from China and could not be achieved without war with China. Evidently she was unaware that, with her stridency on this issue, she undermined the Dalai Lama, who always tried, with great difficulty, to wean radical followers away from suicidal anti-China militancy. On the night of October 1, 1992, Petra Kelly was shot by her lover, the former army general and Green politician Gerd Bastian; he then shot himself. It was a death that would have fit better in a Shakespearean tragedy than in the history of the Federal Republic of Germany.

If one tracks the religious undercurrents in environmentalism around the world and seeks to identify the character of green religiosity, it is instructive to pay attention to the extent to which the environmental movement practices a martyr cult. The murder of the Brazilian Chico Mendes (1944–88) was perhaps the event of this type that was most remembered. Mendes was a rubber tapper, the son of a rubber tapper, and from the 1970s a trade-union leader who vocally defended both the interests of rubber tappers and the Amazon rain forest in which they plied their trade. In the 1980s he became an icon for environmentalists around the world, at that time alarmed by the rapid deforestation of Amazonia. Ranchers, whose business required burning tracts of the forest, arranged for his murder late in 1988. Kelly and Mendes were the most famous green martyrs, but not the only ones.

The teacher Hartmut Gründler burned himself alive in protest against nuclear energy policies on the steps of Hamburg's St. Peter's Church on November 21, 1977. That was the Day of Prayer and Repentance, and the horrific spectacle took place against the background of the SPD (West Germany's Social Democratic Party) congress in Hamburg. His act was rather embarrassing for many German environmentalists. Gründler's name never became well known, very much in contrast to the East German priest Oskar Brüsewitz, who burned himself alive in front of his church in 1976 as a protest against the harassment of the Church in the DDR (and whose example Gründler likely had in mind). In contrast to the Christian-fundamentalist followers of the priest Brüsewitz, most Greens thought voluntary martyrdom made little sense. The only exceptions within Europe were British animal-rights activists, who admired and preserved in memory Olive Perry's 1972 self-immolation, a protest against vivisection.[83] But an animal-rights activist's readiness for violence is an outlier within the environmental movement.

Among the spiritual undercurrents within environmentalism one finds above all a mysticism of life, not of death. In comparison to most other great movements in history, it is notable that in the history of the environmental movement seldom has blood flowed – although conflict has abounded and goals as momentous as saving the Earth theoretically could have legitimized much violence. The legitimization of peaceful resistance, however, has needed no martyrs. If one is searching for the practical meaning of green religiosity, this point deserves special attention.

Notes

1　Significantly, neither in C. Merchant's *Major Problems in American Environmental History* nor in the three-volume *Encyclopedia of World Environmental History* is religion a major theme, although these two otherwise superb overviews of environmental history contain not a few pointers to the significance of religion, especially clearly – and this is also significant – for Native Americans. See C. Merchant, *Major Problems in American Environmental History*, Lexington, KY, D. C. Heath & Co., 1993; and S. Krech, J. R. McNeill, and C. Merchant (eds.), *Encyclopedia of World Environmental History*, 3 vols., New York, Routledge, 2004.

2　H.-U. Wehler, *Deutsche Gesellschaftsgeschichte: Bundesrepublik und DDR, 1949–90*, Munich, Beck Verlag, 2008, p. xiii.

3　For the Anglo-American literature, see the *Bibliography on Religion and the Environment*, accessed from http://energybible.com/spiritual_energy/bibliography.html on March 20, 2012.

4　J. G. Frazer, *The Golden Bough: A Study in Magic and Religion*, New York, Macmillan, 1922.

5　C. J. Glacken, *Traces on the Rhodian Shore: Nature and Culture in Western Thought from Ancient Times to the End of the 18th Century*, Berkeley, University of California Press, 1967.

6　X. de Planhol, *Les fondements géographiques de l'histoire de l'Islam*, Paris, Flammarion, 1968.

7　See Mikhail (Chapter 10) in this volume.

8　N. Luhmann, *Ökologische Kommunikation: Kann die moderne Gesellschaft sich auf ökologische Gefährdungen einstellen?*, Opladen, Westdeutscher Verlag, 1986, p. 68; R. A. Rappaport, *Ecology: Meaning, and Religion*, Richmond, CA, North Atlantic Books, 1979.

9   A. Dieterich, *Mutter Erde: Ein Versuch über Volksreligion*, Darmstadt, *Neudruck Wissenschaftliche Buchgesellschaft*, 1967 [1905], p. 13.

10   S. Krech, *The Ecological Indian: Myth and History*, New York, W. W. Norton, 1999.

11   S. D. Gill, *Mother Earth: An American Story*, Chicago, University of Chicago Press, 1987; W. Arrowsmith and M. Korth, *Die Erde ist unsere Mutter: Die großen Reden der Indianerhäuptlinge*, Munich, Wilhelm Heyne Verlag, 1995, p. 151.

12   M. Harris, *Cannibals and Kings*, New York, Random House, 1977.

13   M. Gadgil and R. Guha, *This Fissured Land: An Ecological History of India*, Delhi, Oxford University Press, 1992, p. 87.

14   H. P. Duerr, "Wir wollen staunen," *Der Spiegel*, October 5, 1996, pp. 222–5 (p. 223).

15   A. Michaels, "Sakralisierung als Naturschutz? Heilige Bäume und Wälder in Nepal," in R. P. Sieferle and H. Breuninger (eds.), *Natur-Bilder: Wahrnehmungen von Natur und Umwelt in der Geschichte*, Frankfurt, Campus Verlag, 1999, pp. 117–36 (p. 132).

16   C. Geertz, "Religion as a Cultural System," in M. Banton (ed.), *Anthropological Approaches to the Study of Religion*, London, Tavistock Publications, 1966, pp. 1–46.

17   C. Geertz, *Islam Observed: Religious Development in Morocco and Indonesia*, Chicago, University of Chicago Press, 1968, p. 9.

18   Geertz, *Islam Observed*, p. 11.

19   L. White, "The Historical Roots of Our Ecological Crisis," *Science* 155/3767, 1967, pp. 1203–7.

20   J. B. McCleary gives a definition of the Beatnik, in contrast to Lynn White, that is without spiritual content: "The predecessor of the hippie, widely considered to be personified by an unkempt appearance, poetry reading, and bongo playing." See J. B. McCleary, *The Hippie Dictionary*, Berkeley, Ten Speed Press, 2004, p. 44.

21   Here, quoted in D. Spring and E. Spring (eds.), *Ecology and Religion in History*, New York, Harper and Row, 1974, pp. 23–8.

22   A. Toynbee, "The Religious Background of the Present Environmental Crisis," in Spring and Spring (eds.), *Ecology and Religion in History*, pp. 137–49.

23   R. Dubos, "Franciscan Conservation versus Benedictine Stewardship," in Spring and Spring (eds.), *Ecology and Religion in History*, pp. 114–36.

24   Y. Tuan, "Discrepancies between Environmental Attitude and Behaviour: Examples from Europe and China in Spring and Spring (eds.), *Ecology and Religion in History*, pp. 91–113.

25   M. Elvin, "Three Thousand Years of Unsustainable Growth: China's Environment from Archaic Times to the Present," *East Asian History* 6, 1993, pp. 7–46.

26   G. Linck, "Die Welt ist ein heiliges Gefäß, wer sich daran zu schaffen macht, wird Niederlagen erleiden: Konfliktaustragung an der Natur während der Umbrüche der chinesischen Geschichte," in J. Calließ, J. Rüsen, and M. Striegnitz (eds.), *Mensch und Umwelt in der Geschichte*, Pfaffenweiler, Centaurus-Verlagsgesellschaft, 1989, pp. 327–51 (p. 350).

27   G. Linck, *Yin und Yang: Die Suche nach Ganzheit im chinesischen Denken*, Munich, Beck, 2000.

28   M. Elvin, *The Retreat of the Elephants: An Environmental History of China*, New Haven, CT, Yale University Press, 2004, p. 324.

29   J. Cronin and R. F. Kennedy, *The Riverkeepers*, New York, Simon and Schuster, 1997, p. 229.

30   Cronin and Kennedy, *The Riverkeepers*, p. 20.

31   K. Kern, "Politische Kultur und Umweltpolitik: Amerikanische Erfahrungen und europäische Perspektiven" ("Political Culture and Environmental Politics: American Experiences and European Perspectives"), in *Mainauer Gespräche 1998, Europas Kulturen und ihr Umgang mit der Natur*, Insel Mainau, Lennart-Bernadotte-Stiftung, 1999, pp. 29–62 (p. 43).

32   Although these regions were the most affected by soil erosion and exhaustion. The thesis of the agricultural historian Avery Craven, that soil erosion was a driver of western expansion, applies mostly for the southern states with their cotton, corn, and tobacco cultures; see A. O. Craven,

Soil Exhaustion as a Factor in the Agricultural History of Virginia and Maryland, 1606–1860, Champaign, IL, University of Illinois Press, 1926; and J. Radkau, *Nature and Power: A Global History of the Environment,* New York, Cambridge University Press, 2008, p. 178.

33   J. Gatta, *Making Nature Sacred: Literature, Religion, and Environment in America from the Puritans to the Present,* New York, Oxford University Press, 2004.

34   R. F. Nash, *Wilderness and the American Mind,* 4th ed., New Haven, CT, Yale University Press, 2001 [1967], p. 35.

35   M. Stoll, *Protestantism, Capitalism, and Nature in America,* Albuquerque, NM, University of New Mexico Press, 1997, p. 198.

36   A. Naess, "Selbst-Verwirklichung: Ein ökologischer Zugang zum Sein in der Welt," in J. Seed et al. (eds.), *Denken wie ein Berg,* Freiburg, Verlag Hermann Bauer, 1993, pp. 33–43 (p. 34).

37   H. Weinzierl,"Naturschutz ist Menschenschutz," in K. Stankiewitz (ed.), *Babylon in Bayern,* Regensburg, Edition Buntehunde, 2004, pp. 133–4 (p. 134).

38   S. Fox, *John Muir and His Legacy: The American Conservation Movement,* Boston, Little, Brown and Co., 1981, p. 107.

39   J. Radkau, *Max Weber: A Biography,* Cambridge, Polity, 2009, p. 427.

40   D. M. Berman and J. T. O'Connor, *Who Owns the Sun? People, Politics, and the Struggle for a Solar Economy,* White River Junction, Chelsea Green, 1996, p. 78.

41   T. R. Wellock, *Critical Masses: Opposition to Nuclear Power in California, 1958–1978,* Madison, University of Wisconsin Press, 1998.

42   D. Brower, *Earth's Sake,* Salt Lake City, UT, Gibbs Smith, 1990 [1957], p. 261.

43   Stoll, *Protestantism, Capitalism, and Nature in America,* p. 151.

44   C. Ponting, *A Green History of the World,* London, Penguin Books, 1991, p. 1; J. Diamond, *Collapse: How Societies Choose to Fail or Succeed,* New York, Viking, 2005, chapter 2.

45   T. L. Hunt and C. P. Lipo, "Ecological Catastrophe, Collapse, and the Myth of 'Ecocide' on Rapa Nui," in P. A. McAnany and N. Yoffee (eds.), *Questioning Collapse: Human Resilience, Ecological Vulnerability, and the Aftermath of Empire,* New York, Cambridge University Press, 2010, pp. 21–44.

46   M. Nicholson later looked back with critical distance at the "Doomsday syndrome" around 1970 and the "orgy of pessimism" at the time; in reality the knowledge of the time about global ecological conditions was still sparse. See M. Nicholson, *The New Environmental Age,* Cambridge, Cambridge University Press, 1987, pp. 194, 193.

47   M. Egan, *Barry Commoner and the Science of Survival: The Remaking of American Environmentalism,* Cambridge, MA, MIT, 2007, p. 79.

48   Egan, *Barry Commoner and the Science of Survival,* p. 104. Chicken Little refers to a different version of the "cry wolf effect": the fearful chicken of an American children's story believes that a thud signals that the sky has fallen, and sets the animals temporarily into panic, only afterward to be held up to ridicule.

49   M. Weber, *Max Weber: Ein Lebensbild,* Munich, Piper, 1989 [1926], p. 605.

50   Radkau, *Max Weber,* p. 447.

51   H. Gruhl, *Überleben ist alles: Erinnerungen,* Frankfurt, Herbig, 1990, p. 217.

52   S. Faroqhi (ed.), *Animals and People in the Ottoman Empire,* Istanbul, Eren, 2010.

53   B. Hüppauf, *Vom Frosch: Eine Kulturgeschichte zwischen Tierphilosophie und Ökologie,* Bielefeld, Transcript, 2011.

54   K. Thomas, *Man and the Natural World: Changing Attitudes in England, 1500–1800,* London, Penguin Books, 1984, p. 109.

55   M. de Montaigne, "An Apology for Raymond of Sebund," in *Essays,* 1588, letter 2, accessed from http://www.gutenberg.org/files/3600/3600-h/3600-h.htm#2H_4_0128 on March 20, 2012.

56   M. Dowie, *Conservation Refugees: The Hundred-Year Conflict between Global Conservation and Native Peoples,* Cambridge, MA, MIT, 2009, p. xi.

57   H. Makowski, *Nationalparke in Deutschland: Schatzkammern der Natur – Kampfplätze des Naturschutzes,* Neumünster, Wachholtz Verlag, 1997, p. 36.

58 W. Haber,"Naturschutz in der Kulturlandschaft – ein Widerspruch in sich?," in Bavarian
    Academy for Nature Conservation and Landscape Management (ed.), *Die Zukunft der
    Kulturlandschaft* 1/8, 2008, pp. 15–25. Heinrich Spanier (advisor for conservation in
    Germany's Federal Ministry for the Environment) to Frank Uekötter, March 22, 2004: "In
    Abdera, the home of Democritus, one had declared frogs as sacred. In the conservation scene,
    I am however not sure whether the pathos of religiosity is only a linguistic gimmick or whether
    it is actually so intended. I tend toward the latter … In this respect, the pathos is also a politi-
    cal program. In any case, the religious dimension of conservation as a theme is worth further
    research in history, the present, and the future" (personal communication, March 22, 2004).

59 D. Foreman, *Confessions of an Eco-Warrior*, New York, Crown Publishers, 1991, pp. 3, 105,
    116, 175, also 19, 51, 73, 90.

60 Foreman, *Confessions of an Eco-Warrior*, p. 39.

61 On this and the following, see the biography authored by Schumacher's daughter: B. Wood,
    Alias Papa: A Life of Fritz Schumacher, Oxford, Oxford University Press, 1985.

62 Wood, *Alias Papa*, p. 320.

63 E. F. Schumacher, *Die Rückkehr zum menschlichen Maß: Alternativen für Wirtschaft und
    Technik*, Reinbeck, Rowohlt, 1977, p. 140. "Appropriate technology" was his updated ver-
    sion of what he at first called "intermediate technology."

64 Wood, *Alias Papa*, p. 352.

65 Wood, *Alias Papa*, p. 342.

66 M. Ishimure, *Paradies im Meer der Qualen*, Frankfurt, Insel Verlag, 1995 [1969], p. 8.

67 Ishimure, *Paradies im Meer der Qualen*, p. 360; I. Ishimure, "Reborn from the Earth Scarred
    by Modernity: Minamata Disease and the Miracle of the Human Desire to Live," *Japan
    Focus*, April 27, 2008, accessed from http://www.japanfocus.org/-Ishimure-Michiko/2732
    on March 20, 2012. One recognizes in this article a general disgust toward many aspects of
    modern civilization, especially in Japan: "When I foolishly switch on the television, the beha-
    viour of the performers, the colour and shape of the things they wear on their body: eve-
    rything is the utmost in vulgarity. We Japanese exceed all global standards on this; when it
    comes to vulgarity, we really are number one, aren't we?"

68 Ishimure, *Paradies im Meer der Qualen*, p. 13.

69 R. Usui,"Die Minamata-Krankheit und die Sprache jenseits des ius talionis: Auf der Suche
    nach der Sprache der Anima im *Paradies im Meer der Qualen* von Michiko Ishimure," in
    H. D. Assmann et al. (eds.), *Grenzen des Lebens – Grenzen der Verständigung*, Würzburg,
    Königshausen and Neumann, 2009, pp. 57–66 (p. 64). I owe Ryiuchiro Usui, who is working
    on a book about Michiko Ishimure, for valuable insights on the significance of this writer.

70 This is the thesis of Ryiuchiro Usui (see previous note).

71 Testimony of leading German Greens in G. Hesse and H. H. Wiebe, *Die Grünen und die
    Religion*, Frankfurt, Athenäum, 1988. Significantly, this volume, much like Charlene
    Spretnak's book, was outdated after only a few years.

72 G. Herzberg and K. Seifert, *Rudolf Bahro – Glaube an das Veränderbare: Biographie*, Berlin,
    Aufbau-Taschenbuch-Verl., 2005, p. 407.

73 Die Grünen im Bundestag, *Parliamentary Faction Meeting of 4/28/1983: Minutes 1983–
    1987*, vol. 1, Düsseldorf, Droste, 2008, p. 108.

74 Die Grünen im Bundestag, *Parliamentary Faction Meeting*, p. 334.

75 C. Spretnak, *Green Politics: The Global Promise*, New York, E. P. Dutton, 1984.

76 Spretnak, *Green Politics*, p. 104 (Bahro's statement).

77 Hesse and Wiebe, *Die Grünen und die Religion*, p. 35.

78 Hesse and Wiebe, *Die Grünen und die Religion*, p. 35.

79 Spretnak, *Green Politics*.

80 M. Sperr, *Petra Karin Kelly: Politikerin aus Betroffenheit*, Munich, C. Bertelsmann, 1993,
    p. 139.

81 A. Schwarzer, *Eine tödliche Liebe: Petra Kelly und Gert Bastian*, Cologne, Kiepenheuer &
    Witsch, 2001, p. 117.

82   Schwarzer, *Eine tödliche Liebe*, p. 118.

83   M. Roscher, *Ein Königreich für Tiere: Die Geschichte der britischen Tierrechtsbewegung*, Marburg, Tectum Verlag, 2009, p. 392.

References

Arrowsmith, W., and Korth, M., *Die Erde ist unsere Mutter: Die großen Reden der Indianerhäuptlinge*, Munich, Wilhelm Heyne Verlag, 1995.

Berman, D. M., and O'Connor, J. T., *Who Owns the Sun? People, Politics, and the Struggle for a Solar Economy*, White River Junction, Chelsea Green, 1996.

Bibliography on Religion and the Environment, accessed from http://energybible.com/spiritual_energy/bibliography.html on March 20, 2012.

Brower, D., *Earth's Sake*, Salt Lake City, UT, Gibbs Smith, 1990 [1957].

Craven, A. O., *Soil Exhaustion as a Factor in the Agricultural History of Virginia and Maryland, 1606–1860*, Champaign, IL, University of Illinois Press, 1926.

Cronin, J., and Kennedy, R. F., *The Riverkeepers*, New York, Simon and Schuster, 1997.

Diamond, J., *Collapse: How Societies Choose to Fail or Succeed*, New York, Viking, 2005.

Dieterich, A., *Mutter Erde: Ein Versuch über Volksreligion*, Darmstadt, Neudruck Wissenschaftliche Buchgesellschaft, 1967 [1905].

Dowie, M., *Conservation Refugees: The Hundred-Year Conflict between Global Conservation and Native Peoples*, Cambridge, MA, MIT, 2009.

Dubos, R., "Franciscan Conservation versus Benedictine Stewardship," in D. Spring and E. Spring (eds.), *Ecology and Religion in History*, New York, Harper and Row, 1974, pp. 114–36.

Duerr, H. P., "Wir wollen staunen," *Der Spiegel*, October 5, 1996, pp. 222–5.

Egan, M., *Barry Commoner and the Science of Survival: The Remaking of American Environmentalism*, Cambridge, MA, MIT, 2007.

Elvin, M., *The Retreat of the Elephants: An Environmental History of China*, New Haven, CT, Yale University Press, 2004.

Elvin, M., "Three Thousand Years of Unsustainable Growth: China's Environment from Archaic Times to the Present," *East Asian History* 6, 1993, pp. 7–46.

Faroqhi, S. (ed.), *Animals and People in the Ottoman Empire*, Istanbul, Eren, 2010.

Foreman, D., *Confessions of an Eco-Warrior*, New York, Crown Publishers, 1991.

Fox, S., *John Muir and His Legacy: The American Conservation Movement*, Boston, Little, Brown and Co., 1981.

Frazer, J. G., *The Golden Bough: A Study in Magic and Religion*, New York, Macmillan, 1922.

Gadgil, M., and Guha, R., *This Fissured Land: An Ecological History of India*, Delhi, Oxford University Press, 1992.

Gatta, J., *Making Nature Sacred: Literature, Religion, and Environment in America from the Puritans to the Present*, New York, r, 2004.

Geertz, C., *Islam Observed: Religious Development in Morocco and Indonesia*, Chicago, University of Chicago Press, 1968.

Geertz, C., "Religion as a Cultural System," in M. Banton (ed.), *Anthropological Approaches to the Study of Religion*, London, Tavistock Publications, 1966, pp. 1–46.

Gill, S. D., *Mother Earth: An American Story*, Chicago, University of Chicago Press, 1987.

Glacken, C. J., *Traces on the Rhodian Shore: Nature and Culture in Western Thought from Ancient Times to the End of the 18th Century*, Berkeley, University of California Press, 1967.

Gruhl, H., *Überleben ist alles: Erinnerungen*, Frankfurt, Herbig, 1990.

Die Grünen im Bundestag, *Parliamentary Faction Meeting of 4/28/1983: Minutes 1983–1987*, vol. 1, Düsseldorf, Droste, 2008.

Haber, W., "Naturschutz in der Kulturlandschaft – ein Widerspruch in sich?," in Bavarian Academy for Nature Conservation and Landscape Management (ed.), *Die Zukunft der Kulturlandschaft* 1/8, 2008, pp. 15–25.

Harris, M., *Cannibals and Kings*, New York, Random House, 1977.

Herzberg, G., and Seifert, K., *Rudolf Bahro – Glaube an das Veränderbare: Biographie*, Berlin, Aufbau-Taschenbuch-Verl., 2005.

Hesse, G., and Wiebe, H. H., *Die Grünen und die Religion*, Frankfurt, Athenäum, 1988.

Hunt, T. L., and Lipo, C. P., "Ecological Catastrophe, Collapse, and the Myth of 'Ecocide' on Rapa Nui," in P. A. McAnany and N. Yoffee (eds.), *Questioning Collapse: Human Resilience, Ecological Vulnerability, and the Aftermath of Empire*, New York, Cambridge University Press, 2010, pp. 21–44.

Hüppauf, B., *Vom Frosch: Eine Kulturgeschichte zwischen Tierphilosophie und Ökologie*, Bielefeld, Transcript, 2011.

Ishimure, M., *Paradies im Meer der Qualen*, Frankfurt, Insel Verlag, 1995 [1969].

Ishimure, M., "Reborn from the Earth Scarred by Modernity: Minamata Disease and the Miracle of the Human Desire to Live," *Japan Focus*, April 27, 2008, accessed from http://www.japan focus.org/-Ishimure-Michiko/2732 on March 20, 2012.

Kern, K., "Politische Kultur und Umweltpolitik: Amerikanische Erfahrungen und europäische Perspektiven" ("Political Culture and Environmental Politics: American Experiences and European Perspectives"), in *Mainauer Gespräche 1998, Europas Kulturen und ihr Umgang mit der Natur*, Insel Mainau, Lennart-Bernadotte-Stiftung, 1999, pp. 29–62.

Krech, S., *The Ecological Indian: Myth and History*, New York, W. W. Norton, 1999.

Krech, S., McNeill, J. R., and Merchant, C. (eds.), *Encyclopedia of World Environmental History*, 3 vols., New York, Routledge, 2004.

Linck, G., "Die Welt ist ein heiliges Gefäß, wer sich daran zu schaffen macht, wird Niederlagen erleiden: Konfliktaustragung an der Natur während der Umbrüche der chinesischen Geschichte," in J. Calließ, J. Rüsen, and M. Striegnitz (eds.), *Mensch und Umwelt in der Geschichte*, Pfaffenweiler, Centaurus-Verlagsgesellschaft, 1989, pp. 327–51.

Linck, G., *Yin und Yang: Die Suche nach Ganzheit im chinesischen Denken*, Munich, Beck, 2000.

Luhmann, N., *Ökologische Kommunikation: Kann die moderne Gesellschaft sich auf ökologische Gefährdungen einstellen?*, Opladen, Westdeutscher Verlag, 1986.

Makowski, H., *Nationalparke in Deutschland: Schatzkammern der Natur – Kampfplätze des Naturschutzes*, Neumünster, Wachholtz Verlag, 1997.

McCleary, J. B., *The Hippie Dictionary*, Berkeley, Ten Speed Press, 2004.

Merchant, C., *Major Problems in American Environmental History*, Lexington, KY, D. C. Heath & Co., 1993.

Michaels, A., "Sakralisierung als Naturschutz? Heilige Bäume und Wälder in Nepal," in R. P. Sieferle and H. Breuninger (eds.), *Natur-Bilder: Wahrnehmungen von Natur und Umwelt in der Geschichte*, Frankfurt, Campus Verlag, 1999, pp. 117–36.

Montaigne, M. de, "An Apology for Raymond of Sebund," in *Essays*, 1588, letter 2, accessed from http://www.gutenberg.org/files/3600/3600-h/3600-h.htm#2H_4_0128 on March 20, 2012.

Naess, A., "Selbst-Verwirklichung: Ein ökologischer Zugang zum Sein in der Welt," in J. Seed et al. (eds.), *Denken wie ein Berg*, Freiburg, Verlag Hermann Bauer, 1993, pp. 33–43.

Nash, R. F., *Wilderness and the American Mind*, 4th ed., New Haven, CT, Yale University Press, 2001 [1967].

Nicholson, M., *The New Environmental Age*, Cambridge, Cambridge University Press, 1987.

Planhol, X. de, *Les fondements géographiques de l'histoire de l'Islam*, Paris, Flammarion, 1968.

Ponting, C., *A Green History of the World*, London, Penguin Books, 1991.

Radkau, J., *Max Weber: A Biography*, Cambridge, Polity, 2009.

Radkau, J., *Nature and Power: A Global History of the Environment*, New York, Cambridge University Press, 2008.

Rappaport, R. A., *Ecology: Meaning, and Religion*, Richmond, CA, North Atlantic Books, 1979.

Roscher, M., *Ein Königreich für Tiere: Die Geschichte der britischen Tierrechtsbewegung*, Marburg, Tectum Verlag, 2009.

Schumacher, E. F., *Die Rückkehr zum menschlichen Maß: Alternativen für Wirtschaft und Technik*, Reinbeck, Rowohlt, 1977.

Schwarzer, A., *Eine tödliche Liebe: Petra Kelly und Gert Bastian*, Cologne, Kiepenheuer & Witsch, 2001.

Sperr, M., *Petra Karin Kelly: Politikerin aus Betroffenheit*, Munich, C. Bertelsmann, 1993.

Spretnak, C., *Green Politics: The Global Promise*, New York, E. P. Dutton, 1984.

Spring, D., and Spring, E. (eds.), *Ecology and Religion in History*, New York, Harper and Row, 1974.

Stoll, M., *Protestantism, Capitalism, and Nature in America*, Albuquerque, NM, University of New Mexico Press, 1997.

Thomas, K., *Man and the Natural World: Changing Attitudes in England, 1500–1800*, London, Penguin Books, 1984.

Toynbee, A., "The Religious Background of the Present Environmental Crisis," in D. Spring and E. Spring (eds.), *Ecology and Religion in History*, New York, Harper and Row, 1974, pp. 137–49.

Tuan, Y., "Discrepancies between Environmental Attitude and Behaviour: Examples from Europe and China," in D. Spring and E. Spring (eds.), *Ecology and Religion in History*, New York, Harper and Row, 1974, pp. 91–113.

Usui, R., "Die Minamata-Krankheit und die Sprache jenseits des ius talionis: Auf der Suche nach der Sprache der Anima im *Paradies im Meer der Qualen* von Michiko Ishimure," in H. D. Assmann et al. (eds.), *Grenzen des Lebens – Grenzen der Verständigung*, Würzburg, Königshausen and Neumann, 2009, pp. 57–66.

Weber, M., *Max Weber: Ein Lebensbild*, Munich, Piper, 1989 [1926].

Wehler, H.-U., *Deutsche Gesellschaftsgeschichte: Bundesrepublik und DDR, 1949–90*, Munich, Beck Verlag, 2008.

Weinzierl, H., "Naturschutz ist Menschenschutz," in K. Stankiewitz (ed.), *Babylon in Bayern*, Regensburg, Edition Buntehunde, 2004, pp. 133–4.

Wellock, T. R., *Critical Masses: Opposition to Nuclear Power in California, 1958–1978*, Madison, University of Wisconsin Press, 1998.

White, L., "The Historical Roots of Our Ecological Crisis," *Science* 155/3767, 1967, pp. 1203–7.

Wood, B., *Alias Papa: A Life of Fritz Schumacher*, Oxford, Oxford University Press, 1985.

Further Reading

Dalton, A. M., and Simmons, H., *Ecotheology and the Practice of Hope*, Albany, NY, SUNY Press, 2011.

Dunlap, T. R., *Faith in Nature: Environmentalism as Religious Quest*, Seattle, University of Washington Press, 2005.

Gottlieb, R., *A Greener Faith: Religious Environmentalism and Our Planet's Future*, New York, Oxford University Press, 2009.

Chapter Twenty-Eight

The Environmentalism of the Poor: Its Origins and Spread

Joan Martinez-Alier

Introduction

Environmentalism is a cultural and social movement concerned with the preservation of nature. The environment and its biodiversity would be defended by the creation of nature parks, which humans were welcome to visit (in small groups) but where they would not be allowed to live. This kind of environmentalism has been called "the cult of wilderness." It has existed for over 100 years. Its recent growth has sometimes been explained by a change in social and cultural values after 1968 in Western countries where economic needs were largely covered, and economic distribution conflicts perhaps receded. Once you had a large house and two cars in the garage, you might become concerned by endangered whales and panda bears. Environmentalism was seen as typical of rich societies and was explained in terms of so-called post-materialist values (Ronald Inglehart's term).[1]

But the "cult of wilderness" is only one variety of environmentalism. From 1990 onward, scholars identified a different kind of environmentalism, "the environmentalism of the poor," focusing on the global south but closely related to the "environmental justice" movement in the US. Based on studies of rural conflicts in India and Latin America, Guha and Martinez-Alier highlighted the relationship between poverty and environmental degradation, arguing that, because the poor rely directly on the land and its natural resources and services, they have a strong motivation to be careful managers of the environment.[2] In addition, and against post-materialist theory, it has been shown that there is no positive correlation between economic wealth and concern for the environment.[3] And against the expectations of "ecological modernization" and "sustainable development" it is clear that the clash between the growth of the economy and the health of the environment continues.

A Companion to Global Environmental History, First Edition. Edited by J.R. McNeill and Erin Stewart Mauldin.
© 2012 John Wiley & Sons, Ltd. Published 2015 by John Wiley & Sons, Ltd.

The concept of the "environmentalism of the poor" was born within the discipline of social history. The thesis does not assert that as a rule poor people feel, think, and behave as environmentalists. This is not so. The thesis is that in the many resource-extraction and waste-disposal conflicts in history and today, the poor are often on the side of the preservation of nature against business firms and the state. This behavior is consistent with their interests and values. The environmentalism of the poor appears mostly at local levels but also at national and international levels in complaints against ecologically unequal exchange and climate injustices, in claims concerning the "ecological debt," and in international court cases asking for compensation for environmental liabilities of foreign companies.

The environmentalism of the poor centers then on social justice, including claims to recognition and participation, and builds on the premise that the fights for human rights and environment are inseparable. From the resistance, new institutions arise. The successful anti-gold-mining movements in Tambo Grande, Peru, and Esquel, Argentina, around 2000, appealed to local democracy and imposed a new institution, the local referendum or public consultation, which allows the expression of values that would otherwise remain hidden.[4] The action by the rubber tappers' union in Xapuri, Acre, in Brazil in the 1980s, led by Chico Mendes against cattle ranchers, resulted in parts of Amazonia being designated as "extractive reserves" combining conservation and the sustainable use of forests.

The environmentalism of the poor relates to actions in situations where the environment is a source of livelihood. It is linked to other values, such as the defence of indigenous territorial rights or claims to the sacredness of particular elements of nature (a mountain, a forest, or even a tree). When livelihood is threatened, those affected will be motivated to act provided that there is a sufficient degree of democracy and they are not suffocated by fear. Environmentalism, in most parts of the world today, as in most eras in history, is more often than not the environmentalism of the poor.

In the environmentalism of the poor as in environmental justice movements in general, it is important to recognize the contributions of women. Bina Agarwal argued that women more often collect water, gather wood, look for medicinal plants, tend to domestic animals, and grow crops, and therefore they have greater knowledge and awareness of their community's direct dependence on the natural environment.[5] This does not imply that women have an empathy with nature denied to men for biological reasons. The argument is based on social roles. In an urban setting, it is women who often take leading positions in environmental justice conflicts (in contrast to labor-union struggles) as regards complaints against waste-dumping, or air and water pollution. Women are often the main actors in environmental conflicts. And they also propose solutions, such as the Kenyan Green Belt Movement founded by the Nobel Prize winner Wangari Maathai in 1977, which sought to plant trees in parts of the country affected by deforestation and soil erosion (and has since broadened its mission).

Causes of Environmental Conflicts

The fundamental clash between economy and the environment comes from two facts. First, population growth. In the twentieth century, global population grew fourfold. It now seems that "peak population" will be reached at about 8.5 billion, by 2045. Second, the social metabolism of industrial economies. Energy cannot be recycled. Therefore, the energy from fossil fuels is used only once, and new supplies must be obtained from "commodity frontiers."[6] Similarly, materials are recycled only in part, therefore even an economy that did not grow would need fresh supplies of iron ore, bauxite, copper, and other materials. The growth in the number of resource-extraction conflicts (and also waste-disposal conflicts, of which the most notable today arises from the production of excess carbon dioxide) is explained by the social metabolism: the modern economy has come to rely on enormous quantities of energy and materials which must come from somewhere, and when used, must go somewhere transformed into wastes. The situation under neoliberal policies is not much different from what it would be under Keynesian social-democratic policies.

Many socioenvironmental conflicts stem from the poor trying to retain environmental resources for their livelihood. Whereas wealthier people of the north have in general lost the idea of the environment as their source of livelihood, the poor and largely rural populations of the south are more connected to the environment, and thus have a more intimate understanding of what is at stake by not managing it carefully.[7] In the north on the other hand, while people consume large quantities of imported energy and materials, and produce increasing amounts of waste, many of the effects are exported elsewhere. Since subsistence production outside the market is often a key aspect of societies in the south, particularly in rural areas, it can be argued that the south also has a greater sense of community, and, with it, a greater connection to the importance of community management.[8]

"Governing the commons"[9] is not a new topic for the social historian. Not only do new institutions arise from environmental conflicts, but also new networks are born. Oilwatch, for example, emerged in 1995, because of numerous protests and direct actions against oil exploration and exploitation by indigenous peoples and local communities in Colombian and Ecuadorian Amazonia and in the Niger delta. This is a "network" of national and regional organizations that acts at the local level so that they can intervene in a united way at the international level in protection of the lives and livelihood of local peoples.[10] It deals with issues of biodiversity, pollution, deforestation, protection of indigenous territorial rights, and global climate change. At the Kyoto conference on climate change in 1997, 200 organizations signed a declaration that helped draw attention to the link between oil extraction and carbon-dioxide emissions, from which 10 years later the Yasuni ITT proposal in Ecuador was born – an arrangement by which the government of Ecuador will be paid to refrain from oil development in Amazonia.[11] The environmentalism of the poor has thus led to international networks being created that have successfully linked the local movements to wider global issues. Other examples include the International Rivers Network supporting local struggles against dams.[12]

The capacity of the global south for spreading the message of the environmentalism of the poor is limited by scarcity of resources. It can be helped by media and communication networks, although these are normally dominated by northern-based actors. Sometimes, as with famous court cases (Chevron Texaco in Ecuador, Shell in Nigeria), the silence is broken. But there remains a lack of knowledge about the many thousands of environmental conflicts around the world that conform to the theory of the environmentalism of the poor, in history and at present.

An environmentalism of the poor can exist in rich countries too. In the US, in the early 1980s, the first social movements appeared among poor and minority communities against disproportionate burdens of pollution and lack of voice on environmental policy implementation. They were rooted in the earlier civil-rights movement, but in some ways echoed earlier movements among populist farmers in US history. This environmental justice movement, as it is called in the US, was initially concerned only by the incidence of pollution in areas inhabited by what in that country are described as "minority populations," or "people of color" – which at the world level are a majority. The movement called attention to the link between pollution, race, and poverty.[13] In the US, asthma is distributed unequally among children in urban environments; it is more common in poor areas, where black and Hispanic populations are concentrated. Research can confirm the existence of such socioenvironmental injustices, thus anticipating the birth of social movements – although injustices do not necessarily lead to action either at present or in history.

This new movement pointed out that the mainstream environmental organizations in the US such as the Sierra Club were staffed predominantly by upper- and middle-class white people, and that they lacked concern for the plight of the poor and ethnic minorities. The fight against so-called "environmental racism" became spectacularly successful in public policy-making when President Clinton in February 1994 issued executive order 12898 asking all departments in the administration to make achieving environmental justice part of their mission, with the Environmental Protection Agency leading the way. Little by little, the environmental justice movement in the US started to collaborate with environmental movements around the world in defense of the poor and the indigenous.[14]

Another contributing factor to the environmentalism of the poor is the strengthening of the international movement for indigenous rights concerned with political rights and cultural preservation. Indigenous territories in many countries (Canada, Australia, Peru, Brazil, India, South Africa, and Namibia) are also at the frontiers of mineral, fossil-fuel, and timber extraction. There are over 370 million indigenous people in Africa, the Americas, Asia, Europe, and the Pacific. Territorial struggles overlap with environmental struggles. The 2007 UN Declaration on the Rights of Indigenous Peoples defends indigenous territories against waste disposal and against resource extraction without prior informed consent. The declaration is an achievement that comes after centuries of social, economic, and environmental injustices.

The Chipko Movement and the Environmentalism of the Poor

The Unquiet Woods was Ramachandra Guha's first book, published in 1989. It has become a classic of peasant studies and socioenvironmental history. This was the first scholarly elaboration of the notion of an "environmentalism of the poor." The book goes over the history of a Himalayan region that has a deep significance for India's culture across many centuries. In Garhwal, peasant resistance in defense of access to the forest took the form of ritual petitions to the king backed by demonstrations; while in Kumaun, there were less structured rebellions and massive arson in pine-plantation forests around 1920–1. This varied history and its protagonists were remembered in the 1970s, when the Chipko movement started.

Forests became state property in India under colonial rule. The main technical advisers were German experts trained in forest economics, or, rather, tree-plantation economics. Their policy was to grow uniform stands of trees as long as it was financially viable to do so, which meant comparing the rate of growth of trees to the rate of interest. In 1849, the experts introduced another factor: the rent obtainable from the land for some years (as pasture, for instance) once the trees were cut and while waiting for the next "crop" of trees. This additional factor would be reason for a shorter rotation. This narrowly economistic training was the kind of management expertise that German foresters had when they arrived in India hired by the British.

India, however, had rich and diverse tropical forests which people depended upon for their subsistence livelihood. The logic of multiple use could not be totally ignored by the administrators. It was obvious that forests in India were used by the local populations for "non-timber" products; they held much biodiversity; and they provided flood control and other "environmental services" (to use today's jargon). In fact, in India the debate on the multifunctionality of forests goes back 150 years. The new institutions of Joint Forest Management since the late 1980s (partly the result of the Chipko movement) have a long intellectual and social history.

While in *The Unquiet Woods* Guha tended to blame German forestry-science for too single-mindedly imposing the logic of plantations over the various uses to which the Himalayan peasantry deployed oak forests, in a more recent book he quotes from Dietrich Brandis, the inspector general of forests between 1864 and 1883, to the effect that village forests should be preserved. The German argued that forests ought to provide free firewood for home consumption or for sale by poor women. He also saw forests as desirable for producing wood for the making of agricultural implements and carts; for bamboo, wood, and grass in thatching, flooring, and fencing; for leaves and branches in the preparation of manure; and for grazing, except in areas that had been closed for forest regeneration.[15] These humane considerations of Brandis notwithstanding, in Garhwal and Kumaun the trend toward the commercial exploitation of uniform tree plantations under state ownership was clear. Demand from the railways was the main factor. Later, already after independence (1947) the demand from other industries such as paper, plywood, and sporting goods (e.g., cricket bats) compounded the logic of plantation forestry, for which untidy forests had to be cleared. The local population resented the felling of trees by outside industries.

Guha's book explains the history of peasant resistance and its varying forms in Garhwal and Kumaun. It also explains the rise of the Chipko movement in the 1970s ("the environmentalism of the poor"), asking whether this was a new environmental movement or only another form of peasant resistance – this time with an explicit ecological content. Finally, Guha provided a running commentary comparing peasant resistance in these two regions to peasant resistance or acquiescence in other countries: the enclosure of forests in western Germany as described by the young Marx; peasant or rural labourers' movements recounted by the British social historians E. P. Thompson and Eric Hobsbawm; everyday forms of resistance among Southeast Asian peasants as analyzed by James Scott; and even peasant movements as presented in rural novels by Balzac. Guha emphasized the world significance of peasant resistance, drawing comparisons far and wide.

Southern Europe and Latin America

While in India Ramachandra Guha is the father of environmental history, outside India he had many followers. In Spain, socioenvironmental history by González de Molina and Ortega Santos on Granada province in Andalusia followed Guha in counting and classifying several kinds of forest crimes as signs of discontent with the disentailment of forests (and shrub areas) in the nineteenth century. While in India forests became state property under colonial rule, in Europe there was bourgeois privatization – a "tragedy of enclosures."[16] The explicit common ground between Ortega Santos and Guha is enclosures leading to the overexploitation of natural resources and the "discovery" of the ecological content within historical struggles over those resources that were managed as commons. In this manner, social and environmental history came together.

In his studies of southern Italy's mountain communities, Marco Armiero has shown how the privatization of land and the juridical attack after the eighteenth century against any form of common access to natural resources inspired harsh resistance from the local poor.[17] To southern Italian peasants, resisting land enclosures meant, above all, defending a particular socioecological asset: forests were not wood "quarries" but rather a complex and cohesive whole of wood, leaves, grasses, fruits, games, and soil, which required local knowledge and multiple property/access rights.

In her recent book on "water enclosures" in Italy, Stefania Barca has shown how, in the nineteenth century, deforestation for commercial purposes combined with river damming for industrial use radically increased environmental vulnerability (floods, landslides, malaria). The costs fell mostly upon local peasant communities.[18] In the twentieth century, electrification spurred a new wave of social costs (mainly dispossession and relocation affecting mountain communities), including apocalyptic inundations caused by hydroelectric dams. Meanwhile, however, the rapid postwar industrialization of the country was also spurring a new kind of environmentalism of the poor: "working class environmentalism."[19] Left with little choice but to migrate into cities to work in highly harzardous jobs, in Italy as elsewhere people started sometimes to organize and collectively struggle for environmental justice, taking corporations to court for environmental crimes related to asbestos or chemical pollution.

In the central Andes (Bolivia, Peru, Ecuador) the historical struggle, as analyzed by Florencia Mallon and others, was not over forests but over agricultural and pastoral land.[20] It pitted the haciendas of colonial origin against the indigenous communities. In the Andes no one explicitly studied the *ecological* content of such struggles – the fights specifically linked with access to natural resources (pastures, water, dung) by poor people which can be called "socioecological distribution conflicts." Until the 1980s there was in fact little socioenvironmental history in the Andes, and indeed in all Latin America. More recently, that has changed dramatically. One excellent monograph, Elinor Melville's *Plague of Sheep*, explains how, in the sixteenth and seventeenth centuries, the Mezquital Valley near Mexico City was turned into a semi-desert while the Otomi people, indigenous to it, nearly disappeared. This was a classic socioenvironmental conflict over land and water.[21]

The GDP of the Poor

The notion of "GDP of the poor" is a new way of making the distinction between market-provisioning and non-market provisioning which Aristotle called respectively chrematistics and oikonomia. The GDP of the poor is not measured in terms of money but in terms of livelihood. In rural contexts, livelihoods depend to a large extent on the non-market provision of products and services (such as water). In energy terms, the peasant economy rests directly on solar energy powering current photosynthesis to a much larger extent than do industrial and urban systems, which are based largely on fossil fuels (photosynthesis from the remote past).

The economic analysis of rural populations immediately reveals the relative importance of non-market inputs, production, and systems of exchange and redistribution. Peripheral markets often play a role (as in the Andean peasants' saying, that there are potatoes for eating and other [varieties of] potatoes for selling), but production and consumption decisions depend on forms of reciprocity and communitarian redistribution.

The analysis of rural livelihoods can rely, if one wishes, on notions of use and investment in natural capital, human capital (knowledge of biodiversity, for instance), and social capital. They can also rely on the notion of availability of resources that gives the capability both to make a living and to make a choice to change the relationships governing the ways in which resources are controlled in society, as in land-reform movements. Economic growth as measured by GDP often implies the destruction of such natural capital, human capital (including the many languages which are being lost), and social capital (as the generalized market system takes over). Economic growth, as conventionally understood, implies the loss of some capabilities, while others are gained. The balance cannot accurately be drawn in money terms as in a profit-and-loss statement or a cost–benefit analysis. It would require a multicriteria approach able to cope with incommensurable values.

Pavan Sukhdev, Pushpam Kumar, and Haripriya Gundimeda advanced the notion of the "GDP of the poor" in their work between 2008 and 2010.[22] If a tribal or poor rural family sees the availability of clean water, fodder, and wood outside the market endangered because of a mining project or a dam, the loss cannot easily be made good. One plastic bottle of water will cost the equivalent of 10–50 percent of the rural daily wage. The products and services from nature which are not counted in terms of GDP are essential to the poor. When they are damaged or destroyed, there are complaints. This inspires the environmentalism of the poor and indigenous.

Two Ecuadorean Women

The environmentalism of the poor is discernible in their own language. Disputes in Ecuador provide two examples. First (in my own translation) I quote the words of a woman of Muisne, Ecuador, regarding the conflict between conservation of mangroves and the development of the shrimp industry as it stood in the late 1990s. She argued in terms of what in the US would be called "environmental justice" against "environmental racism." The coastal population of the province of Esmeraldas in Ecuador is, in its majority, of African descent.

> We have always been ready to cope with everything, and now more than ever, but they want to humiliate us because we are black, because we are poor, but one does not choose the race into which one is born, nor does one choose not to have anything to eat, nor to be ill. But I am proud of my race and of being a conchera[23] because it is my race that gives me strength to do battle in defense of what my parents were, and my children will inherit; proud of being a conchera because I have never stolen anything from anyone, I have never taken anybody's bread from his mouth to fill mine, because I have never crawled on my knees asking anybody for money, and I have always lived standing up.
>
> Now we are struggling for something which is ours, our ecosystem, but not because we are professional ecologists but because we must remain alive, because if the mangroves disappear, a whole people disappears, we all disappear, we shall no longer be part of the history of Muisne, we shall ourselves exist no longer ... I do not know what will happen to us if the mangroves disappear, we shall eat garbage in the outskirts of the city of Esmeraldas or in Guayaquil, we shall become prostitutes, I do not know what will happen to us if the mangroves disappear ... We think, if the camaroneros[24] who are not the rightful owners nevertheless now prevent us and the carboneros[25] from getting through the lands they have

taken, not allowing us to get across the swamps, shouting and shooting at us, what will happen next, when the government gives them the lands, will they put up big "Private Property" signs, will they even kill us with the blessing of the president?

Over the Andes, in Ecuador's rain forest, oil exploration and drilling has been under way since 1964, giving rise to local environmental conflicts. I now summarize an account by Valeria Pacheco (a journalist for Agence France Presse) of a visit to Rumipampa in Orellana, Ecuador, a few days after the court decision against Chevron Texaco of February 14, 2011. Pacheco met a woman who

> has no legal training, and doesn't speak the Spanish that dominates government in Quito but indigenous villager Maria Aguinda helped bring a landmark judgment against US oil giant Chevron for polluting the rain forest she calls home. The diminutive grandmother whose modest home sits near marshes clogged for decades in sticky oil has been at the heart of the David-and-Goliath case, and spoke out after Chevron was slapped last week with a $9.5 billion fine ... "Before I die they have to pay me for the dead animals, and for what they did to the river, and the water and the earth."

It is important to note here that Maria Aguinda puts her grievance in environmental terms, citing the animals, water, and earth. Texaco's environmental impacts, the source of her complaint, became Chevron's problem when Chevron acquired Texaco in 2001. That problem included a lawsuit filed in 1993 by "Maria Aguinda et al." on behalf of some 30,000 residents of Ecuador's rain forest. The settlement, however large, cannot bring back what Aguinda lost.

> Aguinda said she believes her husband and two of their 10 children died from effects of the pollution ... "When Texaco came we never thought they would leave behind such damage, never. Then they began to drill a well and set up burn pits," she said, helped in translation by her son William Grefa. "It changed our life: hunting, fishing, and other food, it's all finished."

Here Aguinda raised issues of damaged health and lost livelihood. This is a classic formulation of the environmentalism of the poor. Texaco undertook cleanup operations, which left Aguinda unsatisfied:

> The operation has done little to improve conditions, Aguinda said. "With the cleanup that Texaco left, the air is just unbearable. I can't live above the oil," groaned Aguinda, who grew visibly irritated talking about the disaster. "If someone comes here from Texaco," he'll get "pepper in his eyes," she winced. A strong petroleum smell permeates Rumipamba, home to nine families, some of whom complain of headaches.

In this journalist's account of Maria Aguinda's predicament, different languages of valuation are deployed side by side, such as livelihood, human right to life and health, and local indigenous territorial rights. These are different from the language of economic valuation of negative externalities appropriate in a court case for damages. These different languages may be used by the same people. There are still other languages available, for instance that of environmental justice against "environmental racism" (as in the Muisne case on mangrove destruction), or that of sacredness. Who has the right, or the power, to impose one particular language of valuation?

The deployment of different valuation languages is not only a strategy to get redress. It responds to deeply ingrained cultural values. Struggles around the protection of forests, rivers, mangroves, wetlands, or biodiversity often appear when poor communities mobilize for the defense of the environment as a source of livelihood; economic growth not only negates ecological processes, however: it also negates the cultural processes that are at the basis of people's valuation of and relationship to the natural world.[26]

In India

In the magazine *Down to Earth* of August 15, 2008, Sunita Narain, the Indian environmental researcher and author, described some environmental struggles in South Asia that exhibit the same awareness that underlies the environmentalism of the poor in Ecuador. She writes:

In Sikkim, bowing to local protests, the government has cancelled 11 hydro-electric projects. In Arunachal Pradesh, dam projects are being cleared at breakneck speed and resistance is growing. In Uttarakhand last month, 2 projects on the Ganga were put on hold and there is growing concern about the rest. In Himachal Pradesh, dams are so controversial that elections were won where candidates said they would not allow these to be built. Many other projects, from thermal power stations to Greenfield mining, are being resisted.

In the early twenty-first century, India has posted remarkable rates of economic growth. Part of that outcome is the result of increased foreign investment, avidly courted by Indian politicians, but not always welcomed by villagers. Narain continues:

The South Korean giant Posco's iron ore mine, steel plant and port are under fire. The prime minister has promised the South Korean premier the project will go ahead by August. But local people are not listening. They don't want to lose their land and livelihood and do not believe in promises of compensation. In Maharashtra, mango growers are up in arms against the proposed thermal power station in Ratnagiri.

In India the struggles occasioned by the poor's resistance to development projects has at times taken violent form. Peasants take matters into their own hands when they judge that the courts or other conventional routes will not yield satisfaction. As Narain puts it:

In every nook and corner of the country where land is acquired, or water sourced, for industry, people are fighting even to death. There are wounds. There is violence. There is also desperation. Like it or not, there are a million mutinies today[27] ... [In Kalinganagar villagers] knew they were poor. But they also knew modern development would make them poorer. It was the same in prosperous Goa, where I found village after village fighting against the powerful mining lobby.[28]

Conventional economics looks at environmental impacts in terms of externalities or lost environmental services which should be internalized into the price system. But one can see externalities (following K. W. Kapp) not as market failures but instead as cost-shifting successes on the part of industrialists and states. These give rise to environmental complaints and movements, to the environmentalism of the poor, in India as in Ecuador.[29]

In Mexico

Defending the environment is, in Mexico, a dangerous activity. Consider the story of Julián Vergara as told by Víctor Toledo. He finds modern peasant environmentalists responding to grievances similar to some of those that ignited the Mexican Revolution in 1910 – Emiliano Zapata's first protest, after all, concerned a sugar company using up communal water in Morelos. Toledo goes back to October 22, 1992 when the press

> published a small note that floated away like water in a river: "Early this morning Julián Vergara, a peasant leader and president of the El Tianguis common-land owners commission, was assassinated by an unknown assailant who shot him in the chest with a shotgun. The deceased was an environmentalist who opposed the felling of the forests in the Acapulco municipality."

Toledo, writing in 2000, continued:

> [A]s often occurs in this country of helplessness and injustice, the memory of Julián Vergara remains buried under the heavy flagstones of our cruel and forgetful time. How many Julián Vergaras will we have to lose in the heroic defence of the forests, springs and rivers of Mexico? I dream of the day when we can reconstruct these disgraceful stories and redeem the icy silence of the hundreds, maybe thousands, of heroic anonymous peasants who risked their lives to preserve the habitat and natural resources of the nation and the world.

Toledo understood peasant resistance to forest clearance as a form of environmentalism of the poor, which he sought to present as an updated version of the revolutionary tradition of Zapata:

> Then we would show that this solidarity with nature, with our fellow humans and future generations, that so many environmentalists around the world eagerly look for, exists already in the collective unconscious and the cultures of many rural settlements, surviving in the face of the most dangerous contamination: that which favors individualism and competition. We would discover that the only difference between the old agrarian martyrs and the new rural defenders of nature is in rhetorical fashions because the "zapatas" of a century ago are today's "environmentalists of the poor."[30]

An additional case is the small Zapotillo dam in Jalisco state. I visited in 2009 because I had read in a local newspaper, *La Jornada*, about this projected dam in the Verde River. It was intended to supply water to the cities of León and Guanajuato as well as other towns. The project involves the flooding of Temacapulín and other small settlements. The affected inhabitants' complaints and demonstrations negate the myth – according to journalist Mario Edgar López Ramírez – that the defence and conservation of the environment are only a luxury of wealthy people. The journalist explained that

> in 2005 the villages of Temacapulín, Acasico and Palmarejo were plagued by small signs of protest, stuck to houses, posts and trees, with inscriptions on them such as, "Dau, the children of Palmarejo will get even with you" (referring to Enrique Dau Flores, former director of the State Water Commission).

Of the 30 million *pesos* destined to buy houses in Temacapulín, only 5 million had been paid out. On February 27 and March 1, 2009, the Guadalajara University radio and television coverage managed to restore the balance of information in favour of the environmentalism of the poor. The interviews with a variety of local people showed their demands for environmental justice: "it is good that they want to bring water to León, but not at the price of throwing us out, not at the cost of flooding our homes, our church, and our ancestors' tombs."[31]

As the last sentence suggests, there is a clash between sacredness and economic values, as in the case of the *concheras* of Ecuador. In the end, if the dam is built, money indemnities will be paid out. Since the people are poor, it won't be a large amount.

There are also interclass cases of environmentalism in Mexico as in other places. Protests against nuclear power stations or the successful court case in 2009 against the Canadian company New Gold who owns the San Xavier Mine on San Pedro Hill in San Luís Potosí have united Mexicans across the socioeconomic spectrum. This open-cast mine has been using cyanide for years in a protected zone. (A successful court decision does not necessarily imply in Mexico that the mine will close down.) There are, also, many examples in Mexico and around the world of an environmentalism – represented by Nature Conservancy, IUCN, and WWF, often financed by companies like Shell and Rio Tinto – that worships nature at its wildest, forgetting about or even displacing the poor and indigenous. There are, at times, collaborations between this environmentalism of the rich and popular environmentalism. The fight to defend mangrove swamps and forests, in Ecuador, in Mexico, in India and elsewhere, could be a shared undertaking. There is in many cases no conflict between the environmentalism of the rich and the environmentalism of the poor. And in those cases where such conflict exists, recognition of broad overlapping interests can often minimize it. One must admit, however, that cases will remain in which ambitions for a given landscape among different groups of environmentalists will be incompatible.

Conclusion

For analytical purposes, modern environmentalism may be divided into three branches (as in a single tree) or three currents (as in a single river). Naturally they overlap to some extent. But they represent different social experiences of nature, different economic relationships with nature, and different intellectual traditions. The first two, in their modern forms, are slightly more than a century old. The third, although identified by scholars only recently, has in one or another form existed for millennia, albeit without explicit reference to the environment as such. They are:

1. The "cult of wilderness," which arises from the love of beautiful landscapes and from deeply held values, not from material interests. This includes the "deep ecology" movement. John Muir is a good representative of this current as one of its early and most eloquent exponents. The cult of wilderness sometimes has religious overtones, as with Muir, but today is backed by the science of conservation biology. For decades it has been fighting a rearguard action to preserve threatened ecosystems because of the products and services that they provide, ranging from aesthetic joy to absorbing carbon dioxide from the atmosphere.
2. The "gospel of eco-efficiency," connected to the "sustainable development" and "ecological modernization" movements. It rests intellectually on the optimism of "Environmental Kuznets Curves" which suppose that the richer a society becomes the more it can afford to be concerned with the environment and the more it will

invest in environmental protection. The gospel seeks ecological modernization, a socioeconomic transition based on two pillars: first, an economic pillar: eco-taxes and markets in emissions permits; second, a technological pillar: support for materials- and energy-saving changes. This is a reformist, managerial movement, and at the same time a research program of worldwide relevance on the energy and material throughput in the economy, and on the possibilities of "delinking" economic growth form its material base.

3. The "environmentalism of the poor" and in general environmental justice movements, which are based on a material interest in the environment as a source and a requirement for livelihood. These movements show more concern with today's poor people than with the rights of other species and of future generations of humans. They are centered on resource-extraction conflicts (over fossil fuels, mining, or biomass) and waste-disposal conflicts. Its protagonists are local populations (indigenous or not), sometimes led by women, whose livelihoods are threatened in rural or urban contexts by environmental impacts.

The third type of environmentalism was not recognized as such until the 1980s and 1990s, as actors in such conflicts over environmental justice have often not used an environmental idiom when expressing their grievances. The conflicts are not only over distribution: they are also about recognition and participation in decision-making about the environment.

The poor are not always environmentalists, and environmentalists are not always poor. There are, however, numerous cases of environmentalism of the poor and of impoverished indigenous communities around the world, indeed now more than ever before. This is because of the increase in the social metabolism, in resource extraction and waste creation, now reaching into the last frontiers.

"Environmentalism of the Poor" were words that Guha and I started to use shortly after we met for the first time in August 1988 in Bangalore. We noticed, then, how his work on the Chipko movement fitted with the work I was starting to do on Latin American environmental movements. In December 1988 Chico Mendes was killed defending the Amazon forest in Brazil. He was a local union leader. The rubber tappers, with some inspiration from the Catholic "Theology of Liberation" (but largely unaware of Gandhi's forms of struggle), developed a technique for opposing the cattle ranchers who came to destroy the forest. They sat down or stood up peacefully, with women and children confronting the armed invaders in a series of *empates* – a Portuguese word to which they gave new meaning (roughly: standoffs). This was a clear case of the environmentalism of the poor. Some years later, after Ken Saro-Wiwa and his companions had been killed in 1995 for complaining about Shell and the Nigerian government, we wrote a book entitled *Varieties of Environmentalism* explaining the theory of the environmentalism of the poor as if we were academic scavengers or decomposers usefully recycling corpses and other cruel historical realities.[32]

Environmental struggles resort sometimes to the language of economic valuation, for instance when compensation for externalities is requested. This is the case in forensic contexts claiming damages. Thus, as we have seen, Chevron Texaco was fined US$ 9.5 billion on February 14, 2011 by a court in Sucumbios, Ecuador, for damage done between 1964 and 1990 in that region. This included not only remediation costs but also an element of reparation for moral damage. In such a case the bottom line is money. In many other cases, the poor and indigenous have tried to stop degradation of the environment by arguing not in terms of economic costs but in terms of rights (territorial rights, human rights), or in terms of sacredness. Languages of valuation are often not translatable into one another.

The study of the environmentalism of the poor arose in social history in India and, to some extent, in Latin America. It is closely related to political ecology, which studies conflicts of ecological distribution, that is the social, spatial, and intertemporal patterns of access to the benefits obtainable from natural resources and from the environment as a life-support system, including its "cleaning up" capacities. The determinants of ecological distribution are in some respects natural (climate, topography, rainfall patterns, minerals, soil quality, and so on). They are clearly, in other respects, social, cultural, political, and technological.

The environmentalism of the poor is an "environmental justice" movement, although it arose independently of the 1980s movement in the US against "environmental racism." The connections between both movements are now closer than ever, both in an analytical and scholarly sense, and in the world of environmental politics.[33] Environmental injustices are not only local, they are also global. Hence the notions of ecologically unequal trade and of ecological debt. First, the exports of raw materials and other products from relatively poor countries are sold at unsustainable rates and at prices which do not include compensation for local or global externalities. Second, rich countries make a disproportionate use of environmental space without payment, and even without recognition of other people's entitlements to such services (particularly, the disproportionate free use of carbon dioxide sinks and reservoirs). The collection of ecological debt would contribute to the "ecological adjustment" which in moral terms the north must make.

The case for a general "win-win" solution (better environment with economic growth) is far from proven. On the contrary, since the global economy is not "dematerializing" in per capita terms, there are increasing local and global conflicts over the sharing of the burdens of pollution (including the enhanced greenhouse effect) and over the access to natural resources.

The environmentalism of the poor, popular environmentalism, livelihood ecology, liberation ecology, and the movement for environmental justice (local and global), emerged historically from the complaints against the appropriation of communal environmental resources and against the disproportionate burdens of pollution. They may help to move society and economy in the direction of social justice and ecological sustainability.

Notes

1   See R. Inglehart, "Public Support for Environment Protection: Objective Problems and Subjective Values in 43 Societies," *Political Science and Politics* 28, 1995, pp. 57–72; R. Inglehart, *Modernization and Postmodernization*, Princeton, NJ, Princeton University Press, 1997; J. Martinez-Alier, "The Environment as a Luxury Good or 'Too Poor to Be Green'?," *Ecological Economics* 13, 1995, pp. 1–10; R. Guha, *Environmentalism: A Global History*, New York, Longman, 2000; S. Brechin, "Objective Problems, Subjective Values, and Global Environmentalism: Evaluating the Postmaterialist Argument and Challenging a New Explanation," *Social Science Quarterly* 80/4, 1999, pp. 793–809.

2   R. Guha, *The Unquiet Woods: Ecological Change and Peasant Resistance in the Himalaya*, Ranikhet, Permanent Black, 1989; and J. Martinez-Alier, *Ecological Economics (Energy, Environment and Society)*, Oxford, Blackwell, 1991.

3 Inglehart, "Public Support for Environment Protection,"; Inglehart, *Modernization and Postmodernization*; R. E. Dunlap and R. York, "The Globalization of Environmental Concern and the Limits of the Post-Materialist Explanation: Evidence from Four Cross-National Surveys," *Sociological Quarterly* 49, 2008, pp. 529–63.

4 D. Schlosberg, *Defining Environmental Justice*, New York, Oxford University Press, 2007; L. Urkidi and M. Walter, "Dimensions of environmental justice in anti-gold mining movements in Latin America," *Geoforum* 42/6, 2011, pp. 686–95.

5 B. Agarwal, "The Gender and Environment Debate: Lessons from India," *Feminist Studies* 18/1, 1992, pp. 119–58.

6 J. W. Moore, "Sugar and the Expansion of the Early Modern World-Economy: Commodity Frontiers, Ecological Transformation, and Industrialization," *Review of the Fernand Braudel Center* 23, 2000, pp. 409–33.

7 I. Davey, "Environmentalism of the Poor and Sustainable Development: An Appraisal," *Journal of Administration & Governance* 4/1, 2009, pp. 1–10; Guha, Environmentalism; J. Martinez-Alier, "Environmental Justice as a Force for Sustainability," in J. Pieterse (ed.), *Global Futures: Shaping Globalization*, New York, Zed, 2000, pp. 148–76; J. Martinez-Alier, *The Environmentalism of the Poor: A Study of Ecological Conflicts and Valuation*, Delhi, Oxford University Press, 2002.

8 Davey, "Environmentalism of the Poor and Sustainable Development."

9 E. Ostrom, *Governing the Commons: The Evolution of Institutions for Collective Actions*, Cambridge, Cambridge University Press, 1990.

10 Oilwatch, *Home page*, n.d., accessed from http://www.oilwatch.org/index.php?lang=en on March 20, 2012.

11 L. Rival, "Ecuador's Yasuní-ITT Initiative: The Old and New Values of Petroleum," *Ecological Economics* 70/2, 2010, pp. 358–65.

12 Davey, "Environmentalism of the Poor and Sustainable Development"; W. M. Adams, *Green Development: Environment and Sustainability in the Third World*, London, Routledge, 2001.

13 R. Bullard, *Confronting Environmental Racism: Voices from the Grassroots*, Boston, South End Press, 1993.

14 R. Bullard, "Environmental Justice: It Is More Than Waste Facility Siting," *Social Science Quarterly* 77, 1996, pp. 493–9; J. Agyeman, R. Bullard, and B. Evans, *Just Sustainabilities: Development in an Unequal World*, London, Earthscan, 2003.

15 R. Guha, *How Much Should a Person Consume? Thinking through the Environment*, Berkeley, University of California Press, 2006, pp. 206–7.

16 A. Ortega Santos, *La Tragedia de los Cerramientos: Desarticulación de la Comunidad en la Provincia de Granada*, Valencia, Fundación Inst. de Historia Social, 2002.

17 M. Armiero, "Seeing Like a Protestor: Nature, Power, and Environmental Struggles," *Left History* 13/1, 2008, pp. 59–76; M. Armiero, *Rugged Nation: Mountains and the Making of Modern Italy, 1860–2000*, Cambridge, White Horse Press, 2011.

18 S. Barca, *Enclosing Water: Nature and Political Economy in Mediterranean Valley, 1796–1916*, Cambridge, White Horse Press, 2010.

19 S. Barca, "Bread and Poison: The Story of Labor Environmentalism in Italy, 1968–1998," in J. Melling and C. Sellers (eds.), *Dangerous Trade: Histories of Industrial Hazards across a Globalizing World*, Philadelphia, Temple University Press, forthcoming.

20 F. E. Mallon, *The Defense of Community in Peru's Central Highlands: Peasant Struggle and Capitalist Transition 1860–1940*, Princeton, NJ, Princeton University Press, 1983.

21 E. Melville, *A Plague of Sheep: Environmental Consequences of the Conquest of Mexico*, Cambridge, Cambridge University Press, 1994.

22 See P. Sukhdev, *The Economics of Ecosystems and Biodiversity: An Interim Report*, 2009, accessed from http://www.teebweb.org/InformationMaterial/TEEBReports/tabid/1278/Default.aspx on March 20, 2012, chapter 3.

23 A woman who collects shellfish for subsistence and for sale.

24 Owners of shrimp farms.
25 Charcoal-makers, who here use mangrove wood.
26 A. Escobar, "Difference and Conflict in the Struggle over Natural Resources: A Political Ecology Framework," *Development* 49/3, 2006, pp. 6–13.
27 Narain is alluding to V. S. Naipaul's 1990 book entitled *India: A Million Mutinies Now.*
28 S. Narain, "Learn to Walk Lightly," *Down to Earth*, August 15, 2008, accessed from http://www.downtoearth.org.in/content/learn-walk-lightly on March 20, 2012.
29 E. Leff, *Ecologia y Capital: Hacia una Perspectiva Ambiental del Desarrollo*, Mexico City, Universidad Nacional Autonoma de Mexico, 1986; J. O'Connor, "Nature, Socialism: A Theoretical Introduction," *Capitalism, Nature, Socialism* 1/1, 1988, pp. 11–38.
30 V. M. Toledo, "El otro zapatismo: Luchas indígenas de inspiración zapatista," *Ecología política* 18, 2000, pp. 11–22.
31 M. E. L. Ramírez, "La presa El Zapotillo: ecologismo de los pobres, cerco informativo y la reacción de Medios UdeG," *La Jornada Jalisco*, March 7, 2009, accessed from http://archivo.lajornadajalisco.com.mx/2009/03/07/index.php?section=politica&article=006n1pol on March 22, 2012.
32 R. Guha and J. Martinez-Alier, *Varieties of Environmentalism: Essays North and South*, London, Earthscan, 1997.
33 R. Nixon, *Slow Violence and the Environmentalism of the Poor*, Cambridge, MA, Harvard University Press, 2011.

References

Adams, W. M., *Green Development: Environment and Sustainability in the Third World*, London, Routledge, 2001.

Agarwal, B., "The Gender and Environment Debate: Lessons from India," *Feminist Studies* 18/1, 1992, pp. 119–58.

Agyeman, J., Bullard, R., and Evans, B., *Just Sustainabilities: Development in an Unequal World*, London, Earthscan, 2003.

Armiero, M., *Rugged Nation: Mountains and the Making of Modern Italy, 1860–2000*, Cambridge, White Horse Press, 2011.

Armiero, M., "Seeing Like a Protestor: Nature, Power, and Environmental Struggles," *Left History* 13/1, 2008, pp. 59–76.

Barca, S., "Bread and Poison: The Story of Labor Environmentalism in Italy, 1968–1998," in J. Melling and C. Sellers (eds.), *Dangerous Trade: Histories of Industrial Hazards across a Globalizing World*, Philadelphia, Temple University Press, forthcoming.

Barca, S., *Enclosing Water: Nature and Political Economy in Mediterranean Valley, 1796–1916*, Cambridge, White Horse Press, 2010.

Brechin, S., "Objective Problems, Subjective Values, and Global Environmentalism: Evaluating the Postmaterialist Argument and Challenging a New Explanation," *Social Science Quarterly* 80/4, 1999, pp. 793–809.

Bullard, R., *Confronting Environmental Racism: Voices from the Grassroots*, Boston, South End Press, 1993.

Bullard, R., "Environmental Justice: It Is More Than Waste Facility Siting," *Social Science Quarterly* 77, 1996, pp. 493–9.

Davey, I., "Environmentalism of the Poor and Sustainable Development: An Appraisal," *Journal of Administration & Governance* 4/1, 2009, pp. 1–10.

Dunlap, R. E., and York, R., "The Globalization of Environmental Concern and the Limits of the Post-Materialist Explanation: Evidence from Four Cross-National Surveys," *Sociological Quarterly* 49, 2008, pp. 529–63.

Escobar, A., "Difference and Conflict in the Struggle over Natural Resources: A Political Ecology Framework," *Development* 49/3, 2006, pp. 6–13.

Guha, R., *Environmentalism: A Global History*, New York, Longman, 2000.

Guha, R., *How Much Should a Person Consume? Thinking through the Environment*, Berkeley, University of California Press, 2006.

Guha, R., *The Unquiet Woods: Ecological Change and Peasant Resistance in the Himalaya*, Ranikhet, Permanent Black, 1989.

Guha, R., and Martinez-Alier, J., *Varieties of Environmentalism: Essays North and South*, London, Earthscan, 1997.

Inglehart, R., *Modernization and Postmodernization*, Princeton, NJ, Princeton University Press, 1997.

Inglehart, R., "Public Support for Environment Protection: Objective Problems and Subjective Values in 43 Societies," *Political Science and Politics* 28, 1995, pp. 57–72.

Leff, E., *Ecologia y Capital: Hacia una Perspectiva Ambiental del Desarrollo*, Mexico City, Universidad Nacional Autonoma de Mexico, 1986.

Mallon, F. E., *The Defense of Community in Peru's Central Highlands: Peasant Struggle and Capitalist Transition 1860–1940*, Princeton, NJ, Princeton University Press, 1983.

Martinez-Alier, J., *Ecological Economics (Energy, Environment and Society)*, Oxford, Blackwell, 1991.

Martinez-Alier, J., "The Environment as a Luxury Good or 'Too Poor to Be Green'?," *Ecological Economics* 13, 1995, pp. 1–10.

Martinez-Alier, J., "Environmental Justice as a Force for Sustainability," in J. Pieterse (ed.), *Global Futures: Shaping Globalization*, New York, Zed, 2000, pp. 148–76.

Martinez-Alier, J., *The Environmentalism of the Poor: A Study of Ecological Conflicts and Valuation*, Delhi, Oxford University Press, 2002.

Melville, E., *A Plague of Sheep: Environmental Consequences of the Conquest of Mexico*, Cambridge, Cambridge University Press, 1994.

Moore, J. W., "Sugar and the Expansion of the Early Modern World-Economy: Commodity Frontiers, Ecological Transformation, and Industrialization," *Fernand Braudel Center* 23, 2000, pp. 409–33.

Narain, S., "Learn to Walk Lightly," *Down to Earth*, August 15, 2008, accessed from http://www.downtoearth.org.in/content/learn-walk-lightly on March 20, 2012.

Nixon, R., *Slow Violence and the Environmentalism of the Poor*, Cambridge, MA, Harvard University Press, 2011.

O'Connor, J., "Nature, Socialism: A Theoretical Introduction," *Capitalism, Nature, Socialism* 1/1, 1988, pp. 11–38.

Oilwatch, *Home page*, n.d., accessed from http://www.oilwatch.org/index.php?lang=en on March 20, 2012.

Ortega Santos, A., *La Tragedia de los Cerramientos: Desarticulación de la Comunidad en la Provincia de Granada*, Valencia, Fundación Inst. de Historia Social, 2002.

Ostrom, E., *Governing the Commons: The Evolution of Institutions for Collective Actions*, Cambridge, Cambridge University Press, 1990.

Ramírez, M. E. L., "La presa El Zapotillo: ecologismo de los pobres, cerco informativo y la reacción de Medios UdeG," *La Jornada Jalisco*, March 7, 2009, accessed from http://archivo.lajornadajalisco.com.mx/2009/03/07/index.php?section=politica&article=006n1pol on March 22, 2012.

Rival, L., "Ecuador's Yasuní-ITT Initiative: The Old and New Values of Petroleum," *Ecological Economics* 70/2, 2010, pp. 358–65.

Schlosberg, D., *Defining Environmental Justice*, New York, Oxford University Press, 2007.

Sukhdev, P., *The Economics of Ecosystems and Biodiversity: An Interim Report*, 2009, accessed from http://www.teebweb.org/InformationMaterial/TEEBReports/tabid/1278/Default.aspx on March 20, 2012.

Toledo, V. M., "El otro zapatismo: Luchas indígenas de inspiración zapatista," *Ecología política* 18, 2000, pp. 11–22.

Urkidi, L., and Walter, M., "Dimensions of environmental justice in anti-gold mining movements in Latin America," *Geoforum* 42/6, 2011, pp. 686–95.

Further Reading

Crosby, A. W., *Ecological Imperialism: The Biological Expansion of Europe, 900–1900*, Cambridge, Cambridge University Press, 1986.

Gerber, J. F., "Conflicts over Industrial Tree Plantations in the South: Who, How and Why?," *Global Environmental Change* 21/1, 2011, pp. 165–78.

Peet, R., and Watts, M., *Liberation Ecologies: Environment, Development, Social Movements*, London, Routledge, 1996.

Peluso, N., *Rich Forests, Poor People: Resource Control and Resistance in Java*, Berkeley, University of California Press, 1992.

Rocheleau, D., Thomas-Slayter B., and Wangari, E. (eds.), *Feminist Political Ecology: Global Perspectives and Local Experience*, London, Routledge, 1996.

Swyngedouw, E., "Globalisation or 'Glocalisation'? Networks, Territories or Rescaling," *Cambridge Review of International Affairs* 17/1, 2004, pp. 25–48.

Swyngedouw, E., "Neither Global nor Local: Glocalization and the Politics of Scale," in K. Cox (ed.), *Spaces of Globalization*, New York, Guilford Press, 1997, pp. 137–66.

Index

A Companion to Global Environmental History, First Edition. Edited by J.R. McNeill and Erin Stewart Mauldin.
© 2012 John Wiley & Sons, Ltd. Published 2015 by John Wiley & Sons, Ltd.

Printed and bound by CPI Group (UK) Ltd, Croydon, CR0 4YY

27/10/2024

14580380-0001